INTERACTING BINARIES

To learn more about the AIP Conference Proceedings, including the
Conference Proceedings Series, please visit the webpage
http://proceedings.aip.org/proceedings

INTERACTING BINARIES

Accretion, Evolution, and Outcomes

Cefalù, Sicily   4 – 10 July 2004

EDITORS
Luciano Burderi
L. Angelo Antonelli
Francesca D'Antona
Gian Luca Israel
INAF-OAR, Astronomical Observatory of Rome
Rome, Italy

Tiziana Di Salvo
University of Palermo
Palermo, Italy

Luciano Piersanti
Amedeo Tornambè
Oscar Straniero
INAF-OAT, Astronomical Observatory of Teramo
Teramo, Italy

SPONSORING ORGANIZATIONS
INAF-OAR - Astronomical Observatory of Rome, Italy
INAF-OAT - Astronomical Observatory of Teramo, Italy
University of Palermo - Department of Physics and Astronomy

Melville, New York, 2005
AIP CONFERENCE PROCEEDINGS ■ VOLUME 797

Editors:

Luciano Burderi
L. Angelo Antonelli
Francesca D'Antona
Gian Luca Israel

INAF-OAR
Astronomical Observatory of Rome
Via di Frascati, 33
00040 Monte Porzio Catone
ITALY

E-mail: burderi@mporzio.astro.it
    antonelli@mporzio.astro.it
    dantona@mporzio.astro.it
    gianluca@mporzio.astro.it

Luciano Piersanti
Amedeo Tornambè
Oscar Straniero

INAF-OAT
Astronomical Observatory of Teramo
Via Mentore Maggini snc
64100 Teramo
ITALY

E-mail: piersanti@te.astro.it
    tornambe@te.astro.it
    straniero@te.astro.it

Tiziana Di Salvo
University of Palermo
Dept. of Physics and Astronomy
Via Archirafi, 36
90123 Palermo
ITALY
E-mail: disalvo@fisica.unipa.it

Technical Editor:

Giuliana Giobbi
INAF-OAR
Astronomical Observatory of Rome
Via di Frascati, 33
00040 Monte Porzio Catone
ITALY
E-mail: giobbi@mporzio.astro.it

Cover: photo courtesy of Maria Teresa Menna.

Authorization to photocopy items for internal or personal use, beyond the free copying permitted under the 1978 U.S. Copyright Law (see statement below), is granted by the American Institute of Physics for users registered with the Copyright Clearance Center (CCC) Transactional Reporting Service, provided that the base fee of $22.50 per copy is paid directly to CCC, 222 Rosewood Drive, Danvers, MA 01923. For those organizations that have been granted a photocopy license by CCC, a separate system of payment has been arranged. The fee code for users of the Transactional Reporting Service is: 0-7354-0286-8/05/$22.50.

© 2005 American Institute of Physics

Individual readers of this volume and nonprofit libraries, acting for them, are permitted to make fair use of the material in it, such as copying an article for use in teaching or research. Permission is granted to quote from this volume in scientific work with the customary acknowledgment of the source. To reprint a figure, table, or other excerpt requires the consent of one of the original authors and notification to AIP. Republication or systematic or multiple reproduction of any material in this volume is permitted only under license from AIP. Address inquiries to Office of Rights and Permissions, Suite 1NO1, 2 Huntington Quadrangle, Melville, N.Y. 11747-4502; phone: 516-576-2268; fax: 516-576-2450; e-mail: rights@aip.org.

L.C. Catalog Card No. 2005933838
ISBN 0-7354-0286-8
ISSN 0094-243X
Printed in the United States of America

CONTENTS

Preface..xiii
Acknowledgments...xv
Photo Gallery..xvii

Population Synthesis for Low and Intermediate Mass Binaries.................1
  L. R. Yungelson

SESSION 1
COMPACT BINARIES IN GLOBULAR CLUSTERS

Interacting X-Ray Binaries in Globular Clusters: 47Tuc vs. NGC 6397........13
  J. E. Grindlay
X-Ray Sources in Globular Clusters of Other Galaxies......................23
  W. H. G. Lewin and F. Verbunt
X-Ray Sources in Globular Clusters...30
  F. Verbunt
X-Ray Binaries in the Globular Cluster 47 Tucanae..........................40
  C. O. Heinke, J. E. Grindlay, P. D. Edmonds, H. N. Cohn, P. M. Lugger,
  F. Camilo, S. Bogdanov, and P. C. Freire
Globular Clusters and Neutron Star Binaries................................46
  F. D'Antona
Formation and Evolution of Compact Binaries with an Accreting
White Dwarf in Globular Clusters...53
  N. Ivanova and F. A. Rasio
Eclipsing Binaries in the Galactic Globular Cluster Omega Centauri.........61
  A. Calamida, G. Bono, R. Buonanno, C. E. Corsi, M. Monelli,
  M. Dall'Ora, L. M. Freyhammer, and A. Munteanu

SESSION 2
MILLISECOND BINARY PULSARS

Millisecond Pulsars in Low-Mass X-Ray Binaries.............................71
  D. Chakrabarty
Indirect Evidence for an Active Radio Pulsar in SAX J1808.4−3658
during Quiescence..81
  S. Campana, P. D'Avanzo, L. Stella, G. L. Israel, G. Marconi, J. Casares,
  R. Hynes, and P. Charles
Inhomogeneous Accretion Flow in X-Ray Binary Pulsars.......................87
  M. T. Menna, L. Burderi, L. Stella, N. Robba, and M. van der Klis
Accretion and Magneto-Dipole Emission in Fast-Rotating Neutron
Stars: New Spin-Equilibrium Lines..95
  T. Di Salvo, L. Burderi, F. D'Antona, and N. Robba

Searching for Optical Counterparts to MSP Companions in Galactic
Globular Clusters 103
    F. R. Ferraro
Temporal Analysis of the Millisecond X-Ray Pulsar SAX J1808.4–
3658 during the 2000 Outburst... 110
    A. Papitto, L. Burderi, T. Di Salvo, M. T. Menna, and L. Stella
General Relativistic Effects on the Evolution of Binary Systems......... 116
    G. Lavagetto, L. Burderi, F. D'Antona, T. Di Salvo, R. Iaria, and
    N. R. Robba

SESSION 3
SN FROM MASSIVE BINARY SYSTEMS AND GRBS

Accretion Power in GRBs 123
    T. Piran
The Quark-Deconfinement Nova Model for Gamma-Ray Bursts 132
    I. Bombaci
Dynamical Evolution of Neutrino Cooled Disks........................... 138
    W. H. Lee
Gamma-Ray Bursts and Afterglow Polarisation 144
    S. Covino, E. Rossi, D. Lazzati, D. Malesani, and G. Ghisellini
The Empirical Grounds of the Supernova/Gamma-Ray Burst
Connection 150
    M. Della Valle
A Few Insights into the Swift Mission 163
    G. Chincarini *(On behalf of the Swift Team)*
The REM Telescope: A Robotic Facility to Promptly Follow-up GRBs
and Cosmic Fast Transients 173
    L. A. Antonelli *(On behalf of the REM Team)*
Rapid GRB Follow-up with the 2-m Robotic Liverpool Telescope 181
    A. Gomboc, M. F. Bode, D. Carter, C. Guidorzi, A. Monfardini,
    C. G. Mundell, A. M. Newsam, R. J. Smith, I. A. Steele, and J. Meaburn

SESSION 4
ACCRETION ON BLACK HOLES AND MICROQUASARS

Black Hole X-Ray Binary Jets 189
    E. Gallo, R. Fender, and C. Kaiser
Black Hole States: Accretion and Jet Ejection 197
    T. Belloni
The Dynamical Fingerprint of Intermediate Mass Black Holes in
Globular Clusters 205
    M. Colpi, B. Devecchi, M. Mapelli, A. Patruno, and A. Possenti
Accretion-Ejection Instability, LFQPO and Lag Structure of
Microquasars... 213
    P. Varnière

Black Hole Hunting in the Andromeda Galaxy 219
    R. Barnard, J. P. Osborne, U. Kolb, and C. A. Haswell
Transient QPOs in the Microquasar XTE J1859+226 225
    P. Casella, T. Belloni, J. Homan, and L. Stella
X-Ray States of Black Hole Binaries in Outburst 231
    R. A. Remillard
Black Hole Formation in X-Ray Binaries: The Case of GRO J1655-40 241
    V. Kalogera, B. Willems, M. Henninger, T. Levin, and N. Ivanova
The 2003 Outburst of the Galactic Microquasar V4641 Sgr (=SAX
J1819.3-2525) 249
    D. Maitra and C. D. Bailyn

SESSION 5
ACCRETION ON WHITE DWARFS AND NOVAE

Magnetic Accretion onto White Dwarfs 257
    G. H. Tovmassian
High-Angular Resolution Imaging of Interacting Binaries 265
    M. Karovska
On the Excitation of Hydrodynamical Turbulence in Accretion Discs 271
    O. A. Kuznetsov
Rapid Oscillations in Cataclysmic Variables, and a Comparison with
X-Ray Binaries 277
    B. Warner and P. A. Woudt
High-Speed Photometry of AM CVn Stars 287
    P. A. Woudt, B. Warner, J. Patterson, and C. Espaillat
Formation of the "Precessional" Spiral Wave in the Cool Accretion
Disk in Semidetached Binaries 295
    D. V. Bisikalo, A. A. Boyarchuk, P. V. Kaygorodov, O. A. Kuznetsov, and
    T. Matsuda
Mass and Angular Momentum Loss during RLOF in Algols 301
    W. Van Rensbergen, C. De Loore, and D. Vanbeveren
RX J0806.3–1527: Ten Years of Phase Coherent Monitoring in the
Optical and X-Ray Bands 307
    G. L. Israel, S. Dall'Osso, V. Mangano, L. Stella, S. Covino, D. Fugazza,
    S. Campana, G. Marconi, S. Mereghetti, and U. Munari
The Supersoft X-Ray Source CAL 83: A Massive White Dwarf 313
    T. Lanz, M. Audard, F. Paerels, and G. A. Telis
Modelling the Evolution of Nova Outbursts 319
    D. Prialnik and A. Kovetz
The First Three Years of the Outburst and Light-Echo Evolution of
V838 Mon and the Nature of its Progenitor 331
    U. Munari and A. Henden

SESSION 6
RAPID VARIABILITY AND SECULAR EVOLUTION OF LMX-RAY BINARIES

**The Accretion Disc Corona in LMXBs: Size and Temperature –
Fundamental Consequences** 339
    M. Bałucińska-Church and M. J. Church
Timing Neutron Stars.. 345
    M. van der Klis
X-Ray Studies of Three Binary Millisecond Pulsars 359
    N. A. Webb, J.-F. Olive, and D. Barret
Echo Tomography of Sco X-1 Using Bowen Fluorescence Lines 365
    J. Casares, T. Muñoz-Darias, I. G. Martínez-Pais, R. Cornelisse,
    P. A. Charles, T. R. Marsh, V. S. Dhillon, and D. Steeghs
**Spectral Changes during Six Years of Scorpius X-1 Monitoring with
BeppoSAX Wide Field Cameras**.. 371
    P. Santolamazza, F. Fiore, L. Burderi, and T. Di Salvo
The Evolution of Low-Mass X-Ray Binaries 377
    H. Ritter
Binary Population Synthesis: Theory and Applications 386
    P. Podsiadlowski, S. Rappaport, E. Pfahl, Z. Han, and M. E. Beer
Optical Spectroscopy of (Candidate) Ultra-Compact X-Ray Binaries 396
    G. Nelemans and P. Jonker
An Absorbed View of a New Class of INTEGRAL Sources..................... 402
    E. Kuulkers
**Exploring the Nature of Weak Chandra Sources near the Galactic
Centre**.. 410
    R. M. Bandyopadhyay, J. C. A. Miller-Jones, K. M. Blundell, F. E. Bauer,
    P. Podsiadlowski, Q. D. Wang, S. Rappaport, and E. Pfahl
**Chemical Composition of Secondary Stars in LMXBs: Implications
on the Progenitors of Black Holes and Neutron Stars** 416
    J. I. González Hernández
Stellar-Mass Black Hole Binaries as ULXs.................................. 422
    S. Rappaport, P. Podsiadlowski, and E. Pfahl
**Young Rotation-Powered Pulsars as Ultraluminous X-Ray Sources in
Star-Forming Galaxies**.. 434
    L. Stella and R. Perna

SESSION 7
SUPERNOVAE TYPE Ia

**Binaries, Cluster Dynamics and Population Studies of Stars and
Stellar Phenomena** 445
    D. Vanbeveren
**Thermonuclear Supernova Models, and Observations of Type Ia
Supernovae** 453
    E. Bravo, C. Badenes, and D. García-Senz

White Dwarf Merging and the Emission of Gravitational Waves............463
    J. Isern, E. García-Berro, J. Guerrero, P. Lorén-Aguilar, and J. A. Lobo
White Dwarfs Undergoing Hydrogen Shell Burning in Single
Degenerate Binary Systems...471
    M. Orio, T. Rauch, E. Leibowitz, and E. Tepedelenlioglu
Galactic Disk Abundance Ratios: Constraining SNIa Stellar Yields...........476
    C. Chiappini
Clues on Type Ia Supernovae Progenitors 482
    L. Piersanti and A. Tornambé
Supernovae Type Ia: An Observational Perspective 491
    J. Danziger
Rotating Type Ia SN Progenitors: Explosion and Light Curves 497
    I. Domínguez, L. Piersanti, E. Bravo, S. Gagliardi, O. Straniero, and
    A. Tornambé

SESSION 8
SECULAR EVOLUTION OF HIGH-MASS X-RAY BINARIES

XMM-Newton EPIC & OM Observations of Her X-1 over the 35 d
Beat Period and an Anomalous Low State 507
    S. Zane, G. Ramsay, M. A. Jimenez-Garate, J. W. den Herder, M. Still,
    P. T. Boyd, and C. J. Hailey
The Properties of the Absorbing and Line Emitting Matter in IGR
J16318-4848 513
    G. Matt, M. Guainazzi, A. Ibarra, and E. Jimenez-Bailon
RXTE Observation of the Low-Mass X-Ray Binary Pulsar GX1+4............519
    T. Kohmura and S. Kitamoto
The Double Pulsar System J0737−3039: News and Views 523
    M. Burgay, N. D'Amico, A. Possenti, A. Lyne, M. Kramer,
    M. McLaughlin, D. Lorimer, D. Manchester, F. Camilo, J. Sarkissian,
    P. Freire, and B. C. Joshi

POSTER SESSION

Phase Resolved Blue Spectroscopy of SS433................................533
    A. D. Barnes, P. A. Charles, J. S. Clark, R. Cornelisse, and C. Knigge
Irradiation Effects in Compact Binaries 537
    M. E. Beer and P. Podsiadlowski
Magnetic Pumping in Accretion Disk Coronae 541
    R. Belmont and M. Tagger
High Resolution Spectroscopy of 4U 1728-34 from a Simultaneous
Chandra-RXTE Observation 545
    A. D'Aí, R. Iaria, T. Di Salvo, G. Lavagetto, N. R. Robba, L. Burderi,
    M. Mendez, and M. van der Klis
Single Stars and Supernovae from Wolf-Rayet Secondaries 549
    L. Dray and C. Tout

The Early Spectral Evolution of Nova Sgr 2004 553
    A. Ederoclite, E. Mason, M. Della Valle, R. Gilmozzi, and R. E. Williams

A Multiple Mass-Ejection by the Symbiotic Prototype Z And During
its 2000-03 Outburst.. 557
    A. Skopal, L. Errico, A. A. Vittone, S. Tamura, M. Otsuka, M. Wolf, and
    V. G. Elkin

What Happens when a Hot Star Shines on a Cool One? 561
    K. Exter, T. Barman, D. Pollacco, V. Pustynski, S. Bell, and I. Pustylnik

Optical Counterpart of the XTE J0929-314 in Quiescence:
Constraints on the Magnetic Field.. 565
    M. Monelli, G. Fiorentino, L. Burderi, F. D'Antona, N. Robba, and V. Testa

Non-Axisymmetric Structure of Accretion Disks around the Neutron
Star in Be/X-Ray Binaries... 569
    K. Hayasaki and A. T. Okazaki

Modeling of Gas Flow Structure in Symbiotic Star Z And in
Quiescent and Active States 573
    E. Y. Kilpio, D. V. Bisikalo, A. A. Boyarchuk, and O. A. Kuznetsov

Fe II Emission Lines of RR Tel during an Obscuration Event 577
    D. Kotnik-Karuza, M. Friedjung, K. Exter, F. P. Keenan, and D. L. Pollacco

LS 5039 / RX J1826.2-1450: A Young Pulsar? 581
    A. Martocchia, C. Motch, and I. Negueruela

A Refined Method for Measuring Jet Speeds 585
    J. Miller-Jones, K. Blundell, and P. Duffy

Time-Delayed Transfer Functions Simulations for LMXBs 589
    T. Muñoz-Darias, I. G. Martínez-País, and J. Casares

BeppoSAX Observations of the X-Ray Binary Pulsar GX 1+4 593
    S. Naik, P. J. Callanan, and B. Paul

Forced Oscillations in Accretion Disks and kHz QPOs...................... 599
    J. Pétri

Gamma Ray Burst Progenitors... 603
    J. Petrovic and N. Langer

The Light Curve of the Companion to PSR B1957+20 607
    M. Reynolds, P. Callanan, A. Fruchter, M. Torres, M. Beer, and R. Gibbons

Spectroscopic Analysis of the Companion to the Binary MSP PSR
J1740−5340 in NGC 6397 611
    E. Sabbi, F. R. Ferraro, R. Gratton, A. Bragaglia, A. Possenti, and
    N. D'Amico

Radius and Temperature Evolution of the White Dwarf in AS 296
during the 1988-1994 Outburst... 615
    A. Siviero and U. Munari

Numerical Simulations of the Thermal Instability Collapse in
Radiation Pressure Dominated Disks...................................... 619
    V. Teresi, D. Molteni, and E. Toscano

On the Mass Distribution of Neutron Star Masses in HMXBs 623
    A. van der Meer, L. Kaper, M. H. van Kerkwijk, and E. P. J. van den
    Heuvel

Creating Ultra-Compact Binaries through Stable Mass Transfer 627
    M. V. van der Sluys, F. Verbunt, and O. R. Pols

X-Ray Modulation from Non-Axisymmetric Structures in Accretion
Disk.. 631
    P. Varnière, E. Blackman, and M. Muno

Time and Spectral Changes of GRS 1915+105 in the ρ Class 635
    G. Ventura, E. Massaro, T. Mineo, G. Cusumano, M. Litterio, M. Feroci,
    P. Casella, and G. Matt

Chandra Localizations of LMXBs: IR Counterparts and their
Properties... 639
    S. Wachter, J. W. Wellhouse, and R. M. Bandyopadhyay

Discovering Interacting Binaries with Hα Surveys 643
    A. Witham, C. Knigge, J. Drew, P. Groot, R. Greimel, and Q. Parker

Spectroscopy and Near-Infrared Photometry of the Helium Nova
V445 Puppis.. 647
    P. A. Woudt and D. Steeghs

Stabilization of Helium Shell Burning by Rotation in Accreting White
Dwarfs 651
    S.-C. Yoon and N. Langer

White Dwarfs with Jets as Non-Relativistic Analogues of Quasars and
Microquasars?.. 655
    R. Zamanov, M. F. Bode, P. Marziani, R. J. Davis, S. P. S. Eyres,
    A. Gomboc, J. Porter, and A. Skopal

List of Participants.. 659
Author Index.. 667

PREFACE

Ten years after the meeting "Evolutionary links in the Zoo of Interacting Binaries", organized at the Osservatorio Astronomico di Roma in Monteporzio in 1993, the Osservatorio di Roma-Monteporzio and the Osservatorio di Teramo, both belonging to the National Institute for Astrophysics (INAF), together with the Department of Physics and Astronomy of the University of Palermo, decided to organize a second meeting to examine the progress made in the field of interacting binaries. This field covers a number of important astrophysical problems, e.g. accretion onto, and physics of compact objects, end stages of stellar evolution and especially Supernovae Type Ia, the puzzle of γ–ray bursts.

Indeed, the last ten years have dramatically increased the quantity and quality of observational data, enlarging the zoo of sources with new compelling exotic systems. Among these new objects we mention the relativistic binary pulsar PSR J0737-3039, the shortest orbital (?) period binary RX J0806.3+1527, the growing class of millisecond X–ray binary pulsars. The models for Type Ia supernovae made interesting steps forward, and there is a growing interest in these objects as standard candles, as they have provided support for an accelerating universe. Finally, the puzzle of γ-ray bursts has been partially solved, at least for a part of them, firmly establishing their extragalactic origin.

Our aim was to give an updated overview of both observations and theory, drawing together the two communities.

We decided to hold the meeting "Interacting Binaries: accretion, evolution and outcomes" in Cefalù (Sicily), birthplace of two of the organizers. This quite private reason resulted in a fascinating setting for discussions, also outside the beautiful meeting place, the "Sala delle Capriate" belonging to the Comune di Cefalù. The meeting was indeed extremely successfull, both in terms of number of participants (about 150) and for the lively discussions. We think we have accomplished a good job, and everybody returned home full of ideas... and work to do.

We wish to thank the Scientific Organizing Committee, and the Local Organizing Committee who worked hard in the months preceeding the meeting, since this was the first international astrophysics conference planned to be held in Cefalú.

We would like to warmly thank the local authorities who enthusiastically supported the Conference: the Comune di Cefalú and the Fondazione Mandralisca. Other Institutions (Provincia Regionale di Palermo, Azienda Autonoma Provinciale Turismo di Palermo, Ente Parco delle Madonie, Azienda Autonoma Soggiorno e Turismo Cefalù) have sponsored the meeting.

The citizens of Cefalù have enjoyed a parallel public outreach program organized during the evenings of the meeting, concluded by an intriguing public conference on the origin and ultimate fate of the Universe, held by the Director of the Osservatorio di Roma Prof. R. Buonanno. The growing interest towards this kind of events is demonstrated by the fact that the Comune di Cefalù and the Fondazione Mandralisca, in coordination with the scientific institutions (the Observatories of Rome and Teramo plus the University of Palermo, Dept. of Physics and Astronomy) decided to organize a new Astrophysics meeting in 2006, and to sponsor a prize for young researchers in this field.

Finally then: see you in Cefalù in Summer 2006!

Luciano Burderi & Francesca D'Antona

ACKNOWLEDGMENTS

The Editors wish to thank the Director of the Astronomical Observatory of Rome, Prof. Roberto Buonanno, the Director of the Astronomical Observatory of Teramo, Prof. Amedeo Tornambè, and Prof. N. R. Robba of the Department of Physical Sciences of the University of Palermo, for the organizational and financial support without which the meeting, and the publication of this volume, would have been impossible.

Photo courtesy of Luigi Stella.

Speech and jokes at the second social dinner.
Photo courtesy of Luigi Stella.

Photo courtesy of Maria Teresa Menna.

Photo courtesy of Maria Teresa Menna.

Photos courtesy of Maria Teresa Menna.

Population synthesis for low and intermediate mass binaries

L. R. Yungelson

Institute of Astronomy, 48 Pyatnitskaya Str., 119017, Moscow, Russia

Abstract. A review of the basic principles of population synthesis for binary stars is presented. We discuss the break-up of low and intermediate mass close binaries over different evolutionary scenarios and, as an example, briefly consider results of the population synthesis for SN Ia.

Keywords: stars: binary, stars:late stages of evolution, supernovae
PACS: 97.80.-d, 97.60.s, 97.60.Bw

INTRODUCTION

Population synthesis for binary stars is a convolution of the statistical data on initial parameters and birthrates of binaries with the evolutionary scenarios for them. The main goals of population synthesis are:

- to understand the descent of binaries of different types from unevolved main-sequence binaries which differ only in the masses of components M_1, M_2, and their separations a and to understand the links between the objects of different types;
- to estimate the incidence of binaries of different types and occurrence rates of certain *events* in them (for instance, explosions of Novae or Supernovae);
- to find the distributions of binaries of different types over parameters like masses of components, orbital periods, luminosity of components, spatial velocities;
- to model 'observed" samples of stars applying selection effects.

Below, we review the basic principles of population synthesis, discuss the break-up of low and intermediate mass close binaries over different evolutionary scenarios and, as an example, consider results of population synthesis studies for Supernovae Ia (SN Ia). We define *low and intermediate mass binaries as objects in which components end their evolution as white dwarfs*.

EVOLUTIONARY SCENARIOS FOR BINARY STARS

The main concept of the population synthesis is that of an "evolutionary scenario", the sequence of transformations of a binary system with given initial (M_{10}, M_{20}, a_0) that it can experience in Hubble time (the term is coined by van den Heuvel and Heise [1]).

For construction of evolutionary scenarios all binaries are separated into "close" and "wide" ones. In close binaries components interact between themselves, the principal form of interaction is mass exchange through Roche lobe overflow. The possibility for interaction of components in a given binary is defined by critical radius R_{cr}, a boundary within which the atmosphere of the evolving star can expand while star

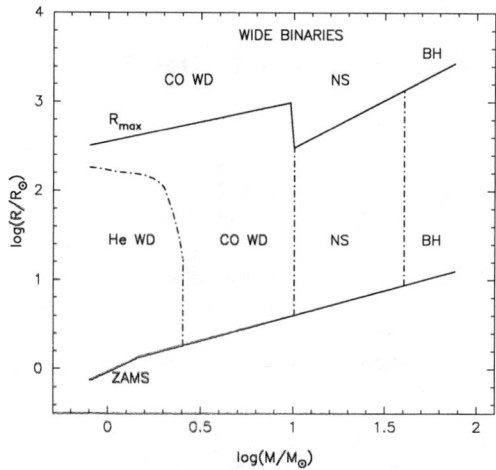

FIGURE 1. Descendants of components of close binaries depending on the radius of star at the instant of RLOF. Close in evolutionary sense stars fill Roche lobes before expanding to R_{max}. For wide binaries the products of evolution are the same as for single stars.

remains a distinctly separate object. If $R_\star > R_{cr}$, stellar matter starts to flow out. Figure 1 sketches ZAMS- and maximum radii of stars and shows descendants of components of binaries depending on their mass and radius at the instant of RLOF (for solar chemical composition stars). The upper mass limit for progenitors of He white dwarfs (WD) is certain to several tenth of $M_\odot$. The boundary between progenitors of WD and neutron stars (NS) is known to about $1 M_\odot$ (it may be different for single stars and components of close binaries). The least certain is the lower cut-off of the masses of progenitors of black holes (BH). It is expected that between 20 and $50 M_\odot$, components of close binaries may produce NS or BH, depending on such parameters as stellar winds, rotation, and magnetic fields [2–4]; the limit of $40 M_\odot$ is sketched in Fig. 1 provisionally.

The data necessary for construction of evolutionary scenarios is provided by stellar evolution computations. The main data are: the rates of mass loss through RLOF or stellar wind (SW) for stars in different evolutionary stages and the nature of the products of evolution; time scales of evolutionary stages; initial–final mass relations; response of stars to accretion; transformations of separations due to mass exchange/loss and angular momentum loss, especially in the common envelope stages.

Evolutionary scenarios for binaries with certain initial parameters may be convolved with star formation rate, binarity rate, distributions of binaries over M_{10}, a_0, $q_0 = M_{20}/M_{10}$ to get birthrates of stars of different types. Convolved with lifetimes this gives number distribution of stars of different kinds at any epoch of the Galactic history, their distributions over basic parameters, and occurrence rates of different events, e. g., SN. Applying selection effects one gets "observed samples" of different binaries. The first numerical population synthesis studies were carried out independently by Kornilov and Lipunov [5] for NS, Dewey and Cordes [6] for radiopulsars, Politano [7] for cataclysmic variables, Lipunov and Postnov [8] for compact magnetized stars.

Binarity rate for stars is not known for sure, it was studied for separate groups of stars only. Studies of selection effects for spectroscopic and visual binaries [9, 10] suggest that binarity rate may be close to 100%.

The studies of spectroscopic and visual binaries [9, 10] suggest that (i) IMF for primary components of binaries may be well approximated by power law $dn/dM \propto M^{-2.5}$; (ii) distribution of binaries is flat in $\log a$. This allows to estimate that among binaries $\sim 40\%$ are close and $\sim 96\%$ are low and intermediate mass ones.

Common envelopes

The crucial problem that plagues all population synthesis studies is uncertainty about mass and angular momentum loss (AML) from the system and especially about *common envelopes* (CE). It is assumed that CE arise when accretor in a close binary is unable to accrete all matter supplied by the donor [11]. The usual assumption is that the envelope of the donor engulfs both components; the latter spiral in, losing energy. This results in reduction of the separation of components and may end either in the merger of components or in the dissipation of CE. This is understood qualitatively, but neither analytical theory nor convincing enough computations of the processes inside common envelopes exist. Therefore, usual approach is based on comparison of the energy released in rapprochement of components and binding energy of CE by applying (most commonly) an equation [12, 13]

$$\frac{M_i(M_i - M_f)}{\lambda R} = \eta_{ce}\left(\frac{M_f m}{2a_f} - \frac{M_i m}{2a_i}\right), \qquad (1)$$

where indexes i and f refer to initial and final masses and separations, m is the mass of accretor, η_{ce} is the efficiency of deposition of energy into common envelope, λ is a function of stellar radius that, in essence, describes how well donor envelope is bound to the core (it varies $\sim 3-4$ times along RG and AGB and depends on the definition of "core" boundary and possible release of thermal energy [14–16]. If whole ensemble of close binaries is considered and Eq. (1) is applied to any system, irrespective to the evolutionary state of components, satisfactory agreement with observations is usually achieved if $\eta_{ce}\lambda \simeq 1-2$.

However, Nelemans and co-authors [17], for instance, have shown that it is impossible to reconstruct evolution of observed close binary helium WD back to main-sequence binaries if Eq. (1) is applied for the stage of unstable mass transfer between components of comparable mass. Instead, it was suggested in [17] to use for this stage an equation based on the angular momentum (J) balance:

$$\frac{\Delta J}{J} = \gamma\left(\frac{\Delta M}{M_d + M_a}\right), \qquad (2)$$

where ΔM is the amount of matter lost by the system. Nelemans and Tout [18] have shown that applying Eq.(2) with $\gamma \simeq 1.5$ it is possible to explain the origin of all known binary WD and pre-CV. Equations (1) and (2) give similar results if masses of components strongly differ [18].

In the case of super-Eddington accretion in the systems with NS or BH accretors formation of common envelopes, perhaps, may be avoided. The energy of the accreted matter released close to the compact object (at r_c) may be sufficient to expel the "excessive" matter from the Roche-lobe surface around the compact object (at $r_{L,c}$) if accretion rate $\dot{M} \leq \dot{M}_{Edd} r_{L,c}/r_c$ (see, e. g., [19]). It is also suggested that common envelopes may be avoided by driving excess of matter by optically-thick winds from the surface of hydrogen-burning accreting white dwarfs [20]. Note, that the model of radiatively driven winds still lacks elaborated theoretical justification, but it is supported by the presence of outflows in many systems with compact components.

If CE formation is avoided as described above, it is usually *assumed* that the matter leaves the system with specific angular momentum of the accreting component. This *assumption* proved to be useful, for instance, for the models of formation of millisecond radiopulsars [21, 22] or ultracompact low-mass X-ray binaries with WD donors [23].

Evolutionary paths of low and intermediate mass close binaries

The character of mass-exchange in a binary depends on the mass ratio of components and the structure of the envelope of Roche-lobe filling star [24, 25]. For stars with radiative envelopes, to the first approximation, mass exchange is stable if $q \lesssim 1.2$; for $1.2 \lesssim q \lesssim 2$ it proceeds in the thermal time scale of the donor; for $q \gtrsim 2$ it proceeds in the dynamical time scale. Mass loss occurs in dynamical time scale if the donor has deep convective envelope. In practice this means that overwhelming majority ($\sim 90\%$) of close binaries pass through one to four stages of common envelope.

A run of population synthesis code (e. g., [26]) for $\sim 1.5 \times 10^6$ close binaries with $0.8 \leq M_{10}/M_\odot \leq 11.4$ produces ~ 600 different scenarios of evolution. In Fig. 2 we present a kind of summary of scenarios for systems that have first unstable RLOF. For description, we denote evolutionary routes in Fig. 2 as "path **N**".

In the tightest systems primaries overflow Roche lobe when they are still in MS. The least massive of these systems ($M_{10} \lesssim 1.5 M_\odot$) evolve under the influence of AML via magnetically coupled stellar wind (MSW) and gravitational waves radiation (GWR). An outcome of evolution may be formation of a W UMa system that merges into rapidly rotating single star (path **1**). About 2% of these systems do not finish their evolution in Hubble time and remain MS-stars or giants. Majority of merger products end their evolution as WD (path **2–3**). Peculiarly, some merger products accumulate mass $\gtrsim 10 M_\odot$ and explode as SN II (path **4**).

In wider systems donors overflow Roche lobes when they have He- or CO-cores. About one third of binaries avoid merger in CE and form pairs consisting of a He- or CO/ONe-WD and a low-mass companion that does not evolve past MS or subgiant stage. The systems remain wide enough to avoid merger due to AML via GWR and/or MSW (path **6**). Systems that merge in CE typically finish their evolution as WD (path **7–8**).

If a binary with a He-core donor does not merge in CE, it first forms a He-star+MS-star pair (path **9**), that may be observed as an sdB star with MS companion. When He-star finishes its evolution, a pair harboring a CO white dwarf and a MS-star appears (path **10**). Similar systems are produced by binaries where RLOF occurs at AGB (path **11**).

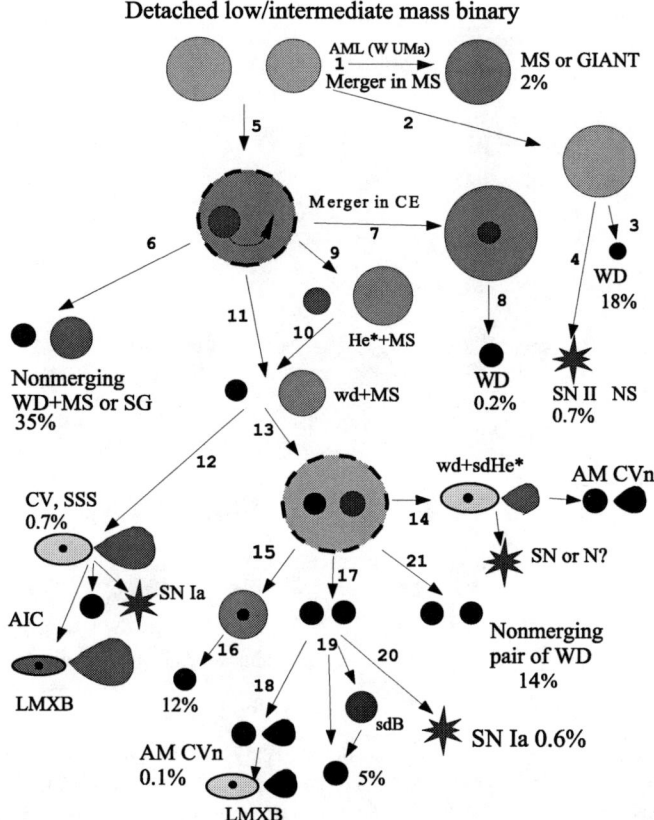

FIGURE 2. Flowchart of the evolution for close binaries that experience unstable first RLOF. The numbers indicate "path" discussed in the text. For some important channels the fraction of all close binaries passing through them is indicated.

The tightest of the latter systems ($a \simeq$ several $R_\odot$) that have low-mass ($\lesssim 1.5 M_\odot$) MS-companions to WD evolve under AML and may form cataclysmic variables (path **12**). If accreted hydrogen burns at the surface of WD stably, the system may appear as supersoft X-ray source (SSS in the plot). The outcome of the evolution of CV is not completely clear. The donor may be disrupted when its mass decreases below several hundredth of $M_\odot$ [27]; accretor may accumulate enough mass to explode as SN Ia (see below); WD may experience an accretion induced collapse into a NS and a low-mass X-ray binary may be formed. Actually, for a considerable fraction of CV evolution is, most probably, frozen when $\dot{M}$ becomes low.

The second common envelope may form when MS-companion to WD overfills its Roche-lobe (path **13**). If the donor has a He-core and system does not merge, an sdB+WD system may emerge (path **14**). If separation of components is sufficiently small, AML via GWR may bring He-star into contact while He is still burn-

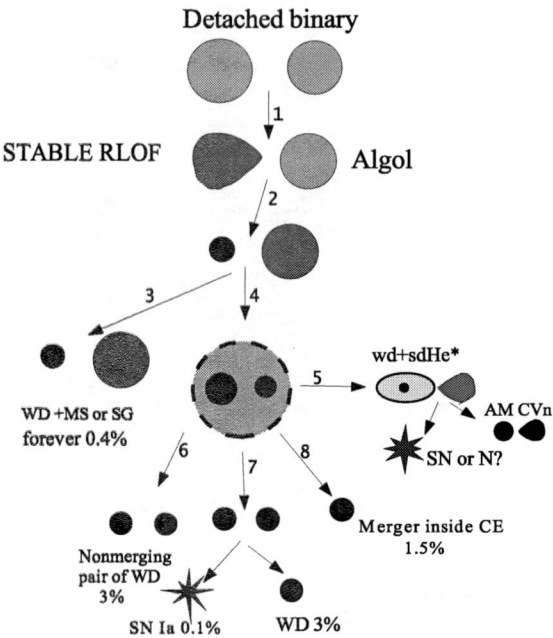

FIGURE 3. Flowchart of the evolution for close binaries that experience stable first RLOF.

ing in its core. If $M_{He}/M_{wd} \lesssim 1.2$ stable mass exchange is possible with typical $\dot{M} \sim 10^{-8}$ $M_\odot$ yr^{-1}[28]. It was suggested that accretion of He onto CO WD at such $\dot{M}$ leads to SN, when accreted He-layer detonates and initiates detonation of C (edge-lit detonation [29]). However, it seems more likely that the lifting effect of rotation reduces the power of He-ignition and a He-nova is produced instead [30]. If the system survives SN or nova explosion, an AM CVn-type system may be formed.

Some 12% of close binaries merge in the second CE and form single WD (path **15–16**). This may create certain excess of massive WD. If the system avoids merger, a pair of WD is formed (path **17**). The closest of them may be brought into contact by AML via GWR. If in a CO+He white dwarf pair conditions for stable mass exchange are fulfilled, an AM CVn system forms (path **18**), see for details [31–33]. About 5% of close binary WD merge (path **19**); it is not yet clear how the merger proceeds and it is possible that for He+He or CO+He pairs a helium star is an intermediate stage [34].

A fraction of close binaries produces merging CO WD pairs ("double-degenerates, DD") with total mass greater than Chandrasekhar mass M_{Ch}(path **20**) that may explode as SN Ia [12, 35, 36]. They are probably the most promising candidates for SN Ia (see below).

Figure 3 shows typical evolutionary scenarios for the systems that stably exchange mass in the first RLOF. It applies to $\sim 10\%$ of all close binaries. In the first RLOF (path **1**) binaries may be identified with Algol-type systems. Algol phase ends by formation of WD+MS-star system directly (path **3**) or with an interfering phase of He-star+MS-star (similarly to (path **9-10**) in Fig. 2). In rare cases WD+MS remain "frozen" in this state (path **3**). RLOF by the former secondary usually results in CE, since $q \gg 1$ (path **4**). If the system did not merge in CE, the donor had a nondegenerate He-core and system is close enough, a semidetached WD+He-star system may arise due to the AML via GWR and in such a system a SN or He-nova may explode (path **5**). In this case formation of an AM CVn star also is possible, like in (path **14**) in Fig. 2. For the systems that avoid merger in CE the following outcomes of evolution become possible: formation of a non-merging in a Hubble time pair of WD (path **6**); formation of a DD that merges due to AML via GWR and possibly produces SN Ia or a massive WD (path **7**). Merger in CE also produces a massive WD (path **8**).

A special case of evolution, that involves binaries with both low/intermediate and high mass components, leads to formation of recycled pulsars. First, in the system with initial $q_0 \sim 10$ the primary becomes a NS. If the system is not disrupted by SN explosion, low-mass secondary after a stage of stable RLOF (when the system is a low-mass X-ray source) becomes a He WD. In another scenario q_0 is larger and, before forming a WD companion to NS, the system passes through CE. In both cases NS are recycled by accretion.

Another peculiar example of evolution involves binaries with $q_0 \sim 1$ and mass of components $\gtrsim (6-7) M_\odot$. If the first mass exchange is stable, initially less massive component accumulates enough mass to explode as SN II. In this case massive white dwarf (former primary) gets a young neutron star companion (see [37, 38] for details).

CANDIDATE SUPERNOVAE IA

Figure 4 summarizes the main evolutionary scenarios that are currently considered to result in formation of candidate SN Ia systems. For completeness, the channel of symbiotic stars is shown, though, they are not "close binaries" strictly. A detailed review of the evolution to SN Ia is recently published in [39]. Here we give a brief outline.

DD-merger scenario (**A**) operates both in young ($\sim 10^9$) yr and old ($\sim 10^{10}$) yr populations [26, 39] and may provide for our Galaxy the rate of SN Ia (ν_{Ia}) comparable to the one inferred from observations [40]: $(4 \pm 2) \times 10^{-3}$ yr^{-1}. The DD-scenario recently got strong support both from the theory and observations. Population synthesis [41] predicted that it may be necessary to study up to ~ 1000 field WD with $V \lesssim 16 \div 17$ for finding a proper merger candidate. This was done within SPY-project and resulted in discovery of the first super-Chandrasekhar DD [$M_t = (1.46 \pm 0.12) M_\odot$] that will merge in 4 Gyr and two DD with mass close to M_{Ch} that will merge in one to two Hubble times [42]. This means that formation of DD candidate SN Ia is quite possible. On the "theoretical" side, SPH calculations of DD-mergers with account for nuclear burning [34] have shown that in the core-disc structure that forms, in the region of interaction the temperature rises rapidly, lifts degeneracy, expands the matter and thus quenches thermonuclear flash. Flame does not propagate to the center, contrary to previously expected (e. g., [43]). Stable accretion from the disk is possible. On the other hand, deposition of the angular momentum into the core spins-up rotation of WD, causes deformation of WD and AML by distorted configuration via GWR and finally makes possible close-to-center explosive ignition of carbon [44, 45].

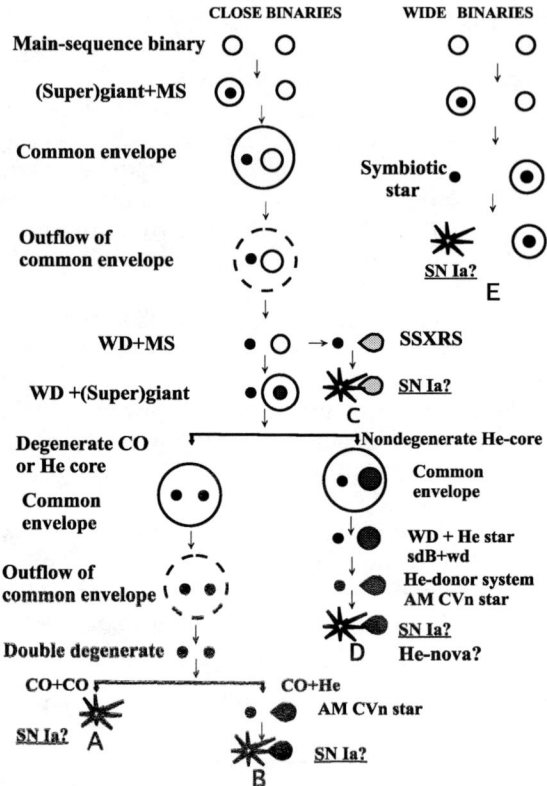

FIGURE 4. Main evolutionary channels resulting in candidate SN Ia systems.

Scenario **B** operates both in young and old populations, but gives a minor contribution to ν_{Ia} since typical total masses of the systems are well below M_{Ch}.

"Single-degenerate" (SD) scenario (**C**) is often considered as the most promising one. Population synthesis estimates lend to it from $\sim 10\%$ [46] to $\sim 50\%$ [47] of the observed ν_{Ia}. The uncertainty in the estimates of ν_{Ia} from this channel stems from still poorly constrained efficiency of mass accumulation in the process of accretion that is accompanied by hydrogen and helium burning flashes [48–50]. The estimates become more favorable if mass transfer may be stabilized by mass and momentum loss from the system. "Observational" problem with this scenario is the absence of hydrogen in the spectra of SN Ia, while it is expected that $\sim 0.15\ M_\odot$ of H-rich matter may be stripped from the companion by the SN shell [51]. Recently discovered SN Ia 2001ic and similar 1997cy with signatures of H in the spectra [52] may belong to the so-called SN 1.5 type or occur in symbiotic systems [53]. Also, under favorable for this scenario conditions, it seems to overproduce supersoft X-ray sources [46]. However, SD-scenario got recently support from discovery of a possible companion to Tycho SN [54].

Scenario **D** may operate in populations where star formation have ceased no more than ~ 1 Gyr ago and produce SN-scale events at the rates that are comparable with the inferred Galactic SN Ia rate. But "by construction" of the model, the most rapidly moving products of explosions have to be He and Ni; this is not observed. The spectra produced by ELD are not compatible with observations of the overwhelming majority of SN Ia [55]. On the theoretical side, it is possible that lifting effect of rotation that reduces effective gravity and degeneracy in the helium layer may prevent detonation [30].

Scenario **E** is the only way to produce SN Ia in a wide system, via accumulation of a He layer for ELD or of M_{Ch} by accretion of stellar wind matter in a symbiotic binary [56]. However, its efficiency is probably low ($\nu_{Ia} \sim 10^{-6}$ yr^{-1}), because of typically low mass of WD and low $\dot{M}$ in symbiotic systems [57].

CONCLUSION

For the conclusion, let us note that, despite evolution of close binary stars is, at least qualitatively, understood and for certain systems population synthesis provides even quantitative agreement with observations, there are some important problems to solve:
- mass and angular momentum loss from the systems (including stellar winds), evolution in common envelopes;
- processes during merger of stars, from main-sequence stars through relativistic objects; evolution of merger products;
- the role of rotation, angular momentum deposition during accretion and associated instabilities;
- mass limits between white dwarfs/neutron stars/black holes.

The author acknowledges financial support from the LOC of the conference and from Russian Foundation for Basic Research (grant no. 03-02-16254).

REFERENCES

1. van den Heuvel, E. P. J., and Heise, J., *Nat*, **239**, 67 (1972).
2. Ergma, E., and van den Heuvel, E. P. J., *A&A*, **331**, L29–L32 (1998).
3. Wellstein, S., and Langer, N., *A&A*, **350**, 148–162 (1999).
4. Fryer, C. L., *ApJ*, **522**, 413–418 (1999).
5. Kornilov, V. G., and Lipunov, V. M., *Soviet Astronomy*, **27**, 334 (1983).
6. Dewey, R. J., and Cordes, J. M., *ApJ*, **321**, 780–798 (1987).
7. Politano, M. J., , Ph.D. thesis, Illinois Univ. at Urbana-Champaign, Savoy (1988).
8. Lipunov, V. M., and Postnov, K. A., *Ap&SS*, **145**, 1–45 (1988).
9. Popova, E. I., Tutukov, A. V., and Yungelson, L. R., *Ap&SS*, **88**, 55–80 (1982).
10. Vereshchagin, S., Tutukov, A., Yungelson, L., Kraicheva, Z., and Popova, E., *Ap&SS*, **142**, 245–254 (1988).
11. Paczyński, B., in *Structure and Evolution of Close Binary Systems*, edited by P. Eggleton, S. Mitton, and J. Whelan, Kluwer, Dordrecht, 1976, p. 75.
12. Webbink, R. F., *ApJ*, **277**, 355 (1984).
13. de Kool, M., *ApJ*, **358**, 189–195 (1990).
14. Han, Z., Podsiadlowski, P., and Eggleton, P. P., *MNRAS*, **270**, 121–130 (1994).
15. Dewi, J. D. M., and Tauris, T. M., *A&A*, **360**, 1043–1051 (2000).
16. Tauris, T., and Dewi, J. D. M., *A&A*, **369**, 170–173 (2001).

17. Nelemans, G., Verbunt, F., Yungelson, L. R., and Portegies Zwart, S. F., *A&A*, **360**, 1011–1018 (2000).
18. Nelemans, G., and Tout, C. A., in *Rev. Mex. Astron. Astrofis. Conf. Ser.*, (2004), pp. 39–40.
19. King, A. R., and Begelman, M. C., *ApJ*, **519**, L169–L171 (1999).
20. Kato, M., and Hachisu, I., *ApJ*, **437**, 802–826 (1994).
21. Tauris, T. M., and Savonije, G. J., *A&A*, **350**, 928–944 (1999).
22. Tauris, T. M., van den Heuvel, E. P. J., and Savonije, G. J., *ApJ*, **530**, L93–L96 (2000).
23. Yungelson, L. R., Nelemans, G., and van den Heuvel, E. P. J., *A&A*, **388**, 546–551 (2002).
24. Tutukov, A. V., Fedorova, A. V., and Yungelson, L. R., *SvAL*, **8**, 365–370 (1982).
25. Hjellming, M. S., and Webbink, R. F., *ApJ*, **318**, 794–808 (1987).
26. Tutukov, A. V., and Yungelson, L. R., *Astronomy Reports*, **46**, 667–683 (2002).
27. Ruderman, M. A., and Shaham, J., *Nat*, **304**, 425–427 (1983).
28. Savonije, G. J., de Kool, M., and van den Heuvel, E. P. J., *A&A*, **155**, 51–57 (1986).
29. Livne, E., and Glasner, A., *ApJ*, **370**, 272–281 (1991).
30. Yoon, S.-C., and Langer, N., *A&A*, **419**, 645–652 (2004).
31. Tutukov, A. V., and Yungelson, L. R., *MNRAS*, **280**, 1035–1045 (1996).
32. Nelemans, G., Portegies Zwart, S. F., Verbunt, F., and Yungelson, L. R., *A&A*, **368**, 939–949 (2001).
33. Marsh, T. R., Nelemans, G., and Steeghs, D., *MNRAS*, **350**, 113–128 (2004).
34. Guerrero, J., García-Berro, E., and Isern, J., *A&A*, **413**, 257–272 (2004).
35. Tutukov, A. V., and Yungelson, L. R., *Nauchn. Informatsii*, **49**, 3 (1981).
36. Iben, I. J., and Tutukov, A. V., *ApJS*, **54**, 331 (1984).
37. Portegies Zwart, S. F., and Yungelson, L. R., *MNRAS*, **309**, 26–30 (1999).
38. Tauris, T. M., and Sennels, T., *A&A*, **355**, 236–244 (2000).
39. Yungelson, L. R., astro-ph/0409677, (2004).
40. Cappellaro, E., and Turatto, M., in *ASSL Vol. 264: The Influence of Binaries on Stellar Population Studies*, 2001, p. 199.
41. Nelemans, G., Yungelson, L. R., Portegies Zwart, S. F., and Verbunt, F., *A&A*, **365**, 491 – 507 (2001).
42. Napiwotzki, R., Christlieb, N., Drechsel, H., Hagen, H.-J., Heber, U., Homeier, D., Karl, C., Koester, D., Leibundgut, B., Marsh, T. R., Moehler, S., Nelemans, G., Pauli, E.-M., Reimers, D., Renzini, A., and Yungelson, L., *The Messenger*, **112**, 25 (2003).
43. Mochkovitch, R., and Livio, M., *A&A*, **236**, 378–384 (1990).
44. Piersanti, L., Gagliardi, S., Iben, I. J., and Tornambé, A., *ApJ*, **598**, 1229–1238 (2003).
45. Piersanti, L., Gagliardi, S., Iben, I. J., and Tornambé, A., *ApJ*, **583**, 885–901 (2003).
46. Fedorova, A. V., Tutukov, A. V., and Yungelson, L. R., *Astron. Lett.*, **30**, 73–85 (2004), astro-ph/0309052.
47. Han, Z., and Podsiadlowski, P., *MNRAS*, **350**, 1301–1309 (2004).
48. Iben, I. J., and Tutukov, A. V., *ApJS*, **105**, 145 (1996).
49. Cassisi, S., Iben, I. J., and Tornambe, A., *ApJ*, **496**, 376 (1998).
50. Piersanti, L., Cassisi, S., Iben, I. J., and Tornambé, A., *ApJ*, **521**, L59–L62 (1999).
51. Marietta, E., Burrows, A., and Fryxell, B., *ApJS*, **128**, 615–650 (2000).
52. Hamuy, M., Phillips, M. M., Suntzeff, N. B., Maza, J., Gonzalez, L. E., Roth, M., Krisciunas, K., Morrell, N., Green, E. M., Persson, S. E., and McCarthy, P. E., *Nat*, **424**, 651 (2003).
53. Chugai, N., and Yungelson, L., *Astron. Lett.*, **30**, 83 – 91 (2004), astro-ph/0308297.
54. Ruiz-Lapuente, P., Comeron, F., Méndez, J., Canal, R., Smartt, S. J., Filippenko, A. V., Kurucz, R. L., Chornock, R., Foley, R. J., Stanishev, V., and Ibata, R., *Nat*, **431**, 1069–1072 (2004).
55. Hoeflich, P., Khokhlov, A., Wheeler, J. C., Phillips, M. M., Suntzeff, N. B., and Hamuy, M., *ApJ*, **472**, L81 (1996).
56. Tutukov, A. V., and Yungelson, L. R., *Astrophysics*, **12**, 521–530 (1976).
57. Yungelson, L., Livio, M., Tutukov, A., and Kenyon, S. J., *ApJ*, **447**, 656 (1995).

SESSION 1

COMPACT BINARIES IN GLOBULAR CLUSTERS

Interacting X-ray Binaries in Globular Clusters: 47Tuc vs. NGC 6397

Jonathan E. Grindlay

Harvard Observatory, 60 Garden St., Cambridge, MA 02138

Abstract. Our deep *Chandra* exposures of 47Tuc and moderate exposures of NGC 6397 reveal a wealth of new phenomena for interacting X-ray binaries (IXBs) in globular clusters. In this (late) Review, updated since the conference, I summarize recent and ongoing analysis of the millisecond pulsars, the compact binaries containing white dwarfs and neutron stars, and the chromospherically active binaries in both globular clusters. Spectral variability analysis enables new insights into source properties and evolutionary history. These binary populations, now so "easily" visible, are large enough that their properties and spatial distributions reveal new hints of compact object formation and binary interactions with their parent cluster. Neutron stars appear overabundant, relative to white dwarfs, in 47Tuc vs. NGC 6397. The IXBs containing neutron stars (i.e., MSPs and qLMXBs), as the most massive and ancient compact binary sample, may trace the protocluster disk in 47Tuc, whereas compact binaries may have been ejected preferentially along the cluster rotation equator during the recent core collapse in NGC 6397.

Keywords: white dwarfs, neutron stars, compact binaries, globular clusters
PACS: 95.85.Nv, 97.30.Qt, 97.60.Gb, 97.60.Jd, 97.80.Gm, 97.80.Jp, 98.20.Gm

INTRODUCTION

Globular clusters continue to delight interacting binary afficionados. Whereas only ~ 30y ago binaries were virtually unknown in globulars and intermediate mass black holes (IMBHs) were thought to be (e.g. Bahcall and Ostriker 1975) the objects responsible for the population of luminous X-ray sources discovered in 4 globular clusters, compact binaries are now known (e.g. Hut et al 1992) to "rule" the dynamics of globular clusters. Still, it is only now becoming clear with the sharp X-ray eye of *Chandra* just how numerous and interactive compact binaries are when in globular clusters. Not only are there the accreting white dwarfs, or cataclysmic variables (CVs), as the dominant population of low luminosity compact X-ray binaries, as well as the (much) smaller population of quiescent low mass X-ray binaries (qLMXBs) – both originally suspected from the original *Einstein* survey (Hertz and Grindlay 1983) – but also the significant population of primordial binaries containing main sequence (and sub-giant) stars detected by their coronal emission as "active binaries" (ABs). More unexpected was the discovery with the first *Chandra*observation of 47Tuc (Grindlay et al 2001a; heraftr GHE01a) that millisecond pulsars (MSPs) are "easily" detected in globulars by their thermal as well as (later recognized for at least 3 of the 19 MSPs with precise locations in47Tuc) non-thermal pulsar wind shock emission. Magnetospheric emission, as from more luminous pulsars, is not generally detected. Other surprises have come from *Chandra* and XMM observations of many other globular clusters (see Verbunt and Lewin 2005 and Verbunt, these proceedings, for recent reviews).

CP797, *Interacting Binaries: Accretion, Evolution, and Outcomes*,
edited by L. Burderi, L. A. Antonelli, F. D'Antona, T. Di Salvo,
G. Luca Israel, L. Piersanti, A. Tornambè, and O. Straniero
© 2005 American Institute of Physics 0-7354-0286-8/05/$22.50

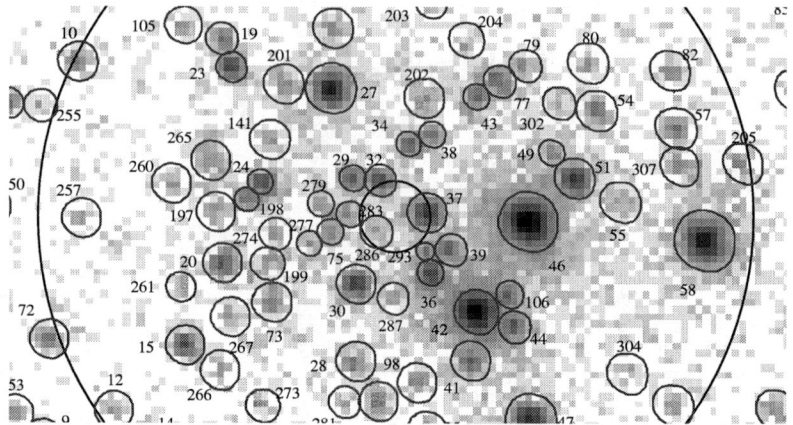

FIGURE 1. *Chandra* image of central region of 47Tuc with circles marking optical core radius (24″) and estimated uncertainty in cluster center position (2.5″). Source W37 is now identified as a qLMXB, leaving only W286 as a contender for a central IMBH.

Here I review the compact binary populations in two particularly interesting, and contrasting, globular clusters, NGC 104 (hereafter 47Tuc) and NGC 6397. I present, or at least touch on, results that suggest the entire suite of low luminosity X-ray sources in globulars – CVs, ABs, MSPs and qLMXBs – are truely interacting X-ray binaries (IXBs), with evidence from each type of source for interactions of the IXB with other cluster members or with the cluster as a whole. Details are given in followup papers.

X-RAY OVERVIEW OF 47TUC

47Tuc remains the "model" globular cluster for a wide range of studies, particularly for IXBs. The very low absorption column $N_H = 1.3 \times 10^{20}$ cm^{-2} allows maximum sensitivity for very soft sources like the thermal emission from the polar caps of MSPs or Polars, and the well-measured cluster dynamics provide a framework for considering sources within the cluster context. Our initial (Mar. 2000) *Chandra* observation (70 ksec, ACIS-I) was presented by GHE01a, and an initial detailed study of the MSPs by Grindlay et al (2002; hereafter GCH02). The rich harvest of IXBs (108 within the central 2.5′×2.5′) merited a deep (4 ×65ksec; Oct. 2002) followup with ACIS-S. Initial results and the overall source catalog are presented by Heinke et al (2005a; hereafter HGE05a, and see Heinke, these Proceedings).

New limits for a central IMBH

One of the new results enabled by the first *Chandra* observation of 47Tuc was the first restrictive X-ray limit for the presence of an IMBH in a globular cluster. Thanks to the detection of hot gas in 47Tuc (the *only* globular in which gas has been detected) by the variable dispersion measure (DM) of its MSPs (Freire et al 2001) and the 0.5″ spatial resolution of *Chandra*, the brightest *Chandra* source with position consistent with being at the precise cluster center allowed a IMBH mass limit of $470 M_\odot$ to be derived assuming Bondi-Hoyle accretion with a (low) efficiency ($\varepsilon \sim 10^{-4}$) advection flow (GHE01a). The limiting source within the $\sim 2.5''$ uncertainty region for the cluster center was source W37, for which the source luminosity was $L_x \sim 1 \times 10^{31}$ erg s^{-1}. This source has since been identified as a probable qLMXB (Heinke et al 2005b; hereafter HGE05b), leaving only source W286 as a contender (W32 is just out of the error circle, but since a IMBH could "wander", it is also a possible candidate and with L_x similar to W37 would yield a similar mass limit; see Fig. 1). The $\sim 10 \times$ lower luminosity of W286 reduces by a factor of ~ 3 the upper limit for IMBH mass (GHE01a),

$$M(IMBH) \lesssim [L_x (0.5\text{-}2.5\text{keV})/(4.5 \times 10^{25} \varepsilon_{-4} T_{100keV})]^{0.5} \sim 150 M_\odot,$$

where L_x is evaluated in the 0.5-2.5 keV band and is 1.2×10^{30} erg s^{-1} for W286 (HGE05a), ε_{-4} is the advection-accretion efficiency in units of 10^{-4}, and T_{100keV} is the advection-accretion temperature in units of 100 keV. The normalization again assumes the ISM in 47Tuc has density 0.1cm^{-3} as suggested by the DM variations of the MSPs. If the MSP winds have evacuated a bubble in the central core of 47Tuc (note that MSP-W = W29, for which the MSP wind is prominent (see below) is very close to the cluster center), then the IMBH mass limit is correspondingly uncertain.

Millisecond Pulsars: 47Tuc-W is not alone?

The initial *Chandra* observation of 47Tuc showed the MSPs to be predominantly soft thermal sources (GHE01a), with 9 of the 15 then located by radio timing detected and several others plausibly detected. Detailed initial studies (Grindlay et al 2002) of the MSP L_x vs. $\dot{E}$ relation showed a significantly flatter dependence ($L_x \propto \dot{E}^{0.5}$) than the linear relation found for the predominantly magnetospheric emission from more luminous pulsars. This work has been extended with an attempt to separate the thermal vs. non-thermal MSP X-ray luminosities vs. $\dot{E}$ (Grindlay 2005) using data from the 2002 deep dataset on 47Tuc. A more detailed study is nearing completion (Bogdanov et al 2005, in preparation; hereafter BGH05).

A key result, first realized at this Cefalu meeting, is that the eclipsing MSP-W, first located from the HST discovery of its optical companion (Edmonds et al 2002), is remarkably similar to the quiescent low mass X-ray binary (qLMXB) and first-discovered accreting millisecond pulsar, J1808-3658. The deep *Chandra* data showed (Bogdanov, Grindlay and van den Berg 2005; hereafter BGvB05) that its X-ray lightcurve shows broad eclipses of its hard flux but no evidence for the sharp and total eclipse expected for the soft thermal component from the NS (Figure 2a). This can be explained by the system geometry shown in Figure 2b: the MSP wind produces a standing shock at (or near) the L1 point, where mass from the secondary is overflowing the Roche lobe, and non-thermal (synchrotron) emission from this shock is then eclipsed by the secondary at binary phases $\phi \sim 0.4$ - 0.6. The longer-rise egress is due to the longer visibility of the

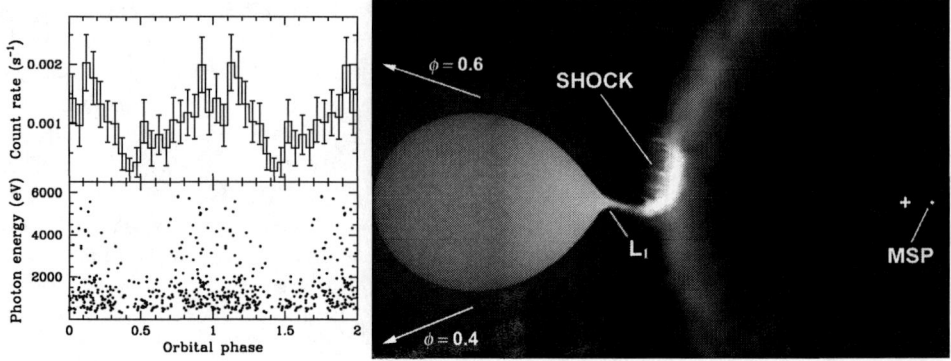

FIGURE 2. *Left:* a) Light curves of MSP-W showing total counts and counts vs. energy folded on the 3.2h binary period. *Right:* b) Sketch of MSP-W to explain eclipse geometry (from BGvB05).

swept-back emission region.

A complete study of the full sample of 19 MSPs with precise locations, now including the new timing positions for MSPs -R and -Y (Freire et al 2005, in preparation), is presented by BGH05 and shows that at least 3 (-W, -J and -O) of the 19 MSPs show significant emission above 2 keV. Certainly for -W, and probably also for the eclipsing systems -J and -O, this is non-thermal (PL) emission from shocked gas excited by the MSP wind. This component for MSP X-ray emission was found (GCH02) to be dominant in the first (and still only) MSP in NGC 6397, N6397-A, which, like that for 47Tuc-W, is probably filling its Roche lobe but not able to accrete. Both have "red straggler" (to red of cluster main sequence) companions, which may indicate mass loss from a sub-giant (N6397-A) or a puffed up envelope for a main sequence star (47Tuc-W). This may be due to the dynamical heating of the companions in their re-exchange. Two of the CVs in NGC 6397 may also have "puffed up" companions given their binary periods vs. inferred companion masses (Taylor et al 2005; hereafter TGE05).

CVs vs. ABs: Distinct spectral variability

Another surprise revealed by the initial 47Tuc observation were the luminosities and spectral hardness of the active binaries. This is extended in the 2002 observation, with ABs making up the largest total population of sources with (likely) optical IDs: extrapolating from the HST coverage used by Edmonds et al (2003) for IDs and extended by HGE05 for initial IDs from 2002 data, the total AB source detections may be as high as 178 vs. 113 for the CVs. Whereas the AB vs. CV X-ray luminosity functions (XLFs) cross at $L_x \sim 10^{30.5}$ erg s^{-1}, with CVs definitely dominant (16:3) for $L_x \gtrsim 10^{31}$ erg s^{-1}, distinguishing these two very different binary populations below this divide is increasingly difficult given the similarly hard spectra (in many cases) and even short-timescale variablity. The latter is revealed by the very similar flare like behaviour of the CV W51 vs. the AB W47, as shown in Fig. 6 of HGE05: both show factor of ~ 20

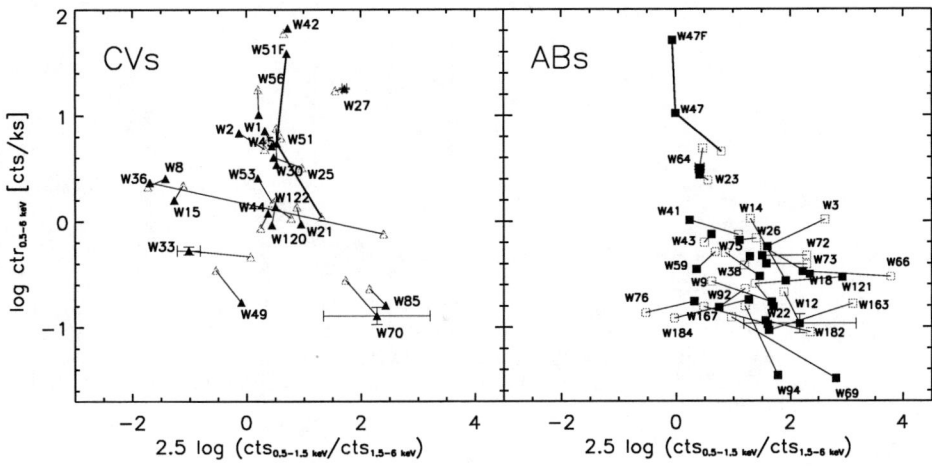

FIGURE 3. X-ray color magnitude diagrams (XCMDs) of *Left:* a) CVs and *Right:* b) ABs in 47Tuc. The 2000 (-I) data (open, green) are connected to the 2002 (-S) data (filled, black) for each source; the large flares for W47 and W51 (*not* removed from the total emission) are connected as a 3rd point ("F").

increases in count rate (0.5-6keV) with rise times ~30min and durations ~2-3h. This is perhaps "typical" for very large flares on BY Dra or RS CVn ABs, but is unprecedented for a CV. It is not clear if this is an accretion instability or how rare such giant CV flares are: W51 had two other smaller flares, each with a ~6×increase and similar timescales, in the third of the four 65ksec obserations in the 2002 data), and only one other CV (W2) had two comparable (~5×increase) closely-spaced flares out of the lightcurves examined for the 22 optically identified CVs for each of the 4 × 65ksec observations.

The flare spectra, however, are very different and may point the way to distinguish ABs from CVs. The AB flare is very hard (vs. the quiescent spectrum), with a Bremsstrahlung fit yielding kT = 169 ±29 keV and absorption column N_H = 7 ×10^{20} cm^{-2} vs. a quiescent spectrum (HGE05) with MEKAL fit of kT = 9 ±1.3 keV and N_H = 11.8 ±1.2 ×10^{20} cm^{-2}. In contrast, the remarkable flare (or is it "super-blobby" accretion?) from the CV W51 is fit with (Brems) kT = 3.5 ±1 keV and N_H = 2 ×10^{20} cm^{-2}, or *softer* than the quiescent emission (HGE05) with (MEKAL) kT = 5.7 ±0.8 keV and N_H = 4.8 ±1.1 ×10^{20} cm^{-2}. Harder spectra are typical of stellar (or solar) flares, where non-thermal processes dominate, whereas an accretion instability and higher $\dot{m}$ for a CV would be expected to be more optically thick and thus softer.

Spectral variability differences for the identified CVs vs. ABs in 47Tuc should be evident in XCMDs for the 2000 vs. 2002 observations (Figure 3). The ACIS-I counts have been transformed to those expected for ACIS-S for the nominal spectra for quiescent emission, taken to be (Brems) kT = 10 keV (CVs) vs. kT = 1 keV (ABs). Comparing sources with at least one of the two measurments $\gtrsim$1ct(0.5-8keV)/ksec for minimal error bars, the 18 CVs show 8:6:4 with positive:negative:uncertain slopes vs. 2:4:0 for the 6 ABs, so the differences are only suggestive.

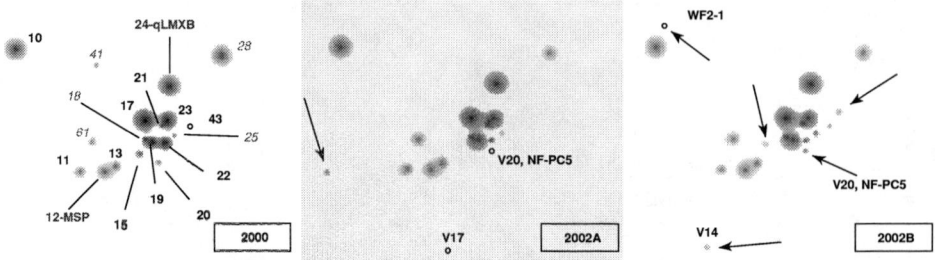

FIGURE 4. *Chandra* images of central region of NGC 6397 from ACIS-I (left; see GHE01b) and -S (middle and right; see GvBB05). Sources are numbered as in GHE01b and new sources are marked. CV source numbers are blue, ABs are black, the MSP and qLMXB are red, and unidentified source *numbers* are magenta. Scale: CVs U17 and U23 are 10.0″ apart. The cluster center is about 1″ from U19/CV2.

VS. NGC 6397

NGC 6397 is the perfect "foil" for 47Tuc: it is core collapsed, with a power law cusp in its core flattening to a core radius with the latest estimate from HST (for main sequence stars) as $r_c = 4.4 \pm 3.2''$ (TGE05) rather than the "perfect" King model with $r_c = 24''$ for 47Tuc. Given the factor of $\sim$2 closer distance of NGC 6397 (2.3kpc), the isothermal core is $\sim$8×smaller in radius than that for 47Tuc and thus for the quoted stellar luminosity density (log $\rho_L \sim$5.68 vs. 4.81 for 47Tuc) has a core mass (and IXB factory) some $\sim$400×smaller than that of 47 Tuc. It is metal-poor ([Fe/H] = -2.0 vs. -0.7 for 47Tuc) and has absolute magnitude -6.63 vs. -9.42 for 47Tuc, suggesting a mass ratio of 13 for constant M/L. Thus it provides a contrast in its dynamical history, metallicity and mass.

Our initial 50 ksec observation of NGC 6397 with ACIS-I in July 2000 (Grindlay et al 2001b; hereafter GHE01b) was followed up with a comparable exposure (2 × 25 ksec) with ACIS-S on May 13 and 15, 2002. Details are reported in Grindlay et al (2005; hereafter GvBB05); highlights for comparison with 47Tuc are reported here. The Wavdetect images of the ACIS-I vs. -S exposures are shown in Figure 4. In the core region shown, sources U15, U20 and U41 are detected only in the ACIS-I observation and 6 new sources (marked with arrows) are detected in one or both of the -S exposures. The Vxx designations mark identifications with variable stars in the cluster reported by Kaluzny and Thompson (2003) and references therein. Open circles mark lower-threshold detections with one either an AB (V20) or possibly a "non-flickerer" (NF-PC5; a probable He WD, see Taylor et al 2001) – both are within the 95% confidence radius ($\sim$0.5″). Additional source details and identifications with HST stars are given in GvBB05 and TGE05. Additional sources of note are outside the field shown (e.g. U60, recently identified with HST by TGE05 as the 9th CV, is along the U24 - U12 line to the SE). Source U28, tentatively classified as a CV by GHE01b from its hard spectrum, is in fact identified by Cool et al (in preparation) as a background edge-on Seyfert galaxy(!) so that the cluster CV total from HST/WFPC2 identifications (TGE05) remains at 9.

Clearly many sources are highly variable. U22 = CV5 decreased its X-ray luminosity by a factor of $\sim$10 and the ABs show large variations. The XCMD for the ACIS-I

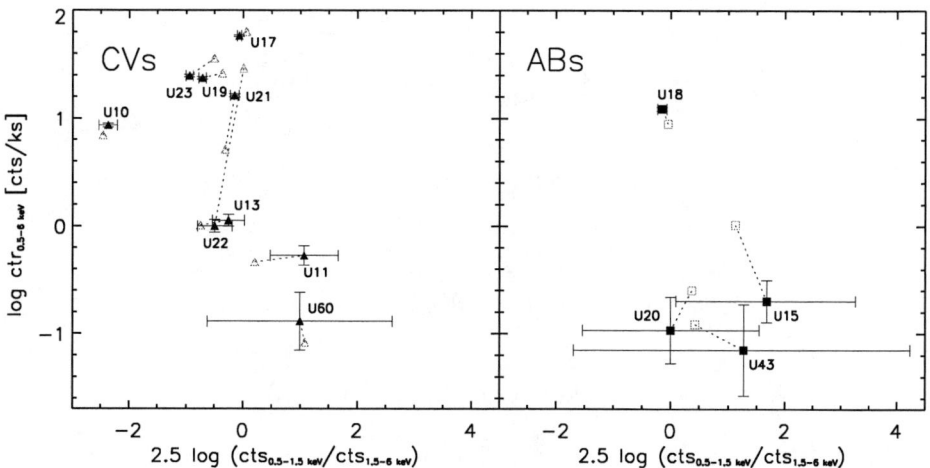

FIGURE 5. X-ray color magnitude diagrams (XCMDs) of *Left:* a) confirmed CVs and *Right:* b) candidate ABs in NGC 6397. Colors and labels for the 2000 vs. 2002 data are as in Fig. 3.

vs. -S spectral variability of the CVs and HST-identified (TGE05) CVs is shown in Figure 5 for comparison with the 47Tuc plot (Fig. 3). Although there are fewer objects, the CVs now all show brighter-softer variations and the ABs (with larger errors) are again brighter-harder. All four CVs (CV1/U23, CV2/U19, CV4/U21 and CV6/U10) with binary periods discovered with HST (TGE05) show significant modulation in the combined *Chandra* data; details are discused by GvBB05.

DO IXBS TRACE CLUSTER FORMATION AND EVOLUTION?

The IXB populations and spatial distributions in both clusters can be compared for constraints on compact object populations and IXB formation/evolution. First, since the CVs and qLMXB/MSP systems are both significantly over-produced, by factors of $\gtrsim 10$, in these clusters vs. the field populations (see GvBB05), both are produced in exchange encounters (primarily) with primordial binaries or, particularly during core collapse for NGC 6397, by tidal capture. NSs are more concentrated in the core and thus favored in IXB production, and lower mass WDs may be more easily expelled than NSs in the core collapse that has occurred in NGC 6397. Thus if both clusters have the same initial mass functions (IMFs) and thus ultimately ratios of white dwarfs (WDs) to neutron stars (NSs), the ratios of IXBs containing each should be similar in both clusters, with perhaps more NS-IXBs expected in NGC 6397.

TABLE 1. Comparison IXB counts: 47Tuc vs. NGC 6397

	Source Type	47Tuc	NGC 6397
Observed:	qLMXBs (NSs)	5	1
	MSPs (NSs)	$\sim 30^*$	~ 2
	CVs (WDs)	~ 30	~ 12
	NSs/WDs in IXBs	~ 1	~ 0.25
Derived:	Γ_c (rel. coll. rate)	~ 3	~ 0.3
	M_{GC} (rel. total mass)	~ 13	~ 1
	(NS/WD) / (Γ_c/M_{GC})	~ 5	~ 0.8

* MSP and CV numbers for both clusters are estimated totals

NSs vs. WDs in IXBs: cluster IMF?

In fact, as we originally suspected (GHE01b), the relative numbers of WDs vs. NSs locked up in IXBs in 47Tuc vs. NGC 6397 are surely not the same. In Table 1 we summarize the observed IXBs for both clusters and derive the ratio of expected IXBs containing WDs vs. NSs *normalized* by the ratio of relative collision number, or rate of IXB production, per unit cluster mass. The scaling for collision number, in a cluster core with density ρ_c and core radius r_c is $\Gamma_c \propto \rho_c^{1.5} r_c^2$, is taken from Verbunt (2003) and Heinke et al (2003). The bottom line is that the NS/WD ratio in IXB systems appears to be enhanced by a factor of ~ 6 in 47Tuc vs. NGC 6397. It is conceivable that the core collapse in NGC 6397 would favor production of WD-IXBs by tidal capture simply because WDs are so much more numerous than NSs in any cluster. However binary "burning" to halt core collapse presumably produces a net loss of binaries in the core (and indeed the ABs appear in Figure 4 to be relatively deficient near the cluster center and have a core radius measured (TGE05) to be significantly larger ($r_c = 9.3 \pm 3.5''$) than the CVs ($r_c = 1.0 \pm 3.5$/arcsec). If the NS/WD ratio difference is due to CV production in core collapse, then the youngest cluster CVs should be in the central core. It is interesting that the three optically faintest CVs (7, 8 and 9 = U11, U13, and U60), all with absolute magnitudes Mv ~ 11.5 - 12 and companion masses thus $\sim 0.1 M_\odot$ (TGE05) are indeed farther out from the cluster center as expected if CVs 1-6 were more recently created.

If CV (vs. LMXB) production is not favored in core collapse, then the lower NS/WD ratio in NGC 6397 points to the underlying NS population being deficient. This could be (partly) due to the lower escape velocity for its reduced cluster mass, but the present mass may be greatly reduced by tidal stripping and so the initial masses are obviously uncertain. More likely, the large metallicity difference suggests a much flatter IMF for 47Tuc and thus NS initial production, as originally suggested by Grindlay (1987) for the luminous galactic globular cluster LMXBs and as now suggested for the pronounced metallicity dependence of extragalactic cluster LMXBs (e.g., Jordan et al 2004).

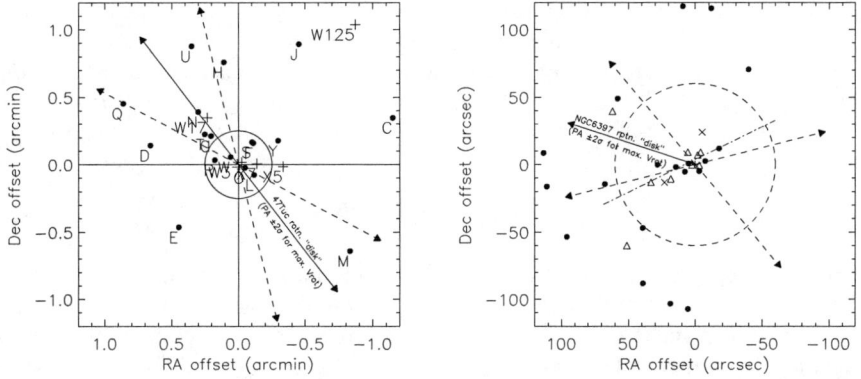

FIGURE 6. IXB sources locations vs. cluster rotation equator for *Left:* a) MSPs (filled dots) and qLMXBs (+) in 47Tuc; circle marks optical core radius, and *Right:* b) all sources in NGC 6397: triangles=CVs, X=MSP and qLMXB, and filled dots= ABs and unIDs. The source "equatorial" plane is fit with the dot-dashed line for sources inside the circle; exterior sources are increasingly background objects.

Interacting binaries vs. cluster rotation?

As noted in GHE01b and as now even more apparent in Figure 4, the IXBs in NGC 6397 appear possibly anisotropic: at radii $\lesssim 1'$ from the cluster center they are predominantly scattered along a NW-SE "line" which is matched by the "line" of 5-6 luminous blue stragglers in the core. With more sources (mostly ABs) now detected in the cluster (Figure 4), this trend is strengthened (e.g. 5 of the 6 "new" sources are roughly along this axis). Very similar cluster core flattening of the *Chandra* source distributions are apparent for the globulars NGC 6440 and NGC 6266 presented by Pooley et al (2002). 47Tuc also shows an apparent anisotropy of its NS-IXBs (MSPs and qLMXBs), as shown by Grindlay (2005) who examined correlations with the cluster proper motion.

A better (or more plausible) possible interpretation, if any of these anisotropies are significant (and simulations are planned) is to consider cluster rotation. Kim, Lee and Spurzem (2004) have found that core collapse and mass segregation are enhanced by cluster rotation. In Figure 6 the NS-IXBs for 47Tuc and the complete IXB sample for NGC 6397 are shown with the cluster rotation equator directions ($\pm 2\sigma$) from the rotation measurements of Gephardt et al (1995) marked. For 47Tuc, the bulk of the MSPs and qLMXBs are within the $\pm 2\sigma$ region around the rotational equator, and for NGC 6397 the flattened core region is at $\sim 2.5\sigma$ from the rotation equator line (but also, given the large uncertainties, $\sim 2.5\sigma$ from the rotation axis!). The position angle (PA) of the rotation velocity equator for NGC 6397 is very uncertain and only measured in integrated light; for 47Tuc the PA is in excellent agreement for both integrated light and individual stellar velocities. Large stellar velocity samples are needed to measure the stellar rotation for NGC 6397 (and NGC 6440 and NGC 6266) to test these alignments.

For 47Tuc, the implication of alignment of the NS-IXBs could be that the NSs, as the oldest objects in the cluster and which may have formed in the proto-cluster disk, have retained their mean angular orbital momentum despite having acquired companions in exchange collisions. Angular momentum alignment would remain fixed in the cluster frame, which was not necessarily the case for anisotropies induced by the cluster proper motion (Grindlay 2005). For NGC 6397, the most natural expectation is that rotational flattening during core collapse has scattered IXBs preferentially in the equatorial plane.

CONCLUSIONS

Interacting binaries in globular clusters are allowing new domains of stellar and binary evolution to be studied: from clues to the formation of the very first massive stars (and NSs), to the oldest and least luminous CVs to the extremes of stellar binaries. They point the way to new dynamical phenomena, including re-re-cycling and alignment processes with cluster angular momentum. High resolution *Chandra* imaging has provided the key for new understanding and new questions.

ACKNOWLEDGMENTS

I thank Maureen van den Berg, Slavko Bogdanov, and Craig Heinke for their many key contributions to the analysis summarized here. This work was supported in part by *Chandra* grant GO2-3059A and HST grant GO-0944.

REFERENCES

1. Bahcall, J. and Ostriker, J. 1975, Nature, 256, 23
2. Bogdanov, S., Grindlay, J. and van den Berg, M. 2005, ApJ, in press (BGvB05)
3. Edmonds, P.D. et al 2002, ApJ, 579, 741
4. Edmonds, P.D., Gilliland, R.L, Heinke, C.O. and Grindlay, J.E. 2003, ApJ, 596, 1177
5. Freire, P. et al 2001, MNRAS, 326, 901
6. Gephardt, K., Pryor, C., Williams, T. and Hesser, J.E. 1995, AJ, 110, 1699
7. Grindlay, J.E. 1987, Proc. IAU Symp. 125, pp. 173 - 185 (Reidel)
8. Grindlay, J.E. Heinke, C., Edmonds, P.D., and Murray, S.S. 2001a, Science, 292, 2290 (GHE01a)
9. Grindlay, J.E., Heinke, C.O., Edmonds, P.D. et al 2001b, ApJ, 563, L53 (GHE01b)
10. Grindlay, J.E., Camilo, F., Heinke, C.O. et al 2002 ApJ, 81, 470 (GCH02)
11. Grindlay, J.E. 2005, Proc. Aspen Workshop on Binary Pulsars, ASP Conf. Proc., in press
12. Heinke, C. et al 2003, ApJ, 508, 501
13. Heinke, C. et al 2005a, ApJ, 625, 796 (HGE05a)
14. Heinke, C. et al 2005b, ApJ, 622, 556 (HGE05b)
15. Hertz, P. and Grindlay, J.E. 1983, ApJ, 275, 105
16. Hut, P. et al 1992, PASP, 104, 981
17. Jordan, A. et al 2004, ApJ, 613, 279
18. Kaluzny, J. and Thompson, I. 2003, AJ, 302, 757
19. Kim, E., Lee, H. and Spurzem, R. 2004, MNRAS, 351, 220
20. Pooley, D. et al 2002, ApJ, 569, 405
21. Taylor, J.M., Grindlay, J.E., Edmonds, P.D. and Cool, A.M. 2001 ApJ, 553, L169
22. Taylor, J.M., Grindlay, J.E., Edmonds, P.D. et al 2005, ApJ, submitted (TGE05)
23. Verbunt, F. 2003, ASP Conf. Ser. 296, 245
24. Verbunt, F. and Lewin, W. 2005, in Compact Stellar X-ray Sources (W. Lewin and M. van der Klis, eds.), Cambridge Press, in press

X-ray sources in globular clusters of other galaxies

Walter H.G. Lewin* and Frank Verbunt[†]

*Massachusetts Institute of Technology, Physics Department Center for Space Research, MA 02139, USA
[†]Astronomical Institute, Postbox 80.000, 3508 TA Utrecht, the Netherlands

Abstract. A large number of X-ray sources in globular clusters of galaxies other than the Milky Way has been found with Chandra. We discuss three issues relating to these sources. The X-ray luminosity function (XLF) of the sources in globular clusters of M31 is marginally compatible with the XLF of globular clusters of the Milky Way. The individual XLFs of a dozen elliptical galaxies, *after correction for incompleteness*, are compatible with one another and show no break; however, the XLF found by adding the individual XLFs of elliptical galaxies has a break at $L_x \simeq 5 \times 10^{38}$ erg s^{-1}. For the moment there is no evidence for a difference between the XLFs of sources inside and outside globular clusters of elliptical galaxies. It is not (yet?) possible to decide which fraction of low-mass X-ray binaries in elliptical galaxies outside globular clusters have formed inside globular clusters.

Keywords: Globular clusters, X-ray sources
PACS: 98.20.Gm, 97.80.Jp

INTRODUCTION

Observations with Chandra of an increasing number of nearby galaxies are revealing a sizable population of bright X-ray sources, many of which are in globular clusters. In elliptical galaxies, as many as half of all bright X-ray sources are in globular clusters; in spiral galaxies a smaller fraction of the bright sources is in globular clusters. The absence of recent star formation in elliptical galaxies implies that the X-ray sources in them are mainly low-mass X-ray binaries. Some questions that have arisen from early research are whether globular clusters do contain black holes? and whether the sources in elliptical galaxies outside globular clusters originate in globular clusters?

Elliptical galaxies are a good target for the study of X-ray sources in globular clusters, because many have 5000 to 10 000 globular clusters. The central galaxy of the Virgo cluster, M 87, has 13500 globular clusters! The X-ray sources detected so far in the faraway galaxies are of necessity *very* bright, typically $L_x > 5 \times 10^{37}$ erg s^{-1}. Early studies indicate that about 4% of the globular clusters contains an X-ray source above this limit, roughly corresponding to 1 source per $5 \times 10^6 L_{\odot,I}$ (Sarazin et al. 2003, Kundu et al. 2003).

We have given an overview of the early papers on this topic in Verbunt & Lewin (2005). Comparison between studies is difficult, as they have different detection limits and compute X-ray luminosities in different energy bands based on different assumed spectra. The study of Kim & Fabbiano (2004) remedies this, and provides a first consistent comparison between a fair number (14) of different elliptical galaxies; we discuss

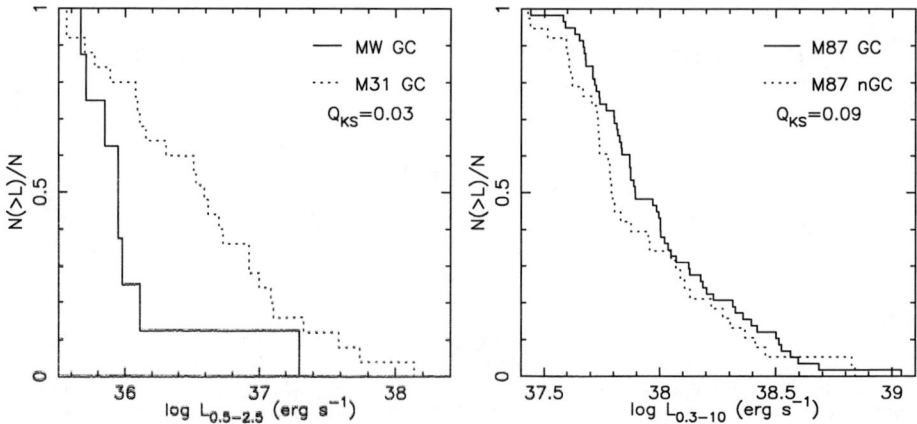

FIGURE 1. *Left: Comparison of the normalized cumulative X-ray luminosity functions (at $L_x > 10^{35.5}$ erg s^{-1}) of globular clusters of our Milky Way and of M 31. The Chandra luminosities given by Di Stefano et al. (2002) were multiplied by 0.46 to convert them to the energy range of ROSAT data from Verbunt et al. (1995). After Verbunt & Lewin (2005). Right: Comparison of the normalized cumulative X-ray luminosity functions (at $L_x > 10^{37.4}$ erg s^{-1}) of M87 sources in globular clusters and outside them. After Jordán et al. (2004). The Kolmogorov-Smirnov probability that the normalized distributions are the same is 0.03 for the Milky Way vs. M 31, and 0.09 for in vs. out globular clusters of M 87.*

this important study in Sect. 3. In Sect. 2 we compare the X-ray luminosities of globular clusters in our galaxy and in the Andromea nebula. In Sect. 4 we discuss the question whether sources outside globular clusters can have formed inside them.

GLOBULAR CLUSTERS IN THE MILKY WAY AND IN M31: DIFFERENT OR THE SAME?

The Andromeda Nebula, a.k.a. M 31, was found with Einstein to have rather more bright X-ray sources in globular clusters than our Milky Way (Van Speybroeck et al. 1979). Since this discovery it has been debated whether this simply reflects the larger number of globular clusters of M 31, or an X-ray luminosity function with a higher fraction of bright sources in M 31 (e.g. Verbunt et al. 1984). The debate continued in the ROSAT and Chandra era (Magnier et al. 1992, Supper et al. 1997, Di Stefano et al. 2002). In Figure 1 we show the cumulative X-ray luminosity functions for globular clusters of our Milky Way and of M 31. A 2-sided Kolmogorov-Smirnov test gives a probability of 3% that both distributions are the same. Thus, the evidence that the luminosity functions are intrinsically different is at the 2-sigma level: suggestive, but not conclusive.

The X-ray luminosity functions of the globular clusters in M 31, and the bulge in M 31 (Kong et al. 2002, 2003) are both roughly compatible with the X-ray luminosity function of elliptical galaxies (see next Section), an indication that the X-ray sources in all these old stellar populations are similar. (With a population of eight persistent sources and 5 transients, the globular cluster system of our galaxy has too few bright X-ray sources to constrain the slope of luminosity function well.)

X-RAY LUMINOSITY FUNCTIONS OF GLOBULAR CLUSTER SYSTEMS OF ELLIPTICAL GALAXIES

Kim & Fabbiano (2003) investigated the incompleteness at the low-luminosity end of the X-ray luminosity function of the elliptical galaxy NGC 1316, and showed that the apparent break in the XLF disappears when appropriate corrections are made. They then investigated XLFs of 13 more elliptical galaxies, and showed that none of these shows a significant break after correction (Kim & Fabbiano 2004, see also Gilfanov 2004). The elliptical galaxies comprise 7 galaxies of the Virgo cluster (at 17 Mpc), 2 of the Fornax cluster (19.9 Mpc), and 5 others (between 11 and 29 Mpc). The effects leading to incomplete detection efficiency at low X-ray luminosities are

- the presence of diffuse emission of the hot interstellar medium in E and S0 galaxies
- the Eddington bias, enhancing the number of sources near the detection threshold
- source confusion
- the larger point-spread function near the detector edge

The X-ray luminosities are determined in the energy range 0.3-8.0 keV; and sources are counted outside the innermost region of each galaxy ($r > 20''$) but within the 25 mag isophote. Corrections for the source numbers and total X-ray luminosity of the galaxy are made on the assumption that they scale as the optical luminosity, i.e. with $r^{1/4}$: this affects the normalization, but not the slope of the luminosity function. It should be noted that Kim & Fabbiano (2003, 2004) do not discriminate between sources within and outside globular clusters, and give the XLFs of all sources related to elliptical galaxies. They find that the XLF of every individual elliptical galaxy is compatible with a power law $N(L_x) \propto L_x^{-\beta}$, with $\beta = 2.0 \pm 0.2$. The variation between the β-values of different ellipticals can be ascribed to small-number statistics. In studies of individual elliptical galaxies it is found that the XLF for sources in globular clusters is the same as the XLF for sources outside globular clusters (Maccarone et al. 2003, Sarazin et al. 2003, Jordán et al. 2004, see Fig. 1). Thus, we may take the overall XLF as a proxy for the XLF of the globular cluster sources.

Since the individual X-ray luminosity functions appear to be the same, they can be added. The added XLF, at luminosities in the 0.3-8.0 keV band $L_x > 6 \times 10^{37}$ erg s^{-1}, of all 14 elliptical galaxies is marginally compatible with a single power law with $\beta = 2.1 \pm 0.1$. A broken power law gives a somewhat better fit, and is shown in Figure 2. The best-fit break luminosity is $(5.6 \pm 1.6) \times 10^{38}$ erg s^{-1}, and the values for β at lower and higher luminosities are 1.8 ± 0.2 and 2.8 ± 0.6, respectively. Remarkably, the break is at about the flux where the luminosity function of AGNs and quasars also shows a break. Whereas background sources typically comprise about 5% of the sources within the 25 mag isophote, they are unlikely to cause the break in the added XLF of the elliptical galaxies, which has been corrected for the background sources (Kim & Fabbiano 2004).

The break is at a luminosity which is about twice the Eddington limit for hydrogen-rich material, and comparable to the Eddington limit for hydrogen-poor material (see

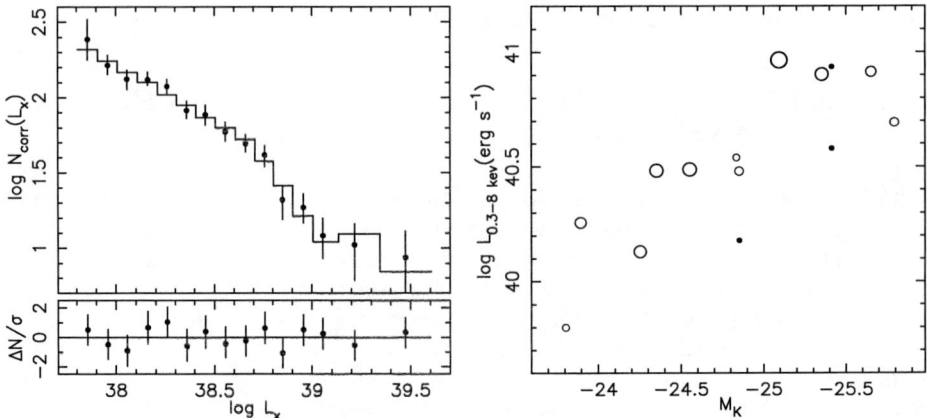

FIGURE 2. *Left: total corrected number of sources in 14 elliptical galaxies as a function of X-ray luminosity in the 0.3-8.0 keV range. Right: corrected total X-ray luminosity of elliptical galaxies as a function of their absolute K-band magnitude M_K. The total X-ray luminosity is computed from the corrected X-ray luminosity function, extrapolated downwards to $L_x = 10^{37}$ erg s^{-1}. The size of the symbols ∘ scales with the number of globular clusters per unit luminosity of the galaxy S_N. • indicates a galaxy for which S_N is not known. (After Kim & Fabbiano 2004.)*

also Kuulkers et al. 2003). From their high X-ray luminosity and soft X-ray spectrum, it has been argued that the sources above the break are accreting black holes (e.g. Angelini et al. 2001). As regards the sources below the break, Bildsten & Deloye (2004) note that the assumption that most very luminous X-ray sources are ultracompact binaries, in which the donor star is a hydrogen-depleted degenerate star, explains the location of the break at the Eddington limit of hydrogen-poor material, the high X-ray luminosities, the source incidence (or birth rate), and the X-ray luminosity function. The model does not explain why sources outside globular clusters, which are less likely to be ultracompact binaries (if our own Milky Way is a guide), have the same X-ray luminosity function. This brings us to the question whether sources outside globular clusters could have formed inside them.

OUTSIDE AND INSIDE GLOBULAR CLUSTERS

In elliptical galaxies it is found that the spatial distribution of X-ray sources outside globular clusters is similar to the spatial distribution of globular cluster X-ray sources (e.g. Jordán et al. 2004 for M 87). Similarly, the X-ray luminosity functions of globular cluster sources and other sources are very similar (Fig. 1). This has led to the revival of a suggestion by Grindlay & Hertz (1985) that all low-mass X-ray binaries, including those now outside globular clusters, were formed inside globular clusters (White et al. 2002).

In the Milky Way and in M 31 there are about 10 bright low-mass X-ray binaries in the disk for each one in a globular cluster. In elliptical galaxies, there is about 1 source outside globular clusters for each one in them. This suggests that the majority of low-mass X-ray binaries in the disk of the Milky Way, M 31 and by generalization in spiral galaxies are formed in the disk, and not in globular clusters (Verbunt & Lewin 2005). It should be noted that we cannot compare fractions inside and outside globular clusters for the Milky Way and M 31 on one hand and elliptical galaxies at the other hand *at the same luminosities*, because there are very few X-ray sources in the Milky Way and M 31 above $L_x \sim 5 \times 10^{37}$ erg s^{-1}, which is the lower limit of detectable sources in ellipticals (see Fig. 1).

The total X-ray luminosity of low-mass X-ray sources in elliptical galaxies scales with the luminosity (hence presumably stellar mass) of the elliptical galaxy (Figure 2). This is expected for sources formed outside globular clusters. It is also expected if the sources are mainly formed in globular clusters, because the number of globular clusters is on average higher in large elliptical galaxies than in small ones. Correlations of the total X-ray luminosity have been made with the specific frequency S_N of globular clusters (i.e. the total number of globular clusters divided by the luminosity of the galaxy). We refer to this here as the "global" specific frequency which includes all globular clusters in the entire galaxy. These correlations greatly suffer from a serious problem related to our lack of knowledge of the global S_N. Reliable values for the number of globular clusters, in general, come from HST-WFPC2 with a very small (5.7 square arc minutes) field of view. This provides only a reliable local value for S_N but not the global value (see the discussion in Section 8.3 of Verbunt & Lewin 2005).

Kim & Fabbiano (2004) note that the correlation between the total X-ray luminosity (of low-mass X-ray sources) and the infrared luminosity of elliptical galaxies has a scatter which is larger than the measurement accuracy. In Fig. 2 we indicate the elliptical galaxies with a large S_N with a large symbol. We see that the elliptical galaxies brighter in X-rays than the average correlation indeed have a tendency to have a higher S_N. This indicates that globular cluster sources contribute significantly to the total X-ray luminosity of the galaxy; to argue that sources outside globular clusters originate inside globular clusters, one would have to show that the total X-ray luminosity (or better, number) of the sources *outside* globular clusters separately scales with specific frequency S_N. In the sample of galaxies studied by Kim & Fabbiano (2004) there is *no* correlation between M_K and S_N (see Fig. 2).

CONCLUSIONS AND OUTLOOK

The conclusions drawn above, i.e. that the X-ray luminosity function of the X-ray sources in old populations is universal and has a break at $L_x \simeq 5 \times 10^{38}$ erg s^{-1}, are based on a study which does not discriminate between sources inside and outside globular clusters. With the increasingly large sample of sources it is possible and necessary to repeat this study for the globular cluster sources separately. This may also help in adressing the question which fraction of the sources outside globular clusters were actually formed inside them. The location of the break suggests that the sources above the break are accreting black holes. The sources below the break may be dominated by ultracompact sources (Bildsten & Deloye 2004).

There are two important areas of further study not discussed above. The first is the question why clusters with a high metallicity have a higher probability of containing

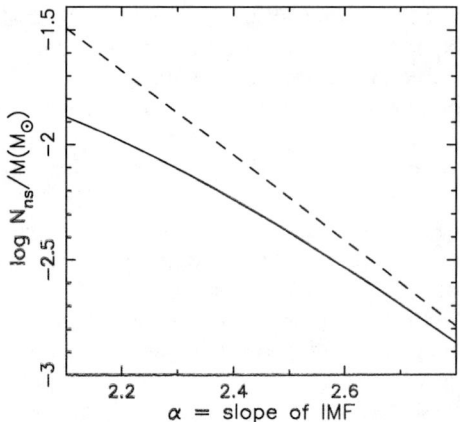

FIGURE 3. *The number of neutron stars per solar mass formed from an initial mass distribution $N(m) \propto m^{-\alpha}$ as a function of the index α. The initial range of stellar masses is taken to be from 0.08 to 80 $M_\odot$, and of neutron-star progenitors from 8 to 80 $M_\odot$. The solid curve is normalized on the initial mass of the cluster, the dashed curve on the current luminous mass, assuming that stars from 0.8 to 8 $M_\odot$ have evolved into white dwarfs.*

an X-ray source, even if the number of collisions in them is the same. An interesting calculation by Jordán et al. (2004) shows that a moderate dependence on metallicity of the slope of the Initial Mass Function leads to a difference in the numbers of neutron stars in metal-rich and metal-poor clusters (per unit mass) which is large enough to explain the observed preference for high-metallicity clusters (see Fig. 3). Briefly, if we write the number of stars at mass $m \equiv M/M_\odot$ as $N(m) = Km^{-\alpha}$, we have for the number of stars with initial mass between m_1 and m_2 (that have evolved into neutron stars) per unit mass between m_a and m_b:

$$\frac{N(m_1, m_2)}{M(m_a, m_b)(M_\odot)} = \frac{2-\alpha}{1-\alpha} \frac{m_1^{1-\alpha} - m_2^{1-\alpha}}{m_a^{2-\alpha} - m_b^{2-\alpha}} \quad (1)$$

Jordán et al. (2004) discuss the number of neutron stas per unit of *initial* mass of the cluster, i.e. they take $m_b = m_2$. This ratio, for values $m_a = 0.08$, $m_b = m_2 = 80$ and $m_1 = 8$, is shown as a solid line in Fig. 3 as a function of α. If α is lower at higher metallicity, i.e. metal-rich clusters have a shallower initial mass function, then metal-rich clusters have a relatively higher number of neutron stars. We may add that this effect is even stronger if we scale not on the initial mass, but on the current luminous mass, i.e. the stars that determine the observed K magnitude. Eq. 1 for $m_2 = 80$ as before, but with $m_b = 0.8$, is shown in Fig. 3 as a dashed line. (Stars in the range $0.8 < m < 8$ have evolved into white dwarfs.) This deserves further study, in particular whether the required dependence of the slope of the IMF on metallicity can be more firmly established.

The second research area is the indication that the probability for a globular cluster to contain an X-ray source increases slower with density than the number of collisions occurring in the cluster (Jordán et al. 2004). The models for formation and evolution of X-ray sources in globular clusters must address these questions.

REFERENCES

1. Angelini, L., Loewenstein, M., and Mushotzky, R., 2001, ApJ 557, L35
2. Bildsten, L. and Deloye, C., 2004, ApJ 607, L119
3. di Stefano, R., Kong, A., Garcia, M., and et al., 2002, ApJ 570, 618
4. Gilfanov, M., 2004, MNRAS 349, 146
5. Grindlay, J. and Hertz, P., 1985, in D. Lamb and J. Patterson (eds.), Cataclysmic Variables and Low Mass X-ray Binaries, pp 79–91, Reidel, Dordrecht
6. Jordán, A., Côté, P., and Ferrarese, L. e. a., 2004, ApJ 613, 279
7. Kim, D.-W. and Fabbiano, G., 2003, ApJ 586, 826
8. Kim, D.-W. and Fabbiano, G., 2004, ApJ 611, 846
9. Kong, A., Di Stefano, R., Garcia, M., and Greiner, J., 2003, ApJ 585, 298
10. Kong, A., Garcia, M., Primini, F., and et al., 2002, ApJ 577, 738
11. Kundu, A., Maccarone, T., Zepf, S., and Puzia, T., 2003, ApJ 589, L81
12. Kuulkers, E., den Hartog, P., in 't Zand, J., Verbunt, F., Harris, W., and Cocchi, M., 2003, A&A 399, 663
13. Maccarone, T., Kundu, A., and Zepf, S., 2003, ApJ 586, 814
14. Magnier, E., Lewin, W., van Paradijs, J., and et al., 1992, A&AS 96, 379
15. Sarazin, C., Kundu, A., Irwin, J., and et al., 2003, ApJ 595, 743
16. Supper, R., Hasinger, G., Pietsch, W., and et al., 1997, ApJ 317, 328
17. van Speybroeck, L., Epstein, A., Forman, W., Giacconi, R., Jones, C., Liller, W., and Smarr, L., 1979, ApJ 234, L45
18. Verbunt, F., Bunk, W., Hasinger, G., and Johnston, H., 1995, A&A 300, 732
19. Verbunt, F. and Lewin, W., 2005, in W. Lewin and M. van der Klis (eds.), Compact stellar X-ray sources, p. in press, Cambridge University Press
20. Verbunt, F., van Paradijs, J., and Elson, R., 1984, MNRAS 210, 899
21. White, N., Sarazin, C., and Kulkarni, S., 2002, ApJ 571, L23

X-ray sources in globular clusters

Frank Verbunt

Astronomical Institute, Postbox 80.000, 3508 TA Utrecht, the Netherlands

Abstract. Observations with BeppoSAX, RXTE and Chandra suggest that many of the bright X-ray sources in globular clusters have ultrashort binary periods. This is remarkable as such systems are not easily formed. With accurate optical astrometry of HST images, the large numbers of low-luminosity X-ray sources discovered with Chandra can be classified as quiescent low-mass X-ray binaries, pulsars, cataclysmic variables, and magnetically active binaries. The number of cataclysmic variables is found to scale with the number of close stellar encounters.

Keywords: Globular clusters, X-ray sources
PACS: 98.20.Gm, 97.80.Jp

INTRODUCTION

The known populations of X-ray sources in globular clusters have grown with the increasing sensitivity of X-ray detectors, from bright low-mass X-ray binaries discovered in the 1970s (UHURU, OSO-7, Ariel-V), to faint low-mass X-ray binaries and cataclysmic variables in the 1980s (Einstein, HEAO-2), recycled radio pulsars in the 1990s (ASCA, ROSAT), and magnetically active close binaries in the 2000s (Chandra). In this review I summarize the new knowledge gained in the last years about the bright X-ray sources, and discuss in particular the possibility that many of these have ultrashort orbital periods (Sect. 2). Then I explain how various new technical possibilities of Chandra, XMM and HST expand our knowledge of the faint sources (Sect. 3). Some considerations on formation rates end the paper.

THE BRIGHT SOURCES: NEUTRON STARS ACCRETING AT HIGH RATES

We now know 13 bright X-ray sources ($L_{0.5-2.5 keV} > 10^{35.5}$ erg s^{-1}, say) in 12 globular clusters of our Galaxy (Table 1). Most were discovered in the 1970s with the first X-ray satellites; relatively recent additions are the source in Terzan 6, discovered in the ROSAT All Sky Survey, and the two sources into which the bright X-ray emission from NGC 7078 – known since the 1970s – was resolved by Chandra. X-ray bursts were detected from some sources soon after discovery, and some sources (Terzan 1 and Terzan 5) were initially detected only during bursts. It required the large temporal and spatial coverage by the BeppoSAX Wide Field Cameras and the RXTE All Sky Monitor to find that all but one of the bright sources in globular clusters are X-ray bursters, and thus are neutron stars accreting from a companion. The 13th source, in NGC 7078, may be a neutron star as well.

TABLE 1. *List of the thirteen bright X-ray sources in globular clusters with the year of discovery, year of the first observation of a burst, absolute magnitude in B (or U,J, as indicated), orbital period, and indications whether the source is a transient (T), and whether the source has properties associated with an ultrashort period (U) or with a normal period (N) in its optical magnitude (O), X-ray spectrum (X), or burst maximum (B). See text for explanation. Detailed references are given in Verbunt & Lewin (2004).*
[a] *Luminous X-ray emission from NGC 7078 was already found in 1974.* [b] *or its alias at 13.2 m.*

cluster	disc	burst	M_λ	P_b	TOXB
NGC 1851	1975	1976	5.6B		UUU
NGC 6440	1975	1999	3.7B		T−N−
NGC 6441	1974	1987	2.4B	5.7hr	−NN
NGC 6624	1974	1976	3.0B	11.4m	UUU
NGC 6652	1985	1998	5.6B		UUU
NGC 6712	1976	1980	4.5B	20.6m[b]	UUU
NGC 7078-1[a]	2001		0.7B	17.1hr	−−−
NGC 7078-2[a]	2001	1990	3.1U		−−U
Terzan 1	1981	1981			T−−−
Terzan 2	1977	1977			−NU
Terzan 5	1981	1981	1.7J		T−U−
Terzan 6	1991	2003		12.36h	T−N−
Liller 1	1976	1978			T−−−

Remarkably, of the five orbital periods currently known, two are ultrashort, at periods of 11.4 m and 20.6 m (or its alias 13.2 m). I return to this below. Five of the 13 sources are transients, in the sense that their luminosity at times has dropped below $\sim 10^{34}$ erg s^{-1}.

Recently, two remarkable illustrations of the danger of identifications without accurate positions have been found. The first is the discovery that the bright ultraviolet source in NGC 6652, with a period of 43.6 m (Deutsch et al. 2000), is *not* the optical counterpart of the bright X-ray source in this cluster, but of a second, relatively faint X-ray source, presumably a faint low-mass X-ray binary (Heinke et al. 2001). The second is the discovery that the single faint source significantly detected in a Chandra observation of Terzan 1 is *not* the quiescent counterpart of the transient in this cluster, but a different source (see Figure 1). A warning that the optical centers of globular clusters are not always as accurate as advertized is provided by the study of Terzan 6: the hitherto assumed optical center was probably offset from the actual center by the influence of a bright star outside the cluster. A re-determination of the cluster center, in which each star gets the same weight irrespective of its brightness, places the center very close to the accurately determined position of the X-ray source (in 't Zand et al. 2003).

With three ROSAT observations of NGC 6624, van der Klis et al. (1993) showed that the 11.4 m period of the X-ray source in this cluster decreases, in contradiction of the model in which the donor delivering matter to the neutron star is a white dwarf. The negative period derivative is confirmed by Chou & Grindlay (2001). Some hesitance in accepting this derivative as intrinsic may remain, since the differences between observed periods as a function of time and a quadratic fit look irregular. Also, it is possible that acceleration of the binary in the cluster potential affects its observed period derivative: the optical counterpart of the X-ray source is very close to the cluster center (0.6±0.3″, King et al. 1993). Nonetheless, one must now seriously consider the possibility that the 11.4 m period is decreasing rather than increasing.

Are there models which can explain this? To discuss this, we note first that the

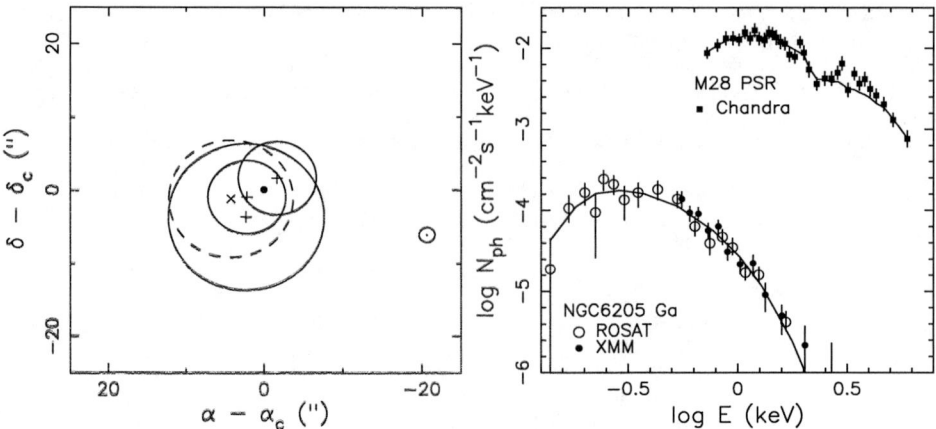

FIGURE 1. *Left:* Accurate positions for the bright source in Terzan 1 were determined with EXOSAT (× with dashed error circle, from Parmar et al. 1989) and with three independent ROSAT observations (+ with solid error circles, redetermined by the author after Johnston et al. 1995). Coordinates are with respect to the cluster center. The bright source had switched off in 1999 (Guainazzi et al. 1999). The position of the one source detectable in a short Chandra observation, indicated as the small error circle on the right, is not compatible with the position of the previously bright source (Wijnands et al. 2002). *Right:* Comparison of the X-ray spectrum of a quiescent low-mass X-ray binary (NGC6205 Ga, below left) and a radio pulsar (in M28, up right). Note the sensitivity of ROSAT to very soft photons. Data from ROSAT and XMM from Gendre et al. (2003); from Chandra from Becker et al. (2003).

formation of the bright X-ray binaries in globular clusters is thought to follow two main paths: either a tidal capture of a closely passing star by a single neutron star, or an exchange encounter in which a neutron star encounters a binary and swaps place with one binary star. If a low-mass main-sequence star is captured, and starts transferring mass to the neutron star, the orbit will decrease until the donor star becomes degenerate, after which the orbit expands again. The minimum period reached this way is near 70-80 m, and shorter periods cannot be explained this way. If a more massive main-sequence star is captured, and evolves into a giant before mass transfer starts, the mass transfer is unstable and a spiral-in of the neutron star into the envelope of the giant may bring it into a close orbit with the core of the giant. This must have happened in the past, when stars with a mass sufficiently high to induce unstable mass transfer still existed in globular clusters: it may take several billion years before the core of the giant, cooled into a white dwarf, comes into contact under the influence of gravitational radiation and starts mass transfer. Such ultrashort period binaries can therefore be observed today (Rasio et al. 2000). If a neutron star collides directly with a giant, the envelope will be expelled, and the core may enter an orbit around the neutron star. A white-dwarf donor of mass M_c fills its Roche-lobe at an orbital period of roughly $P_b \sim 1^m (M_\odot/M_c)$; thus ultrashort periods are possible, but the period must lengthen with time as the mass of the white dwarf decreases (Verbunt 1987). Tutukov et al. (1985) note that ultrashort periods can also be reached in binaries in which a star starts to transfer mass to its companion after it has already evolved somewhat, and depleted its core from hydrogen. The donor

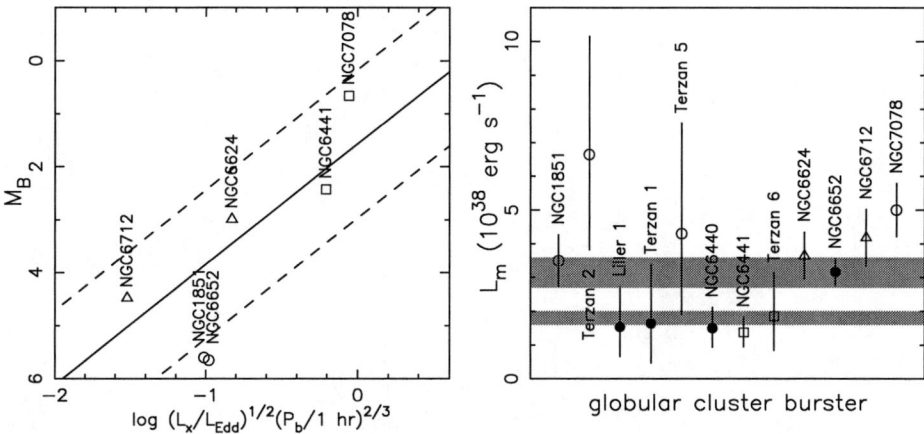

FIGURE 2. *Left: Absolute B magnitude of optical counterparts to X-ray sources in globular clusters as a function of the X-ray luminosity (expressed in Eddington luminosity $2 \times 10^{38} erg\, s^{-1}$) and orbital period (in hours). Data from Deutsch et al. (2000) and Heinke et al. (2001), see Table 1. The solid and dashed lines give the average relation and spread around the average for X-ray binaries in the galactic disk, as derived by Van Paradijs & McClintock (1994). Right: maximum observed luminosities L_m during X-ray bursts, compared with the Eddington limits for hydrogen-rich ($X = 0.73$) and hydrogen-poor ($X = 0$) matter (lower and upper gray band, respectively). After Kuulkers et al. (2003). In both figures: □ normal period, △ ultrashort period, O period not known (set at 1 hr for the left plot). A filled symbol indicates that the peak flux was determined for an ordinary burst, and may be lower than the Eddington limit.*

then becomes degenerate at smaller mass, and shorter orbital period. Tutukov et al. do a limited number of calculations and find a minimum period of about 20 m; Podsiadlovski et al. (2002) show that periods as short as 5 m can be reached this way. This scenario may be called 'magnetic capture'.

Van der Sluys et al. (these proceedings, see also Van der Sluys et al. 2004) note that none of the relevant computations of magnetic capture done by Pylyser & Savonije (1988) reach such extremely short periods, and investigate in some detail which initial conditions (initial binary period and donor mass) lead to ultrashort orbital periods *within a Hubble time*. They find that only binaries in a very small range of initial orbital periods and in a small range of donor masses evolve to ultrashort periods within the Hubble time, and that each of these binaries spends only a small fraction of its evolution at ultrashort periods. They conclude that only an extremely small fraction (< 0.001) of a population of X-ray binaries at any time can be expected to be at ultrashort periods. Whereas the observation of one such system amongst the thirteen globular cluster sources could be explained by magnetic capture, the observation of two or more demands another explanation.

In this context it is worthy of note that there are good indications that many of the bright X-ray sources in globular clusters have ultrashort periods, as first pointed out by Deutsch et al. (1996). As seen in Table 1, two of five orbital periods known are ultrashort. Indirect evidence for ultrashort periods is provided by three observations. First, it is known that the optical light of bright low-mass X-ray binaries is dominated

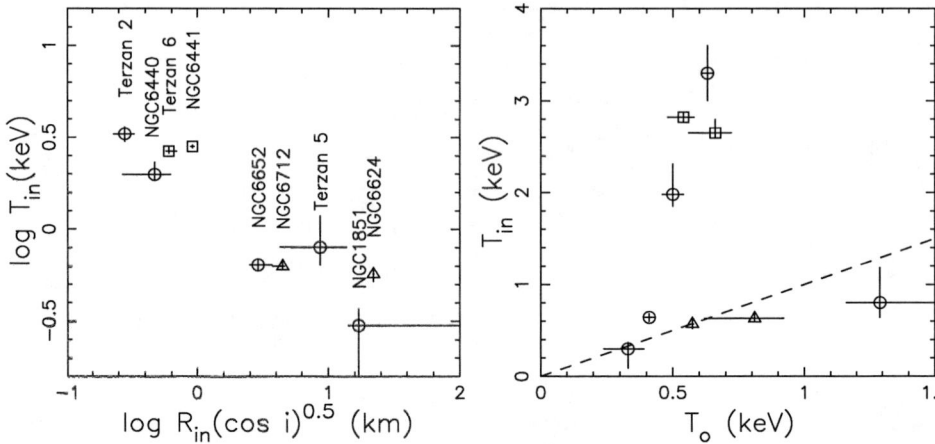

FIGURE 3. *Fitted values of model parameters of X-ray spectra of X-ray sources in globular clusters. Shown are the blackbody temperatures T_{in} of the photons leaving the inner disk as a function of the projected inner disk radius $R_{in}\sqrt{\cos(i)}$ (left) and of the blackbody temperatures T_o of photons entering the Comptonization region (right). After Sidoli et al. (2001); NGC 6652 added from Parmar et al. 2001, Terzan 5 added from Heinke et al. (2003). Symbols as in Fig. 2.*

by reprocessing of X-rays that impinge on the accretion disk. Thus, if the disk is small and/or if the X-ray luminosity is small, one expects a low optical luminosity. Van Paradijs & McClintock (1994) derive that the expected absolute visual magnitude scales with disk size (via orbital period P_b) and X-ray luminosity L_x roughly as $M_V \propto P_b^{2/3} L_x^{1/2}$. Figure 2 shows that the transition between systems with normal periods (i.e. periods above 80 m) and those with ultrashort periods lies near $M_B = 3$. The B magnitudes of the sources in NGC 1851 and NGC 6652 thus suggest ultrashort periods.

The second indication is the maximum luminosity reached in a burst. Kuulkers et al. (2003) have analyzed an extensive data set of X-ray bursts of sources in globular clusters, and note that the maximum luminosity reached in radius expansion bursts, when the source is assumed to reach the Eddington limit, corresponds to the Eddington limit for hydrogen-poor material in the two systems with ultrashort periods, and to that for hydrogen-rich material in the source with a normal period in NGC 6641. The maximum luminosities in the other cluster sources then suggest ultrashort periods in NGC 1851, NGC 6652, NGC 7078-2, and Terzan 2 (Fig. 2).

A third indication comes from the X-ray spectra. A model which is often used – but debated as to its physical correctness – is that in which a sum of blackbody spectra from annuli of the accretion disk is combined with a Comptonized spectrum. The parameters are the inner radius and inner temperature of the disk, and the temperature of the photons entering the Comptonized region, characterized by an optical depth and electron temperature. The best-fit parameters appear to separate the ultrashort systems in NGC 6624 and NGC 6712, for which the inner radius is compatible with a neutron star radius and for which the photons entering the Comptonization region have a temperature compatible with those leaving the disk, from the systems with normal periods in

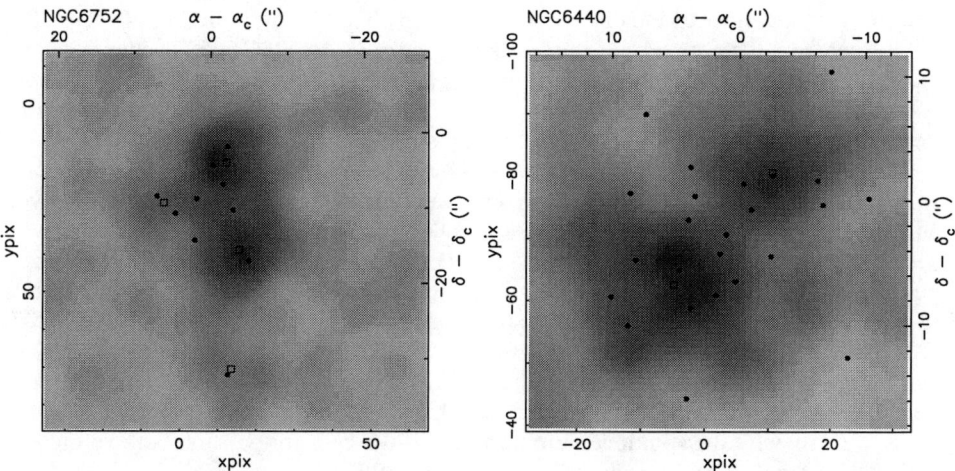

FIGURE 4. *Smoothed grey scale images of NGC 6752 and NGC 6440 as obtained with ROSAT, with the positions of the sources derived from the ROSAT data (□) and the much more numerous positions of the sources detected with Chandra (●), illustrating the dramatic increase in resolution and sensitivity. The uncertain direction ('boresight') of the ROSAT observation allows a shift between the ROSAT and Chandra image of NGC 6440 by 5″. ROSAT data from Verbunt & Johnston (2000) and Verbunt et al. (2000); Chandra positions from Pooley et al. (2002a,b).*

NGC 6441 and Terzan 6, for which the inner disk radii are too small, and the photons entering the Comptonization region too soft, for the model to be physically viable (Sidoli et al. 2001). This suggests that the sources in NGC 1851, NGC 6652 and Terzan 5 also have an ultrashort period, and those in NGC 6440 and Terzan 2 normal periods (Sidoli et al. 2001, Parmar et al. 2001, Heinke et al. 2003).

Collating these three observational indicators, Verbunt & Lewin (2004) find that the sources in NGC 1851 and NGC 6652 probably, and those in NGC 7078-2 and Terzan 2 possibly, have ultrashort periods (Table 1). The source in NGC 6440 possibly has a normal period. The indications are contradictory in Terzan 2 and absent in NGC 7078-1 and Liller 1. Thus, certainly 2, probably 4, and possibly 6 or 8 of the thirteen bright X-ray sources in globular clusters have ultrashort periods!

VARIOUS TECHNIQUES TO STUDY THE DIM SOURCES

CCDs have a better **spectral resolution** than proportional counters, and as a result the spectral energy distribution of X-ray sources is better determined from observations with Chandra and XMM than with ROSAT (Fig. 1). If a source has a soft spectrum, with a characteristic blackbody temperature of $\lesssim 0.3\,\mathrm{keV}$ say, and an X-ray luminosity $\gtrsim 10^{32}\mathrm{erg\,s^{-1}}$, then it is almost certain to be a neutron star accreting from a companion. With this method an increasing number of faint accreting neutron stars has been discovered in globular clusters, showing that the number of such sources is more than ten times higher than the number of bright sources (Pooley et al. 2003). In general, accreting neutron stars are close to the cluster center, within about 2 core radii, but of order 20% are further from the core. For example, the bright source in NGC 6652 is about six core radii from the center of that cluster (Heinke et al. 2001). With the accuracy of Chandra positions, the error in the position of a source with respect to the cluster center is now dominated by the uncertainty in the cluster center.

The Chandra telescope has a much better **spatial resolution** than ROSAT, and as a result many more faint sources have been discovered in clusters (Figure 4). The improved accuracy of the source positions allows a fairly safe identification of X-ray sources with accurately localized radio pulsars on the basis of positional coincidence alone. A dozen pulsars has thus been detected in 47 Tuc without ambiguity; two more sources are blends of two pulsars each. A single pulsar has been detected in each of the clusters NGC 6397, NGC 6752 and M 28 (Grindlay et al. 2002, D'Amico et al. 2002, Becker et al. 2003). The latter source, already detected with ASCA and ROSAT (Saito et al. 1997, Verbunt 2001), is very bright, and shows that the total X-ray luminosity in the 0.5-2.5 keV range scales directly with the spindown luminosity, as is the case for pulsars (both young and recycled) in the disk of the galaxy (Verbunt et al. 1996).

The **positional accuracy** of Chandra and of the Hubble Space Telescope allows much more secure identification of optical counterparts to the X-ray sources than earlier ROSAT data. To take full advantage of this opportunity, one must align the coordinates from Chandra and HST. From the absolute position of a Chandra source (with accuracy $0.6''$, say) and of the HST objects (with accuracy of $1''$, say; but occasionally as big a $3''$) we must look for counterparts in the HST image in a circle with a radius of $\sqrt{0.6^2 + 1^2} = 1.2''$, or occasionally even larger. On the other hand, once we have securely identified at least one Chandra source with an HST object, we can align the coordinate frames with this match, and the error circles for the remaining sources are much reduced: for sufficiently bright X-ray sources the relative positions are accurate to $\lesssim 0.1''$, the relative optical positions are accurate to $\lesssim 0.1''$, and thus the radius of the search circle now is $\sqrt{0.14^2 + \sigma_m^2}''$, where σ_m is the accuracy of the match.

In the case of a rare source type, like pulsars, identifications are secure even with the larger error circles of the absolute astrometry; to be useful for optical identifications, the pulsar radio position must be tied to the optical coordinate frame. A beautiful example of this is the identification of the pulsar in NGC6397 with a Chandra source, on the basis of the binary period, detected both in the radio and in the optical (Ferraro et al. 2001). If such a direct match is not available, one may improve the HST absolute positions by connecting its coordinate system to absolute astrometry. For this purpose, the USNO CCD Astrograph Catalog (UCAC: Zacharias et al. 2000) turns out to be excellent, as shown by Bassa et al. (2004) for NGC 6121. Bassa et al. first identify 91 stars in an ESO 2.2m Wide Field Imager observation with stars in UCAC to determine the coordinates of the ESO WFI image, and then identify about 200 stars in the HST WFPC2 image with stars on the ESO WFI image to determine the coordinates of the HST frame with a 1 σ accuracy of $0.12''$ both in right ascension and in declination. This allows identification of three optical objects with Chandra sources, which are then used to align the Chandra frame to the HST frame, with an accuracy of about $0.14''$.

Once we have optical counterparts, we can complete the **classification** of the sources. From the X-ray data alone we can classify the quiescent low-mass X-ray binaries (from

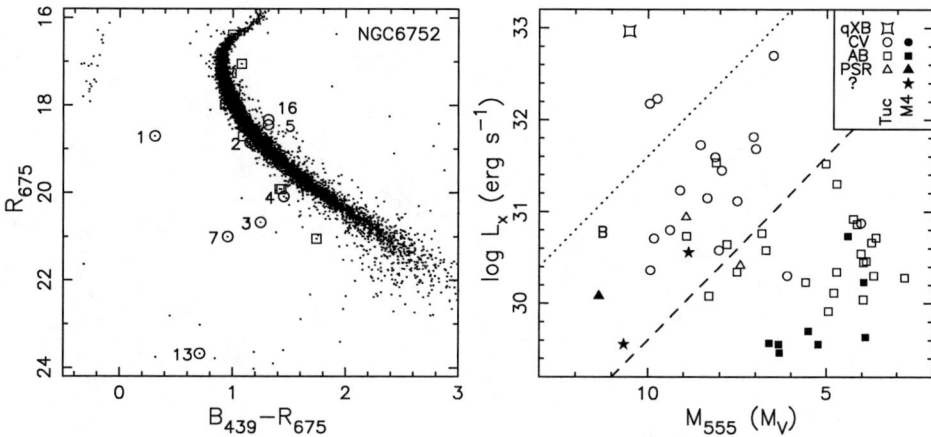

FIGURE 5. *Left: Colour – magnitude diagram for NGC6752. Optical counterparts for X-ray sources are indicated ⊙ and labelled with the Chandra source number; other stars within the X-ray error circles, but presumably not the counterparts are indicated □. From a re-analysis by Bassa (private communication) after Pooley et al. (2002a). Right: The X-ray luminosity as a function of absolute visual magnitude for various source types in 47 Tuc (open symbols) and M4 (filled symbols). The dashed line roughly separates the cataclysmic variables (CV) from the active binaries (AB). Other source types are quiescent low-mass X-ray binaries (qXB), and companions to millisecond pulsars (PSR). Sources which may be either a cataclysmic variable or an active binary are indicated with a star. A 'B' indicates the position of an X-ray source (X2) in M4, which may be a background quasar. The dotted line has L_x hundred times the dashed line. Data from Edmonds et al. (2003), Grindlay et al. (2001) and Bassa et al. (2004).*

their spectrum and luminosity, see Fig. 1) and the pulsars (from positional coincidence). In the optical colour-magnitude diagram, cataclysmic variables are bluer than the main sequence, and active binaries are more luminous than the main-sequence (Fig. 5). Cataclysmic variables have a higher X-ray to optical flux ratio than magnetic binaries. In a diagram of X-ray luminosity versus absolute visual magnitude, the line

$$\log L_{0.5-2.5\,keV}\,(\mathrm{erg/s}) = -0.4 M_V + 33.6$$

roughly separates cataclysmic variables from active binaries (Fig. 5). The variability of cataclysmic variables can cause confusion, if data are combined that were not taken simultaneously. Also, background quasars may maskerade as cataclysmic variables in the L_x-M_V diagram, as may be the case for X2 in M4, tentatively classified as a cataclysmic variable by Bassa et al. (2004), but identical to the blue source, probably a quasar, discovered by Bedin et al. (2003; Bedin & King, private communication). This source is indicated 'B' in Fig. 5. The dotted line, a factor 100 in X-ray luminosity above the dashed line, roughly gives an upper limit to the X-ray luminosities of cataclysmic variables in the galactic disk (based on ROSAT data, Verbunt et al. 1995). Remarkably, two cataclysmic variables in 47 Tuc are well above this line (Grindlay et al. 2001).

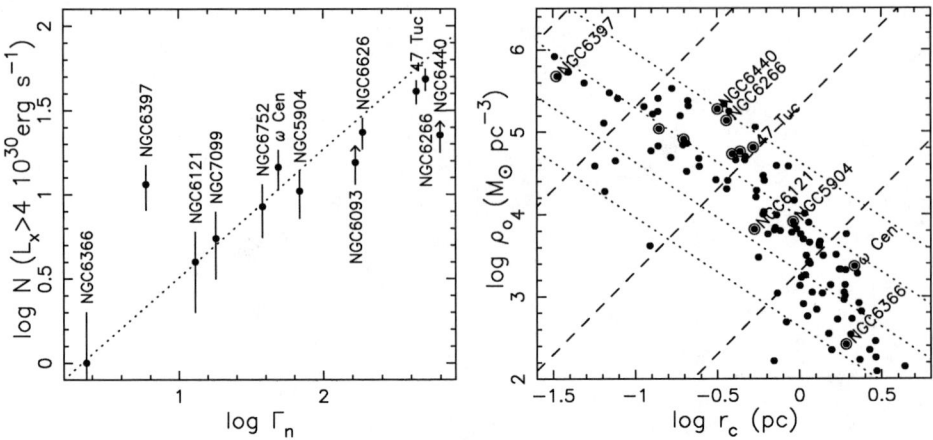

FIGURE 6. *Left: The number of X-ray sources detected with Chandra in globular clusters above a threshold limit, as a function of a normalized collision number Γ_n. An arrow indicates a lower limit. The dashed line indicates the best fit: $N \propto \Gamma_n^{0.74}$. After Pooley et al. (2003). Right: central density of globular clusters as a function of core radius, with lines of constant collision number $\Gamma \propto \rho_o^{1.5} r_c^2$ (dotted, steps of factor 10) and of constant destruction number $\gamma \propto \rho_o^{0.5}/r_c$ (dashed, steps of factor 10). Clusters from the left panel are encircled, and (some) indicated with their name. After Verbunt (2003).*

SOME CONSIDERATIONS ON FORMATION RATES

If (quiescent) low-mass X-ray binaries and cataclysmic variables are formed in globular clusters mainly through stellar encounters, their numbers should scale with the collision number Γ of the cluster. Γ scales approximately with the central density ρ_o and core radius r_c of the cluster as $\Gamma \propto \rho_o^{1.5} r_c^2$ (e.g. Verbunt 2003). Pooley et al. (2003) show that this predictions holds rather well for the X-ray sources with $L_x > 4 \times 10^{30}$ erg s^{-1} (in the 0.5-2.5 keV band, Fig. 6). More precisely, with $N \propto \Gamma^{\alpha}$, they find $\alpha = 0.74 \pm 0.36$, allowing a direct proportionality, but suggesting that N increases slower than Γ. Since most of these sources are actually cataclysmic variables (see e.g. Fig. 5), this indicates that most cataclysmic variables in globular clusters are *not* evolved from primordial binaries, but rather formed via stellar encounters. NGC 6397 is an interesting cluster in this respect, as it has a rather larger number than predicted from its Γ of binaries with neutron stars (one quiescent low-mass X-ray binary, one radio pulsar) and cataclysmic variables. Pooley et al. (2003) suggest that the cluster was more massive in the past, and lost single stars from its outskirts in passages close to the galactic center; binaries escape this stripping process as they reside in the cluster core. The problem with this explanation is that NGC 6397 shows no excess of active binaries in it core. Perhaps the explanation of this is that the active binaries are destroyed in close encounters. The rate at which a single binary undergoes an encounter is roughly $\gamma \propto \rho_o^{0.5}/r_c$ (Verbunt 2003), and NGC 6397 has the highest γ of the globular clusters studied so far (Fig. 6). Since active binaries tend to be wider than cataclysmic variables, they would be more in danger of a destructive encounter.

A wonderful outlook on future work is provided by the study of bright X-ray sources in the globular clusters of M 87: Jordán et al. (2004) show that the probability for a cluster to house a bright X-ray source scales slower with ρ_o than Γ. They argue that this may be due to destruction of sources in the highest-density clusters.

REFERENCES

1. Bassa, C., Pooley, D., Homer, L., et al. 2004, ApJ 609, 755
2. Becker, W., Swartz, D., Pavlov, G., et al., 2003, ApJ 594, 798
3. Bedin, L., Piotto, G., King, I., and Anderson, J., 2003, AJ 126, 247
4. Chou, Y. and Grindlay, J., 2001, ApJ 563, 934
5. D'Amico, N., Possenti, A., Fici, L., et al. 2002, ApJ 570, L89
6. Deutsch, E., Anderson, S., Margon, B., and Downes, R., 1996, ApJ 472, L97
7. Deutsch, E., Margon, B., and Anderson, S., 2000, ApJ 530, L21
8. Edmonds, P., Gilliland, R., Heinke, C., and Grindlay, J., 2003, ApJ 596, 1177
9. Ferraro, F., Possenti, A., D'Amico, N., and Sabbi, E., 2001, ApJ 561, L93
10. Gendre, B., Barret, D., and Webb, N., 2003, A&A 403, L11
11. Grindlay, J., Heinke, C., Edmonds, P., and Murray, S., 2001, Science 292, 2290
12. Grindlay, J., Camilo, F., Heinke, C., Edmonds, P., Cohn, H., and Lugger, P., 2002, ApJ 581, 470
13. Guainazzi, M., Parmar, A., and Oosterbroek, T., 1999, A&A 349, 819
14. Heinke, C., Edmonds, P., and Grindlay, J., 2001, ApJ 562, 363
15. Heinke, C., Edmonds, P., Grindlay, J., LLoyd, D., Cohn, H., and Lugger, P., 2003, ApJ 590, 809
16. in 't Zand, J., Hulleman, F., Markwardt, C., et al. 2003, A&A 406, 233
17. Johnston, H., Verbunt, F., and Hasinger, G., 1995, A&A 298, L21
18. Jordán, A., Côté, P., and Ferrarese, L. et al. 2004, ApJ 613, 279
19. King, I., Stanford, S., Albrecht, R., et al., 1993, ApJ 413, L117
20. Kuulkers, E., den Hartog, P., in 't Zand, J., et al. 2003, A&A 399, 663
21. Parmar, A., Stella, L., and Giommi, P., 1989, A&A 222, 96
22. Parmar, A., Oosterbroek, T., Sidoli, L., Stella, L., and Frontera, F., 2001, A&A 380, 490
23. Podsiadlowski, P., Rappaport, S., and Pfahl, E., 2002, ApJ 565, 1107
24. Pooley, D., Lewin, W., Homer, L., 2002a, ApJ 569, 405
25. Pooley, D., Lewin, W., Verbunt, F., 2002b, ApJ 573, 184
26. Pooley, D., Lewin, W., Anderson, S., et al., 2003, ApJ 591, L131
27. Pylyser, E. and Savonije, G., 1988, A&A 191, 57
28. Rasio, F., Pfahl, E., and Rappaport, S., 2000, ApJ 532, L47
29. Saito, Y., Kawai, N., Kamae, T., et al., 1997, ApJ (Letters) 477, L37
30. Sidoli, L., Parmar, A., Oosterbroek, T., et al. 2001, A&A 368, 451
31. Tutukov, A., Fedorova, A., Ergma, E., and Yungelson, L., 1985, Pis'ma Astron. Zh. 11, 123
32. van der Klis, M., Hasinger, G., Verbunt, F., et al. 1993, A&A 279, L21
33. van der Sluys, M., Verbunt, F., and Pols, O., 2004, A&A in press, astro-ph/0411189
34. van Paradijs, J. and McClintock, J., 1994, A&A 290, 133
35. Verbunt, F., 1987, ApJ 312, L23
36. Verbunt, F., 2001, A&A 368, 137
37. Verbunt, F., 2003, in G. Piotto, G. Meylan, S. Djorgovski, and M. Riello (eds.), New horizons in globular cluster astronomy, pp 245–254, ASP Conf. Ser. 296
38. Verbunt, F. and Johnston, H., 2000, A&A 358, 910
39. Verbunt, F. and Lewin, W., 2004, in W. Lewin and M. van der Klis (eds.), Compact stellar X-ray sources, p. in press, Cambridge University Press
40. Verbunt, F., Bunk, W., Hasinger, G., and Johnston, H., 1995, A&A 300, 732
41. Verbunt, F., Kuiper, L., Belloni, T., Johnston, H., et al. 1996, A&A 311, L9
42. Verbunt, F., van Kerkwijk, M., in 't Zand, J., and Heise, J., 2000, A&A 359, 960
43. Wijnands, R., Heinke, C., and Grindlay, J., 2002, ApJ 572, 1002
44. Zacharias, N., Urban, S., Zacharias, M., et al., 2000, AJ 120, 2131

X-ray Binaries in the Globular Cluster 47 Tucanae

C O. Heinke[*,†], J. E. Grindlay[†], P. D. Edmonds[†], H. N. Cohn[**],
P. M. Lugger[**], F. Camilo[‡], S. Bogdanov[†] and P. C. Freire[§]

[*]*Northwestern University, 2131 Tech Drive, Evanston, IL 60208*
[†]*Harvard University, 60 Garden Street, Cambridge, MA 02138*
[**]*Indiana University, Dept. of Astronomy, Swain West 319, Bloomington, IN 47405*
[‡]*Columbia Astrophysics Laboratory, 550 West 120th Street, New York, NY 10027*
[§]*NAIC, Arecibo Observatory, HC03 Box 53995, PR 00612*

Abstract. *Chandra* observations of globular clusters provide insight into the formation, evolution, and X-ray emission mechanisms of X-ray binary populations. Our recent (2002) deep observations of 47 Tuc allow detailed study of its populations of quiescent LMXBs, CVs, MSPs, and active binaries (ABs). First results include the confirmation of a magnetic CV in a globular cluster, the identification of 31 additional chromospherically active binaries, and the identification of three additional likely quiescent LMXBs containing neutron stars. Comparison of the X-ray properties of the known MSPs in 47 Tuc with the properties of the sources of uncertain nature indicates that relatively few X-ray sources are MSPs, probably only ∼30 and not more than 60. Considering the ∼30 implied MSPs and 5 (candidate) quiescent LMXBs, and their canonical lifetimes of 10 and 1 Gyr respectively, the relative birthrates of MSPs and LMXBs in 47 Tuc are comparable.

Keywords: pulsars, X-ray binaries, globular clusters
PACS: 95.85.Nv,97.60.Gb,97.80.Jp,98.20.Gm

INTRODUCTION

Globular clusters are efficient factories for the production of X-ray binaries from populations of primordial binaries (see papers by D'Antona et al. and Ivanova et al., this volume). Although the nature of bright ($L_X > 10^{36}$ ergs s^{-1}) X-ray sources in globular clusters as low-mass X-ray binaries containing neutron stars has been long established [1, 2], the nature of faint X-ray sources has been more difficult to determine. At least four types of low-luminosity ($L_X = 10^{30-34}$ ergs s^{-1}) X-ray systems are now known to exist in globular clusters: cataclysmic variables [CVs; 3], LMXBs in quiescence [qLMXBs; 3, 4], millisecond radio pulsars [MSPs; 5], and chromospherically active main-sequence binaries [ABs; 6]. A recent review of X-ray sources in globular clusters can be found in Verbunt and Lewin [7,see also Verbunt, these proceedings].

The advent of the *Chandra* X-ray Observatory has allowed detailed study of the populations of faint X-ray sources in globular clusters [8], especially in the dense and nearby cluster 47 Tuc [9]. The large MSP population in 47 Tuc [10] has been detected in X-rays with *Chandra*, allowing studies of the MSPs' X-ray luminosities and spectra [11, 12]. The precise positions have allowed identification of optical counterparts to 58% (45 of 77) of the X-ray sources detected within a deep *HST* imaging field [13, 14]. Within 47 Tuc, 22 CVs and 29 ABs have been unambiguously identified, in addition to two qLMXBs and 17 MSPs.

A major mystery associated with globular clusters is the formation mechanism of millisecond pulsars. The logical progenitors of MSPs, the LMXBs, appear at first glance to be far too small in numbers to produce the estimated numbers of MSPs [15]. The number of MSPs in 47 Tuc has been estimated at >200 [10], suggesting ~10000 MSPs in the Galactic globular cluster system. However, only 13 bright LMXBs are known in our globular cluster system. For typically assumed lifetimes of 10 and 1 Gyr respectively, the MSP birthrate is two orders of magnitude higher than the LMXB birthrate. It is possible that many MSPs in globular clusters were born in the distant past from intermediate-mass X-ray binaries, and/or that X-ray heating greatly reduces the lifetimes of LMXBs [16]. However, the problem cannot be regarded as conclusively solved via these theoretical mechanisms at the present time. Here we discuss observations of 47 Tuc which may reduce this discrepancy in lifetimes by two orders of magnitude.

CHANDRA OBSERVATIONS OF 47 TUC

Following our successful 70 ksec *Chandra* observation of 47 Tuc in 2000 [9], we obtained a deeper 280 ksec observation in late 2002, spread over more than a week to constrain source variability. These observations used the backside-illuminated ACIS-S3 chip for maximal low-energy sensitivity. Within the 2.79' half-mass radius of 47 Tuc, 300 sources were detected in the 2002 observations [Figure 1; Grindlay, these proceedings; 17].

Comparing the positions of the new X-ray sources with unusual objects in 47 Tuc, we find that an additional 31 active binaries, identified from *HST* variability studies [18], can be unequivocally identified with *Chandra* X-ray sources. None of the new CV candidates suggested by [19] are identified with X-ray sources. Eighty-seven X-ray sources show clear variability, on timescales ranging from hours and weeks (within the 2002 observations) to years (comparing the 2002 and 2000 observations) and decades (comparing the 2002 detections with the ROSAT results, [20]). Based on the radial distribution of the detected sources, roughly 70 of them are probably background sources.

Simple spectral fits to the individual sources find that the majority can be well-fit by absorbed thermal plasma models (VMEKAL in XSPEC) using the cluster metallicity. This is expected for CVs and ABs (though some of the brightest CVs require complicated multi-component spectra). However, the known qLMXBs and MSPs are generally not well-fit by thermal plasma models [17, 12]. There are three additional sources that are unusually soft and bright when compared to other CVs and ABs.

One, W37, shows eclipses with a 3.087 hour period and strong variations in N_H, with an X-ray spectrum generally best described by a hydrogen-atmosphere model appropriate for neutron star surfaces [21]. The other two, X4 and W17, have X-ray spectra best described by a hydrogen-atmosphere model with a strong power-law component. X4 also shows short-term variability, which may be from either component. The hydrogen-atmosphere models give implied radii consistent with 10-13 km for all three sources, indicating that all three are probably quiescent LMXBs. However, the relative strength of the power-law components (60-65% of the 0.5-10 keV unabsorbed flux) and low luminosities ($\sim 5 \times 10^{31}$ ergs s^{-1}) for two, and high absorption for the third, indicate

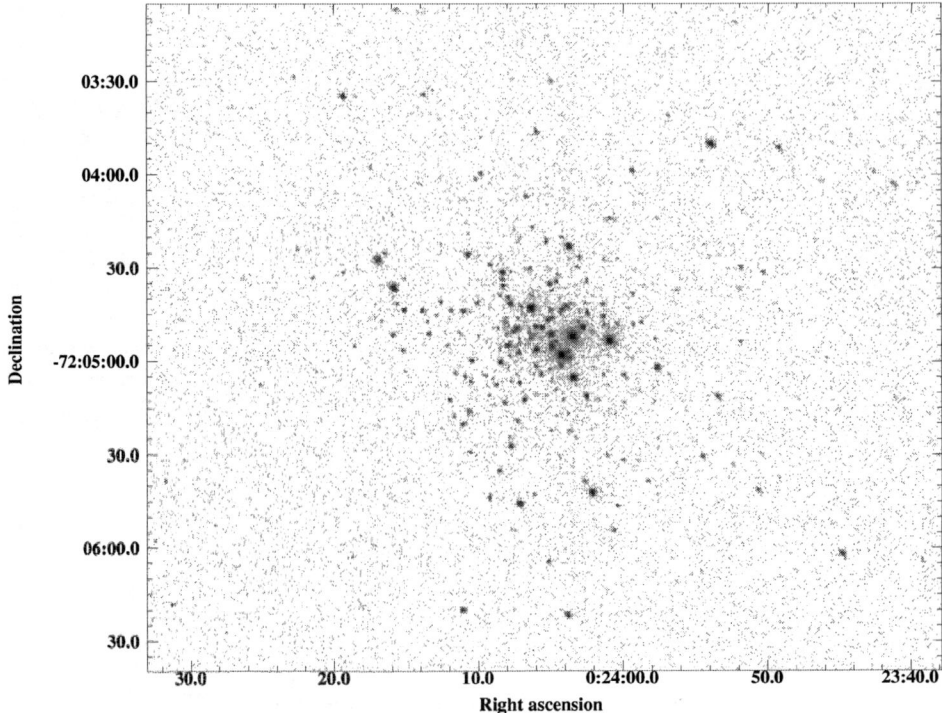

FIGURE 1. *Chandra* X-ray image of 47 Tuc, 0.3-6 keV energy range. The corners of this view are located at the half-mass radius.

that qLMXBs such as these would probably not have been identified in other globular clusters [22].

Cataclysmic variables in globular clusters seem to have higher X-ray fluxes, and lower optical fluxes, than CVs in the rest of the Galaxy [14]. This has led to speculation that the accretion flow onto CVs in globular clusters may be generally controlled by the magnetic field of the white dwarf, since such systems (DQ Her and AM Her systems) have little or no disk and relatively high X-ray/optical flux ratios [23]. However, only one DQ Her-type system has been suggested in a globular cluster [X9 in 47 Tuc, 9]. Our new *Chandra* observations identify a large-amplitude sinusoidal modulation with a 4.7-hour period from the CV X10, and a very soft blackbody component to its X-ray spectrum, identifying this CV as an AM Her, or polar. We also identify an N_H column above the cluster value for 12 of the 22 CVs, possibly indicating a DQ Her nature. Period searches are underway to confirm this possibility (Grindlay et al. 2005, in prep.).

MILLISECOND PULSARS IN 47 TUC

Of the 22 known MSPs in 47 Tuc, 17 have known positions, 16 through radio timing [10, 24], and one through matching an X-ray source to a variable optical source with an orbital period and phase that match those of 47 Tuc-W [25]. The MSPs G and I, separated by only $0.1''$, are detected as a single source in the *Chandra* image. The MSPs F and S, separated by only $0.74''$, are detected as a single, extended source. All other MSPs with known positions are clearly detected in this image.

Comparing the independently detected MSPs' radio pseudoluminosities [10] with their X-ray luminosities, no correlation is seen. This implies that MSPs with lower radio pseudoluminosities will have X-ray luminosities similar to those of the known MSPs. The X-ray emission is probably mostly from the hot polar caps of the neutron stars [11], while the radio emission originates higher in the magnetosphere. Gravitational bending assures that we will see virtually all MSPs in the X-ray, regardless of radio beaming fractions [12]. Thus, nearly all MSPs in 47 Tuc should be detected among our X-ray sources, with X-ray luminosities between 2×10^{30} and 2×10^{31} (30-350 counts).

We can compare the X-ray properties (X-ray "colors", variability, and spectral fits) of the unknown sources in this luminosity range to those of the known MSPs, CVs and ABs. We show in Figure 2 the distributions of MSPs, good candidate MSPs, ABs, and unknown sources in an X-ray color-color plot. (Likely background sources, located farther than $100''$ from the center of 47 Tuc, are also indicated, and are somewhat harder on average.) The distributions of MSPs and ABs are seen to be significantly different, and the unknown source distribution is more similar to the ABs than the MSPs.

We compare the numbers of MSPs, ABs, CVs, and background sources with each property (variability, spectral agreement with a VMEKAL model, colors within one of three classes) to the total number of unknown sources with that property. We assume that the ratio of unknown members of a class (e.g. ABs) to the known members of that class is the same (within binomial statistics) for each property. That is, if the number of ABs not yet identified is 60% of the number of identified ABs, and 20 known ABs are variable, then $\sim$12 unknown variable sources are probably ABs. We fit the numbers of unknown sources with each property using the numbers of known sources, to determine the ratios of unknown to known sources in each class. We can thus estimate the number of unidentified MSPs as 7^{+10}_{-7}. These unidentified MSPs must include the five known MSPs without known positions, suggesting that few MSPs remain undetected. The 95% single-sided upper limit on the total number of MSPs is 42; varying our choice of information to include gives a range of upper limits up to 56. Because our CV sample is X-ray selected, we cannot provide a similar constraint upon the number of CVs.

Can this estimate be missing a significant number of radio MSPs? Radio-faint MSPs could be X-ray faint also. However, there is no evidence of such a correlation among the known MSPs. Submillisecond pulsars or pulsars in very tight orbits could escape radio detection, but since these would be relatively energetic they would tend to be X-ray brighter, not fainter, than the known MSPs. The most likely way for MSPs to remain hidden is to be completely enshrouded in ionized gas from a nondegenerate companion (see Grindlay, these proceedings), similar to (but more extreme than) 47 Tuc W [10, 12]. This would obscure the radio signal, and the shock from the pulsar would also emit X-rays, altering the X-ray emission spectrum. However, 47 Tuc W is rather

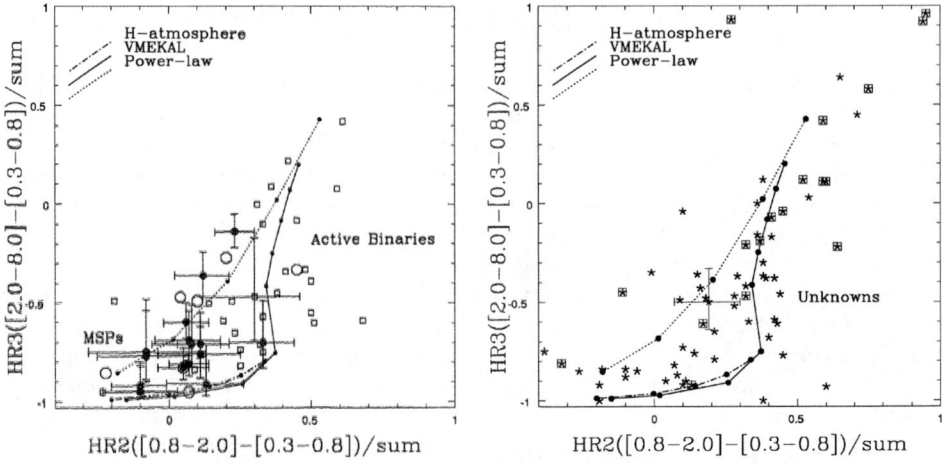

FIGURE 2. X-ray color-color plots showing sources with 30-350 counts in 47 Tuc. Left: Locations of MSPs (large filled dots), candidate MSPs (open octagons), and ABs (open squares). Right: Locations of sources of unknown nature (stars), and the subset of unknown sources located beyond 100″ from the center of the cluster (stars enclosed by squares); the latter are likely dominated by background sources. Model spectral tracks are plotted in both, with small dots representing (lower left to upper right): H-atmosphere for 75, 100, 125, 150, 175 eV; VMEKAL thermal plasma for 0.4, 0.5, 0.7, 1, 2, 3, 5, 10 and 30 keV; and a power-law with photon index 3, 2.5, 2, 1.5, 1. Error bars are plotted for all MSPs, and a few other representative faint sources. Increasing N_H moves sources up and to the right.

luminous for an MSP (and since the X-rays from the shock would need to obscure the thermal component, any "hidden" MSPs would be similarly luminous). There are very few X-ray sources of unknown nature in 47 Tuc of similar luminosity as 47 Tuc W. We conclude that unless there exists a class of MSPs that are very different from the MSPs we know, there are probably not more than 60 MSPs in 47 Tuc. This estimate agrees with independent constraints from *HST* identifications of X-ray sources [14] and integrated radio continuum flux measurements [26].

We extrapolate from this constraint in two directions. Two lines of evidence suggest that of order 10% of the neutron stars in a dense cluster are recycled into MSPs. Six of the eighty known MSPs in globular clusters have companions of masses $\sim 0.1-0.2\,M_\odot$, yet are eclipsing. All of these systems with optical identifications are probably MSP–main sequence binaries, suggesting binary exchange of a white dwarf for a main sequence star and putting the pulsar on a path to further recycling. Since 7% of these pulsars were doubly recycled, the recycling rate for neutron stars in dense clusters is likely to be of the same order. Recent cluster simulations [27] also suggest a neutron star recycling rate of 5-15%, higher for denser clusters. Thus we can extrapolate a total number of ~ 300 neutron stars in 47 Tuc (a rough, but empirical estimate). This is significantly less than has been previously predicted, and helps to resolve the neutron star retention problem in globular clusters [28].

The other direction is toward relative LMXB/MSP birthrates. These studies of 47 Tuc have increased the inferred number of qLMXBs and decreased the inferred number of MSPs, producing a ratio of $\sim$30/5=6 (compare to a ratio of $\sim$1000 from [15]). If qLMXBs can be counted among the progenitors of MSPs as a stage in the outburst cycle of LMXBs, and LMXB/qLMXBs live for $\sim$1 Gyr, then there is no longer an MSP birthrate problem in 47 Tuc, and in globular clusters generally.

ACKNOWLEDGMENTS

We thank the Penn State team for the ACIS_EXTRACT software, and D. Lloyd for his neutron star atmosphere models. C. H. was supported in part by *Chandra* grant GO2-3059A, and by the Lindheimer fund at Northwestern University.

REFERENCES

1. W. H. G. Lewin and P. C. Joss, "X-Ray Bursters and the X-Ray Sources of the Galactic Bulge," in *Accretion-Driven Stellar X-ray Sources*, 1983, p. 41.
2. J. E. Grindlay, P. Hertz, J. E. Steiner, S. S. Murray, and A. P. Lightman, *ApJL*, **282**, L13–L16 (1984).
3. P. Hertz and J. E. Grindlay, *ApJ*, **275**, 105–119 (1983).
4. F. Verbunt, R. Elson, and J. van Paradijs, *MNRAS*, **210**, 899–914 (1984).
5. Y. Saito, N. Kawai, T. Kamae, S. Shibata, T. Dotani, and S. R. Kulkarni, *ApJL*, **477**, L37 (1997).
6. C. D. Bailyn, J. E. Grindlay, and M. R. Garcia, *ApJL*, **357**, L35–L37 (1990).
7. F. Verbunt and W. H. G. Lewin, (2004), astro-ph/0404136.
8. D. Pooley, W. H. G. Lewin, S. F. Anderson, and et al., *ApJL*, **591**, L131–L134 (2003).
9. J. E. Grindlay, C. Heinke, P. D. Edmonds, and S. S. Murray, *Science*, **292**, 2290–2295 (2001).
10. F. Camilo, D. R. Lorimer, P. Freire, A. G. Lyne, and R. N. Manchester, *ApJ*, **535**, 975–990 (2000).
11. J. E. Grindlay, F. Camilo, C. O. Heinke, P. D. Edmonds, H. Cohn, and P. Lugger, *ApJ*, **581**, 470–484 (2002).
12. S. Bogdanov, J. E. Grindlay, C. O. Heinke, F. Camilo, and W. Becker, *ApJ* (2005), to be submitted.
13. P. D. Edmonds, R. L. Gilliland, C. O. Heinke, and J. E. Grindlay, *ApJ*, **596**, 1177–1196 (2003).
14. P. D. Edmonds, R. L. Gilliland, C. O. Heinke, and J. E. Grindlay, *ApJ*, **596**, 1197–1219 (2003).
15. S. R. Kulkarni, R. Narayan, and R. W. Romani, *ApJ*, **356**, 174–183 (1990).
16. E. Pfahl, S. Rappaport, and P. Podsiadlowski, *ApJ*, **597**, 1036–1048 (2003).
17. C. O. Heinke, J. E. Grindlay, P. D. Edmonds, H. N. Cohn, P. M. Lugger, F. Camilo, S. Bogdanov, and P. C. Freire, *ApJ, submitted* (2004).
18. M. D. Albrow, R. L. Gilliland, T. M. Brown, P. D. Edmonds, P. Guhathakurta, and A. Sarajedini, *ApJ*, **559**, 1060–1081 (2001).
19. C. Knigge, D. R. Zurek, M. M. Shara, and K. S. Long, *ApJ*, **579**, 752–759 (2002).
20. F. Verbunt and G. Hasinger, *A&A*, **336**, 895–901 (1998).
21. D. A. Lloyd, (2003), astro-ph/0303561.
22. C. O. Heinke, J. E. Grindlay, P. D. Edmonds, *ApJ, in press* (2004).
23. J. E. Grindlay, "High Resolution Studies of Compact Binaries in Globular Clusters with HST and ROSAT," in *IAU Symp. 174: Dynamical Evolution of Star Clusters*, 1996, p. 171.
24. P. C. Freire, F. Camilo, D. R. Lorimer, A. G. Lyne, R. N. Manchester, and N. D'Amico, *MNRAS*, **326**, 901–915 (2001).
25. P. D. Edmonds, R. L. Gilliland, F. Camilo, C. O. Heinke, and J. E. Grindlay, *ApJ*, **579**, 741–751 (2002).
26. D. McConnell, A. A. Deshpande, T. Connors, and J. G. Ables, *MNRAS*, **348**, 1409–1414 (2004).
27. N. Ivanova, K. Belczynski, J. M. Fregeau, and F. A. Rasio, (2003), astro-ph/0312497.
28. E. Pfahl, S. Rappaport, and P. Podsiadlowski, *ApJ*, **573**, 283–305 (2002).

Globular Clusters and Neutron Star Binaries

Francesca D'Antona

INAF, Osservatorio di Roma, I-00040 Monteporzio, Italy

Abstract. I summarize some aspects of the research at OAR on the neutron star population in Globular clusters. In particular I discuss the evolutionary stage of the interacting millisecond pulsar binary PSR J1740-5340 in the cluster NGC 6397 and point out that the donor component can not be a recently acquired low mass star. The precise mass derived for this star helps to constrain the loss of angular momentum during the radio–ejection phase which this system is traversing. I then discuss the problem of the large number of neutron stars present in globular clusters and show that it might be due to an atypical initial mass function, which is necessary to understand the chemical anomalies shown by the low mass stars evolving today.

Keywords: stars:globular clusters, millisecond pulsars, binaries, star formation in globular clusters
PACS: 98.20.Gm, 97.60.Gb, 97.80.Jp

INTRODUCTION

One of the most intriguing questions about neutron stars (NSs)in Globular Clusters (GCs) is their high number: there are more than 50 millisecond pulsars (MSP), to be compared with about the same number in the whole Galaxy, in spite of the fact that GCs represent only a small fraction of the galactic mass. There is of course a common wisdom answer to this problem, namely that GCs are the best place to produce binary systems containing a NS even a long time after its formation: capture and exchange interactions in binaries may be the dominant mechanism which allows recycling, and they provide a new opportunity to the NS to be revealed as a pulsar. Nevertheless, it is not straightforward to account for the large number of cluster NS, due to the fact that the kick velocity at supernova explosion provides quite a small retention factor. In addition, in this talk we point out an evolutionary indication that the exchange interaction rate may have been overestimated. This problem will be discussed with the help of results from the modeling of the secular evolution of binaries containing neutron stars, which our group is presently investigating, and I will show how an explanation for the high number of NSs may come from an entirely different field of investigation in GC research, namely from the chemical composition of GC stars and their anomalies.

MODELING THE EVOLUTION OF SYSTEMS CONTAINING MILLISECOND PULSARS

Four years ago we started at OAR a new project for computing the evolution of interacting binaries with a compact component, putting together the classic evolutionary code ATON (originally by Mazzitelli (1989) [20]), which follows the donor evolution with the input physics revised by Ventura et al. (1998) [31], with a general relativistic code that describes the behavior of the neutron star. The first results of the application of this code to the evolution of the neutron star through mass and angular momentum accretion are described in Lavagetto et al. (2004) [19]. Having this aim in mind, we studied in detail the possible physical conditions encountered in the presence of mass transfer. In particular, Burderi et al. (2001) [2] envisioned that mass accretion could be prevented by "radio–ejection", that is the sweeping effects of the energy outflow from a rapidly spinning NS undergoing a radio pulsar phase [18, 25, 26]. As soon as the accreting plasma moves out beyond the light cylinder radius (where an object corotating with the NS attains the speed of light), the NS generates magneto dipole radiation and relativistic particles, whose pressure may expel the matter overflowing the Roche lobe. In [2] we determined the dependence of this mass ejection mechanism on the parameters of the system, and have shown that it naturally provides an explanation for the values of the mass and rotation of the observed millisecond pulsars (MSPs) (Thorsett & Chakrabarty 1999 [29]; Tauris & Savonije 1999 [28]). In order for the radio–ejection phase to be initiated, it is necessary that the system suffers a phase of detachment, so that the radio pulsar can be switched on. Interestingly enough, a case for radio–ejection was –possibly– identified a short time after the publication of the Burderi et al. (2001) suggestion, namely the case of PSR J1740–5340.

The case of PSR J1740-5340

As soon as D'Amico et al. (2001) [6] discovered the millisecond radio pulsar PSR J1740.5340 in the globular cluster NGC 6397 —at an orbital period of 32.5 hr and with a spin period P_{spin}=3.65s— whose radio signal was randomly occulted during the orbital period, it was suddenly clear to Luciano Burderi and me that we were probably witnessing the just proposed radio–ejection phase! We were in the presence of a quasi–main sequence star losing mass towards its neutron star companion, but not accreting it. The models built up for the donor star confirmed that the system could be in radio--ejection (Burderi et al. 2002 [3]). Further determination of the dynamical mass of the companion resulted in a well constrained $0.3 M_\odot$ [17, 22].

Conservative mass transfer as presented in [3] leads to a mass of about $0.45 M_\odot$ at the present stage of evolution. Nonetheless, it is quite possible to obtain the correct donor mass and evolutionary stage by adjusting the specific angular momentum losses from the system, starting from the time at which radio–ejection begins [10]. Figure 1 shows examples of this modellization: it appears that reproduction of the location of the optical component of PSR J1740-5340 and of its present mass requires that the phase of conservative mass accretion which is necessary to accelerate the neutron star is followed by a phase during which mass is lost from the system with the specific angular momentum of the donor.

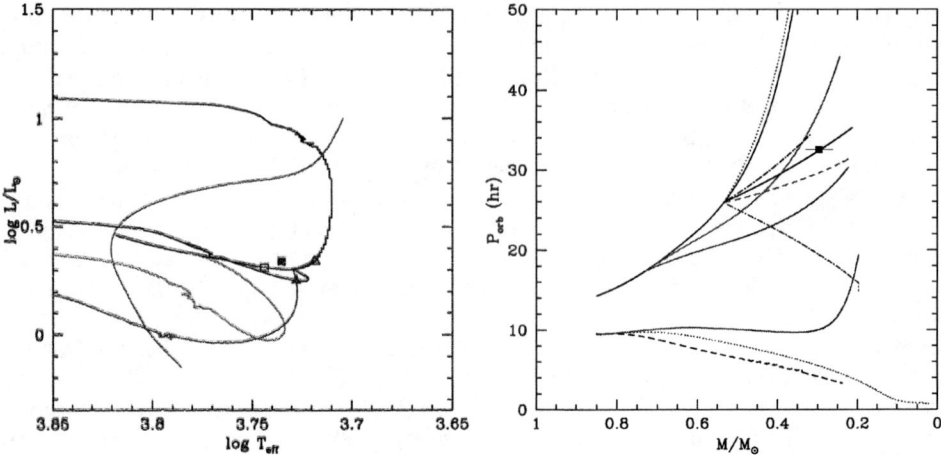

FIGURE 1. On the left we show the HR diagram with the location of the optical companion of PSR J1740-5340 (full square) and some evolutionary tracks which pass through its location. All tracks start close to the turnoff of a $M=0.85 M_\odot$, when hydrogen is almost fully depleted in the central regions. The (blue) continuous line evolving to high luminosities is the conservative evolution considered in [3]. The (black) curve, in which the full triangle indicates the orbital period of 32.5hr, shows the evolution obtained by assuming that, starting at a period of ~ 26hr, the mass is lost from the system with the specific angular momentum of the donor star. This same evolution is shown in the right side of the figure in the orbital period versus donor mass plane, as the (black) line which passes through the present donor mass and orbital period of PSR J1740-5340, gauging the angular momentum loss. Several other choices of the loss of angular momentum are also shown. The lines at the bottom of this figure represent evolutions starting below the bifurcation period.

Is the optical star in PSR J1740-5340 a recently acquired companion?

Notice also that the HR diagram location of the donor, close to the main sequence and luminous as the turnoff stars, is a strong constraint against the interpretation of this star as a recently acquired low mass companion, totally independent from the evolution which spun up the neutron star. This is also confirmed by the CN cycled chemistry of the donor envelope, observed by Ferraro et al. (2002) [12]. Hansen et al. (2003) [15] estimate a binary exchange lifetime $\sim 10^9$yr in the core of NGC 6397, and propose that the binary containing PSR J1740-5340 was previously hosting a low mass white dwarf, remnant of the star which spun up the pulsar, and thereafter the white dwarf was lost in an exchange interaction, resulting in a millisecond pulsar with a close main sequence companion. But, in this case, the optical component should have a mass close to $0.8M_\odot$, and not $0.3M_\odot$. A main sequence component of $0.3M_\odot$ would be intrinsically much dimmer, and the pulsar irradiation is by far too small to be the responsible of the luminosity. Therefore, it is possible that the donor in PSR J1740-5340 may have not been the original secondary of the system, but, even if it has been recently exchanged, it must have begun mass transfer soon after the dynamical interaction, and it is the star whose mass transfer has accelerated the neutron star. Such constraint must be taken into account when discussing the dynamical evolution of the binary population.

In the right part of Figure 1 the bottom lines represent the orbital period versus mass evolution of systems which start their binary mass exchange when the secondary is less evolved, that is, they remain below the bifurcation period [30, 11]: notice that the evolution which leads to the orbital period of PSR J1740-5340 must have begun very close to the bifurcations period: this seems to be an important ingredient of many evolutions to millisecond pulsars: for example, it may be a necessary ingredient to explain the concentration of short orbital periods ($\leq$ 10d) in the MSPs of globular clusters, and for explaining the LMXBs at ultrashort periods $P \simeq 40m$ (e.g. Nelson & Rappaport 2003 [21]) and in the end we must find a consistent framework to explain it.

THE EARLY STAGES OF FORMATION OF GLOBULAR CLUSTERS

Hints on the neutron stars (and possibly black hole) population of globular clusters come from an entirely different field of research: the observations and modellization of the star to star abundance variations in the spectra for a large fraction of globular clusters. For a general and updated summary of this problem see Gratton et al. (2004) [13]. The abundance anomalies involve elements which are processed through CNO and hot CNO cycles, and are commonly attributed to some kind of pollution from the ejecta of massive asymptotic giant branch stars (AGB), whose envelopes are CNO processed by hot bottom burning. This hypothesis was put forward more than 20 years ago by Cottrell and Da Costa (1981) [5] and D'Antona et al. (1983) [7], but only a few years ago it was reappreciated: on the observational side it was made clear that the chemical anomalies were present at the turnoff and on the subgiant branch, stars in which they can not be inputed to any 'in situ' mechanism; on the theoretical side, it was clearly shown that hot bottom burning in low metallicity AGBs could activate the ON cycle depleting oxygen and producing some of the observed anomalies (Ventura et al. 2001 [32]).

The most popular —although debated!— scenario hypothesizes a first prompt star formation, during which all the massive and intermediate mass stars are born, together with a more or less relevant fraction of the low mass stars evolving today. There follows a *second stage* of star formation, at an age from ~ 70 to ~ 200Myr, during which other low mass stars are born *directly* from the winds lost at low velocities from the massive AGB stars. Part of this scenario, developed during the latest two years, was made clear when D'Antona et al. (2002) [8] have shown that the helium enrichment in the ejecta of AGBs may be a long sought solution to at least part of the so called second parameter problem, that is the wildely differing morphology of the horizontal branches (HB) in clusters which have about the same age and chemistry. Along this line of thought, Vittoria Caloi

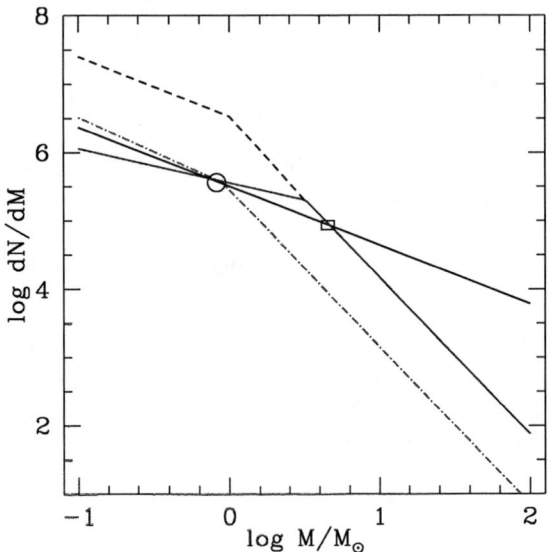

FIGURE 2. Sketch of possible IMFs for NGC2808. A normal IMF, which follows Salpeter's slope at $M \geq 0.9 M_\odot$ and is flatter at smaller masses, would either pass through the open circle representing the population of the HB red clump (at mass $\simeq 0.8 M_\odot$, dash-dotted line), or through the rectangle at 4– $5 M_\odot$ (dashed-continuous line). An IMF passing through both points would have a shape $dN/dM \propto M^{-1}$, or it must be discontinuous, with a small number of low mass stars compared to the larger mass stars. Notice that the normal IMF, passing only through the open circle, predicts a number of netron stars and black holes much smaller than the IMF anchored to the rectangle.

and I [9] have developped a model able to explain the complex bimodal morphology of the HB in the cluster NGC 2808, reconstructing the initial mass function needed to account for the population of the red part of the HB, which is attributed to the first burst of star formation, and for the blue part of the HB, which is described as the helium enriched descendants of the second stage of star formation.

The description provides two points of the initial mass function (IMF), which indeed depict a very atypical shape (Figure 2). In order to have enough stars in the blue part of the HB, we would need a *very flat* IMF. This could imply that the GC intermediate mass stars could be born preferentially in the central regions of the cluster, or could be dynamically segregated there. Evidence for clustering of massive stars in the central parts of young clusters is accumulating [23], [27], [1]. It is possible that the ejecta of the intermediate mass stars gave birth to the second generation, and the external parts of the cluster, containing all the other first generation low mass stars, were lost, leaving the present proportion of first and second generation stars. The IMFs shown in Fig. 2 can be considered then simple examples of what actually should be obtained by considering the proper dynamic evolution of the cluster, including the loss of low mass stars from it.

If the IMF of the first generation, which we can pin down at the intermediate masses $\sim 4-6 M_\odot$, can be extrapolated to the larger masses, which explode as supernovae and leave a neutron star or black hole remnant, the total number of compact remnants is about a factor 5–7 larger than if we consider a standard IMF. Although the numbers provided by [9] should be taken as helpful indication, and certainly not as predictions, we see that this large number of remnants helps to solve the problem that GCs harbour a large number of millisecond pulsars (MSPs). The richest GCs (in particular 20 MSPs have been discovered in the cluster 47 Tuc [4]) may contain more than ~ 1000 NSs – Pfahl et al. 2003 [24]– a number embarassingly large, close to the total number which is predicted by a normal IMF (although the revision of the NS numbers by Heinke et al. 2005, in these Proceedings brings this figure down to a minimum of ~ 300 NS). Therefore, most NSs should remain in the cluster, implying a high retention factor, at variance with the characteristic kick speeds of single radiopulsar in the Galaxy. This in fact is an average $<v> \sim 250-300$ Km/s [14], much larger than the escape velocity from typical clusters (~ 25 Km/s). Pfahl et al. (2003) [24] have shown that the retention factor is between 1% and 8%, even including consideration of binariety, but they point out that this value would have been much smaller if they had taken into account a number of other effects. The retention problem becomes much less severe if the IMF of NS progenitors follows the constraint we obtain at the intermediate mass stars, necessary to interpret the blue HB morphology.

ACKNOWLEDGMENTS

I thank all the friends and coworkers which have contributed to the ideas and work presented here, and especially Luciano Burderi, Vittoria Caloi, Tiziana Di Salvo, Paolo Ventura, and the phD students Giuseppe Lavagetto and Anamaria Teodorescu. Work on MSPs at the INAF-OAR is supported by the Italian Minister of Research (MIUR) under the Cofin 2003 national program.

REFERENCES

1. Brocato, E., Castellani, V., Di Carlo, E., Raimondo, G., & Walker, A. R. 2003, AJ, 125, 3111
2. Burderi, L., et al. 2001, ApJ, 560, L71
3. Burderi, L., D'Antona, F. & Burgay, M. 2002, ApJ, 574, 325
4. Camilo, F., Lorimer, D. R., Freire, P., Lyne, A. G., & Manchester, R. N. 2000, ApJ, 535, 975
5. Cottrell, P. L. & Da Costa, G. S. 1981, ApJ, 245, L79
6. D'Amico, N., Lyne, A. G., Manchester, R. N., Possenti, A., & Camilo, F. 2001, ApJ, 548, L171
7. D'Antona, F., Gratton, R., & Chieffi, A. 1983, Memorie della Societa Astronomica Italiana, 54, 173
8. D'Antona, F., Caloi, V., Montalbán, J., Ventura, P., & Gratton, R. 2002, A&A, 395, 69
9. D'Antona, F. & Caloi, V. 2004, ApJ, 611, 871
10. Di Salvo, T., Burderi, L. & D'Antona, F. 2005, in preparation
11. Ergma, E. 1996, A&A, 315, L17

12. Ferraro, F.R., Possenti, A., D'Amico, N., Sabbi, E. 2001, ApJ 561, L93
13. Gratton, R., Sneden, C. & Carretta, E. 2004, Annu. Rev. A &A, 42, 385
14. Hansen, B. M. S. & Phinney, E. S. 1997, MNRAS, 291, 569
15. Hansen, B.M.S., Kalogera, V., & Rasio, F.A. 2003, ApJ, 586, 1364
16. Heinke, C.O., et al. 2005, in these Proceedings
17. Kaluzny, J., Rucinski, S.M. & Thompson, I.B. 2002, ApJ, 574, 325
18. Illarionov, A., & Sunyaev, R. 1975, A&A, 39, 185
19. Lavagetto, G., Burderi, L., D'Antona, F., Di Salvo, T., Iaria, R., Robba, N.R. 2004, MNRAS, 348, 73
20. Mazzitelli, I. 1979, A&A, 79, 251
21. Nelson, L.A. & Rappaport, S. 2003, ApJ, 598, 431
22. Orosz J.A. & van Kerkwijk, M. H. 2003 A&A, 397, 237
23. Panagia, N., Romaniello, Scuderi & Kirshner 2000, ApJ 539, 197
24. Pfahl, E., Rappaport, S. & Podsiadlowski, P. 2003, ApJ, 573, 283
25. Ruderman, M., Shaham, J., & Tavani, M. 1989, ApJ, 336, 507
26. Shaham, J., & Tavani, M. 1991, ApJ, 377, 588
27. Stolte, A., Grebel, E. K., Brandner, W., & Figer, D. F. 2002, A&A, 394, 459
28. Tauris, T. M., & Savonije, G. J. 1999, A&A, 350, 928
29. Thorsett, S. E., & Chakrabarty, D. 1999, ApJ, 512, 288
30. Tutukov, A. V., Fedorova, A. V., Ergma E., & Yungelson, L. R. 1985, *Soviet Astron. Lett.*, **11**, 123
31. Ventura P., Zeppieri A., Mazzitelli I., D'Antona F., 1998, A&A, 334, 953
32. Ventura, P., D'Antona, F., Mazzitelli, I., & Gratton, R. 2001, ApJ Letters, 550, L65

Formation and evolution of compact binaries with an accreting white dwarf in globular clusters.

N. Ivanova* and F.A. Rasio*

Department of Physics and Astronomy, 2145 Sheridan Rd, Evanston, IL 60208, USA

Abstract. The population of compact binaries in dense stellar systems is affected strongly by frequent dynamical interactions between stars and their interplay with the stellar evolution. In this contribution, we consider these effects on binaries with a white dwarf accretor, in particular cataclysmic variables and AM CVns. We examine which processes can successfully lead to the creation of such X-ray binaries. Using numerical simulations, we identify predominant formation channels and predict the expected numbers of detectable systems. We discuss also why the distribution of cataclysmic variables has a weaker dependence upon the cluster density than the distribution of quiescent low-mass X-ray binaries and why dwarf nova outbursts may not occur among globular cluster cataclysmic variables.

Keywords: binaries: close – binaries: general – globular clusters: general – globular cluster: individual (NGC 104, 47 Tucanae) – stellar dynamics.
PACS: 97.8.-d; 98.20.Gm

INTRODUCTION

In the past few years, substantial progress has been made in optical identification of *Hubble Space Telescope* counterparts to *Chandra* X-ray sources in several globular clusters (GCs). Valuable information was obtained for the population of cataclysmic variables (CVs), chromospherically active binaries and quiescent low-mass X-ray binaries (qLMXBs) (Edmonds et al. 2004, Heinke et al. 2003b, Haggard et al. 2004). For the first time we can compare populations of such binaries in GCs and in the Galactic field, as well as calculate the rates of formation of these binaries and their population characteristics. In particular, as many as 22 CVs have been already identified in 47 Tuc, posing a number of interesting questions: e.g., their ratio of X-ray flux to optical flux is higher than in field CVs (Edmonds et al. 2003) and they do not show dwarf nova outbursts, which are common for field CVs (Shara et al. 1996). Also, it has been found that the specific incidence of harder X-ray sources (primarily CVs) depends more weakly on density than that of qLMXBs (Heinke et al. 2003a).

The present population of close binaries with a white dwarf (WD) accretor in GCs has not necessarily evolved from primordial binaries, since it can be influenced by dynamical encounters. In this contribution we study these processes by combining an advanced binary population code and careful treatment of all dynamical interactions with a simplified dynamical cluster background model. We first consider what mechanisms are likely to produce a CV or an AM CVn and then compare our predictions with the results of numerical simulations. We also discuss how the formation of mass-transfer systems with

CP797, *Interacting Binaries: Accretion, Evolution, and Outcomes,*
edited by L. Burderi, L. A. Antonelli, F. D'Antona, T. Di Salvo,
G. Luca Israel, L. Piersanti, A. Tornambè, and O. Straniero
© 2005 American Institute of Physics 0-7354-0286-8/05/$22.50

a WD accretor is different from the formation of these with systems with a neutron star (NS) accretor.

BINARIES IN A DENSE CLUSTER

There are several ways to destroy a primordial binary in a globular cluster. If the binary is in a dense region, dynamical encounters play a significant role in its evolution. For instance, there is a high probability to "ionize" a soft binary as a result of a dynamical encounter. In the case of a hard binary, destruction can happen through a physical collision during the encounter. Additionally, a primordial binary can be destroyed through an evolutionary merger (most important for hard binaries) or following to a SN explosion. Overall, we find that, if a typical cluster initially had as many as 100% of its stars in primordial binaries, the binary fraction at an age of 10-14 Gyr will be only $\leq 10\%$ (Ivanova et al. 2005).

The typical formation scenario for CVs in the field (low density environment) involves the common envelope (CE) evolution. The minimum initial orbital period for the progenitor binary is about 100 days and the primary mass has to be smaller than 8 $M_\odot$. In the core of a GC with core density $\rho_c \sim 10^5$ pc^{-3}, a binary with such a period will experience a dynamical encounter before its primary leaves the main sequence. The primordial channel for the CV formation could succeed therefore only if this binary entered the dense cluster core after the CE event. The contribution of the primordial channel depends mainly on the cluster half-mass relaxation time, which regulates how fast binaries will segregate into the central dense region. The situation is even more striking for the formation of AM CVn. In this case, the standard formation channel requires the occurrence of two CE events, and the primordial binary is expected to be even wider.

Dynamical formation of CVs.

There are two main types of dynamical encounters that lead to the dynamical formation of a binary consisting of a main sequence star (MS) and a WD: exchange interactions, and physical collisions between a red giant (RG) and a MS star. However, only a fraction of the dynamically formed MS-WD binary systems is capable of becoming a CV. We will consider first the post-encounter binary evolution of a dynamically formed MS-WD binary (to find out which post-encounter parameters favor CV formation), and then determine which encounters give these specific parameters.

A close MS-WD binary looses its angular momentum through magnetic braking (MB) and gravitational wave (GW) emission. The orbital shrinkage due to the synchronization of the MS star with the orbital motion can be neglected: less than a few percent of the total orbital angular momentum is required to spin-up a MS. A post-exchange hard binary has an average eccentricity $e \sim 0.7$, following a thermal distribution, and a post-collision binary has eccentricity $e \geq 0.4$ (Lombardi et al. 2005).

Depending on the MB prescription – the standard MB (Skumanich 1972, Pylyser & Savonije 1988) or the reduced MB (Ivanova & Taam 2003) – in a dynamically formed eccentric binary with $e \approx 0.7$, MB is at most comparable to GW emission (see Fig. 1).

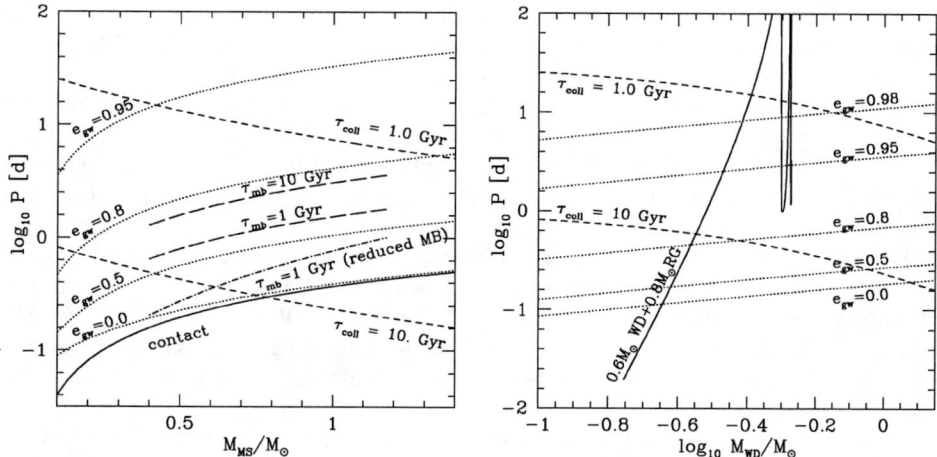

FIGURE 1. The fate of post-encounter binaries. **Left panel:** MS-WD binaries where the primary is a WD of 0.6 $M_\odot$. P is the post-encounter orbital period and M_{MS} is the mass of a MS secondary. The short-dashed lines show the binary periods for collision times of 1 Gyr and 10 Gyr. The dotted lines delineate the binaries that will shrink within 1 Gyr due to gravitational wave emission, for different post-exchange eccentricities. The long-dashed lines delineate the binaries that will shrink within 1 Gyr and 10 Gyr for the standard magnetic braking prescription, and dash-dotted line – for reduced magnetic braking. Below the solid line the binary is in contact. **Right panel:** WD-WD binaries where the primary is a WD of 0.6 $M_\odot$. P is the post-encounter orbital period and M_{WD} is the mass of a WD secondary. The dashed and dotted lines are the same as in the left panel. The solid line shows binaries that can be formed through a collision between a WD of 0.6 $M_\odot$ and a RG of 0.8 $M_\odot$ ($\alpha_{CE}\lambda = 1$).

In order to become a CV within 1 Gyr, a binary should have a post-encounter period ~ 1 day (the separation is then $\approx 10 R_\odot$), or $P \sim 10$ day and very high eccentricity, $e \geq 0.96$. It is not clear which post-exchange binaries will dominate in producing CVs. At first glance, in the encounter with a hard binary, the post-exchange separation is comparable to the separation in the pre-encounter binary. Therefore, considering the collision time, the formation of a wider binary through exchange is more likely. On the other hand, it is rather unlikely to produce such a very high eccentricity through an exchange.

In the case of a MS-WD binary with $P \geq 10$ days and a moderate eccentricity $e \leq 0.8$, dynamical interactions that occur after binary formation cannot harden such a binary significantly. Let us consider a binary consisting of a 1 $M_\odot$ MS and a 0.6 $M_\odot$ WD. Even if each hardening encounter could reduce the orbital separation by 50%, the hardening of this binary from 10 days to 1 day (at this period MB starts to be efficient) will take about 20 Gyr. In contrast to hardening, another mechanism is important: *eccentricity pumping*. The mean time between successive collisions $\tau_{coll} \leq 1$ Gyr and therefore a binary can experience many encounters. If the acquired eccentricity $e \geq 0.95$, the binary can shrink through GW emission even if its initial period is bigger than 10 days.

As the post-exchange separation in a binary that becomes a CV is comparable to the separation in the pre-encounter binary, the pre-encounter binary has to be very tight.

It is not likely that, in the case of a tight binary (with moderate eccentricity), the pre-encounter binary was a MS-MS binary or a MS-WD binary: an encounter with so close a binary will likely lead to a physical collision rather than to an exchange (Fregeau et al. 2004). This restriction, together with $\tau_{\text{coll}} \sim$ few Gyr for pre-collision binaries of $P \leq 1$ day, predicts that most of the post-exchange CVs can be formed in a three-body encounter where the pre-encounter binary had $P \geq 3$ days and the post-encounter binary has high eccentricity $e \geq 0.8$ (or this eccentricity was increased in subsequent encounters). No hardening is expected.

For binaries formed through MS-RG collisions, the binary separation can be estimated using the standard CE prescription with $\alpha_{\text{CE}}\lambda = 1$ (Iben & Livio 1993). The consideration of a RG inner structure through its evolution predicts that only collision of a MS star with a RG with core mass $M_{\text{core}} \leq 0.3 M_\odot$ or with a giant at the core helium burning stage with $M_{\text{core}} \approx 0.6 M_\odot$ will lead to the formation of a tight enough binary (see also Fig. 1, right panel).

A WD of $0.3\ M_\odot$ cannot be formed unless it evolved through a CE event or a physical collision. A post-CE binary for this WD is very hard and has $\tau_{\text{coll}} \geq 10$ Gyr, so the number of single $0.3 M_\odot$ WDs is negligible. This restricts the exchange formation channel for CVs with a low-mass WD: all CVs with a low mass WD companion must be formed either through a CE event (in a primordial binary or in a dynamically formed binary with $P \sim 10 - 100$ days), or as a result of a physical collision.

Dynamical formation of AM CVn systems.

Let us consider first a hard post-exchange WD-WD binary. As we described previously, the typical eccentricity is $e \sim 0.7$, the separation is comparable to the pre-exchange separation; there is no post-exchange hardening. The main difference with MS-WD binaries is that post-exchange WD-WD binary periods that will allow a binary to evolve to mass transfer (MT) are several times smaller (see Fig. 1, right panel). The collision time for pre-encounter binaries is about 10 Gyr, therefore making the formation an AM CVn binary through three-body exchange (with subsequent GW decay) rather unlikely.

A variant for this channel is the dynamical formation of a relatively wide MS-WD binary that will subsequently evolve through CE. The collisional time for these relatively wide MS-WD binaries is less than 1 Gyr and therefore a significant fraction of such binaries may participate in some other encounter (destruction, hardening or eccentricity pumping) before the CE occurs.

The second important channel is again a physical collision, in this case of a single WD with a RG (see Fig. 1, where we show possible outcomes of a such a collision).

We therefore expect that only a post-CE system can become an AM CVn, where the post-CE system could be from a primordial binary, a post-collision binary, or a dynamically formed binary.

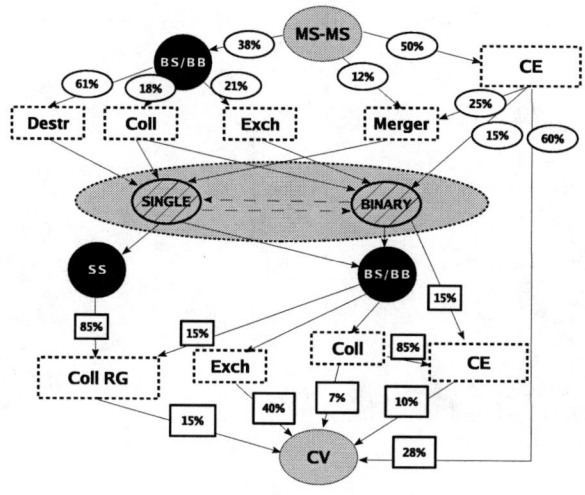

FIGURE 2. Main formation channels of CVs in a typical cluster.

NUMERICAL RESULTS

To study stellar evolution in a dense system, we use a Monte Carlo approach which couples one of most advanced binary population synthesis codes currently available (Belczynski et al. 2002), a simple model for the cluster, and a small N-body integrator for accurate treatment of all relevant dynamical interaction processes (Fregeau et al. 2004). The complete description of the method is provided in Ivanova et al. 2004. Here we treated physical collisions with a simple CE prescription with post-collision $e = 0$, though in a more realistic treatment the post-collision binary parameters should depend on the impact parameter (c.f. Lombardi et al. 2004).

We studied a "typical" cluster, starting with $N = 10^6$ stars and initial binary fraction 100%. This high primordial binary fraction is needed in order to match the observed binary fraction in GC cores today (Ivanova 2005). The core density $\rho_c = 10^{4.7} M_\odot$ pc^{-3}, one-dimensional velocity dispersion $\sigma_1 = 10$ km/s, escape speed from the cluster $v_{esc} = 43$ km/s, and half-mass relaxation time $t_{rh} = 1$ Gyr. We also considered a 47 Tuc-type cluster, characterized by $\rho_c = 10^{5.2} M_\odot$ pc^{-3}, $\sigma_1 = 11$ km/s, $v_{esc} = 60$ and $t_{rh} = 3$ Gyr. We adopt the broken power law IMF of Kroupa (2002) for single primaries, a flat mass-ratio distribution for secondaries and the distribution of initial binary periods constant in logarithm between contact and 10^7 d. The mass of the cluster at 11 Gyr is $\sim 2 \times 10^5 M_\odot$.

In Fig. 2 we show the main formation channels for CVs that operate in a typical cluster at 11 Gyr. The most important channel for CV formation is through an exchange encounter – it provides 40% of CVs, and also 7% of CVs are in binaries that experienced a merger during the last three-body encounter; 15% of CVs were formed as a result of a physical collision between a RG and a MS star; in 10% the CE occured in a dynamically formed binary; and 28% of CVs are provided by the primordial channel. In total, the number of post-CE systems and post-exchange systems is about the same.

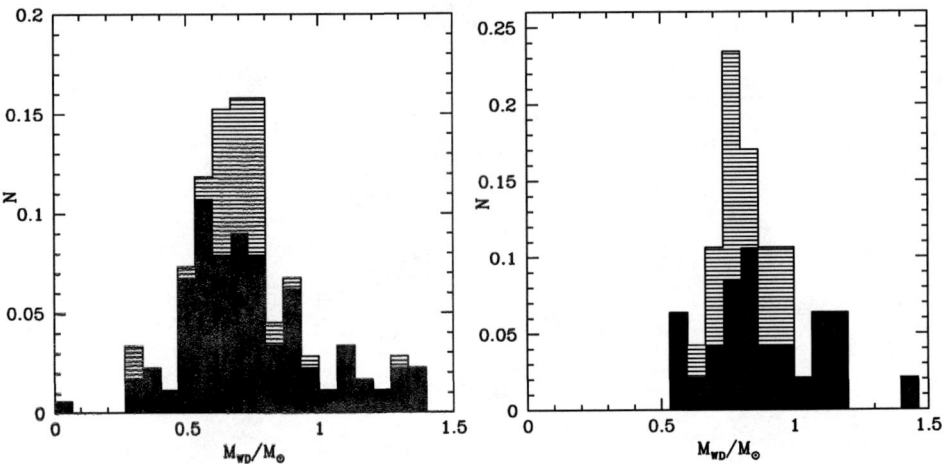

FIGURE 3. CVs (left) and AM CVns (right) in a cluster core: the accreting WD-mass distribution. The solid filled area corresponds to dynamically formed binaries; the hatched area to systems formed directly from primordial binaries.

Typical participants of a successful physical collision (leading to CV formation) are a MS star of $0.3 - 0.9\ M_\odot$ and a RG of about $1 - 1.7 M_\odot$ with a core around $0.3 M_\odot$ or He core burning giant with core mass around $0.5 M_\odot$. CVs formed this way are similar to post-CE CVs from primordial binaries, the typical post-exchange period is about 0.2 day.

For CVs formed through an exchange, there are two types of successful three-body encounters: (a) a single, relatively heavy WD of about $0.7 - 1.4 M_\odot$ and a MS-MS binary with a total mass usually $\leq 1 M_\odot$; (b) a single relatively massive MS star of around the turn-off mass and a MS-WD or WD-WD binary. In most cases the WD-MS or WD-WD binary was not a primordial binary, but a dynamically formed binary. The number of successful encounters between MS star and WD-WD binary is relatively small, and no successful four-body encounter occurred. A post-exchange binary typically has a WD heavier than a post-CE binary. As a result, the distribution of WD masses in CVs is more populated at the high mass end (see Fig. 3) compared to the field population, which has two very well distinguished peaks at ~ 0.4 and $\sim 0.5\ M_\odot$ and exponentially decreases with WD mass (Willems, priv. comm.). As expected, most CVs with a heavy WD were formed dynamically.

In Fig. 4 we show the main formation channels for AM CVns that occur in a typical cluster. As predicted, the main difference with CV formation is that there are only post-CE channels. The role of physical collisions for AM CVns is slightly more important than for CVs, although this is only for the relative fraction. The number of CVs formed by physical collisions is about 3 times higher than the number of AM CVns formed by physical collisions. At the moment of a physical collision, in 2/3 of cases, the participants are a RG of $0.9 - 1.4 M_\odot$ and a relatively massive MS star with well

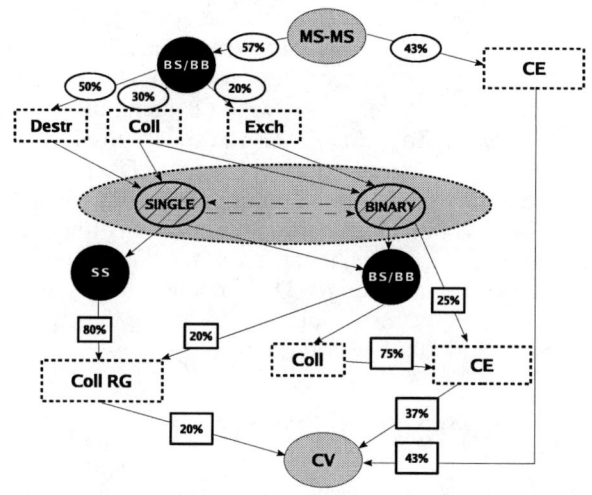

FIGURE 4. Main formation channels of AM CVns in a typical cluster.

developed He core. The AM CVns we see are the result of MT that started when the donor was at the late MS stage, stripped down to the He core (the mass of the He donor in this system is typically about $0.01 M_\odot$), or the system has passed through another CE event. About a third of physical collisions occur between a RG and heavy WD. Overall, this channel provides AM CVns with accretor masses at present (11 Gyr) from $0.55 M_\odot$ to $1.4 M_\odot$, compared to the primordial channel where masses of accretors are mainly distributed between $0.55 M_\odot$ and $0.9 M_\odot$ (Fig. 3; see also Nelemans et al. 2001).

The exchange formation channel for AM CVns involves an exchange encounter and a CE event. A single WD acquires a relatively massive companion that becomes later a RG or, with a 0.6-0.8 $M_\odot$ MS star, a second encounter occurs, with either a physical collision between MS stars or exchange with a more massive MS.

The total numbers of CVs and AM CVns present in a typical cluster are not very different from these in field population — they are comparable (per whole cluster population) and only 2-3 times larger in the core than in the field (per unit mass). Dynamical formation is responsible only for 60%-70% of CVs in the core, but there is also a fraction of CVs that never entered the core. The GC core density variation therefore does not play a significant role, in contrast to the case of NS binaries, where almost systems were formed dynamically (Ivanova et al. 2004) and the numbers have a direct strong dependence on the cluster collision rate (Pooley et al. 2003). We expect to have about 1 CV per $200-300 M_\odot$ in the core of a typical cluster and about 1 CV per $400-600 M_\odot$ in a 47 Tuc type cluster (about 100 CVs in the core of 47 Tuc, in quite reasonable agreement with observations, Edmonds et al. 2003). The number of AM CVns is typically 3-4 times smaller than the number of CVs.

One of the remaining questions is why CVs in GCs do not show nova outbursts. There are many differences between the binary properties of GC and field CV populations. From our simulations, we see that an accreting WD in a GC CV is typically more massive than in field, and the MS donor is less massive. A CV therefore is characterized by a smaller MT rate, $\dot{M} \sim 10^{-11} M_\odot \mathrm{yr}^{-1}$, which is much smaller than the typical value for dwarf novae ($\sim 10^{-9} M_\odot \mathrm{yr}^{-1}$). Nevertheless, we should mention this this value is large enough to provide $L_X \sim 10^{33} \mathrm{ergs\ s}^{-1}$, which is about the maximum observed luminosity for CVs in, e.g., 47 Tuc. Small MT rates also possibly explain why the optical fluxes from these objects are unexpectedly low in UV (Edmonds et al. 2003). According to the disk instability model which is currently used to describe dwarf novae cycles (Frank, King & Raine 2002), the condition for instability to occur is described in terms of a critical MT rate. For instance, there is a critical accretion rate below which a CV disk is cold and stable. The accreting WD did not necessarily evolve through CE. This possibly results in a higher magnetic field WD. A higher than usual magnetic field could also help suppress disk instabilities even further.

ACKNOWLEDGMENTS

We thank B. Willems and C. Heinke for helpful discussions. NI was supported by a *Chandra Theory* grant, FAR acknowledges support from NASA ATP Grant NAG5-12044.

REFERENCES

1. Belczynski, K., Kalogera, V., & Bulik, T. 2002, ApJ, 572, 407
2. Edmonds, P. D., Gilliland, R. L., Heinke, C. O., & Grindlay, J. E. 2003, ApJ, 596, 1177
3. Edmonds, P. D., Kahabka, P., & Heinke, C. O. 2004, ApJ, 611, 413
4. Frank, J., King, A., & Raine, D. J. 2002, Accretion Power in Astrophysics: Third Edition, by Juhan Frank, Andrew King, and Derek J. Raine. Cambridge University Press, 2002, 398 pp.
5. Fregeau, J. M., Cheung, P., Portegies Zwart, S. F., & Rasio, F. A. 2004, MNRAS, 352, 1
6. Haggard, D., Cool, A. M., Anderson, J., Edmonds, P. D., Callanan, P. J., Heinke, C. O., Grindlay, J. E., & Bailyn, C. D. 2004, ApJ, 613, 512
7. Heinke, C. O., Grindlay, J. E., Lugger, P. M., Cohn, H. N., Edmonds, P. D., Lloyd, D. A., & Cool, A. M. 2003a, ApJ, 598, 501
8. Heinke, C. O., Grindlay, J. E., Edmonds, P. D., Lloyd, D. A., Murray, S. S., Cohn, H. N., & Lugger, P. M. 2003b, ApJ, 598, 516
9. Iben, I. J. & Livio, M. 1993, PASP, 105, 1373
10. Ivanova, N., Fregeau, J.M., & Rasio, F.A. 2004, to appear in Binary Radio Pulsars, ASP Conf. Series, ed. F.A. Rasio & I.H. Stairs
11. Ivanova, N., Belczynski, K., Fregeau, J.M., & Rasio, F.A. 2005, MNRAS in press
12. Ivanova, N. & Taam, R. E. 2003, ApJ, 599, 516
13. Kroupa, P. 2002, Science, 295, 82
14. Lombardi, J.C., Dooley, K., Proulx, Z., Ivanova, N., Rasio, F. 2005, in preparation
15. Nelemans, G., Portegies Zwart, S. F., Verbunt, F., & Yungelson, L. R. 2001, A&A, 368, 939
16. Pooley, D., et al. 2003, ApJ, 591, L131
17. Pylyser, E. & Savonije, G. J. 1988, A&A, 191, 57
18. Skumanich, A. 1972, ApJ, 171, 565
19. Shara, M. M., Bergeron, L. E., Gilliland, R. L., Saha, A., & Petro, L. 1996, ApJ, 471, 804

Eclipsing binaries in the galactic globular cluster Omega Centauri.

A.Calamida*, G.Bono*, R.Buonanno*, C. E. Corsi*, M.Monelli*, M.Dall'Ora*, L.M.Freyhammer[†] and A.Munteanu**

*INAF-Osservatorio Astronomico di Roma, Via Frascati 33, 00040, Monte Porzio Catone, Italy
[†]Royal Observatory of Belgium, Ringlaan 3, B-1180 Brussels, Belgium
**Univeritat Politècnica de Catalunya, Spain

Abstract. We present a photometric survey aimed at the identification of variable stars in the galactic globular cluster ω Cen. Our photometric catalogue is based on $\sim$ 400 exposures of the central region of the cluster(Fov $\sim$ 15'×15') collected with the Danish Telescope. The catalogue has an accuracy of about $\sim$ 0.02 mag at B $\sim$ 18 mag. We present here preliminary results concerning the possible identification of two variable stars and an eclipsing binary candidate in ω Cen.

Keywords: binaries: eclipsing - globular clusters: general
PACS: 97.80 - 98.20

INTRODUCTION

ω Cen is the most massive Galactic Globular Cluster (M $\sim 5 \times 10^6 M_\odot$, Meylan et al. 1995) and the only one which clearly shows a well-defined spread in metallicity. According to recent estimates based on sizable samples of evolved Red Giant and Sub-Giant stars, the metallicity distribution shows three peaks around [Fe/H] = -1.7, -1.5 and -1.2, together with a tail of metal-rich stars approaching [Fe/H] $\sim$ 0.5 (Norris et al. 1996; Hilker et al. 2004; Pancino et al. 2004). Even though ω Cen presents properties which still need to be properly understood, its stellar content is a gold mine to investigate several open problems concerning stellar evolution and its dependence on the metallicity. This could apply not only to red giant stars, but also to RR Lyrae stars, hot HB stars and binary stars. Accurate stellar masses and radii can be determined by analyzing the light curves and radial velocity curves of eclipsing binaries. Moreover, the comparison between theoretical models and data can supply fundamental constraints on basic parameters, such as age, distance and helium abundance for the cluster binaries (Niss et al. 1978, Paczynsky 1997). The identification and the analysis of binary stars in ω Cen is therefore very useful to provide an indipendent determination of distance and age. Thompson et al. (2001), using a detached eclipsing binary, OGLE 17, has estimated a distance modulus of $(m-M)_V$ =14.05±0.11 for this cluster. Moreover, estimates of binary stars' helium abundance in ω Cen could be helpful to investigate the origin of the helium enhancement that some of the different stellar populations in the cluster seem to show (Piotto et al. 2005; Lee et al. 2005).

The knowledge of globular clusters' binary fraction is also important for better understanding their dynamical evolution: the gravitational binding energy of the binaries, actually, through encounters with single stars, can be extracted and converted in kinetic energy, and halt the collapse of the cluster core. The efficiency of this process strongly depends on the binary fraction in the cluster. The only accurate estimate of a cluster binary frequency has been recently provided by Albrow et al. (2001), who estimated an overall binary frequency of $\sim$ 13-14% for the globular cluster 47 Tucanae, from HST observations of the cluster's dense core (c = log $r_t/r_c \sim$ 2.03).

ω Cen could be particularly useful in the study of binary frequencies in clusters; ithas an half-mass relaxation time (t $\sim \times 10^{11}$ yrs) longer than the cluster age and, therefore, it should have preserved a large fraction of primordial binaries, which are expected to have periods ranging from 200 to 4000 days. This cluster has a low central concentration (c $\sim$ 1.61), with a central density of log$\rho_0 \sim$ 3.12 $L_{V_\odot}/pc^3$, that makes it a good laboratory to search for binaries in its core. A preliminary estimate of a binary frequency of about 3-4% was given in Mayor et al. (1996). This estimate is pretty low but it means that we expect to find at least $\sim$ 3 - 4$\times 10^4$ binaries in this cluster.

The Cluster AgeS Experiment (CASE) has up to now identified 30 eclipsing binaries, plus about 30 contact binaries (mostly short period, P < 1day) and about 70 SX Phoenicis (pulsating Blue Stragglers) stars in the central part of ω Cen.

The main purpose of our study is to search for eclipsing binaries, short period variables like SX Phe, and long period variables, thanks to the long time base-line of our dataset.

OBSERVATIONS AND DATA REDUCTION

B, V and Gunn i images of ω Cen were collected during the years 1995, 1996, 1998 and 1999 with the Danish 1.54m Telescope at ESO, La Silla. We have a total of $\sim$ 5000 observations which cover a 15$\times$15 arcmin^2 field centered on the cluster. The seeing conditions were good, ranging from 1" to 1.5".

The data have been reduced with DAOPHOTII/ALLSTAR (Stetson 1987), using an IDL procedure to accurately select the best point-spread function (PSF) stars across each frame. All the frames have been geometrically transformed to a common reference catalogue for each year separately, using transformations determined with DAOMATCH/DAOMASTER.

We present here some preliminary results for the 1995 observations, which catalogue includes $\sim 2 \times 10^5$ stars and has a typical accuracy of $\sim$ 0.02 mag at B = 18 mag.

DATA ANALYSIS

Three-color time-series CCD photometry (B, V, i) was performed during the year 1995, covering the central field for 9 nights and single-channel (B) photometry for the central 9'$\times$9' field (F0) in two separate nights. Every 15-20 minutes each star is covered with a B,V,i color cycle. This means that stars with pulsation periods larger than 0.04 days could in theory be detected. Opposed to this, each F0 image has been acquired instead every 3-4 minutes to search for short period variables, such as SX Phe stars. We present the analysis of just a small portion of our 1995 dataset, focusing on data from field F3 (Fig. 1).

FIGURE 1. Mosaic of all 1995 images of ω Cen: the box is the F3 pointing (9'×9').

Identification of variables

To identify variable stars we adopted an approach similar to the Welch & Stetson (1993) algorithm. This method correlates variations in the photometric bands and estimates a variability index W:

$$W = \sqrt{\frac{n}{n-1}} \sum_{i=1}^{n} \prod_{j=1}^{3} (\delta m_{ij})$$

where m_{ij} are the magnitudes, i=1,...n, the number of measurements, and j=1,2,3, the number of photometric bands. In our case, we determined the normalized magnitude residuals for the three bands of F3 pointing:

$$\delta m_{ij} = \frac{m_{ij} - \bar{m}_j}{\sigma_{m_{ij}}}$$

where $\sigma_{m_{ij}}$ is the estimated uncertainties for each magnitude determination and $\bar{m}_j$ is the weighted mean in the j band. The magnitude residual are plotted against each other in Fig.2. For the selection we adopted the criterion:

$$\sum_{i=1}^{n} \frac{|\delta m_{ij}|}{n} \geq 4$$

in one of the three bands. We obtained $\sim$ 1000 candidates as variable stars, a list that was reduced to 17 candidates after a more accurate analysis (see Fig.3).

Additional searches for variables were performed by applying Fourier analysis based on the Scargle periodogram and least-square fitting of the unevenly sampled light curves

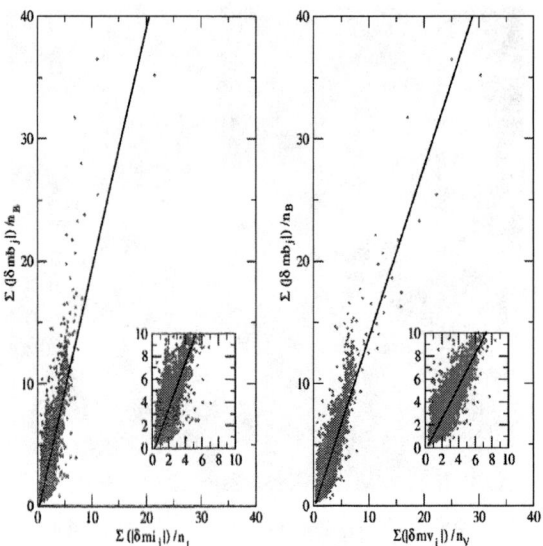

FIGURE 2. The normalized magnitude residuals for the B band versus the I band (a); B band versus the V band (b). See text for details.

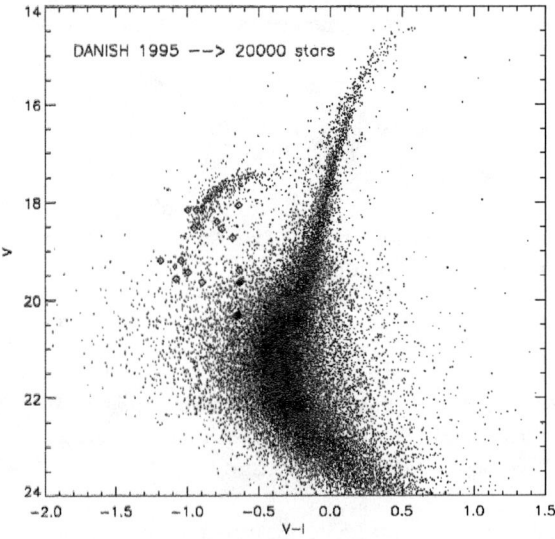

FIGURE 3. (V,V-I) CMD of ω Cen. The 17 candidate variables stars are marked with diamonds symbols. Two of the three stars we selected to perform a preliminary period analysis are marked with filled diamonds(see text).

to harmonic functions. The Lomb-Scargle analysis (Woodward 1994) is a completely different method of spectral analysis with respect to the Fourier analysis, not only because it deals with unevenly spaced data, but also because it differs conceptually from the latter. The Lomb-Scargle method actually evaluates sines and cosines only at instants of time corresponding to the data. Moreover, the exact expression of the Lomb normalized periodogram (spectral power as a function of frequency) ensures that it is equivalent to the equation that one would obtain by estimating the spectral power at a given frequency by linear least-squares fitting to the model. It weights the data on a "per point" basis instead of on a "per time interval" basis, such as the Fourier method does, when the sampling can lead the latter approach into erroneous biased results. Thus this method, complementary to the classical Fourier method, can help us obtain a more reliable criterion for variable stars detection. This analysis, as well as selection our criteria above, can be considered preliminary as it was intended only as a study of which spectral analysis methods are able to provide a better and more reliable tools for identification and characterization of variable stars in general, and eclipsing variables in particular. We report here this preliminary period analysis performed on three selected candidates.

Example 1: a long-period pulsating variable

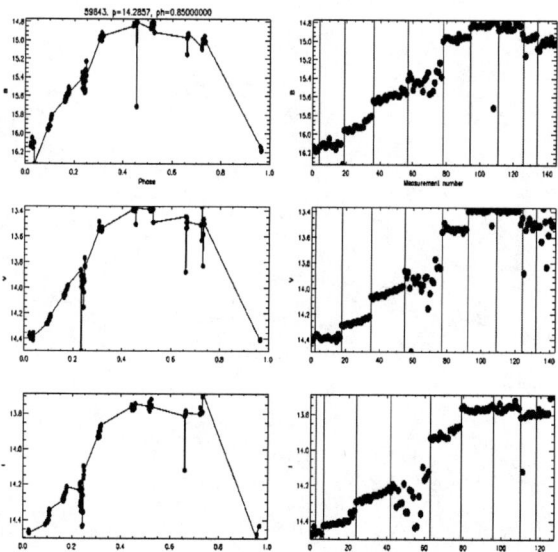

FIGURE 4. From top to bottom: B,V,i light curves of a RV Tauri type variable. Left panels show the light curves phased with a period of 14.26 days, while the right panels show the subsequent measurements with individual nights delimited by vertical bars. Note the non-photometric observing conditions on night #5 (points 42-63). This variable is V1 of Kaluzny et al. (2004): the estimated period by Kaluzny is about 29.34 days, that means our estimate is a period alias.

Example 2: a short-period pulsating variable

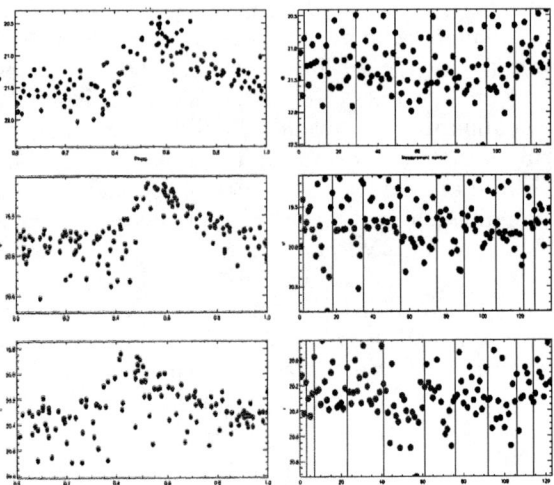

FIGURE 5. Like Fig.4, but for an SX Phe candidate with a period of 0.067 days. This is V199 SX Phe variable identified by Kaluzny et al. 2004.

Example 3: an eclipsing binary

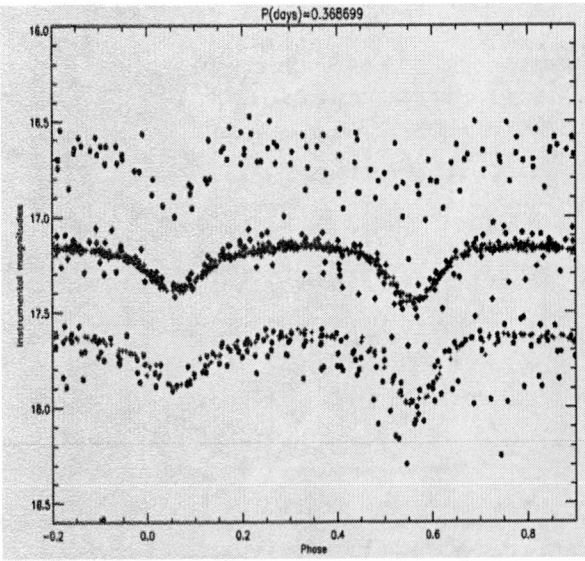

FIGURE 6. From top to bottom follows instrumental light curves in the i, V (diamond symbols) and B bands for the known eclipsing binary V205 (Kaluzny et al. 2004). BV light curves from Kaluzny et al. (2004) are superimposed with "+" symbols and our own data were added magnitude offsets to match these magnitudes. The light curves are phased with the orbital period.

CONCLUSIONS

We presented preliminary results concerning the detection of intrinsic variables and binary stars in ω Cen. Even though, the current photometric database has not been properly calibrated, we already detected several variables identified by OGLE. In the near future we plan to extend the same analysis to the other three fields and to include new time series data collected in the same bands with the same telescope. The entire dataset will allow us to detect variables and binaries with periods of the order of years, if any. To speed up the variable identification and the period analysis we are also developing new algorithms, since the entire catalogue will include more that 10^8 measurements. The homogeneous approach to the reduction of these data will provide the unique opportunity to investigate transient phenomena (mode transition, Bono et al. 1995; Bono et al. 2003) in regular variables and in long-period variables and to detect long secondary modulation such as the Blazhko effect (LaCluyze, A. et al. 2004).

ACKNOWLEDGMENTS

This work was supported by the Belgian Fund for Scientific Research (FWO) in the framework the project "IAP P5/36" of the Belgian Federal Science Policy, by MIUR/PRIN 2003 in the framework of the project: "Continuity and Discontinuity in the Galaxy Formation", and by Danish NSRC in the framework of the project: "Structure and evolution of stars: new insight from eclipsing binaries and pulsating stars".

REFERENCES

1. Albrow M. D. et al., *ApJ*, 559 1060 (2001)
2. Bono, G.; Castellani, V.; Stellingwerf, R. F., *ApJ*, 445 145 (1995)
3. Bono, G.; Petroni, S.; Marconi, M. ,"Interplay of Periodic", in *Cyclic and Stochastic Variability in Selected Areas of the H-R Diagram*, edited by C.Sterken, ASP, San Francisco, 2003, pp. 71
4. Harris, W. E., *yCat.7195*, OH (1996)
5. Hilker, M., Kayser, A., Richtler, T., Willemsen, P., *A&A*, 422 9 (2004)
6. Kaluzny,J.et al., *A&AS*, 125 343 (1997)
7. Kaluzny,J. et al., *A&A*, 424 1101 (2004)
8. LaCluyze, A. et al., *AJ*, 127 1653 (2004)
9. Lee, Y.W. et al., *ApJ*, 621 57 (2005)
10. Mayor, M., *ASPC*, 90 190 (1996)
11. Meylan, G.; Mayor, M.; Duquennoy, A.; Dubath, P., *A&A*, 303 761 (1995)
12. Norris,J. E.; Freeman, K. C.; Mighell, K. J., *ApJ*, 462 241 (1996)
13. Norris, J.E., *ApJ*, 612 25 (2004)
14. Niss, B., Jorgensen, H. E., Laustsen, S ., *A&AS*, 32 387 (1978)
15. Paczynski, B., "Space Telescope Science Institute Series",in *The Extragalactic Distance Scale*, edited by M. Livio, Cambridge: Cambridge University Press, 1997, pp. 273
16. Pancino, E.,"Carnegie Obs. Astrophysics Ser.", in *Origin and Evolution of the Elements*, edited by McWilliam & M. Rauch, Cambridge: Cambridge Univ. Press, 2004, pp. 45
17. Stetson,P.B., *PASP*, 99 191 (1987)
18. Thompson, I.B., Kaluzny, J., Pych, W., Burley, G. et al., *AJ*, 121 3089 (2001)
19. Welch, D. L.; Stetson, Peter B., *AJ*, 105 1813 (1993)
20. Woodward, R. C. Jr., Scherb F., Roesler F. L., Oliversen, R. J., *Icarus*, 111 45 (1994)

SESSION 2

MILLISECOND BINARY PULSARS

Milliseconds Pulsars in Low-Mass X-Ray Binaries

Deepto Chakrabarty

Department of Physics and Center for Space Research, Massachusetts Institute of Technology, Cambridge, MA 02139, USA

Abstract. Despite considerable evidence verifying that millisecond pulsars are spun up through sustained accretion in low-mass X-ray binaries (LMXBs), it has proven surprisingly difficult to actually detect millisecond X-ray pulsars in LMXBs. There are only 5 accretion-powered millisecond X-ray pulsars known among more than 80 LMXBs containing neutron stars, but there are another 11 "nuclear-powered" millisecond pulsars which reveal their spin only during brief, thermonuclear X-ray bursts. In addition, 2 of the accretion-powered pulsars also exhibit X-ray burst oscillations, and their unusual properties, along with the absence of persistent pulsations in most LMXBs, suggest that the magnetic fields in many LMXBs may be hidden by accreted material. Interestingly, the nuclear-powered pulsars offer a statistically unbiased probe of the spin distribution of recycled pulsars and show that this distribution cuts off sharply above 730 Hz, well below the breakup spin rate for most neutron star equations of state. This indicates that some mechanism acts to halt or balance spin-up due to accretion and that submillisecond pulsars must be very rare (and are possibly nonexistent). It is unclear what provides the necessary angular momentum sink, although gravitational radiation is an attractive possibility.

INTRODUCTION

Since the discovery of the first millisecond radio pulsar (Backer et al. 1982), it has been believed that these old, weak-field ($\sim 10^8$ G) pulsars were spun up to millisecond periods by sustained accretion onto neutron stars in low-mass X-ray binaries (NS/LMXBs; Alpar et al. 1982). It is known from accretion torque theory (e.g., Ghosh & Lamb 1979) and from observations of strong-field ($> 10^{11}$ G) accreting neutron stars (Bildsten et al. 1997) that steady disk accretion onto a magnetized neutron star will lead to an equilibrium spin period

$$P_{\rm eq} \sim 1\,{\rm s}\left(\frac{B}{10^{12}\,{\rm G}}\right)^{6/7}\left(\frac{\dot M}{10^{-9} M_\odot\,{\rm yr}^{-1}}\right)^{-3/7}, \qquad (1)$$

where B is the pulsar's surface dipole magnetic field strength and $\dot M$ is the mass accretion rate. Also, most neutron stars in LMXBs are old and have weak magnetic fields, based both on the occurrence of thermonuclear X-ray bursts (which require $B < 10^{10}$ G; see Joss & Li 1980) and on their X-ray spectral (Psaltis & Lamb 1998) and timing properties (see, e.g., van der Klis 2000). Thus, as long as we accept that sustained accretion onto neutron stars can somehow attenuate their strong birth fields down to $\sim 10^8$ G strengths (see, e.g., Bhattacharya & Srinivasan 1995), it is natural to expect neutron stars in LMXBs to be spinning at millisecond periods. (Note that although the general trend of

Equation (1) is expected to extend down to 10^8 G, the precise power-law dependences are likely to be modified; see Psaltis & Chakrabarty 1999).

A robust prediction of this model is that the neutron stars in LMXBs should be X-ray pulsars, since a 10^8 G field should be strong enough to truncate the Keplerian accretion disk and channel its flow onto the neutron star's magnetic poles. However, the detection of accretion-powered millisecond X-ray pulsars proved elusive for nearly two decades, with a series of X-ray missions failing to detect millisecond pulsations from NS/LMXBs down to stringent upper limits on the pulsed fraction (e.g., Vaughan et al. 1994). The launch of the *Rossi X-Ray Timing Explorer* (*RXTE*; Bradt, Rothschild, & Swank 1993; Jahoda et al. 1996) in December 1995 finally provided an instrument with sufficient flexibility and sensitivity to detect SAX J1808.4−3658, the first accretion-powered millisecond X-ray pulsar (Wijnands & van der Klis 1998; Chakrabarty & Morgan 1998), as well as several additional examples (see Table 1). A key element proved to be the highly flexible pointing and scheduling ability of *RXTE*. The Galactic X-ray sky is highly variable, with some sources lying dormant for years and only intermittently becoming active in X-ray emission. The combination of having both the *RXTE* All Sky Monitor to determine when an X-ray transient becomes active and the ability to rapidly repoint the main *RXTE* instruments at a newly active source were crucial in enabling the discovery of millisecond X-ray pulsars.

Along the way, *RXTE* detected other classes of rapid X-ray variability which also point to millisecond spins for these neutron stars but which also raised a number of new questions about the underlying physics.

RAPID VARIABILITY IN ACCRETING NEUTRON STARS

RXTE has identified three distinct classes of millisecond variability in accreting neutron stars:

- **Kilohertz quasi-periodic oscillations (kHz QPOs)** (van der Klis et al. 1996; see van der Klis 2000 for a review): These are pairs of relatively high-Q peaks in the X-ray intensity power spectrum whose frequencies drift by hundreds of hertz (200–1200 Hz) as the source intensity varies, but whose separation frequency (typically a few hundred Hz) remains roughly constant (see example in left panel of Figure 1). This approximate separation frequency has a unique, reproducible value for each of the over 20 NS/LMXBs in which this phenomenon is observed. These oscillations are believed to arise in the inner accretion disk flow, where the dynamical time scale is of order milliseconds.

- **X-ray burst oscillations** (Strohmayer et al. 1996; see Strohmayer & Bildsten 2005 for a review): These are nearly coherent millisecond oscillations observed only during thermonuclear X-ray bursts (which typically last ~ 10 s). The observed frequencies drift by a few Hz over ~ 5 s during the bursts, asymptotically reaching a maximum frequency that is unique and reproducible for each of the 13 sources in which this phenomenon is observed (see right panel of Figure 1). These maximum frequencies lie in the 270–619 Hz range. The oscillation (at least at the burst onset) is understood as a temperature anisotropy caused by the nuclear burning, with

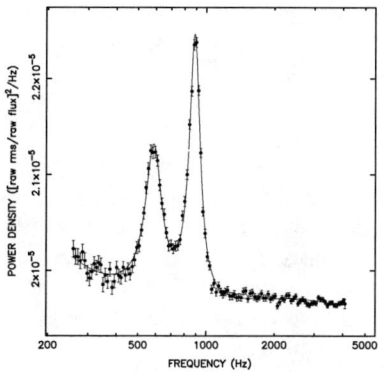
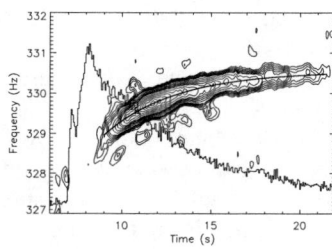

FIGURE 1. *Left:* Power spectrum of X-ray intensity showing kHz QPO pair in Sco X-1. Adapted from van der Klis et al. (1997). *Right:* X-ray burst oscillations in 4U 1702−43. The solid histogram shows the X-ray intensity history of the X-ray burst. The contours show the Fourier power level as a function of frequency and time, indicating a drifting oscillation starting at 328 Hz ($t = 7$ s) and ending at 330.5 Hz. Adapted from Strohmayer & Markwardt (1999).

the frequency drift interpreted as angular momentum conservation in a cooling, decoupled burning layer on the stellar surface (Strohmayer et al. 1997; Cumming & Bildsten 2000). Since their discovery, it was suspected that the millisecond oscillations were somehow tracing the stellar spin, but until recently there was some question about whether the oscillations might be a harmonic of the spin. A list of these sources is given in Table 1.

- **Persistent accretion-powered pulsations** (Wijnands & van der Klis 1998; see Wijnands 2004 for a review): These are the objects originally expected by the recycling hypothesis, NS/LMXBs whose persistent (non-burst) emission contains coherent millisecond pulsations. Oddly, all five of the these pulsars (see Table 1) are in soft X-ray transients: accreting systems that lie dormant for years, with intermittent outbursts of accretion lasting about a month (these outbursts are understood to arise from an accretion disk instability; see Frank, King, & Raine 2002). Moreover, all five also have very short orbital periods[1] and very low mass accretion rates. Also, the pulsed amplitude of these systems is about 6 percent, while the upper limit on the pulsed amplitude for most NS/LMXBs is 1 per cent or less, raising the question of what is different about this group of 5 pulsars: why is it so difficult to find persistent millisecond pulsations in most NS/LMXBs? It should be noted that none of the millisecond X-ray pulsars has been detected as a radio pulsar.

While both kHz QPOs and X-ray burst oscillations are often observed in the same

---

[1] As an aside, I also note that there are five LMXBs whose orbital periods are within 90 seconds of 42 minutes: three of the millisecond X-ray pulsars, the strong-field pulsar 4U 1626−67, and the quiescent LMXB NGC 6652B. This excess is statistically significant, and there is no selection effect for finding this period. What is so special about this particular orbital period?

source, until 2002 neither phenomena had ever been observed in accretion-powered millisecond pulsar. Since it was only in this latter class that the neutron star spin was definitively known, the relationship between these three classes of variability was not clear. Still, some patterns were clear. In particular, the separation frequency $\Delta\nu_{kHz}$ organized the burst oscillation sources into two groups: the slow oscillators (with $\nu_{burst} <$ 400 Hz) all have $\nu_{burst} \approx \Delta\nu_{kHz}$, while the fast oscillators (with $\nu_{burst} > 500$ Hz) all have $\nu_{burst} \approx 2\Delta\nu_{kHz}$. The photospheric radius expansion properties of the X-ray bursts also divide along these lines, with fast oscillations occurring preferentially in radius expansion bursts (Muno, Galloway, & Chakrabarty 2004). Since both $\Delta\nu_{kHz}$ and ν_{burst} are reproducible characteristics of a given source and the most likely mechanism for such a stable frequency is the stellar spin, there was considerable debate as to whether it is $\Delta\nu_{kHz}$ or ν_{burst} that is the fundamental spin frequency.

This question was finally settled by the detection in kHz QPOs in two accretion-powered X-ray pulsars in 2002 and 2003. In the rapid rotator SAX J1808.4−3658, $\Delta\nu_{kHz}$ is half the 401 Hz spin frequency (Wijnands et al. 2003); while in the slow rotator XTE J1807−294, $\Delta\nu_{kHz}$ roughly equals the 190 Hz spin frequency (Markwardt et al. 2005), verifying the odd phenomenology described above. (The possibility that the spin frequency in SAX J1808.4−3658 is actually 200.5 Hz was excluded by the very stringent non-detection of pulsations at this frequency; see Morgan et al. 2005.) In addition, X-ray burst oscillations were also detected in the 401 Hz pulsar SAX J1808.4−3658 (Chakrabarty et al. 2003) and the 314 Hz pulsar XTE J1814−338 (Strohmayer et al. 2003); in both cases, ν_{burst} was equal to the spin frequency.

These new observations lead to three conclusions:

- X-ray burst oscillations directly trace the neutron star spin (and are not higher harmonics of a fundamental), and may thus be thought of as nuclear-powered pulsations.
- The kHz QPO separation frequency $\Delta\nu_{kHz}$ is sometimes roughly the spin frequency (in slow rotators) and sometimes roughly half the spin frequency (in fast rotators). The origin of these QPOs is uncertain, but the underlying mechanism clearly has some coupling to the stellar spin.
- Most neutron stars in LMXBs are indeed spinning at millisecond periods, but for some reason only a small fraction of them are visible as persistent, accretion-powered millisecond pulsars.

This last point is puzzling, since one would expect a 10^8 G accreting neutron star to be a pulsar. There have been several possible explanations discussed for why most NS/LMXBs are not pulsars. One possibility is that the most of the NSs have magnetic fields that are too weak to channel the accretion flow, although there is evidence for radio pulsars with fields much weaker than 10^8 G. Another possibility is that the neutron stars in non-pulsing LMXBs are surrounded by a scattering medium that attenuates pulsations (Brainerd & Lamb 1987; Kylafis & Phinney 1989). Indeed, Titarchuk, Cui, & Wood (2003) have recently argued that the data support this argument, although there is not yet a consensus on this point (Heindl & Smith 1998; Psaltis & Chakrabarty 1999). A third possibility is that gravitational self-lensing might attenuate the pulsations (Wood, Ftaclas, & Kearney 1988; Meszaros, Riffert, & Berthiaume 1988). In the next section, I will discuss recent evidence that suggests that magnetic field strength may indeed be the relevant factor.

A RANGE OF LMXB MAGNETIC FIELD STRENGTHS?

It is instructive to compare the behavior of the X-ray burst oscillations observed in the pulsar SAX J1808.4−3658 with those observed in the non-pulsing LMXBs. In SAX J1808.4−3658, strong millisecond oscillations around the 401 Hz spin frequency were observed during 4 X-ray bursts in 2002, with very similar characteristics in each burst (Chakrabarty et al. 2003). An example is shown in Figure 2. First, a rapidly drifting oscillation (increasing from 397 to 403 Hz) was detected during the burst rise. Second, no oscillations were detected during the radius expansion phase of the burst (typical of other burst oscillation sources as well). Finally, a strong oscillation reappeared during the cooling phase of the burst, at a constant frequency nearly equal to the spin frequency (which was known precisely from the pre-burst persistent pulsations), but exceeding it by one part in 70000.

This observed frequency drift demonstrates that this is a similar phenomena as the burst oscillations observed in other (non-pulsing) neutron stars, although the oscillations in SAX J1808.4−3658 have some very unusual traits: the drift time scale is an order of magnitude faster than in the other neutron stars (compare right panel of Figure 1), and the maximum oscillation frequency is reached during the burst rise, inconsistent with angular momentum conservation in a cooling, contracting burning shell. In fact, as evident in Figure 2, the oscillation *overshoots* the spin frequency during the burst rise. The rapid, overshooting drift probably indicates that SAX J1808.4−3658 has a stronger magnetic field than the other burst oscillation sources (Chakrabarty et al. 2003), since a sufficiently strong field will suppress rotational shearing in the burning layer and may act as a restoring force. Strohmayer et al. (2003) also interpreted the frequency evolution of the burst oscillations in the pulsar XTE J1814−338 as evidence for a stronger than normal magnetic field in that LMXB.

This magnetic field argument is particularly appealing given that only SAX J1808.4−3658 and XTE J1814−338, among a total of 13 burst oscillation sources (and two of only five systems out of a total of over 80 NS/LMXBs), show persistent pulsations in their non-burst emission. The absence of persistent pulsations in most of these systems suggests that they lack a sufficiently strong magnetic field for the accretion flow to be magnetically channeled. Indeed, it has been proposed that the absence of persistent pulsations in most NS/LMXBs is due to diamagnetic screening of the neutron star magnetic field by freshly accreted material, which would occur above a critical value of $\dot{M}$ at a few percent of the Eddington rate (Cumming, Zweibel, & Bildsten 2001). In this context, it is interesting to note that all five of the persistent millisecond X-ray pulsars like at the low end of the $\dot{M}$-distribution for LMXBs. Thus, while all the NS/LMXBs may have underlying surface field strengths of $\sim 10^8$ G (see, e.g., Psaltis & Chakrabarty 1999), it may be that the non-pulsing LMXBs have *effective* field strengths that are much lower due to screening by accreted material. Presumably, the underlying field would emerge when the accretion eventually halts; thus, this idea is

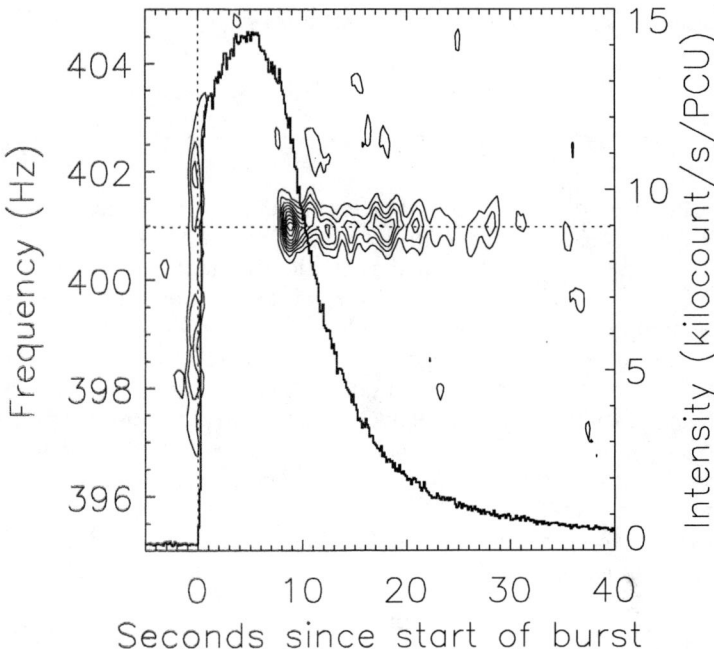

FIGURE 2. X-ray burst oscillation in the pulsar SAX J1808.4−3658. The histogram shows the X-ray intensity during the thermonuclear X-ray burst. The contours show the Fourier power levels as a function of frequency and time. The horizontal dotted line indicates the (known) pulsar spin frequency. A rapidly drifting oscillation is detected during the burst rise and overshoots the spin rate. A stationary oscillation near the spin rate is detected in the burst tail. Adapted from Chakrabarty et al. (2003).

still consistent with the absence of millisecond radio pulsars with fields much weaker than 10^8 G.

THE SPIN DISTRIBUTION OF RECYCLED PULSARS

Having observationally verified that millisecond pulsars are spun up or "recycled" in LMXBs, it is interesting to ask what the underlying spin distribution of recycled pulsars is, and whether there is any limit to the recycling process. One might expect the distribution of spin frequencies to simply reflect the range of equilibrium spins corresponding to the magnetic field strength distribution in LMXBs. However, if the effective field strength of most NS/LMXBs is considerably below 10^8 G as discussed above, then the resulting boundary layer accretion onto the neutron star might be capable of spinning up the pulsar to submillisecond periods. Certainly, a strict upper limit on the neutron

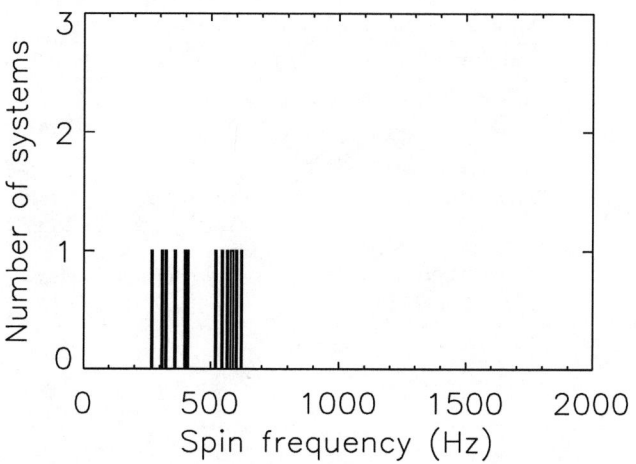

FIGURE 3. The spin frequency distribution of nuclear-powered millisecond X-ray pulsars[2]. There is a sharp drop off in the population at spins above 730 Hz. *RXTE* has no significant selection effects against detecting oscillations as fast as 2000 Hz, making the absence of fast rotators extremely statistically significant. Based on results from Chakrabarty et al. (2003).

star spin rate is given by the centrifugal breakup limit, up to 3 kHz depending upon the neutron star equation of state (Cook, Shapiro, & Teukolsky 1994; Haensel, Lasota, & Zdunik 1999).

Although the substantial known population of millisecond radio pulsars in principle provides an ideal probe of the spin distribution of recycled pulsars, severe observational selection effects have historically made it difficult to make a statistically accurate estimate. However, the burst oscillation sources (nuclear-powered pulsars) are an ideal probe: they are bright and easily detected throughout the Galaxy, their signals are short-lived enough to avoid modulation losses due to orbital Doppler smearing, and *RXTE* has no significant selection effects against detecting oscillations as fast as 2 kHz.

The spin frequencies (see Table 1) of the 13 known nuclear-powered millisecond pulsars are plotted in Figure 3[2]. (The three non-bursting, accretion-powered millisecond pulsars are omitted to keep the sample unbiased.) The spins are consistent with a uniform distribution within the observed 270–619 Hz range. The absence of any pulsars at lower frequencies is not surprising, since such low equilibrium spin rates would require somewhat higher magnetic field strengths, which would then suppress the thermonuclear X-ray bursts necessary for burst oscillations. However, the absence of spins above 619 Hz is extremely significant, given that there is no significant loss of *RXTE* sensitivity out

---

[2] Omitted is the newly-discovered, slow (45 Hz) nuclear-powered pulsation in EXO 0748−676, reported two months after the Cefalu meeting (Villareal & Strohmayer 2004).

TABLE 1. Millisecond Pulsars in X-Ray Binaries

Object	Spin Frequency (Hz)	Orbital Period
ACCRETION-POWERED PULSARS		
XTE J0929−314	185	43.6 min
XTE J1807−294	191	41 min
XTE J1814−338	314	4.27 hr
SAX J1808.4−3658	401	2.01 hr
XTE J1751−305	435	42.4 min
NUCLEAR-POWERED PULSARS[2] (Burst Oscillations)		
EXO 0748−676	45	3.82 hr
4U 1916−05	270	50 min
XTE J1814−338	314	4.27 hr
4U 1702−429	330	?
4U 1728−34	363	?
SAX J1808.4−3658	401	2.01 hr
SAX J1748.9−2021	410	?
KS 1731−260	524	?
Aql X-1	549	19.0 hr
X1658−298	567	7.11 hr
4U 1636−53	581	3.8 hr
X1743−29	589	?
SAX J1750.8−2900	601	?
4U 1608−52	619	?

to 2 kHz. Under the simple assumption of a uniform distribution out to some maximum value, the observed distribution yields an maximum spin frequency of 730 Hz (95% confidence; Chakrabarty et al. 2003). This limit is consistent with the fastest known millisecond radio pulsar, PSR B1937+21, which has $P_{spin} = 641$ Hz. This limit is also well below the breakup frequency for nearly all equations of state for rapidly rotating neutron stars. Recent radio pulsar surveys, in which selection effects are accounted for, are independently finding similar evidence for a maximum spin frequency around 700 Hz, as reported at this meeting (McLaughlin et al. 2005).

It is thus clearly demonstrated observationally that the population of pulsars with spins above 730 Hz must drop off dramatically (although not necessarily to zero: the existence of submillisecond pulsars is *not* excluded, but such objects are evidently at best very rare). It remains unclear what causes this drop off. Magnetic spin equilibrium can only account for the observed distribution if the entire sample has surface magnetic field strength $\sim 10^8$ G. However, as noted above, fields this strong should be dynamically important for the accretion flow and lead to persistent millisecond pulsations, making it difficult to understand the lack of pulsations in most NS/LMXBs and instead suggesting a wider range of magnetic field strengths.

Alternatively, several authors have shown that gravitational radiation can carry away substantial angular momentum from accreting neutron stars, driven by excitation of an r-mode instability in the neutron star core (Wagoner 1984; Andersson, Kokkotas, & Stergioulas 1999), or by a rotating, accretion-induced crustal quadrupole moment (Bildsten 1998), or by large (internal) toroidal magnetic fields (Cutler 2002). These losses can balance accretion torques for the relevant ranges of spin and $\dot{M}$, and the predicted gravitational radiation strengths are near the detection threshold for, e.g., the planned Advanced LIGO detectors (Bildsten 2003). Such a detection would rely on operating a narrow-band search in a known frequency range, but this approach has the advantage of searching for a persistent signal that can be integrated, unlike the transient signals like those produced by binary mergers.

It will be of considerable interest to measure the shape of the spin distribution more accurately, in order to determine whether we are observing a gradual fall off, an abrupt cutoff, a pile up, etc. — at present, the data cannot distinguish between these possibilities, and thus cannot meaningfully discriminate between theoretical models. In the long run, it is likely to be the radio systems that provide the best measurement of the distribution, based simply on ease of detectability and the available instrumentation.

ACKNOWLEDGMENTS

It is a pleasure to thank Lars Bildsten, Duncan Galloway, Jake Hartman, Scott Hughes, Andrew King, Miriam Krauss, Fred Lamb, Craig Markwardt, Nergis Mavalvala, Ed Morgan, Mike Muno, Dimitrios Psaltis, Saul Rappaport, Tod Strohmayer, Michiel van der Klis, and Rudy Wijnands for useful discussions and collaborations. I also thank Luciano Burderi and Hale Bradt for inviting me to Cefalu to give this review talk.

REFERENCES

1. Alpar, M. A., Cheng, A. F., Ruderman, M. A., & Shaham, J. 1982, Nature, 300, 728
2. Andersson, N., Kokkotas, K. D., & Stergioulas, N. 1999, ApJ, 516, 307
3. Backer, D. C., Kulkarni, S. R., Heiles, C. E., Davis, M. M., & Goss, W. M. 1982, Nature, 300, 615
4. Bhattacharya, D. & Srinivasan, G. 1995, in *X-Ray Binaries*, ed. W. H. G. Lewin et al. (Cambridge: Cambridge U. Press), 495
5. Bildsten, L. 1998, ApJ, 501, L89
6. Bildsten, L. 2003, in ASP Conf. Ser. Vol. 302, *Radio Pulsars*, ed. M. Bailes, D. J. Nice, & S. E. Thorsett (San Francisco: ASP), 289
7. Bildsten, L. et al. 1997, ApJS, 113, 367
8. Bradt, H. V., Rothschild, R. E., & Swank, J. H. 1993, A&AS, 97, 355
9. Brainerd, J. & Lamb, F. K. 1987, ApJ, 317, L33
10. Chakrabarty, D. & Morgan, E. H. 1998, Nature, 394, 346
11. Chakrabarty, D., Morgan, E. H., Muno, M. P., Galloway, D. K., Wijnands, R., van der Klis, M., & Markwardt, C. B. 2003, Nature, 424, 42
12. Cook, G. B., Shapiro, S. L., & Teukolsky, S. A. 1994, ApJ, 421, L117
13. Cumming, A. & Bildsten, L. 2000, ApJ, 544, 453
14. Cumming, A., Zweibel, E., & Bildsten, L. 2001, ApJ, 557, 958
15. Cutler, C. 2002, Phys. Rev. D, 66, 4025
16. Frank, J., King, A. R., & Raine, D. 2002, *Accretion Power in Astrophysics*, 3rd ed. (Cambridge: Cambridge U. Press)
17. Ghosh, P. & Lamb, F. K. 1979, ApJ, 234, 296
18. Haensel, P., Lasota, J. P., & Zdunik, J. L. 1999, A&A, 344, 151
19. Heindl, W. A. & Smith, D. M. 1998, ApJ, 506, L35
20. Jahoda, K., Swank, J. H., Giles, A. B., Stark, M. J., Strohmayer, T., Zhang, W., & Morgan, E. H. 1996, Proc. SPIE, 2808, 59

21. Joss, P. C. & Li, F. K. 1980, ApJ, 238, 287
22. Kylafis, N. & Phinney, E. S. 1989, in *Timing Neutron Stars*, ed. H. Ögelman & E. P. J. van den Heuvel (Dordrecht: Kluwer), 731
23. Markwardt, C. B. et al. 2005, in preparation
24. McLaughlin, M. A. et al. 2005, in ASP Conf. Ser. Vol. 328, *Binary Radio Pulsars*, ed. F. A. Rasio & I. H. Stairs (San Francisco: Astron. Soc. Pacific), 43 (astro-ph/0404181)
25. Meszaros, P., Riffert, H., & Berthiaume, G. 1988, ApJ, 325, 204
26. Morgan, E. H. et al. 2005, in preparation
27. Muno, M. P., Galloway, D. K., & Chakrabarty, D. 2004, ApJ, 608, 930
28. Psaltis, D. & Chakrabarty, D. 1999, ApJ, 521, 332
29. Psaltis, D. & Lamb, F. K. 1998, in *Neutron Stars and Pulsars*, ed. N. Shibazaki et al. (Tokyo: Univ. Acad. Press), 179
30. Strohmayer, T. & Bildsten, L. 2005, in *Compact Stellar X-Ray Sources*, ed. W. H. G. Lewin & M. van der Klis (Cambridge: Cambridge U. Press), in press (astro-ph/0301544)
31. Strohmayer, T. E. & Markwardt, C. B. 1999, ApJ, 516, L81
32. Strohmayer, T. E., Jahoda, K., Giles, A. B., & Lee, U. 1997, ApJ, 486, 355
33. Strohmayer, T. E., Markwardt, C. B., Swank, J. H., & in 't Zand, J. J. 2003, ApJ, 596, L67
34. Strohmayer, T. E., Zhang, W., Swank, J. H., Smale, A., Titarchuk, L., Day, Charles, & Lee, U. 1996, ApJ, 469, L9
35. Titarchuk, L., Cui, W., & Wood, K. S. 2003, ApJ, 576, L49
36. van der Klis, M. 2000, ARA&A, 38, 717
37. van der Klis, M., Swank, J. H., Zhang, W., Jahoda, K., Morgan, E. H., Lewin, W. H. G., Vaughan, B., & van Paradijs, J. 1996, ApJ, 469, L1
38. van der Klis, M., Wijnands, R. A. D., Horne, K., & Chen, W. 1997, ApJ, 481, L97
39. Vaughan, B. A. et al. 1994, ApJ, 435, 362
40. Villareal, A. R. & Strohmayer, T. E. 2004, ApJ, 614, L121
41. Wagoner, R. V. 1984, ApJ, 278, 345
42. Wijnands, R. 2004, in *The Restless High-Energy Universe*, ed. E. P. J. van den Heuvel et al. (Elsevier), in press (astro-ph/0309347)
43. Wijnands, R. & van der Klis 1998, Nature, 394, 344
44. Wijnands, R., van der Klis, M., Homan, J., Chakrabarty, D., Markwardt, C. B., & Morgan, E. H. 2003, Nature, 424, 44
45. Wood, K. S., Ftaclas, C., & Kearney, M. 1988, ApJ, 324, L63

Indirect evidence for an active radio pulsar in SAX J1808.4–3658 during quiescence

S. Campana, P. D'Avanzo*, L. Stella, G.L. Israel, G. Marconi[†], J. Casares**, Rob Hynes[‡] and Phil Charles[§]

*INAF-Osservatorio Astronomico di Brera, Via Bianchi 46, I–23807, Merate (LC), Italy
[†]INAF-Osservatorio Astronomico di Roma, Via Frascati 33, I–00040, Roma, Italy
**Instituto de Astrofisica de Canarias, 38200 La Laguna, Tenerife, Spain
[‡]Astronomy Department, Univ. of Texas at Austin, 1 University Station C1400, Austin, TX 78712
[§]Department of Physics and Astronomy, University of Southampton, Southampton SO17 1BJ, UK

Abstract. Millisecond radio pulsars are neutron stars that have been spun-up by the transfer of angular momentum during the low-mass X–ray binary phase. The transition from an accretion-powered to a rotation-powered pulsar takes place on evolutionary timescales at the end of the accretion process, however it may also occur sporadically in systems undergoing transient X–ray activity. We have obtained the first optical spectrum of the low mass transient X–ray pulsar SAX J1808.4–3658 in quiescence. Similar to the black widow millisecond pulsar B1957+20, this X–ray pulsar shows a large optical modulation at the orbital period due to an irradiated companion star. Using the brightness of the companion star as a bolometer, we conclude that a very high irradiating luminosity, a factor of ~ 100 larger than directly observed, must be present in the system. This most likely derives from a rotation-powered neutron star that resumes activity during quiescence.

INTRODUCTION

SAX J1808.4–3658 was the first-discovered low mass X–ray binary (LMXB) transient showing coherent pulsations. This confirmed unambiguously the long-sought connection between LMXBs and millisecond radio pulsars (Bhattacharya & van den Hevuel 1991; Tauris & van den Hevuel 2004). The detection of coherent X–ray pulsations during outbursts testifies that the neutron star possesses a magnetic field of $B \sim 10^8 - 10^9$ G, sufficient for a small magnetosphere to form (Psaltis & Chakrabarty 1999; Menna et al. 2003; Di Salvo & Burderi 2003). Unlike persistent LMXBs, SAX J1808.4–3658 is a transient system, i.e. it is active in X–rays only for short intervals lasting a few months (outbursts) followed by quiescent periods of years.

During quiescence LMXB transients are very faint in X–rays (5-6 orders of magnitude less than in outburst) usually with luminosities of $10^{32} - 10^{33}$ erg s^{-1}. Transient systems, therefore, represent a unique laboratory for the study of compact objects in accretion regimes that are inaccessible to persistent sources. Most neutron star transients are characterized by a quiescent X–ray spectrum consisting of a soft component, usually ascribed to the cooling neutron star which has been heated during outbursts (Brown et al. 1998; Rutledge et al. 1999), and a hard power-law tail (with photon index $\Gamma \sim 1-2$) the nature of which is still debated (Campana & Stella 2000). Several mechanisms have been put forward to explain these spectral components, ranging from accretion disks (in different flavours such as advection-dominated or convection-dominated disks, disks stopping at the magnetosphere, etc.; Narayan et al. 1997; Blandford & Begelman 1999; Igumenshchev et al. 2003) to emission from the interaction between the relativistic wind from a re-activated radio pulsar with matter outflowing from the companion star (Stella et al. 1994; Campana et al. 1998b). XMM-Newton carried out the first detailed study of SAX J1808.4–3658 in quiescence: SAX J1808.4–3658 was dimmer than any other neutron star transient (with a 0.5–10 keV luminosity of 5×10^{31} erg s^{-1} at a distance of 2.5 kpc, in't Zand et al. 2001) and exhibited only a hard power-law component ($\Gamma = 1.4^{+0.6}_{-0.3}$, 90% confidence level) in its quiescent X–ray spectrum (Campana et al. 2002). These results were confirmed by a subsequent XMM-Newton observation (Wijnands 2002).

In the optical, SAX J1808.4–3658 is dim during quiescence (mean $R \sim 20.9 \pm 0.1$; Homer et al. 2001) and brightens considerably in outbursts ($R \sim 16.2 \pm 0.2$; Wang et al. 2001). The optical light curve in outburst and quiescence is modulated at the orbital period and it is in anti-phase with the X–ray light curve, likely indicating that irradiation of the companion star plays a crucial role in spite of the low X–ray luminosity. This is unlike other quiescent transients. The mass function derived from X–ray data ($4 \times 10^{-5} M_\odot$, Chakrabarty & Morgan 1998) and the requirement that the companion fills its Roche lobe led to the conclusion that it must be a rather low mass star, possibly a brown dwarf (Bildsten & Chakabarty 2001). A white dwarf companion is ruled out because it would not fill its Roche lobe. Homer et al. (2001) proposed that the bulk of the optical emission in quiescence arises from the internal energy release of a remnant disc and the orbital modulation from the varying contribution of the heated face of the companion star. Burderi et al. (2003) noted that the required irradiating luminosity needed to match the optical flux, however, is a factor $10 - 50$ higher than the quiescent X–ray luminosity of SAX J1808.4–3658. These authors proposed an alternative scenario, in which the irradiation is due to the rotational energy emitted in the form of a relativistic particle wind from the fast spinning neutron star, which switched to the rotation-powered regime during quiescence. Their results are in agreement with the weak constraints from the optical magnitudes by Homer et al. (2001).

DATA ANALYSIS

We obtained data on SAX J1808.4–3658 in quiescence with the ESO-VLT (UT4 Yepun) during two half nights on July 12-13 2002. We carried out I band photometry with 3 min exposures with FORS2 (pixel size of $0''.126$/pixel and a field of view of $6'.8 \times 6'.8$) over one orbital period. Spectroscopy of the same target was performed using the low resolution grism 600RI (centered at 6780 Å with a resolution of 55 Å/mm) covering 5120–8450 Å and a $1''$ slit with 3 min spectra over four orbital periods. Data reduction was done in MIDAS to remove the bias level, flat-field.

The region around the optical companion of SAX J1808.4–3658 is crowded and poor seeing conditions complicated the analysis (varying between $1.5 - 2''$). We take advantage of previous ESO-VLT images (obtained during quiescence in 1999 with seeing $\sim 0.5''$) to de-blend our data (these magnitudes were $V = 21.82 \pm 0.03$, $R =$

21.63 ± 0.04 and $I = 21.08 \pm 0.04$). Our I photometry shows a dimmer source ($I = 21.5 \pm 0.1$) and a clear modulation at the 2.01 hr orbital period with a semi-amplitude 0.2 ± 0.04 mag (65% in flux). We also re-calculated the modulation semi-amplitude in V (0.13 ± 0.06) and R (0.39 ± 0.09) using the Homer et al. (2001) data. The I-band light curve shows a clear maximum at phase 0.52 ± 0.05 and a single minimum at phase 0.02 ± 0.05 (based on the precise X–ray ephemerides Chakrabarty & Morgan 1998). This is a clear indication of emission from an irradiated companion (e.g. Charles & Coe 2004) and it argues against emission from the impact point between the gas stream from the companion and an accretion disk (the hot spot) since this has maximum at phase $0.8 - 0.9$.

We obtained spectra at 3-min intervals over four orbital periods. We selected spectra taken with seeing better than $1.6''$ (due to poor seeing conditions), collecting a total of 51 min of good data. Wavelength calibrations used HeArNe arc lamp observations. Second-order flexure effects were corrected using night-sky emission lines. This correction was always < 0.3 Å. Spectra were corrected for slit-losses according to Diego (1985). We also account for the contaminating stars, estimating their relative contribution in a $1''$ slit on the good-seeing VLT images and interpolating to the spectral range. Errors were tracked along these processes resulting in a 0.1 mag error. A weak Hα emission line is visible in the spectrum (equivalent width $EW = 10.3 \pm 3.7$ Å, 68% confidence level, and $FWHM = 44.0 \pm 6.3$ Å, see insert in Fig. 1).

MODELLING THE DATA

What is the cause of the optical emission during quiescence? Likely candidates are emission from the companion star and/or the disk. In order to fit within the Roche lobe of a 2.01 hr binary the companion mass has to be less than $0.17\ M_\odot$. In the model of Bildsten & Chakrabarty (2001) the most likely companion is a $0.05\ M_\odot$ brown dwarf bloated by irradiation to fill its Roche lobe ($0.13\ R_\odot$). The maximum intrinsic optical luminosity from the companion for any of the models by Bildsten & Chakrabarty (2001) is $\sim 3 \times 10^{31}$ erg s^{-1} (corresponding to a star temperature of 4800 K for a distance of 2.5 kpc). This is too low a luminosity to account for the observed optical flux, which is a factor of > 10 brighter. We therefore turn to the accretion disk as a possible source. Assuming that the quiescent X–ray luminosity is powered by accretion we can infer the expected mass inflow rate and derive the corresponding optical luminosity (including irradiation), which fails to account for what we see by more than a factor of 100. Disk models may be envisaged with a much higher mass accretion rate together with a truncation radius (fine) tuned to avoid optical and soft X–ray violation of observed data. However, this kind of models still require some additional ingredient to explain the large optical phase modulation. The emission from the pulsar could itself extend to the optical and extrapolation of the power law X–ray flux (assuming the XMM-Newton observation found the X–rays in a similar state) could account for about 30% of the optical luminosity. But this also could not explain the observed orbital phase modulation.

It is instructive to compare the properties of SAX J1808.4–3658 with those of the *black widow* pulsar PSR B1957+20 (Fruchter et al. 1988), which consists of a 1.6 ms radio pulsar irradiating its white dwarf companion (orbital period of 9.16 hr) with a rotational energy of 10^{35} erg s^{-1}. X- and γ-rays are generated in an inter-binary shock front, which causes ablation and heating of the companion (Phynney et al. 1988; Arons & Tavani 1994). An X–ray nebula has recently been revealed around PSR B1957+20 confirming this scenario (Stappers et al. 2003), and the orbital modulation is large with an R-band semi-amplitude > 4 mag (Callanan et al. 1995). A similar system is the eclipsing millisecond radio pulsar PSR J2051–0827, consisting of a 4.5 ms pulsar orbiting its very low mass companion ($\sim 0.03\ M_\odot$) every 2.4 hr (Stappers et al. 1996). Radio eclipses as well as a ~ 3.3 mag optical modulation have been observed in PSR J2051–0827 (Stappers et al. 2001), however no X–ray observations are available.

Inspired by this analogy and following Burderi et al. (2003), we attempt to account for the optical and X–ray spectra (even if not close in time) as well as the V, R and I modulations of SAX J1808.4–3658 with an irradiated star plus the contribution of the shock front. We fit the data by using the irradiating luminosity ($L_{\rm irr}$), the fractional luminosity difference between the heated and the cold face of the companion (f) and interstellar absorption (A_V) as free parameters (e.g. Chakrabarty 1998). We obtain a good fit to all the available data (reduced $\chi^2 = 0.7$ with 57 degrees of freedom[1], see Fig. 1). In particular, the required irradiating luminosity is $L_{\rm irr} = (4^{+3}_{-1}) \times 10^{33}$ erg s^{-1} (90% confidence level for three free parameters, i.e. $\Delta\chi^2 = 6.3$). The best fit fraction is $f = 0.65 \pm 0.10$ resulting in temperature difference at the two faces of the companion star of about 1000 ± 300 K. This temperature difference is similar to the one observed in PSR J2051–0827. The amplitude modulation of SAX J1808.4–3658 is instead smaller than in the case of PSR J2051–0827 and PSR B1957+20. This could be due to some remaining contamination due to our de-blending process, or it could be an inclination effect. For the most strongly modulated PSR J1957+20, the inclination is $\sim 70°$. PSR J2051–0827 has an inclination of $\sim 40°$.

The estimated absorption is $A_V = 1.0 \pm 0.5$, in line with previous estimates. Optical emission from the shock front accounts for about 15% of the total emission[2]. In the fit we assumed that all the irradiating luminosity is re-emitted by the star. Therefore the irradiating luminosity we derived represents only a lower limit since a relativistic particle wind could have rather different effects from those of X–ray irradiation in terms of effective albedo. Moreover, we note that evidence of this large irradiating luminosity comes from the equivalent width of the $H\alpha$ line. If the line comes from reprocessing, one can roughly infer an irradiating luminosity of $\gtrsim 3 \times 10^{33}$ erg s^{-1}. This can be estimated by taking the line flux and increasing it by the fraction of emitted flux intercepted by the star ($R_*^2/4a^2$, with R_* the companion star radius taken equal to the Roche lobe radius and a the orbital separation) and by the fraction of energy re-emitted in $H\alpha$ (e.g. Hynes et al. 2002). We assumed a conservative value of 0.3 (see the discussion in Hynes et al. 2002).

---

[1] The conversion efficiency of rotational energy into X–rays for SAX J1808.4–3658 is $\lesssim 10^{-3}$, a factor of 10 higher than PSR 1957+20 (Stappers et al. 2003), as expected due to geometrical reasons.

[2] We fit our data with an irradiated star plus the extrapolation to the optical of the X–ray tail. The presence of a disk is not required by our fit. We tried in any case to fit with an irradiated star plus a disk model finding similar results for the star parameters and a large disk inner radius ($\gtrsim 10^9$ cm).

CONCLUSIONS

The required irradiating luminosity is in all cases large ($\sim 4 \times 10^{33}$ erg s^{-1}), indeed much larger than the observed X–ray luminosity in quiescence; neither accretion-driven X rays, nor the intrinsic luminosity of the companion star or disk are able to account for it (see also Burderi et al. 2003). The only source of energy available within the system is then the rotational energy of the neutron star. In order to have such a large spin-down luminosity one needs a neutron star magnetic field $\bar{B} \gtrsim 6 \times 10^7$ G (the companion star albedo might be larger than zero and the neutron star emission may also be partially beamed; this value is smaller than the one estimated by Burderi et al. 2003, since they did not account for the contamination from nearby stars). The required magnetic field is well in the range inferred from X–ray observations during the outbursts (see above). Prospects for directly observing SAX J1808.4–3658 pulsing in the X–ray band are rather hard since with XMM-Newton we collected less than 300 counts in 30 ks (Campana et al. 2002). Given the crowding around SAX J1808.4–3658 in the optical band, it would be difficult to detect an $H\alpha$ nebula like that around PSR B1957+20. On the other hand a search in the radio band looking for a millisecond radio pulsar would require searches at high frequencies to overcome the effects of free-free absorption (Campana et al. 1998a; Ergma & Antipova 1999; Burgay et al. 2003; Burderi et al. 2003) and a favorable orientation of the radio beam.

REFERENCES

1. Arons, J., Tavani, M. 1994, ApJS, 90, 797
2. Bhattacharya, D., van den Heuvel, E. P. J. 1991, Phys. Rep. 203, 1
3. Bildsten, L., Chakabarty, D. 2001, ApJ, 557, 292
4. Blandford, R. D., Begelman, M. C. 1999, MNRAS, 303, L1
5. Brown, E. F., Bildsten, L., Rutledge, R. E. 1998, ApJ, 504, L95
6. Burderi, L., et al. 2003, A&A, 404, L43
7. Burgay, M., et al. 2003, ApJ, 589, 902
8. Callanan, P. J., van Paradijs, J., Rengelink, R. 1995, ApJ, 439, 928
9. Campana, S., Colpi, M., Mereghetti, S., Stella, L., Tavani, M. 1998a, A&A Rev., 8, 279
10. Campana, S., Stella, L. 2000, ApJ, 541, 849
11. Campana, S., et al. 1998b, ApJ, 499, L65
12. Campana, S., et al. 2002, ApJ, 575, L15
13. Chakrabarty, D. 1998, ApJ, 492, 342
14. Chakrabarty, D., Morgan, E. H. 1998, Nat, 394, 364
15. Charles, P. A., Coe, M. J. 2004, in *Compact Stellar X-Ray Sources*, eds. W.H.G. Lewin and M. van der Klis, (Cambridge University Press), in press
16. Diego, F. 1985, PASP, 97, 1209
17. Di Salvo, T., Burderi, L. 2003, A&A 397, 723
18. Ergma, E., Antipova, J. 1999, A&A, 343, L45
19. Fruchter, A. S., Stinebring, D. R., Taylor, J. H. 1988, Nat, 333, 237
20. Homer, L., Charles, P. A., Chakrabarty, D., Zyl, L. 2001, MNRAS, 325, 1471
21. Hynes, R. I., et al. 2002, MNRAS, 330, 1009
22. Igumenshchev, I. V., Narayan, R., M. A. Abramowicz, M. A. 2003, ApJ, 592, 1042
23. in't Zand, J. J. M., et al. 2001, A&A, 372, 916
24. Menna, M. T., Burderi, L., Stella, L., Robba, N., van der Klis, M. 2003, ApJ, 589, 503
25. Narayan, R., Garcia, M. R., McClintock, J. E. 1997, ApJ, 478, L79
26. Phinney, E. S., Evans, C. R., Blandford, R. D., Kulkarni, S. R. 1988, Nat, 333, 832
27. Psaltis, D., Chakrabarty, D. 1999, ApJ, 521, 332

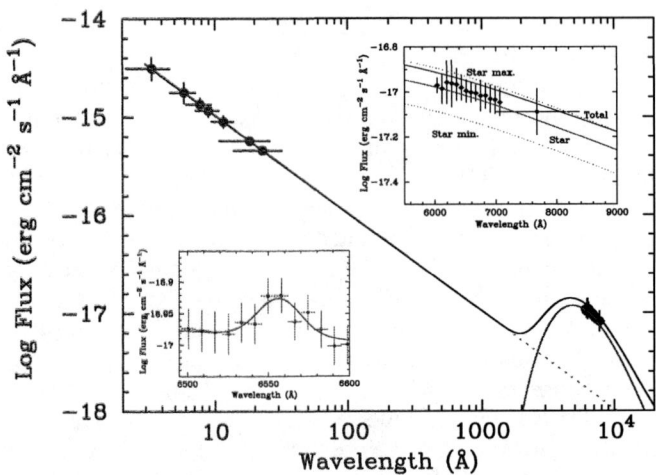

FIGURE 1. Central panel: X-ray to optical (VLT) spectrum of SAX J1808.4−3658. The data are corrected for absorption ($A_V = 1.0$) using standard extinction curve. Dots shows the XMM-Newton (X-rays) and VLT (optical spectrum plus I photometry) data. The continuous line shows the overall best fit for an irradiating luminosity of 4×10^{33} erg s^{-1}. The dotted line shows the contribution from shock emission between the relativistic pulsar wind and matter outflowing from the companion. The power law tail has a photon index of $\Gamma = 2.0$, this provides a slightly better fit to the data with respect to a $\Gamma = 1.5$ model (in that case the required irradiating luminosity is higher). The lower continuous line shows the mean contribution from the bloated ($0.13\ R_\odot$) companion star. This spectrum together with the optical V, R and I modulation has been fitted with an irradiated star plus shock emission model. Free parameters were the ratio between the 'hot' and the 'cold' star surface and absorption. A downhill simplex method for the search of the minimum has been adopted. Errors were computed in the 3-dimension space with $\Delta\chi^2 = 6.3$. Upper insert: Optical spectrum with the best fit model spectrum (upper continuous line). The mean star spectrum (lower continuous line) together with the minimum and maximum spectra (dotted lines) are also shown. Lower insert: particular of the broad ($FWHM = 44.0 \pm 6.3$ Å) $H\alpha$ line. The line equivalent width is $EW = 10.3 \pm 3.7$ Å.

28. Rutledge, R. E., et al. 1999, ApJ, 514, 945
29. Stappers, B. W. et al. 1996, ApJ, 465, L119
30. Stappers, B. W. et al. 2001, ApJ, 548, L183
31. Stappers, B. W. et al. 2003, Science, 299, 1372
32. Stella, L., et al. 1994, ApJ, 423, L47
33. Tauris, T. M., van den Heuvel, E. P. J. 2004, in *Compact Stellar X-Ray Sources*, eds. W.H.G. Lewin and M. van der Klis, (Cambridge University Press), in press
34. Wang, Z., et al. 2001, ApJ, 563, L61
35. Wijnands, R. 2003, ApJ, 588, 425

Inhomogeneous Accretion Flow in X-ray Binary Pulsars

M. T. Menna*, L. Burderi*, L. STella*, N. Robba[†] and M. van der Klis**

I.N.A.F. - Osservatorio Astronomico di Roma, via Frascati 33, I–00040 Monteporzio Catone (Roma), Italy
[†]*Dipartimento di Scienze Fisiche ed Astronomiche, Università di Palermo, via Archirafi 36, I–90123 Palermo, Italy*
**Astronomical Institute Anton Pannekoek, University of Amsterdam, Kruislaan 403, 1098 SJ Amsterdam, The Netherlands*

Abstract. We analyze the power spectrum of SAX J1808.4-3658, the first accreting millisecond binary pulsar discovered, to look for a broadening in the wings of the harmonic line, in analogy to what had been previously found in some high mass X-ray binaries. We indeed detect a broadening at the base of the 401 Hz peak, which is consistent with the convolution of the low frequency noise present in the power spectrum with the harmonic line. We interpret this as the result of a coupling between a fraction of the aperiodic and periodic variability, suggesting that at least part of the noise originates in a region close to the neutron star surface at the magnetic poles.

Keywords: pulsars: individual (SAX J1808.4-3658) – accretion – stars: magnetic fields – stars: neutron – X-rays: stars
PACS: 97.80.Jp, 97.60.Jd

INTRODUCTION

The main characteristic of the power spectra of X-ray pulsators is the presence of harmonic lines due to the flux modulation caused by the neutron star spin. Moreover the power density spectra of low mass X-ray binaries show broad band noise components originating from aperiodic variability, which, being typically located at low frequncies, is often referred to as red noise. Other features often present in power spectra are broad peaks, quasi periodic oscillations, bumps and wiggles (see *e.g.* [18] for a review). All this features are not unique to low mass X-ray binaries, but can be found also in in high mass X-ray binaries, containing X-ray pulsators and black hole candidates (see *e.g.* [2]).

In the case of high mass X-ray binaries, if the noise is produced, at least in part, by matter funneled by the magnetic field lines and accreting onto the magnetic polar caps of the neutron star, a modulation of that part of the aperiodic signal at the spin period would be expected (harmonic coupling). In the power spectra this would appear as a broadening of the wings of the harmonic lines due the convolution of each harmonic line with the red noise [3]. This effect has been indeed found in the power spectra of some X-ray pulsators ([4, 12]), indicating that aperiodic variability is modulated at the spin period of the neutron star. More recently, a coupling between the period and aperiodic variability of Hercules X-1 has been reported [15].

SAX J1808.4-3658 was the first millisecond accretion powered X-ray pulsar discovered by RXTE [17]. It exhibited a spin frequncy of ~ 401 Hz and a periodic modulation at 2.01 hours testifying to the presence of a low mass companion orbiting the neutron star with a mass function of the secondary star of $3.8 \times 10^{-5} M_\odot$ [5]. RXTE observed the source in outburst from the 11th of April to the 6th of May. The X-ray spectrum remained stable, featuring a power law shape with a photon index ~ 2 in the 15-100 keV band and a high energy cut-off above about 100 keV ([7, 9]). The power spectrum showed a red noise component with a knee frequecy $v_{knee} \sim 1$ Hz.

This high frequncy harmonic line makes SAX J1808.4-3658 the ideal candidate to investigate the presence of harmonic coupling, in fact in this case, $v_{spin} \gg v_{knee}$ which means that the typical duration of the aperiodic variations is several spin cycles. If, on the contrary the spin frequancy is too low, *i.e.* $v_{spin} \ll v_{knee}$ the broadening of the harmonic line's wings is so wide that it blends in the red noise itself ([4, 12]).

MODEL

Analytic derivations of the power spectrum of a signal consisting of aperiodic variability, a periodic modulation and a coupling of the two was described by [4] and [12]. Our aim is to use an analytic function to fit the power spectrum of the SAX J1808.4-3658 to derive the entity (if present) of the coupling. In line with [4], we adopt randomly occurring shots as a convenient description of the aperiodic variability giving rise to the red noise (shot noise model). In this model the aperiodic variability is produced by a random superposition (in the light curve) of shots. These instabilities might originate in the accretion flow onto the compact object. We note that we can apply this model without loss of generality, as the coupling effect is independent of the mathematical model adopted to describe the red noise.

For a detailed derivation of the analytic formula used in our fit see [14], in the following we briefly summarize some useful definitions and results. The power spectra of the signal described above, taking into account the finite duration T of the light curves and adopting the normalization of Leahy [13] is of the form:

$$P(v) = P_0 + P_{RN} + P_{HL} + P_{CPL} + P_{WN} \tag{1}$$

The first term is the power at zero frequency and is only dependent on the total intensity of the signal, the second term is the red noise term (RN), the third describes the harmonic line, and the coupling term P_{CPL} consists in a rescaled versions of the red noise placed at the harmonic frequencies. The scaling factor is $r = P_{CPL}/P_{RN} = \frac{1}{4}\left(\frac{c_1}{(1-c_1)}\right)^2 \left(1 + \frac{\lambda_{DF}}{(1-c_1)^2 \lambda_{LC}}\right)^{-1}$, where λ_{DF} and λ_{LC} are respectively the constant arrival mean rate of diffuse and localized shots and c_1 is defined by the modulation function of the harmonic signal $M(t) \propto c_1 \cos(2\pi v_0 t)$. The white noise component P_{WN}, induced by counting statistics has, adopting the normalization of Leahy, an expected value of 2. Because of detector dead-time effects (*e. g.* [16]), the value is usually slightly less than 2 and thus it will be determined by the fit of the power spectra, as described in the next section.

We define the coupling fraction $\eta = \lambda_{LC}/(\lambda_{LC} + \lambda_{DF})$ as the ratio of the arrival rates of the localized shots over the total shot rate. Furthermore we define the pulsed fraction $\mathscr{P}_f = (M(t)_{MAX} - M(t)_{MIN})/M(t)_{MAX} = 2c_1$, where $M(t)_{MAX}$ and $M(t)_{MIN}$

are respectively the maximum and the minumum of the modulating function over one period. With these definitions $0 \leq \mathscr{P}_f \leq 1$ and we have $c_1/(1-c_1) = \mathscr{P}_f/(2-\mathscr{P}_f)$. Moreover, as $M(t) \geq 0$ always, we have $1/2 \leq (1-c_1) \leq 1$ and $1 \leq 1/(1-c_1)^2 \leq 4$. Therefore the following relation must hold:

$$\frac{4\sqrt{r}}{\kappa_1 + 2\sqrt{r}} \leq \mathscr{P}_f \leq \frac{4\sqrt{r}}{\kappa_2 + 2\sqrt{r}} \qquad (2)$$

being $\kappa_1 = \left[1 + \left(\frac{1-\eta}{\eta}\right)\right]^{-1/2}$ and $\kappa_2 = \left[1 + 4\left(\frac{1-\eta}{\eta}\right)\right]^{-1/2}$ In the following section we will determine r from the fit of the power spectra and with the above relation we will constrain the values of the pulsed fraction in relation to the fraction of coupled shots.

Moreover we will be able to derive a relation between the diffuse component of the signal, as defined above, and the pulsed fraction. Averaging the signal only over the shot noise process we define the "observed" pulsed fraction

$$\mathscr{P}_{f\,\text{obs}} = (\langle I(t)\rangle_{\text{MAX}} - \langle I(t)\rangle_{\text{MIN}})/(\langle I(t)\rangle_{\text{MAX}} - I_{\text{bck}})$$

as the ratio of the difference of the averaged maximum and minimum intensity, over the difference of the averaged maximum and background intensity. We can also define the diffuse fraction as the ratio of the diffuse component over the sum of the diffuse and localized components as $\xi = \frac{\langle I_{\text{DF}}\rangle}{\langle I_{\text{DF}}\rangle + \langle I_{\text{LC}}\rangle}$. With a litte algebra (see [14]) we have $\xi = 1 - \frac{\mathscr{P}_{f\,\text{obs}}}{\mathscr{P}_f}$. We will see in the next section how the constraints on the pulsed fraction will have a bearing also on the fraction of the diffuse component.

DATA ANALYSIS

We used the observations of SAX J1808.4-3658 performed during the April-May 1998 outburst by the Rossi X-Ray Timing Explorer (RXTE) Proportional Counter Array (PCA) [11]. We used data collected with $\sim 122\mu s$ time resolution and 64 PHA channels which were available for all observations.

We first corrected photons arrival times for satellite motion with respect to the solar system barycenter using the BeppoSax position for the source [10] $R.A. = 18h08m29s$, $Dec = -36° 58.6'(J2000)$. We then corrected for the orbital motion of the neutron star by using the orbital parameters derived by [5]. For our analysis we rebinned the data into bins of $\Delta t = 488.28125\mu s$ (corresponding to a Nyquist frequency of 1024 Hz) and we used the events from all 64 PHA channels. For each observation, we calculated consecutive FFTs over time intervals of $T = 64s$ corresponding to power spectra with a frequency range of 0.015625-1024 Hz. These power spectra were then averaged to produce a single power spectra for each observation, 16 in all.

In our analysis we selected and analyzed only the observations in which the source count rate was highest (between ~ 800 and $\sim 500 cts/s$) and approximately constant, that is, 6 consecutive observations between April 11 and 20. This was necessary because the simple model of the aperiodic variability described in the previous section assumes that the characteristics of the signal do not vary with time. Examining the power spectra of

TABLE 1. Results of the fit

	Parameter	Value	Error (90% confidence level)
WN		1.994	±0.001
RN	A_1	1.7	$\pm 6 \times 10^{-1}$
	p_1	1.1×10^1	$\pm 0.6 \times 10^1$
	p_2	2.6	±1
	p_3	1.6×10^{-1}	$\pm^{8}_{9} \times 10^{-2}$
	τ_1	3.45×10^{-1}	$\pm^{4}_{5} \times 10^{-3}$
	α_1	3.3	$\pm^{3}_{2} \times 10^{-1}$
	τ_2	6.5×10^{-2}	$\pm^{1}_{0.5} \times 10^{-2}$
	α_2	2.0	±0.2
	τ_3	3.9×10^{-3}	$\pm 1. \times 10^{-3}$
	α_3	1.5	±0.1
	A_2	7.3×10^1	±3
	A_3	3.4×10^{-3}	$\pm 1 \times 10^{-3}$
Coupling	r	0.002	±0.001

each observation we consistently find an almost constant value for the red noise intensity and shape.

The shape of the red noise in SAX J1808.4-3658 was found to be reasonably well described by the sum of three empirical functions: $f_{RN}(v) = \sum_{i=1}^{3} p_i f_i(v)$ where: $f_i(v) = [1 + (2\pi \tau_i v)^{\alpha_i}]^{-1}$ and p_i indicates the normalization of each of them.

In order to fit the spectrum efficiently, we logarithmically rebinned the spectrum so that the number n_i of original bins averaged in the ith new bin follows $n_i = \text{int}(1.02^i)$, where $\text{int}(x)$ is the integer part of x. The region around the coherent modulation peak ($400.8 < v < 401.2$) was left at the original Fourier resolution, while the intervals $352.37 < v \leq 400.8$ and $401.2 \leq v < 449.64$, in which the peak broadening would be more important, were (uniformly) rebinned by a factor of 31 (see Fig. 1). The region around the second harmonic of the modulation was excluded (see above). We used a fitting function:

$$F(v) = A_1 f(v) + A_2 g(v - v_0) + A_3 f(v - v_0) + A_4$$

where $g(v) = W_T^2(v)$ and $f(v) = p_1 f_1(v) + p_2 f_2(v) + p_3 f_3(v)$ as defined above. Thus $A_1 f(v) = P_{RN}$, $A_2 g(v - v_0) = P_{HL}$, $A_3 f(v - v_0) = P_{CPL}$ and $A_4 = P_{WN}$. In order to check the robustness of our fit we compared different fitting procedures and we conclude that the region that affects the fit is only the narrow region around 401 Hz and a significant coupling is present independently of the detailed fitting procedure. The values determined for the f_i and A_1, A_2, A_3 parameters are listed in Tab. 2 along with their errors. From the best fit parameters A_3 and A_1 we derive $r = P_{CPL}/P_{RN} = 0.005$. An F-test calculates the probability that the $\Delta \chi^2$ variation is due to a chance improvement $P = 8.38 \times 10^{-6}$. With the derived value of r, we plot in figure (2) the two constraints defined by equation (2) in the $\eta - \mathcal{P}_f$ plane. The area between the curves defines the allowed region for the parameters. It can be seen that for any allowed value of the pulsed fraction a fraction of shots between $0.8 - 100\%$ must be modulated at the spin period

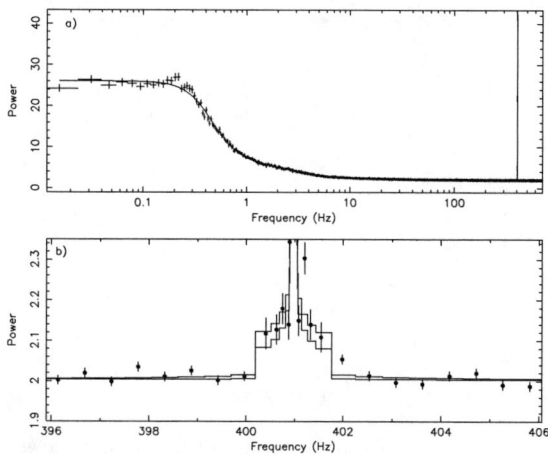

FIGURE 1. The power spectrum of SAX J1808.4-3658. Upper panel shows the measured broad band power spectrum and the simultaneous modelling of the red noise and harmonic components (solid line). The lower panel shows an expanded view of the region around the first fundamental harmonic: data have been binned by a factor 35 to increase the signal to noise ratio. Both the fit with and without coupling with the red noise have been plotted. The lower line is without coupling.

(coupled). Moreover, even in the most favourable case of 100% localized shots, the modulation has to be $> 17\%$ in order to account for the broadening detected. It can be noted that, in our discussion, we have assumed an identical amplitude for the localized and diffuse shots, that is, we have assumed that the blobs originating in the disk survive intact to the polar caps. It might be argued that the magnetic field just scoops off a fraction of each blob. We examined, for example, a case in which just 10% of each blob is threaded to the neutron star surface. In this scenario we obtain that nothing much changes in the favourable case of total coupling (100%), whilst for the maximum value of the pulsed fraction (100%), a minimum fraction of shots greater than 45% must be modulated at the spin period. That is, if just a part of each blob is scooped off, then more blobs must be funneled to observe the same effect.

A different constraint can be derived comparing the "observed" pulsed fraction value $\mathscr{P}_{f\,\mathrm{obs}}$ with its theoretical value $\mathscr{P}_f$ as derived from eq. 6 above. To compute the value of the "observed" pulsed fraction we have folded the data corrected for the orbital motion modulo the spin period and obtained a pulse profile. Adopting the background value for this observation of 130 c/s [17] we found $\mathscr{P}_{f\,\mathrm{obs}} \sim 12\%$. We can note from fig. (2)

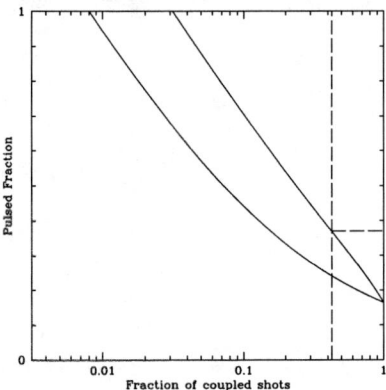

FIGURE 2. The $\mathcal{P}_f$ vs η relation. The dashed vertical line indicates a lower limit on the fraction of coupled shots $\eta \geq 0.42$ derived from the constraints of Fig. 3. This, in turn, poses an upper limit on the pulsed fraction $\mathcal{P}_f \leq 0.37$.

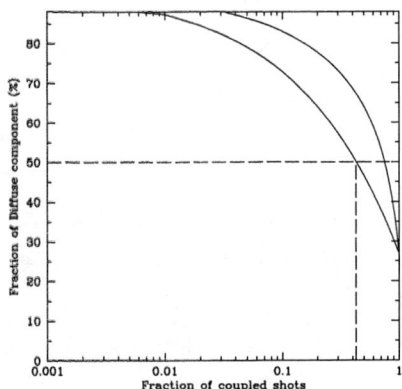

FIGURE 3. The ξ vs η relation. The horizontal line indicates an upper limit for the fraction of diffuse component $\xi \leq 50\%$. This, in turn, poses a lower limit on the fraction of coupled shots $\eta \geq 0.42$, as shown by the dashed vertical line.

that the pulsed fraction $0.17 \leq \mathcal{P}_f \leq 1$ and, recollecting the definition of the fraction of diffuse component $\xi = 1 - \frac{\mathcal{P}_{f\,obs}}{\mathcal{P}_f}$ we can pose a constraint on ξ, $1 - \frac{\mathcal{P}_{f\,obs}}{1} \leq \xi \leq 1 - \frac{\mathcal{P}_{f\,obs}}{\mathcal{P}_{f\,min}}$, and so $0.294 \leq \xi \leq 0.88$. We can now plot in fig (3) the diffuse fraction ξ, versus the fraction of coupled shots η.

DISCUSSION

We have shown the presence of a broadening at the base of the harmonic line. This can be interpreted as the result of a coupling between a fraction (between 0.8% and 100%) of the aperiodic and periodic variability in SAX J1808.4-3658. Moreover we have demonstrated that a fraction between $\sim 30\%$ and $\sim 90\%$ of the total emission arises from a component that is not modulated by the lighthouse effect (diffuse component).

Some more stringent constraints can be derived if we assume that the diffuse component originates from an accretion disk and the localized component derives from accretion onto the neutron star surface. As the spin period of SAX J1808.4-3658 is known, we can compute the corotation radius for this system i.e. the radius at which the speed of a particle rigidly rotating with the neutron star equals the local Keplerian value, $R_{co} = 1.5 \times 10^6 m^{1/3} P_{-3}^{2/3}$ cm, where m is the NS mass in units of solar masses and P_{-3} is the NS spin in milliseconds. In our case we get $R_{co} = 30.8$ km assuming a mass of $1.4 M_\odot$. For the accretion not to be centrifugally inhibited the inner rim of the disc must be located inside the corotation radius. Therefore we have $R_{disc} \leq R_{co}$. In the standard scenario of disk accretion onto magnetized neutron stars, the accreting matter forms a disk whose inner radius is truncated at the magnetosphere by the interaction of the accretion flow with the magnetic field of the NS. In this case the magnetospheric radius r_m is a fraction $\phi \leq 1$ (an expression for ϕ can be found in Burderi et al. 1998[1]; for $L \sim 10^{36}$ ergs/s we get $\phi \sim 0.3$) of the Alfvén radius R_A defined as the radius at which the energy density of the (assumed dipolar) NS magnetic field equals the kinetic energy density of the spherically accreting (free falling) matter:

$$R_A = 2.23 \times 10^6 R_6^{-2/7} m^{1/7} \mu_{26}^{4/7} \varepsilon^{2/7} L_{37}^{-2/7} \text{ cm} \qquad (3)$$

(see, e.g., [8]), where R_6 is the NS radius, R_{NS}, in units of 10^6 cm, m is the NS mass in solar masses, μ_{26} is the NS magnetic moment in units of 10^{26} G cm^3, ε is the ratio between the specific luminosity and the specific binding energy ($L = \varepsilon \times GM\dot{M}/R_{NS}$, G is the gravitational constant, M is the NS mass, $\dot{M}$ is the accretion rate), and L_{37} is the accretion luminosity in units of 10^{37} erg/s, respectively. In this context we have $R_{disc} = r_m$. Since the accretion disc is virialized, its luminosity (averaged over the shot process) is $\langle I_{DF} \rangle = 0.5 \, \varepsilon \, GM\dot{M}/r_m$. The total luminosity is $\langle I_{DF} \rangle + \langle I_{LC} \rangle = \varepsilon \, GM\dot{M}/R_{NS}$. Therefore we have $\xi = 0.5 \, R_{NS}/r_m$. A lower limit for the inner rim of the accretion disk is $r_m \geq R_{NS}$ and therefore we get the constraint $\xi \leq 0.5$. This constraint is indicated as a dashed horizontal line in fig. 3. This immediately gives a lower limit on the fraction of coupled shots $\eta > 42\%$ indicated ad the dashed vertical line in figures 2 and 3. This in turn limits the value of the pulsed fraction below 0.37 as indicated in fig. 2 by the dashed horizontal line. An upper limit can be derived from the constraint derived in the previous paragraph $\xi \geq 0.294$ that gives $r_m \leq 1.7 R_{NS}$. Adopting as a typical NS radius $R_{NS} \sim 10$ km, we get $r_m \leq 17$ km, that is consistent with the requirement that accretion is not centrifugally inhibited. Moreover adopting $\phi = 0.3$ we can derive an upper limit for the magnetic field of the NS. Adopting the explicit espression for Alfvén

---

[1] $\phi = 0.21 \alpha^{4/15} n_{0.615}^{8/27} m^{-142/945} [(L_{37}/\varepsilon)^{8/7} R_6^{8/7} \mu_{26}^{5/7}]^{4/135}$, see the text for the definition of the symbols.

radius in $r_m \leq 1.7 R_{NS}$ we obtain $\mu_{26} \leq 5.1\, \phi_{0.3}^{-7/4} R_6^{9/4} m^{-1/4} \varepsilon^{-1/2} L_{37}^{1/2}$, where $phi_{0.3}$ is the ϕ factor in units of 0.3. Taking $m = 1.4$, $\varepsilon = 1$ and $L_{37} = 0.1$ that are appopriate for our source, we get $\mu_{26} \leq 1.5 R_6^{9/4}$. This is well within the range $(1-5) \times 10^{26}$ G cm^3 that are the lower and upper limits derived by [5] and di [6] respectively.

This scenario implies that the aperiodic variability is generated in a region affected by the lighthouse modulation, close to the neutron star surface. Our result is in line with the coupling of the red noise discovered analyzing the power spectra of some high mass X-ray binaries ([4, 12]). Our present result constrains some of the models proposed for the origin of the aperiodic variability. Indeed a scenario like the one proposed by [1] in which the aperiodic variability is generated by a shot noise process associated with magnetospheric flares, caused by magnetic reconnection all around the neutron star magnetosphere, cannot work for the origin of the aperiodic variability in SAX J1808.4-3658 in case the red noise component were coupled with the periodic modulation. On the other hand, inhomogeneities could arise in the accretion flow during the plasma penetration in the magnetosphere (*e.g.* via Kelvin-Helmholtz instabilities, see [4] for a more extensive discussion). The subsequent impact of this inhomogenous accretion flow funneled by the magnetic field lines onto the magnetic caps appears to be a viable mechanism for the origin of the red noise.

ACKNOWLEDGMENTS

This work was partially supported by a grant from the Italian Ministry of University and Research (Cofin 2001021123_002).

REFERENCES

1. J. J. Aly, and J. Kuijpers, *Astronomy and Astrophysics*, **227**, 473 (1990).
2. T. Belloni, G. Hasinger, *Astronomy and Astrophysics*, **230**, 103 (1990).
3. L. Burderi, N. R. Robba, G. Cusumano, *Adv.SpaceRes.*, **13**, 291 (1993).
4. L. Burderi, N. R. Robba, N. La Barbera, M. Guainazzi, *Astrophysical Journal*, **481**, 943 (1997).
5. D. Chakrabarty, E. H. Morgan, *Nature*, **394**, 346 (1998).
6. T. Di Salvo, L. Burderi,*Astronomy and Astrophysics*, **397**, 723 (2003).
7. M. Gilfanov, et al. *Astronomy and Astrophysics*, **338**, L83 (1998).
8. S. Hayakawa, *Phys. Rep.*, **121**, 317 (1985)
9. W. A. Heindl, and D. M. Smith, *Astrophysical Journal*, **506**, L35 (1998).
10. J. J. in 't Zand, et al.,*Astronomy and Astrophysics*, **331**, L25 (1998)
11. K. Jahoda, J. H. Swank, A. B. Giles, M. J. Stark, T. Strohmayer, W. Zhang, E. H. Morgan, *Proc. SPIE*, **2808**, 59 (1996).
12. D. Lazzati, L. Stella, *Astrophysical Journal*, **476**, 267 (1997).
13. D. A. Leahy, W. Darbro, R. F. Elsner, M. C. Weisskopf, S. Kahn, P. G. Sutherland, J. E. Grindlay, *Astrophysical Journal*, **266**, 160L (1983).
14. M. T. Menna, L. Burderi, L. Stella, N. R. Robba, M. van der Klis, *Astrophysical Journal*, **589**, 503 (2003).
15. D. Moon, S. S. Eikenberry, *Astrophysical Journal*, **552**, L135 (2001).
16. M. van der Klis, in "Fourier Techniques," in *X-ray Timing, Timing Neutron Stars*, edited by H. Ogelman and E. van den Heuvel, Kluwer Academic Publishers, Dordrecht, 1989, p. 27.
17. R. Wijnands, M. van der Klis, *Nature*, **394**, 344 (1998).
18. R. Wijnands, M. van der Klis, *Astrophysical Journal*, **514**, 939 (1999).

Accretion and Magneto-Dipole Emission in Fast-Rotating Neutron Stars: New Spin-Equilibrium Lines

Tiziana Di Salvo[*], Luciano Burderi[†], Franca D'Antona[**] and Natale Robba[*]

[*]*Dipartimento di Scienze Fisiche ed Astronomiche, Universitá di Palermo, via Archirafi 36, 90123 Palermo, Italy.*
[†]*Dipartimento di Fisica, Universitá degli Studi di Cagliari, SP Monserrato-Sestu km 0, 7, Monserrato (CA) 90042, Italy.*
[**]*Osservatorio Astronomico di Roma, via Frascati 33, 00040 Monteporzio Catone (Roma), Italy.*

Abstract. Although most of the proposed equations of state predict minimum spin periods well below one millisecond if more than about 0.3 solar masses are accreted onto a low magnetized neutron star through a Keplerian accretion disc, the spin periods of the recently discovered accreting millisecond pulsars all cluster in the quite narrow range between 1.7 and 5.4 ms; these spin periods are uncomfortably higher than the theoretical predictions. We propose and discuss here the possibility that emission due to a magneto-dipole rotator is relevant even during the accretion phase in fast-spinning neutron stars; this mechanism is able to explain the quite long spin periods observed in both low mass X-ray binaries and millisecond radio pulsars.

Keywords: accretion discs – stars: neutron stars — X-ray: stars — X-ray: binaries — X-ray: general
PACS: 97.60.Gb; 97.60.Jd

INTRODUCTION

Low-mass X-ray binaries (hereafter LMXBs) consist of a neutron star, generally with a weak magnetic field (less than $10^9 - 10^{10}$ Gauss), accreting matter from a low-mass (≤ 1 $M_\odot$) companion star. Most of these systems are X-ray transients; these are usually found in a quiescent state, with luminosities in the range $10^{31} - 10^{33}$ ergs/s. On occasions they exhibit outbursts, with peak luminosities between 10^{36} and 10^{38} ergs/s (see Campana et al. 1998 for a review). Adopting typical values for the conversion efficiency of the accreting matter energy into X-rays during outbursts and quiescent states, the inferred variations in the mass accretion rate in the proximity of the central source are typically a factor of 10^5.

These sources are closely related to millisecond radio pulsars since it is widely accepted that millisecond radio pulsars are the outcomes of the evolution of LMXBs: this is the so-called recycling scenario (see e.g. Bhattacharya & van den Heuvel 1991 for a review). Indeed, the low magnetic field of millisecond radio pulsars (usually lower than 10^9 Gauss) and the fact that these are often found in binary systems with very light (≤ 0.3 $M_\odot$) companion stars suggest that millisecond radio pulsars are old neutron stars, spun-up to periods of the order of milliseconds by a previous phase of accretion of matter and angular momentum. Although widely accepted, up to few years ago there was no proof that neutron stars can be spun-up to millisecond periods by the accretion of matter because of the lack of coherent pulsations in the progenitors, the neutron stars in LMXBs. The situation dramatically changed in 1998, when the first accreting millisecond pulsar, SAX J1808.4–3658, has been discovered. SAX J1808.4–3658 is a transient LMXB with an orbital period $P_{\rm orb} \simeq 2$ h showing coherent X-ray pulsations at ~ 2.5 ms (Wijnands & van der Klis 1998; Chakrabarty & Morgan 1998; see also Papitto et al. 2005). More recently, five other LMXBs have been discovered to be millisecond X-ray pulsars. All of them are X-ray transients and compact systems (orbital periods shorter than few hours). These are: XTE J1751–305 ($P_{\rm spin} \sim 2.3$ ms, $P_{\rm orb} = 42$ min), XTE J0929–314 ($P_{\rm spin} \sim 5.4$ ms, $P_{\rm orb} = 43$ min), XTE J1807–294 ($P_{\rm spin} \sim 5.2$ ms, $P_{\rm orb} = 40$ min), XTE J1814–338 ($P_{\rm spin} \sim 3.2$ ms, $P_{\rm orb} = 4.3$ hr), and finally the most recently discovered (on Dec 2, 2004) and the fastest among the accreting millisecond pulsars, IGR J00291+5934 ($P_{\rm spin} \sim 1.67$ ms, $P_{\rm orb} = 2.45$ hr; see Wijnands 2005, and references therein, for a review of the observational properties of all the accreting millisecond pulsars).

Indeed two of the main physical parameters of a neutron star are its spin period and magnetic field; these parameters determine most of the observed characteristics of a LMXB as well as the evolution of the neutron star spin itself. It is therefore useful to have a measure (or at least some constraints) on these parameters.

The presence and intensity of a magnetic field in LMXBs is an important question to address. The connection between LMXBs and millisecond radio pulsars indicates that neutron stars in LMXBs have magnetic fields of the order of $B \sim 10^8 - 10^9$ Gauss. In this case, the accretion disc in LMXBs should be truncated at the magnetosphere, where the disc pressure is balanced by the magnetic pressure exerted by the neutron star magnetic field. If the neutron stars in LMXBs have magnetic fields similar to those of millisecond radio pulsars, then the accretion disk should be truncated quite far (depending on the accretion rate) from the stellar surface, and the magnetic field should affect the accretion process, as indeed happens for the accreting millisecond pulsars. To constrain the magnetic field in LMXBs, Di Salvo & Burderi (2003) have developed a method to put an upper limit to the strength of the magnetic field of neutron stars for which the spin period and the X-ray luminosity during X-ray quiescent periods are known. This is obtained using simple considerations about the position of the magnetospheric radius during quiescent periods. We applied this method to the accreting millisecond pulsar SAX J1808.4–3658, which shows coherent X-ray pulsations at a frequency of ~ 400 Hz and a quiescent X-ray luminosity of $\sim 5 \times 10^{31}$ ergs/s, and found that $B \leq 5 \times 10^8$ Gauss in this source. Combined with the lower limit inferred from the presence of X-ray pulsations, this constrains the SAX J1808.4–3658 neutron star magnetic field in the quite narrow range $(1-5) \times 10^8$ Gauss. Similar considerations applied to the case of Aql X–1 and KS 1731–260 give neutron star magnetic fields lower than $\sim 10^9$ Gauss. Di Salvo & Burderi (2003) also show that residual accretion very unlikely powers the weak X-ray luminosity during quiescent phases in SAX J1808.4–3658.

The neutron star spin frequency is generally not easily directly observed in LMXBs, but it can be known from the so-called type-I X-ray burst oscillations. During these bursts nearly-coherent oscillations are sometimes observed, the frequencies of which are in the rather narrow range between 300 and 600 Hz (see van der Klis 2000; Strohmayer 2001 for reviews). This frequency is interpreted as the neutron star rotation frequency (see e.g. Chakrabarty et al. 2003), due to a hot spot in an atmospheric layer of the rotating neutron star. More indirectly, we can obtain information about the spin frequency of the neutron star from the so-called kHz QPOs. Many LMXBs show rich time variability both at low and at high frequencies, in the form of noise components or quasi periodic oscillations (QPOs). In particular, QPOs at kilohertz frequencies (kHz QPOs), with frequencies ranging from a few hundred Hz up to 1200 – 1300 Hz, have been observed in the emission of about 20 LMXBs (see van der Klis 2000 for a review). Usually two kHz QPO peaks ("twin peaks") are simultaneously observed, the difference between their centroid frequencies being in the range 250–350 Hz (usually similar, but not exactly identical, to the corresponding nearly-coherent frequency of the burst oscillations, or half that value). In a few cases, the recently discovered accreting millisecond pulsars, we can directly measure the spin frequency of the neutron star from coherent X-ray pulsations. From all these direct and indirect measures of the spin period of neutron stars in LMXBs, we infer that periods of fast-spinning neutron stars all cluster in the range 1.6 – 5.5 ms.

This poses a problem to address, that is the apparent absence of very fast-spinning neutron stars. Most of the equations of state proposed for ultra-dense matter predict minimum spin periods well below one millisecond if more than 0.1 – 0.3 solar masses have been accreted onto a low magnetized neutron star through a Keplerian accretion disc (e.g. Burderi et al. 1999; Lavagetto et al. 2004). However, the minimum spin period of a neutron star (or in a LMXB or in a millisecond radio pulsar) detected up to date is about 1.5 ms (Backer et al. 1982), uncomfortably higher than the theoretical predictions. It is unknown whether there is a natural brake for the spin-up process, or if it continues until the mass shedding limit is reached at submillisecond periods. However, the lack of very fast-spinning neutron stars in both the progenitors (the LMXBs) and the final end-products of the evolution (the millisecond radio pulsars) supports the idea that some mechanism limits accretion torques in spinning up millisecond pulsars. The most popular mechanisms invoked to limit the spin frequency of neutron stars are: i) gravitational radiation losses (e.g. Wagoner 1984; Bildsten 1998; Andersson et al. 1999), or ii) accretion centrifugally inhibited by the onset of a 'propeller' phase (Illarionov & Sunyaev 1975, but see also Rappaport et al. 2004). Both these proposed mechanisms, however, show some problems when applied to the evolution of these systems. Indeed, the virial theorem sets stringent limits on the fraction of matter that can be ejected during a propeller phase, the typical ejection efficiency is $\sim 50\%$, and this is often not enough to prevent the formation of very fast spinning neutron stars. Similarly, gravitational wave emission can balance the torque due to accretion, although with some ad hoc assumptions (Ushomirsky, Cutler, & Bildsten 2000), and might explain the observed spin periods of spun-up neutron stars.

We propose in this paper that emission due to a magneto-dipole rotator is relevant even during the accretion phase in millisecond X-ray pulsars; we show that this mechanism is able to explain the observed relatively long spin periods of neutron stars.

DISK-MAGNETIC FIELD INTERACTION

The position of the magnetospheric radius (R_M, the radius at which the accretion disk is probably truncated) is determined by the instantaneous balance of the pressure exerted by the accretion disc and the pressure exerted by the neutron star magnetic field. It turns out that R_M is a fraction $\phi \leq 1$ of the Alfvèn radius R_A determined equating the ram pressure of matter in free fall, spherically accreting onto the neutron star, with the magnetic pressure:

$$R_M = \phi R_A \simeq 3.48 \times 10^6 \, \phi \, m^{-1/7} \mu_{26}^{4/7} \dot{M}_{-10}^{-2/7} \text{ cm}, \qquad (1)$$

where m is the neutron star mass in Solar masses, μ_{26} is the neutron star magnetic moment, μ, in units of 10^{26} Gauss cm^3, and $\dot{M}_{-10}$ is the accretion rate in units of 10^{-10} M$_\odot$yr^{-1}. Equation (1) shows that as $\dot{M}$ decreases, R_M expands. Accretion of matter occurs when the magnetospheric radius is inside the corotation radius R_{CO}, the radius at which the Keplerian angular frequency of the orbiting matter is equal to the neutron star spin: $R_{CO} = 1.5 \times 10^6 \, m^{1/3} P_{-3}^{2/3}$ cm, where P_{-3} is the neutron star spin period in milliseconds. According to the standard theory, the accreting matter transfers its specific angular momentum (the Keplerian angular momentum at the magnetospheric radius) to the neutron star, therefore spinning it up. This process goes on until the pulsar reaches the equilibrium spin frequency, that is the Keplerian frequency at R_M. The spin equilibrium condition is therefore $R_M = R_{CO}$. The maximum equilibrium frequency is attained when the co-rotation radius is at the neutron star surface, $R_M = R_{CO} = R_{NS}$.

Accretion onto a spinning magnetized neutron star is centrifugally inhibited once R_M expands beyond the corotation radius R_{CO}. In this case the accreting matter could in principle be ejected from the system: this is called propeller phase (Illarionov & Sunyaev 1975).

Finally, if R_M further expands beyond the light-cylinder radius (where an object corotating with the neutron star attains the speed of light, $R_{LC} = 4.8 \times 10^6 P_{-3}$ cm), the neutron star becomes generator of magneto-dipole radiation and relativistic particles (i.e. a radio pulsar). Indeed, a common requirement of all the models of the emission mechanism from a rotating magnetic dipole is that the space surrounding the neutron star is free of matter up to R_{LC}. In this case, we expect the pulsar to lose energy (and angular momentum) according to the Larmor formula: $L_{PSR} = 2/(3c^3)\mu^2\Omega^4$, where Ω the neutron star spin frequency. The switch-on of the magneto-dipole rotator in transient LMXBs during X-ray quiescent periods (when the accretion rate is so low that the space inside the light-cylinder radius is most probably free of matter) has been invoked to explain the optical counterpart of the accreting millisecond pulsar SAX J1808.4–3658 during X-ray quiescence. Measures in the V band show an unexpectedly large optical luminosity, inconsistent with both intrinsic emission from the companion star and X-ray reprocessing in optical (Homer et al. 2001). To explain these data we have proposed that a magnetic dipole rotator is active in this source during quiescence and its bolometric luminosity (given by the Larmor's formula) powers the reprocessed optical emission. The resulting optical luminosity and colors are perfectly in agreement with observations (Burderi et al. 2003).

MAGNETO-DIPOLE EMISSION AND SPIN-EQUILIBRIUM LINES

As mentioned above, during the accretion phase the neutron star in a LMXB is accelerated by the transfer of mass and angular momentum through a Keplerian accretion disk. During the LMXB phase, the companion star loses about $1 M_\odot$, and the conservation of angular momentum tells us that a small fraction of this mass ($\sim 0.1 - 0.3 \, M_\odot$, see e.g. Lavagetto et al. 2004) is enough to bring a non-rotating neutron star to its minimum period. Therefore, at the end of the accretion phase we expect to find the neutron star very close to spin equilibrium, that is given by the condition: $R_M = R_{CO}$. From this condition we can derive how the equilibrium period depends on the mass accretion rate and the magnetic field of the neutron star:

$$P_{-3} = 5.23 \phi^{3/2} m^{-5/7} \mu_{26}^{6/7} \dot{M}_{-10}^{-3/7} \text{ for } R_M = R_{CO} > R_{NS}$$

$$P_{-3} = 0.54 R_6^{3/2} m^{-1/2} \text{ for } R_M = R_{CO} = R_{NS}, \tag{2}$$

where R_6 is the neutron star radius in units of 10^6 cm. This condition gives the equilibrium line plotted as a solid line in Figure 1, where we have assumed an average accretion rate of $\dot{M} = 10^{-10} \, M_\odot/\text{yr}$, and a typical value $\phi = 0.2$ (see e.g. Burderi et al. 2002). It is clear that magnetic field strengths below 5×10^8 Gauss will give equilibrium periods below 1.5 ms.

This standard scenario presents a few problems, the most important of which is the lack of a population of submillisecond pulsars, that is expected as the outcome of LMXB evolution: despite thoroughly searched, no pulsars with spin periods below 1.5 ms have been discovered up to now. On the contrary most of the observed spin periods cluster in the range 2 – 5 ms. To solve these problems we propose that another ingredient should be added to this scenario. As we mentioned above, the neutron star switches on as a magneto-dipole rotator when the space up to the light cylinder radius is free of matter, and in this case it will appear as a radio pulsar. However, there is no reason to exclude that the magneto-dipole rotator is emitting energy and angular momentum even in presence of matter. We expect it will emit the same amount of energy, although the pulsar will not be observed as a radio pulsar, and most of the energy emitted will be promptly absorbed by the matter surrounding the neutron star. In this case, the spin equilibrium condition is given by the balance between the accretion torque and the radiation torque. The accretion torque can be written as:

$$\tau_{acc} = j_K(R_M) \dot{M} = \dot{M} \sqrt{GM R_M} \sim 1.56 \times 10^{32} \phi^{1/2} m^{6/14} \mu_{26}^{2/7} \dot{M}_{-10}^{6/7} \text{ dyne cm},$$

where $j_K(R_M)$ is the specific angular momentum at the magnetospheric radius (we are assuming here that all the angular momentum of the accreting matter at the accretion radius is transferred to the neutron star). The expression above is valid as far as the magnetospheric radius is outside the neutron star surface, and this is true for magnetic moments stronger than a critical value given by: $\mu_{26,\text{crit}} = 0.072 \phi^{-7/4} m^{1/4} R_6^{7/4} \dot{M}_{-10}^{1/2}$. For magnetic moments below this value the accretion torque can be written as

$$\tau_{acc} = \dot{M} \sqrt{GM R_{NS}} \sim 0.73 \times 10^{32} R_6^{1/2} m^{1/2} \dot{M}_{-10} \text{ dyne cm}.$$

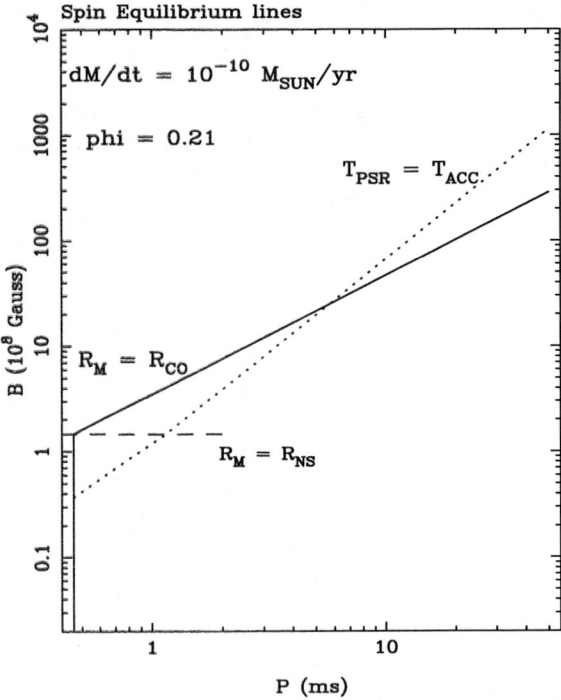

FIGURE 1. Spin Equilibrium lines in the magnetic field strength vs. spin period diagram, calculated for a mass accretion rate of 10^{-10} M$_\odot$/yr and a typical value of 0.21 for the ϕ parameter. The solid line represents the standard equilibrium condition $R_M = R_{CO}$, while the dotted line represent the new equilibrium condition proposed in this paper $\tau_{acc} = \tau_{psr}$. The dashed line indicates the magnetic field at which the magnetospheric radius reaches the neutron star radius at the given $\dot{M}$; below this line no pulsations can therefore be observed in accreting neutron stars.

The radiation torque can be written as:

$$\tau_{psr} = (2/3c^3)\mu^2\Omega^3 \sim 6.12 \times 10^{31} \mu_{26}^2 P_{-3}^{-3} \text{ dyne cm}.$$

The equilibrium condition, $\tau_{acc} = \tau_{psr}$, therefore implies:

$$P_{-3} = 0.73 \phi^{-1/6} m^{-1/7} \mu_{26}^{4/7} \dot{M}_{-10}^{-2/7} \text{ for } R_M > R_{NS}$$

$$P_{-3} = 0.94 R_6^{-1/6} m^{-1/6} \mu_{26}^{2/3} \dot{M}_{-10}^{-1/3} \text{ for } R_M = R_{NS} \qquad (3)$$

The resulting spin equilibrium line is thus plotted as a dotted line in Figure 1. As it is easy to see, for a given magnetic field (less than a few 10^9 Gauss) the equilibrium period is now much larger than with the standard spin equilibrium line. In particular a magnetic field of 5×10^8 Gauss, will give an equilibrium period around 2 ms. For high magnetic field strengths ($> 10^9$ Gauss in the example shown in Fig. 1), the torque due to the emission of the magneto-dipole rotator does not have any effect on the equilibrium period anymore. Finally, for magnetic field strength below the dashed line (where the magnetospheric radius reaches the neutron star radius) no pulsations can be observed from a neutron star accreting at the given $\dot{M}$ (10^{-10} $M_\odot$/yr in the example of Figure 1). This naturally explains why usually LMXBs do not show coherent pulsations, unless they are quite faint or with relatively high magnetic fields.

DISCUSSION

In this paper we propose that the effects of lost of energy and angular momentum due to a magneto-dipole rotator cannot be neglected in fast rotating neutron stars even during accretion phases. Indeed in fast pulsar, the emission calculated according to the Larmor formula gives an emitted power that can easily be comparable to the X-ray luminosity due to accretion. We have shown that, if these effects are included in the calculation of the maximum spin frequency of a neutron star in a LMXB, the neutron star cannot reach very fast (submillisecond) periods, thus making theoretical predictions in agreement with the observed periods of millisecond pulsars. We present in this paper new spin-equilibrium lines for accreting neutron stars.

That a magneto-dipole rotator emits energy (and angular momentum) according to the Larmor formula even in presence of matter inside the light-cylinder radius (although in this case the emission is very efficiently absorbed by the surrounding matter and therefore not observable) is not only plausible but also, in a sense, "observed". The discovery of the relativistic double pulsar system J0737-3039 (Burgay et al. 2003) shows that this is possible. In this system, in fact, the magnetosphere of pulsarA penetrates the magnetosphere of the companion, pulsarB (Lyne et al. 2004). Despite of the influence of J0737-3039A's energy flux on the magnetosphere of J0737-3039B (which causes orbital modulation of the flux density and the pulse shape of J0737-3039B), J0737-3039B is still losing energy and angular momentum according to the radio-pulsars' theory. Note that, if indeed a magneto-dipole rotator loses energy and angular momentum even during accretion phases, then this mechanism must be considered in the calculations of the spin equilibrium lines for a neutron star. We have shown that the spin equilibrium periods calculated in this way are much more consistent with observations.

Including this new ingredient in the calculation of the spin-equilibrium periods of accreting neutron stars naturally explains the observed relatively long spin period of LMXBs (measured directly for millisecond pulsars, or, indirectly, from burst oscillations) and millisecond radio pulsars, which all cluster in the range between 1.5 and 6 ms, and in particular explains the lack of very fast spinning (submillisecond) neutron stars. Our model also predicts, in given conditions, the possibility to observe a spin-down of the neutron star even during accretion. A possible spin-down episode, with a $\dot{\nu} \simeq -9.2 \times 10^{-14}$ Hz/s, during an accretion phase has been reported for the millisecond pulsar XTE J0929-314 (Galloway et al. 2002). If this spin-down rate can be entirely ascribed to magneto-dipole emission, then a spin frequency of 185 Hz requires a magnetic field of about 4×10^9 Gauss. In these conditions, the radiation torque caused by magneto-dipole emission will be dominant with respect to the accretion torque if the source luminosity, during the spin-down episode, was less than 1.510^{36} ergs/s, and this is true if the source distance is about 5 kpc.

ACKNOWLEDGMENTS

This work was partially supported by the Ministero della Istruzione, della Universitá e della Ricerca (MIUR).

REFERENCES

1. Andersson N., Kokkotas K., Schutz B.F., 1999, ApJ, 510, 846
2. Backer D.C., Kulkarni S.R., Heiles C., Davis M.M., Goss W. M., 1982, Nature, 300, 615
3. Bhattacharya D., van den Heuvel E.P.J., 1991, Physics Report, 203, 1
4. Bildsten L., 1998, ApJ, 501, L89
5. Burderi L., Di Salvo T., D'Antona F., Robba N.R., Testa V., 2003, A&A, 404, L43
6. Burderi L., Di Salvo T., Stella L., et al., 2002, ApJ, 574, 930
7. Burderi L., Possenti A., Colpi M., Di Salvo T., D'Amico N., 1999, ApJ, 519, 285
8. Burgay M., D'Amico N., Possenti A., et al., 2003, Nature, 426, 531
9. Campana S., Colpi M., Mereghetti S., Stella L., Tavani M., 1998, A&A Rev., 8, 279
10. Chakrabarty D., & Morgan E.H., 1998, Nature, 394, 346
11. Chakrabarty D., Morgan E.H., Muno M.P., Galloway D.K., Wijnands R., van der Klis M., Markwardt C., 2003, Nature, 424, 42
12. Di Salvo T., & Burderi L., 2003, A&A, 397, 723
13. Galloway D.K., Chakrabarty D., Morgan E.H., Remillard R.A., 2002, ApJ, 576, L137
14. Homer L., Charles P.A., Chakrabarty D., van Zyl L., 2001, MNRAS, 325, 1471
15. Illarionov A.F., & Sunyaev R.A., 1975, A&A, 39, 185
16. Lavagetto G., Burderi L., D'Antona F., Di Salvo T., Iaria R., Robba N. R., 2004, MNRAS, 348, 73
17. Lyne A.G., Burgay M., Kramer M., et al., 2004, Science, 303, 1153
18. Papitto A., Menna M. T., Burderi L., Di Salvo T., D'Antona F., Robba, N. R., 2005, ApJ, 621, L113
19. Rappaport S.A., Fregeau J.M., Spruit H., 2004, ApJ, 606, 436
20. Strohmayer T.E., 2001, in Astrophysical Sources of Gravitational Radiation for Ground-based Detectors, astro-ph/0101160
21. van der Klis M., 2000, ARA&A, 38, 717
22. Ushomirsky G., Cutler C., Bildsten L., 2000, MNRAS, 319, 902
23. Wagoner R. V., 1984, ApJ, 278, 345
24. Wijnands R., 2005, To appear in Nova Science Publishers (NY) volume "Pulsars New Research", astro-ph/0501264
25. Wijnands R., & van der Klis M., 1998, Nature, 394, 344

Searching for optical counterparts to MSP companions in Galactic Globular Clusters

F.R. Ferraro

Dipartimento di Astronomia, Università di Bologna, Via Ranzani 1, 40127 Bologna (ITALY)

Abstract. We are carring out a programme which exploits the advantages offered by the current generation of astronomical instruments (from the ground and from space) in a coordinate effort to understand the formation mechanism and evolution of Millisecond Pulsar (MSP) binary systems in Globular Clusters. In this framework I present a flower of the most recent results obtained by our group.

Keywords: Globular clusters; Millisecond pulsars; Dynamics and stellar evolution.
PACS: 98.20.Gm; 97.10.Zr; 97.20.Rp

BINARY BY-PRODUCTS IN GCS

Ultra-dense cores of Galactic Globular Clusters (GCs) are very efficient "kilns" for generating exotic objects, such as low-mass X-ray binaries, cataclysmic variables, millisecond pulsars (MSPs), blue stragglers (BSS), etc. Most of these objects are thought to result from the evolution of various kinds of binary systems originated and/or hardened by stellar interactions. The nature and even the existence of binary by-products can be strongly affected by the cluster core dynamics, thus serving as a diagnosticis of the dynamical evolution of GCs.

In this framework, we started a long-term project aimed to study the possible link between the dynamical evolution of clusters and the evolution of their stellar population. Indeed, among the possible collisional by-product zoo, MSPs are invaluable probes to study cluster dynamics: they form in binary systems containing a neutron star (NS) which is eventually spun up through mass accretion from the evolving companion. Despite the large difference in total mass between the disk of the Galaxy and the GC system, about 50% of the entire MSP population has been found in the latter. This is not surprising since in the Galactic disk MSPs can only form through the evolution of primordial binaries, while in GC cores dynamical interactions can lead to the formation of several different binary systems, suitable for recycling NS.

THE SURPRISING COMPANION TO THE MSP IN NGC6397

In this specific sector, we have recently done an outstanding discovery which promises to open a new perspective in understanding the MSP formation mechanisms in GCs: we found that the companion to the binary MSP (PSR J1740-5340) in NGC6397 is a peculiar, bright variable star.

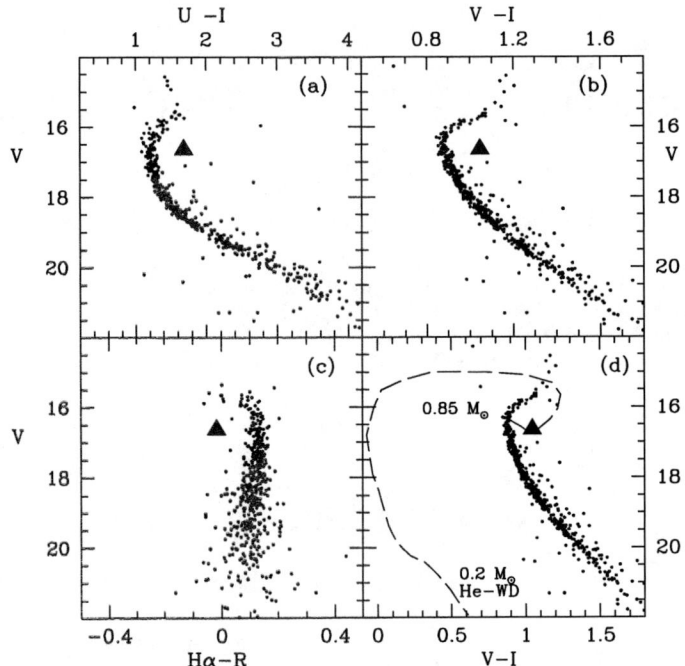

FIGURE 1. The optical counterpart to the MSP companion in NGC6397. Multiband Color Magnitude Diagram for stars detected in a region around the nominal position of the MSP. The optical counterpart of the MSP companion is marked with a large filled triangle in all panels. In *Panel (d)* the evolution path suggested by Burderi et al. (2002) is shown.

PSR J1740-5340 was identified by D'Amico et al. (2001a) during a systematic search for MSP in GCs carried out with the Parkes radio telescope. The pulsar displays eclipses at a frequency of 1.4 GHz for more than 40% of the 32.5 hr orbital period and exhibits striking irregularities of the radio signal at all orbital phases (D'Amico et al 2001b). This suggests that the MSP is orbiting within a large envelope of matter released from the companion, whose interaction with the pulsar wind could be responsible for the modulated and probably extended X-ray emission detected with CHANDRA (Grindlay et al. 2001, 2002). By using high resolution multiband HST observations and the position of the MSP inferred from radio timing, we identified a bright variable star (hereafter COM J1740-5340), as the optical counterpart to MSP companion (Ferraro et al. 2001, see Figure 1), whose optical modulation nicely agrees with the orbital period of the MSP itself.

This is the first example ever observed of a MSP companion whose light curve is dominated by ellipsoidal variations, suggestive of a tidally distorted star, which almost completely fills (and is still overflowing) its Roche lobe. In the classical framework of the MSP recycling scenario, the usual companion to a binary MSP should be either a white dwarf (WD) or a very light $(0.01 - 0.03 M_\odot)$ almost exhaused star. None of these scenarios can be applied to COM J1740-5340: it is too luminous to be a white dwarf ($V \sim 16.5$, comparable to the turn-off stars of NGC6397); moreover its mass ($M \sim 0.2 M_\odot$), recently constrained by radial velocity observations, is too high for a very light stellar companion. As a consequence, a wealth of intriguing scenarios have flourished in order to explain the nature of this binary (see Orosz & van Kerkwijk 2003, Grindlay et al. 2002 for a review). In particular, Burderi et al. (2002) suggested that the position of COM J1740-5340 in the Color Magnitude Diagram (CMD) is consistent with the evolution of an (slightly) evolved Sub Giant Branch (SGB) star orbiting the NS and loosing mass. The future evolution of this system will generate a He-WD/MSP pair (see *panel (d)* in Figure 1). COM J1740-5340 could be a star acquired by exchange interaction in the cluster core or alternatively the same star that spun up the MSP and still overflowing its Roche lobe. The latter case suggests the fashinating possibility that PSR J1740-5340 is a new-born MSP, the very first observed just after the end of the recycling process.

Thanks to the unusual brightness of the companion ($V \sim 16.5$), this system represents an unique laboratory to study the formation mechanism of binary MSP in GCs allowing unprecedented detailed spectroscopic observations. In this contest, we are coordinating a spectro-photometric programme at ESO-Telescopes. In particular a first set of high resolution spectra have been acquired at the VLT with UVES. From these data we obtained a number of interesting results:

- The determination of the radial velocity curve has allowed an accurate measure of the mass ratio of the system ($M_{PRS}/M_{COM} = 5.83 \pm 0.13$) which suggests a mass of $M_{COM} \sim 0.25 M_\odot$ by assuming $M_{PSR} \sim 1.4 M_\odot$ (see Ferraro et al 2003a).
- The H_α emission from the system was already noted by Ferraro et al (2001, see also *Panel (c)* in Figure 1) and fully confirmed by the high-resolution spectra (Sabbi et al 2003a). In particular, the complex structure of the H_α line suggests the presence of a matter stream going from the companion toward the NS. Note that because the flux of radiation from the pulsar the material would never reach the NS surface, creating a cometary-like gaseous tail which feeds the presence of (optically thin) hydrogen gas outside the Roche lobe.
- The unexpected detection of strong He I absorption lines implies the existence of a region (at $T > 10,000 K$), significantly hotter than the rest of the star (Ferraro et al 2003a, Sabbi et al 2003a,b). The intensity of the He I line correlates with the orbital phase, suggesting the presence of a region on the companion surface, heated by the millisecond pulsar flux.
- COM J1740-5340 has been found to show a large rotation velocity ($V \sin i = 50 \pm 1 K m s s^{-1}$). The derived abundances are found fully compatible with those of normal unperturbed stars in NGC 6397, with the exception of a few elements (Li, Ca, and C). In particular, the lack of C suggests that the star has been peeled down to regions where incomplete CNO burning occurs (Sabbi et al. 2003b), favouring a scenario where the companion is a SGB star which has lost most of its mass (see also Ergma & Sarna 2003).

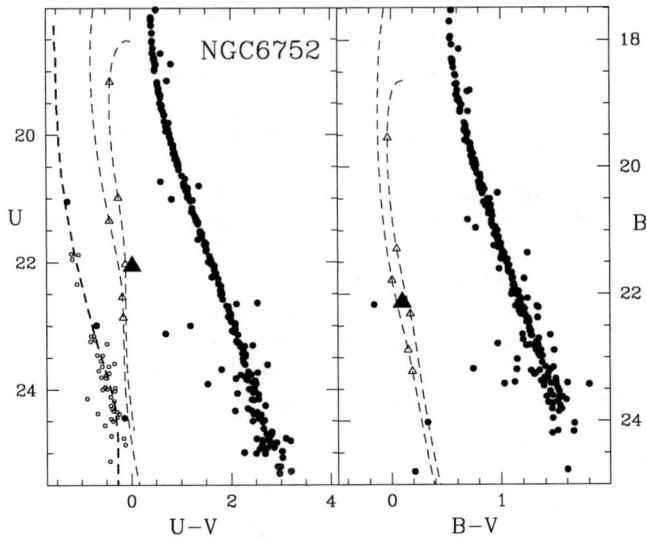

FIGURE 2. The optical counterpart to the MSP companion in NGC6752. $((U, U-V))$ and $((B, B-V))$ CMDs for the stars identified in a region of $80'' \times 80''$ centered at the nominal position of the PSR J1911-5958A. The optical counterpart to the pulsar companion (COM J1911-5958A) is marked with a *large filled triangles*. The *heavy dashed line* is the CO-WD cooling sequence from Wood (1995); the two *light dashed lines* are the cooling tracks for He WD masses 0.197 and 0.172 $M_\odot$ from Serenelli et al (2002).

THE CASE OF NGC6752

Another interesting object has captured our attention: the binary MSP recently discovered in the outskirts of the nearby globular cluster NGC 6752 PSR J1911-5958A (hereafter PSR-A) (D'Amico et al. 2001). It is located quite far away (at about $6'$) from the cluster optical center. Indeed PSR-A is the more off-centered pulsar among the sample of 41 MSPs whose position in the corresponding cluster is known.

By using deep high-resolution multiband images taken at the ESO *Very Large Telescope* we recently identified the optical binary companion to PSR-A (COM-PSR-A, Ferraro et al 2003b). The object turns out to be a blue star whose position in the CMD is consistent with the cooling sequence of a low mass ($M \sim 0.17 - 0.20 M_\odot$), low metallicity Helium WD (He-WD) (see Figure 2, see also Bassa et al 2003). The anomalous position of PSR-A with respect to the globular cluster center ($\sim 6'$) suggested that this system has been recently (< 1 Gyr) ejected from the cluster core as the result of a strong dynamical interaction. On the other hand, the cluster and its MSP population present a

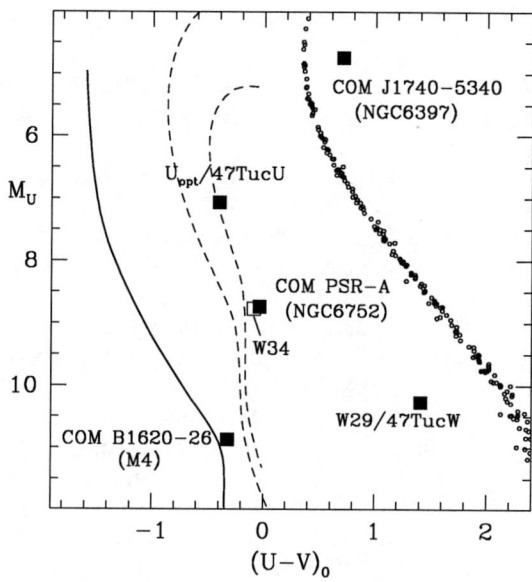

FIGURE 3. Optical counterparts to the MSP companion in GCs. All the optical counterparts to MSP companion detected so far in GCs are plotted (as large fille squares) in the $(M_U, U-V)$ plane. The cooling sequences for He-WD (from Serenelli et al. 2002) and for the CO-WD (from Wood 1995) are plotted. Main sequence stars of NGC6752 are also plotted as reference (see Ferraro et al 2003b).

number of additional anomalies:

- NGC6752 hosts also the second most displaced MSP seen in a GC, the isolated MSP PSR J1911-6000C.
- In the plane of the sky, the positions of the other 3 known MSPs in the cluster (PSR-B, D, E, all isolated pulsars, see D'Amico et al. 2001a, 2002) are close to the cluster center, as expected on the basis of mass segregation in the cluster. D'Amico et al. (2002) showed that PSR-B and E are experiencing a large acceleration with a line-of-sight component a_l directed toward the observer (see also Ferraro et al 2003c).
- By using a proper combination of high-resolution HST/WFPC2 and ground-based wide-field images, we have recently demonstrated that the brightness profile of the cluster cannot be properly reproduced by a simple King model, suggesting that NGC 6752 is experiencing a post-core-collapse bounce (Ferraro et al 2003c).

All these evidence would suggest the presence of a local perturber(s) of the pulsar dynamics, such as a binary black hole of intermediate $(100-200 M_\odot)$ mass, as recently proposed by Colpi et al. (2002,2003).

PHOTOMETRIC PROPERTIES OF MSP COMPANION IN GCS

The companion to PSR-A is the second He-WD which has been found to orbit a MSP in GCs. Curiously both objects have been found to lie on the same mass He-WD cooling sequence.

Since the first discovery of an optical counterpart to a MSP companion in GC, U_{opt}, the companion to PSR J0024−7203U in 47 Tuc (Edmonds et al 2001), the zoo of the optical MSP counterparts in GCs is rapidly increasing. Fig. 3 shows a comparison of the photometric properties of the available optical identifications of MSP companions hosted in GCs. Note that we include also an additional potential MSP companion (W34 in 47 Tuc) discussed by Edmonds et al. (2003). Two among the five sources seem really peculiar: the bright object in NGC6397 (which is as luminous as the turn off stars and shows quite red colours) and the faint W29 in 47 Tuc, which is also too red to be a He-WD (Edmonds et al 2002). Indeed, U_{opt} and COM-PSR-A are found to lie nearly on the same mass He-WD cooling sequence and W34 in 47 Tuc curiously shares the same photometric properties of COM-PSR-A. Indeed, if confirmed as a MSP companion, W34 would be the third He-WD companion orbiting a MSP in GGCs roughly located on the same-mass cooling sequence. If further supported by additional cases, this evidence could confirm that a low mass $\sim 0.15 - 0.2\ M_\odot$ He-WD orbiting a MSP is the favoured system generated by the recycling process of MSPs in GCs (Rasio et al 2000).

ACKNOWLEDGMENTS

It is a pleasure to thank Giacomo Beccari, Elena Sabbi, Andrea Possenti, Nichi D'Amico and the many other friends who are collaborating to this project. This research was supported by the *Agenzia Spaziale Italiana* (ASI) and by the *Ministero della Istruzione dell'Università e della Ricerca* (MIUR).

REFERENCES

1. Bassa, C. G., Verbunt, F., van KerKwijkm M. H., & Homer, L. 2003, A&A, 409, L31
2. Burderi, L.,D'Antona, F., Burgay, M., 2002, ApJ, 574, 325
3. Bassa, C. G., Verbunt, F., van KerKwijkm M. H., & Homer, L. 2003, A&A, 409, L31
4. Colpi, M., Possenti, A., & Gualandris, A. 2002, ApJ, 570, L85
5. Colpi, M., Mapelli, M., & Possenti, A. 2003, ApJ, 599, 1260
6. D'Amico, N., Lyne, A. G., Manchester, R. N., Possenti, A., & Camilo, F. 2001a, ApJ, 561, L171
7. D'Amico, N., Possenti, A., Manchester, R. N., Sarkissian, J., Lyne, A. G., & Camilo, F. 2001b, ApJ, 561, L89
8. D'Amico, N., Possenti, A., Fici, L., Manchester, R. N., Lyne, A. G., Camilo, F., & Sarkissian, J. 2002, ApJ, 570, L89
9. Edmonds, P. D., Gilliland, R. L., Heinke, C. O., Grindlay, J. E., & Camilo, F. 2001, ApJ, 557, L57
10. Edmonds, P. D., Gilliland, R. L., Camilo, F., Heinke, C. O., & Grindlay, J. E. 2002, ApJ, 579, 741
11. Edmonds, P. D., Gilliland, R. L., Heinke, C. O, & Grindlay, J. E. 2003 ApJ, 596, 1197
12. Ergma, E., Sarna, M.J., 2003, A&A, 399, 237
13. Ferraro, F. R., Possenti, A., D'Amico, N., & Sabbi, E. 2001, ApJ, 561, L93
14. Ferraro, F. R., Sabbi, E., Gratton, R., Possenti, A., D'Amico, N., Bragaglia, A., Camilo, F., 2003a, ApJ, 584, L13
15. Ferraro, F. R., Possenti, A., Sabbi., E.,D'Amico, N., 2003b, ApJ, 596, L211

16. Ferraro, F. R., Possenti, A., Sabbi., E., Lagani, P., Rood, R. T., D'Amico, N. & Origlia, L., 2003c, ApJ, 595, 179
17. Grindlay, J. E., Heinke, C. O., Edmonds, P. D., Murray, S. S.& Cool, A. 2001, ApJ, 563, L53
18. Grindlay, J. E., Camilo, F., Heinke, C. O., Edmonds, P. D., Cohn, H., Lugger, P., 2002, ApJ, 581, 470
19. Orosz, J.A., van Kerkwijk, M. H., 2003, A&A, 397, 237
20. Rasio, F.A., Pfahl, E.D., Rappaport, S., 2000, ApJ, 532, L47
21. Sabbi, E., Gratton, R., Ferraro, F. R., Bragaglia, A., Possenti, A., D'Amico, N., Camilo, F., 2003a, ApJ, 589, L41
22. Sabbi, E., Gratton, R., Bragaglia, A.,Ferraro, F. R., Possenti, A., Camilo, F., D'Amico, N., 2003b, A&A, 412, 829
23. Serenelli, A. M., Althaus, L. G., Rohrmann, R. D., & Benvenuto, O. G. 2002, MNRAS, 337, 1091

Temporal Analysis of the Millisecond X-ray Pulsar SAX J1808.4-3658 During the 2000 Outburst

Alessandro Papitto*, Luciano Burderi[†], Tiziana Di Salvo**, Maria Teresa Menna* and Luigi Stella*

*Osservatorio Astronomico di Roma, via Frascati 33, 00040 Monteporzio Catone (Rome, Italy)
[†]Univ.degli Studi di Cagliari, Dip.di Fisica, SP Monserrato-Sestu km 0.7, 09042 Monserrato, Italy
**Dipartimento di Scienze Fisiche e Astronomiche, Universitá di Palermo, via Archirafi 36, 90123, Palermo, Italy

Abstract. We report a temporal analysis of the millisecond X-ray Pulsar SAX J1808.4-3658 during the 2000 outburst, observed with RXTE.

The observed maximum luminosity was approximately a factor of ten lower than in the other outbursts exhibited by the source, and this low flux level forced us to use a technique based on the χ^2 obtained with an epoch folding search to discriminate between different possible orbital solutions, in order to correct the data for the orbital motion.

In the subsequent searches for periodicities we clearly detected the 401Hz pulsation in at least two observations, but in the faintest the pulsed fraction varied from 20 % ca. to the absence of signs of coherent pulsation at all, while the measured flux remained at an almost constat level. This erratic behaviour is discussed in the context of the centrifugal inhibition of accretion.

Keywords: pulsars:general – pulsars: individual(SAX J1808.4–3658) – stars:magnetic fields – stars: neutron – X-rays: binaries
PACS: <Missing classification>

INTRODUCTION

The X-ray transient SAX J1808.4-3658 was discovered in September 1996 when it exhibited an outburst detected by the BeppoSAX Wide Field Cameras (in't Zand et al. 1998). The source had a maximum luminosity of $\sim 2.5 \times 10^{36} erg\ s^{-1} (3-25 keV)$ and also showed three type-I X-ray bursts that led to the identification of the compact object as a neutron star (NS) and to the derivation of a distance of about $2.5 kpc$ (in 't Zand et al. 1998;2001).

The source was found in outburst again in 1998 April, when the high temporal resolution of the Proportional Counter Array (PCA) on board the *Rossi X-ray Timing Explorer* (RXTE) made it possible for Wijanands & van der Klis (1998) to discover coherent 401 Hz pulsations, making this source the first known accretion-driven millisecond pulsar. The peak luminosity was about $3.5 \times 10^{36} erg\ s^{-1}$ in the $3-150 keV$ band. Timing analysis performed on data collected over the period 1998 April 11-18 made it possible for Chakrabarty & Morgan (1998, hereafter CM98) to determine the orbital parameters of the system, such as the ~ 2 hr orbital period.

SAX J1808.4-3658 was detected in outburst for the third time on the 2000 January 21st (van der Klis et al. 2000), and observed again with RXTE/PCA, but this time at a flux level of about a tenth of the previously observed fluxes. Nevertheless it is worth noting that in the days preceeding the detection the source could not be observed due to solar constrainsts. The maximum luminosity reached by the source was $\sim 2.5 \times 10^{35} erg\, s^{-1} (3-25 keV)$ on the 2nd of February, but its proximity to the Sun at that time made it impossible to determine the exact moment of the start of the outburst, so that the peak luminosity could have been significantly higher. The behaviour of the X-ray luminosity, towards the end of the 2000 outburst, was highly erratic with dramatic variations by a factor ~ 20 on time scales of 5 hours and by an even larger factor (up to 1000) on time scales of a few days (Wijnands et al. 2001).

This kind of behaviour was again exhibited by the same source in the later stages of the 2002 October outburst (Wijnands 2004). In the first few weeks of this outburst the light curve resembled very nearly the 1998 one, with a similar peak luminosity and a steady decay that after about two weeks became steeper; five days later the flux raised again, reaching about one tenth of the peak level. Then the source entered in a low level activity state, with luminosity swings similar to those of the 2000 outburst, suggesting on one hand that a true outburst may have occurred, although unobserved, in 2000, and on the other that this low level activity state could be typical for this source.

OBSERVATIONS

Throughout this work we used the Target of Opportunity pubblic domain data of the PCA on board RXTE (Bradt et al. 1993). In particular, we analyzed data from the PCA (Jahoda et al. 1996), which is composed of a set of five xenon proportional counters operating in the $2-60$ keV energy range with a total effective area of $\sim 7000\, cm^2$. We considered all the available PCA observations of SAX J1808.4-3658 taken during the 2000 outburst covering the period January 21st - March 1st, focusing then our attention on the two brighter observations of this outburst (February 2nd and 8th), in which the spin modulation was strong enough to be analyzed. As we had to perform a temporal analysis on these data, first of all we reported all the events arrival times to the solar system barycenter, using the JPL's ephemerides DE200. In order to extract the light curves we considered data taken in Standard 2 Mode (129 channels; 16s in resolution), while event mode data $E\_125US\_64M\_0\_1S$ (64 channels; $122\mu s$ resolution) were used to perform temporal analysis. The data were background subtracted and analyzed according to the RXTE Cook Book recipes, using FTOOLS ver.5.2.

ORBITAL MOTION

The orbital motion of the source relative to the observer, determines a Doppler shift of the spin ferquency of the NS. In other terms the difference between the emission and the arrival times of the X-ray photons depends on the position of the source with respect to the line of nodes of the binary system. Neglecting any second order term in expanding $(t_{em} - t_{arr})$ the relation between emission and arrival times is:

$$t_{em} = t_{arr} - \frac{a\sin i}{c} \sin\left[\frac{2\pi}{P_{orb}}(t_{arr} - T^*)\right] \qquad (1)$$

where the t_{arr}'s are the photon arrival times at the line of barycenter of the Solar System, the t_{em}'s are the emission times referred to barycenter of the binary system, $\frac{a \sin i}{c}$ is the projected semimajor axis of the orbit, and T^* is the time of ascending node passage. Therefore with an observational estimate of the orbital parameters $a \sin i$, P_{orb} and T^* we can correct the arrival times of the photons emitted by the compact object for the delays induced by the orbital motion.

The high temporal resolution RXTE data of SAX J1808.4–3658 span a period of $\Delta T_{data} \sim 1.8$ yr, from April 1998 to February 2000, during which SAX J1808.4–3658 performed $N_{max} \sim 8000$ orbital cycles and $n_{max} \sim 22$ billion spin cycles. In order to perform a timing analysis over the whole ΔT_{data} we must unambigously associate n, i.e. the number of elapsed spin cycles since the 1998 outburtst, to any given t_{arr}. Unfortunately, even neglecting any error on the spin period estimate P_{spin}, induced by the errors in the orbital correction, this association is not possible as the uncertainity on the CM98 estimate of the spin period ($\sigma_{Pspin} \sim 5 \times 10^{-12}$) is at least one order of magnitude larger than the one needed to allow it ($\sigma_{Pspin} \leq P_{spin}/n_{max} \sim 1 \times 10^{-13}$ s.), thus ruling out the possibility of making an overall timing analysis over ΔT_{data}.

However, the large value of N_{max}, that becomes even larger considering the 2002 outburst data too, suggested us to develop a method of timing of the orbital period P_{orb} in order to improve the orbital period estimate of CM98. Our idea is to estimate the time of passage at a given point in the orbit, e.g. the ascending node T^*, for any given orbital cycle N, and to fit these times T^*s vs the corresponding N to improve the estimate of P_{orb} and to give a value (or an upper limit) for the orbital period derivative $\dot{P}_{orb}$.

Our strategy to estimate the values of T^* relies on the fact that in correcting the photon arrival times with equation 1, using a certain set of orbital parameters, the more they are far from the 'real' ones, the more the resulting pulse profiles obtained with an Epoch Folding will be smeared, hence with a smaller pulse fraction.

The experimental estimate of each T^* can be thus achieved by correcting the light curves of interest with different values of T^*, choosing as our estimate of this parameter the one that gives the maximum χ^2 in an epoch folding search.

This technique is discussed in details in Papitto et al. 2005 and its application to the three outburst of SAX J1808.4–3658, allowed us to determine an unique orbital solution valid over the entire $5\,yr$ period (1998 april - 2002 october) for which high temporal resolution data were available. We revised the previous estimate of the orbital period, $P_{orb} = 7249.1569(1)\,s$, and reduced the corresponding error by 1 order of magnitude with respect to that previuosly reported in CM98. Moreover we could get to the first constraint on the orbital period derivative, $-6.6 \times 10^{-12} < \dot{P}_{orb} < +0.8 \times 10^{-12}\,s\,s^{-1}$. These orbital parameters allowed us to produce, via a folding technique, pulse profiles at any given time.

PULSE PROFILES DURING 2000 FEBRUARY 8 OBSERVATION

We focused our attention on the two brighter observations of the 2000 outburst in which the count rate was high enough to allow an analysis of the pulse profiles.

We initially performed an epoch folding search on these data dividing each cycle in

5 phase bins and a Gaussian fit of the resulting χ^2 vs. spin period curve resulted gave an estimate of $P_{SPIN} = 0.002493919711(2)s$. The average pulse profiles, obtained by folding the light curves of interest around this value of the spin period and dividing each spin cycle into 20 phase bins, are shown in Figure 1; the average pulse fractions, evaluated as $pf = (I_{MAX} - I_{MIN})/I_{MAX}$, with I_{MAX} and I_{MIN} maximum and minimum count rate respectively, were of $(11.5\% \pm 0.5\%)$ (February 2) and $(10.4\% \pm 0.7\%)$ (February 8) (left panel of Fig.2).

While in the February 2 observation the pulsating actvity remained stable and clear troughout its whole duration, in the February 8 one it showed an intermittent behaviour, without any appreciable simultaneous variation in the X-ray flux. Considering $200s$ time intervals and evaluating the pf on the folded pulse profiles as before, we detected pulse fraction variations from $\sim 20\%$ to the absence of significant coherent pulsation at all, regardless of a nearly constant luminosity. The overall behaviour of the pulse fraction and luminosity is showed in the right panel of Fig.2.

In order to search for possible correlations between the pulse fraction and the X-ray flux, we performed a Pearson's linear correlation test on the roughly linearly increasing part of this template. This test measures the strenght of a supposed linear relationship between two variables in terms of a coefficient $r = cov(X,Y)/(\sigma_X \sigma_Y)$, $|r| \leq 1$. We found $r = 0.80$ for the points representing pulse fraction and time and $r = -0.18$ for luminosity and time, which mean, for $N = 12$ points, that the probability the pulsed fraction is not linearly correlated with time is less than 0.2%, while the the probability that luminosity is not linearly correlated with time is more than 50%.

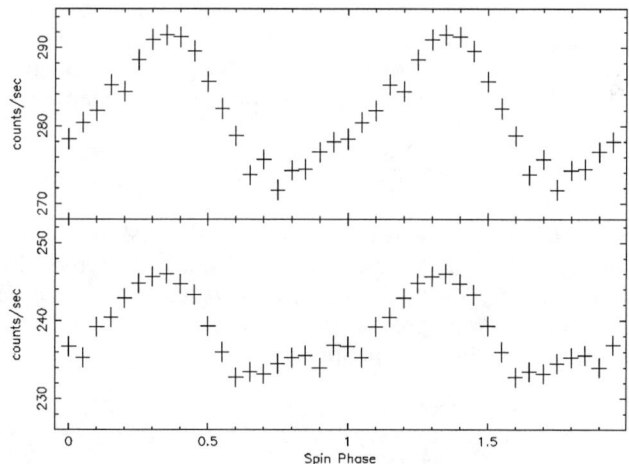

FIGURE 1. Average pulse profiles of the entire February 2 (top) and February 8 (bottom) observations. The light curves were folded around $P_{SPIN} = 2.493919711ms$ while every cycle was divided in 20 phase bins.

Finally we searched for any spectral variations that may be associated with the disappearance of pulsations, but the source remained quite hard troughout the observations, with a behaviour reminiscent of that of an atoll sources.

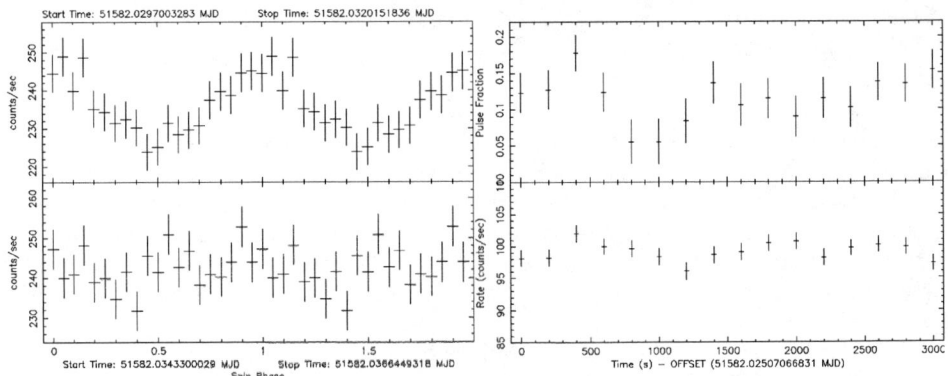

FIGURE 2. (Left):Pulse profiles in two different $200\,s$ intervals of the February 8 observation. The folded period is $P_{spin} = 2.493919711ms$, obtained from an Epoch Folding Search, as discussed in text. The top panel is referred to the third interval, while the bottom one is referred to the fifth. (Right): Pulsed fraction vs. time (top panel) and corresponding count rate $(2-60keV)$ (bottom panel) for the February 8 observation.

DISCUSSION

In the standard picture of disc accretion onto a neutron star of radius R_{NS}, whose magnetic field is strong enough to disrupt the disc at some magnetospheric radius $R_M > R_{NS}$, the total X-ray energy output can be divided in two terms:

$$L_X = \frac{GM\dot{M}}{2R_M} + GM\dot{M}\left[\frac{1}{R_{NS}} - \frac{1}{2R_M}\right]$$

The first represents the emission from matter in the disc that gradually spirals in, down to the magnetospheric radius, while the second arises from matter infall onto the NS polar caps. A fraction f of the second term, depending mainly on the geometry of accretion path and on scattering processes near the surface, will be actually pulsed. It is then possible to express the pulsed fraction as:

$$pf = \frac{f \times GM\dot{M}\left[\frac{1}{R_{NS}} - \frac{1}{2R_M}\right]}{\frac{GM\dot{M}}{2R_M} + GM\dot{M}\left[\frac{1}{R_{NS}} - \frac{1}{2R_M}\right]} = f \times \left(1 - \frac{R_{NS}}{2R_M}\right)$$

Expressing the magnetospheric radius as a roughly constant fraction of the Alfven radius $(R_A \sim \dot{M}^{-2/7})$, we then expect the pulse fraction to be a slowly decreasing function of $\dot{M}$ and an increasing function of f.

During 2000, February 8 observations SAX J1808.4-3658 showed strong variations in pulsed fraction, and in particular a sharp cut-off of the pulsating activity, while the X-ray flux, and hence the mass accretion rate, remained at an almost constant level. According to the simple picture we outlined before, these variations can be accounted only by changes in the parameter f or by a marginal decrease in the mass accretion rate that triggers a different accretion state.

The first hypotesis involves rather polar caps size than scattering environment variations, as no significant spectral evolution was detected during the disappearance of pulses. It has to be noted that fluctuations in the polar caps extent and structure were also invoked by C.Markwardt (IB2004), in explaining the apparent alternating spin up/down beahviour, exhibited by the same source during the October 2002 outburst.

The intermittent behaviour of pulsations can even be interpreted according to the propeller scenario (Illarionov & Sunyaev 1975), in which the accretion of matter onto the NS surface is inhibited by the onsetting of a centrifugal barrier, when the mass accretion rate drops below a critical value and the magnetospheric radius equals the corotation radius. Subsequently Spruit & Tamm (1993) have pointed out how the condition $R_M = R_{CO}$ is energetically insufficient to unbind all the incoming matter from the gravitational influence of the NS, and then that there must be an intermediate accretion state, for $R_M \sim R_{CO}$, in which mass accretion is neither allowed in the standard way nor halted by the centrifugal barrier, but rather proceeds in a cyclic way. RXTE observed the source again approximately 40 hours after the Feb8 observation, but this time was below the detection threshold ($few \times 10^{33} ergs \, s^{-1}$); we can therefore interpret the disappearance of pulsations as an intermediate phase of the onsetting of centrifugal barrier, in which the pulses starts to become less clear, but the luminosity did not had the expected cut off still. However a detailed coverage of these luminosity swings is missing and we could not even study the possibility of a connection between these pulsed fraction variations and the orbital motion, as already observed by Kirsch et al. (2004) on another source of this class, XTE J1807-294.

REFERENCES

1. Bradt, H., Rotschild, R.E. & Swank, J.H. 1993, *A&AS*, 97, 355.
2. Chakrabarty, D. & Morgan, E.H. 1998, *Nature*, 394, 346 (CM98).
3. Ilarionov, A.F. & Sunyaev, R.A. 1975, *A&A*, 39,185I.
4. in't Zand, J.J.M. et al. 1998, *A&A*, 331, L25.
5. in't Zand, J.J.M. et al. 2001, *A&A*, 372, 916.
6. Jahoda, K. et al. 1996, Proc.SPIE 2808, 59.
7. Kirsch, M.G.F. et al. 2004, *A&A*, 423L, 9
8. Papitto, A., Menna, M.T., Burderi, L., Di Salvo, T., D'Antona, F. & Robba, N.R. 2005, *ApJL*, 621.
9. Spruit, H.C. & Tamm, R.E. 1993, *ApJ*, 402, 593S.
10. van der Klis, M. et al. 2000, *IAU Circ.*, 7358, 3.
11. Wijnands, R. & van der Klis, M. 1998, *Nature*, 394, 344.
12. Wijnands, R., et al. 2001, *ApJ*, 560, 892.
13. Wijnands, R. 2004, in AIP Conf.Proc. 714, X-Ray Timing 2003: Rossi and beyond, ed. P.Kaaret, F.K.Lamb & J.H. Swank (Melville: AIP), 209

General relativistic effects on the evolution of binary systems

G. Lavagetto*, L. Burderi[†], F. D'Antona[†], T. Di Salvo*, R. Iaria* and N. R. Robba*

*Dipartimento di Scienze Fisiche ed Astronomiche, Università di Palermo, via Archirafi n.36, 90123 Palermo, Italy.
[†]Osservatorio Astronomico di Roma, Via Frascati 33, 00040 Monteporzio Catone (Roma), Italy.

Abstract. When a radio pulsar brakes down due to magnetodipole emission, its gravitational mass decreases accordingly. If the pulsar is hosted in a binary system, this mass loss will increase the orbital period of the system. We show that this relativistic effect can be indeed observable if the neutron star is fast and magnetized enough and that, if observed, it will help to put tight constraints to the equation of state of ultradense matter. Moreover, in Low Mass X-ray Binaries that evolve towards short periods, the neutron star lights up as a radio pulsar during the "period gap". As the effect we consider contrasts the orbital period decay, the system spends a longer time in this phase. As a consequence, the neutron star can survive this phase only if it is non-supramassive. Since in such bianries $\sim 0.8 M_\odot$ can be accreted onto the neutron star, short period ($P \leq 2$ h) millisecond X-ray pulsars like SAX J1808.4-3658 can be formed only if either a large part of the accreting matter has been ejected from the system, or the equation of state of ultradense matter is very stiff.

INTRODUCTION

Neutron stars (hereafter NSs) are the densest objects we can observe directly: their density can exceed by an order of magnitude the nuclear density and therefore their binding energy is non-negligible. Constraining the equation of state (EOS) of matter at supernuclear densities is of great interest. Anyway, only a few constraints that are sufficiently model-independent have been found from radio and X–ray observations. Moreover, these constraints did not allow to rule out most of the EOSs proposed in the literature [1].

Millisecond radio pulsars (MSPs) are some of the fastest spinning NSs known to date, and their timing behavior is measurable with great precision thanks to their stability. Of particular interest are millisecond pulsars hosted in binary systems: in particular, timing studies of NS-NS binaries allowed us to test general relativity to an unprecedented precision (see for example [2]).

These pulsars are thought to originate from Low Mass X-ray binaries (LMXBs) once accretion stops [3]. LMXBs in which the mass transfer is driven by orbital angular momentum losses evolve towards short orbital periods, typically below 2 h. It should be expected that these LMXBs evolve through a "period gap" similar to the gap found in the distribution of cataclysmic binaries [4].

Here we present a mechanism that, if observed in binary MSPs, could allow to put strong constraints on the EOS of NSs. We will show also how this mechanism affects the evolution of shrinking LMXBs, allowing us to make some predictions on such systems.

CONSTRAINING THE EOS WITH BINARY RADIO PULSAR TIMING

In a NS, the variation gravitational mass M_G per unit time depends on the variation of both the baryonic mass M_B and the angular momentum J of the star [5]:

$$\dot{M}_G = \Phi \dot{M}_B + \frac{\omega}{c^2} \dot{J}. \quad (1)$$

where ω is the NS spin frequency, and Φ is the energy needed to bring a unit mass from infinity to the surface of the star. Although a pulsar does not lose matter ($\dot{M}_B = 0$, it loses angular momentum via magnetodipole radiation, and therefore loses gravitational mass. This will have effects on the orbital evolution of the system. ăWe obtain for the evolution of the orbital period P [6]:

$$\frac{\dot{P}}{P} = -2 \frac{\dot{M}_G}{M_c} \frac{q}{1+q} + \frac{\dot{P}_{GW}}{P} \quad (2)$$

where $q = M_c/M_G$ and $\dot{P}_{GW}$ is the orbital period derivative due to the emission of gravitational waves, that is in general negative. As the gravitational mass of the NS decreases, the period of the binary system widens, yielding a positive contribution to the orbital period derivative, opposite to the contribution of gravitational waves emission. Since the effect of gravitational mass loss is $\propto P$, while the effect of gravitational waves emission is $\propto P^{-5/3}$ [7], the former effect will be dominant in systems with large enough orbital periods (say $P \geq 6$ h). To a good approximation we can write:

$$\dot{M}_G = \frac{I}{c^2} \omega \dot{\omega} \quad (3)$$

where I is the momentum of inertia of the NS. In a binary MSP it is often possible to measure both the spin frequency and its derivative with high precision. The orbital period derivative depends then only on measured quantities (the spin frequency ω and its derivative $\dot{\omega}$), on the masses of the two stars and of the momentum of inertia of the neutron star (see equation 2). We can extract information on the two masses from the mass function $f(M)$, that is measurable in binary MSP with very good precision. Using it, we can impose constraints on the momentum of inertia of the NS. Since the momentum of inertia depends strongly on the EOS of the NS [8], the detection of this effects will allow us to discriminate between various EOSs on a solid observational basis. Suppose, for example, that a system with an orbital period $P = 8$ h, a spin period $P_s = 2$ ms, $\dot{P}_s = 3 \times 10^{-19}$ and a mass function of $0.005 M_\odot$ is observed, and that the orbital period derivative has been measured to be $+2.5 \times 10^{-14}$. In figure 1, we plot the values of the masses of the two stars that are compatible with the value of the orbital period derivative wee obtained, in the hypothesis that the NS is governed by EOS A (the pure neutron EOS by Pandharipande [9]) or by EOS BBB (a realistic EOS by Baldo,

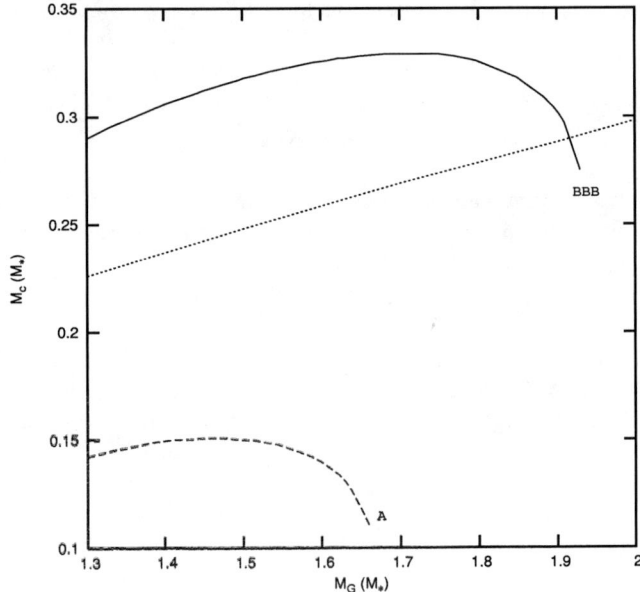

FIGURE 1. Allowed values of the mass of the primary versus the mass of the companion (both in solar masses) for the system described in the text. The two lines are for NSs with EOS A (dashed line) and NSs with EOS BBB (solid line). The dotted line indicates the lower limit on the companion mass obtained if the mass function is $5 \times 10^{-3} M_\odot$.

Bombaci and Burgio, [10]) respectively. The mass function requires that the values of the two masses should be above the dashed line in figure 1. Therefore, this particular system would allow us to rule out EOS A on an almost model-independent basis. This could be a potentially very powerful method for constraining the EOS of NS with very precise measurements and with an almost model-independent method.

Which is, between the known millisecond binary radio pulsars, the best candidate for detecting such an effect? The most promising object we found is PSR J0218+4232, which is constituted of a NS spinning at 2.3 ms in orbit with a white dwarf companion, with an orbital period of 2 days. The mass ratio is measured to be 0.13 ± 0.04 [11].

According to equation (2), we find that the orbital period derivative of the system is $\dot{P}_{orb} = 2.5 \times 10^{-14} I_{45}$, where I_{45} is the moment of inertia of the NS in units of 10^{45} g cm^2. The variation in the orbital period can be derived from the measure of the periastron time delay. The effect of an orbital period derivative on the periastron arrival time is given by:

$$\Delta T_{per} \simeq 0.5 \frac{\dot{P}}{P} \Delta T_{obs}^2. \qquad (4)$$

We find that we will need $118/I_{45}^{1/2}$ years of observation to detect a delay of 1 second. However, if a pulsar with an higher spin-down energy will be discovered, this effect might become observable in a much shorter observation time.

EVOLUTION OF LOW MASS X-RAY BINARIES TOWARDS SHORT ORBITAL PERIODS

LMXBs that transfer mass due to angular momentum losses evolve towards short orbital periods. As the period becomes smaller and smaller, the companion becomes fully convective (typically at an orbital period $\sim$ 3 h), magnetic braking is thought to stop, the companion recovers thermal equilibrium and the mass transfer ceases. The binary system will continue to shrink due to gravitational waves emission, until it reaches an orbital period $\sim$ 2 h, when mass transfer resumes.

What happens to the NS during this gap in the accretion process? It will light up as a millisecond radio pulsar since accretion has stopped. It will moreover be spinning very fast, as a considerable amount of matter has been accreted [12]. Two general relativistic effects are relevant during the gap:

1. The widening of the system due to the loss of gravitational mass from the pulsar (see the preceding section). This widening has the effect of increasing the duration of the detached phase of the system. This increase can vary strongly depending on the spin-down energy of the primary and on q (see equation 2).
2. The silent collapse to a black hole if the pulsar is supramassive (i.e. its mass exceeds the maximum non-rotating mass), once it loses enough angular momentum.

The NS survives the gap only if the collapse time (e.g. the time it takes for it to collapse to a black hole, T_c) is larger than the gap time (e.g. the time the system needs to cross the period gap, T_g). Else, the NS will collapse to a black hole before the mass transfer resumes.

We can define the collapse time via the equation:

$$J_{\text{in}} - J_{\text{crit}} = -\int_0^{T_c} \dot{J} dt \qquad (5)$$

where J_{in} is the angular momentum at the beginning of the detached phase, J_{crit} is the critical angular momentum below which the star collapses, and $\dot{J}$ is the angular momentum lost during the pulsar phase. Since we know that $\dot{J} = \dot{E}/\omega$, using the relativistic formulation of the energy released from a rotating dipole [13] we obtain

$$\dot{J} = -\frac{2}{3c^3}\mu^2\omega^3 \left(\frac{f}{N^2}\right)^2 \qquad (6)$$

where μ is the magnetic dipole moment of the NS and

$$N = (1-2\chi)^{1/2}, \quad \chi = \frac{GM_G}{c^2 R}$$
$$f = \frac{3}{8}\chi^{-3}\left[\log N^2 + 2\chi(1+2\chi)\right]. \qquad (7)$$

where R is the equatorial radius of the NS. On the other hand, the time the system spends in the gap, i.e. without accreting, is defined by the equation

$$\Delta P_{\text{gap}} = -\int_0^{T_g} \dot{P} dt \qquad (8)$$

where ΔP_{gap} is the amplitude of the period gap and $\dot{P}$ is defined in equation (2). In most situations $\Delta P_{\text{gap}} \sim 1$ h, and the mass of the companion is $\leq 0.25 M_\odot$. Integrating these two equations we find that, over a vast range of EOSs and of initial conditions, if the NS is supramassive and an even weak magnetic field ($\sim 10^8$ G), $T_g > T_c$. This means that LMXBs that host a NS and have a period shorter than 3 h cannot be supramassive. This is true, obviously, only if the binary system was a NS-main sequence binary: in systems where the companion is a white dwarf, and that evolve from short periods towards long periods and do not evolve though a pulsar phase. For instance, the surface magnetic field of the first millisecond X-ray pulsar discovered, SAX J1808.4-3658 [14], has been estimated to be in the range $(1-5) \times 10^8$ G [15]. If this system has a MS companion, so that it evolved from longer orbital periods, it cannot be supramassive, as it survived the period gap. The primary will not be a supramassive NS only if one of the following holds:

1. A relevant part of the accreting matter has been ejected from the system, and the mass transfer has therefore been non-conservative for most of the binary evolution.
2. The maximum non-rotating mass of the NS is very high, $\geq 2M_\odot$ (i.e. the EOS of the NS is very stiff).

REFERENCES

1. Stergioulas, N., *Living Reviews in Relativity*, **1**, 8 (1998).
2. Burgay, M., D'Amico, N., Possenti, A., Manchester, R. N., Lyne, A. G., Joshi, B. C., McLaughlin, M. A., Kramer, M., Sarkissian, J. M., Camilo, F., Kalogera, V., Kim, C., and Lorimer, D. R., *Nature*, **426**, 531 (2003).
3. Bhattacharya, D., and van den Heuvel, E. P. J., *Phys. Rep.*, **203**, 1–124 (1991), URL http://cdsads.u-strasbg.fr/cgi-bin/nph-bib_query?bibcode=1991%PhR...203....1B&db_key=AST.
4. Spruit, H. C., and Ritter, H., *A&A*, **124**, 267–272 (1983).
5. Bardeen, J. M., *ApJ*, **162**, 71 (1970).
6. Esposito, L. W., and Harrison, E. R., *ApJ*, **196**, L1 (1975).
7. Lorimer, D. R., *Living Reviews in Relativity*, 4, 5 (2001).
8. Cook, G. B., Shapiro, S. L., and Teukolsky, S. A., *ApJ*, **424**, 823 (1994).
9. Arnett, W. D., and Bowers, R. L., *ApJS*, **33**, 415 (1977).
10. Baldo, M., Bombaci, I., and Burgio, G. F., *A&A*, **328**, 274–282 (1997).
11. Bassa, C. G., van Kerkwijk, M. H., and Kulkarni, S. R., *A&A*, **403**, 1067–1075 (2003).
12. Lavagetto, G., Burderi, L., D'Antona, F., Di Salvo, T., Iaria, R., and Robba, N. R., *MNRAS*, **348**, 73–82 (2004).
13. Rezzolla, L., and J. Ahmedov, B., *MNRAS*, **352**, 1161–1179 (2004).
14. Wijnands, R., and van der Klis, M., *Nature*, **394**, 344–346 (1998).
15. Di Salvo, T., and Burderi, L., *A&A*, **397**, 723–727 (2003).

SESSION 3

SN FROM MASSIVE BINARY SYSTEMS AND GRBS

Accretion Power in GRBs

Tsvi Piran

*Racah Institute for Physics, The Hebrew University, Jerusalem, 91904, Israel
and Theoretical Astrophysics, Caltech, Pasadena, CA 91125, USA*

Abstract. I discuss the implication of the temporal structure of GRBs to the nature of their inner engine. I argue that the temporal strucutre shows that GRBs must involve internal shocks (or another kind of internal interaction within a relativistic outflow). To produce these internal shocks GRB inner engines must vary on a time scale of a fraction of a second and, on the other hand, they should be active for the whole duration of the burst, namely for several dozen of seconds. This implies that from the point of view of the central engine GRBs are a "quasi steady state" phenomenon. Accretion onto a newly formed black hole is the most likely mechanism that can satisfy these conditions and can power GRBs. I discuss the implication of accretion models of massive disks around black holes to GRB modelling.

Keywords: Gamma-ray bursts, accretion
PACS: 97.10.Gz, 98.70.Rz

INTRODUCTION

Our understanding of Gamma-Ray Bursts (GRBs) was revolutionized during the last ten years. According to the generally accepted Fireball model (see e.g. [1, 2] for recent reviews) the gamma-rays and their subsequent multiwavelength afterglow are produced when the kinetic energy of ultra-relativistic flow is dissipated. Most current observations, from prompt emission to late afterglow, from γ-rays via X-ray optical and IR to radio, are consistent with this model.

Today there is a clear evidence that (at least some) long duration GRBs are associated with Supernovae [3, 4, 5, 6, 7, 8]. This has been suggested by the Collapsar model [9, 10, 11] and confirmed by the association of long duration GRBs with type Ic supernovae [7, 8]. The origin of short GRBs is not clear but merging binary neutron stars [12] are possible candidates. In both cases it is believed that the GRB's "inner engine" involve accretion onto a newborn black hole, even though the details of the acceleration and collimation of the relativistic outflow are not understood.

I begin with summarizing the constrains on the "inner engines" of GRBs that arise from the temporal structure of GRBs. Sari and Piran [13] have shown that variable GRBs can be generated only by internal interactions [1] within the flow. To produce internal shocks the central engine must produce a long and variable wind. Kobayashi et al [14] have shown that the observed internal shocks light curve reflects almost directly the temporal activity of the inner engine. This is the best direct evidence on what is

---

[1] This interaction is usually considered as a collisionless shock. However the exact nature of the interaction is unimportant for most of our arguments.

happening at the center of the GRB. I review the arguments leading to these conclusions.

I then discuss additional observational results [15, 16, 17] and a theoretical toy model [18] that explains these observations within the internal shocks paradigm. The implication of these results is that the inner engine must be active for a very long time scale (the duration of the burst) compared to its own dynamical time scale (the dynamical scale of a typical compact object – black hole or a neutron star – is less than a millisecond). The most natural engine that can satisfy these conditions is a rather massive compact accretion disk around a newly formed compact object. In this case the observed time of the burst is the accretion time of the disk, while the fluctuation time is the dynamical time scale of the disk. The energy requirement suggests that the disk contains about $0.1 M_\odot$ and such a massive disk can arise only during the formation process of the compact object.

I conclude with a discussion of the implications of accretion theory to the black hole-accretion disk model for the inner engine of GRBs. I show that this theory implies that long GRBs should be produced by a Collapsar while short one by merging neutron star binaries. I refer the reader to [2] for a recent detailed review on other properties of GRBs.

TIME SCALES IN GRBS - OBSERVATIONS

Most GRBs are highly variable. Fig. 1 depicts the light curve of a typical variable GRB (GRB920627)[2]. The variability time scale, δt, as determined by the width of the peaks is much shorter (in some cases by a more than a factor of 100) then T, the duration of the burst. Variability on a time scale of milliseconds is seen in some long bursts [17].

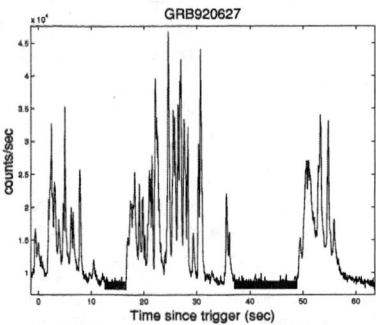

FIGURE 1. The light curve of GRB920627. The total duration of the burst is 52sec, while typical pulses are 0.8sec wide. Two quiescent periods lasting ~ 10 seconds are marked by horizontal solid bold lines.

A comparison of the pulse width distribution and the pulse separation, Δt, distribution, reveals an excess of long intervals [15, 16]. These long intervals can be classified as quiescent periods [19], relatively long periods of several dozen seconds with no activity. When excluding the quiescent periods Nakar and Piran [15, 16] find that both distributions are lognormal with a comparable parameters: The average pulse interval,

---

[2] About 15% of all GRBs are not variable. Clearly the arguments that follow do not apply to this subgroup

$\bar{\Delta t} = 1.3 sec$ is larger by a factor 1.3 then the average pulse width $\bar{\delta t} = 1 sec$. One also finds that the pulse widths are correlated with the preceding interval [15, 16].

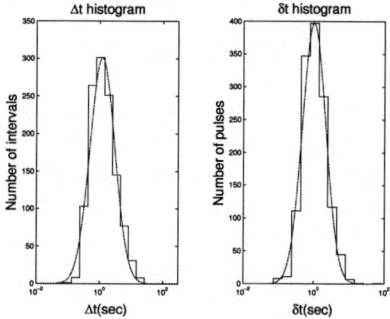

FIGURE 2. The pulse width distribution (right) and the distribution of intervals between pulses (left) (from [16]).

The results described so far are for long bursts. The variability of short ($T < 2sec$) bursts is more difficult to analyze. The duration of these bursts is closer to the limiting resolution of the detectors. Still most ($\sim 66\%$) short bursts are variable with $\delta t/T < 0.1$ [17].

TIME SCALES IN GRBS - THEORY

Consider a spherical relativistic emitting shell with a radius R, a width Δ and a Lorentz factor Γ. This can be a whole spherical shell or a spherical like section of a jet whose opening angle θ is larger than Γ^{-1}. Because of relativistic beaming an observer would observe radiation only from a region of angular size Γ^{-1}. Photons emitted by matter moving directly towards the observer (point A in Fig. 3) will arrive first. Photons emitted by matter moving at an angle Γ^{-1} (point D in Fig. 3) would arrive after $t_{ang} = R/2c\Gamma^2$. This is also the time of arrival of photons emitted by matter moving directly towards the observer but emitted at $2R$ (point C in Fig. 3). Thus, $t_{rad} \approx t_{ang}$ [13, 20].

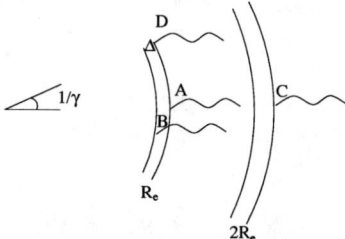

FIGURE 3. Different time scale from a relativistic expanding shell in terms of the arrival times (t_i) of various photons: $t_{ang} = t_D - t_A$, $t_{rad} = t_C - t_A$ and $T_\Delta = t_B - t_A$.

At a given point particles are continuously accelerated and emit radiation as long as the shell with a width Δ is crossing this point. The photons emitted at the front of this shell will reach the observer a time $T_\Delta = \Delta/c$ before those emitted from the rear (point

B in Fig. 3). In fact photons are emitted slightly longer as it takes some time for the accelerated photons to cool. For most reasonable parameters the cooling time is much shorter from the other time scales [21] and I ignore it hereafter.

Light curves are divided into two classes according to the ratio between T_Δ and $t_{ang} \approx t_{rad}$. The emission from different angular points smoothes the signal on a time scale t_{ang}. If $T_\Delta \leq t_{ang} \approx t_{rad}$ the resulting burst will be smoothed with a width $t_{ang} \approx t_{rad}$. Sari and Piran [13] have shown that for external shocks $\Delta/c \leq R/c\Gamma^2 \approx t_{rad} \approx t_{ang}$. External shocks can produce only smooth bursts!

A necessary condition for the production of a variable light curve is that $T_\Delta = \Delta/c > t_{ang}$. This can be easily satisfied within internal shocks (see Fig 4). Consider an "inner engine" emitting a relativistic wind active over a time $T_\Delta = \Delta/c$ (Δ is the overall width of the flow in the observer frame). The source is variable on a scale L/c. The internal shocks will take place at $R_s \approx L\Gamma^2$. At this place the angular time and the radial time satisfy: $t_{ang} \approx t_{rad} \approx L/c$. Internal shocks continue as long as the source is active, thus the overall observed duration $T = T_\Delta$ reflects the time that the "inner engine" is active. Note that now $t_{ang} \approx L/c < T_\Delta$ is trivially satisfied. The observed variability time scale in the light curve, δt, reflects the variability of the source L/c. While the overall duration of the burst reflects the overall duration of the activity of the "inner engine".

Numerical simulations [14] have shown that not only the time scales are preserved but the source's temporal behavior is reproduced on an almost one to one basis in the observed light curve. I return to this point in the next section where I describe a simple toy model that explains this result.

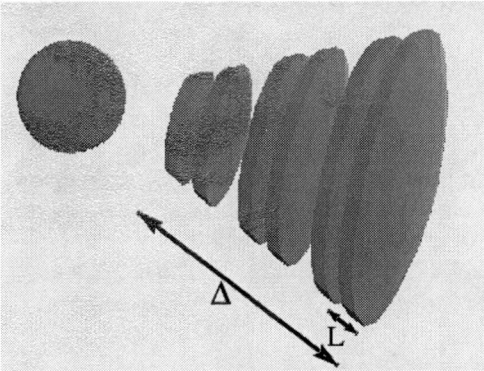

FIGURE 4. The internal shocks model (from [22]). Faster shells collide with slower ones and produce the observed γ rays. The variability time scale is L/c while the total duration of the burst is Δ/c.

AN INTERNAL SHOCKS TOY MODEL

The discovery [15, 16] that the distribution of pulse widths and pulse separations are comparable and that there is a correlation between the pulse width and the preceding interval provides an independent evidence in favor of the internal shocks model. Furthermore it suggests that the different shells emitted by the internal engine are most likely "equal energy" rather than "equal mass" shells.

Both features arise naturally within the internal shocks model [18] in which both the pulse duration and the separation between the pulses are determined by the same parameter. We outline here the main arguments showing that. Consider two shells with a separation L. The slower outer shell Lorentz factor is $\Gamma_1 = \Gamma$ and the inner faster shell Lorentz factor is $\Gamma_2 = a\Gamma$ ($a > 2$ for an efficient collision), both in the observer frame. The shells' are ejected at t_1 and $t_2 \approx t_1 + L/c$. The collision takes place at a radius $R_s \approx 2\Gamma^2 L$ (Note that R_s does not depend on Γ_2). The emitted photons from the collision will reach the observer at time (omitting the photons flight time, and assuming transparent shells):

$$t_o \approx t_1 + R_s/(2c\Gamma^2) \approx t_1 + L \approx t_2 . \tag{1}$$

The photons from this pulse are observed almost simultaneously with a (hypothetical) photon that was emitted from the "inner engine" together with the second shell (at t_2). This explains why Kobayashi et al [14] find numerically that for internal shocks the observed light curve replicates the temporal activity of the source.

Consider now four shells emitted at times t_i ($i = 1, 2, 3, 4$) with a separation of the order of L between them. Assume that there are two collisions: between the first and the second shells and between the third and the fourth shells. The first collision will be observed at t_2 while the second one will be observed at t_4. Therefore, $\Delta t \approx t_4 - t_2 \approx 2L/c$. Now assume a different collision scenario, the second and the first shells collide, and afterward the third shell takes over and collide with them (the forth shell does not play any roll in this case). The first collision will be observed at t_2 while the second one will be observed at t_3. Therefore, $\Delta t \approx t_3 - t_2 \approx L/c$. Numerical simulations [18] show that more then 80% of the efficient collisions follows one of the two scenarios described above. Therefore one can conclude that:

$$\Delta t \approx L/c . \tag{2}$$

Note that this result is independent of the shells' masses.

The pulse width is determined by the angular time (ignoring the cooling time): $\delta t = R_s/(2c\Gamma_s^2)$ where Γ_s is the Lorentz factor of the shocked emitting region. If the shells have an equal mass ($m_1 = m_2$) then $\Gamma_s = \sqrt{a}\Gamma$ while if they have equal energy ($m_1 = am_2$) then $\Gamma_s \approx \Gamma$. Therefore:

$$\delta t \approx \begin{cases} R_s/2a\Gamma^2 c \approx L/ac & \text{equal mass,} \\ R_s/2\Gamma^2 c \approx L/c & \text{equal energy.} \end{cases} \tag{3}$$

The ratio of the Lorentz factors a, determines the collision's efficiency. For efficient collision the variations in the shells Lorentz factor (and therefore a) must be large.

It follows from Eqs. 2 and 3 that for equal energy shells the Δt-δt similarity and correlation arises naturally from the reflection of the shells initial separation in both variables. However, for equal mass shells δt is shorter by a factor of a than Δt. Since a has a large variance this would wipes off the Δt-δt correlation. This suggests that equal energy shells are more likely to produce the observed light curves.

IMPLICATIONS OF THE INTERNAL SHOCKS MODEL TO THE INNER ENGINE

The results presented above show that within the internal shocks model the GRB light curve follows the activity of the internal engine. This, in turn, implies that the internal engine must have a rather short typical time scale (as implied by the variability time scale) of about 1 msec or less and a long over all activity time scale (as implied by the duration of the bursts) of several dozen seconds or even a thousand seconds in some cases. The first time scale suggest that GRBs involve a compact object, most likely a black hole. The second time scale indicates activity much longer than the typical dynamical time scale of a compact object and suggest that the "inner engine" involves accretion onto a black hole, with the duration of the burst being the accretion time.

IMPLICATIONS FROM ACCRETION THEORY

Several scenarios could lead to a black hole - massive accretion disk system. This could include mergers (NS-NS binaries [12, 23], NS-BH binaries [24] WD-BH binaries [25], BH-He-star binaries [26]) and models based on "failed supernovae or "Collapsars" [9, 10, 11]. Narayan, Piran and Kumar [27] (hereafter NPK01) have shown that accretion theory suggests that from all the above scenarios only Collapsars could produce long bursts and only NS-NS (or NS-BH) mergers could produce short bursts. I outline here the essence of their argument.

NPK01 considered a generic accretion model of a GRB in which a mass, $m_{d\odot}$, goes into orbit around a relativistic star of mass $3m_\odot$. The orbiting mass is initially inserted into a torus at a radius $r_{out}R_S$, where R_S is the Schwarzschild radius of the central star: $R_S = 2GM/c^2 = 8.85 \times 10^5 m_3$ cm. Starting from the initial toroidal configuration, the mass spreads out by viscosity and becomes an accretion flow extending from $r = 1$ (the horizon of the central black hole) to $r \sim r_{out}$.

If the accretion disk is larger than a few tens or hundreds of Schwarzschild radii, the accretion will proceed via a convection-dominated accretion flow (CDAF) in which most of the matter escapes to infinity rather than falling onto the black hole (see e.g. [28]) this poses a problem for models in which "large" ($r > 100$) disks arise. Ball, Narayan and Quataert [29] estimate the mass accretion rate in the CDAF is estimated to be $\dot{m}$

$$\dot{m} = 3.39 \times 10^4 m_3^{-1} m_d r_{out}^{-5/2}. \qquad (4)$$

The accretion time scale is not simply equal to $m_d/\dot{m}$. The reason is that much of the mass in a CDAF actually flows out of the system rather than into the central black hole. To estimate the accretion time, t_{acc}, [27] use the turbulent velocity of the connective blobs. They find:

$$t_{acc} \approx 4.17 \times 10^{-4} \alpha_a^{-1} m_3 r_{out}^{3/2} \text{ s}, \qquad (5)$$

where α_a is the dimensional parameter of the accretion viscosity. This scaling estimates agrees with numerical simulations of CDAFs [30]. One can notice immediately that long accretion times require large disks and hence if the duration of the burst is determined by the duration of the accretion the accretion torus must be large. However, in this case the mass, $m_{acc} = t_{acc}\dot{m}$, accreted by the black hole, are very small:

$$\begin{aligned} m_{acc} &= m_d, & r_{out} &\leq 14.1\alpha_a^{-1}m_d, \\ &= 14.1\alpha_a^{-1}m_d r_{out}^{-1}, & r_{out} &> 14.1\alpha_a^{-1}m_d. \end{aligned} \quad (6)$$

When r_{out} is large, the accreted mass is much less than m_d. The reason is that the bulk of the mass is ejected from the system, flowing out at $r \sim r_{out}$. The energy for the ejection is provided by convective energy flux from the interior of the flow. [27] find that CDAF is possible only if:

$$r_{out} > 118\alpha_a^{-2/7} m_3^{-1} m_d^{3/7}. \quad (7)$$

If the mass in the accretion flow is initially at a radius greater than the above limit, then the flow will become a CDAF at $r = r_{out}$, and for all $r < r_{out}$.

Within the inner region of the disk, very near the black hole the temperatures a high enough so that the disk can cool by emitting thermal neutrinos [31]. This neutrino-dominated accretion flow (NDAF) is efficient. An NDAF behaves like a thin accretion disk and use the basic theory of thin disks to estimate the accretion time and the accretion rate [31, 27]:

$$\begin{aligned} t_{acc} &= 2.76 \times 10^{-2} \alpha_a^{-6/5} m_3^{6/5} r_{out}^{4/5} \text{ s}, \\ \dot{m} &= 36.2 \alpha_a^{6/5} m_3^{-6/5} m_d r_{out}^{-4/5}. \end{aligned} \quad (8)$$

In a cooling-dominated thin disk, very little mass is expected to be lost to outflows, so nearly all the mass in the disk to be accreted by the star, i.e.

$$m_{acc} \approx m_d. \quad (9)$$

These results assume that the disk is gas pressure dominated. Slightly different resutls are obtained if degeneracy pressure takes over the gas pressure [27]. DiMatteo, Narayan and Perna [32] considered with the details of the neutrino transport within the inner parts of the disk. They find that the disk can become optically thick to neutrinos in its most inner parts. If the neutrinos are sufficiently trapped then energy advection becomes once more important.

NS-NS and BH-NS merger models, with $(r_{out}, m_d) = (10, 0.1)$ and $(10, 0.5)$ (see [31]), are well inside the NDAF zone and, according to our calculations, are capable of producing GRBs. However, this is only if the black hole is small (few $M_\odot$). If the black hole is larger than $\sim 10 M_\odot$ its Schwarszchild radius becomes too large and there is not enough "room" for an NDAF solution around it. Moreover, the neutron star in this case is swallowed whole by the BH and it is not tidally disrupted to create an accretion disk. On the other hand, unless the viscosity is very small, such disks cannot produce long bursts lasting hundreds or even tens of seconds. This suggests that NS-NS mergers and BH-NS mergers with smaller BH masses produce short GRBs but not long ones.

Other merger models, specifically the BH-WD and the BH-He star merger models, would appear not to be viable GRB engines. As the secondaries in these systems are not compact, they would form accretion flows with large values of r_{out}. For instance, [31] estimate $r_{out} \sim 3000$ for a BH-WD binary and $r_{out} \sim 5000$ for a BH-He star binary. At these radii, the accretion flow will be a very extended CDAF and hardly any mass will be accreted. Although the time scales of these models are consistent with long bursts, the extremely small value of m_{acc} suggests that these models do not produce GRBs of any kind.

In binary mergers, one can expect that a certain fixed amount of mass is instantaneously input into the accretion flow. The Collapsar model [11] corresponds to a different scenario in which mass is steadily fed over a period of time by fallback from the supernova explosion. [11] show that the time scale of the GRB is set by the physics of fallback rather than by accretion. Further, the time scales they obtain are consistent with observations of long GRBs. While the time scale may be set by fallback, the efficiency of the burst still depends on the nature of the post-fallback accretion. Efficient accretion, where most of the fallback material reaches the black hole, is possible only if r_{out} is small and falls within the NDAF zone. If Collapsars have a distribution of r_{out} and m_d, then only those systems that have $r_{out} \leq 100\alpha^{-2/7}m_3^{-1}$ will make bursts. Systems with larger angular momentum, and hence larger r_{out}, will form CDAFs and will eject most of the mass.

CONCLUSIONS

I cannot provide a recipe for a GRB "inner engine". However I can list the specifications of this engine (for a long variable GRB). It must satisfy the following conditions:

- It should accelerate $\sim 10^{51}$ ergs to a variable relativistic flow with $\Gamma > 100$.
- It should collimate this flow, with a varying degree of collimation (up to 1°).
- It should be active from several seconds up to several hundred seconds (according to the duration of the observed burst).
- It should vary on a time scale of a fraction of a second (corresponding to the duration of a typical pulse within the burst).
- Different shells of matter should have a comparable energy and their different Lorentz factors should arise due to a modulation of the accelerated mass.
- At times the engine should stop for several dozen seconds (resulting in a quiescent periods).

The natural model that satisfies these conditions is an accretion disk around a new born black hole. Arguments based on accretion theory [27] suggest that:

- Long bursts are produced by a Collapsar in which the duration of the activity of the disk depends on the time that matter fallback on the disk from the supernova explosion.
- Short bursts are produced by Neutron star binary mergers (or black hole-neutron star mergers) an in those the duration of the bursts is dictated by the duration of the accretion of the disk onto the black hole.

ACKNOWLEDGMENTS

This research was supported by a grant from the US-Israel Binational Science Foundation.

REFERENCES

1. T. Piran, Physics Reports, **314**, 575 (1999).
2. T. Piran, *Rev. Mod. Phys*, **76**, 1143 (2004)..
3. T. Galama et al., Nature, **395**, 70, (1998).
4. Bloom, J. S. et al., Nature, **401**, 453 (1999).
5. Reichart, D. E. 1999, ApJ, **521**, L111, (1999).
6. Galama, T. J. et al., ApJ, **536**, 185 (2000).
7. K. Z. Stanek, T. Matheson, P. M. Garnavich et al., *Ap. J. Lett.,*, **591**, L17–L20 (2003).
8. J. Hjorth, J. Sollerman, P. Møller et al., *Nature,*, **423**, 847–850 (2003).
9. S. E. Woosley, Ap. J., **405**, 273 (1993).
10. B. Paczynski, Ap. J. Lett., **494**, L45 (1998).
11. A. I. MacFadyen and S. E. Woosley, Ap. J., **524**, 262 (1999).
12. D. Eichler, M. Livio, T. Piran, and D. N. Schramm, Nature, **340**, 126 (1989).
13. R. Sari and T. Piran, Ap. J., **485**, 270 (1997).
14. S. Kobayashi, T. Piran, and R. Sari, Ap. J., **490**, 92 (1997).
15. E. Nakar and T. Piran, astro-ph/0103011, GRBs in the Afterglow Era, Eds. E. Costa, F. Fronteira and K. Hjorth, (Springer) (2001).
16. E. Nakar, and T. Piran, *Mon. Not. RAS,*, **331**, 40–44 (2002).
17. E. Nakar, and T. Piran, *Mon. Not. RAS,*, **330**, 920–926 (2002).
18. E. Nakar, and T. Piran, *Ap. J. Lett.,*, **572**, L139–L142 (2002).
19. E. Ramirez-Ruiz, and A Melroni, MNRAS, **320**, K25 (2001).
20. E. E. Fenimore, C. D. Madras, and S. Nayakshin, Ap. J., **473**, 998, (1996).
21. R. Sari, R. Narayan and T. Piran, Ap. J., **473**, 204, (1996).
22. R. Sari, PhD thesis, (1998)
23. R. Narayan, B. Paczynski, and T. Piran, *Ap. J. Lett.,*, **395**, L83–L86 (1992).
24. B. Paczynski, *Acta Astronomica*, **41**, 257–267 (1991).
25. C. L. Fryer, S. E. Woosley, M. Herant, and M. B. Davies, *Ap. J.,*, **520**, 650–660 (1999).
26. C. L. Fryer, and S. E. Woosley, *Ap. J. Lett.,*, **502**, L9–+ (1998).
27. R. Narayan, T. Piran and P. Kumar, Ap. J., **557**, 949, (2001).
28. J. M. Stone, J. E. Pringle, and M. C. Begelman, *Mon. Not. RAS,*, **310**, 1002–1016 (1999).
29. Ball, G. H., Narayan, R., & Quataert, E. *Ap. J.,*, **552**, 221, (2001).
30. I. V. Igumenshchev, and M. A. Abramowicz, *Ap. J. Supp.,*, **130**, 463–484 (2000).
31. R. Popham, S. E. Woosley, and C. Fryer, *Ap. J.,*, **518**, 356–374 (1999).
32. Di Matteo, T., Perna, R., & Narayan, R. *Ap. J.,*, **579**, 706 (2002).

The Quark-Deconfinement Nova model for Gamma-Ray Bursts

I. Bombaci

Dipartimento di Fisica "E. Fermi", Università di Pisa & INFN, Sezione di Pisa, largo B. Pontecorvo, 3, I-56127, Pisa, Italy

Abstract. We report on a new model which is able to explain how a gamma-ray burst (GRB) can take place days or years after a supernova explosion. We show that above a threshold value of the gravitational mass a pure hadronic star ("neutron star") is metastable to the conversion into a quark star (hybrid star or strange star), *i.e.* a star made at least in part of deconfined quark matter. The stellar conversion process can be delayed if finite size effects at the interface between hadronic and deconfined quark matter phases are taken into account. A huge amount of energy, on the order of $10^{52} - 10^{53}$ ergs, is released during the conversion process and can produce a powerful gamma-ray burst. The delay between the supernova explosion generating the metastable neutron star and the new collapse can explain the delay inferred in GRB 990705 and in GRB 011211.

Keywords: Gamma rays: Gamma Ray Bursts. Stars: Neutron Stars, Strange Stars. Dense Matter: equation of state, quark matter
PACS: 98.70.Rz; 97.60.Jd; 26.60.+c; 25.75.Nq; 21.65.+f

INTRODUCTION

In the last few years, the evidence for a physical connection between supernova (SN) explosions and long-duration Gamma Ray Bursts (GRBs) has mounted [1, 2, 3, 4, 5, 6, 7]. In particular, the detection of X-ray spectral features in the X-ray afterglow of some GRBs [1], has given evidence for a possible time delay between the SN explosion and the associated GRB. In the case of GRB 990705, GRB 020813, and GRB 011211, it has been possible to estimate the time delay between the two events. For GRB 990705 the supernova explosion is evaluated to have occurred a few years before the GRB [2, 11], a few months before the burst in the case of GRB 020813 ([7]), while for GRB 011211 about four days before the burst [5].

This possible delayed SN-GRB connection demands for a two-step explosion mechanism. The first event, in this scenario, is the supernova explosion which forms a compact stellar remnant, *i.e.* a neutron star (NS); the second catastrophic event is associated with the NS and it is the energy source for the observed GRB. The two-step scenario outlined above, poses severe problems for most of the current theoretical models for the central energy source (the so-called *central engine*) of GRBs. The main difficulty of all these models is to understand the origin of the second "explosion", and to explain the long time delay between the two events.

In the so-called *supranova* model [12] for GRBs the second catastrophic event is

---

[1] The statistical significance of the detection of these spectral lines is of some concern [8, 9, 10].

the collapse to a black hole of a *supramassive* neutron star, *i.e.* a fast rotating NS with a baryonic mass M_B above the maximum baryonic mass $M_{B,max}$ for non-rotating configurations. In this model, the time delay between the SN explosion and the GRB is equal to the time needed by the fast rotating newly formed neutron star to get rid of angular momentum and to reach the limit for instability against quasi-radial modes where the collapse to a black hole occurs [13]. In the following, we report on some recent research [14, 15] which try to make a connection between GRBs and the quark-deconfinement phase transition in the core of neutron stars.

NEUTRON STARS: HADRONIC OR QUARK STARS?

In a simple and conservative picture the core of a neutron star is modeled as a uniform fluid of neutron rich nuclear matter in equilibrium with respect to the weak interaction (β-stable nuclear matter). However, due to the large value of the stellar central density and to the rapid increase of the nucleon chemical potentials with density, hyperons (Λ, Σ^-, Σ^0, Σ^+, Ξ^- and Ξ^0 particles) are expected to appear in the inner core of the star. Other *exotic* phases of hadronic matter such as a Bose-Einstein condensate of negative pion (π^-) or negative kaon (K^-) could be present in the inner part of the star.

Quantum Chromodynamics (QCD) predicts a phase transition from hadronic matter to a deconfined quark phase to occur at a density of a few times nuclear matter saturation density. Consequently, the core of the more massive neutron stars is one of the best candidates in the Universe where such deconfined phase of quark matter (QM) could be found. Since β-stable hadronic matter posses two conserved "charges" (*i.e.*, electric charge and baryon number) the quark-deconfinement phase transition proceeds through a mixed phase over a finite range of pressures and densities according to the Gibbs' criterion for phase equilibrium [16]. At the onset of the mixed phase, quark matter droplets form a Coulomb lattice embedded in a sea of hadrons and in a roughly uniform sea of electrons and muons. As the pressure increases various geometrical shapes (rods, plates) of the less abundant phase immersed in the dominant one are expected. Finally the system turns into uniform quark matter at the highest pressure of the mixed phase [17]. Compact stars which possess a "quark matter core" either as a mixed phase of deconfined quarks and hadrons or as a pure quark matter phase are called [18] *Hybrid Stars* (HyS). In the following of this paper, the more *conventional* neutron stars in which no fraction of quark matter is present, will be referred to as *pure Hadronic Stars* (HS).

A complementary manifestation of quark matter in compact stars is the possible existence of a new family of compact stars consisting completely of a deconfined mixture of *up* (*u*), *down* (*d*) and *strange* (*s*) quarks (together with an appropriate number of electrons to guarantee electrical neutrality) satisfying the so-called Bodmer–Witten hypothesis [19]. Such compact stars have been called *strange stars* [20, 21] (SS) and their constituent matter [22] as *strange quark matter* (SQM). Presently there is no unambiguous proof about the existence of strange stars, however, a sizable amount of observational data collected by the new generations of X-ray satellites, is providing a growing body of evidence for their possible existence [23, 24, 25]. In the following, we will refer to hybrid stars and strange stars collectively as quark stars (QS).

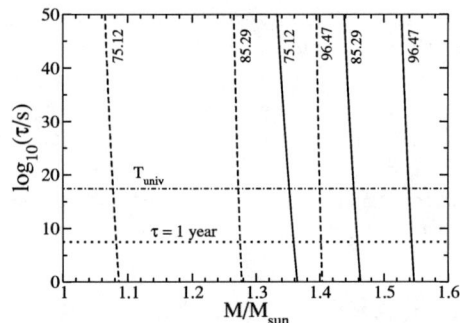

FIGURE 1. Nucleation time as a function of the gravitational mass of the hadronic star. Solid lines correspond to a value of $\sigma = 30$ MeV/fm^2 whereas dashed ones are for $\sigma = 10$ MeV/fm^2. The nucleation time corresponding to one year is shown by the dotted horizontal line. The different values of the bag constant (in units of MeV/fm^3) are plotted next to each curve. The hadronic phase is described with the GM1 model for the equation of state

NUCLEATION OF QUARK MATTER IN HADRONIC STARS

In cold ($T = 0$) bulk matter the quark-hadron mixed phase begins at the *static transition point* defined according to the Gibbs' criterion for phase equilibrium

$$\mu_H = \mu_Q \equiv \mu_0, \qquad P_H(\mu_0) = P_Q(\mu_0) \equiv P_0 \qquad (1)$$

where $\mu_H = (\varepsilon_H + P_H)/n_{b,H}$ and $\mu_Q = (\varepsilon_Q + P_Q)/n_{b,Q}$ are the chemical potentials for the hadron and quark phase respectively, ε_H (ε_Q), P_H (P_Q) and $n_{b,H}$ ($n_{b,Q}$) denote respectively the total (*i.e.*, including leptonic contributions) energy density, the total pressure and baryon number density for the hadron (quark) phase.

Consider now the more realistic situation in which one takes into account the energy cost due to finite size effects in creating a drop of deconfined QM in the hadronic environment. As a consequence of these effects, the formation of a critical-size drop of QM is not immediate and it is necessary to have an overpressure $\Delta P = P - P_0$ with respect to the static transition point. Thus, above P_0, hadronic matter is in a metastable state, and the formation of a real drop of QM occurs via a quantum nucleation mechanism. A sub-critical (virtual) droplet of deconfined QM moves back and forth in the potential energy well separating the two matter phases on a time scale $v_0^{-1} \sim 10^{-23}$ seconds, which is set by the strong interactions. This time scale is many orders of magnitude shorter than the typical time scale for the weak interactions, therefore quark flavor must be conserved during the deconfinement transition. We will refer to this form of deconfined matter, in which the flavor content is equal to that of the β-stable hadronic system at the same pressure, as the Q*-phase. Soon afterwards a critical size drop of QM is formed the weak interactions will have enough time to act, changing the quark flavor fraction of the deconfined droplet to lower its energy, and a droplet of β-stable SQM is formed (hereafter the Q-phase).

In our scenario, we consider a pure HS whose central pressure is increasing due to spin-down or due to mass accretion, *e.g.*, from the material left by the supernova explosion, or from a companion star. As the central pressure exceeds the threshold value P_0^* at the static transition point, a virtual drop of quark matter in the Q*-phase can be formed in the central region of the star. As soon as a real drop of Q*-matter is formed, it will grow very rapidly and the original HS will be converted to and hybrid star or to a strange star, depending on the details of the equation of state (EOS) [2] for quark matter employed to model the phase transition.

To calculate the nucleation time τ, *i.e.*, the time needed to form the first critical droplet of deconfined QM in the hadronic medium, one can use the relativistic quantum nucleation theory [27] (for more details, see [14, 15]). The nucleation time τ, can be calculated for different values of the stellar central pressure P_c. In Fig. 1, we plot τ as a function of the gravitational mass M_{HS} of the HS corresponding to the given value of the central pressure, as implied by the solution of the Tolmann-Oppeneimer-Volkov equations [28] for the pure HS sequences. Each curve refers to a different value of the bag constant and surface tension σ. As we can see, from the results in Fig. 1, a metastable hadronic star can have a mean-life time many orders of magnitude larger than the age of the universe $T_{univ} = (4.32 \pm 0.06) \times 10^{17}$ s. As the star accretes a small amount of mass (of the order of a few per cent of the mass of the sun), the consequential increase of the central pressure leads to a huge reduction of the nucleation time and, as a result, to a dramatic reduction of the HS *mean-life time*.

To summarize, pure hadronic stars having a central pressure larger than the static transition pressure for the formation of the Q*-phase are metastable to the "decay" (conversion) to a more compact stellar configuration in which deconfined QM is present (HyS or SS). These metastable HS have a *mean-life time* which is related to the nucleation time to form the first critical-size drop of deconfined matter in their interior [3]. We define as *critical mass* M_{cr} of the metastable HS, the value of the gravitational mass for which the nucleation time is equal to one year: $M_{cr} \equiv M_{HS}(\tau=1 \text{ yr})$. Pure hadronic stars with $M_H > M_{cr}$ are very unlikely to be observed. M_{cr} plays the role of an *effective maximum mass* for the hadronic branch of compact stars. While the Oppenheimer–Volkov maximum mass [28] $M_{HS,max}$ is determined by the overall stiffness of the EOS for hadronic matter, the value of M_{cr} will depend in addition on the bulk properties of the EOS for quark matter and on the properties at the interface between the confined and deconfined phases of matter (*e.g.*, the surface tension σ).

In Fig. 2, we show the mass-radius (MR) curve for pure HSs within the GM1 model for the EOS of the hadronic phase, and that for hybrid or strange stars for different values of the bag constant B. The configuration marked with an asterisk on the hadronic MR curves represents the HS for which the central pressure is equal to P_0^* and $\tau = \infty$. The full circle on the HS sequence represents the critical mass configuration, in the case $\sigma = 30$ MeV/fm^2. The full circle on the HyS (SS) mass-radius curve represents the hybrid

---

[2] For the hadronic phase we use an EOS model based on a relativistic lagrangian of hadrons interacting via the exchange of σ, ρ and ω mesons. Particularly, we use the Glendenning–Moszkowski (GM) EOS [26]. For the quark phase we use the phenomenological EOS of ref. [22] which is based on the MIT bag model for hadrons. The parameters here are: the mass m_s of the strange quark, the bag constant B and the QCD structure constant α_s. In the present work, we take $m_u = m_d = 0$, $m_s = 150$ MeV and $\alpha_s = 0$.
[3] The actual *mean-life time* of the HS will depend on the mass accretion or on the spin-down rate which modifies the nucleation time via an explicit time dependence of the stellar central pressure.

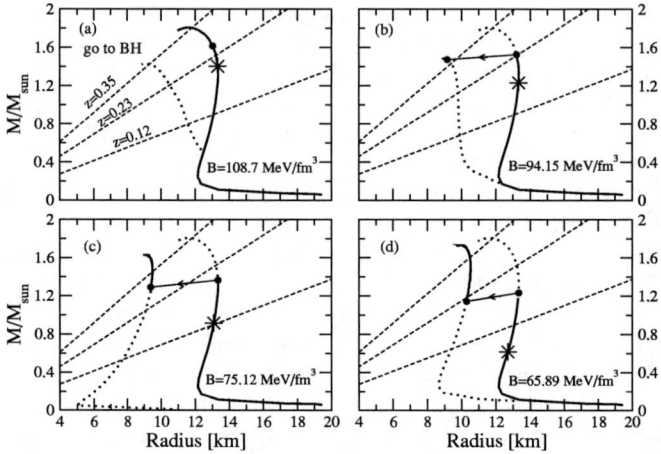

FIGURE 2. Mass-radius relation for pure HS described within the GM1 model and for HyS or SS configurations for several values of the bag constant. The configuration marked with an asterisk represents the HS for which the central pressure is equal to P_0^* and $\tau = \infty$. The conversion process of the HS, with a mass equal to M_{cr}, into a final QS is denoted by the full circles connected by an arrow. In all the panels $\sigma = 30$ MeV/fm^2. The dashed lines show the gravitational red shift deduced for the X-ray compact sources EXO 0748-676 ($z = 0.35$) and 1E 1207.4-5209 ($z = 0.12 - 0.23$).

(strange) star which is formed from the conversion of the hadronic star with $M_{HS}=M_{cr}$. We assume that during the stellar conversion process the total number of baryons in the star (*i.e.* the stellar baryonic mass) is conserved. Thus the total energy liberated in the stellar conversion is given [29] by the difference between the gravitational mass of the initial hadronic star ($M_{in} \equiv M_{cr}$) and that of the final hybrid or strange stellar configuration $M_{fin} \equiv M_{QS}(M_{cr}^b)$ with the same baryonic mass: $E_{conv} = (M_{in} - M_{fin})c^2$.

The stellar conversion process starts to populate the new branch of quark stars (see Fig. 2). Long term accretion on the quark star can next produce stars with masses up to the limiting mass $M_{QS,max}$ for the QS sequence.

The delayed stellar conversion process, described so far, represents the second "explosion" – the *Quark Deconfinement Nova* (QDN) – in the two-step scenario proposed in ref. [14] to explain the delayed SN-GRB connection. The total energy liberated during the stellar conversion is in the range [15] $0.5-1.7 \times 10^{53}$ erg. This huge amount of energy will be mainly carried out by the neutrinos produced during the stellar conversion. Near the surface of a compact stellar object, due to general relativity effects, the efficiency of the neutrino-antineutrino annihilation into e^+e^- pairs is strongly enhanced with respect to the Newtonian case [30], and it could be as high as 10%. Thus the total energy deposited into the electron-photon plasma can be of the order of $10^{51}-10^{52}$ erg.

The strong magnetic field of the compact star will affect the motion of the electrons and positrons, and in turn could generate an anisotropic γ-ray emission along the stellar magnetic axis. This picture is strongly supported by the analysis [31] of the early optical afterglow for GRB 990123 and GRB 021211. Moreover, it has been recently shown [32] that the stellar magnetic field could influence the velocity of the "burning front" of hadronic matter into quark matter. This results in a strong geometrical asymmetry of the forming quark matter core along the direction of the stellar magnetic axis, thus providing a suitable mechanism to produce a collimated GRB. Other anisotropies in the GRB could be generated by the rotation of the star.

In summary, the present scenario [14, 15] implies, as a natural consequence a two step-process which is able to explain the inferred delayed connection between SN explosions and GRBs, giving also the correct energy to power GRBs. It is important to stress that the second explosion (QDN) take place in a "baryon-clean" environment due to the previous SN explosion. Moreover, it is possible to have different time delays between the two events since the *mean-life time* of the metastable hadronic star depends on the value of the stellar central pressure. Thus the present model is able to interpret a time delay of a few years (as inferred [2, 11] for GRB 990705), of a few months (as in the case [7] of GRB 020813), of a few days (as deduced [5] for GRB 011211), or the nearly simultaneity of the two events (as in the case [6] of SN2003dh and GRB 030329).

REFERENCES

1. J.S. Bloom et al., Nature **401**, 453 (1999).
2. L. Amati et al., Science **290**, 953 (2000).
3. L.A. Antonelli et al., Astrophys. Jour. **545**, L39 (2000).
4. L. Piro et al., Science **290**, 955 (2000).
5. J.N. Reeves et al., Nature **414**, 512 (2002).
6. J. Hjorth et al, Nature **423**, 847 (2003).
7. N.R. Butler et al., Astrophys. Jour. **597**, 1010 (2003).
8. R.E. Rutledge and M. Sako, Mon. Not. RAS **339**, 600 (2003).
9. K.N. Borozdin and S.P. Trudolyubov, Astrophys. Jour. **583**, L61 (2003).
10. J.N. Reeves et al. Astron. and Astrophys. **403**, 463 (2003).
11. D. Lazzati, G. Ghisellini, L. Amati, F. Frontera, M. Vietri and L. Stella, Astrophys. Jour. , **556**, 471 (2001).
12. M. Vietri and L. Stella, Astrophys. Jour. **507**, L45 (1998).
13. B. Datta, A.V. Thampan and I. Bombaci, Astron. and Astrophys. **334**, 943 (1998).
14. Z. Berezhiani, I. Bombaci, A. Drago, F. Frontera and A. Lavagno, Astrophys. Jour. **586**, 1250 (2003).
15. I. Bombaci, I. Parenti and I. Vidaña, Astrophys. Jour. **614**, 314 (2004).
16. N.K. Glendenning, Phys. Rev. D **46**, 1274 (1992).
17. H. Heiselberg, C.J. Pethick and E.F. Staubo, Phys. Rev. Lett. **70**, 1355 (1993).
18. N.K. Glendenning, Compact Stars, Springer Verlag (1996).
19. A.R. Bodmer, Phys. Rev. D**4**, 1601 (1971); E. Witten, Phys. Rev. D**30**, 272 (1984).
20. C. Alcock, E. Farhi and A. Olinto, Astrophys. Jour. **310**, 261 (1986).
21. P. Haensel, J.L. Zdunik and R. Schaefer, Astron. and Astrophys. **160**, 121 (1986).
22. E. Farhi and R.L. Jaffe, Phys. Rev. D**30**, 2379 (1984).
23. I. Bombaci, Phys. Rev. C**55**, 1587 (1997).
24. K.S. Cheng, Z.G. Dai, D.M. Wai and T. Lu, Science **280**, 407 (1998).
25. X.-D. Li, I. Bombaci, M. Dey, J. Dey and E.P.J. van den Heuvel, Phys. Rev. Lett. **83**, 3776 (1999).
26. N.K. Glendenning and S.A. Moszkowski, Phys. Rev. Lett. **67**, 2414 (1991).
27. K. Iida and K. Sato, Prog. Theor. Phys. **1**, 277 (1997); Phys. Rev. C**58**, 2538 (1998).
28. J.R. Oppenheimer and G.M. Volkoff, Phys. Rev. **55**, 374 (1939).
29. I. Bombaci and B. Datta, Astrophys. Jour. **530**, L69 (2000).
30. J.D. Salmonson and J.R. Wilson, Astrophys. Jour. **517**, 859 (1999).
31. B. Zhang, S. Kobayashi and P. Meszaros, Astrophys. Jour. **595**, 950 (2003).
32. G. Lugones, C.R. Ghezzi, E.M. de Gouveia Dal Pino and J.E. Horvath, Astrophys. Jour. , **581**, L101 (2002).

Dynamical evolution of neutrino cooled disks

William H. Lee

Instituto de Astronomía, UNAM, Apdo. Postal 70–264, Cd. Universitaria, México DF 04510

Abstract. The evolution of accretion disks in the so-called hypercritical regime, where the main source of cooling is neutrino emission, is relevant for the study of gamma ray burst (GRB) central engines. For short bursts, which may arise from compact binary merger remnant disks, no external agent feeds the disk, and when the initial mass supply is exhausted no further energy release is possible. For long bursts, possibly arising from massive rotating core collapse, the infalling envelope may supply the disk with matter and allow the energy release to continue for some time. We give here a general overview of the conditions in such disks, and present detailed calculations of their structure and evolution (for the particular case of disks arising from mergers), taking into account the effects of neutrino opacities.

Keywords: accretion — dense matter — hydrodynamics — gamma rays: bursts
PACS: 95.30.Lz, 95.30.Tg, 95.85.Pw, 95.85.Ry, 97.10.Gz, 97.60.Jd, 98.70.Rz

INTRODUCTION

The nature of GRB progenitors is a problem that is still under intense investigation, eight years after the discovery of the first optical afterglow in 1997 [13]. A connection between long GRBs and supernovae has been established in several instances, and it is now widely believed that they are in general related. For the class of short bursts there is less information to go on, because there are as yet no afterglow observations. However, violent accretion onto newborn black holes seems like a possible scenario that could drive these events. In the case of short bursts they could arise from compact object mergers [3](independent detection of a gravitational wave signal coincident with a GRB would be a wonderful development), and for long bursts the ingredients in the collapsar model could explain the association between massive stars and GRBs[12]. In either case, the GRB is presumably powered by high accretion rates from a disk onto a black hole, in a regime far above the classical Eddington limit for electromagnetic radiation to play a role in cooling. We have performed in the past 3D hydrodynamics calculations of black hole–neutron star mergers [6, 9] to determine the possible outcome of these events. More recently, we have used these as input for a new study [10, 11] in which our goal is to study the long–term evolution of such disks, in the context of GRB central engines. We report here on our latest results.

All the calculations we discuss have been carried out in cylindrical coordinates (r,z) assuming azimuthal symmetry. We use a Lagrangian Smooth Particle Hydrodynamics (SPH) code, which naturally follows the evolution of the gas without the need for a pre–determined grid. The self–gravity of the gas is neglected, and the interacion with the central mass is through a Newtonian potential $\Phi \propto -1/r$. The accretion of matter is modeled by the presence of an absorbing boundary at the Schwarszchild radius $r_{Sch} = 2GM_{BH}/c^2$.

PHYSICS INPUT

Although quite a bit of information regarding the structure of disks in the present regime can be obtained analytically [15, 14, 7], a full solution requires a numerical approach, because the gas is in an intermediate region concerning the equation of state, in many ways (see also ref.[19] for a 3D numerical approach on shorter timescales than those discussed here). First, the temperature ($\simeq 10^{11}$ K) and density ($\simeq 10^{12}$ g cm^{-3}) are such that photodisintegration is important, leading to a composition dominated by free neutrons and protons. In some regions α particles may be formed, and the energy equation needs to take into consideration the corresponding energy input/output to follow the fluid evolution correctly. The transition from one to the other is quite rapid[18], so that when helium nuclei are present, the electron fraction is essentially $Y_e = 1/2$. Second, when photodisintegration takes place, the density is so high that neutronization occurs, lowering the electron fraction below 1/2, in some cases dramatically. The gas then consists of free nucleons, plus a mixture of electrons and positrons (because the temperature is above the threshold for pair creation). The nucleons are to great accuracy just an ideal non–degenerate Boltzmann gas. If the electron gas were fully degenerate, electrons would dominate over positrons, and one could approximately say that the pressure from this part was that of a fully degenerate Fermi gas at $T = 0$. On the other hand, if the gas was far from degeneracy, one could approximate it as a mixture of ultra-relativistic electron/positron pairs, with a pressure $\propto T^4$. In reality, neither of these conditions is met, and one must solve the thermodynamics in greater detail to obtain meaningful results. The degeneracy parameter, defined as μ_e/kT, where μ_e is the electron chemical potential, is approximately 2–4. We have used an equation of state which takes these effects into account, and is valid for relativistic Fermions with arbitrary degeneracy [2].

The density is so high within the disk that photons are completely trapped. The only possible source of cooling (other than advection) is neutrino emission. The most important reactions are electron/positron capture onto free n/p and pair annihilation, which we have included in our calculations. The rates for the first are taken from detailed nuclear theory calculations in tabular form [8], and the latter from fitting formulae [4]. We also consider nucleon–nucleon bremmsstrahlung and plasmon decay, although their contribution to the total cooling is negligible. Since the optical depth to neutrinos is larger than unity in some regions and in most of the cases we have studied, not all the energy transferred to neutrinos will be lost from the system. A full calculation with adequate neutrino transport is currently beyond the scope of this work, but as a first approximation we simply suppress the cooling rate by a factor $\exp(-\tau_\nu)$. Thus the opaque regions will be unable to cool, while those that remain transparent will [11].

Concerning the neutronization, the solution for the composition depends on whether the fluid is optically thick or thin to its own neutrino emission [1]. The main contribution to the neutrino opacities is from coherent scattering off free neutrons and protons, and α particles. We use the standard expressions for these (e.g., for scattering of free nucleons the cross section is $\sigma = (1/4)\sigma_0(E_\nu/m_e c^2)^2$, where E_ν is the neutrino energy and $\sigma_0 = 1.76 \times 10^{-44}$ cm^2) [20], thus assuming that the emergent neutrino spectrum is not modified, and do not take into account neutrino absorption by nucleons (which could lead to heating and the driving of winds or other outflows from the surface of the disk).

The scattering cross section is energy dependent, and we assume that the mean neutrino energy is comparable with the electron Fermi energy (in accordance with our noting that the dominant cooling term is that coming from electron/positron capture onto n/p, and that the electrons are somewhat degenerate).

If neutrinos are trapped within the gas, one can solve for the composition by imposing equilibrium in the chemical potentials: $\mu_e + \mu_p = \mu_n$, having set the chemical potential of the neutrinos equal to zero. This condition, plus charge neutrality, allows one to obtain the electron fraction Y_e as a function of density and temperature, and applies in the regime where $\tau_\nu \gg 1$. On the other hand, if the density is low enough that neutrinos can escape as soon as they are produced, there is no equilibrium and the composition is determined by the condition of equal capture rates of electrons and positrons onto protons and neutrons. The result for Y_e is valid if $\tau_\nu \ll 1$. We have performed an interpolation by means of optical depth–dependent weights of the two solutions to yield a first approximation to the composition in the disk. This is necessary because the surface where $\tau \approx 1$ is roughly coincident with the isodensity surface with $\log\rho(\text{g cm}^{-3}) = 11$, which is in most cases within our disk.

EFECTS OF OPACITY

The most important consequence of the disk being opaque to its own emission is that the cooling timescale is lengthened considerably. The disk initially has an internal energy reservoir, $E_{int} \simeq 10^{53}$ erg. If this were radiated assuming transparency, the luminosity would exhibit a power–law decay (with the index depending somewhat on the strength of the viscosity), and the duration would be $\simeq 50$ ms. If, on the other hand, the material is opaque, and if the accretion timescale is long enough, the neutrino luminosity remains approximately constant, over a cooling timescale $t_{cool} \simeq E_{int}/L_\nu$, and $L_\nu \simeq$ few $\times 10^{52}$ erg s^{-1}. This occurs for an effective viscosity parameter $\alpha \leq 0.1$, since $t_{cool} > t_{acc}$ in such a situation, where t_{acc} is the accretion timescale. Otherwise the disk is drained of matter faster than it can cool, and the luminosity drops away on an accretion timescale. Opaqueness, then, combined with a moderate efficiency for angular momentum transport, can produce a high neutrino luminosity for approximately one second. This is comparable to the duration of a short GRB, and thus becomes relevant in this context. Note that this mechanism can work on even longer timescales, if one assumes that the central disk is fed externally by its surroundings, as is the case in the popular collapsar model invoked for long bursts.

A second consequence of the opaqueness of the material is related to the instantaneous structure of the disk. When the optical depth rises above unity, the internal energy that would otherwise be lost remains in the fluid, and increases the pressure. This causes not only a rise in the disk's scale height, but in its dependence on the radius. In the outer parts, the pressure scale height H (which can be calculated for example as $P/\nabla P$) roughly scales as $H/R \propto R$. This is the regime in which the disk cools efficiently, and thins as one approaches the central mass. In the inner regions this changes to $H/R \simeq$ constant: the disk effectively inflates because of the rise in pressure. This is apparent when inspecting the scale height H, and also the density as a function of radius in the equatorial plane, $z = 0$. At the transition radius where $\tau_\nu = 1$, the profile changes

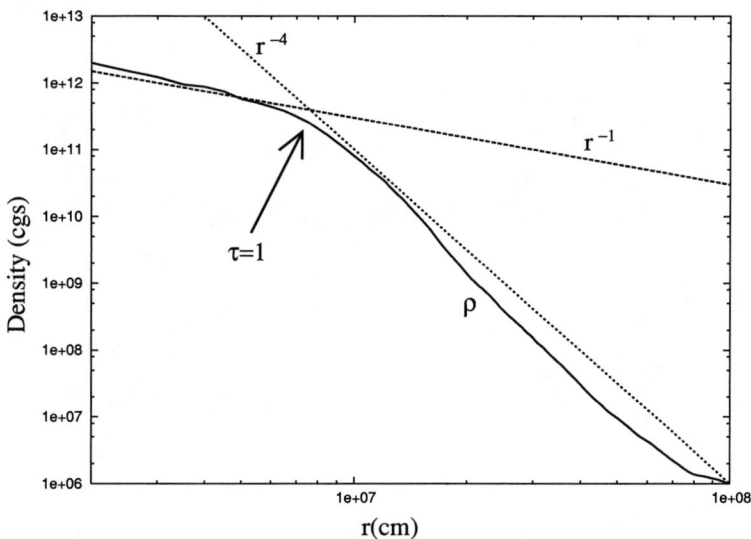

FIGURE 1. Density profile (solid line) in the equatorial plane for a dynamical calculation with viscosity parameter $\alpha = 0.001$, 50 ms after the start of the calculation. The transition from the transparent to the opaque regime is indicated, as are reference power laws showing the change in disk structure.

abruptly from $\rho \propto r^{-5}$ (transparent) to $\rho \propto r^{-1}$ (opaque), or, in terms of surface density, from $\Sigma \propto r^{-4}$ to $\Sigma \simeq$ constant.

A final aspect involes the change in composition which occurs as the optical depth rises. For the transparent regime, the electron fraction is determined by equality in the capture rates of electrons and positrons, as mentioned above. The solution shows a trend in which Y_e increases with decreasing density and temperature, and thus with increasing radius (the disk is more tenuous and cool at greater distances from the central mass). This gives $dY_e/dr > 0$ and $ds/dr > 0$, implying stability with regards to convection. In the opaque regime, Y_e is fixed by equilibrium among the chemical potentials, and the solution yields $dY_e/dr < 0$ and $ds/dr < 0$. This immediately means that there is convective instability, as occurs below the accretion shock in the instants following core collapse supernovae and the formation of a proto–neutron star [5]. In addition to the velocity field generated by the presence of viscosity through the stress tensor, one has convective overturn within the gas, on a timescale roughly given by H/c_s, where c_s is the local sound speed. The implications for the structure of the disk need to be examined carefully, since it has been noted that convection may transport angular momentum inward in an accretion disk (rather than outward, as is the case with a standard viscosity), and thus reduce or suppress the actual accretion onto the central object [17].

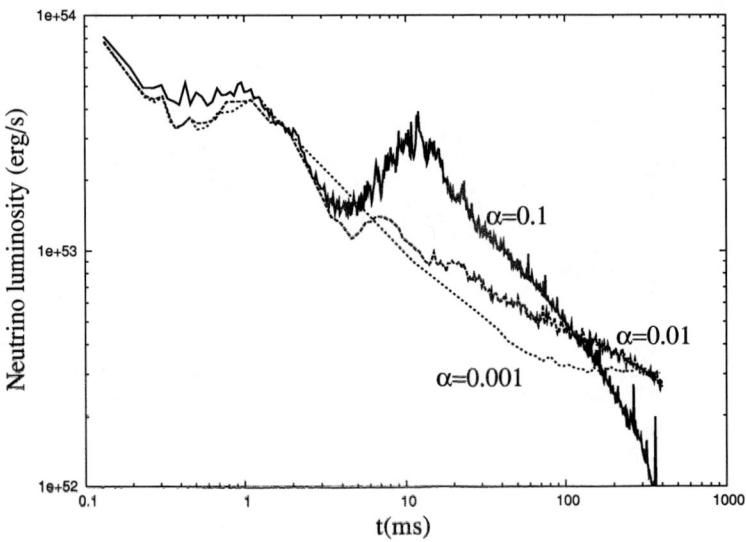

FIGURE 2. Neutrino luminosity as a function of time for three dynamical calculations using different magnitudes for the viscosity α.

ENERGETICS AND GAMMA RAY BURSTS

The physics input we have included in our code allows us for the first time to present realistic estimates concerning the structure, evolution and energetics of disks in these systems over hundreds of dynamical timescales. The neutrino luminosity $L_\nu \approx$ few $\times$ 10^{52} erg s^{-1} for approximately one second, depending on the assumed strenght of the viscosity. This energy release could in principle power a relativistic fireball that could give rise to a GRB. The effect of neutrino opacities is crucial, since it allows the luminosity to remain roughly constant for a limited time interval (the power of an outflow driven by $\nu\bar{\nu}$ annihilation would scale as L_ν^2, so a steep power–law decay in L_ν would make such a scenario unlikely). Likewise, a strong neutrino–driven wind will have a stronger impact over a longer time if its energy input remains nearly steady. The neutronization in the inner regions of the disk is intense, and in principle neutron–rich outflows are possible from the region closest to the central mass [16]. We plan to investigate these matters in depth based on our calculations in future work.

ACKNOWLEDGMENTS

It is a pleasure to acknowledge collaboration and discussions with E. Ramirez-Ruiz and D. Page. Financial support was provided by CONACyT (36632E).

REFERENCES

1. Beloborodov, A.M. 2003, ApJ, 588, 931
2. Blinnikov, S.I., Dunina–Barkovskaya, N. V. & Nadyozhin, D. K. 1996, ApJS, 106, 171
3. Eichler, D., Livio, M., Piran, T. & Schramm, D. N. 1989, Nature, 340, 126
4. Itoh, N., Hayashi, H., Nishikawa, A. & Kohyama, Y. 1996, ApJS, 102, 411
5. Janka, H.-Th., Müller, E. 1996, A&A, 306, 167
6. Kluźniak, W., Lee, W. H. 1998, ApJ, 494, L53
7. Kohri, K., Mineshige, S. 2002, ApJ, 577, 311
8. Langanke, K. & Martínez–Pinedo, G. 2001, Atomic Data and Nuclear Data Tables, 79, 1
9. Lee, W.H., 2001, MNRAS, 328, 583
10. Lee, W.H., Ramirez–Ruiz, E. 2002, ApJ, 577, 893
11. Lee, W.H., Ramirez–Ruiz, E. & Page, D. 2004, ApJ, 608, L5
12. MacFadyen, A.I., Woosley, S.E. 1999, ApJ, 524, 262
13. Metzger, M.R., et al. 1997, Nature, 387, 878
14. Narayan, R., Piran, T., Kumar, P. 2001, ApJ, 557, 949
15. Popham, R., Woosley, S. E., Fryer, C. 1999, ApJ, 518, 356
16. Pruet, J., Woosley, S.E., Hoffman, R.D., 2003, ApJ, 586, 1254
17. Quataert, E., Gruzinov, A. 2000, ApJ, 545, 842
18. Qian, Y.–Z. & Woosley, S.E. 1996, ApJ, 471, 331
19. Setiawan, S., Ruffert, M. & Janka, H.-Th. 2004, MNRAS, 352, 753
20. Shapiro, S. A. & Teukolsky, S. L. 1983, *Black Holes, White Dwarfs and Neutron Stars*, Wiley Interscience, NY

Gamma-Ray Bursts and Afterglow Polarisation

S. Covino*, E. Rossi†, D. Lazzati**, D. Malesani‡ and G. Ghisellini*

INAF / Brera Astronomical Observatory, V. Bianchi 46, 22055, Merate (LC), Italy
†*Max Planck Institute for Astrophysics, Garching, Karl-Schwarzschild-Str. 1, 85741 Garching, Germany*
**JILA, University of Colorado, 440 UCB, Boulder, CO 80309-0440, USA*
‡*International School for Advanced Studies (SISSA-ISAS), via Beirut 2-4, I-34014 Trieste, Italy*

Abstract. Polarimetry of Gamma-Ray Burst (GRB) afterglows in the last few years has been considered one of the most effective tool to probe the geometry, energetic, dynamics and the environment of GRBs. We report some of the most recent results and discuss their implications and future perspectives.

Keywords: polarisation, gamma-ray burst
PACS: 95.30.Gv, 98.70.Rz

INTRODUCTION

Polarimetry has always been a niche observational technique. It may be difficult to apply, requiring special care for the instruments, data reduction and analysis. Indeed, for real astronomical sources, where often the polarisation degree is fairly small at the level of a few per cent, the signal to noise required to derive useful information has to be very high. However, the amount of information that can be extracted by a polarised flux is also very high, since polarisation is an expected feature of a large number of physical phenomena of astronomical interest. This is particularly true for unresolved sources as GRB afterglows, where polarimetry offers one of the best opportunity to infer on the real geometry of the system. In particular, time resolved polarimetry can in principle give fundamental hints on the jet luminosity structure and on the evolution of the expanding fireball. This would provide reliable tools to discriminate among different scenarios. Finally, it has been recently realised that polarimetry of GRB afterglows can offer a direct way to study the physical condition of the Inter-Stellar Medium (ISM) around the GRB progenitor. GRB polarimetry, thus, becomes a powerful probe for gas and dust in cosmological environments, a valuable research field by itself.

In the following of this contribution we want to briefly comment on the most recent advancement in the field and discuss the likely future perspectives that are now open by the advent of the GRB dedicated Swift satellite with its unprecedented rapid localisation capabilities [1].

SYNCHROTRON AND BEAMING?

The first pioneeristic attempts, culminated with the successful observation of a $\sim 1.7\%$ polarisation level in GRB 990510 [2, 3], were driven by the hypothesis that the afterglow

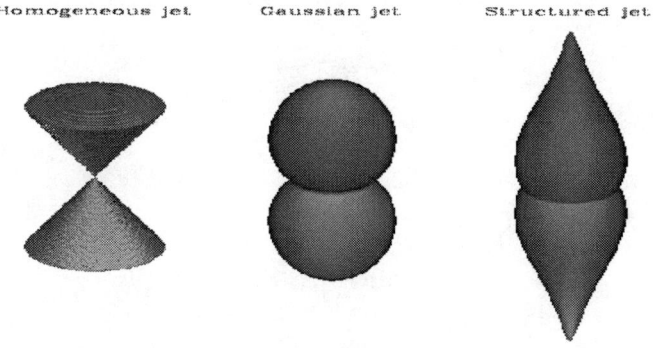

FIGURE 1. Possible different jet structures. From Rossi et al. [15].

emission were due to synchrotron radiation [4, 5, 6]. GRB 990510 was also a perfect case for testing the hypothesis of a geometrically beamed fireball. Indeed, the detection of an achromatic break in the optical light curve [7, 8], together with the observed degree of polarisation, gave support to this scenario. Shortly after this result, it was realised that a jetted ultra-relativistic outflow would produce a characteristic time evolution of the polarisation degree and position angle [9, 10]. The detailed shape of the polarisation curves depends on the dynamical evolution. Testing this model against data is thus a powerful diagnosis of the geometry and dynamics of the fireball.

A large number of polarimetric observations has been carried out since GRB 990510. A review of these data has been compiled by Covino et al. [11] and Björnsson [12]. However, until recently, the detection of a low level of polarisation required strong observational efforts. This prevented a satisfactory time coverage of the afterglow decay and, in turn, a convincing test for the model predictions.

HOMOGENEOUS, STRUCTURED AND MAGNETISED JETS

Lacking strong observational constraints, an improvement of the reference models was achieved considering more physical descriptions for the GRB afterglow jets. In the basic model the energy distribution is homogeneous, making the jet a single entity. More complex beam and magnetic field patterns (Fig. 1), reflecting a physically more plausible scenario, were studied in several papers [13, 14, 15] showing that the light curve is barely affected by this parameter, while the polarisation and position angle evolution changes substantially, providing a further diagnostic tool Fig. 2.

The universal structured jet model predicts that the maximum of the polarisation curve is at the time of the break in the light curve. The position angle remains constant throughout the afterglow evolution. On the contrary, the homogeneous jet model requires two maxima before and after the light curve break and, more importantly, the position angle shows a sudden rotation of 90° between the two maxima, roughly simultaneously to the break time of the light curve. At early and late time the polarisation should be

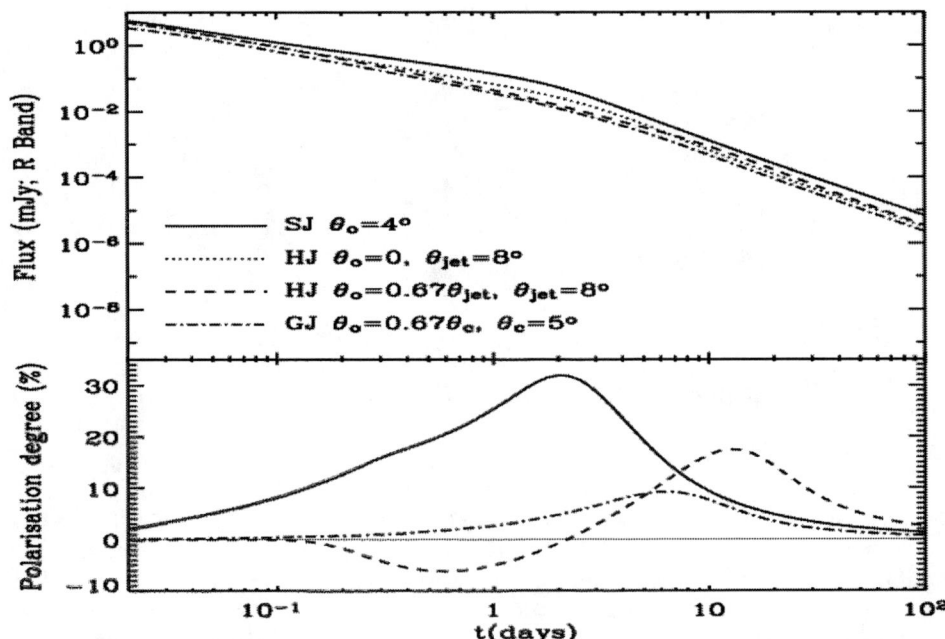

FIGURE 2. Light curve and polarisation evolution for different jet structures. SJ stands for structured jet, HJ homogeneous jet, GJ for Gaussian jet. The figure shows the similarity of the predicted light curves for the various models while the polarisation changes considerably. Negative polarisation degrees mark a 90° rotation for the position angle. From Rossi et al. [15].

essentially zero (Fig. 2).

This last result is substantially modified if it is assumed that a large-scale magnetic field is driving the fireball expansion. The topics has been widely discussed in the context of polarimetry by Granot & Königl [13], Lazzati et al. [14] and [15]. Magnetised jets can be both homogeneous and structured. We do not discuss here the details of this recent research branch. However, we note that, at early times, a large-scale ordered magnetic field produces a non negligible degree of polarisation, contrary to the purely hydrodynamical models. Polarimetry may therefore be the most powerful available diagnostic tool to investigate the fireball energy content and its early dynamical evolution.

Dust Induced Polarisation

The observed low polarisation level from GRB afterglows is often comparable to the expected polarisation induced by dust. Dust grains are known to behave like a dichroic, possibly birefringent, medium [16]. Significant amounts of dust are expected to lie close to the GRB site, as a consequence of the observation of a supernova (SN) component in a few GRBs. The measured polarisation will be modified by the propagation of radiation

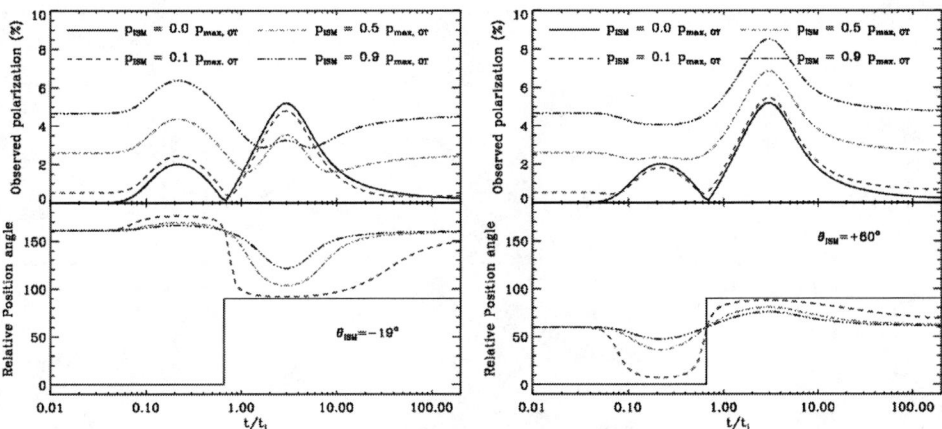

FIGURE 3. Assuming as a reference a typical polarisation curve with a homogeneous jet, the presence of some dust along the line of sight deeply modify the observed time evolution if the dust-induced polarisation is comparable to the intrinsic one, as it seems to be the rule for GRB afterglow at least at rather late time after the high-energy event [11]. Depending on the relation between the position angle of the dust-induced polarisation and of the intrinsic GRB afterglow polarisation, the typical shape of the curve can be removed or even enhanced. From Lazzati et al. [16].

through dusty media. This effect is, contrary to the intrinsic afterglow polarisation, wavelength dependent. The different wavelength dependence open the interesting possibility to study the polarisation signature from the afterglow to study the physical characteristics of dust in cosmological environments: probably the only way to study dust close to star formation regions at high redshift. Even assuming that dust properties close to GRB formation sites are comparable to what we know in the Milky Way (MW), it is important to take into account this component once information from time evolution polarimetry are derived. The superposition of the intrinsic time evolution to dust-induced components for the GRB host galaxy and the MW may substantially alter the expected behavior (Fig. 3).

OBSERVATIONS VS. THEORY

So far, a rather satisfactory coverage of the polarisation evolution of a GRB afterglow has been obtained for three events only: GRB 021004 [16, 17, 18, 19], GRB 030329 [20, 21], and GRB 020813 [22, 14]. However, firm conclusions from the analysis could have been derived for the last case only. GRB 021004 and GRB 030329 showed some remarkable similarities given that their light curves were characterised by a large number of "bumps" or rebrightenings. Several different possibilities has been proposed to model the irregularities in the light curve invoking clumping in the external medium [23]; a more complex and not axi-symmetric energy distribution in the fireball [18] or delayed energy injections [19]. It was soon clear [16] that the standard models for polarisation could not be applied in these conditions, since they are all derived in cylindrical

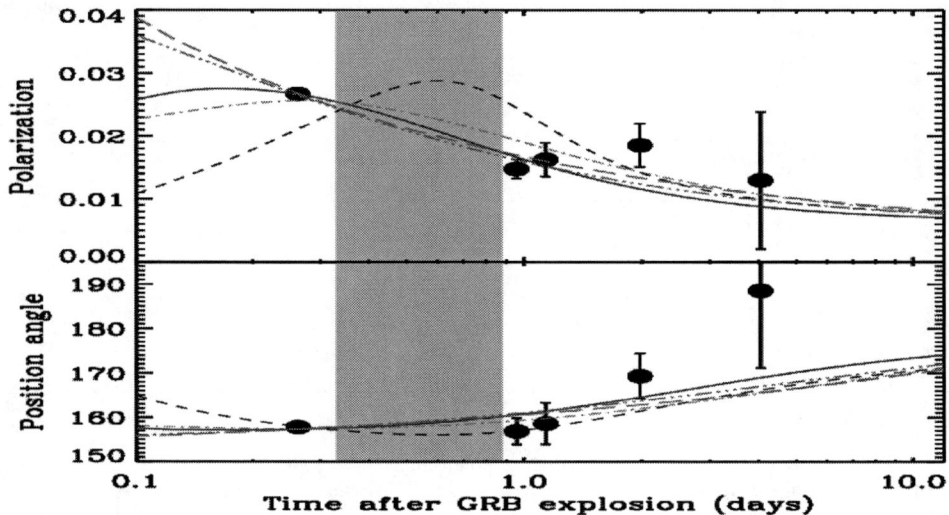

FIGURE 4. Polarisation data for GRB 020813 [22]. Different curves refer to different models. From Lazzati et al. [14].

symmetry. Even for GRB 030329, for which a remarkable dataset was obtained [20], no convincing explanation of the polarization and light-curve erratic behaviors has so far been obtained. It is not clear yet to what extent GRB 021004 and GRB 030329 belong to the same population of long GRBs. It is argued however that the failed detection of this erratic behavior in other afterglows (such as GRB 020813) is not due to a coarser sampling of the light curve.

GRB 020813 was the best case for model testing. Its light curve was remarkably smooth [24], in several optical/infrared bands, and a break in the light curve was clearly singled out. A few polarimetric observations have been carried out providing for the first time polarisation data before and after the light curve break time [22]. Lazzati et al. [14] applied to this event a more quantitative approach not limited, as usually done in the past, to the bare qualitative search of features in the polarisation curve (i.e. rotation of the position angle, etc.). A formal analysis was carried out, taking into account the GRB host galaxy and MW dust induced polarisation and the intrinsic GRB afterglow polarisation. All current jet models were considered, including homogeneous and structured jets, with and without a coherent magnetic field. The dataset, did not allow us to strictly derive a best fitting model. The main result was to rule out the basic homogeneous jets model at a confidence larger than 3σ, mainly because of the lack of the predicted 90° position angle rotation. Again the role of the MW dust induced polarisation is significant. All magnetized models and structured jets fit satisfactorily the data, the ambiguity being mainly due to the lack of early time measurement, i.e. where magnetised or not magnetised models mostly differ (see Fig. 4).

The debate is still far from being settled. Recently, for GRB 030226 Klose et al. [25] a quite low upper limits ($\sim 1\%$) was reported, in rather strict coincidence with the break time, therefore close to the maximum for the polarisation curve if we assume a structured jet model. With one only measurement it is difficult to draw firm conclusions, since this null polarisation measurement may well be due to dust induced polarisation superposed destructively to the intrinsic, if any, GRB afterglow polarisation.

It is finally worth, even though tautological, to report that, as soon as Swift will be fully operational, distributing routinely prompt localisations, a new era will be open even for GRB polarimetry. It will allow us to carry out more stringent tests to the available models and therefore strictly constraint geometry, energetics and dynamics of the fireball.

REFERENCES

1. Gehrels, N., Chincarini, G., Giommi, P., et al. 2004, ApJ 611, 1005
2. Covino S., Lazzati D., Ghisellini G., et al. 1999, A&A 348, 1
3. Wijers R.A.M.J., Vreeswijk P.M., Galama T.J., et al. 1999, ApJ 523, 177
4. Paczyński B., Rhoads J.E. 1993, ApJ 418, 5
5. Mészáros P., Rees M.J. 1997, ApJ 476, 232
6. Sari R., Piran T., Narayan R. 1998, ApJ 497, 17
7. Israel G.L., Marconi G., Covino S., et al. (1999), A&A 348, 5
8. Harrison F.A., Bloom J.S., Frail D.A., et al. (1999), ApJ 523, 121
9. Ghisellini G., Lazzati D. (1999), MNRAS 309, 7
10. Sari R. (1999), ApJ 524, 43
11. Covino S., Ghisellini G., Lazzati D., Malesani D. 2004, ASP Conf. Ser. 312, 169
12. Björnsson G. (2003), astro-ph/0302177
13. Granot J., Königl A. (2003), ApJ 594, 83
14. Lazzati D., Covino S., Gorosabel J.R., et al. (2004), A&A 422, 121
15. Rossi E.M., Lazzati D., Salmonson J.D., Ghisellini G. (2004), MNRAS 354, 86
16. Lazzati D., Covino S., di Serego Alighieri S., et al. (2003), A&A 410, 823
17. Rol E., Wijers R.A.M.J., Fynbo J.P.U. et al. (2003), A&A 405, 23
18. Nakar E., Oren Y. (2004), ApJ 602, 97
19. Björnsson G., Gudmundsson E.H., Jóhannesson G. (2004), ApJ 615, 77
20. Greiner J., et al. (2003), Nature 426, 157
21. Klose S., Palazzi E., Masetti N., et al. (2004), A&A 420, 899
22. Gorosabel J., Rol E., Covino S., et al. (2004), A&A 422, 113
23. Lazzati D., Rossi E., Covino S., Ghisellini G., Malesani D. (2002), A&A 395, 5
24. Covino S., Malesani D., Tavecchio F. et al. (2003), A&A 404, 5
25. Klose S., Greiner J., Rau A. et al. (2004b), AJ 128, 1942

The Empirical Grounds of the Supernova/Gamma-Ray Burst Connection

Massimo Della Valle

INAF/Osservatorio Astrofisico di Arcetri, Largo E. Fermi 5, Firenze

Abstract. For decades Supernovae have been believed to be among the most energetic phenomena occurring in the Universe after the Big Bang. In the recent years observations from both satellites and ground based telescopes have allowed to discover that comparable amounts of energy are released by Gamma-Ray Bursts and that these two classes of events have a deep connection. In this article we review the observational status of the supernova/gamma-ray burst connection and highlight the open questions.

Keywords: Supernovae, Gamma Ray Bursts
PACS: 97.60.Bw, 98.70.Rz

INTRODUCTION

Gamma Ray Bursts (GRBs) are sudden and powerful flashes of gamma-ray radiation which occur randomly in the sky at the rate of about one per day (as observed by the BATSE instrument). The distribution of the durations at MeV energies ranges from 10^{-3} s to about 10^3 s and is clearly bimodal [59]. The bimodality is also apparent from the spectral properties: long bursts ($T > 2$ s) tend to be softer than the short ones. In the discovery paper [57] pointed out the lack of evidence for a connection between GRBs and Supernovae (SNe) (as proposed by Colgate (1968) [17]) and the origin and the distance scale of GRBs remained a mystery for almost three decades. Thanks to observations with *Beppo*SAX [9], detections of X-ray, optical and radio afterglows of Gamma-Ray Bursts [18, 105, 37] have established that they originate at cosmological distances[81], thus settling the dispute on the GRB distance (and energetics) scale. It was clear, since the beginning, that GRBs involve the production of huge amounts of energy ($\sim 10^{53}$ ergs after assuming their emission to be isotropic) roughly comparable to the binding energy of a neutron star. Therefore, regardless of the exact model for GRBs, it appeared likely that some of them might be associated with the collapse of massive stars [112, 85]. The paper is organized as it follows: in section 2 we discuss the circumstantial evidences for the existence of a link between GRBs and the death of massive stars, in sections 3, 4 and 5 we review the SN/GRB associations so far established; in section 6 we estimate the rate of Hypernovae and in section 7 we discuss the possible existence of a lag between the SN explosion and the associated gamma-ray event. Finally in section 8 we present the conclusions and highlight the open problems. Reviews on this topic have been recently published [21, 73, 98].

2. CIRCUMSTANTIAL EVIDENCES

Before 2003 the existence of a connection between SNe and GRBs were supported by several facts, nevertheless none of them was really conclusive.

1) SN 1998bw was the first SN discovered spatially and temporally coincident with a GRB (GRB 980425; [43]). Unexpectedly, SN 1998bw was discovered not at cosmological distances, but in the nearby galaxy ESO 184-G82 at $z = 0.0085$. This implied that GRB 980425 was underenegetic by 4 orders of magnitudes with respect to typical "cosmological GRBs". Moreover, the absence of a conspicuous GRB afterglow contrasted with the associated SN, which had expansion velocities a factor 3-4 larger than those of normal Ib/c SNe and was characterized by a peak luminosity of $\sim 10^{43}$ ergs s^{-1} (for a distance to SN 1998bw of ~ 40 Mpc). This is about 10 times brighter than typical SNe Ib/Ic [15], therefore suggesting that a large amount of ^{56}Ni must have

been synthesized in SN explosion [55, 113]. The theoretical modeling of the lightcurve and spectra suggests that SN 1998bw can be well reproduced by an extremely energetic explosion of an envelope-stripped star, with a C+O core of about ~ 10 $M_\odot$, which originally was $\sim 40 M_\odot$ on the main sequence. This picture is consistent with the radio properties of SN 1998bw, which can be explained as due to the interaction of a mildly relativistic ($\Gamma \sim 1.6$) shock with a dense circumstellar medium [60, 111] due to a massive progenitors that has entirely lost its H envelope.

Höflich et al. (1999)[50] presented an alternative picture based on the hypothesis that all SNe-Ic are the results of aspherical explosions. In this case the apparent luminosity of the SN may vary up to 2 mag, according to different combinations of the geometry of the explosion and line of sight of the observer. This result can explain the high luminosity at maximum of 1998bw, without calling for a dramatic overproduction of ^{56}Ni (~ 0.2 $M_\odot$ ^{56}Ni) and would allow SN 1998bw to have an explosion energy ($\sim 2 \times 10^{51}$ ergs) similar to that of 'normal' core-collapse supernovae. Maeda et al. (2002)[70], after analyzing the line profiles in late time spectra of SN 1998bw, also give some support to the idea that SN 1998bw was the product of an asymmetric explosion viewed from near the jet direction (and characterized by high kinetic energy, of $\sim 10^{52}$ ergs). The idea that Hypernovae and more generally SNe-Ib/c can be produced by asymmetric explosions is supported by polarimetric observations of core-collapse SNe (e.g. [109, 65]), which seem to indicate that the degree of polarization increases along the SN-types sequence: II–Ib–Ic (i.e. with decreasing the envelope mass).

However, the association between two peculiar astrophysical objects such as GRB 980425 (very faint gamma-ray emission, unusual afterglow properties) and SN 1998bw (overluminous SN characterized by unusual spectroscopic features) was believed only suggestive, rather than representative, of the existence of a general SN/GRB connection.

2) the light curves of many afterglows show rebrightenings that have been interpreted as emerging supernovae outshining the afterglow several days or weeks after the GRB event ([6, 114] and references therein). However, since other explanations such as dust echoes [27] or thermal re-emission of the afterglow light [110] could not be ruled out, only spectroscopic observations during the rebrightening phase could remove the ambiguity. Indeed spectroscopic features of SNe are unique, being characterized by

FWHM ~ 100 Å (see section 4).

3) the detection of star–formation features in the host galaxies of GRBs [25, 38] has independently corroborated the existence of a link with the death of massive stars. For example, Christensen et al. (2004)[14] have found that the GRB hosts are galaxies with a fairly high (relative to the local Universe) star formation of the order of $10\,M_\odot\,\mathrm{yr}^{-1}/L^\star$ (see also [63]). Also the location of the GRBs within their host galaxies seems consistent with the regions that contain massive stars [8].

4) Some GRB afterglows have shown absorption features at velocities of a few $\times 10^3$ km/s that has been interpreted as the result of the interaction with the stellar winds originating from the massive progenitors [13, 82].

3. SN 2002lt/GRB 02121

One of the first opportunities to carry out spectroscopic observations during a GRB afterglow rebrightening arrived in late 2002 [22]. GRB 021211 was detected by the HETE–2 satellite [19] ă, allowing the localization of its optical afterglow [35] and the measurement of the redshift z=1.006 [107]. Fig. 1 shows the result of the late-time photometric follow-up, carried out with the ESO VLT–UT4, together with observations collected from literature. A rebrightening is apparent, starting ~ 15 days after the burst and reaching the maximum ($R \sim 24.5$) during the first week of January. For comparison, the host galaxy has a magnitude $R = 25.22 \pm 0.10$, as measured in late-time images. A spectrum of the afterglow + host obtained 27 days after the GRB, during the rebrightening phase is reported in in the rest frame of the GRB (red solid line). The spectrum of the bump is characterized by broad low-amplitude undulations blueward and redward of a broad absorption, the minimum of which is measured at ~ 3770 Å (in the rest frame of the GRB), whereas its blue wing extends up to ~ 3650 Å. The comparison with the spectra of SN 1994I, and to some extent also of SN 1991bg and SN 1984L (fig. 2 in [22]) supports the identification of the broad absorption with a blend of the Ca II H and K absorption lines, the blueshifts corresponding to the minimum of the absorption and to the edge of the blue wing imply velocities $v \sim 14\,400$ km/s and $v \sim 23\,000$ km/s, respectively. The exact epoch when the SN exploded depends crucially on its rising time to maximum light. SN 1999ex, SN 1998bw and SN 1994I (the best documented examples of type-Ic SNe) reached their B-band maximum ~ 18, 16 and 12 days after the explosion [51]. In Fig. 1 we have added the light curve of SN 1994I (dereddened by $A_V = 2$ mag) to the afterglow and host contributions, after applying the appropriate K-correction (solid line). As it can be seen, this model reproduces well the shape of the observed light curve. A null time delay between the GRB and the SN explosions is required by our photometric data, even if a delay of a few days is also acceptable given the uncertainties in the measurements. It is interesting to note that SN 1994I, the spectrum of which provides the best match with the observations, is a typical type-Ic [29] event rather than a bright *Hypernova* as the ones proposed for association with long GRBs [43, 99, 53, 71].

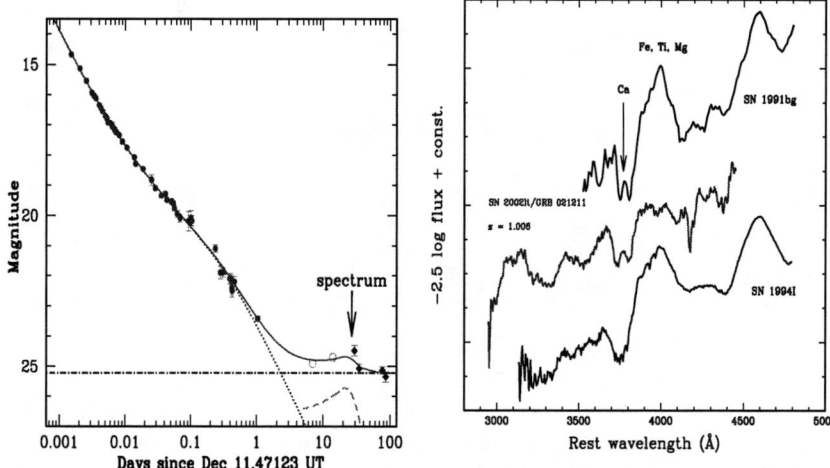

FIGURE 1. Left panel. Light curve of the afterglow of GRB 021211. Filled circles represent data from published works [35, 67, 86], open circles are converted from HST measurements [39], while filled diamonds indicate our data; the arrow shows the epoch of our spectroscopic measurement. The dotted and dot-dashed lines represent the afterglow and host contribution respectively. The dashed line shows the light curve of SN 1994I reported at z = 1.006 and dereddened with $A_V = 2$ (from [64]). The solid line shows the sum of the three contributions. **Right panel.** Spectrum of the afterglow+host galaxy of GRB 021211 (middle line), taken on Jan 8.27 UT (27 days after the burst). For comparison, the spectra of SN 1994I (type Ic, bottom) and SN 1991bg (peculiar type Ia, top) are displayed, both showing the Ca absorption. Plots from [22, 23]

4. THE "SMOKING GUN": GRB 030329/SN 2003DH

The peculiarity of the SN 1998bw/GRB 980425 association and the objective difficulties to collect data for SN 2002lt (4 h to get one single spectrum) prevented us from generalizing the existence of a SN/GRB connection, although both cases were clearly suggestive. The breakthrough in the study of the GRB/SN association arrived with the bright GRB 030329. This burst, discovered by the HETE satellite, was found at a redshift $z = 0.1685$ [47], close enough to allow detailed photometric and spectroscopic studies. SN features were detected in the spectra by several groups (e.g [99, 53, 74]) and the associated SN (SN 2003dh) looked striking similar to SN 1998bw (Fig. 2). The gamma-ray and afterglow properties of this GRB were not unusual among GRBs, therefore, the link between GRBs and SNe was eventually established to be general, likely concerning all "classical" and "long", cosmological GRBs.

The modeling of the early spectra of SN 2003dh [77] has shown that SN 2003dh had a high explosion kinetic energy, $\sim 4 \times 10^{52}$ ergs (in spherical symmetry). However, the light curve derived from fitting the spectra suggests that SN 2003dh was not as bright as SN 1998bw, ejecting only ~ 0.35 $M_\odot$ of ^{56}Ni. The progenitor was a massive main sequence star of $\sim 35-40 M_\odot$. The spectral analysis of the nebular-phase emission lines

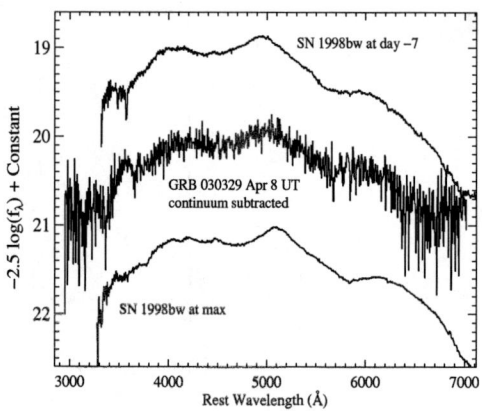

FIGURE 2. Spectrum of April 8 with the smoothed spectrum of April 4 scaled and subtracted. The residual spectrum shows broad bumps at approximately 5000 and 4200 Å (rest frame), which is similar to the spectrum of the peculiar Type Ic SN 1998bw a week before maximum light [87]. Plot from [99].

carried out by Kosugi et al. (2004) [58] suggests that the explosion of the progenitor of the GRB 030329 was aspherical.

5. GRB 031203/SN 2003LW: AN (ALMOST) TWIN OF GRB 980425/SN 1998BW

GRB 031203 was a 30s burst detected by the INTEGRAL burst alert system [80] on 2003 Dec 3. At $z = 0.1055$, it was the second closest burst after GRB 980425. The burst energy was extremely low, of the order of 10^{49} erg, well below the "standard" reservoir $\sim 10^{51}$ erg of normal GRBs (e.g. [36]). Only GRB 980425 and XRF 020903 were less energetic. In this case, a very faint NIR afterglow could be discovered, orders of magnitude dimmer than usual GRB afterglows [71]. A few days after the GRB, a rebrightening was apparent in all optical bands [4, 101, 16, 41]. The rebrightening amounts to $\sim 30\%$ of the total flux (which is dominated by the host galaxy), and is coincident with the center of the host galaxy to within $0.1''$ (~ 200 pc). For comparison, we plot in Fig. 3 the VRI light curves of SN 1998bw (solid lines; from [43]), placed at $z = 0.1055$ and dereddened with $E_{B-V} = 1.1$. Even after correcting for cosmological time dilation, the light curve of SN 2003lw is broader than that of SN 1998bw, and the latter lightcurve requires an additional stretching factor of ≈ 1.1 to match the R and I data points. The R-band maximum is reached in ~ 18 (comoving) days after the GRB. Assuming a light curve shape similar to SN 1998bw, which had a rise time of 16 days in the V band, our data suggest an explosion time nearly simultaneous with the GRB. However, given that SN 2003lw was not strictly identical to SN 1998bw, and as we lack optical data in the days immediately following the GRB, a lag of a few days cannot be ruled out. A precise determination of the absolute magnitude of the SN is made difficult

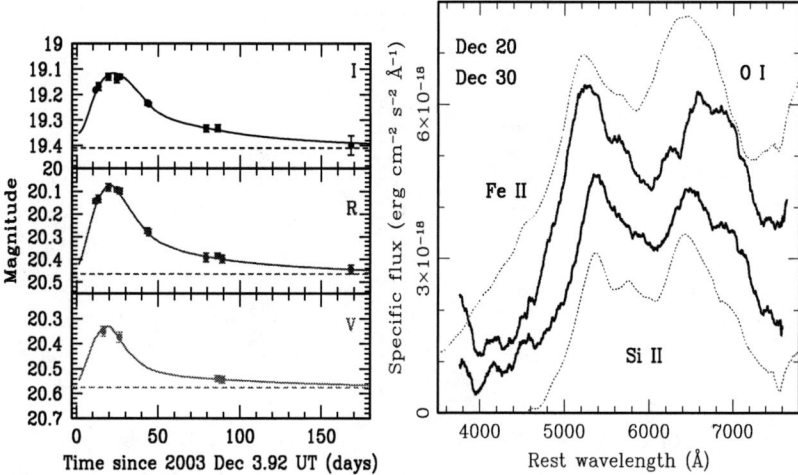

FIGURE 3. Left panel. Optical and NIR light curves of GRB 031203 (circles). The solid curves show the evolution of SN 1998bw ([43, 79]), rescaled at z = 0.1055, stretched by a factor 1.1, extinguished with E(B-V)=1.1, and brightened by 0.5 mag. The dashed lines indicate the host galaxy contribution. The vertical dotted lines mark the epochs of our spectra. **Right panel.** Spectra of SN 2003lw, taken on 2003 December 20 and 30 (solid lines), smoothed with a boxcar filter 250 Å wide. Dotted lines show the spectra of SN 1998bw (from [87]), taken on 1998 May 9 and 19 (13.5 and 23.5 days after the GRB, or 2 days before and 7 days after the V-band maximum, respectively), extinguished with E(B-V)= 1.1 and a Galactic extinction law ([12]). The spectra of SN 1998bw were vertically displaced for presentation purposes.

by the uncertain extinction. Based on the Balmer ratios of the host galaxy we derive the average combined Galactic and host extinction to be $E_{B-V} \approx 1.1$. Given the good spatial coincidence of the SN with the center of the host, such value is a good estimate for the SN extinction. With the assumed reddening, SN 2003lw appears brighter than SN 1998bw by 0.5 mag in the V, R, and I bands. The absolute magnitudes of SN 2003lw are hence $M_V = -19.75 \pm 0.15$, $M_R = -19.9 \pm 0.08$, and $M_I = -19.80 \pm 0.12$. Fig. 3 shows the spectra of the rebrightening on 2003 Dec. 20 and Dec. 30 (14 and 23 rest-frame days after the GRB), after subtracting the spectrum taken on 2004 Mar. 1 (81 rest-frame days after the GRB, [100]). The spectra of SN 2003lw are remarkably similar to those of SN 1998bw obtained at comparable epochs (shown as dotted lines in Fig. 3, see [71] for details). Both SNe show very broad absorption features, indicating large expansion velocities, this makes SN 2003lw another example of Hypernova.

6. RATES OF SNE IB/C, HYPERNOVAE AND GRBS

The measurement of the SN rate is based on the control-time methodology [115] which implies the systematic monitoring of galaxies of known distances and the use of appro-

TABLE 1. Hypernovae

SN	vel (km/s)	References
1997dq	958	[78]
1997ef	3539	[28]
1998bw	2550	[43]
1999as	36000	[52]
2002ap	632	[76, 33]
2002bl	4757	[30]
2003bg	1320	[31]
2003dh	46000	[99, 53]
2003jd	5635	[32, 75]
2003lw	30000	[71]
2004bu	5549	[34]

priate templates for the lightcurves of each SN type (see [11] for bias and uncertainties connected with this procedure). Unfortunately all Hypernovae reported in Tab. I have not been discovered during time 'controled' surveys, then any attempt to derive an absolute value of the rate of Hypernovae should be taken with great caution. One possibility is to compute the frequency of occurence of all SNe-Ib/c and Hypernovae in a limited distance sample of objects. From the Asiago catalogue we have extracted 91 SNe-Ib/c, 8 of which are Hypernovae, with $cz < 6000$ km/s. This velocity threshold is suitable to make the distance distributions of 'normal' Ib/c and Hypernovae, statistically indistinguishable (KS probability=0.42). After assuming that the host galaxies of both 'normal' SNe Ib/c and Hypernovae have been efficiently (or inefficiently) monitored by the same extent (and excluding 1998bw that has been discovered because of the gamma emission associated with the GRB 980425) one can infer that Hypernovae are about $\sim 8\%$ of the total number of SNe Ib/c (normal Ib/c+Hypernovae). Since Hypernovae can be brighter than normal SNe-Ib/c, their discovery may be favored, therefore it could be that thay are over-represented in the SN sample and then 8% should be regarded as an upper limit for their frequency of occurence. The rate of type Ib/c SNe is about 2×10^{-3} yr^{-1} for a "Milky-Way-like" galaxy [11], therefore, one can estimate the rate of Hypernovae to be less than $\sim 2 \times 10^{-4}$ yr^{-1}. These figures have to be compared with the 'canonical' rate of GRB of 2×10^{-7} yr^{-1} estimated for the 'average' galaxy, after rescaling for the beaming factor fb^{-1} of ~ 500 [36] or ~ 75 [49, 88]. These data implies that the ratio GRB/Hypernovae spans the range $\sim 0.5 \div 0.07$, which makes hard to believe that every Hypernova is able to produce a gamma-ray event. This finding is consistent with the radio properties exhibited by 2002ap [3], which do not support the association of this Hypernova with a GRB.

7. SN-GRB TIME LAG

Several authors have reported the detection of Fe lines in GRB X-ray afterglows [89, 1, 92]. If valid (see [96] for a critical view) these observations would have broad implications for both GRB emission models and would strongly link the GRBs with the SN explosions. For example Butler et al. (2003) [10] have reported the detection in

TABLE 2. Supernova-Gamma Ray Burst time lag. A negative time lag indicates that the SN explosion precedes the GRB.

GRB	SN	$+\Delta t (days)$	$-\Delta t (days)$	References
GRB 980425	1998bw	0.7	−2	[55]
GRB 000911	bump	1.5	−7	[62]
GRB 011121	2001kw	0	−5	[7]
			a few	[44]
GRB 021211	2002lt	1.5	−3	[22]
GRB 030329	2003dh	2	−8	[56]
			−2	[75]
GRB 031203	2003lw	0	−2	[71]

a Chandra spectrum of emission lines which intensity and blueshift would imply that a supernova occurred > 2 months prior to the γ event. These kind of observations can be accommodated in the framework of the *supranova* model [106], where a SN is predicted to explode months or years before the γ burst. In Tab. II we have reported the estimates of the lags between the SN explosions and the respective GRBs, as measured by the authors of the papers. After taking the data of Tab. II at their face value, one can conclude that all SN/GRB associations, so far established, give strong support to the idea that SNe and GRBs occur simultaneously (e.g. [69]) and pose severe temporal constraints to those models where the SN is expected to occur before [106] or after [95] the gamma-ray event.

8. CONCLUSIONS AND OPEN QUESTIONS

Data presented in previous sections provide robust empirical grounds to the idea that some types of core-collapse SNe are the progenitors of long-duration GRBs. On the other hand, the existence of SN/GRB associations poses intriguing questions which have not yet been answered.

1. *Which kind of SNe are connected with long-duration GRBs and XRFs?*

Evidence based on the associations between SN 1998bw/GRB 980425, SN 2003dh/ GRB 030329, and SN 2003lw/GRB 031203 would suggest that the parent SN population of GRBs is formed by the bright tail of Hypernovae, that is, by a sub-class of type-Ib/c SNe which are characterized by high/moderate luminosity peaks ($M_B \sim -19.5 \div -17.0$) and high expansion velocity of the ejecta (~ 30000 km/s). However, there is growing evidence that the SN population associated with GRBs is formed by an heterogeneous class of progenitors including dim Hypernovae and standard Ib/c events ([22]; [40](see Fig. 4); [102]; [66](see Fig. 5); [72], [91]).

Possible associations between GRBs and other types of core-collapse SNe (type IIn) have been claimed in the past [46, 103, 94] on the basis of spatial and temporal coincidences. However, in a recent study, Valenti et al. (2005) [104] have not confirmed the association with type-IIn to be statistically significant. Currently the best evidence for the case of an association between a Supernova IIn and a gamma ray burst has been provided by Garnavich et al. (2003)[44] who find that the color evolution of the bump associated with GRB 011121, is consistent with the color evolution of an underlying SN

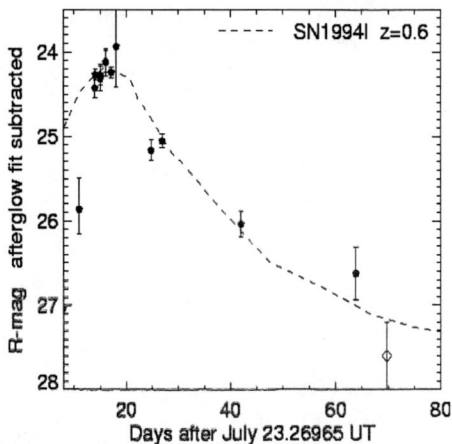

FIGURE 4. The lightcurve of the bump associated with XRF 030723 compared with the B lightcurve of SN 1994I redshifted to z=0.6 (data from [93]). Plot from [40]

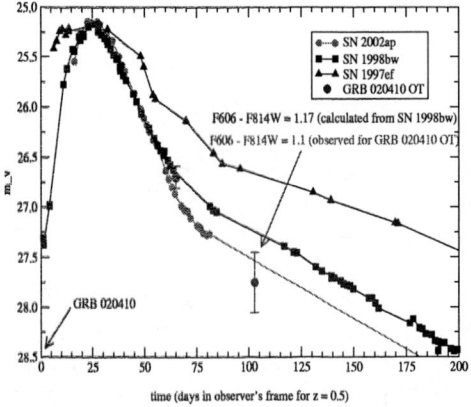

FIGURE 5. Lightcurve of the SN associated with GRB 020410 compared with SN 1997ef [55], 2002ap ([42, 86]) and SN 1998bw ([43, 79]). Plot from [66].

(2001ke) strongly interacting with a dense circumstellar gas due to the progenitor wind.

2. What are the most frequent gamma events in the Universe?

GRB 031203 was quite similar to GRB 980425, albeit more powerful. Both events consisted in a single, underenergetic pulse. Their afterglows were very faint or absent in the optical, and showed a very slow decline in the X rays. Lastly, they were both accompanied by a powerful Hypernova. Therefore, GRB 980425 can no longer be considered as a peculiar, atypical case. Both bursts were so faint, that they would have been easily missed at cosmological distances. Since the volume they sample is $10^5 \div 10^6$ times smaller than that probed by classical distant GRBs, the rate of these events could be dramatically larger, perhaps they are the most common GRBs in the Universe. However we are still left with the question of whether or not these bursts belong to a different local population of γ-bursts [5, 97] or they are typical cosmological bursts observed off-axis ([84, 26, 113, 48]). Naively one may expect that the spectroscopic and photometric similarities exhibited by SNe 1998bw, 2003dh and 2003lw, may indicate a common origin for the associated GRBs (in spite of dramatic differences in their γ-energy budgets and afterglow properties) rather than suggesting two distinct classes of progenitors.

3. *Are the "red bumps" always representative of the signatures of incipient SNe?*

Or can some of them be produced by different phenomena as dust echoes ([27]) or thermal re-emission of the afterglow light ([110])? To date, only for GRB 021211/SN 2002lt ([22]) a spectroscopic confirmation was obtained. On the other hand, Garnavich et al. (2003)[44], Fynbo et al. (2004)[40] did not find clear SN spectroscopic features in the bumps of GRB 011121 and XRF 030723.

4. *Is the lack of an optical bump indicative of the lack of a supernova?*

Or rather do long GRBs have a heterogeneous class of progenitors including, besides SNe of different magnitudes at maximum, also merging between compact objects (e.g. [2, 24]).

We note that the lightcurve of the afterglow associated to GRB 030329 ([74, 68]) did not show the bump which is believed to be caused by the emerging SN.

5. *What causes some small fraction of SNe Ib/c to produce observable GRBs, while the majority do not?*

With the obvious exceptions of SN 1998bw, SN 2003dh and SN 2003lw, none of the Hypernovae reported in Tab. I have been associated with GRBs by direct observations (the association GRB 971115/SN 1997ef, for example, has been proposed by Wang & Wheeler (1998)[108] on the basis of spatial and temporal coincidences). The situation is even more intriguing if one considers that Hypernovae are only a small fraction of 'normal' SNe-Ibc (less than 10%) and thus only a tiny fraction of SNe-Ibc, about $\sim 0.7\% \div 5\%$ seems to be able to produce GRBs. This may imply that the evolution leading a star to produce a GRB requires very special circumstances (rotation?, binary interaction? see [90, 83]) other than being 'only' a massive star. As an alternative, one can argue that most SNe Ib/c produce GRBs but we are able to detect only those which are view at very small angles relative to the jet direction. For larger angles we see XRFs, and at even larger angles, relative to the jet direction, we observe only the SN event [61, 20]

With an expected rate of discovery of about 2 event/week the *Swift* satellite [45] will allow the GRB community to obtain in the next $3 \div 4$ years an accurate spectroscopic classification for dozens SNe associated with GRBs and to provide conclusive answers to several of the above questions.

ACKNOWLEDGMENTS

The author wishes to thank Nino Panagia and Daniele Malesani for useful comments and discussions.

REFERENCES

1. Antonelli, L.A., Piro, L., Vietri, M., et al. 2000, ApJ, 545, L42
2. Belczynski, K., Bulik, T., Rudak, B. 2002, ApJ, 571, 394
3. Berger, E., Kulkarni, S. R., Chevalier, R.A. 2002, ApJ, 577, L5
4. Bersier, D., et al. 2004, GCN Circ. 2544
5. Bloom, J. S., Kulkarni, S. R., Harrison, F., Prince, T., Phinney, E. S., Frail, D. A. 1998, ApJ, 506, L105
6. Bloom, J. S., Kulkarni, S. R., Djorgovski, S. G. et al. 1999, Nature, 401, 453
7. Bloom, J. S.; Kulkarni, S. R.; Price, P. A. et al. 2002b, ApJ, 572, L45
8. Bloom, J. S., Kulkarni, S. R., Djorgovski, S. G. 2002a, AJ, 123, 1111
9. Boella,G., Butler,R.C., Perola,G.C.,et al. 1997, A&AS, 122, 299
10. Butler, N.R., Marshall, H.L., Ricker, G.R. Vanderspek, R.K., Ford, P.G., Crew, G. B., Lamb, D.Q., Jernigan, J.G. 2003, ApJ, 597, 1010
11. Cappellaro, E., Turatto, M., Benetti, S., Tsvetkov, D. Y., Bartunov, O. S., Makarova, I. N. 1993, A&A, 268, 472
12. Cardelli, J. A., Clayton, G. C., & Mathis, J. S. 1989, ApJ, 345, 245
13. Chevalier, R.A., Li, Z. 2000, ApJ, 536, 195
14. Christensen, L., Hjorth, J., Gorosabel, J. 2004, A&A, 425, 913
15. Clocchiatti, A.; Wheeler, J. C. 1997, ApJ, 491, 375
16. Cobb, B. E., Baylin, C. D., van Dokkum, P. G., Buxton, M. M. & Bloom, J. S. 2004, ApJ, 608, L93
17. Colgate, S. 1968, Can. J. Phys., 46, 476
18. Costa, E., Frontera, F., Heise, J., et al. 1997, Nature, 387, 783
19. Crew, G. B., Lamb, D. Q., Ricker, G. R., et al. 2003, ApJ, 599, 387
20. Dado, S., Dar, A., De Rujula, A. 2004, A&A, 422, 381
21. Dar, A. 2004, Astrophysics Workshop, Vulcano, Italy, May 24-29, 2004; astro-ph/0405386
22. Della Valle M., Malesani, D., Benetti, S. et al. 2003, A&A, 406, L33
23. Della Valle, M. Marziani, P., Panagia, N. 2005, in the Proced. of the GRB Conference, Rome, Sept. 2004
24. Della Valle M., Malesani, D., Benetti, S. et al. 2004, in the Proceedings of the 2003 GRB Conference (Santa Fe, 2003 Sep 8-12), eds. Fenimore, E., Galassi, M., p. 403
25. Djorgovski, S. G., Kulkarni, S. R., Bloom, J. S., Goodrich, R., Frail, D. A., Piro, L.; Palazzi, E. 1998, ApJ, 508, L17
26. Eichler, D., Levinson, A. 1999, ApJ, 521, L117
27. Esin, A. & Blandford, R. 2000, ApJ, 534, L151
28. Filippenko, A.V. 1997, IAUC, n. 6783
29. Filippenko, A.V., Barth, A.J., Matheson, T. et al. 1995, ApJ, 450, L11
30. Filippenko, A.V., Leonard, D.C., Moran, E.C. 2002, IAUC, n.7845
31. Filippenko, A.V. & Chornock, R. 2003, IAUC, n. 8084
32. Filippenko, A.V., Foley, R.T., Swift, B. 2003, IAUC, n. 8234
33. Foley, R.J., Papenkova, M.S., Swift, B. J. et al. 2003, PASP, 115, 1220
34. Foley, R.J., Wong, D.S., Moore, M. & Filippenko, A.V. 2004, IAUC, n. 8353
35. Fox, D. W., Price, P. A., Soderberg, A. M., et al. 2003, ApJ, 586, L5
36. Frail, D. A., Kulkarni, S. R., Sari, R. et al. 2001, ApJ, 562, L55
37. Frail, D.A., Kulkarni, S.R., Nicastro, S.R. et al. 1997, Nature, 389, 261
38. Fruchter, A.S., Thorsett, S. E., Metzger, M.R. et al. 1999, ApJ, 519, L13
39. Fruchter, A. S., Levan, A., Vreeswijk, P. M., Holland, S. T. & Kouveliotou, C. 2002, GCN Circ, 1781
40. Fynbo, J., Sollerman, J., Hjorth, J. et al. 2004, ApJ, 609, 962
41. Gal-Yam, A., Moon, D.S., Fox, D.B. et al. 2004, ApJ, 609, L59
42. Gal-Yam, A., Ofek, E.O., Shemmer, O. 2002, MNRAS, 332, L73

43. Galama, T.J., Vreeswijk, P.M., van Paradijs, J., et al. 1998, Nature, 395, 670
44. Garnavich, P. M., Stanek, K. Z., Wyrzykowski, L. et al. 2003, ApJ, 582, 924
45. Gehrels, N., Chincarini, G., Giommi, P., et al. 2004, ApJ, 611, 1005
46. Germany L., Reiss, D. J. Sadler, E.M., Schmidt, B.P. Stubbs, C. W. 2000, ApJ, 533, 320
47. Greiner et al. 2003, GCN 2020
48. Guetta, D., Perna, R., Stella, L., Vietri, M. 2004, ApJ, 615, L73
49. Guetta, D., Piran, T., Waxman, E. 2005, ApJ, 619, 412
50. Höflich, P.; Wheeler, J. C.; Wang, L. 1999, ApJ, 521, 179
51. Hamuy, M. 2003, in "Core Collapse of Massive Stars", ed. C.L. Fryer, Kluwer, Dordrecht; (astro-ph/0301006)
52. Hatano, K., Branch, D., Nomoto, K., Deng J.S., Maeda, K., Nugent P., Aldering, G. 2001, 198th BAAS, 33, p. 838
53. Hjorth J., Sollerman, J., Moller, P. et al. 2003, Nature,423, 847
54. Iwamoto, K., Nakamura, T., Nomoto, K. et al. 2000, ApJ, 534, 660
55. Iwamoto, K., Mazzali, P.A., Nomoto, K., et al. 1998, Nature, 395, 672
56. Kawabata, K. S., Deng, J., Wang, L. et al. 2003, ApJ, 593, L19
57. Klebesadel, R.W., Strong, I.B., Olson, R. A. 1973, ApJ, 182, L85
58. Kosugi, G. Mizumoto, Y. Kawai, N. et al. 2004, PASJ, 56, 61
59. Kouveliotou, C. et al. 1993, ApJ, 413, L101
60. Kulkarni, S. R., Frail, D. A., Wieringa, M. H. et al. 1998, Nature, 395, 663
61. Lamb, D. Q., Donaghy, T. Q., Graziani, C. 2005, ApJ, 620, 355
62. Lazzati, D., Covino, S., Ghisellini, G. et al. 2001, A&A, 378, 996
63. Le Floc'h, E., Duc, P.-A., Mirabel, I. F. et al. 2003, A&A, 400, 499
64. Lee, M.G.; Kim, E. Kim, S.C.; Kim, S. L.; Park, W.; Pyo, T.S. 1995, JKAS, 28, 31L
65. Leonard, D.C., Filippenko, A.V., Barth, A.J.& Matheson, T. 2000, ApJ, 536, 239
66. Levan, A., Nugent, P., Fruchter, A. et al. 2005, ApJ, sub-mitted, astro-ph/0403450
67. Li, W., Filippenko, A. V., Chornock, R. & Jha, S. 2003, ApJ, 586, L9
68. Lipkin, Y. M., Ofek, E. O., Gal-Yam, A., et al. 2004, ApJ, 606, 381
69. MacFadyen, A. I., Woosley, S.E. 1999, ApJ, 524, 262
70. Maeda, K., Nakamura, T., Nomoto, K., Mazzali, P., Patat, F., Hachisu, I. 2002, ApJ, 565, 405
71. Malesani D. Tagliaferri, G., Chincarini, G. et al.2004, ApJ, 609 L5
72. Masetti, N., Palazzi, E., Pian, E. et al. 2003, A&A, 404, 465
73. Matheson, T. 2005, in the proceedings of "Supernovae as Cosmological Lighthouses", Padua, 2004, ASP conference Series; astro-ph/0410668
74. Matheson, T.; Garnavich, P. M.; Stanek, K. Z. et al. 2003a, ApJ, 599, 394
75. Matheson, T., Challis, P. & Kirshner, R. 2003b, IAUC, n. 8234
76. Mazzali et al., 2002, ApJ, 572, L61
77. Mazzali, P., Deng, J., Tominaga, N. et al. 2003, ApJ, 599, L95
78. Mazzali, P. A., Deng, J., Maeda, K., Nomoto, K., Filippenko, A. V., Matheson, T. 2004, ApJ, 614, 858
79. McKenzie, E.H., Schaefer, B. E. 1999, PASP, 111, 964
80. Mereghetti, S., & Götz, D. 2003, GCN Circ. 2460
81. Metzger, M.R., Djorgovski, S.G., Kulkarni, S. R. et al. 1997, Nature, 387, 878
82. Mirabal, N., Halpern, J. P., Chornock, R., Filippenko, A.V., Terndrup, D. M., Armstrong, E., Kemp, J., Thorstensen, J. R., Tavarez, M., Espaillat, C. 2003, ApJ, 595, 935
83. Mirabal, I, F. 2004, RMxAC, 20, 14
84. Nakamura, T. 1999, ApJ, 522, L101
85. Paczyński, B. 1998, ApJ, 494, L45
86. Pandey, S.B., Anupama, G.C., Sagar, R., Bhattacharya, D., Castro-Tirado, A.J., Sahu, D.K., Parihar, P., Prabhu, T.P. 2003, A&A, 408, L21
87. Patat, F., Cappellaro, E., Danziger, I.J., et al. 2001, ApJ, 555, 900
88. Piran, T. 2005, in the Proced. of the GRB Conference, Rome, Sept. 2004
89. Piro, L., Costa, E., Feroci, M. et al. 1999, ApJ, 514, L73
90. Podsiadlowski, P., Mazzali, P., Nomoto, K., Lazzati, D., Cappellaro, E. 2004, ApJ, 607, L17
91. Price, P.A., Kulkarni, S.R., Schmidt, B.P. et al. 2003, ApJ, 584, 931
92. Reeves, J. N., Watson, D., Osborne, J. P. et al. 2002, Nature, 416, 512

93. Richmond, M. W., Van Dyk, S. D., Ho, W., et al. 1996, AJ, 111, 327
94. Rigon, L., Turatto, M., Benetti, S., Pastorello, A., Cappellaro, E., Aretxaga, I., Vega, O., Chavushyan, V., Patat, F., Danziger, I. J. Salvo, M. 2003, MNRAS, 340, 191
95. Ruffini, R., Bianco, C.L., Fraschetti, F., Xue, S., Chardonnet, P. 2001, ApJ, 555, L117
96. Sako, M., Harrison, F., Rutledge, R. 2005, ApJ, in press, astro-ph/0406210
97. Soderberg, A. M., Frail, D. A., Wieringa, M. H. 2004, ApJ, 607, L13
98. Stanek, K.Z. 2005, in First International Workshop on "Stellar Astrophysics with the World Largest Telescopes", Torun, Poland, 7-10 September 2004. astro-ph/0411361
99. Stanek, K. Z., Matheson, T., Garnavich, P. M. et al. 2003, ApJ, 591, L17
100. Tagliaferri, G., et al. 2004, IAU Circ. 8308
101. Thomsen, B., Hjorth, J., Watson, D. et al. 2004, A&A, 419, L21
102. Tominaga, N., Deng, J., Mazzali, P., Maeda, K., Nomoto, K., Pian, E., Hjorth, J.,Fynbo, J. 2004, ApJ, 612, L105
103. Turatto M., Suzuki, T., Mazzali, P., Benetti, S., Cappellaro, E., Danziger, I.J., Nomoto, K., Nakamura, T., Young, T.R., Patat, F. 2000, 534 L57
104. Valenti, S., Cappellaro, E., Della Valle, M., Frontera, F., Guidorzi, C., Montanari, E. 2005, in the Proced. of the GRB Conference, Rome, Sept. 200
105. van Paradijs, J., Groot, P.J., Galama, T., et al. 1997, Nature, 386, 686
106. Vietri, M. & Stella, L. 1998, ApJ, 507, L45
107. Vreeswijk, P. M., Fruchter, A., Hjorth, J., & Kouveliotou, C. 2002, GCN Circ, 1785
108. Wang L. and Wheeler J.C., 1998, ApJ,504, L87
109. Wang, L., Howell, D. A., Höflich, P., Wheeler, J. C. 2001, ApJ, 550, 1030
110. Waxman, E. & Draine, B. T. 2000, ApJ, 537, 796
111. Weiler, K.W., Panagia, Nino, Montes, M.J., Sramek, R,A. 2002, ARA&A, 40, 387
112. Woosley, S. 1993, ApJ, 405, 273
113. Woosley, S. E., Eastman, R.G., Schmidt, B.P. 1999, ApJ, 516, 788
114. Zeh, A., Klose, S., Hartmann, D. H. 2004, ApJ, 609, 952
115. Zwicky, F. 1938, ApJ, 88, 529

A few insights into the Swift Mission

Guido Chincarini
On behalf of the Swift Team

guido@merate.mi.astro.it

Abstract. I shall give a very brief summary of the status of the Swift mission after launch. Except for a failure of the active cooling of the CCD of the X-ray telescope XRT), all the instrumentation after activation is working according to specification. The failure of the active cooling did not affect the performance of the instrument. The Burst Alert Telescope is performing in a superlative mode. Remarkable is the capability of the spacecraft in pointing the target soon after the trigger. This enables the XRT to start observations in less than 2 minutes after the trigger and gather data during the tail of the prompt and at the very beginning of the afterglow. In addition to the image of the first light with XRT, I shall show the X-ray light curve of one of the many GRBs we observed.

Keywords: gamma rays: bursts, astronomical and space-research instrumentation: X- and γ-ray telescopes and instrumentation
PACS: 98.70.Rz, 95.55.Ka

INTRODUCTION

At the time of this writing, the communication I gave during the meeting is largely obsolete. The Swift satellite [1] was launched on November 20th 2004 from Cape Kennedy and by now not only we did activated all the instruments, but we also had a reasonable number of calibration observations, we detected a number of Gamma Ray Bursts and we made some analysis and interpretation. I asked therefore permission to the Editors to modify somewhat my presentation and make short summary of the instrumentation describing the status after launch and activation.

The observations we obtained so far may be slightly in the future improved due to the still ongoing calibration. Of these only a small part has been submitted for publication so that, while I will show some results, I will avoid to touch upon interesting yet unpublished data. The main point is that the satellite since the very beginning had a good performance, albeit a few unforeseen drawbacks, according to expectation. The instrumentation is detecting GRBs and other sources on a regular basis.

In Section II we will discuss some of the characteristics of the Instrumentation (for a detailed discussion see [1]). In section III I will illustrate a few GRBs discovered by Swift.

THE MISSION

Thanks to the experience of previous missions, BeppoSAX in particular, related to the detection and study of GRBs we could design a multi-wavelengths observatory, Swift, capable not only of pointing the target soon after detection, but also to communicate in real time with the ground, so as to enable the observations with ground-based facilities immediately. The global concept of the mission is indeed based on fast reaction and immediate communication. This is accomplished through: 1) technical design and related implementation, innovative instrumentation and spacecraft solutions. In addition the mission uses a communication network that allows at any time, according to the specific needs, the download and upload of data and commands and 2) a global, worldwide, collaboration and organization of the astronomical community which actively relates to the spacecraft and instrumentation as a team. In advance to the robotic performance of the satellite, the science team quickly reacts as soon as a GRB is detected. Naturally, the staff of highly expert engineers and scientists who control the satellite is fundamental to the operation.

The primary goal of the mission is the detection and study of the different classes of GRBs. On the other hand, with a field of view of the Burst Alert Telescope (BAT) covering about 1/8 of the sky, Figure 1, the instrument will also detect other events in the sky flaring at high energies, but will also perform a deep survey of the sky at high energies. This will be a unique collection of data both for Galactic and extragalactic (absorbed AGN for instance) objects.

The Burst Alert Telescope (BAT) is a rather sensitive instrument, Figure 2 with a coded mask working for imaging, in the energy range 15 to 150 keV and with a non–coded response up to 500 keV. The sensitivity down to 15 keV has the advantage to continue the response of the X-ray Telescope (XRT), albeit with lower sensitivity, to higher energies. Indeed, thanks to the fast moving spacecraft, we will have the quasi-simultaneity between the detection of the last part of the burst and the beginning of the softer X ray photons by the XRT, this is sensitive in the energy range 0.2–10 keV.

The spacecraft must be capable of moving on the target, according to pre–launch simulations, and depending on the location of the GRB on the detector, in a time range of about 10 to 100 s. We can now compare this expectation with the real operation, where we must also take into account for the time it takes to the on boad software to compute the location of the burst.

The performance is excellent and according to expectation. Following the detection of the burst GRB050128, XRT started observations after 108 s of the trigger and with the burst GRB050219a XRT was already observing 92 s after the BAT trigger.

The operations to be carried out, in automatic mode, soon after the BAT trigger and before the Narrow Field Instruments are on the target are: a) evaluation of the detection and trigger by the on board computer, b) integration and estimate of the location of the burst, c) command to the spacecraft to slew and communicate to the ground of the characteristics of the burst detected, d) transmission via TDRSS (see

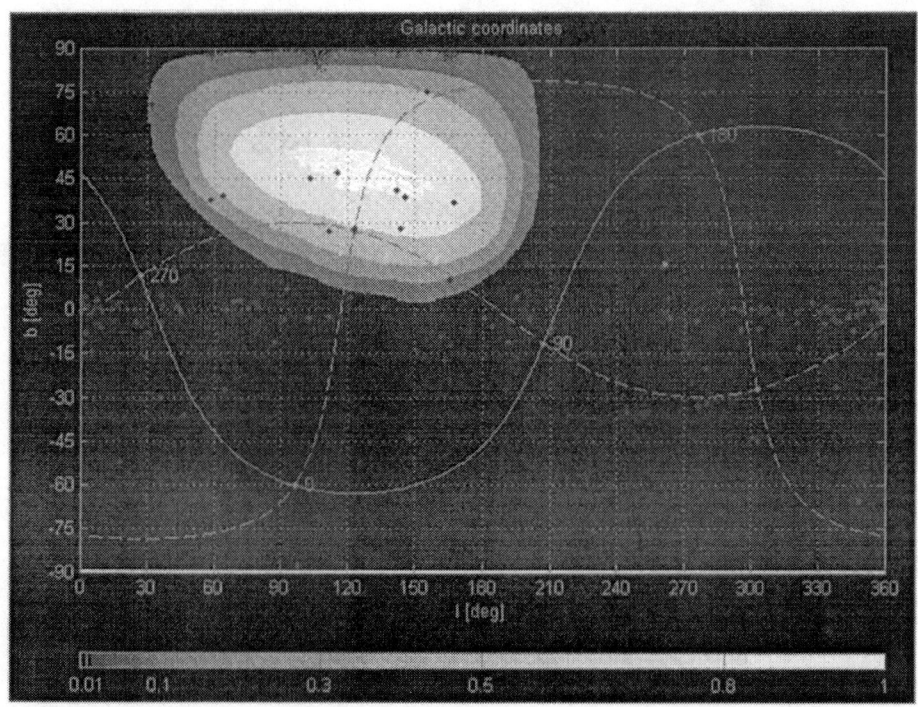

FIGURE 1. The field of view of the Burst Alert Telescope projected on the sky. The on-board catalogue has a list of known sources, in the figure some of these are plotted and concentrated along the galactic plane, so as to avoid giving a false alert on obvious and well know high energy sources.

section III) of the coordinates of the detected object to the ground, e) stop on target and start observation and start of new observations with the narrow field instruments.

The trigger algorithm, as devised and written by the Los Alamos group headed by Ed Fenimore, relays on a very large number of criteria and is based on the HETE–2 experience. It is essentially related to the excess of detector count rate over the background and constant sources. The trigger time is a very well-defined time in the operational sense. Physically it also corresponds, or at least it is very close, to the onset of the phenomenon.

The coupling of BAT and XRT, with a fast slewing and setting on target, is probably one of the most important assets for the science we will get out of this mission on GRBs. In the same way the Ultraviolet–Optical Telescope (UVOT) is enlarging the observational band to the ultraviolet and optical bands on all targets removing any constraint such as daily cycle and cloudy and clear weather. The time it takes to the spacecraft to slew and point the narrow field instrument on the position estimated by the on-board computer following the analysis of the BAT observations is, as stated above, very close to the expectations. The sequence depicted in Figure 3 and due to Scott Barthelemy refers to the GRB 050117. Immediately following the trigger, BAT

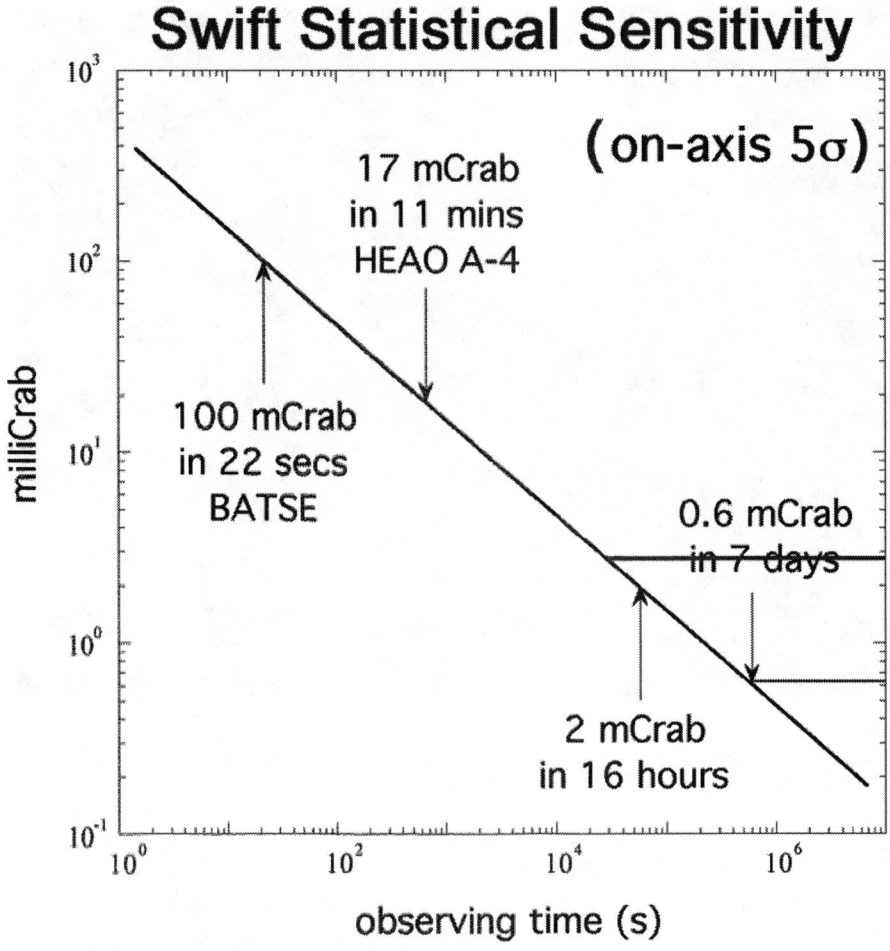

FIGURE 2. The sensitivity of the BAT instrument as a function of the integration time.

continues to collect data on the field and, at the same time, alerts via TDRSS the Mission Operation Center and the community via GCN (Gamma–Ray Burst Coordinates Network, [2]). BAT gets the image and measure the centroid. In the case we are considering it took 81 seconds to carry out this operation. The detection occurred just before the spacecraft entered the South Atlantic Anomaly (SAA). The spacecraft begins to slew to the target after 116 s and sets pointing the position measured by BAT. At this point the narrow band instruments start the observations.

The accuracy of the BAT coordinates depends on the intensity of the source. It is generally better than 4–5 arcmin, and we are now as accurate as about 1 arcmin. This is not sufficient to point large telescopes for spectroscopy but it is good enough to image

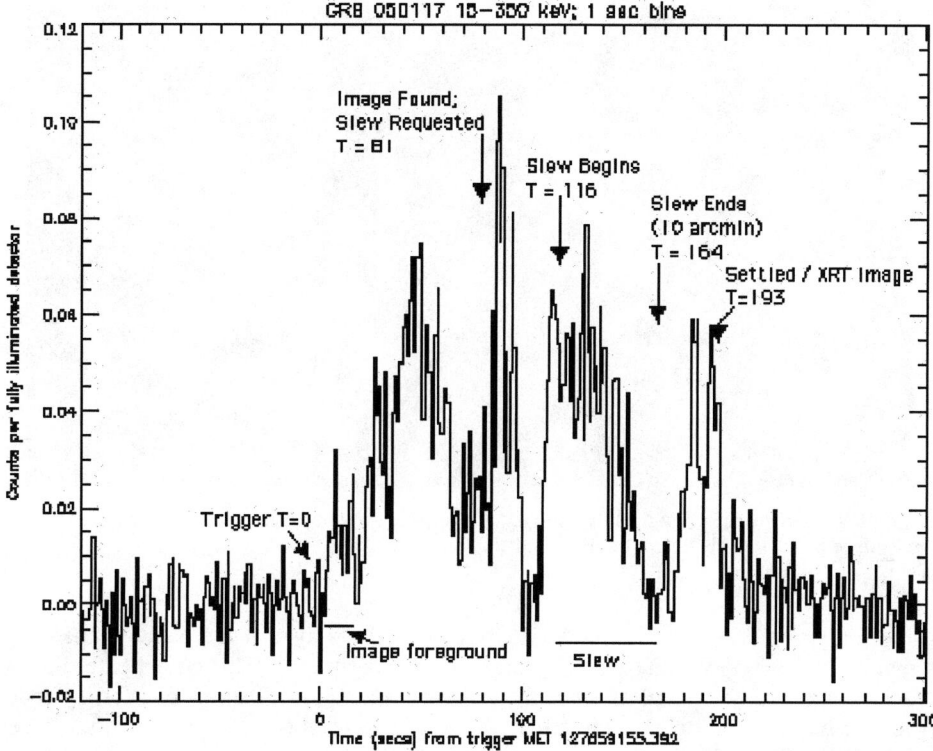

FIGURE 3. A time sequence of the counts detected by BAT in occasion of the Burst GRB050117. While crossing the SAA the XRT telescope does not collect data. Courtesy of Scott Barthelemy.

the field not only with small telescopes but, in most cases, also with medium size and large telescopes.

Naturally, in order to identify the object and to carry out spectroscopic observations, a ground-based telescope needs a second image to detect source variability and therefore the GRB, unless, due to the brightness of the burst the identification is immediately recognized. It is because of this that not only all of these observations complement each other, but that the XRT telescope also plays a fundamental role in the identification of the GRB afterglow. Indeed, XRT will measure and communicate to the ground, within minutes, the position with an accuracy of a few seconds. Independently of the UVOT detection, if the image is too faint (below the 20th magnitude) or very red the UVOT can not detect it, a large telescope can identify the source and start spectroscopy.

Soon after the cooling system of the CCD of the XRT was turned on, the system failed. The causes of this failure are not yet known, even if we suspect the instrument was hit by a strong protons current due to a solar flare. The consequence is that the CCD is being cooled only by the passive cooler. Test showed that, by using the passive

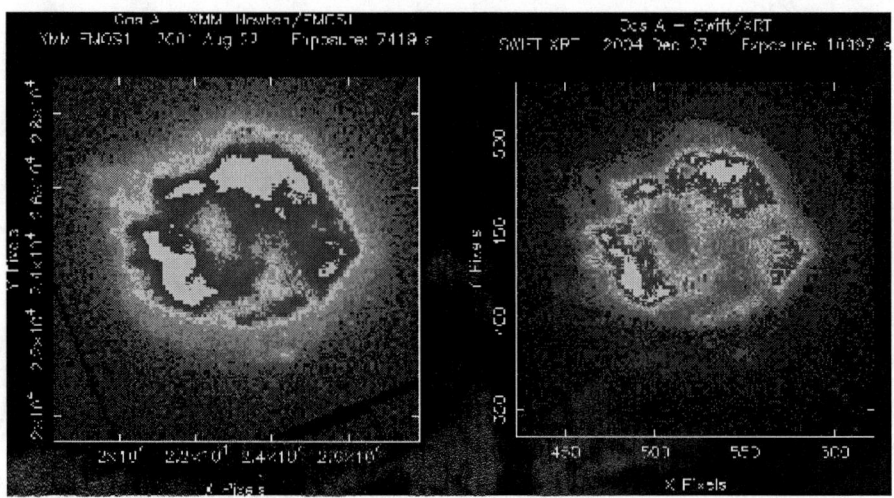

FIGURE 4. On the right first light image of CAS A obtained with the XRT telescope and on the left the same image obtained with XMM. Montage courtesy by Paolo Giommi.

cooler only, we must operate the spacecraft in such a way that the CCD temperature remains always below about −55 degrees during observations. This complicate slightly the planning and the operation, since the passive cooler should be kept out, for instance, from the Earth radiation and any other source of radiation (heat source). On the other hand, all the tests which were carried out after this failure showed that the instruments performs well, as far as sensitivity, spectral and spatial resolution is concerned according to specification.

Figure 4 shows the first light image obtained with XRT and a similar image of Cas A obtained with XMM–Newton. Naturally, XMM–Newton has a larger collecting area but the compromise which was made for the focus adjustment, indeed a strategy we planned for the JETX SXG mission, slightly favors XRT. The PSF is as measured on the ground and the focus of the instrument is as expected, see Figure 5.

This is particularly important for the science and the signal-to-noise for point-like sources. Figure also shows that the mirror module was not affected by the launch and that the heating system works as expected (being the optics maintained at about −20 Celsius with a very small gradient in order to avoid distortions in the image).

As mentioned above, it is one of the key assets of this mission: to get to the burst soon after the discovery to follow the light curve with an unprecedented accuracy and statistics, and also to get any eventual spectral feature in detail. Finally the good quality of the instrument and the low background of the orbit make possible the discovery and study of serendipitous sources in addition to allowing, over the years, the making of a good survey in the soft X ray where with long integration we should be able to reach a flux of about $3 \; 10^{-15}$ erg cm^{-2} s^{-1}.

The third instrument on Board of Swift has been activated only recently and it is now

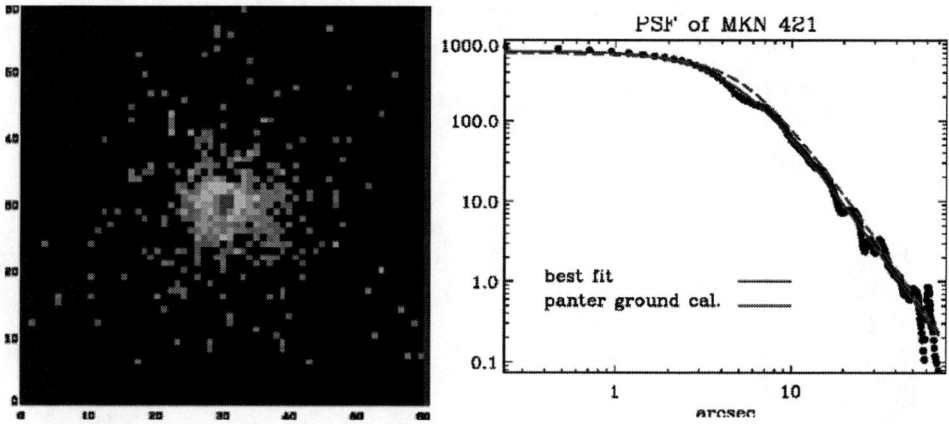

FIGURE 5. The Point Spread Function (PSF) obtained from in flight calibration of XRT matches perfectly with the PSF measured at the PANTER facility during the ground-based end-to-end calibrations. The focus adjustment and related strategy, that is a temperature regulated telescope and slightly on axis out of focus setting in order to maximize the corrected field, we devised for the mirrors worked out perfectly.

being calibrated.

Due to a focusing inconvenient it remains above the resolution expected but again, this does not change much the accuracy of the astrometry or the photometric properties of the instrument.

Fundamental to the strategy of the mission is the communication system which uses the TDRSS for uploading and downloading of the urgent information (trigger time and position of the burst for instance) and commands, and of the Malindi ASI Ground Segment which is used to upload regularly the program or commands and to download all the data. The record of Malindi in relation to the Swift mission is that over more than 1000 passes we missed only two contact, an unmatched record that make reliability an asset of the ground station. This high efficiency is due in large part to the work of Luca Salotti and the Malindi staff. In addition to the various automatisms we have the mission relay on an excellent and collaborating team of persons who immediately reduce the data and take decisions on the best way to proceed with the analysis and with the activity of the spacecraft. It is because of such an organization that the mission is being working at its optimum. In fact we are able to alert immediately the ground based observatories on most of the bursts. Such organization and trigger system is important not only for the research activity we are carrying out with Swift but for all the related activity, such as the high energy sky survey and the galactic or extragalactic Target of Opportunity which are bound to give us new insights into the secrets of Nature.

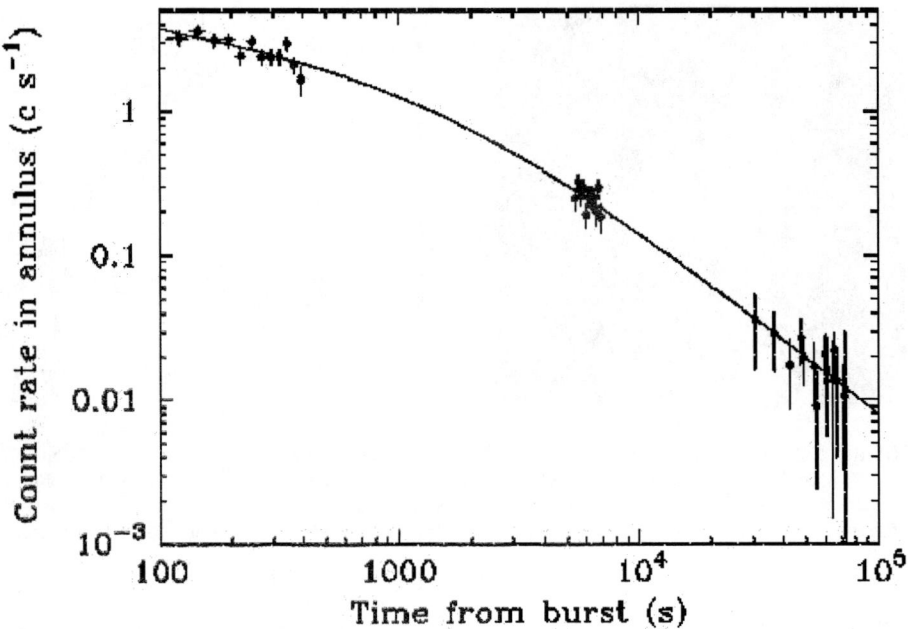

FIGURE 6. The XRT light curve of GRB050128. Courtesy of Sergio Campana.

SOME PRELIMINARIES

The sensitivity of BAT covers a softer range of energies than the BATSE experiment. However we expect, as shown in a detailed analysis by Sakamoto, that the type of bursts we detect are the same as those detected by BATSE. The effectiveness of the fast pointing spacecraft is shown by the Burst light curve shown in Figure 6, [3]. For this burst, XRT started gathering data after 108 seconds from the trigger so that it was possible to detect the change of slope which occurred after 1500 seconds. Hereafter, a very soft decline at the beginning of the light curve the brightness fades more rapidly, from a slope of $\delta_1 \sim 0.3$ we pass to a slope of $\delta_2 \sim 1.3$. The Fluence of the Burst was of about $4.5\ 10^{-6}$ erg cm^{-2}.

We should also mention the burst detected by BAT on the 15th of February 2005. This burst lasted for about 10 seconds and was followed by XRT for about 1 week for a total exposure of 43604 seconds allowing, at the same time, testing XRT capabilities in deep exposures, Giommi 2005. The counterpart has been detected in the visual by UVOT and in the near infrared by the UKIRT Telescope, Tanvir et al., 2005, GCN 3031, while no radio counterpart has been found. This burst happens to be an XRF with a very low fluence and a peak emission, by definition, in the soft energy range. This shows the capability of BAT to detect XRF as well.

In addition to the many new contributions the Swift mission will give to astronomy in

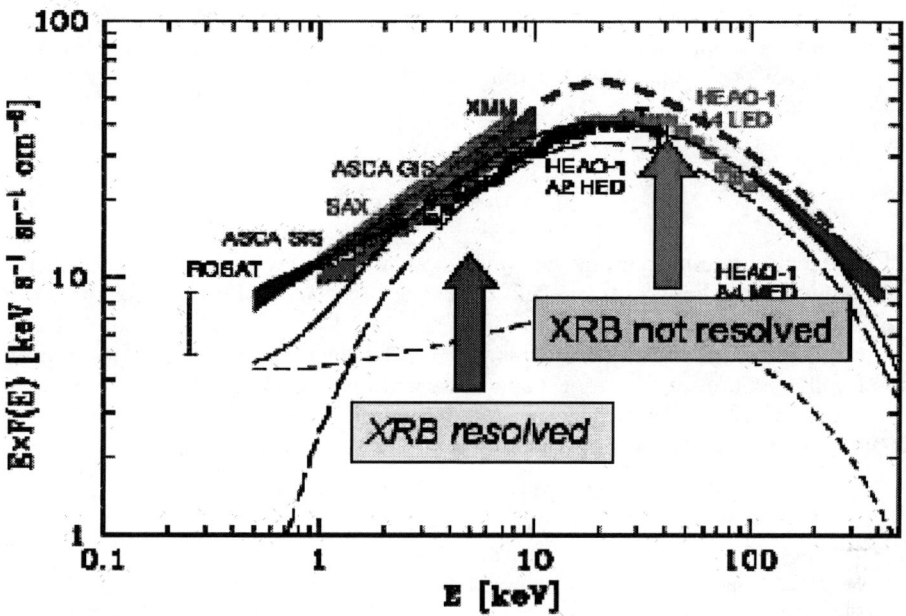

FIGURE 7. The X ray spectrum background. Graphic courtesy of Daniele Spiga.

the various fields of endeavors, the high energy survey will be a major contribution, thanks to the unique sensitivity of the BAT instrument in the 15 âĂŞ 200 keV range and the complete coverage of all the sky.

As it is shown in Figure 7 the understanding of the X ray sky in the soft energy band is nicely explained, thanks also to the deep observations carried out by the Chandra satellite on the Lockman Hole [4] and Chandra Deep Fields ([5, 6] among many others). It is not known yet the identity of the population contributing to the high energy part of the spectrum. We have difficulties under various scenarios and we certainly know that AGN cannot make it all by themselves unless we call into play obscured AGN. Indeed Rosati et al., 2002 [7] , following the analysis of the Chandra 1 Million second exposure field concluded that "there is still a not negligible population of faint hard sources to be discovered with better sensitivity at high energies. Such sources are expected to generate the 30 keV bump in the XRB spectrum" and the search continues. It is also straightforward to estimate the contribution given by the accretion of matter into Black Holes (BH) in the Universe, assuming the population of BHs in the Universe is inferred from the population of spheroidal galaxies. We should be able to come close to the observed background, but it seems that the best way is to observe and count, at least partially, the objects that generate such a background. Swift may be able to do just and give also a better measure of the Energy content of the Universe.

We started this interesting journey with a well-performing machine and an excellent team. The Principal Investigator (PI) Neil Gehrels is paying attention that all duties are performed, commitments honored and the community served as to make a real progress in science. Hopefully everything will happens, as they did so far, according to expectations.

Acknowledgments

I am deeply grateful to all the persons who made this experience possible, management, engineering, support teams and colleagues. The International collaboration has been a enriching experience. The mission, developed by a NASA-lead collaboration, has been supported by NASA, ASI and PPARC. I finally would like to thank all my closer collaborators for the work we have carried out together. Finally I would like to thank Angelo Antonelli and all the Editors for their help and patience. Sergio Campana for suggestions and comments on the draft and Dino Fugazza for help.

REFERENCES

1. Gehrels, N., Chincarini, G., Giommi, P., et al. 2004, ApJ 611, 1005
2. Barthelemy, S.D., et al., 2000, in *"Gamma Ray Bursts"*, eds. R. M. Kippen, R. S. Mallozzi and G. J. Fishman, AIP (New York), p. 731
3. Campana, S., Antonelli, L. A., Chincarini, G., et al., 2005, ApJ, 625, L26
4. Hasinger, G., et al., 2001, A&A, 365, L45
5. Giacconi, R., et al., 2001, ApJ, 551, 624
6. Moretti, A., et al., 2003, ApJ, 588, 696
7. Rosati, P., et al., 2002, ApJ, 566, 667

The REM Telescope: a Robotic Facility to Promptly Follow-up GRBs and Cosmic Fast Transients.

L. Angelo Antonelli

on behalf of the REM Team

INAF-Osservatorio Astronomico di Roma, Italy

Abstract. The REM (Rapid Eye Mount) Telescope is a fully robotic fast-slewing 60-cm telescope. It has been primarily designed and realized to follow-up the early phases of GRBs afterglow detected by space-borne experiments such as SWIFT, HETE II, INTEGRAL, AGILE. Since June 2003, REM is operating in the La Silla Observatory (Chile). The Telescope is equipped with a high throughput Infrared Camera (REMIR) and an optical imaging spectrograph (ROSS) which are simultaneously fed by a dichroic. This allows to collect high S/N data in an unprecedented large spectral range ($0.45 - 2.3\,\mu m$) on a telescope of this size. The wide band covered, the very fast pointing capability (60 degrees in 5 seconds) and the full robotization make REM the ideal experiment for observation of Gamma Ray Bursts and other fast transients. The REM observatory is an example of a versatile and agile facility necessary to complement large telescopes in fields in which rapid response and/or target pre-screening are necessary. This paper describes the main characteristics and operation modes of the REM observatory and gives an overview of preliminary results obtained during the Science Verification Phase.

Keywords: Gamma Ray Bursts, Afterglows, Robotic Telescopes
PACS: 95.55.Fw, 98.70.Rz

INTRODUCTION

In the last decade, multiwavelenght observations of Gamma Ray Bursts (GRBs)(e.g. [1, 2, 3] and prompt observations by optical robotic Earth–based facilities (e.g [4]), showed that a very fast reaction is the winning strategy to study fast transient phenomena, such as GRBs. Following this strategy, new space-borne experiments, such as SWIFT, HETE II, INTEGRAL, AGILE, have been developed in the last years in order to detect GRBs (or any other transient phenomena at high energies) and promptly distribute coordinates of the detected event. In such a scenario, has become of fundamental importance for the astronomical community to develop new automatic ground-based facilities in order to react to satellites triggers faster than a traditional "human operated" astronomical observatory.

In June 2003, after 2.5 years from the Kick Off meeting, a team of Italian, Irish and French scientists and engineers enjoyed the first light of the REM (Rapid Eye Mount) telescope and its instrumentation at ESO Observatory of La Silla (Chile) [5]. This is a fully robotic fast-slewing telescope conceived to follow up the early afterglow of GRBs detected by Space-borne-alert systems [6, 7].

The ambitious goal of the REM project is to discover and permit the study of the the most distant astronomical sources ever observed so far. Such a goal is related to

FIGURE 1. *Left panel* The REM Telescope and instruments are pointing the sky. *Right panel* The first building just in front is the "dome" hosting the REM Telescope. It is located at the ESO observatory of La Silla (Chile) ("Notre Dome de La Silla")

the detection and observation of GRBs afterglows. About a 30% of the observed GRBs do not show any optical afterglow. At least for a part of them it could be that Ly-α absorption dumps all the light at optical wavelengths because they explode in high-z galaxies. Ly-α absorption falls in the REM Infrared Arm wavelength range for sources with redshifts between 8 and 15. A burst in this z-range could be detected by REM provided it is bright enough (where enough means $K < 15$, a figure expected for this kind of event if observed promptly). The Astrometry, to a few tenth of an arcsec, is made available by REM in a time-scale of tens of second allowing one to observe the transient with larger area telescopes when it is still very bright. As a consequence, via REM, a 8-mt class telescope equipped with suitable IR spectrographs, could collect a high resolution high S/N spectrum of a source at $z > 10$, even within or beyond the range of the expected red-shift of re-ionization ($8 < z < 20$).

THE REM DOME

The REM dome is a concrete building with sliding roof located within the ESO La Silla Observatory area (2400 m a.s.l.). The dome is entirely robotically controlled via a PLC system that governs its opening and closing, its external functions such as the local meteo station, the air conditioning and temperature control system and the emergency illumination system. The situation in the dome can be monitored in real time via webcams the images of which, once stored, provide also visual data base of the activity occurred in the dome in presence or absence of human operators.

THE TELESCOPE.

The REM telescope is a Ritchey-Chretien system with a 60 cm f/2.2 primary and a overall f/8 focal ratio mounted in an alt-azimuth mount in order to provide stable Nasmyth focal stations, suitable for fast motions. REM has two Nasmyth focal stations:

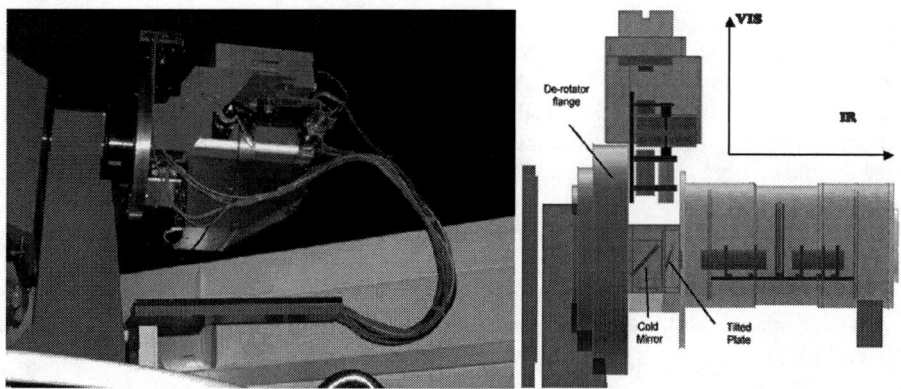

FIGURE 2. The de-rotated nasmyth focal station hosts both the instruments REM-IR and ROSS. These instruments are mounted onto the instrument flange, REM-IR along the nasmyth optical axis and ROSS orthogonal to such axis. The specific dichroic used for beam repartition has been customly designed in house and manufactured under the direct supervision of the REM technical team. A tilted plate allows to dither IR images without moving the telescope.

one is idle and the other hosts the instrumental flange. The image at the Nasmyth focal station in use is de-rotated and the instrumental flange is designed to receive a load of 250 Kg to be compared to the actual estimated total weight of the instrumentation of about 70 Kg. In the working focal station is located a dichroic, working at 45 degrees in the f/8 convergent beam, to split the beam in order to feed both instruments of the REM telescope: the REM–IR camera and the ROSS Spectrograph.

The Telescope has been manufactured by Teleskoptechnik Halfmann Gmbh in Augsburg (Germany). Mirrors are made by Carl Zeiss AG (Germany) and they are coated with protected silver to maximize reflection efficiency in such a large (0.45 - 2.3 μm) wavelength range. The altitude and azimuth motors made by ETEL allows a maximum speed of 12 deg/sec on both axis while the Heidenain encoders (237 steps per arcsec) allows excellent pointing, slewing and tracking precisions. The mechanical structure has been designed with stiffness in mind, because of the fast motion, but also taking into account the background contamination at NIR wavelength.

THE REM IR CAMERA (REMIR).

The REMIR camera is a fully cryogenic NIR ($0.9 - 2.3\,\mu m$) camera designed and developed with high throughput as major goal ([8, 9]). The camera has a focal reducer scheme in order to reform a white pupil in a cold environment for Lyot-stop positioning. A filter wheel with 10 positions is located at the reformed pupil allowing one to insert filters and grisms for slit-less spectroscopy or polarimeters in a parallel beam. The camera changes the focal ratio from f/8 to f/5.3 providing a plate-scale of 64.4 as/mm that allows one to position a 9.9 x 9.9 am FOV on a 512x512 (18 μm pitch) HgCdTe chip of the HAWAII series produced by Rockwell.

The Filter for the REM-IR camera are standard high performance IR filters. The J, H and Ks units have been ordered in the framework of the international Consortium lead by University of Hawaii (PI A. Tokunaga). Beside these standard NIR bands REMIR's filter wheel is also hosting a 1 μm filter (z'), a narrow band H2 filter and a dispersing element (grism) to perform slitless spectroscopy also at IR wavelengths.

The whole camera train is mounted in a dewar, designed and manufactured by IRLabs (Tucson, AZ), and operated in a cool environment. The chip working temperature is 77 K and will be guaranteed at the detector location and at the cold stop position. The optical train is kept at a temperature of about 100-120 K in order to save cooling power. The cryogenics, initially supported by a Stirling-Cycle cryo-pump, is now provided by a traditional system adopting liquid nitrogen. A view of REMIR and ROSS instruments is given in Fig. 2.

THE ROSS SPECTROGRAPH.

REM Nasmyth A is also hosting the slitless spectrograph ROSS with an orthogonal development relative to the REMIR optical axis. The idea of ROSS Spectrograph [10] originated from the consideration that optical/NIR burst spectral distributions may evolve so rapidly that truly simultaneous multi-band observations are required. Multiplexing the spectral response with rotating photometric filters may not be satisfactory at the expected evolution time-scale of the phenomena.

The spectrograph consists of a fore-optics which images a pupil at the location of the dispersing element and re-maps the focal plane onto the detector unit. The selected detector head is a commercial Apogee AP47 camera hosting a Marconi 47-10 1K x 1K 13 μm pitch CCD. The plate scale of the REM telescope (43 arcsec/mm) matches properly with the specifications and allows one to cover a 9.54 x 9.54 am^2 with a scale of 0.56 as/px.

The dispersion is obtained by insertion at the pupil location of an Amici Prism 66 mm long. The prism spreads the 0.45-0.95 μm wavelength range on 60 pixels, allowing the recording of 30 2-pixels bins along the range. The optical quality of ROSS is very good. The Amici prism is accompanied by classical V,R,I imaging filters and as a possible upgrade we are considering a double Wollaston Polarimeter. The ROSS mechanics and optical mountings have been manufactured in house (University of Perugia and INFN laboratories in Perugia).

INSTRUMENT PERFORMANCES

The overall transmission of the REMIR and ROSS optics have been measured during the acceptance tests at gOlem laboratories in Merate. We summarize the transmission in Fig. 3. In this figure we show the QE of the ROSS CCD and the REM-IR FPA, the reflection/transmission of the dichroic. The long-dashed line represents the convolution of all this component with the transmission of the optics and of either the Amici prism, on the optical side, or the ordinary J,H,Ks filter transmission. We can see from the figure that even in the most unfortunate circumstance (the J band) REM transmission always

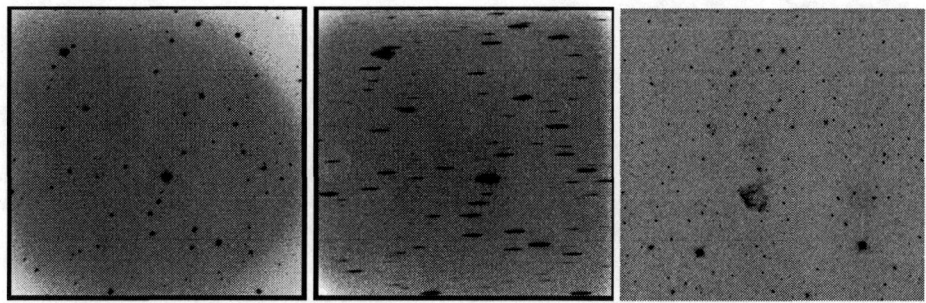

FIGURE 3. (*Left and Mid panel*) The same field observed by ROSS with R filter and Amici Prism. Amici prism spreads the 0.45-0.95 μm wavelength range on 60 pixels. (*Right panel*). REMIR H band observaions of the plerion around SGR0526 showing the good quality of infrared images.

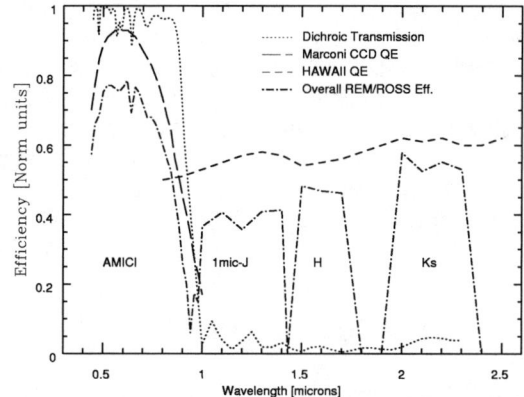

FIGURE 4. Overall efficiency of the Dichroic, Optical and NIR Camera

exceeds 30%.

The regime in which REMIR is operated is mainly sky limited and RON and DARK have little effect on the performance. The limiting magnitudes computed during Science Verification Phase observations are reported in table 1.

TABLE 1. Limiting magnitudes as a function of the S/N (5 or 3) and of the passband, for the different integration times obtained during the REM Science Verification Phase.

$T_{int.}$	J 5σ	J 3σ	H 5σ	H 3σ	K 5σ	K 3σ
10 sec	16.1	16.7	15.9	16.4	15.0	15.5
30 sec	17.2	17.8	16.8	17.4	16.3	16.8
600 sec	18.2	18.9	17.5	18.1	17.2	17.8

THE OBSERVING SOFTWARE

The REM Observing Software (REMOS, [11]) package is divided in several modules each managing a distinct task: communication between the REM community and the telescope, communication between orbital facilities and the telescope, security checks for the source of information, checks for visibility of the target and scientific priority with respect to the current (if any) REM target, full log of operations, etc. The whole system is designed to be extremely modular and with a high level of reliability and, in principle, the various sub-processes do not even need to run on the same computer. The system, therefore, can be trained to automatically react and configure to face with a large set of possible hardware failures.

Connection to external alert sources is one of most fundamental tasks performed by REMOS. GCN and IBAS networks provide alerts for SWIFT, HETEII and Integral, respectively. The code has been developed with some capabilities in recovering network generating warning messages and attempting to establish a new connection.

Observing activity is managed through a scheduler and processing software REM-SCHDLR and REMOBS. REMSCHDLR is the process that computes the schedule for secondary science targets. In the REM context each observation can be requested generating an Observing Block (OB), i.e. a bunch of data defining a logic unit of observations characterised by the same instrumental set up for both the NIR and optical cameras. Once OBs are scheduled REMOBS takes care of all operations required to drive the observations generating warning messages to the REMIR and ROSS subsystems, creating log files, etc.

THE REDUCTION SOFTWARE

The on-line IR image reduction pipeline [12] contains also the artificial intelligence for transient detection and decision making about transient information handling. Images are processed on a fast bi-processor computer, different from the single processor computer devoted to handle camera operation. As soon as images are written by the camera control software at the end of the observations they are sent over the network to the bi-proc that reduces them through the usual procedure used for NIR images, i.e. subtraction of a median "sky'" frame, obtained through a raw median of the single jittered images, flat-fielding, re-centring and average of the single images to produce a final scientific image. The astrometry is then computed on each image using a list of object extracted from the 2MASS catalogue, that is resident on the computer. A standard photometry software (SExtractor) is finally run to produce the photometric catalogue of the objects in the field. The core of the REMIR software is the transient detection algorithm named DECOAR. DECOAR is a decision tree that is run whenever a new catalogue is generated. For each instance, a grade is assigned to possible transient candidates and when only one of them is left with a grade above threshold, it is marked as a safe transient detection and released to REMOS for further management (e.g alert to larger telescope, release of a GCN, etc.). DECOAR is also able to manage REMIR in order to modify observation strategy if the GRB is not detected or if it is too bright.

The Ross Control Software (RCS) has been designed to operate ROSS as a stand alone device. RCS conists in three main programs running on two PCs. RCSClient and RCSServer run the single processor as a demon on RossPC and handle the observations with ROSS. The real time image processing in the ROSS spectrograph runs on a second bi-processor machine, different from the REMIR one, called RossBP. The pipeline is an automatic, fast image analysis procedure launched at each new acquisition in case of a GRB observation. The main goals of this procedure are: determination of ROSS spectrograph field astrometry, correcting for pointing/derotator uncertainties; search associations with catalogued objects and isolate possible OT candidates. Sub-arcsec field astrometry determination is actually needed in order to locate GRB counterparts candidates by catalogue comparison.

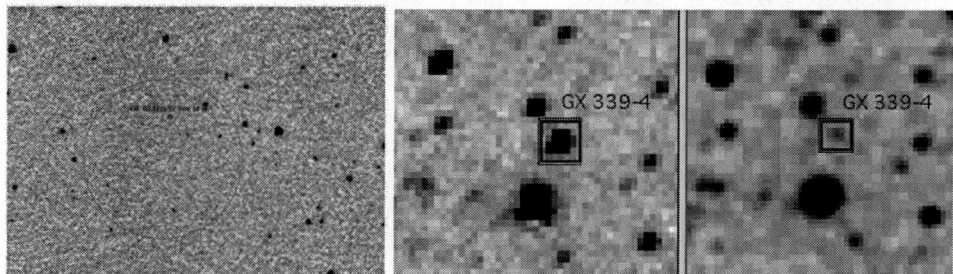

FIGURE 5. *Left panel* REMIR H band observation of GRB 040511 performed 12 hours after the burst, due to late time observation only an upper limit could be derived. Overimposed there is the HETE II error circle and the Optical Transient position. *Mid and Right panel* The galactic black hole candidate GX 334-9 observed by REM in Ks band during a high state compared with an archive image (from 2MASS) in the same band.

THE FIRST YEAR OF REM OBSERVATIONS.

REM has been primarily designed to follow up the early phases of NIR/optical afterglow of gamma ray bursts detected by dedicated satellites such as HETE-II, INTEGRAL and in particular SWIFT. During 2004, REM observed many GRBs detected by both HETE-II and INTEGRAL. Unfortunately, in all cases events happened during La Silla daytime so the fully automatic reaction was not tested and only upper limits on the afterglow emission could be derived. SWIFT (launched on November 20, 2004) is now providing more triggers (about 100/yr) and in the first months of 2005 we had the possibility to observe a couple of GRB fields in less then 30 sec but no afterglows were detected. One year of REM observations of GRBs are reported in table 2.

As all other robotic facilities dedicated to GRB science, for considerable amount of time REM remains idle in the sense that it does not have any GRB transient to point to. During such idle phase REM serves the community as a fast pointing NIR imager particularly suitable for multi-frequency monitoring of highly variable and transient sources. Among the obvious applications of REM idle time we find AGN and variable stars multi-frequency monitoring. Some Key-programs of interest for the REM-team have been identified and the related preparatory work has been initiated. During 2004 REM has been used in association with INTEGRAL to monitor Galactic Black Hole

TABLE 2. GRBs observed by REM from Dec 2003 to May 2005 and reported in GCN circulars by the REM Team. The large part of these GRBs happened during La Silla day-time and were observed many hours after the burst, so only upper limit could be derived. In two cases only REM had the opportunity to promptly react to a GRB trigger (39s and 56s after trigger) but in both cases no afterglow were detected and only upper limits could be derived. **Notes:** *n.a.*: not available (i.e. bad acquisition or only one filter was used) ; *n.o.*: instrument not operative.

GRB	$T - T_{burst}$ hrs	V mag	R mag	I mag	J mag	H mag	K mag	Trigger	GCN #
GRB 031203	6.14	n.o.	n.o.	n.o.	>16.5	>15.0	>14.5	INTEGRAL	2466,2471
GRB 040106	11	>19.3	>18.7	>18.2	n.o.	n.o.	n.o.	INTEGRAL	2511
GRB 040223	16.6	n.a.	n.a.	>17.0	n.a.	>15.5	>15.0	INTEGRAL	2531
GRB 040511	12	n.o.	n.o.	n.o.	>18.6	>18.2	>16.6	HETE II	2600
GRB 040903	4.67	n.o.	n.o.	n.o.	n.a.	>15.2	n.a.	INTEGRAL	2697
GRB 041223	12	n.a.	>18.0	n.a.	n.o.	n.o.	n.o.	SWIFT	2907
GRB 050128	3	>18.2	>18.2	>17.9	n.a.	>17.0	n.a.	SWIFT	2999, 3004
GRB 050209	2.4	>19.1	>19.1	>18.5	n.a.	n.a.	n.a.	HETE II	3017
GRB 050306	28.9	n.a.	>17.0	n.a.	>15.4	>15.2	>15.0	SWIFT	3076
GRB 050408	8.2	>18.5	>18.3	>17.9	>18.4	>17.4	>16.8	SWIFT	3205
GRB 050412	39 sec	>18.3	>18.0	>17.6	>16.0	>15.3	>15.1	SWIFT	3236
GRB 050416B	1.1	n.o.	n.o.	n.o.	>16.0	>15.4	>15.0	SWIFT	3281
GRB 050509C	56 sec	>19.1	>19.0	>18.5	n.a.	>15.5	n.a.	HETE II	3400, 3427

Candidates or Low- and High-mass X-ray binaries, flaring stars, variable stars, star forming regions and Blazars monitoring.

ACKNOWLEDGMENTS

The REM Project was funded by Italian Ministry for Instruction and Research through COFIN 2000 and PRIN2002, and by CNAA and INAF. ROSS was funded by Italian Space Agency (ASI).

REFERENCES

1. Costa, E., Frontera, F., Heise, J., et al. 1997, Nature, 387, 783
2. van Paradijs, J., Groot, P.J., Galama, T., et al. 1997, Nature, 386, 686
3. Frail, D.A., Kulkarni, S.R., Nicastro, S.R. et al. 1997, Nature, 389, 261
4. Ackerlof C., Mc Kay, E., 1999, IAUCirc., 7100
5. Chincarini, G., Zerbi F.M., Antonelli, L.A., et al., 2003, The Messenger, 113, 40.
6. Antonelli, L. A., Zerbi, F. M., Chincarini, G., et al., 2003, Mem. SAIt, 74, 304
7. Zerbi, F.M., Chincarini, G., Antonelli, L.A., et al., 2004, SPIE, 5492, 108
8. Vitali, F., et al, 2003, SPIE, 4841, 627
9. Conconi, P., Cunniffe, R., D'Alessio, F., et al, 2004, SPIE, 5492, 109
10. Tosti, G., Bagaglia, M., Campeggi, C., et al., 2004, SPIE 5492, 183
11. Covino, S., Stefanon, M., Zerbi, F.M., et al, 2004, SPIE 5492, 110
12. Testa, V., Antonelli, L.A., di Paola, A., et al., 2004, SPIE 5496, 94

Rapid GRB Follow-up with the 2-m Robotic Liverpool Telescope

Andreja Gomboc[*,†], Michael F. Bode[*], David Carter[*], Cristiano Guidorzi[*], Alessandro Monfardini[*], Carole G. Mundell[*], Andrew M. Newsam[*], Robert J. Smith[*], Iain A. Steele[*] and John Meaburn[**]

[*]*Astrophysics Research Institute, Liverpool John Moores University, UK*
[†]*University in Ljubljana, Slovenia*
[**]*University of Manchester, UK*

Abstract. We present the capabilities of the 2-m robotic Liverpool Telescope (LT), owned and operated by Liverpool John Moores University and situated at Observatorio Roque de Los Muchachos (ORM), La Palma. Robotic control and scheduling of the LT make it especially powerful for observations in time domain astrophysics including: (i) rapid response to Targets of Opportunity: Gamma Ray Bursts, novae, supernovae, comets; (ii) monitoring of variable objects on timescales from seconds to years, and (iii) observations simultaneous or coordinated with other facilities, both ground-based and from space. Following a GRB alert from the Gamma Ray Observatories HETE-2, INTEGRAL and Swift we implement a special over-ride mode which enables observations to commence in about a minute after the alert, including optical and near infrared imaging and spectroscopy. In particular, the combination of aperture, site, instrumentation and rapid response (aided by its rapid slew and fully-opening enclosure) makes the LT excellently suited to help solving the mystery of the origin of optically dark GRBs, for the investigation of short bursts (which currently do not have any confirmed optical counterparts) and for early optical spectroscopy of the GRB phenomenon in general. We will briefly describe the LT's key position in the RoboNet-1.0 network of robotic telescopes.

Keywords: Gamma Ray Bursts, Afterglows, Robotic Telescopes
PACS: 95.55.Fw, 98.70.Rz

EARLY GRB AFTERGLOWS IN OPTICAL AND INFRA-RED

Bright Optical Afterglows of Long GRBs

GRB990123 is the famous case of a prompt optical flash. Although detected more than 5 years ago it is still one of the very few (so far, only 4) optical afterglows (GRB 990123, 021004, 021211, 030418) that were detected a few minutes after the gamma ray burst. However, early GRB afterglows are essential in the study of extreme physics of ultra-relativistic flows and shocks and provide unique probes of the circum-burst medium and the nature of GRB progenitors.

In general, optical afterglows fade according to a power law $F \sim t^\alpha$ with power law index α between -0.6 and -2.3 [1]. But in some cases, such as GRB 021004, dense photometric coverage revealed additional light curve structure on short timescales as well as colour changes in early optical afterglows. According to different models such fluctuations could be related to phenomena in jets, circum-burst medium or renewed activity of the central engine [2], [3].

Among the most important issues regarding the origin and overall energetics of GRBs are achromatic breaks observed in some light curves a few hours to a few days after the GRB. Interpreted as due to beaming, they indicate jet opening angles of a few degrees and have led to evidence that most GRBs have a standard energy reservoir [4]. In a number of cases though, the exact time of the break is controversial due to insufficiently dense sampling of the light curve.

Optically Dark GRBs

Contrary to X-ray afterglows, which are detected in most well localized GRBs, optical afterglows were so far observed in approximately half of them. These missing afterglows are usually referred to as "optically dark". Although small robotic telescopes provide fast magnitude upper limits, these usually do not go deep enough to be conclusive. It is still an open question whether dark optical afterglows are merely observationally overlooked or not detected due to some of their properties. Their non-detection could be, for example, due to the fact that they are intrinsically faint or initially bright but very rapidly fading. Other explanations include heavy dust obscuration in host galaxies or their position at high redshifts of $z = 5 - 10$. The latter possibility is extremely interesting since GRBs at such high redshifts would provide a unique cosmological probe of star formation and galaxy evolution in the early universe. To solve the mystery of dark bursts, rapid observations in infrared wavelengths are of key importance.

Afterglows of Short GRBs

For short GRBs no unambiguous afterglow has been detected so far and their nature remains an enigma. The favourite model is a binary neutron star - neutron star or neutron star - black hole merger, although no evidence is available to prove or disprove this. It has been predicted [5] that their early optical afterglows may be much fainter than those of long GRBs: 20 – 23 magnitude initially and fading to 25 – 28 magnitude after a day, indicating that rapid and deep follow-up in the optical may again play a crucial role.

X-ray Flashes

A special class of objects, which seem to be related to GRBs are X-Ray Flashes (XRFs): they are similar to GRBs in many characteristics of their prompt emission with the main difference that they peak in X-rays instead of gamma rays. In the optical, several afterglows (XRF 020903, XRF030723 [6], [7]) and host galaxies (XRF 011030, XRF 020427 [8]) were detected. More observations in the near future will hopefully contribute to clarifying the XRF and GRB similarities and relationship and, on the other hand, also the issue of their differences: i. e. whether they differ due to different total energy, structure of the jet, progenitor's size, redshift, etc.

FIGURE 1. The Liverpool Telescope at ORM, La Palma, Canary Islands.

In view of the many open issues briefly outlined here and the fact that only a handful of optical afterglows were observed within a few minutes or even an hour after the GRB, it is obvious that early multi-colour optical and infrared photometry and spectroscopy can provide valuable information and help better to understand the nature of GRBs and their afterglows.

THE LIVERPOOL TELESCOPE

The Liverpool Telescope (LT, see Fig. 1) is situated at Roque de los Muchachos on La Palma in the Canaries and is operated by Liverpool John Moores University as a National facility under the auspice of PPARC. One of its noticeable characteristics is the clam-shell enclosure, which is fully opening. The benefits of such an enclosure are the minimization of the dome seeing, fast thermal equilibrium (reached in less than 30 min) and, particularly important in GRB follow-up, the short response time, since there is no need to wait for the dome to slew. Drawbacks are potential for windshake and the fact that in case of the enclosure breakdown, the telescope is totally exposed to weather conditions. Windshake is minimised by the 'stiff' structure of the telescope design and the enclosure has battery back-up in the event of power failure.

Other telescope specifications are: 2-m primary mirror, final focal ratio f/10, altitude-azimuth design, image quality < 0.4" on axis, pointing < 2arcsec rms, slew rate of 2°/sec, five instrument ports (4 folded and one straight-through, selected by a deployable, rotating mirror in the AG Box within 30s) and robotic (unmanned) operation with automated scheduler.

At present, instrumentation (Table 1) includes Optical and Infrared cameras. A prototype low resolution spectrograph will be commissioned in 2005 and a higher resolution spectrograph is being developed for 2006.

The telescope began science operations in January 2004 and after the enclosure hydraulics upgrade in summer 2004, the LT is entering fully robotic mode. It is expected

TABLE 1. The Liverpool Telescope instrumentation

RATCam Optical CCD Camera -	2048×2048 pixels, 0.135"/pixel, FOV 4.6'×4.6', 8 filter selections (u', g', r', i', z', B, V, ND2.0) - from LT first light, July 2003
SupIRCam 1 - 2.5 micron Camera - (with Imperial College)	256×256 pixels, 0.4"/pixel, FOV 1.7'×1.7', Z, J, H, K' filters - from late 2004
Prototype Spectrograph - (with University of Manchester)	49, 1.7" fibres, 512×512 pixels, R=1000; $3500 < \lambda < 7000$ Å - from 2005
FRODOSpec Integral field double beam spectrograph - (with University of Southampton)	R=4000, 8000; $4000 < \lambda < 9500$ Å - from 2006

that the telescope will operate in a fully autonomous way (without human intervention) from the beginning of 2005.

Science programmes running on the LT are diverse, but due to robotic control [9] and automated scheduling [10], [11] the LT is especially suited for:

- rapid response to Targets of Opportunity (GRBs, novae, supernovae, comets);
- monitoring of variable objects on timescales from seconds to years (follow-up of the ToO, active galactic nuclei, gravitational lenses etc.);
- observations simultaneous or coordinated with other facilities, both ground-based and from space;
- condition (e.g. seeing, photometricity) dependent or time critical (e.g. binary phase) observations.

RAPID GRB FOLLOW UP OBSERVATIONS WITH THE LIVERPOOL TELESCOPE

The GRB programme on the LT takes advantage of the telescope's robotic control: following receipt of GRB alert from the GCN, the Robotic Control System interrupts ongoing observations, applies over-ride mode and starts with GRB observations according to the basic GRB follow-up strategy, which presented in simple terms proceeds through following steps:

- slew to the position given in GCN alert;
- start optical imaging in about 1 min;
- try to identify the optical transient by comparison with the USNO catalogue or by image subtraction;
- if no candidate afterglow in the optical is detected, continue observations in the infrared;
- if an optical transient is reliably identified:
  - continue with multicolour imaging with intervening spectroscopy (if the optical transient is brighter than magnitude 15 for prototype spectrograph and 19

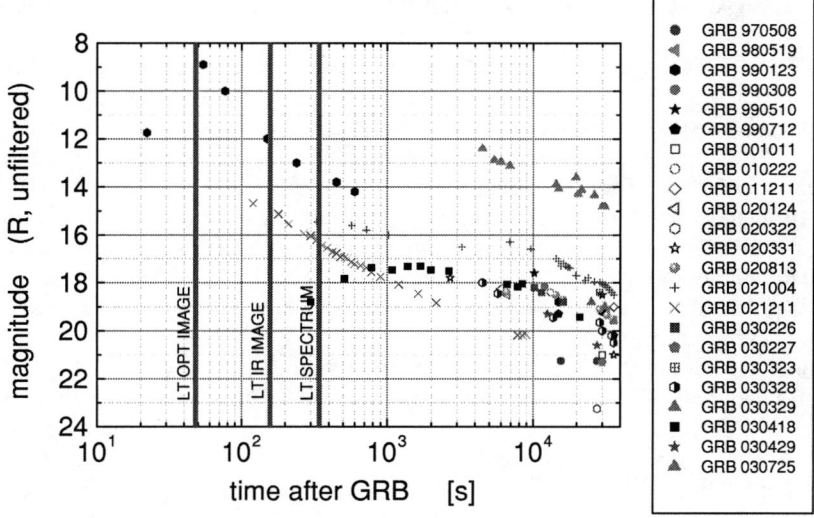

FIGURE 2. Optical afterglows detected in first minutes and hours after the GRB. Vertical lines indicate typical times of first observations with LT instrumentation. (Magnitudes and times are taken from GCN Circulars and are intended for illustrative purpose only.)

for FRODO spectrograph);
- issue a GCN Circular;
- trigger larger facilities.

The advantages of the LT in comparison with smaller robotic telescopes in GRB follow-up are larger aperture and deeper observations, the number of filters, infrared imaging and the possibility of early spectroscopy. These make it particularly suitable for study of:

- afterglows of short GRBs,
- afterglows of optically dark GRBs,
- prompt optical flashes,
- early optical GRB spectrometry, and
- statistical properties of GRBs and their afterglows.

With approximately 25 percent of GRBs occurring at night over La Palma and 70° maximal zenith distance observable by the LT, we expect to follow-up 1 in 6 GRBs immediately following the alert. We plan to monitor GRB afterglows at later stages depending on their scientific significance and in collaboration with other facilities, including the Faulkes Telescopes as part of RoboNet-1.0.

ROBONET-1.0 NETWORK OF ROBOTIC TELESCOPES

RoboNet-1.0 is a project to use a network of three large robotic telescopes: the LT and two Faulkes Telescopes, which are almost exact clones of the LT. The Faulkes Telescope North (FTN) is situated at Maui in Hawaii and has been operating since the end of 2003. Faulkes Telescope South (FTS), situated at Siding Spring, Australia, achieved first light in September 2004. They are financed mainly by the Dill Faulkes Educational Trust and are intended for use by UK schools. They are usually in the remote control mode (operated through the Telescope Management Centre in Liverpool or from schools), but can also operate in the fully robotic mode, identical to the LT's. Most of the observing time is intended for use by UK school children, but some time is available also to the research community.

This is the core of the RoboNet-1.0 project, which is funded by the UK PPARC and includes members of 10 UK university teams in Cardiff, Exeter, Hertfordshire, Leicester, Liverpool JMU, Manchester, MSSL, QUB, St. Andrews and Southampton. Funds were approved for acquisition of time on FTN and FTS for observations on extra-solar planets and GRBs (in the latter case, 275 hours over next 2.5 years, which covers most of Swift's expected lifetime). The principal technological aim of the project is to integrate LT, FTN and FTS into a global network of telescopes to act as a single instrument. The primary research areas of the project, including GRBs, will greatly benefit from the increased sky and time coverage provided by such a network.

ACKNOWLEDGMENTS

The Liverpool Telescope is funded via EU, PPARC, JMU grants and the benefaction of Mr. A. E. Robarts. AG and CG acknowledge the receipt of the Marie Curie Fellowship from the EU and MFB is grateful to the UK PPARC for provision of a Senior Fellowship.

REFERENCES

1. J. Greiner, http://www.mpe.mpg.de/jcg/grbolc.html.
2. D. Bersier et al., *ApJ*, **584**, L43–L46 (2003).
3. D. Lazzati et al., *A&A*, **396**, L5–L9, (2002).
4. D.A. Frail et al., *ApJ*, **562**, L55–L58 (2001).
5. A. Panaitescu et al., *ApJ*, **561**, L171–L174 (2001).
6. A. M. Soderberg et al., GCN Notice 1554 (2002).
7. D. B. Fox et al., GCN Notice 2323 (2003).
8. J.S. Bloom et al., *ApJ*, **599**, 957–963 (2003).
9. S. N. Fraser and I. A. Steele, "Object oriented design of the Liverpool Telescope Robotic Control System," in *Advanced Telescope and Instrumentation Control Software II*, edited by Lewis, Hilton, Proceedings of the SPIE, 2002, **4848**, pp. 443–454.
10. I. A. Steele and D. Carter, "Control software and scheduling of the Liverpool Robotic Telescope," in *Telescope Control Systems II*, edited by Lewis, Hilton, Proceedings of the SPIE, 1997, **3112**, pp. 223–233.
11. S. N. Fraser and I. A. Steele, "Robotic telescope scheduling: the Liverpool Telescope experience," in *Ground-based Telescopes*, edited by Oschmann, Jacobus M., Jr., Proceedings of the SPIE, 2004, **5493**, pp. 331–340.

SESSION 4

ACCRETION ON BLACK HOLES

AND MICROQUASARS

Black hole X-ray binary jets

Elena Gallo*, Rob Fender[†] and Christian Kaiser[†]

*Astronomical Institute 'Anton Pannekoek', University of Amsterdam, Kruislaan 403, 1098 SJ, Amsterdam, the Netherlands
[†]School of Physics and Astronomy, University of Southampton, Hampshire SO17 1BJ Southampton, United Kingdom

Abstract. Relativistic jets powered by stellar mass black holes in X-ray binaries appear to come in two types: steady outflows associated with hard X-ray states and large scale discrete ejections associated with transient outbursts. We show that the broadband radio spectrum of a 'quiescent' stellar mass black hole closely resembles that of canonical hard state sources emitting at four orders of magnitude higher X-ray levels, suggesting that a relativistic outflow is being formed down to at least a few 10^{-6} times the Eddington X-ray luminosity. We further report on the discovery of a low surface brightness radio nebula around the stellar black hole in Cyg X-1, and discuss how it can be used as an effective calorimeter for the jet kinetic power

Keywords: X-ray binaries; Jets, outflows and bipolar flows
PACS: 97.80.Jp; 98.58.Fd

ASTROPHYSICAL RELEVANCE

The production of jets, collimated bipolar outflows with relativistic velocities, appears to be a common consequence of accretion of material onto black holes on all mass scales. However, despite decades of study, we still lack a comprehensive theory that might account for the mechanism(s) of jet production, acceleration and collimation, and address the issue of the coupling between the outflow of matter and the accretion flow (e.g. [1]). The advantage of studying relativistic jets powered by stellar mass objects is simply given by their rapid variability: as the physical timescales associated with the jet formation are thought to be set by the accretor's size, and hence mass, then by observing stellar mass black holes in Galactic X-ray binary systems (BHXBs) on timescales of days to decades we are probing the time-variable jet:accretion coupling on timescales of tens of thousands to millions of years or more for supermassive black holes at the centres of active galactic nuclei (AGN). Even though such jets are expected to be highly radiatively inefficient – being adiabatic expansion the dominant cooling process – in the recent years it has become apparent that they may carry away a significant (if not the dominant) fraction of the liberated accretion power in low luminosity systems, possibly acting as a major source of energy and entropy for the interstellar medium.

RADIO STATES OF BLACK HOLE X-RAY BINARIES

Historically, the key observational aspect of X-ray binary jets lies in their radio emission ([20], [33]); in BHXBs, different radio properties are associated with different X-ray spectral states (see [10], [30] and Remillard, in these Proceedings, for recent reviews).

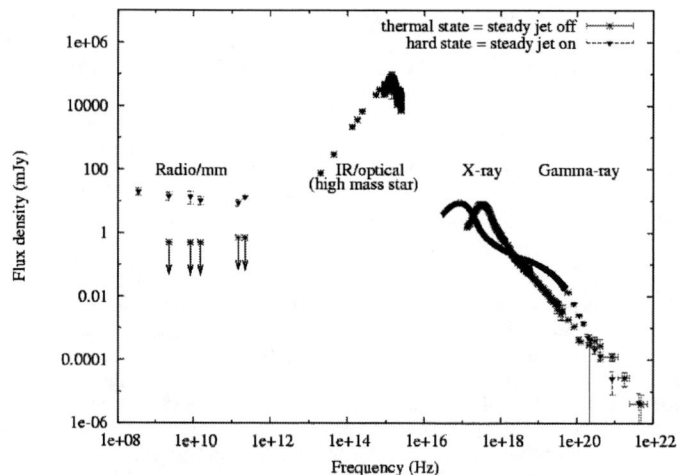

FIGURE 1. Spectral energy distribution of the prototypical 10 $M_\odot$ BH in the high mass X-ray binary Cygnus X-1. When the X-ray spectrum (above $\sim 10^{18}$ Hz) is dominated by a hard power-law component (triangle-points), the system is persistently detected in the radio band. The radio-mm spectrum is flat, due to an inhomogeneous, partially self-absorbed steady jet resolved on milliarcsec-scales [41]. Above a critical X-ray luminosity, the disc contribution becomes dominant (in units of νF_ν), while the hard X-ray power law softens (star-points). In this 'thermal dominant' state the radio emission is quenched by a factor up to about 50 with respect to the hard state. Adapted from [43].

This is illustrated schematically in Figure 1, which shows the spectral energy distribution, from radio to γ-ray wavelengths, of the (prototypical) stellar mass black hole in Cygnus X-1 over different accretion regimes.

While accreting gas at relatively low rates (below a critical X-ray luminosity of a few per cent of the Eddington X-ray luminosity, L_{Edd}), BHXB systems emit the bulk of their radiation in form of a hard X-ray power law component that cuts off at a few hundreds of keV, traditionally interpreted as due to Comptonization of disc photons in a rarefied electron/positron plasma ([42], [44]). In terms of radio properties this *hard* X-ray state is *radio active*, associated with persistent emission and a flat or slightly inverted spectrum which extends up to near-IR (possibly optical) frequencies, thought to be synchrotron in origin ([12] and ref. therein). The outflow nature of the synchrotron-emitting (and thus relativistic) plasma is inferred by brightness temperature arguments, leading to minimum linear sizes for the emitting region that often exceed the typical orbital separations, and thus making it unconfinable by any known component of the binary. This, together with the fact that BHXBs in the hard state display persistent radio emission despite being inevitably subject to expansion losses, imply the presence of a continuously replenished relativistic plasma that is flowing out of the system. The flat radio spectrum would arise in a *steady* partially self-absorbed conical jet, becoming progressively more transparent at lower frequencies as the matter travels away from the launching site ([2], [21]). The collimated nature of these outflows is less certain, as it requires direct imaging to be proven. Milliarcsec-resolution observations of Cyg X-1 in the hard X-ray state have confirmed the jet interpretation of the flat radio-mm spectrum in this system, imaging an extended structure extending to about 15 milliarcsec (∼30 A.U. at 2 kpc), and with an opening angle of less then 2°([41]).

Further indications for the existence of collimated outflows in the hard state of BHXBs come from the stability in the orientation of the electric vector in the radio polarization maps of GX 339−4 over a two year period ([7]). This constant position angle, being the same as the sky position angle of the large-scale, optically thin radio jet powered by GX 339−4 after its 2002 outburst ([17]), clearly indicates a favoured ejection axis in the system. Finally, the milliarcsec scale jet of the (somewhat peculiar) BH candidate GRS 1915+105 ([9]; [15]) in the hard state supports the association of hard X-ray states of BHXBs with steady, partially self-absorbed jets.

It is worth mentioning that some authors propose a jet interpretation (rather than a Comptonizing 'corona') for the X-ray power law which dominates the spectrum of BHXBs in the hard/quiescent (see next Section) state ([28], [27]). In this model, depending on the location of the frequency above which the jet synchrotron emission becomes optically thin to self-absorption and the distribution of the emitting particles, a significant fraction – if not the whole – of the hard X-ray photons would be produced in the inner regions of the steady jet, by means of optically thin synchrotron and synchrotron self-Compton emission.

Above a few per cent of L_{Edd}, BHXBs enter the so called *thermal dominant* X-ray state (starred points in Figure 1), during which the power output is dominated by a thermal component with typical temperatures of about 1 keV, interpreted as the clear signature of a geometrically thin optically thick accretion disc ([39]) extending very close to the central hole. No core radio emission is detected while in the thermal dominant state: the radio fluxes are *quenched* by a factor up to about 50 with respect to the hard X-ray state ([14], [6]), probably corresponding to the physical disappearance of the steady jet. This has been taken as a strong arguments in favour of magneto-hydro-dynamic jet formation in geometrically thick accretion flows ([31]).

Additionally, we observe a second variety of radio jets powered by BHXBs: hard-to-thermal X-ray state transitions appear to be associated with arcsec scale (thousands of A.U.) synchrotron-emitting plasmons moving away from the binary core with highly relativistic velocities ([33], [10] and ref. therein). Unlike milliarcsec scale steady jets, such discrete ejection events display optically thin synchrotron spectra and rapidly decaying fluxes. We shall refer to them as *transient* jets.

RELATIVISTIC OUTFLOWS IN 'QUIESCENCE'

What are the required conditions for a steady jet to exist? We wonder especially whether the steady jet survives in the very low luminosity, *quiescent* X-ray state (with $L_X \lesssim 10^{33.5}$ erg sec^{-1}, i.e. below a few $10^{-5}L_{Edd}$). In such a regime, very few systems have been detected in the radio band, mainly because of sensitivity limitations on the existing telescopes. Among them, the faintest is V404 Cygni, hosting a 12 $M_\odot$ black hole ([38]) and emitting in the X-rays at a few $10^{-6}L_{Edd}$ ([26]). Given the quite large degree of uncertainty about the overall structure of the accretion flow in quiescence (see [30] and ref. therein), it has even been speculated that the total power output of a quiescent BH could be dominated by a radiatively inefficient outflow ([18], [13]) rather than by the local dissipation of gravitational energy in the accretion flow. It is therefore of primary importance to establish the nature of radio emission from quiescent BHXBs.

Radio observations (using the Westerbork Synthesis Radio Telescope, WSRT) of this system, performed on 2002 December 29 (MJD 52637.3) at four frequencies, over the interval 1.4–8.4 GHz, have provided us with the *first* broadband radio spectrum of a quiescent stellar mass black hole. We measured a mean flux density of 0.4 mJy, consistent with that reported by [22], and a flat/inverted spectral index $\alpha = 0.09 \pm 0.19$ (such that $S_\nu \propto \nu^\alpha$). WSRT observations performed one year earlier, at 4.9 and 8.4 GHz, resulted in a mean flux density of 0.5 mJy, confirming the relatively stable level of radio emission from V404 Cyg on a year time-scale; even though the spectral index was not well constrained at that time, the measured value was consistent with the later one.

Synchrotron emission from a partially self-absorbed relativistic outflow of plasma seems to be the most likely explanation for the flat radio spectrum. Optically thin free-free emission as an alternative is ruled out, on the basis that far too high mass loss rates would be required in order to sustain the observed radio flux: even taking into account geometrical effects, such as outflow collimation and/or clumpiness, the mass loss rates can not be lower than 10^{-3} times the Eddington rate (assuming a 10 per cent efficiency in converting mass into light), *i.e.* still far too high for a sub-10^{-5} Eddington BH to produce any observable radio emission (see [16] for details).

The collimated nature of this outflow remains to be proven; based on brightness temperature arguments and the 5.5-hour time-scale variability detected at 4.9 GHz, we conclude that the angular extent of the radio source is constrained between 0.01 at 1.4 GHz and 10 mas at 4.9 GHz (at a distance of 4 kpc; [23]). These arguments led us to suggest that a relativistic jet is being formed in the quiescent state of V404 Cyg, and probably in BHXBs between a few 10^{-6} and a few per cent of L_{Edd} (were the collimated jet is actually resolved), strengthening the notion of 'quiescence' as a low luminosity version of the canonical hard X-ray state ([16]).

A UNIFIED PICTURE

The question remains whether the steady and transient jets of BHXBs have a different origin or are somewhat different manifestations of the same phenomenon. [11] have addressed this issue, showing that: i) the power content of the steady and transient jets are consistent with a monotonically increasing function of L_X; ii) the measured bulk Lorentz factors of the transient jets are systematically higher than those inferred for the steady jets. Based upon these arguments, a unified model for the jet/accretion coupling in BHXBs has been put forward. The key idea is that, as the disc inner boundary moves closer to the hole (hard-to-thermal state transition), the escape velocity from the inner regions increases. As a consequence, the steady jet bulk Lorentz factor rises sharply, causing the propagation of an *internal shock* through the slower-moving outflow in front of it. Eventually, the result of this shock is what we observe as a post-outburst, optically thin radio plasmon. For a thorough description of the model, we refer the reader to

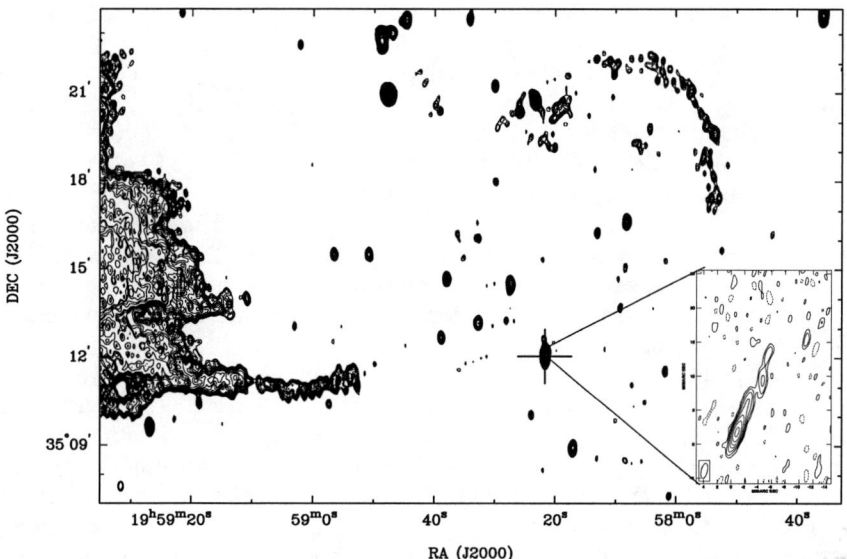

FIGURE 2. Westerbork Synthesis Radio Telescope 1.4 GHz of the field of view of the BHXB and Galactic jet source Cyg X-1: the arcmin scale, semi-ring-like structure northeast of the binary core (marked by a cross) seems to draw an edge between the bright HII region west of Cyg X-1 and the direction of the Cyg X-1 milliarcsec scale jet, shown in the inset (VLBA map from [41]); the average monochromatic flux of the ring is of about 0.08 mJy/beam. We interpret this structure as the result of a strong shock that develops at the location where the collimated jet impacts on the ambient ISM (Gallo, Fender, Kaiser & Russell, in preparation).

Belloni, in these Proceedings.

DISCOVERY OF A JET-POWERED RADIO NEBULA IN CYG X-1

The importance of BHXB jets for the overall energetics and dynamics of the accretion process, and furthermore as a potentially major source of energy input into the galactic interstellar medium (ISM), has yet to be well quantified. For both the steady and transient jets we are forced to make assumptions about the spectral extent (as the jet high frequency emission is generally blocked by the accretion disc or the companion star) and radiative efficiency, basing our estimates on, for example, assumptions of equipartition for which there is little a priori justification. Alternatively, we can constrain the jets' power content by looking at their (gradual or abrupt) interaction with the surrounding medium. As in AGN, the total energy associated with radio lobes and termination shocks was, and to a certain extent remains, the safest way to estimate the jet *power* × *lifetime* product ([3]). In the case of Galactic stellar mass BHs, arcmin-scale radio lobes are associated with two hard state sources in the galactic centre, 1E 1740.9-2942 and GRS 1758-258 ([34], [37]). [4] identified two IRAS sources with flat spectrum symmetric about GRS1915+105 (see also [36]) but argued that a possible association with the arcsec scale jets in these system seemed inconclusive. [25] suggest instead that the two IRAS regions would be the actual jets' impact sites; applying a fluid dynamical model developed for AGN jets ([24]) to this stellar mass system, they conclude that the time-averaged energy transport rate in the jet may be as low as 10^{36} erg s^{-1} (i.e. at least one order of magnitude than what inferred for the discrete ejecta;[14]). If correct, this association would place GRS 1915+105 at the same distance of the two IRAS sources, i.e. at 6.5 kpc (rather than the 12 kpc estimated by [19]), casting doubts even on its 'superluminal' nature. Finally, we have recently witnessed the dynamic formation of arcmin-scale decelerating radio *and* X-ray lobes, following an outburst of the transient BHXB XTE J1550-564 ([5]). These results suggest that radio lobe formation is a common occurrence which might be associated with many more sources and has not been found to date due to low signal-to-noise.

Such considerations led us to look for extended radio nebulae around the more promising (i.e. radio-loud) binary systems powering jets with the WSRT. Observations performed in May 2003 at 1.4 GHz resulted in the *discovery of an arcmin scale semi-ring-like radio nebula around the Galactic jet source Cyg X-1*, presented in Figure 2 (Gallo, Fender, Kaiser & Russell, in preparation). The structure appears to be perfectly aligned with the collimated jet resolved on milliarcsec scale. We note that previous attempts to look for lobes around Cyg X-1 ([29]) resulted in several 'interesting structures' at 1.4 GHz, whose association with Cyg X-1 could not be proven yet. Interestingly, Martí and collaborators (1996) already talked of a 'suggestive shell appearance of the structures' around Cyg X-1, which led them to investigate the possibility of a weak supernova remnant interpretation, eventually excluded due to too low surface brightness.

Modelling the ring of Cyg X-1: jet power and lifetime

Following the self-similar model developed by [24] for extragalactic jet-cocoon systems, we have interpreted the semi-ring radio structure of Cyg X-1 as the result of a strong shock that develops at the location where the collimated jet (resolved on milliarcsec scales) impacts on the ambient ISM. The jet particles inflate a radio lobe which is over-pressured with respect to the surrounding medium, thus the lobe expands sideways forming the observed bow shock that emits bremsstrahlung radiation – *hypothesis which needs to be confirmed through approved optical and deep 90 cm observations of this field*. The very pressure in the cocoon is responsible for keeping the jet confined. Following their formalism, we assume:

1. that the jet and the shocked ISM are in pressure balance;
2. that the rate of energy input is constant and given by the average jet power, Q_0;
3. that the rate of mass transport along the jet is constant and supplied by the bulk kinetic energy of the jet;
4. that the shock expands into an atmosphere of constant mass density ρ_0.

This model implicitly requires the bow shock to be self-similar and a roughly constant jet direction (which seems to be the case here, as the measured proper motion of Cyg X-1 rules out large velocity kicks; [32]).

Knowing the ring monochromatic luminosity, we are able to derive the density of the ionized particles in the ring from the expression of the bremsstrahlung emissivity ε_ν[1] for a pure hydrogen gas emitting at a typical temperature of $T \simeq 10^4$ K (below which the ionization fraction becomes negligible, and above which the cooling time becomes too short). The measured luminosity density, $L_{1.4 \text{ GHz}} \simeq 4.8 \times 10^{17}$ erg s^{-1} Hz (estimated assuming a distance of 2 kpc to Cyg X-1), equals the product $(\varepsilon_\nu \times V)$, where the source volume V is given by the beam area times the measured ring thickness: $V \simeq 2.0 \times 10^{53}$ cm^3. For $T \simeq 10^4$ K and a Gaunt factor $g \simeq 6$, we derive a particle density n_e of the ionized particles of ~ 24 cm^{-3}. The *total* particle density in the bow shock region will be a factor $1/x$ higher though, where x is the ionization fraction. At $\sim 10^4$ K, $x = 0.019$ ([40]), resulting in a particle density n_t of 1260 cm^{-3}.

For a strong shock in a mono-atomic gas, the velocity of the bow shock, v_{bow}, depends on the temperature T of the shocked gas as: $v_{bow} = \sqrt{(16 k_B / 3 m_p) T}$.

For 10^4 K, $v_{bow} \simeq 2.1 \times 10^6$ cm sec^{-1}, justifying the strong shock assumption. In [24] the length L of the jet within the cocoon grows with the time in such a way that: $t = (L/c_1)^{(5/3)} \times (\rho_0/Q_0)^{(1/3)}$ (where the factor c_1 depends on the thermodynamical properties of the jet material and on the aspect ratio; here $c_1 \simeq 1.5$). By writing the time derivative of the above equation, we obtain $t = \frac{3}{5}(L/v_{bow})$, resulting in a jet lifetime of $\sim 0.2 \times (\sin\theta)^{-1}$ Myr, where θ is the jet angle to the line of sight. This value has to be compared with the estimated age of the progenitor of the black hole in Cyg X-1, of a few Myr ([32]). For $t \simeq 0.3$ Myr (obtained with $\theta = 35°$; [35]), and adopting a mass density ρ_0 of the un-shocked material that is ~ 4 times lower than that of the shocked material, we obtain an average jet power Q_0 of a few 10^{35} erg s^{-1} , which would be a significant fraction of the measured 0.1-200 keV X-ray power of Cyg X-1 at the peak of the hard X-ray state ([8]). The results presented here clearly illustrate that finding and measuring jet-powered nebulae in stellar mass black holes offers an alternative and valuable method to address the debated issue of the jet power content in these systems.

ACKNOWLEDGMENTS

The Westerbork Synthesis Radio Telescope (WSRT) is operated by the ASTRON (Netherlands Foundation for Research in Astronomy) with support from the Netherlands Foundation for Scientific Research (NWO). Raffaella Morganti and Tom Oosterloo are gratefully acknowledged for their support in the analysis of the Cyg X-1 data.

REFERENCES

1. Blandford R. D., 2001, in Progress of Theoretical Physics Supplement, astro-ph/0110394
2. Blandford R. D., Königl A., 1979, ApJ, 232, 34
3. Burbidge G. R., 1959, ApJ, 129, 849
4. Chaty S. et al. , 2001, A&A, 366, 1035
5. Corbel S. et al. , 2002, Science, 298, 196

---

[1] $\varepsilon_\nu = 6.8 \times 10^{-38}\ g(\nu,T)\ T^{(-1/2)}\ n_e^2\ \exp(h\nu/k_B T)$ erg cm^{-3} sec^{-1} Hz^{-1}

6. Corbel S. et al. , 2001, ApJ, 554, 43
7. Corbel S. et al. , 2000, A&A, 359, 251
8. Di Salvo T., Done C., Zycki P. T., Burderi L., Robba N. R., 2001, ApJ, 547, 1024
9. Dhawan V., Mirabel I. F., Rodríguez L. F., 2000, ApJ, 543, 373
10. Fender R. P., 2005, to appear in 'Compact stellar X-ray sources', ed. Cambridge, astro-ph/0303339
11. Fender R. P., Belloni T., Gallo E., 2004, MNRAS, 355, 1105
12. Fender R. P., 2001, MNRAS, 322, 31
13. Fender R. P., Gallo E., Jonker P. G., 2003, MNRAS, 343, L99
14. Fender R. P. et al. , 1999, ApJ, 519, L165
15. Fuchs Y. et al. , 2003, A&A, 409, L35
16. Gallo E., Fender R. P., Hynes R. I., 2005, MNRAS, 356, 1017
17. Gallo E., Corbel S, Fender R. P., Maccarone T. J., Tzioumis A. K., 2004, MNRAS, 347, L52
18. Gallo E., Fender R. P., Pooley G. G., 2003, MNRAS, 344, 60
19. Greiner J., Cuby J. G., McCaughrean J., 2001, Nature, 414, 522
20. Hjellming P. M., Han X., 1995, in 'X-ray Binaries', ed. Camdridge, p. 308
21. Hjellming P. M., Johnston K. J., 1988, ApJ, 328, 600
22. Hjellming P. M., Rupen M. P., Mioduszewski A. J., Narayan R., 2000, ATel 54
23. Jonker P. G., Nelemans G., 2004, MNRAS, 354, 355
24. Kaiser C. R., Alexander P., 1997, MNRAS, 286, 215
25. Kaiser C. R. et al. , 2004, ApJ, 612, 332-341
26. Kong A. K. H., McClintock J. E., Garcia M. R., Murray S. S., Barret D., 2002, ApJ, 570, 277
27. Markoff S., Nowak M. A., 2004, ApJ, 609, 972
28. Markoff S., Falcke H., Fender R., 2001, A&A, 372, L25
29. Martí J.,,.306, Rodriguez L. F., Mirabel I. F., Paredes J. M., 1996, A&A, 306, 449
30. McClintock J. E., Remillard R. A., 2005, to appear in 'Compact stellar X-ray sources', ed. Cambridge, astro-ph/0306213
31. Meier D. L., 2001, ApJ, 548, L9
32. Mirabel; I. F., Rodrigues I., 2003, Science, 300, 1119
33. Mirabel I. F., Rodríguez L. F., 1999, ARA&A, 37, 409
34. Mirabel I. F., Rodríguez L. F., Cordier B., Paul J., Lebrun F., 1992, Nature, 358, 215
35. Orosz J., 2002, in Proceedings IAU Symposium No. 212, ed. K.A. van der Hucht, A. Herrero & C. Esteban
36. Rodríguez L. F., Mirabel I. F., 1998, A&A, 340, L47
37. Rodríguez L. F., Mirabel I. F., Martí J., 1992, ApJ, 401, L15
38. Shahbaz T. et al. , 1994, MNRAS, 271, L10
39. Shakura N. I., Sunyaev R. A., 1973, A&A, 24, 337
40. Spitzer L., 1978, 'Physical Processes in the Interstellar Medium', ed. J. Wiley & Sons
41. Stirling A. M. et al. , 2001, MNRAS, 327, 1273
42. Sunyaev R. A., Titarchuk L. G., 1980, A&A, 86, 121
43. Tigelaar S. P., Fender R. P., Tilanus R. P. J., Gallo E., Pooley G. G., 2004, MNRAS, 352, 1015
44. Titarchuk L. 1994, ApJ, 434, 570

Black Hole States: Accretion and Jet Ejection

T. Belloni

INAF-Osservatorio Astronomico di Brera, Via E. Bianchi 46, I-23807 Merate, Italy

Abstract. The complex spectral and timing properties of the high-energy emission from the accretion flow in black-hole binaries, together with their strong connection to the ejection of powerful relativistic jets from the system, can be simplified and reduced to four basic states: hard, hard-intermediate, soft-intermediate and soft. Unlike other classifications, these states are based on the presence of sharp state transitions. I summarize this classification and discuss the relation between these states and the physical components contributing to the emitted flux.

Keywords: accretion: accretion disks – black hole physics – stars: oscillations – X-rays: binaries
PACS: 95.75.Wx, 95.85.Nv, 97.10.Gz, 97.60.Lf, 97.80.Jp

INTRODUCTION

Since the launch of RossiXTE, our knowledge of the high-energy emission of Black Hole Binaries (BHB) has increased enormously, leading to a new, complex picture which is difficult to interpret. At the same time, a clear connection between X-ray and radio properties has been found (see Fender 2005). In the following, I present briefly the state paradigm that is now emerging, based on a large wealth of RossiXTE data from bright transient sources and its connection with jet ejection (see Belloni et al. 2005; Homan & Belloni 2005; Fender, Belloni & Gallo 2004). I discuss this in general terms, in the attempt to simplify the picture as much as possible.

BLACK HOLE STATES AND JET EJECTION

The results of detailed timing and color/spectral analysis of the RossiXTE data of bright BHBs have evidenced a very wide range of phenomena which are difficult to categorize. Nevertheless, it is useful to identify distinct states. Based on the variability and spectral behavior and the transitions observed in different energy bands, we consider the following states in addition to a quiescent state (see Homan & Belloni 2005; Belloni et al. 2005; Casella et al. 2004,2005 for the description of the different QPO types):

- Low/Hard State (LS): this state is associated to relatively low values of the accretion rate, i.e. lower than in the other bright states. The energy spectrum is hard and the fast time variability is dominated by a strong ($\sim$30% fractional rms) band-limited noise. Sometimes, low frequency QPOs are observed. The characteristic frequencies detected in the power spectra follow broad-range correlations (see Belloni, Psaltis & van der Klis 2002). In this state, flat-spectrum radio emission is observed, associated to compact jet ejection (see Gallo, Fender & Pooley 2003; Fender, Belloni & Gallo 2004).

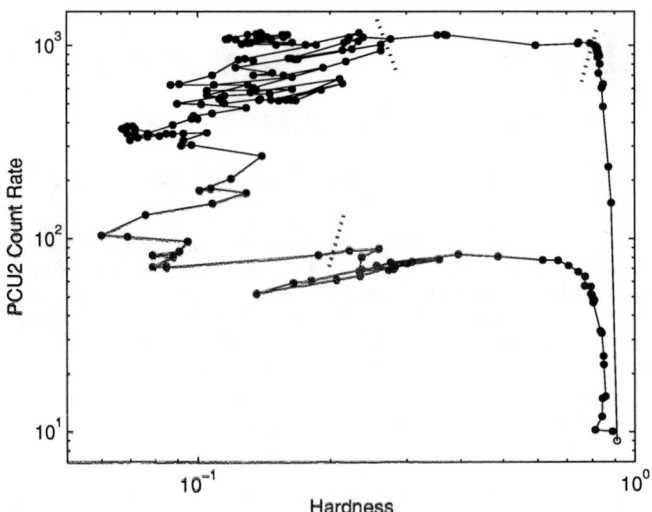

FIGURE 1. Hardness-Intensity diagram of the 2002/2003 outburst of GX 339-4 as observed by the RXTE PCA. The gray lines mark the state transitions described in the text. The inset on the lower right shows the general time evolution of the outburst along the 'q'-shaped pattern. From Belloni et al. (2005).

- Hard Intermediate State (HIMS): in this state, the energy spectrum is softer than in the LS, with evidence for a soft thermal disk component. The power spectra feature band-limited noise with characteristic frequency higher than the LS and usually a rather strong 0.1-15 Hz type-C QPO (see e.g. Casella et al. 2005). The frequencies of the main components detected in the power spectra extend the broad correlations mentioned for the LS. The radio emission shows a slightly steeper spectrum (Fender, Belloni & Gallo 2004). Just before the transition to the SIMS (see below), Fender, Belloni & Gallo (2004) suggested that the jet velocity increases rapidly, giving origin to a fast relativistic jet.
- Soft Intermediate State (SIMS): here the energy spectrum is systematically softer than the HIMS. The disk component dominates the flux. No strong band-limited noise is observed, but transient type-A and type-B QPOs, the frequency of which spans only a limited range. No core radio emission is detected.
- High/Soft State (HS): the energy spectrum is very soft and strongly dominated by a thermal disk component. Only weak power-law noise is observed in the power spectrum. No core radio emission is detected (see Fender et al. 1999; Fender 2005).

This simplified classification needs to be supported by a picture of the time evolution in a transient source. The states described above are defined also in terms of their transitions, which need to be taken into account. A sketch of the evolution of the 2002/2003 outburst of GX 339-4 is ideal to show these transitions (see Homan & Belloni 2005; Belloni et al. 2005). Figure 1 shows the outburst in a Hardness-Intensity Diagram

(HID): the x-axis shows the X-ray hardness and the y-axis the detected count rate (from Belloni et al. 2005). The outburst can be described in the following way.

- The system starts its outburst as a weak hard source (the bottom right in the HID): the first detection with the RXTE/PCA (a factor of ~ 10 lower than the first point in Fig. 1), indicates an X-ray flux a factor of ~ 10 higher than in quiescence (Homan et al. 2005). The flux increases with time, while the X-ray colors indicate that the spectrum is still hard and only mildly softening: the source moves upward in the right branch. During this period, the infrared flux was seen to correlate strongly with the X rays, indicating that both components are connected, possibly originating from a jet (Homan et al. 2005). This branch corresponds to the LS, when flat-spectrum radio emission is observed (see Fender et al. 1999; Fender, Belloni & Gallo 2004). The characteristic frequencies of the different noise components in the power spectrum increase with time. The energy spectrum is roughly described by a cutoff-power law.

- At the top of the right branch, GX 339-4 moves left and enters a horizontal branch, corresponding to the HIMS. The transition is shown by the dotted line. The precise position of this line corresponds to a marked change in the IR/X correlation (Homan et al. 2005). The drifting to the left is the result of two causes: the appearing of a thermal disk component and the steepening of the power-law component. In GRS 1915+105, recent Integral/RXTE observations (Rodriguez et al. 2004) indicate that in this state two hard components are at work (see also Zdziarski et al. (2001). In the power spectra, the characteristic frequencies continue to increase so that the transition in the timing domain appears to be rather smooth. As GX 339-4 moves left in the HID, the radio spectrum steepens slightly and as the source approaches the 'jet line' (see below) the velocity of the jet outflow increases (Fender, Belloni & Gallo 2004).

- When the source, moving left, passes the second dotted line, a very sharp transition is observed. The variation of X-ray hardness is small, indicating that the energy spectrum does not change by a large amount. However, the power spectrum changes abruptly: the strong band-limited noise and type-C QPO disappear and a sharp type-B QPO appears. The source has entered the SIMS. After this transition, the source remains for a long time in the upper left quadrant. The spectrum is soft and dominated by the thermal disk component, and the power spectrum shows complex and variable features. Transient type-B QPOs are observed, but always when the hardness is comparable to that of the first detection. The sharp transition, observed in other systems as well (see Homan & Belloni 2005; Casella et al. 2004) corresponds to the crossing of the 'jet line' (in the terminology of Fender, Belloni & Gallo 2004). It is in correspondence to the crossing of the jet line that the strong radio flare is observed (Gallo et al. 2004) and a fast relativistic jet is launched. This is also observed in other systems such as GRS 1915+105, XTE J1859+226 and XTE J1550-564.

- After entering the SIMS, the source remains for a long time (about five months) in the upper left quadrant, moving in a complex fashion (see Fig. 1). The spectrum is soft and dominated by the thermal disk component, and the power spectrum shows very complex and varying features. Transient type-B QPOs are observed,

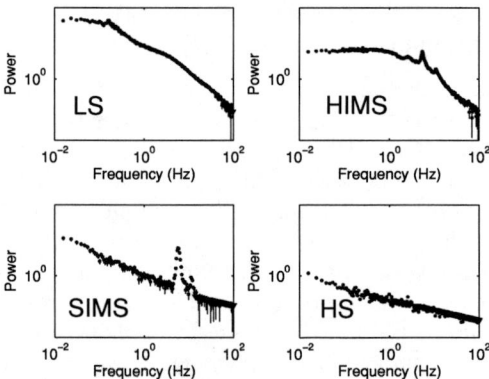

FIGURE 2. Examples for power spectra of GX 339-4 corresponding to the four states described in the text (adapted from Homan & Belloni 2005).

as well as other weak features. However, no strong band-limited noise component is observed. The properties of the SIMS are complex, and in other systems such as XTE J1859+226 fast transitions to and from the HIMS (i.e. between type-A/B and type-C QPOs) are observed (Casella et al. 2004), corresponding to the crossing of the jet line. For that source, there is evidence that subsequent radio flares are observed in correspondence of these transitions (see Casella et al., this volume). During the SIMS, observed radio emission corresponds only to the previous jet ejections (Fender, Belloni & Gallo 2004).

- Once the count rate of GX 339-4 goes below a few hundred counts per second per PCU, it enters the HS. The spectrum remains soft, the flux decreases with time, and the power spectrum only features a weak power-law component. Even integrating all the HS observations, no other timing component appears. No radio emission is observed in the HS (see Fender 2005).
- Below $\sim$100 cts/s/PCU, the source hardens again and enters the HIMS (third dotted line in Fig. 1). Notice that for three observations the source is found back in the range of hardness corresponding to the HS, and indeed HS properties are observed at those times in the power spectrum. No clear instance of SIMS is seen, although one observation at hardness similar to the bright SIMS points does show a clear 1 Hz QPO which could be a low-frequency version of the type-B ones, which were observed around 6-7 Hz at high flux.
- As GX 339-4 continues to harden, it moves back to the LS with a smooth transition. The points in the HID reach almost the same position where they started the outburst.

This picture, albeit simplified, provides a useful framework within which one can work on a modeling of the properties of the accretion flow and the ejection of collimated jets. This classification differs from that proposed by McClintock & Remillard (2005), being based on the presence of sharp state transitions rather than on a characterization in terms of spectral and timing parameters.

PHYSICAL STATES

In order to understand the scheme outlined above, we should consider the main components that are observed from the system.

Spectral components

There is amounting indication that three main spectral continuum components contribute to the observed high-energy flux (I deliberately ignore additional components such as emission lines and Compton reflection). The first is the thermal thin disk, observed at energies below 10 keV. This component is probably present in the LS, but its temperature is in most cases too low to be observable (see e.g. the case of XTE J1118+480; Frontera et al. 2001; McClintock et al. 2001). In the HIMS it contributes a small (varying) degree to the X-ray flux, while it is dominant it in the SIMS and the HS. This component appears to be rather quiet and does not contribute much to the observed fast variability. There is evidence that the innermost radius of this thin disk, as evaluated from the energy spectrum, decreases from large values in the LS to smaller values in the HS, although absolute measurements of this radius are plagued by a number of problems (see e.g. Merloni, Fabian & Ross 2000).

The second component is much harder and is usually fitted with a thermal Comptonization from $\sim$100 keV electrons or with its approximation as a power-law with a high-energy cutoff. This is the component that is observed in the LS and which is clearly associated to the strong band-limited noise typical of this state. In recent years, the possibility that this component originates directly from the jet has been proposed and fits to the broad-band spectra of LS sources have been attempted (see e.g. Markoff et al. 2003). It is not observed in the HS.

The third component is modeled with a steep power law, observed in the HS and which has been reported also for a variety of spectra the state attribution of which is uncertain (see Grove et al. 1998) but in some cases probably corresponding to the HIMS. This component was detected in GRS 1915+105, which is always observed in the HIMS (Zdziarski et al. 2001) and attributed to non-thermal Comptonization. For the SIMS, there is some evidence that no high-energy cutoff is observed with HEXTE, indicating that this component is present (Homan et al., in prep.). Interestingly, Integral/RXTE observations of GRS 1915+105 reported by Rodriguez et al. (2004) show that the combined timing/spectral properties can be explained with the simultaneous presence of both hard components, one with a high-energy cutoff and one without. Therefore, the scenario that is emerging for the high-energy component is one with a thermal component in the LS and a non-thermal one in the HS, while in the HIMS, which is intermediate between these two, both components are present. It is therefore possible that in the top branch in Fig. 1 (HIMS) the thermal hard component decreases and the non-thermal hard component appears, although it is not clear how gradual these changes are. After the jet line is crossed, however, the thermal hard component disappears.

Radio emission and jets

Fender, Belloni & Gallo (2004) proposed a unified scheme in which the compact flat-spectrum jet emission and the fast relativistic jet component originate from the same outflow, which is accelerated to high Lorentz factors as the source approaches the jet line. Indeed, core radio emission is observed only in the LS (down to quiescence level, see Gallo, Fender & Hynes 2004) and HIMS. The remaining two states, SIMS and HS, do not show radio emission.

Noise and QPO components

As shown above, the properties of the fast time variability are complex, but it is useful to examine their behavior in relation to the states described above. Two states, LS and HIMS, show strong band-limited noise and type-C QPOs; the frequencies of these power-spectrum components follow correlations that suggest there is a one-to-one correspondence between components in the two states (see Belloni, Psaltis & van der Klis 2002). Interestingly, these frequencies vary over a large range in the states where large variations in the inferred inner disk radius are observed. Their energy spectrum is hard (see e.g. Rodriguez et al. 2004; Casella et al. 2004). The two soft states, HS and SIMS, while displaying a number of transient feature such as type-A/B QPOs and other weak components difficult to classify (see Casella et al. 2004,2005; Belloni et al. 2005), never show strong band-limited noise components such as those described above. The QPOs in these states, in particular type-B QPOs, vary over a much smaller range of frequencies compared to the type-C ones. Their spectrum is also hard (Casella et al. 2004). The few instances of high frequency QPOs in BHBs were all observed in the SIMS (see e.g. Morgan, Remillard & Greiner 1997; Homan et al. 2001,2003; Cui et al. 2000; Remillard et al. 1999).

DISCUSSION

Above, we discussed the presence of four states of BHBs, which are identified in terms of the presence/absence of three spectral components, two or more power-spectral components, and a radio-emitting jet with varying Lorentz factor.

Before the 90's, when only relatively sparse data were available, two BHB states were clearly defined: a low/hard state and a high/soft state. The general picture that can be drawn from the properties discussed above is now the following. In the hard state, the source is jet-dominated, in the sense that the power in the jet is probably larger than that in the accretion (see Fender, Gallo & Jonker 2003). The dominant component in the X-ray range is the thermal hard component, which is possibly associated to the jet itself. The geometrically thin, optically thick, accretion disk is very soft and has a

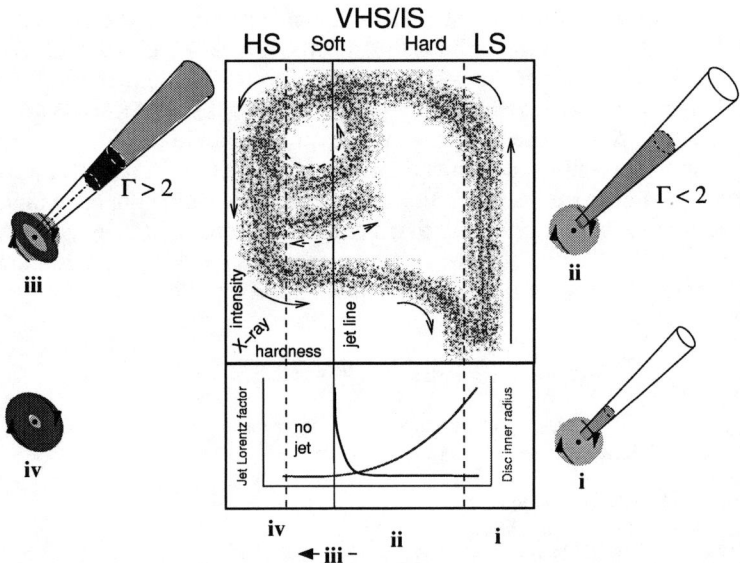

FIGURE 3. Schematic view of the accretion/ejection coupling in BHBs. The top panel shows an idealized HID, the bottom panel shows the Γ factor of the jet and the value of the inner accretion disk radius (see text). The pictures on the side sketch the structure of the three components in the system: jet, disk, and corona. From Fender, Belloni & Gallo (2004).

varying inner radius, so that its contribution to the X-ray emission changes. This state is characterized by strong band-limited noise and type-C QPOs, whose characteristic frequencies show clear correlations between themselves and with spectral parameters (see Wijnands & van der Klis 1999; Psaltis, Belloni & van der Klis 1999; Belloni, Psaltis & van der Klis 2002; Markwardt, Swank & Taam 1999; Vignarca et al. 2003). This state includes the LS and HIMS: the latter is associated to small accretion disk radii, faster jet ejection, steeper energy spectra and higher characteristic frequencies. In the HIMS, the corona component (see below) starts contributing to the X-ray flux (Zdziarski et al. 2001; Rodriguez et al. 2004).

In the soft states (HS and SIMS), the jet is suppressed (jet radio and X-ray components are not observed). The flux is dominated by the accretion disk component, which now has a higher temperature and a small inner radius, possibly coincident with the innermost stable orbit around the black hole, but an additional steep power-law component, with no evidence of a high-energy cutoff up to $\sim$1 MeV is visible, which we associate here (generically) to a corona. In the power spectra, no strong band-limited noise component is observed, and transient QPOs of type A/B are observed, with frequencies above a few Hz and not much variable.

This picture corresponds to the schematic outburst evolution shown in Fig. 3 (from Fender, Belloni & Gallo 2004). In addition to the top horizontal branch in Figs. 1 and 3, transitions across the jet line, which separates the two major states described above, can be seen also on short time scales. In XTE J1859+226, very fast transitions are observed between type-B QPOs at 6 Hz and type-C QPOs at 8 Hz, i.e. between the SIMS and the HIMS with high characteristic frequencies (Casella et al. 2004). These transitions have been seen at times corresponding to major radio activity (Casella, this volume). This association resembles that seen in GRS 1915+105 (Klein-Wolt et al. 2002; Fender & Belloni 2004). These short-term transitions should be studied in more detail. Although timing analysis of the fast variability of BHBs can give us direct measurements of important parameters of the accretion flow, up to now we do not have unique models that permit this. Recent results show that a clear association can be made between type-C QPO, strong band-limited noise and the presence of a relativistic jet. In the framework of unifying models, these results could play an important role.

REFERENCES

1. T. Belloni, D. Psaltis, and M. van der Klis, *ApJ*, 572, 392 (2002).
2. T. Belloni, et al., *A&A*, submitted (2005).
3. P. Casella, et al., *A&A*, 426, 587 (2004).
4. P. Casella, et al., *A&A*, submitted (2005).
5. P. Cui, et al., *ApJ*, 535, L123 (2000).
6. R. P. Fender, in *Compact Stellar X-ray Sources*, Cambridge Univ. Press, in press (2005).
7. R.P. Fender, et al., *ApJ*, 519, L165 (1999).
9. R.P. Fender, and T. Belloni, *ARA&A*, 42, 317 (2004).
9. R.P. Fender, E. Gallo, and P. Jonker *MNRAS*, 343, L99 (2003).
10. R.P. Fender, T. Belloni, and E. Gallo, *MNRAS*, 355, 1105 (2004).
11. F. Frontera, et al., *ApJ*, 561, 1006 (2001).
12. E. Gallo, R.P. Fender, and G. Pooley, *MNRAS*, 344, 60 (2003).
14. E. Gallo, et al., *MNRAS*, 347, L52 (2004).
14. E. Gallo, R.P. Fender, and R. Hynes, *ApJ*, 611, L125 (2004).
15. J.E.. Grove, et al., *ApJ*, 500, 899 (1998).
16. J. Homan, and T. Belloni, in in*From X-ray Binaries to Quasars: Black Hole Accretion on All Mass Scales*, edited by T.J. Maccarone, R.P. Fender & L.C. Ho, Kluwer, in press (2005).
17. J. Homan, et al., *ApJS*, 132, 377 (2001).
18. J. Homan, et al., *ApJ*, 586, 1262 (2003).
19. J. Homan, et al., *ApJ*, in press (astro-ph/0501371) (2005).
20. M. Klein-Wolt, et al., *MNRAS*, 331, 745 (2002).
21. S. Markoff, et al., *A&A*, 397, 645 (2003).
22. J.E. McClintock, and R.A. Remillard, in *Compact Stellar X-ray Sources*, Cambridge Univ. Press, in press (2005).
23. J.E. McClintock, et al., *ApJ*, 555, 477 (2001).
24. A. Merloni, A.C. Fabian, R.R. Ross, *MNRAS*, 313, 193 (2000).
25. E.H. Morgan, R.A. Remillard, and J. Greiner, *ApJ*, 482, 993 (1997).
26. C.B. Markwardt, J.H. Swank, and R.E. Taam, *ApJ*, 513, L37 (1999).
27. D. Psaltis, T. Belloni, and M. van der Klis, *ApJ*, 520, 262 (1999).
28. R.A. Remillard, et al., *ApJ*, 522, 397, (1999).
29. J. Rodriguez, et al., *ApJ*, 615, 416 (2004).
30. F. Vignarca, et al., *A&A*, 397, 729 (2003).
31. R. Wijnands, and M van der Klis, *ApJ*, 514, 939 (1999).
32. A. Zdziarski, et al., *ApJ*, 554, L48 (2001).

THE DYNAMICAL FINGERPRINT OF INTERMEDIATE MASS BLACK HOLES IN GLOBULAR CLUSTERS

M. Colpi*, B. Devecchi*, M. Mapelli†, A. Patruno* and A. Possenti**

Department of Physics G. Occhialini, University of Milano Bicocca, Piazza della Scienza 3, Milano, I 20126 Italy
†*SISSA/ISAS, Via Beirut 4, Trieste, I 34014 Italy*
***INAF, Cagliari Observatory, Poggio dei Pini, Strada 54, Cagliari, I 09012 Italy*

Abstract. A number of observations hints for the presence of an intermediate mass black hole (IMBH) in the core of three globular clusters: M15 and NGC 6752 in the Milky Way, and G1, in M31. However the existence of these IMBHs is far form being conclusive. In this paper, we review their main formation channels and explore possible observational signs that a single or binary IMBH can imprint on cluster stars. In particular we explore the role played by a binary IMBH in transferring angular momentum and energy to stars flying by.

Keywords: Black holes; pulsars; binaries; globular clusters
PACS: 97.60.Ls; 97.60.Jd; 97.80.Jp

INTRODUCTION

A number of different observations suggest that large black holes (BHs) may exist in nature, with masses between $20M_\odot - 10^4 M_\odot$. Heavier than the stellar-mass BHs born in core-collapse supernovae ($3M_\odot - 20M_\odot$; [22]), these intermediate mass black holes (IMBHs) are expected to form in dense stellar systems through complex dynamical processes. Globular clusters thus become prime sites for their search. Recently, Gebhardt, Rich, & Ho [12] suggested the presence of an IMBH of $2^{+1.4}_{-0.8} \times 10^4 M_\odot$ in the globular cluster G1, in M31, to explain its kinematics and surface brightness profile. Gerssen et al. [13] indicate the presence of an IMBH of $1.7^{+2.7}_{-1.7} \times 10^3 M_\odot$ in the galactic globular cluster M15, based on HST kinematical data. An additional puzzling observation comes from the exploration of the globular cluster NGC 6752 with the discovery of 5 millisecond pulsars (MSPs) showing unusual accelerations or locations [5].

NGC 6752 hosts in its core two MSPs (PSR-B and PSR-E) with very high negative spin derivatives that, once ascribed to the overall effect of the cluster potential well, indicate the presence of $\sim 1000 M_\odot$ of under-luminous matter enclosed within the central 0.08pc [9]. NGC 6752 in addition hosts two MSPs with unusual locations: PSR-A, a canonical binary pulsar with a white dwarf companion [5, 10], holds the record of being the farthest MSP ever observed in a globular cluster, at a distance of ≈ 3.3 half mass radii, and PSR-C, an isolated MSP that ranks second in the list of the most offset pulsars known, being at a distance of 1.4 half mass radii from the gravitational center of the cluster [5]. Colpi, Possenti & Gualandris [4] first conjectured that PSR-A was propelled into the halo in a flyby between the binary MSP and a binary stellar-mass BH or a binary IMBH present in the core of NGC 6752. Colpi, Mapelli & Possenti (CMP03 hereon; [3]) proposed later that the position of PSR-A could also be explained as an ejection following a dynamical encounter of a non-recycled neutron star in a binary, by a single IMBH, prompted by the evidence of under-luminous matter in the core of NGC 6752 [9]. The interaction considered was a flyby between the binary pulsar PSR-A and the IMBH having within its sphere of influence a cusp star bound on a Keplerian orbit. CMP03 carried on an extensive analysis of binary-binary encounters with IMBHs, single or in binaries, to asses the viability of their scenario, indicating that IMBHs are best targets for imprinting the necessary thrust to PSR-A at a rate compatible with its persistence in the halo against dynamical friction. Ejection of PSR-A from the core to the halo following exchange interactions off binary stars can not be excluded, but as pointed out by Colpi et al. [4], the binary parameters of PSR-A and its evolution make this possibility remote, and call for fine tuning conditions on binary evolution.

All three these observations, hinting for an IMBH interpretation of the data, are far from being conclusive as regard to the nature of their dark component. Numerical studies by Baumgardt et al. [1] have in fact shown that kinematical features observed in G1 and M15 can be explained if dark low-mass remnants reside in their cores, without need of an exotic IMBH. Also for NGC 6752, the underluminous matter found can be associated to a cluster of compact stars [9]. Thus, other signs of an IMBH should be explored in order to asses the reliability of such interpretations.

On theoretical ground the existence of IMBHs in globular clusters, single or in binaries, has been advanced by several authors (see van der Marel for a review [29]), but the difficulty in finding a clear formation pathway remains. Recently, Portegies Zwart et al. [26, 24] suggested that IMBHs find their formation channel in young star clusters sufficiently dense to become vulnerable to unstable mass segregation [25, 11]. Through the runaway collision of a single heavy star off other stars, a giant stellar object is expected to form that collapses into an IMBH. If this holds true in globular clusters at the time of their formation, there is freedom to believe that gas-dynamical processes, as such suspected to occur in young metal rich star clusters, were at work early in the cluster lifetime.

It has also been speculated that IMBHs in globular clusters may form, alternatively, through binary-single or binary-binary gravitational encounters and mergers among light BHs during a far more advanced stage of cluster evolution [20]. In clusters a few billions years old, the heaviest stars are stellar compact remnants, i.e., neutron stars and black holes. Despite neutron stars form in larger numbers (for any reasonable initial mass function), BHs likely outnumber neutron stars, since they experience weaker natal kicks (due to their larger inertia) and are thus easily retained inside the cluster. The end-result of stellar and dynamical evolution is a dense core of stellar-mass BHs, some bound to stars in binaries. Sinking further by unstable mass segregation these BHs decouple dynamically from the system and get caught in binaries through exchange interactions among BHs [16, 28]. These binaries, in the high density environment of a mass-segregated core, experience frequent interactions, initiating a process of hardening that may proceed until gravitational wave emission drives the evolution of the binaries toward coalescence into a single more massive BH. The process may repeat leading to a larger IMBH [20].

The hardening of binaries via gravitational encounters can however find sudden halt if the BH binary is light enough to experience ejection: since the interactions that produce hardening produce recoil, a sizable fraction of BHs can be ejected so that the core of BH remnants evaporates and dissolves [16, 28, 27]. Single as well as binary BHs leave the cluster, emptying the core of all its BHs but a small number. Thus, it is from a delicate balance between hardening and recoil that sequences of encounters among BHs can drive the core into a state with no BHs or with a few, bound preferentially in binaries. Miller & Hamilton [20] and CMP03 showed, from simple considerations, that a minimum initial seed mass is required (around $50M_\odot$) for the BH to remain in the cluster and grow up to several hundred solar masses, through hardening and coalescence. But a closer and more detailed inspection of sequences of binary-single scattering events by Gultekin, Miller & Hamilton [15] have revealed that in order to avoid ejection, a larger seed BH should exists in situ of $\sim 300M_\odot$, at the onset of dynamical evolution to remain safely inside the cluster and grow further in mass, avoiding ejection. There might be also the possibility that BHs propelled away from the core remain bound to the cluster living in the halo. These BHs may return back by dynamical friction after almost all the central BHs have been expelled by recoil. Since there is no unique outcome for the fate of BHs in globular clusters, various scenarios remain open for investigation. A globular cluster may host:

- a *single* IMBH, with mass $> 300M_\odot$ formed in situ following a runaway merger among heavy stars, at the time of cluster formation. This IMBH may subsequently grow up to $10^3 - 10^4 M_\odot$ or more, by dynamical process, on the core relaxation time. This IMBH may capture stars via relaxation processes or gravitational interactions and be surrounded by a small cusp of stars.
- a *binary* IMBH of mass $50M_\odot - 300M_\odot$ with a BH companion of similar mass or lighter then $10M_\odot$. This binary may form dynamically via exchange interactions and mergers among BHs. The large cross section that such a binary has implies that it is relatively shortlived since close encounters with stars can cause its hardening up to coalescence.
- a *stellar-mass binary* BH can be present composed of two BHs, relic of the most massive stars, perhaps ejected in the halo, that returned back to the core by dynamical friction after BH core evaporation.

SINGLE INTERMEDIATE MASS BLACK HOLES

If a single IMBH of mass $\gtrsim 10^3 M_\odot$ is present in a globular cluster, it can influence stars passing by in various ways. A case of interest occurs when the IMBH is not strictly single, but is surrounded by a swarm of stars, i.e., a small cusp. This cusp is likely to be unstable to ejections, captures and relaxation processes. It is however not implausible that at least a tightly bound star is present. Given the high fraction of binary stars in the core of globular clusters that form dynamically, the interaction of these binaries with the IMBH may bring to three potential observational signatures:

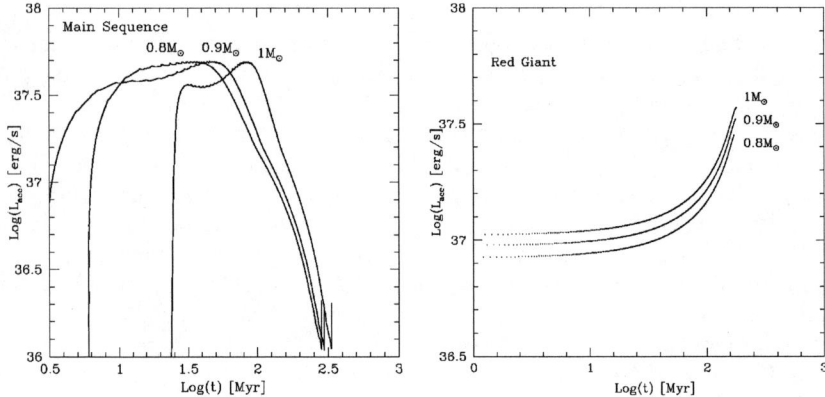

FIGURE 1. Luminosity versus time from accretion onto an IMBH of $1000M_\odot$. The donor is a light star of 0.8, 0.9 and $1M_\odot$, respectively. Left (right) panel refers to accretion when the donor is on the main sequence (red giant phase).

(i) *Flybys* involving the IMBH and its cusp star. As discussed in CMP03, a flyby may explain the ejection from the core into the cluster halo of binary stars, such as PSR-A.
(ii) *Ionization of the binaries*, either interacting with the IMBH and its cusp star (CMP03) or experiencing the tidal field of the IMBH itself [23]. As discussed recently by Pfahl [23], ionization by the tidal field can bring one binary component into a bound-disruption orbit, releasing the other on an hyperbolic orbit. The IMBH in this case may reveal its presence flaring in X-ray when swallowing the tidal debris of the disrupted star every $\sim$1-10 Myr, depending on the details of the capture process.
(iii) *Accretion*. Binary-single or binary-binary encounters with the IMBH can deliver a star on a close orbit around the IMBH. After circularization, a phase of mass transfer can initiate, similarly to what happen in X-ray binaries hosting a stellar-mass BH.

Accretion onto an IMBH

We have explored accretion onto an IMBH considering a low mass donor star, evolving along the main sequence and red giant branch. The donor star is modeled using an updated version of the evolutionary code of Eggleton [7] and of Webbink, Rappaport & Savonije [30]. Figure 1 shows the run of the luminosity versus time: mass transfer via Roche lobe overflow leads, in both cases, to luminosities $\gtrsim 10^{37}$ erg s^{-1}. These correspond to mean accretion rates that are low enough to fulfill the condition for variable mass transfer in a thermal-viscous unstable thin disk [6]. Thus, we find that IMBHs accreting from low mass donors should undergo limit-cycle behavior and appear as transient X-ray sources. The signs that would distinguish an IMBH from a stellar-mass BH in a low-mass X-ray binary should thus be searched in the spectral and timing properties: a softer black body component and longer timescale variabilities may be the distinguish-

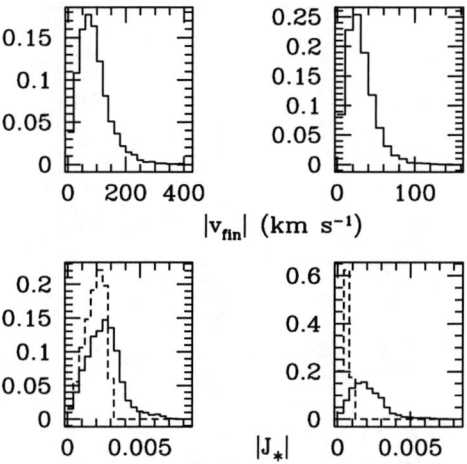

FIGURE 2. Post-encounter velocity and angular distribution (in modulus) of cluster stars scattering off a binary IMBH of mass $[100M_\odot, 50M_\odot]$, separation of 10 AU (left panels) and 100 AU (right panel), and eccentricity $e = 0.7$.

ing features [19]. We can not exclude a priori that bright X-ray sources seen in ellipticals, having globular clusters as optical counterparts, host accreting IMBHs [8, 21].

BINARY INTERMEDIATE MASS BLACK HOLES

A binary IMBH can imprint large recoil velocities to stars flying by. When not too hard and massive, the binary IMBH can also transfer angular momentum to the stars, perturbing them away from dynamical equilibrium. Whether this effect influences only few stars or a sizable number is still unexplored and under our current study (Mapelli et al. 2005 in preparation). We here report preliminary results obtained running single-binary encounters between a binary IMBH and cluster stars.

Supra-thermal stars and angular momentum alignment?

Our aim is to address the following questions: (i) Are stars heated to supra-thermal energies, i.e. to energies in excess of their dispersion values without escaping from the globular clusters? (ii) Is there direct transfer of angular momentum from the binary IMBH to the stars? (iii) Do we observe alignment of the stellar orbit in the direction of the angular momentum of the binary?

Figure 2 shows the post-encounter velocity and angular momentum distributions (in modulus) of stars that impinge on a binary IMBH of mass $[100M_\odot, 50M_\odot]$, and orbital separation a of 10,100 AU, respectively. The hard (softer) binary, with $a = 10$

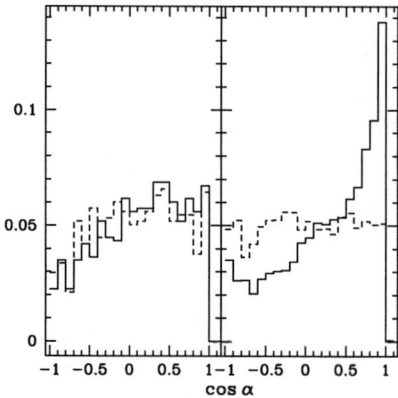

FIGURE 3. The distribution of the angle α between the binary IMBH angular momentum and the orbital angular momentum of the incoming star. The binary IMBH is as in Figure 2. Solid (dashed) line indicates the post-encounter (pre-encounter) distributions. Note that there is an overabundance of corotating stars when the binary IMBH has $a = 100$ AU (right panel).

AU (100 AU), tends to produce stars with velocities above (below or closer to) the escape speed ($\simeq 40\,\mathrm{km\,s^{-1}}$). Respectively, 15% and 80% of the stars are scattered to supra-thermal energies. Angular momentum is exchanged when the binary IMBH has a higher orbital angular momentum relative to that of the incoming stars. The right panel of Figure 3 shows the most favorable case of angular momentum alignment (for the binary IMBH with $a = 100$ AU): we find that alignment involves a sizable fraction of stars ($\sim 70\%$), so introducing an anisotropy in their equilibrium energy and angular momentum distributions. Supra-thermal stars are also those absorbing the largest angular momentum.

A binary IMBH tightens rapidly via binary-single flybys and the bulk of the bound supra-thermal stars with excess angular momentum are produced over a time comparable to the IMBH hardening life, typically of $\sim 10^7 - 10^8$ yrs for a cluster such as NGC 6752. Propelled into the halo, these stars mainly return to equilibrium within a few core radii after a time comparable to the half mass relaxation time $\sim 10^9$ yrs. Thus, their signature may last longer than the hardening process of the binary IMBH, but shorter than the cluster lifetime [1].

Convolving the statistical results of our simulations with projection effects, we find that few hundreds stars should display signs of supra-thermal motion via Doppler line shift or proper motion. Considering that only a fraction of these stars will be detectable, the remaining being white dwarfs, neutron stars or faint stars, a statistically significant identification of such non-equilibrium stars seems to be very difficult. Angular momen-

---

[1] The binary IMBH keeps hardening at a lower peace when the separation falls below one astronomical unit: when in this regime, stars scattering off the binary IMBH leave the cluster being ejected with velocities far above the escape speed.

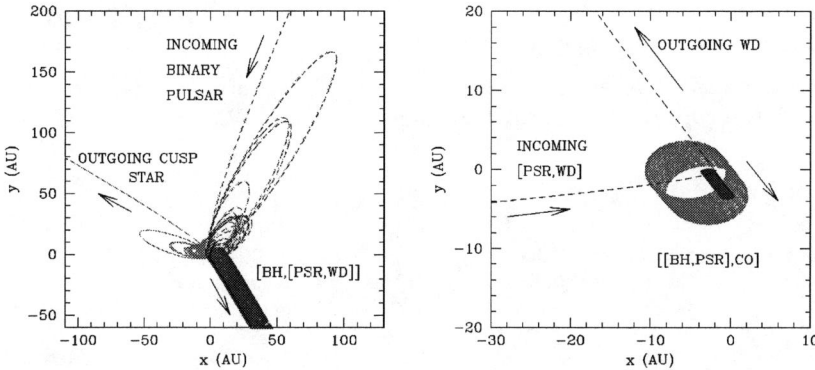

FIGURE 4. A stable [17] triplet (left panel) resulting from the exchange of the cusp star of $1M_\odot$ revolving around a $100M_\odot$ IMBH with a binary MSP (as PSR-A in NGC 6752), and a triplet (right panel) resulting from the ionization of the binary MSP impinging onto a $1000M_\odot$ IMBH and its cusp star. The coalescence timescale due to emission of gravitational waves for the inner binary in the first triplet is 1.7×10^{11}yr, while the BH-PSR binary in the second system should coalesce in 6.4×10^7yr.

tum alignment requires a rather massive binary composed of two large BHs, for transfer to be effective. Thus, angular momentum alignment, induced by a binary IMBH, may remain visible over a time comparable to the cluster lifetime only if a new binary IMBH form via dynamical processes involving other BHs after every coalescence of the progenitor binary. This would generate families of stars with different orientation angles, and characteristic lifetimes, each for every binary that has formed.

Given the perturbative action that an IMBH has on stars, studies on the overall globular cluster evolution as those from Baumgardt et al. [2] can shed further light into the equilibrium properties that a cluster with an IMBH displays.

Millisecond pulsars around IMBHs

As discuss in CMP03, binary IMBH are catalysts for the formation of triplets, resulting from binary-binary interactions. The encounter of a binary pulsar with a binary IMBH can create an extraordinary system: a millisecond pulsar-IMBH-BH triplet (see CMP03) or a star-pulsar-IMBH hierarchical system. Here we focus on an IMBH with a cusp star. Figure 4 shows the formation of a stable triplet through an exchange (left) and ionization of a binary MSP (right).

We are now performing an extensive analysis of binary-single and binary-binary encounters with single and binary IMBH to determine the rate of formation, destruction and coalescence of these systems. The dynamical capture often deliver the captured pulsar (or compact object in general) on a rather tight and eccentric orbit: by looking at the distributions of separation and eccentricities of the triplets found we will be able to determine also the rate of neutron star coalescence by gravitational waves onto the

IMBH, in our nearby universe. These events could have a profound impact for the gravitational waves astronomy.

Acknowledgments We are grateful to Stein Sigurdsson for enlightening discussion. This work was supported in part by the grant PRIN03-MIUR.

REFERENCES

1. Baumgardt, H., Makino, J., McMillan, S. & Portegies Zwart, S. 2003, ApJ, 582, L21
2. Baumgardt, H., Makino, J.& Hut, P. 2004, astro-ph/0410597, to appear in ApJ
3. Colpi, M., Mapelli, M. & Possenti, A. 2003, ApJ, 599, 1260
4. Colpi, M., Possenti, A. & Gualandris, A. 2002, ApJL, 570, 85
5. D'Amico, N., Possenti, A., Fici, L., Manchester, R. N., Lyne, A. G., Camilo, F. & Sarkissian, J. 2002, ApJL, 570, 89
6. Dubus, G., Lasota, J., Hameury, J. & Charles, P. 1999, MNRAS, 303, 139
7. Eggleton, P. P. 1971, MNRAS, 151, 351
8. Fabbiano, G. & White, N. E. 2004, astro-ph/0307077, to appear in "Compact Stellar X-ray Sources" book in preparation for Cambridge University Press (ed s., W. Lewin & M. van der Klis)
9. Ferraro, F. R., Possenti, A., Sabbi, E. & D'Amico, N. 2003, ApJL, 596, 211
10. Ferraro, F. R., Possenti, A., Sabbi, E., Lagani, P., Rood, R. T., D'Amico, N. & Origlia, L. 2003, ApJ, 595, 179
11. Fraitag, M., Rasio, F. A. & Baumgardt, H. 2005, astro-ph/0503129
12. Gebhardt, K., Rich, R. M. & Ho, L. C. 2002, ApJL, 578, 41
13. Gerssen, J., van der Marel, R. P., Gebhardt, K., Guhathakurta, P., Peterson, R. C. & Pryor, C. 2003, AJ, 125, 376
14. Gerssen, J., van der Marel, R. P., Gebhardt, K., Guhathakurta, P., Peterson, R. C. & Pryor, C. 2002, AJ, 124, 3270
15. Gultekin, K. Miller, M. C. & Hamilton, D. P. 2004, astro-ph/0402532, submitted to ApJ
16. Kulkarni, S. R., Hut, P. & McMillan, S. 1993, Nature, 364, 421
17. Mardling, R. A. & Aarseth, S. J. 2001, MNRAS, 330, 232
18. Miller, M. C. 2003, astro-ph/0306173
19. Miller, J. M., Fabian, A. C. & Miller, M. C. 2004, ApJ, 607, 931
20. Miller, M. C. & Hamilton, D. P. 2002, MNRAS, 330, 232
21. Mushotzky, R. 2004, astro-ph/0411040, Prog.Theor.Phys.Suppl. 155, 27-44
22. Orosz, J. A. 2002, astro-ph/0209041
23. Pfahl, E. 2005, astro-ph/0501326, submitted to ApJL
24. Portegies Zwart, S. F. 2004, astro-ph/0406550, Lecture note to appear in "Joint Evolution of Black Holes and Galaxies" of the Series in High Energy Physics, Cosmology and Gravitation. IOP Publishing, Bristol and Philadelphia, 2005, eds M. Colpi, V. Gorini, F. Haardt and U. Moschella.
25. Portegies Zwart, S. F., Baumgardt, H., Hut, P., Makino, J. & McMillan, S. L. W. 2004, Nature, 428, 724
26. Portegies Zwart, S. F. & McMillan, S. L. W. 2002, ApJ, 576, 899
27. Portegies Zwart, S. F. & McMillan, S. L. W. 2000, ApJL, 528, 17
28. Sigurdsson, S. & Hernquist, L. 1993, Nature, 364, 423
29. van der Marel, R. P. 2004, Coevolution of Black Holes and Galaxies, from the Carnegie Observatories Centennial Symposia. Published by Cambridge University Press, as part of the Carnegie Observatories Astrophysics Series. Edited by L. C. Ho
30. Webbink, R. F., Rappaport, S. & Svonije, G. J. 1983, ApJ, 270, 678

Accretion-Ejection Instability, LFQPO and lag structure of microquasars

P. Varnière

Department of Physics & Astronomy, Rochester University, Rochester NY 14627-0171

Abstract. The Accretion-Ejection Instability (AEI) has been proposed to explain the low frequency Quasi-Periodic Oscillation (QPO) observed in low-mass X-Ray Binaries. Its frequency, typically a fraction of the Keplerian frequency at the disk inner radius, is in the right range indicated by observations. I will briefly review the basic of the instability. In a second part I will focus on a simple model to explain the complex lag structure between the hard and soft X-ray observed in microquasars such as GRS 1915+105.

Keywords: xray binaries – observation
PACS: 95.75.-z, 95.85.Nv, 97.80.Jp

INTRODUCTION: A BRIEF PRESENTATION OF THE AEI AND RELATION TO OBSERVATION

The Accretion-Ejection Instability ([8]) is a spiral instability similar to the galactic spiral but driven by magnetic stress instead of self-gravity. This instability affects the inner region of the disk when the plasma $\beta = 8\pi p/B^2$ is of the order of one, i.e. there is equipartition between the gas and magnetic pressure. It forms a quasi-steady spiral pattern rotating in the disk at a frequency of the order of a few times the orbital frequency at the inner edge of the disk. This spiral density wave is coupled with a Rossby vortex that it creates at the corotation radii (corotation between the spiral wave and the gas in the disks). This coupling permits the energy and angular momentum to be stored at the corotation radius, allowing accretion to proceed in the inner part of the disk. Contrary to the magnetic turbulence (such as the Magneto Rotational Instability) based accretion models, the AEI does not heat-up the disk as the energy is transported by waves and not deposited locally. The Rossby vortex twists the foot points of the field lines. In the presence of a low density corona this torsion will propagate as an Alfven wave transporting energy and angular momentum stored in the vortex. This will put energy into the corona where it might power a wind or a jet ([10]).

In order to apply the AEI to the phenomena occurring in microquasars we have used two observables: the presence of jet and the characteristic of the Quasi-Periodic Oscillation (QPO). We have already compared some of the properties of the AEI with observations, mainly the relation between the inner radius of the disk from spectral fit and the QPO frequency ([7, 9]). Moreover, a proof of principle that the AEI is able to modulate the flux from the disk has been shown in [11].

THE COMPLEX LAG STRUCTURE

RXTE has provided us with a better picture of the temporal behavior of X-ray binaries. Using such techniques as Fourier Transform (FT), lag and coherence computation, we obtain an even better picture of the phenomena. The lag between the low-energy (2 - 5 keV) and the high-energy (5 - 20 keV) is generally associated with Inverse-Compton of soft photons producing hard photons. Surprisingly the lags appear to change sign during an observation and also between observations. This is inconsistent with the Inverse-Compton explanation. In GRS 1915+105 (e.g. [2, 4]) and XTE J1550-564 (e.g. [14, 3]), this unusual behavior of the time lag has been reported during outburst and/or the radio-loud state. Namely, the sign of the QPO's time lag changed over a single observation whereas the sign of its first harmonics' time lag stayed the same.

The next section shows simultaneous radio and X-ray data from GRS 1915+105 taken from the plateau/hard-steady χ state ([1, 6]). We use this data to gain insight into the relation between radio/jet and the X-ray timing properties of the system. In the following section we show the behavior of the FT in the case of an absorbed sinusoid. In the last section we make use of this simple, zeroth order, model to explain the complex lag structure observed in GRS 1915+105 and XTE J1550-564 and see what we can infer about those systems.

THE CASE OF GRS 1915+105

[6] studied the hard state (χ state in the classifications by [1], or radio plateau) using simultaneous X-ray and radio observations. Figures 7 and 8 of [6] shows the different temporal behavior of the source depending on the radio loudness. Figure 8 shows how the temporal properties (QPO frequency and phase lag at the QPO frequency) correlate with the different components of the X-ray flux, namely the total flux, the thermal/disk flux and the power-law flux. By looking carefully at the plots, two populations can be distinguished (the triangle and the cross). This distinction is more apparent in the graph showing the lag. On this figure we see that for a QPO frequency higher than about two hertz the QPO frequency appears to be correlated with the total flux and the power-law flux (which in fact dominates the total flux). This applies for most of the low-mass X-ray binaries. For a QPO frequency lower than 2Hz, this QPO frequency no longer correlates with any of the X-ray fluxes. In fact all of the frequencies below 2 Hz appear at a similar flux level for both the thermal and power-law flux, i.e. the cluster of triangles is very narrow. These points also represent the ones with a high radio flux (the triangles represent the radio-loud state) as is seen in fig 1. These radio-loud points are also the only ones to exhibit a positive phase lag. Concerning this lag, there is also another difference besides the change of sign between the radio-loud and the radio-quiet state. It also seems that there is no correlation between the lag and any of the X-ray fluxes. But, depending on wherever the source is radio-loud or radio-quiet, the "clusters" of points appear to be perpendicular to each other.

In the radio-loud case, the temporal behavior of the source is modified for quasi constant X-ray fluxes. These modifications are a function of the radio flux. Figure 7 of [6] show the evolution of the temporal properties such as the QPO frequency, the phase lag, the coherence and the ratio of low-frequency power as a function of the radio flux at 15.2 GHz. Once again the radio-loud and radio-quiet points are well separated. The separation occurs at a radio flux of about 60 mJy. By looking in more detail at the first plot (QPO frequency - radio flux) we see that a QPO frequency less than two hertz is always associated with a radio flux of more than 60 mJy. These same QPOs have a positive phase lag and show much less coherence than the QPOs in the radio quiet state. Moreover, the phase lag which seems totally uncorrelated with the radio flux when it is less than 60 mJy, appears to be correlated with the higher radio fluxes. In the graph of the ratio of low-frequency power as function of the radio flux the possible correlation seems to reverse during the transition between radio-quiet and radio-loud.

Either these QPOs (less than 2Hz, more than 2Hz) arise from a different mechanism (e.g. one related with the jet and the other one not) or there is a threshold in radio flux above which new phenomena appear in addition to the QPO mechanism. This could cause a modification of the temporal behavior of the source, especially relevant to the lag which seems to become proportional to the radio flux. We will focus on this last possibility. The presence of two different unrelated mechanisms seems highly improbable because of the smooth transition in QPO properties as function of time (see for example figure 6 of [6]). However, before exploring the possible origin for the change in temporal properties we will look at the lag definition and its computation through FTs.

FOURIER TRANSFORM AND PHASE LAG

Definition of lag and coherence

We will briefly go over the definition of the lag as presented by [13]. Suppose that $x_1(t_k)$ and $x_2(t_k)$ represent the X-ray flux in two energy bands (soft and hard) at time t_k. We note $X_1(v_j)$ and $X_2(v_j)$ as their FTs at the frequency v_j:

$$X(v_j) = \frac{1}{2\pi} \int x(t_k) e^{-iv_j t_k} dt_k \qquad (1)$$

The time lag between the hard and soft X-ray is then defined as:

$$\delta t(v_j) = \frac{1}{2\pi v_j} \times arg(X_1^\star(v_j) X_2(v_j)) = \frac{arg(X_2) - arg(X_1)}{2\pi v_j} \qquad (2)$$

Modification of one of the two phases (lowering the phase of the soft band or increasing the phase of the hard band) could induce the lag to change sign. More generally, any change in the phase of one of the bands, caused by internal or external phenomena, could lead to a sign change in the lag.

The coherence is a measure of how much of a signal f can be predicted knowing a signal h. In our case it means how much of the high energy flux can be predicted knowing the low energy flux. If the two signals are related then the coherence is high; the maximum equals one, which corresponds to the case where there is a linear transformation to go from one to the other.

Lag: a simple derivation

One can reproduce the observed behavior of the lag and harmonics using simple assumptions about the initial profile. The idea is to compare the Fourier representation of an initial profile (here a constant plus a cosine) taken to be the hard X-ray, to a modified profile taken to be the soft X-ray. We will then compute the lag between them and show a simple way to match the observed lag behavior.

If we compute the FT of a sinusoid function we obtain the frequency, amplitude and phase. In order to make it similar to data we take the sinusoidal profile superimposed on a constant background and add a small amount of random noise to it. By using the FT we still find the frequency, amplitude and phase. Now take into account the case where some part of the modulated emission does not arrive to the observer but a part of it is "absorbed/obscured" by a media located in the system. This would make a profile similar to the one of figure 15 of [5]. We computed the FT using an input profile of unity plus a sinusoid with rms amplitude $rms = 0.14$ minus a Gaussian profile of amplitude γ centered to reproduce a profile like the one from the figure 15 of [5][1].

By doing the FT on this signal we obtain different parameters for the sinusoid. Depending on the amount of "absorption" we can obtain a smaller value for the amplitude but the striking feature is the effect on the phase. We observe a change in the phase. Moreover, the sign of the phase difference is not the same for the fundamental and its first harmonics. If we take the formula for phase lag and say that only the low energy/soft X-rays are absorbed and not the hard one, we can compute the phase lag which appears as a consequence of the absorption of only part of the signal. Doing so reproduces the observed phase characteristics: a different sign for the fundamental and first harmonics. In addition, if the absorption is turned on, it creates a change in the sign of the lag. This comes from the fact that the FT adjusts the data with a shifted sinusoid, creating a phase difference. We propose that this is the origin of the changing sign of the lag presented in the previous section. This will also decrease the coherence between the two bands as a new signal is added to only one band. This happens without changing the primary physical phenomena that produce the emission in the two bands. The above results can be easily illustrated even using two sinusoidal signals with a $\pi/2$ phase between them:

$$\cos\theta + \varepsilon \sin\theta = \frac{1}{\cos\phi} \cos(\theta - \phi), \tan\phi = \varepsilon \quad (3)$$

The presence of a second, small, sinusoidal signal with a phase lag of $\pi/2$ and an amplitude ε is enough to create an "apparent" phase lag of $\phi = \mathrm{atan}(\varepsilon)$, which is about ε the amplitude of the perturbation.

If we now take into account the presence of a small harmonic to the QPO and compare the result from the FT to that with the the same signal absorbed, the effect on the phase is even more striking. Indeed the induced phase lag between the real data and the absorbed

---

[1] In the first step of this work we searched for which type of profiles are able to reproduce the lag structure. In a second step we tried to find similar profiles in observations. The paper [5] has this "absorbed"-like profile we use as an example here.

one does not have the same sign at the fundamental vs. the first harmonics. This could be at the origin of the observed phenomena.

APPLICATION TO MICROQUASAR OBSERVATIONS

Using the above argument it appears that the use of an FT can lead to an incorrect interpretation of the lag in the presence of an absorption which depend on the energy band. To use this idea for the observations of GRS 1915+105 presented in [6], we need to find what may produce the "absorbed" part of the QPO modulation. This has to be related to the jet, either to have the same origin, or be a consequence of it. In the following we will assume that the QPO modulation is created by a hot spiral, for example [12], and we are just interested in further absorption/modulation of this already existing modulation.

Suppose that the basis of the jet/corona gets "between" the observer and the spiral during one orbit of the spiral in the disk. This is enough to "absorb" a part of the flux modulation, especially if it happens when the spiral is "behind" the black hole and therefore near the maximum of the modulation. This simple model is able to explain both the occurrence of changing sign lag and its relation with the jet. In the same way it can also explain the fact that absorption is energy dependent, which makes the coherence drop. In fact, anything located inside the inner radius of the disk that can absorb a small part of the flux coming from the hot spiral could explain the changing sign of the lag and the complex behavior of the harmonics. But this needs to be related to the radio flux and therefore to the jet mechanism.

The first way to check this idea is to look at the QPO profile and see if there is an energy dependent departure from a sinusoidal signal. [5] show for the low-frequency QPO that there is indeed a departure from a sinusoid, which seems compatible with an absorption feature. This kind of analysis is difficult and rarely done for QPOs because of the lack of photons at these timsescale. Another way to check the same properties is to see how the value of the lag depends on the energy band chosen. Using the idea of an energy dependent absorption we see that the negative lag will be more important between the lower energy band (say, $2 - 4$keV) and the highest possible band available than between two high energy bands. It seems possible to have a change of the sign of the lag if we look to high enough energies (for example using INTEGRAL data).

This simple model can also be used with observational data to gain insight into the geometry near the black hole. The pulse shape of the QPO in different energy bands can allow us to constrain the relative geometry of the absorption region with respect to the emitting region (QPO origin), and also the column density of the absorber. We will test several mechanisms that could lead to this "absorbed-like" profile and compare them with observational data.

CONCLUSIONS

This letter shows how absorption can modify the X-ray signal and give rise to an apparent change in the phase lag between the hard and soft photons. The model is phenomenological, and future simulation work is needed to yield more quantitative predictions that can be compared with observational data and thereby giving access to the geometry in the inner part of the disk. Indeed, with numerical simulation we intend to probe the relative geometry of the QPO emission region with respect to the absorbing media by using the shape of the QPO pulse. The use of RXTE and INTEGRAL data together with numerical simulations of the absorption of a "hot-spot" orbiting in the disk will further test this idea.

ACKNOWLEDGMENTS

PV is supported by NSF grants AST-9702484, AST-0098442, NASA grant NAG5-8428, HST grant, DOE grant DE-FG02-00ER54600, the Laboratory for Laser Energetics and the french GDR PCHE.

PV thanks Michel Tagger, Eric Blackman, Jason Maron, Jerome Rodriguez and Mike Muno for all the discussions, helpful comments on the paper and data.

REFERENCES

1. Belloni, T., Klein-Wolt, M., Méndez, M. & van der Klis, M.; van Paradijs, J. 2000, A&A, 355, 271
2. Cui, W. 1999, ApJ, 524, 59
3. Cui, W., Zhang, S.N. & Chen, W. 2000, ApJ, 531, 45
4. Lin, D., Smith, I.A., Liang, E.P. & Bottcher, M. 2000, ApJ, 543, 141
5. Morgan, E.H., Remillard, R.A. & Greiner, J. 1997, ApJ, 482, 993
6. Muno, M., Remillard, R., Morgan, E., Waltman, E., Dhawan, V., Hjellming, R., Pooley, G. 2001, ApJ, **556**, 515
7. Rodriguez, J., Varnière, P., Tagger. M., and Durouchoux, P., 2002, A&A, **387**, 487-496.
8. Tagger, M., and Pellat, R., 1999, A&A, **349** 1003 (**TP99**)
9. Varnière, P., Rodriguez, J. and Tagger. M., 2002, A&A, **387**, 497-506.
10. Varnière, P. and Tagger. M., 2002, A&A, **394**, 329-338.
11. Varnière, P., Muno, M., M. Tagger & A. Frank, proceeding of French society of Astronomy and Astrophysics, 2003.
12. Varnière, P., Muno, M. & Blackman, E., in preparation.
13. Vaughan, B.A. & Nowak, M.A., 1997, ApJ, **474**, L43-46.
14. Wijnands,R., Homan, J. & van der Klis, M., 1999, ApJ, **526**, L33-36.

Black hole hunting in the Andromeda Galaxy

R. Barnard*, J. P. Osborne†, U. Kolb* and C. A. Haswell*

The Open University, Milton Keynes, MK7 6AA, UK
†The University of Leicester, Leicester, LE1 7RH, UK

Abstract. We present a new technique for identifying stellar mass black holes in low mass X-ray binaries (LMXBs), and apply it to XMM-Newton observations of M31. We examine X-ray time series variability seeking power density spectra (PDS) typical of LMXBs accreting at a low accretion rate (which we refer to as Type A PDS); these are very similar for black hole and neutron star LMXBs. Galactic neutron star LMXBs exhibit Type A PDS at low luminosities (10^{36}–10^{37} erg/s) while black hole LMXBs can exhibit them at luminosities $>10^{38}$ erg s^{-1}. We propose that Type A PDS are confined to luminosities below a critical fraction of the Eddington limit, l_c that is constant for all LMXBs; we have examined a sample of black hole and neutron star LMXBs and find they are all consistent with $l_c \sim 0.1$ in the 0.3–10 keV band. We present luminosity and PDS data from 167 observations of X-ray binaries in M31 that provide strong support for our hypothesis. Since the theoretical maximum mass for a neutron star is 3.1 M$_\odot$, we therefore assert that any LMXB that exhibits a Type A PDS at a 0.3–10 keV luminosity greater than 4×10^{37} erg s^{-1} is likely to contain a black hole primary.

Keywords: X-rays: general — X-rays: binaries — Galaxies: individual: M31 — black hole physics — Methods: data analysis
PACS: 95.85.Nv, 97.10.Gz, 97.10.Ri, 97.60.Lf, 97.80.Jp, 98.56.Ne

INTRODUCTION

Van der Klis [1] showed that the power density spectra (PDS) of low mass X-ray binaries (LMXBs) with neutron star primaries at low accretion rates are strikingly similar to those of black hole LMXBs in their low accretion rate states. The fractional rms variability is high ($\sim$30–50%) and the PDS are well described by a broken power law that changes in spectral index, γ, from $\sim$0 to $\sim$1 at frequencies higher than a certain break frequency; the break occurs at 0.01–1 Hz. We will refer to these as Type A PDS. At higher accretion rates, the rms variability is only a few percent, and the PDS are characterised by a power law with $\gamma \sim$1–1.5. We will refer to these as Type B PDS. Van der Klis (1994) proposed that the transition between Type A and Type B PDS occurs when the accretion rate exceeds a critical fraction of the Eddington limit (f_c) that is constant for all LMXBs. He suggested $f_c \sim$1%, as an order of magnitude estimate.

We propose here a new diagnostic for identifying stellar-mass black holes in LMXBs. Since we cannot observe the accretion rate, $\dot{m}$, directly, we must use the lumosity to trace the evolution of the PDS with $\dot{m}$. We define l as L/L_{Edd}, where L is the luminosity and L_{Edd} is the Eddington limit, and $l_c = L_c/L_{\text{Edd}}$ so that Type A PDS are exhibited by LMXBs when $l < l_c$ and Type B PDS are exhibited when $l > l_c$. Throughout this work we use the Eddington limit for uniform spherical accretion of hydrogen. However, the picture is obviously more complicated. Nevertheless, if the accretion discs of black hole and neutron star LMXBs are generally similar, then l_c should be similar also.

Initially, we assume that l_c is constant for all LMXBs, allowing black hole LMXBs to exhibit Type A PDS at considerably higher luminosities than neutron star LMXBs. The Eddington limit for hydrogen rich gas is $\sim 1.3\times 10^{38}$ $(M_1/M_\odot)$ erg s^{-1}, where M_1 is the mass of the primary; hence the maximum luminosity for a neutron star exhibiting a Type A PDS, L_c^{NS} is determined by the maximum mass of a neutron star. This allows us to classify the primaries of LMXBs that exhibit Type A PDS at luminosities higher than L_c^{NS} as black hole candidates. We will assume a maximum neutron star mass of 3.1 $M_\odot$, as is generally accepted.

Here, we use XMM-Newton observations of globular cluster X-ray sources in M31 and results from the literature on Galactic LMXBs to find an empirical value of l_c. We then apply our technique to four XMM-Newton observations of the 63 brightest X-ray sources in the central region of M31, before focusing on RX J0042.4+4112, an X-ray source in the vicinity of M31 that appears to exhibit both Type A and Type B PDS, allowing us to estimate its mass.

OBTAINING AN EMPIRICAL VALUE FOR L_C IN THE 0.3–10 KEV BAND

In practise, we do not observe the full bolometric luminosity of an X-ray source and can only obtain its luminosity in a given energy band. We have been using the XMM-Newton and Chandra X-ray observatories to identify black hole LMXBs in external galaxies, and these have an energy range of $\sim$0.3–10 keV. We therefore estimate l_c in the 0.3–10 keV band, i.e. $L_c^{0.3-10keV}/L_{Edd}$, since it is directly applicable to our observations. The energy spectra of neutron star and black hole LMXBs are similar at low accretion rate; hence l_c should scale to different energy bands in a similar way for all LMXBs.

Estimating l_c from XMM-Newton observations of 14 globular cluster LMXBs in M31

Of the 63 X-ray source that we studied in the central region of M31, 14 were associated with globular clusters by Kong et al. [2]. There are thirteen bright X-ray sources in Galactic globular clusters, and twelve of these have been identified as neutron star LMXBs, while the thirteenth has not been classified [3, and references within]. Hence the 14 globular cluster X-ray sources in our sample are expected to be LMXBs containing $\sim$1.4 $M_\odot$ neutron stars. However, note that Angelini et al. [4] have reported a possible globular cluster black hole binary in the elliptical galaxy NGC 1399, with a 0.3–10 keV luminosity of 5×10^{39} erg s^{-1}.

Figure 1 shows the observed luminosities of the 14 globular cluster X-ray sources for each of the observations that were selected; flat PDS are consistent with Type B, but not Type A. Quoted uncertainties in the luminosity are based on 90% confidence limits on the best fit parameters used to model the spectrum. Each X-ray source is consistent with the hypothesis that Type A PDS are exhibited at lower luminosities than Type B PDS, within errors. Furthermore, twelve are consistent with $1.0 \leq (L_c/10^{37} \text{ erg s}^{-1}) \leq 2.6$,

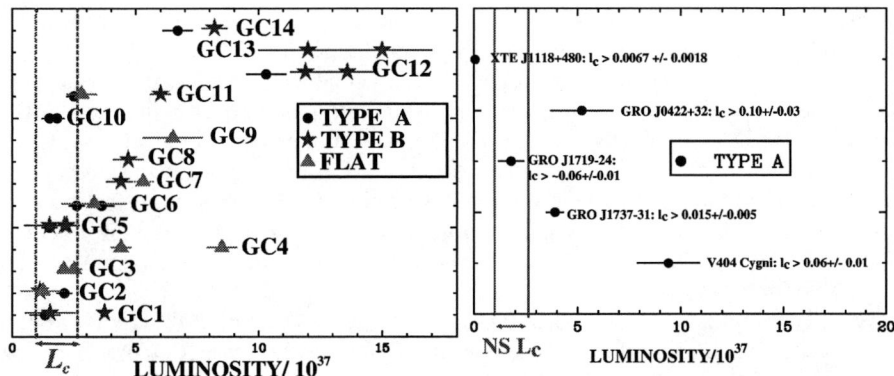

FIGURE 1. *LEFT:* The 0.3–10 keV luminosities of 14 globular cluster X-ray sources in XMM-Newton observations of M31; the errors in the luminosity are given by 90% errors in the parameters of the best fit spectral model. The dashed, vertical lines represent the range of L_c, the luminosity of transition between Type A and Type B PDS for these sources (presumed 1.4 M$_\odot$ neutron stars). Bo 153 and Mita 299 are likely to be multiple bright X-ray sources, but are possible black holes. *RIGHT:* Luminosities of Type A PDS from 5 Galactic black hole LMXBs; lower limits to l_c are given for each object.

assuming a distance to M31 of 760 kpc [5]; this corresponds to $l_c = 0.10 \pm 0.04$ if the primaries in these twelve sources are 1.4 M$_\odot$ neutron stars. Bo 153 and Mita 299 exhibit Type A PDS at luminosities of (10.3 ± 0.8) and $(6.7 \pm 0.6) \times 10^{37}$ erg s^{-1} respectively. These systems would be consistent with our hypothesis if they contained black hole binaries, or if they were composed of two or more bright X-ray sources [like M15, see 6].

l_c for a Galactic neutron star LMXB

4U 1705-44 is a Galactic LMXB that exhibits X-ray bursts, and hence contains a neutron star [7]. It exhibited a Type A PDS in the faintest of four EXOSAT observations, and a Type B PDS in the next faintest; the respective 1–11 keV fluxes were 1.3×10^{-9} and 1.8×10^{-9} erg cm^{-2} s^{-1} [7, 8]. Hence, an accurate distance would yield a tight constraint on l_c. The distance to 4U 1705-44 has been estimated using X-ray bursts as standard candles [see 9]; Christian and Swank [10] find a distance of 11 kpc from Einstein data, while Cornelisse et al. [11] obtain a distance of 8.6 kpc using data from BeppoSAX. If we assume that the distance lies between these two values, $l_c = 0.10 \pm 0.04$.

l_c for Galactic black hole LMXBs

Of the eighteen confirmed Galactic black hole X-ray binaries, three are high mass X-ray binaries (HMXBs), and are persistently bright, and the rest are transient LMXBs. In general outbursts last several months and the X-ray luminosity can increase by a factor of 10^7; outbursts are on average separated by years of quiescence [12, 3]. The outbursts are hysteretic in that the transition from the low/hard state to the high/soft state during the rise of the outburst occurs at a higher luminosity than transition from the high/soft state to the low/hard state in decay [e.g. 13, 14]; hence, estimates of l_c in black hole LMXBs were restricted to those that have been observed during the rise of the outburst, and the subset of outbursts where the transition from low/hard state to high/soft state were not made [see 15, for a review]. Unfortunately, most X-ray observations of Galactic black hole LMXBs during outburst have only covered the decay phase, and are hence unsuitable for this work.

Two of the confirmed black hole LMXBs, GX 339−4 [16] and XTE J1550−564 [17], have recently been caught during the rise of the outburst by the RXTE-ASM, allowing monitoring of the entire outburst by the main instruments of RXTE. Both systems exhibited spectral transitions at bolometric luminosities of ∼20%. This corresponds to 0.3–10 keV luminosities of ∼10% Eddington, i.e. l_c ∼0.1 in the 0.3–10 keV band.

Additionally, nine black hole LMXBs have exhibited outbursts where they remained in the low/hard state [for a review see 15]. Five of these have published Type A PDS, distances and mass estimates; we have used published results to obtain the corresponding minimum value of l_c in the 0.3–10 keV band. The results are presented in Fig. 1; the mass of GRS 1737−31 is given by the mass range of known black holes. We note that Esin et al. [18] find that the SED of GRO 0422+32 at the peak of its outburst is consistent with an accretion rate of ∼ $\dot{m}_{crit}$; this is interesting because we find the 0.3–10 keV luminosity of GRO 0422+32 at that time to be 10±3% Eddington. Results from the other four black hole LMXBs are also consistent with l_c ∼0.1 in the 0.3–10 keV band, although they cannot constrain l_c.

Our empirical value of l_c

The data are all consistent with l_c ∼0.1 for neutron star and black hole LMXBs in the 0.3–10 keV band. Using a maximum neutron star mass of 3.1 M$_\odot$, we classify LMXBs that exhibit Type A PDS at $>4\times10^{37}$ erg s^{-1} as likely black hole LMXBs. So far we have identified 11 candidates in the core of M31, including RX J0042.3+115 [19] and CXOM31 J004303.2+411528 [20].

THE BRIGHT X-RAY POPULATION OF THE CENTRAL REGION OF M31

The 0.3–10 keV luminosities and PDS of the 63 brightest X-ray sources were analysed for each of the four XMM-Newton observations of the central region of M31. The sample is likely to be dominated by X-ray binaries, and since the field of view is dominated by the bulge, they are most likely to be LMXBs. The sample was selected on the criterion that the average 0.3–10 keV EPIC-pn intensity was greater than 0.02 count s^{-1} in at least one of the four observations. For each observation of each source, a PDS was made of the combined EPIC, background subtracted, 0.3–10 keV lightcurve;

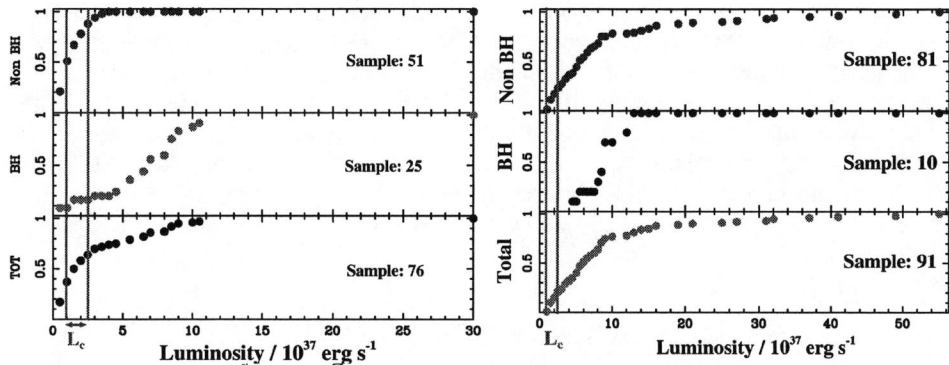

FIGURE 2. Cumulative distribution functions (CDFs) vs. 0.3–10 keV luminosity for Type A PDS (left) and Type B PDS (right) for the brightest X-ray sources in four XMM-Newton observations of the central region of M31. The bottom panel shows the CDF for the whole sample, and the sample is divided into black hole candidates and non-black hole candidate in the middle and top pannels respectively. The vertical lines represent the reange of L_c for a 1.4 $M_\odot$, derived from Fig. 1.

each PDS was averaged over many intervals, which were divided into 64 or 128 bins of 5.2 s duration. Luminosities were obtained from the unabsorbed 0.3–10 keV flux given by best fit models to EPIC-pn spectra. From this sample, all known foreground objects and background AGN identified by Kong et al. [2] were filtered out, as were sources that were resolved into multiple X-ray sources by Chandra. Furthermore, any observation where the classification of the PDS was ambiguous was also rejected.

Of the 167 observations accepted, 76 exhibited Type A PDS and 91 exhibited Type B or flat PDS. X-ray sources that exhibited Type A PDS at luminosities higher than 4×10^{37} erg s^{-1} in at least one of the four observations were classed as black hole candidates, the rest were classed as non-black holes. We note that these black hole identifications are neither certain nor complete.

Cummulative distribution function (CDFs) of Type A PDS vs luminosity are presented in Fig. 2; the CDFs for the non-black hole population is shown in the top panel; the black hole CDF is shown in the middle panel, and the CDF for the whole sample is shown in the bottom panel. The vertical lines indicate the range of transition L_c obtained from the globular clusters in Fig. 1, assuming 1.4 $M_\odot$ neutron star primaries. We see that the total CDF may be broken into two populations: $\sim$0–4 and $\sim$6–30 $\times$ 10^{37} erg s^{-1}; these populations are similar to the non-BH and BH populations we obtain for $l_c \sim 0.1$. Also, 50% of the 51 non-BH Type A PDS were seen at luminosities below 1.0×10^{37} erg s^{-1}, despite the expected strong bias towards high luminosity Type A PDS due to improved statistics.

The CDFs vs. luminosity of Type B and flat PDS for the non-black hole, black hole and total populations of X-ray sources are presented in right hand panels of Fig. 2. As expected, the majority ($\sim$80%) are observed at luminosities higher than the range of L_c values obtained from the globular cluster population, and none are observed below the lower limit to L_c; a break occurs around 10^{38} erg s^{-1}, further indicating that most of the sources contain neutron stars. We do not expect all Type B PDS to occur above l_c, since some of the sources are likely to be transient, and hence hysteretic. Most significantly, none of our black hole candidates exhibit Type B PDS below 4×10^{37} erg s^{-1}; this provides strong support for the idea that Type B PDS are observed at higher luminosities than Type A PDS, and suggests that these black hole candidates are not hysteretic, and are perhaps persistently bright. We note however, that only 10 Type B PDS were observed in our black hole candidates.

Our results show that ~70 LMXBs are consistent with $l_c \sim 0.1$ for neutron star and black hole LMXBs.

ACKNOWLEDGMENTS

This work was funded by PPARC. R.B. would like to thank W. Clarkson, R. Cornelisse, T. Maccarone, J. Wilms and D. Gelino for useful discussions.

REFERENCES

1. M. Van der Klis, *ApJs*, **92**, 511–519 (1994).
2. A. H. K. Kong, M. R. Garcia, F. A. Primini, S. S. Murray, R. DiStefano, and J. E. McClintock, *ApJ*, **577**, 738–756 (2002).
3. J. in't Zand, F. Verbunt, J. Heise, A. Bazzano, M. Cocchi, R. Cornelisse, E. Kuulkers, L. Natalucci, and P. Ubertini, *To appear in "The Restless High-Energy Universe" (2nd BeppoSAX Symposium)*, eds. E.P.J. van den Heuvel, J.J.M. in 't Zand & R.A.M.J. Wijers, Nucl. Instrum. Meth. B Suppl. Ser, astro-ph/0403120 (2004), astro-ph/0403120.
4. L. Angelini, M. Loewenstein, and R. F. Mushotzky, *ApJL*, **557**, L35–L38 (2001).
5. S. van den Bergh, *The galaxies of the Local Group*, Cambridge University Press, Cambridge Astrophysics Series Series, vol no: 35 (2000), URL http://adsabs.harvard.edu/cgi-bin/nph-bib_query?bibcode=2000g%1g....conf.....V&db_key=AST.
6. N. E. White, and L. Angelini, *ApJL*, **561**, L101–L105 (2001).
7. A. Langmeier, M. Sztajno, G. Hasinger, J. Truemper, and M. Gottwald, *ApJ*, **323**, 288–293 (1987).
8. A. Langmeier, G. Hasinger, and J. Truemper, *ApJL*, **340**, L21–L23 (1989).
9. E. Kuulkers, P. R. den Hartog, J. J. M. in't Zand, F. W. M. Verbunt, W. E. Harris, and M. Cocchi, *A&A*, **399**, 663–680 (2003).
10. D. J. Christian, and J. H. Swank, *ApJ Si*, **109**, 177+ (1997), URL http://adsabs.harvard.edu/cgi-bin/nph-bib_query?bibcode=1997A%pJS..109..177C&db_key=AST.
11. R. Cornelisse, E. Kuulkers, J. J. M. in't Zand, F. Verbunt, and J. Heise, *A&A*, **382**, 174–177 (2002), URL http://adsabs.harvard.edu/cgi-bin/nph-bib_query?bibcode=2002A%%26A...382..174C&db_key=AST.
12. X. Chen, J. H. Swank, and R. E. Taam, *ApJL*, **477**, L41+ (1997), URL http://ukads.nottingham.ac.uk/cgi-bin/nph-bib_query?bibcode=1%997ApJ...477L..41C&db_key=AST.
13. S. Miyamoto, S. Kitamoto, K. Hayashida, and W. Egoshi, *ApJL*, **442**, L13–L16 (1995).
14. T. J. Maccarone, and P. S. Coppi, *MNRAS*, **338**, 189–196 (2003).
15. C. Brocksopp, R. M. Bandyopadhyay, and R. P. Fender, *New Astronomy*, **9**, 249–264 (2004).
16. A. A. Zdziarski, M. Gierlinski, J. Mikolajewska, G. Wardzinski, D. M. Smith, B. A. Harmon, and S. Kitamoto, *MNRAS, in press, astro-ph/0402380* (2004), astro-ph/0402380.
17. J. Rodriguez, S. Corbel, and J. A. Tomsick, *ApJ*, **595**, 1032–1038 (2003).
18. A. A. Esin, R. Narayan, W. Cui, J. E. Grove, and S. Zhang, *ApJ*, **505**, 854–868 (1998).
19. R. Barnard, J. P. Osborne, U. Kolb, and K. N. Borozdin, *A&A*, **405**, 505–511 (2003).
20. R. Barnard, U. Kolb, and J. P. Osborne, *A&A*, **423**, 147–153 (2004).

Transient QPOs in the microquasar XTE J1859+226

P. Casella*,†, T. Belloni*, J. Homan** and L. Stella†

*INAF - Osservatorio Astronomico di Brera
via E. Bianchi 46, I–23807 Merate (LC), Italy
†INAF - Osservatorio Astronomico di Roma
Via di Frascati, 33, I–00040 Monte Porzio Catone (Roma), Italy
**Center for Space Research, Massachusetts Institute of Technology
77 Massachusetts Avenue, Cambridge, MA 02139, USA

Abstract. We present an analysis of RXTE data of the microquasar XTE J1859+226, covering the 1999 outburst of this source. We found three main different types of low frequency QPOs with well defined time-lags, coherence and intensity properties. Rapid transitions between different power spectral shapes were observed, always involving a narrow 4.5-6 Hz QPO, and their link with the count rate was studied. Furthermore, the frequency of this QPO was found to be linearly correlated with the count rate.

Keywords: Black holes physics; stars: oscillations; stars: individual: XTE J1859+226
PACS: 97.10.Gz; 97.60.Lf; 97.80.Jp

INTRODUCTION

Low-frequency quasi-periodic oscillations (LFQPOs) with frequencies ranging from a few mHz to $\sim$10 Hz are a common feature in black hole candidates (BHCs). Low-frequency QPOs have been detected in virtually all observed BHCs (see [1]). Usually they are associated with the spectrally hard and intermediate states [2, 3]: they appear together with a flat-top noise component, and on a time scale of days their frequency often correlates with the source count rate (see e.g. [4, 5]). However, in the Ginga observations of the bright transient GS 1124–683 two different types of low frequency QPOs were identified: one associated with a flat-top noise component and one to a steep noise component [6]. The first type showed a strong dependence on count rate, while the other had a rather stable frequency. These two QPOs had centroid frequencies in the 1-10 Hz range and were clearly related to the two PDS 'flavors' of very high state observed in this system [7].

In the RXTE data of XTE J1550–564 two different types of QPOs were reported by [8] and [9]: a broad one (type-A), with a quality factor Q (the QPO frequency divided by the QPO FWHM) of less than 3, and a narrower one (type-B), with a Q larger than 6. Both QPOs had a centroid frequency around 6 Hz and were associated with a weak red-noise component. The more common QPO-type associated with a flat-top noise component had already been observed in the same source and was dubbed 'type-C' by [10]. It is relatively coherent (Q $\gtrsim$ 10) and has a variable centroid frequency (in the range 0.1 - 10 Hz). The strong flat-top noise component has an rms in the range $\sim$10–40%. While type-C QPOs are observed in many systems, the other two are less

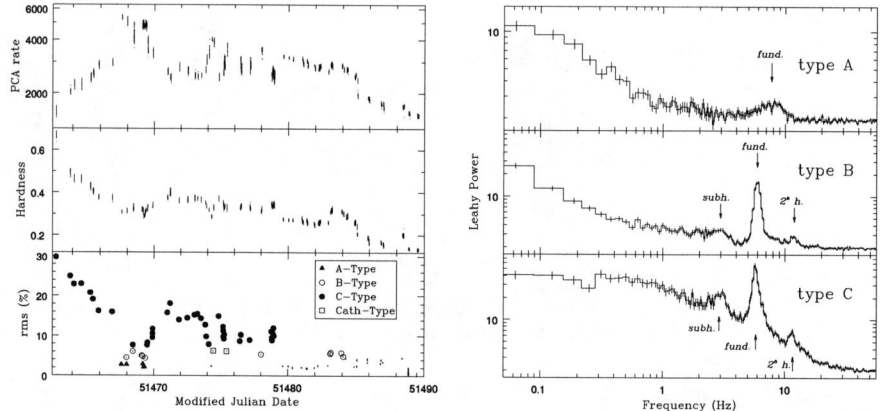

FIGURE 1. *Left*: RXTE/PCA light curve (PCUs 0 and 2, *top panel*), hardness ratio [(7-15 keV)/(2-7 keV), *middle panel*] and total rms (0.03-64 Hz, *bottom panel*) of XTE J1859+226 during the first 25 days of its 1999 outburst. The time resolution for the light- and hardness curves is 16 s. *Right*: Examples of type A, B and C QPOs from our XTE J1859+226 observations. QPO and harmonics peaks are indicated. The Poisson noise was not subtracted.

common. In addition to XTE J1550–564, type-B QPOs were also observed in GX 339–4 (e.g. [11, 12]), GRS 1739–278 [13] and possibly in 4U 1630–47 [14], while type-A QPOs were observed in GX 339–4 [12], H1743-322 [15] and possibly in 4U 1630–47 [14, 16]. Furthermore, in the light of the A-B-C classification, the two QPOs observed in GS 1124–683 (see above) can be tentatively identified with types B and C, although a detailed analysis of Ginga data is necessary to confirm this association.

The soft X-ray transient XTE J1859+226 was discovered on 1999 October 9 with the RXTE All Sky Monitor [17]. After an initial hard rise the source softened at its peak intensity on October 16 and continued to soften for almost two months, when a secondary hard plateau took place. This was rather similar to the behavior observed in many X-ray transients, suggesting a common scenario for the evolution of these objects. Both low-frequency ($\sim 1 - 4$ Hz and ~ 6 Hz) and high-frequency ($\sim 150 - 187$ Hz) QPOs have been reported [18, 19].

Here we present the results of a timing analysis of the 1999 outburst of XTE J1859+226, focusing on the transitions between the three different types of low-frequency QPOs we found and their link with count rate. For further details on the analysis see [20].

OBSERVATIONS

We analyzed 129 RXTE/PCA observations made during the 1999 outburst of the black-hole candidate XTE J1859+226, between MJD 51462 (Oct 11, 1999) and 51626 (Mar 23, 2000). In Figure 1 (left panel) we show the 2-60 keV light curve of the first 25 days

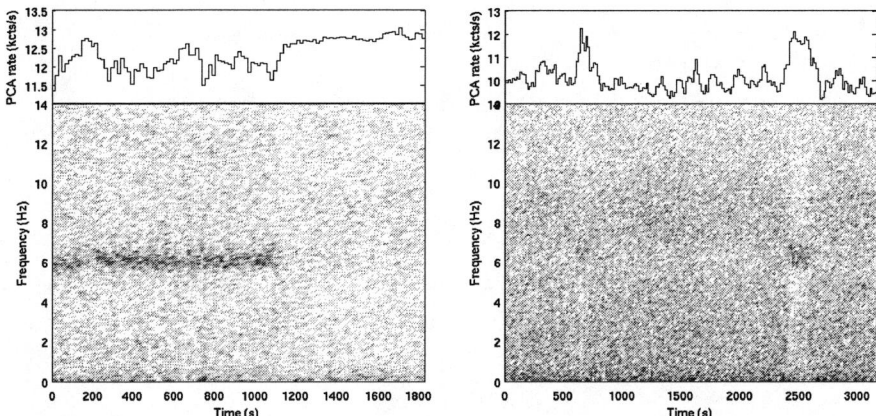

FIGURE 2. Light curves and dynamical PDS for observations 40124-01-13-00 (*left panel*) and 40124-01-14-00 (*right panel*). The lower power value of the type-A QPO peaks with respect to the type-B render them invisible in the dynamical power spectra of the left panel. However, these are clearly seen in the total average PDS.

of the outburst (in which QPOs were observed), the hardness ratio and the integrated 0.03-64 Hz fractional rms of the 2-15 keV light curves. QPOs were detected in many of the observations, with frequencies ranging from $\sim$1 to $\sim$9 Hz. Three main types could be distinguished, which, based on their phase lag and coherence properties, could be associated with type-A, -B, and -C QPOs. In two observations the type B showed a peculiarly evident subharmonic peak, as a result of a higher coherence. We dubbed this sub-type "Cathedral" type. The type identification in each observation is marked in the left-bottom panel of Figure 1. Example power spectra of each type are shown in the right panel of Figure 1.

FAST TRANSITIONS

In some observations, the dynamical PDS showed rapid (on a time scale of a few tens of seconds) transitions between different power spectral shapes. In all cases, the transitions involve type-B QPOs. In Figure 2 we show two examples of different behaviors. In the first half of the observation 40124-01-13-00 (left panel), when the light curve was highly variable, the PDS was of type-B (with a QPO frequency $\sim$ 6 Hz). Simultaneously with the rise observed in the light curve after $\sim$1100 s from the start, the PDS showed a sharp transition to a type-A shape with a QPO frequency of $\sim$ 8 Hz (not visible in the gray scale representation). In the second part of the observation, the light curve was much less variable and had a higher mean count rate. Notice that a brief interval with the same characteristics (type-A QPO, higher flux) was seen $\sim$200 s into the light curve, again with very sharp in and out transitions. For observation 40124-01-14-00 (right panel) the behavior was different: when the source flux was low the power spectrum showed

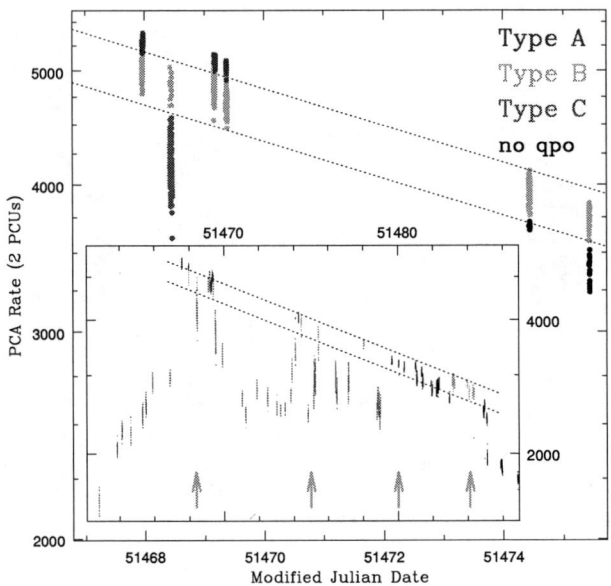

FIGURE 3. Light curve of the observations showing transitions between two different QPO types (bin size 16s). The two parallel lines have been drawn so as to intersect the transition points. In the inset, the lines have been extended to the part of the outburst where QPOs appear and four arrows have been added in correspondence of the observed radio flares [21]. Different grayscales indicate different QPO types. Black points correspond to data in which no QPOs were found.

a type-C shape (~ 8.7 Hz), while during the two peaks in count-rate, when it reached values close to those of the first half of the previous observation, the PDS was of type-B (~ 6.4 Hz). The transitions were again very sharp.

It is worth remarking that in all observed cases the transitions involve type-B QPOs. When the source showed a type-B PDS and underwent a fast transition to a *lower* count rate, the PDS changed to type-C; when on the other hand the transition was to a *higher* count rate, the PDS changed to type-A. In the same way, fast transitions from type-A to type-B and from type-C to type-B always involved a decrease and increase in count rate, respectively. Direct transitions between types A and C were not observed.

In Figure 3 we plot a light curve of all observations in which fast transitions were observed. From this figure it is evident that the count rate at which the transitions occur follow an exponential trend (with a $\tau \sim 28$ days). Type-B QPOs appear in a narrow count rate range, as can be seen in the inset of the same figure, where we show the entire portion of the outburst where low frequency QPOs were detected.

There is a strong evidence of a threshold triggering the type-B QPO superimposed to the roughly exponential decay of the outburst. This confirms the existence of a second parameter (as it was proposed by [9]) in addition to the mass accretion rate, the latter being probably responsible for the long time-scale evolution of the source, and therefore

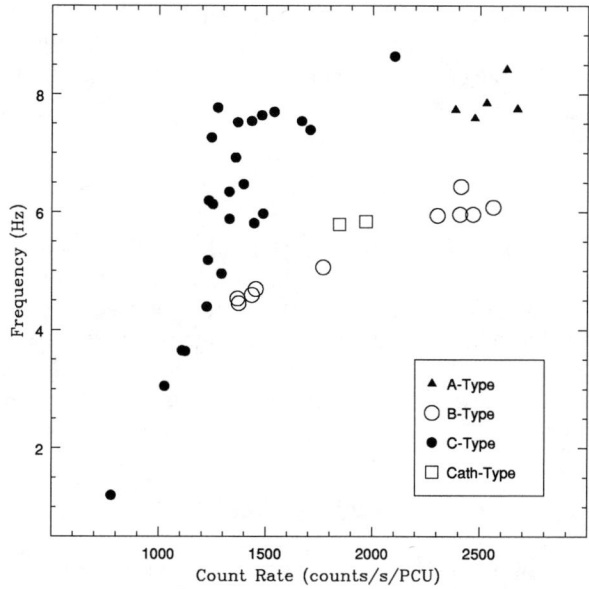

FIGURE 4. Count rate vs. centroid frequency for all observed QPOs.

for the exponential decay. Furthermore, as one can see from the arrows in the inset of Figure 3 (see caption), three out of four radio flares happen in correspondence of the presence of the type B QPO [21]. The association is only putative, mainly because of the incomplete sampling of radio observations. However, it is in line with the recent general scheme proposed by [22] which identifies a "jet line" in the Hardness-Intensity diagram corresponding to the launch of the relativistic jet and to the transition between two different "flavors" of the Intermediate State (see also Belloni, this volume).

FREQUENCY-RATE CORRELATION

In Figure 4, we plot count rate vs. QPOs centroid frequency. Both type-C and -B QPO frequencies appear to be linearly correlated with the count rate, with different slopes, while type-A QPOs are clustered at high count rate and frequency values. The correlation of the type-C frequency with the count rate was already known (see e.g. [4, 5]). For the first time we discovered that type-B QPO also follow a flatter but similar correlation. The dependence of the frequency on the count rate is weaker than that of the type-C QPOs, and as a result of this the type-B QPOs appear in a narrow range of frequencies near ~ 6 Hz. This QPO in fact has been seen at present only at high count rates. However, [12] found a ~ 1 Hz QPO in the outburst decay of the black hole candidate GX 339-4, with properties very similar to those of type-B QPOs. A 1

Hz QPO with similar properties was also observed in XTE J1650-500 (Rossi et al., in preparation). If the identification of these QPOs with the type-B QPOs will be confirmed, this would extend the range of the frequency-count rate correlation to the bottom-left region of the diagram in Figure 4.

CONCLUSION

From the large data set presented here, the classification of low-frequency QPOs into three main types is strengthened and extended to another black-hole candidate. A coherent, general scheme is emerging, in which transitions between different power spectral shapes, which happen when the source is at its highest flux values, play an important role. It is clear that while the long time-scale evolution of the source is driven by the mass accretion rate, a second (still unknown) parameter appears to determine state transitions, which are thus the key to the understanding of the physics of the accretion flow.

REFERENCES

1. van der Klis, M., 2004, to appear in "Compact Stellar X-ray Sources", eds. W.H.G. Lewin and M. van der Klis, Cambridge University Press, Cambridge
2. van der Klis, M., 1995, in "X-ray binaries", eds. W.H.G. Lewin, J. van Paradijs, E.P.J. and van den Heuvel, Cambridge University Press, Cambridge, p. 252
3. McClintock, J. E., & Remillard, R. A., 2004, to appear in "Compact Stellar X-ray Sources", eds. W.H.G. Lewin and M. van der Klis, Cambridge University Press, Cambridge (astro-ph/0306213)
4. Cui, W., 1998, in "High Energy Processes in Accreting Black Holes", eds Poutanen J. & Svensson R., (Graftavallen, Sweden), June 1998 (astro-ph/9809408)
5. Reig, P., Belloni, T., van der Klis, M., et al. 2000, ApJ, 541, 883
6. Takizawa, M., Dotani, T., Mitsuda, K., et al. 1997, ApJ, 489, 272
7. Miyamoto, S., Iga, S., Kitamoto, S., & Kamado, Y., 1993, ApJ, 403, L39
8. Wijnands, R., Homan, J., & van der Klis, M., 1999, ApJ, 526, L33
9. Homan, J., Wijnands, R., van der Klis, et al. 2001, ApJS, 132, 377
10. Remillard, R. A., Sobczak, G. J., Muno, M. P., & McClintock, J. E., 2002, ApJ, 564, 962
11. Miyamoto, S., Kimura, K., Kitamoto, S., et al. 1991, ApJ, 383, 784
12. Belloni, T., Homan, J., Casella, P., et al., 2005, submitted to A&A
13. Wijnands, R., Méndez, M., Miller, J. M., et al. 2001, MNRAS, 328, 451
14. Tomsick, J. A., & Kaaret, P., 2000, ApJ, 537, 448
15. Homan, J., Miller, J. M., Wijnands, R., et al., 2004, submitted to ApJ (astro-ph/0406334)
16. Dieters, S. W., Belloni, T., Kuulkers, E., et al. 2000, ApJ, 538, 307
17. Wood, A., Smith, D. A., Marshall, F. E., Swank, J., 1999, IAUC 7274
18. Cui, W., Shrader, C. R., Haswell, C. A., & Hynes, R. I., 2000, ApJ, 535, L123
19. Focke, W. B., Markwardt, C. B., Swank, J. H., & Taam, R. E., 2000, in "Rossi2000: Astrophysics with the Rossi X-ray Timing Explorer", Greenbelt, March 2000, E104
20. Casella, P., Belloni, T., Homan, J., & Stella, L., 2004, A&A, 426, 587
21. Brocksopp, C., Fender, R. P., McCollough, M., et al., 2002, MNRAS, 331, 765
22. Homan, J., & Belloni, T., 2004, in Proc. of "From X-ray binaries to quasars: Black hole accretion on all mass scales", (Amsterdam, July 2004), eds. T. Maccarone, R. Fender, L. Ho

X-ray States of Black Hole Binaries in Outburst

R. A. Remillard

The MIT Kavli Institute for Astrophysics and Space Research

Abstract. We continue to probe the properties of stellar-size black holes and the physics of black-hole accretion using bright X-ray transients. Progress has been made in the recognizing that the three states of active accretion are related to different physical elements that may contribute radiation: the accretion disk, a jet, and a compact corona. Each of these states offers potential applications for investigation via general relativity in the regime of strong gravity. High-frequency QPOs are especially interesting in this regard, as the evidence mounts for their interpretation as stationary 'voice-prints' that may constrain black-hole mass and spin.

Keywords: Physics of black holes, Black holes
PACS: 04.70.-s, 97.60.Lf

RE-DEFINING X-RAY STATES

It has been proposed recently that X-ray states of black hole binary systems should be redefined in terms of quantitative criteria that utilize both X-ray energy spectra and power density spectra (PDS) [22]. The goals of this effort are to capture the essential elements of historical state descriptions, to incorporate the complexity of black hole behavior revealed in extensive monitoring campaigns with the *Rossi* X-ray Timing Explorer (*RXTE*), and to correct misleading terminology.

The redefinition of X-ray states utilizes four criteria: f_{disk}, the ratio of the disk flux to the total flux (both unabsorbed) at 2-20 keV; the power-law photon index (Γ) at energies below any break or cutoff; the integrated rms power in the PDS at 0.1–10 Hz (r; expressed as a fraction of the average source count rate); and the integrated rms amplitude (a) of a quasi-periodic oscillation (QPO) detected in the range of 0.1–30 Hz.

It had been known for decades that the energy spectra of outbursting black holes often exhibit composite spectra consisting of two broadband components [43]. There is a multi-temperature accretion disk [36, 19, 10] with a characteristic temperature near 1 keV. This modified form of the black-body function is easy to recognize. Usually, it is seen in combination with a quasi-power-law component, which may be modified by disk reflection [5, 33] and/or a cutoff energy (e.g. 100 keV) or a spectral break to a steeper power law. Many black hole systems further exhibit a broad Fe emission line (e.g. [34, 25]) and atomic absorption edges that are primarily due to the cold interstellar medium. In some cases there are also absorption features due to hot gas that is local to the binary system (e.g. [18]).

There is a need to draw attention to those times when black hole radiation is dominated by the heat from the inner accretion disk, while there are no obvious temporal features that signify unexpected oscillations or instabilities that complicate the picture. Toward this end, the thermal state (formerly the "high/soft" state) is defined by the following three conditions: (1) the disk contributes more than 75% of the total unabsorbed flux at

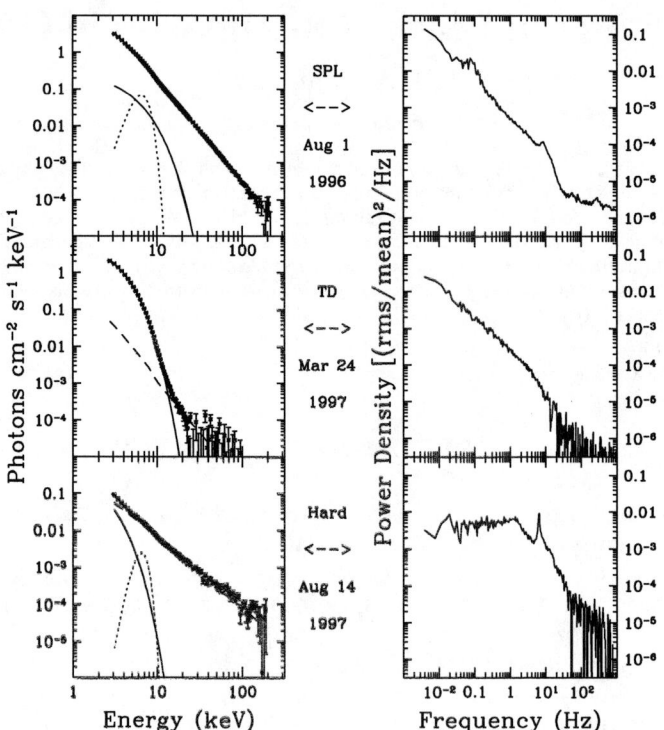

FIGURE 1. Examples of the 3 states of active accretion for the black hole binary GRO J1655-40. Left panels show the energy spectra, with model components attributed to thermal-disk emission (solid line), a power-law continuum (dashes) and a relativistically broadened Fe K–$alpha$ line (dotted). Power-law components for the SPL and hard states are distinguished by different values of the photon index (i.e. slope). The PDS are shown in the right panels. A strong, band-limited continuum characterizes the hard state, while QPOs and the absence of the intense, broad continuum are usually seen in the SPL state.

2–20 keV, i.e. $f > 0.75$, (2) there are no QPOs present with integrated amplitude above 0.5% of the mean count rate, i.e. $a_{max} < 0.005$, and (3) the integrated power continuum is low, with $r < 0.06$.

In principle, the normalization constant for the thermal component may allow numerical estimates of the radius of the inner accretion disk, if the source distance and disk inclination are accurately known [49]. However, such estimates depend on disk models computed under general relativity (GR), with careful attention to the inner disk boundary condition and to effects of radiative transfer [23]. Improved disk models may one day lead to trustworthy results, and this could lead to measures of the dimensionless spin parameter when the mass is well constrained via dynamical measures of stellar motion in the binary system.

There are other occasions when the spectrum from an outbursting black-hole system shows a much greater contribution from an X-ray power-law component. Observations with *CGRO*–OSSE were particularly valuable in showing that there were two forms of these non-thermal spectra [12]. Spectral fits for the power-law component yield clusters of Γ values: one near 1.7 (hard state, with an exponential decrease beyond ~ 100 keV) and the other near 2.5 (steep power law, with no apparent cutoff). In each case, the corresponding PDS also shows a distinct departure from the thermal state. An illustration of the thermal state and the two non-thermal states is given in Fig. 1.

In the hard X-ray state, the accretion-disk component is either absent or it is modified in the sense of appearing comparatively cool and large. The hard state has been clearly associated with the presence of a steady type of radio jet [7, 4]. Transitions to either the thermal state (e.g. GX339-4; [8, 3]) or a steep power law (SPL) state (e.g. Cyg X-1; [50]) effectively quenches the radio emission. This crucial, multi-frequency advancement helps to demonstrate that the three X-ray states of black hole binaries in outburst represent accretion systems that are very different in terms of physical elements, geometry, and energy-transport mechanisms. Hindsight analysis of X-ray data correlated with radio signatures of the steady jet allows a definition of the hard state, again based on three X-ray conditions: (1) $f < 0.2$, i.e. the power-law contributes at least 80% of the unabsorbed 2–20 keV flux, (2) $1.5 < \Gamma < 2.1$ (for power-law, cutoff power law, or broken power-law (using Γ_1), as appropriate), and (3) the PDS yields $r > 0.1$.

Current research on the hard state now addresses more detailed and physical questions such as the jet energetics and ejection velocities ([7]; see also Gallo et al., these proceedings), and the whether the X-ray spectrum represents synchrotron radiation, inverse Compton emission at the base of the jet, or a collimated outflow related to an advection-dominated accretion flow (ADAF) [6, 20, 9, 47].

The SPL component was first linked to the power-law "tail" found in the thermal state, and it was widely interpreted as inverse Compton radiation from a hot corona somehow coupled to the accretion disk. The picture became more complicated when X-ray QPOs were first detected with Ginga for two sources: GX339-4 (6 Hz) and X-ray Nova Muscae 1991 (3–8 Hz) [27, 26]. The QPOs, the high luminosity, and the strength of the power-law component prompted the interpretation that the QPOs signified a new black hole state, labeled as the "very high" state. *RXTE* observations later showed that X-ray QPOs from black-hole binaries are much more common than had been realized [45].

As noted above, *CGRO* observations have shown that the SPL may extend to photon energies as high as 800 keV [12, 44]). This forces consideration of non-thermal Comptonization models [11, 48]. The QPOs impose additional requirements for an oscillation mechanism that must be intimately tied to the electron acceleration mechanism (in the inverse Compton scenario), since the QPOs are fairly coherent ($v/\Delta v \sim 12$; [32]) and are strongest above 6 keV. Despite a wide range in SPL luminosities, the SPL tend to dominate as the luminosity approaches the Eddington limit (e.g. Fig. 1). Furthermore, the occasions of high-frequency QPOs at 100-450 Hz in 7 black hole binaries almost always coincide with a strong SPL spectrum [22]. Overall, the many fundamental differences between the thermal and SPL properties (see Fig. 1) force us to reject alternative state descriptions that unify thermal and SPL observations under a single "soft" state.

The results from intense monitoring campaigns for black-hole transients with *RXTE* motivated the redefinition of the very high state as the "SPL state" [22], with conditions: (1) $\Gamma > 2.4$, (2) $r < 0.15$, and (3) either $f < 0.8$ while a QPO (0.1 to 30 Hz) is present in the PDS (with $a > 0.01$), **or** $f < 0.5$ with no QPOs (i.e. the disk contributes less than half of the flux). Generally, the accretion disk remains visible in the X-ray spectrum in the SPL state. There may be modifications to the thermal parameters (T_{in}, R_{in}) during SPL episodes at high luminosity. In such cases, the disk appears unusually small and hot, and this is a likely artifact of radiative transfer effects that occur when the disk is viewed through a compact corona with moderate optical depth [17].

This quantitative framework for the three states of active accretion in black-hole binary systems is intended to tie X-ray states to each of the three broad-band spectral components that have a distinct physical origin and/or radiation mechanism. These states appear to demonstrate a capacity for quasi-stability, but observations exhibit their inherently transient nature. Clearly, the parameter ranges chosen for the three X-ray states leave substantial room for intermediate conditions. Conversely, we abandon the effort to pigeon-hole every observation into a well-defined state ; this perspective merely honors the complexity of black hole outbursts as seen in the *RXTE* era.

TEMPORAL EVOLUTION OF X-RAY STATES

The temporal evolution of X-ray states for the case of GRO J1655-40 (1996-1997 outburst) is shown in Fig. 2. The ASM light curve is displayed in the top panel, while the unabsorbed flux derived from spectral fits to *RXTE* pointed observations are shown in the bottom panel. Here, the X-ray state is also represented via the choice of plotting symbol: thermal (x), hard (solid square), steep power-law (solid triangle), and any intermediate type (circle). In this particular case, the two intermediate cases exhibit properties that lie between the thermal and SPL definitions.

The state assignments utilize spectral parameters obtained via the modeling prescriptions of Sobczak et al. 1999 [38]. There are a few differences here: all *RXTE* observations are considered (adding programs 10261 and 20187), newer versions of PCA response models are used, the analyses are restricted to PCU #2, and a model with broken power-law (rather than a simple power-law) is used for the beginning of the outburst, up to and including 1996 June 20. The observations are grouped into 67 data intervals, combining short observations that occur on the same day. However, the last three *RXTE* pointings (beginning 1997 Aug 29) are then excluded, since the source flux is below 2 mCrab and the uncertainties in the spectral parameters is large. These are very likely to be additional samples of the hard state. The statistics of the 64 state assignments shown in Fig. 2 are listed in Table 1.

The temporal evolution of X-ray states for XTE J1550-564 (1998-1999 outburst) is shown in Fig. 3, with content and state representations analogous to Fig. 2. For this source spectral modeling efforts follow Sobczak et al. 2000 [39]. The re-analysis again targets PCU #2, a broken power-law model is used when it improves the fit significantly, and this happens frequently for observations before MJD 51140. Some observations on the same day are grouped together, with a net of 201 data intervals, and the statistics

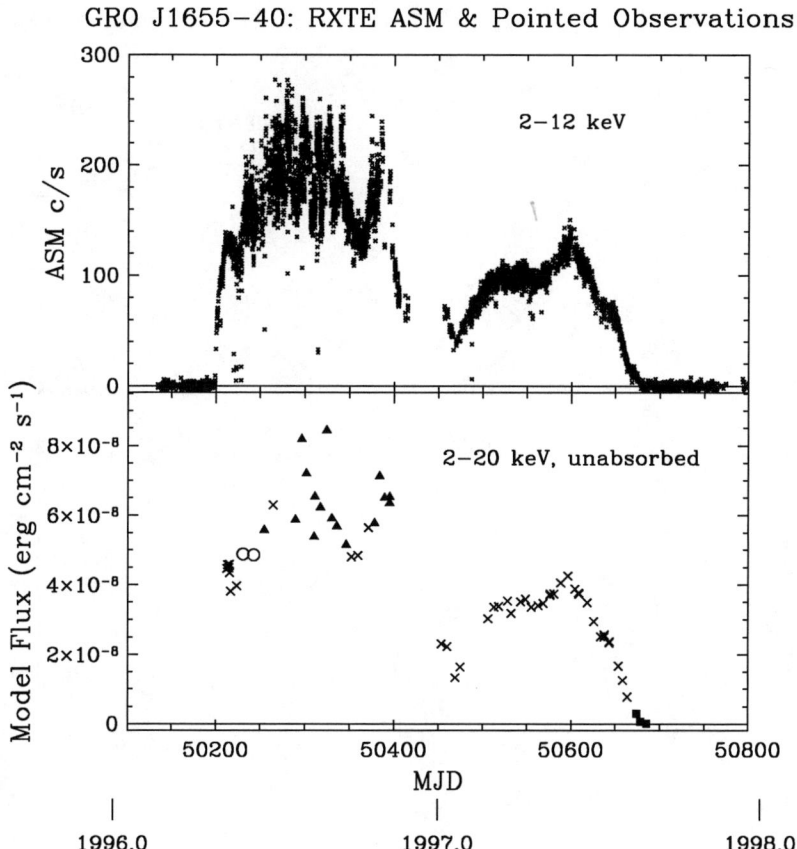

FIGURE 2. X-ray state evolution during the 1996-1997 outburst of GRO J1655-40. The top panel shows the ASM light curve. The bottom panel shows the model flux from PCA pointed observations. Here the symbol type denoted the X-ray state: thermal (x), hard (solid square), steep power-law (solid triangle), and any type of intermediate state (circle).

of state assignments are included in Table 1. There are a relatively large number of intermediate states encountered during this outburst of XTE J1550-564, especially in the MJD range 51050–51140. The latter cases exhibit properties that lie between the SPL and hard states (see [22] for more detailed discussions) and they coincide with the appearance of "C" type QPOs [32].

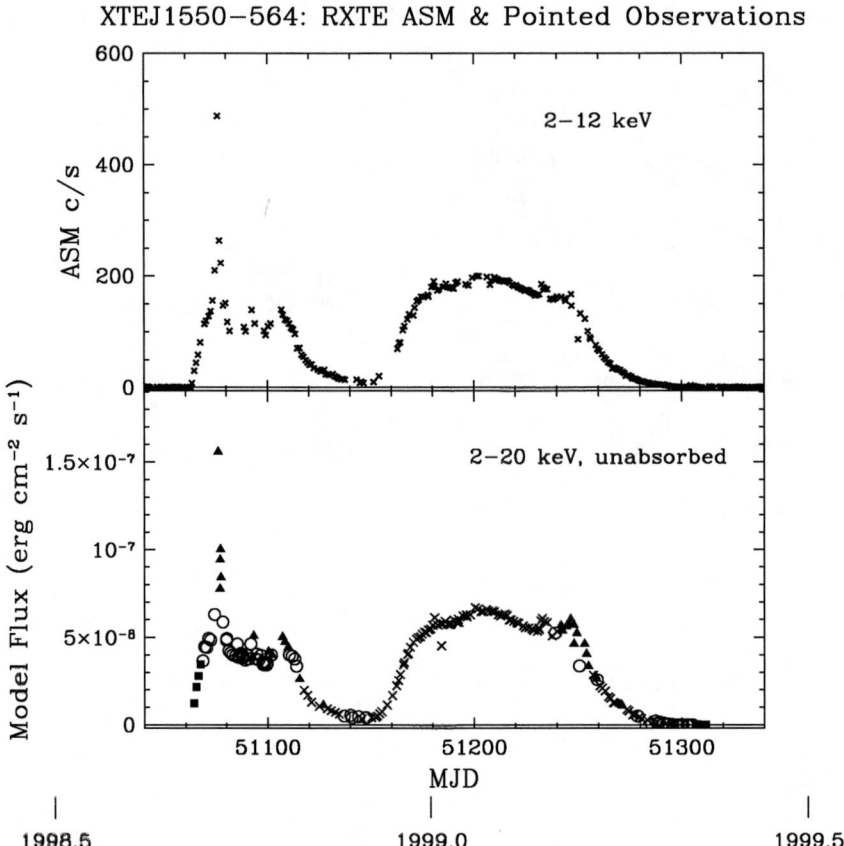

FIGURE 3. X-ray state evolution during the 1998-1999 outburst of XTE J1550-564. The top panel shows the ASM light curve. The bottom panel shows the model flux from PCA pointed observations, again with the X-ray state denoted as: thermal (x), hard (solid square), steep power-law (solid triangle), intermediate (circle).

The 1998-1999 outburst of XTE J1550-564 was followed by successively weaker outbursts in 2000, 2001, 2002, and 2003. The outburst of 2000 again shows multi-state spectral evolution, but the three weaker outbursts appear entirely constrained to the hard state (e.g. [2]).

TABLE 1. Statistics of X-ray State Classifications

	Data intervals	Thermal state	SPL state	Hard state	Intermediate states
GRO J1655-40 (1996-1997)	64	43	16	3	2
XTE J1550-564 (1998-1999)	201	102	30	7	62

HIGH FREQUENCY QPOS FROM BLACK HOLE BINARIES

High-frequency QPOs (HFQPOs; 40-450 Hz) have been detected thus far in 7 black-hole binaries or candidates. These are transient subtle oscillations, with $a \sim 1\%$ [28, 29, 13, 41, 42, 31, 14, 15, 30]. Frequently, one must select photon energy bands above 6 keV or above 13 keV in order to gain significant detection. Furthermore, for statistical reasons most detections require efforts to group observations with similar spectral and/or timing characteristics. All of the current detections for black-hole binaries are all displayed in Fig. 4. HFQPOs above 100 Hz generally occur during the SPL state [22].

Four sources (GRO J1655-40, XTE J1550-564, GRS 1915+105, and H1743-322) exhibit pairs of QPOs that have commensurate frequencies in a 3:2 ratio. These HFQPOs have frequencies above 100 Hz, and they generally occur during the SPL state [22]. GRS 1915+105 shows an additional, slower QPO pair (41 and 67 Hz) [28, 42] that occurs in thermal states with moderate to high luminosity,

In many cases, there is a substantial range in X-ray luminosity within a set of observations that contribute to a particular HFQPO, or a particular pair of commensurate HFQPOs in a given source (e.g. [31]). This supports the conclusion that HFQPO frequency systems are a stable timing signatures inherent to the accreting black hole. This is an important difference from the the kHz QPOs in neutron-star systems, which exhibit large variations in frequency. Moreover, the frequency stability of such a fast oscillations seems to suggest that HFQPOs may represent an invaluable means to probe black hole mass and spin via GR theory.

Commensurate HFQPO frequencies can be seen as a signature of an oscillation driven by some type of resonance condition. In fact, it has been proposed by Abramowicz & Kluzniak [1] that QPOs could represent a resonance in the coordinate frequencies given by GR for motions around a black hole under strong gravity (see [24]). Earlier work had used GR coordinate frequencies and associated beat frequencies to explain QPOs with variable frequencies in both neutron-star and black-hole systems [40].

The "parametric resonance" concept hypothesizes enhanced emissivity from accreting matter at a radius where two of the three coordinate frequencies (i.e. azimuthal, radial, and polar) have commensurate values that match (either directly or via beat frequencies) the observed QPOs. For the cases with known black hole mass, the value of the dimensionless spin parameter (a_*) can be determined via the application of this resonance model if the correct pair of coordinate frequencies can be identified. In fact, reasonable values ($0.25 < a_* < 0.95$) can be derived from the observed HFQPOs for either 2:1 or 3:1 ratios in either orbital:radial or polar:radial coordinate frequencies [31].

The driving mechanism that would allow accretion blobs to grow and survive at the resonance radius has not been specified, and it is known that there are severe damping

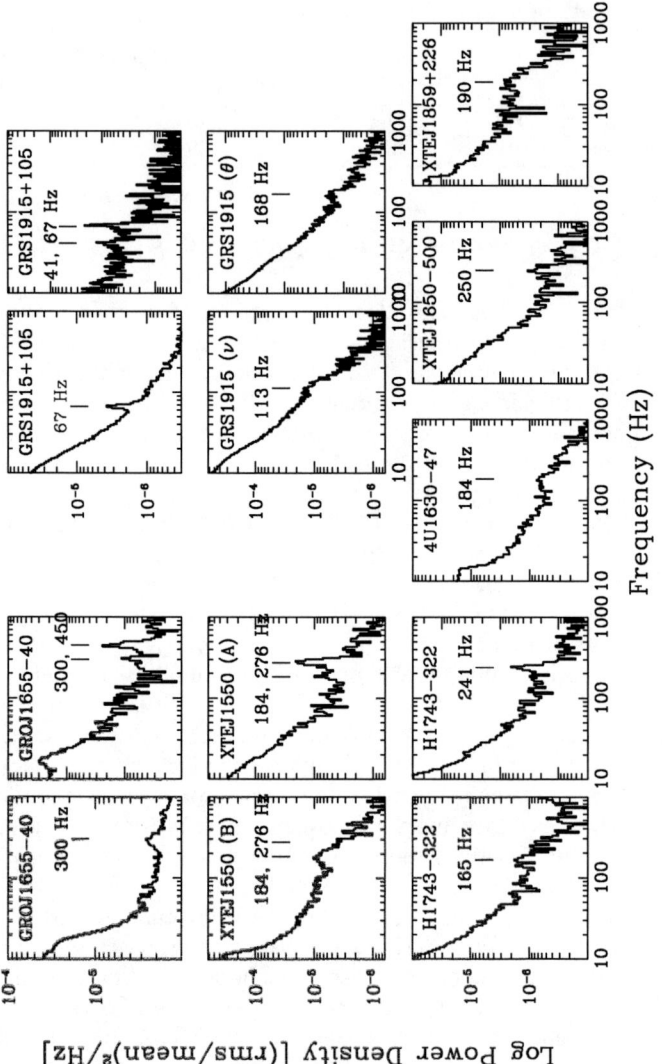

FIGURE 4. High frequency QPOs (40-450 Hz) seen in 7 black hole binary systems. Four sources show pairs of HFQPOs with central frequencies that occur in a 3:2 ratio. GRS 1915+105 shows an additional pair at 41 and 67 Hz. It should be noted that we generally detect one QPO at a time, and QPOs involved in the 3:2 ratio tend to occur alternately, rather than simultaneously.

forces in the inner accretion disk [21]. On the other hand, ray-tracing calculations under GR [37] show that the putative blobs could indeed produce the HFQPO patterns, and that the choice of $3 \times \nu_0$ versus $2 \times \nu_0$ for the stronger QPO is governed by the angular width of the accreting blob. Clearly, there more work is needed to investigate this resonance model.

A possible alternative scenario is to extend the models for "diskoseismic" oscillations to include non-linear effects that might drive some type of resonant oscillation. Diskoseismology treats the inner disk as a resonance cavity in the Kerr metric ([16, 46]). Normal modes have been derived for linear perturbations, and the extension of this theory would be both very interesting and difficult. Another alternative is to consider accretion models that deviate from a thin disk geometry, e.g. an accretion torus and its oscillation modes under GR [35].

For three systems that show HFQPO pairs with frequencies in 3:2 ratio, we have the fortune of black hole mass estimates. In most types of GR oscillations, including the coordinate frequencies discussed above, the oscillation frequency scales with black-hole mass as M^{-1}, with additional dependence on the dimensionless spin parameter and the possibly the radius where the oscillations originate. Surprisingly, these three cases yield a relationship between HFQPO frequency and black hole mass that is consistent with a simple M^{-1} relationship [22]: $\nu_0 = 931 M^{-1}$, where HFQPOs are seen at frequencies of $2 \times \nu_0$ and $3 \times \nu_0$. This result suggests that these black holes have similar values of the dimensionless spin parameter. Furthermore, these results offer strong encouragement for efforts to interpret black-hole HFQPOs via GR theory.

ACKNOWLEDGMENTS

This work was supported by the NASA contract to MIT for the ASM and EDS instruments on *RXTE*. Special thanks are extended to Jeff McClintock and Jeroen Homan for many helpful discussions.

REFERENCES

1. M. A. Abramowicz, and W. Kluzniak, *A&A*, **374**, L19–L20 (2001).
2. T. Belloni, A. P. Colombo, J. Homan, S. Campana, and M. van der Klis, *A&A*, **390**, 199–204 (2002).
3. S. Corbel, et al., *A&A*, **359**, 251–268 (2000).
4. V. Dhawan, I. F. Mirabel, and L. F. Rodriguez, *Ap. J.*, **543**, 373–385 (2000).
5. C. Done, and S. Nayakshin, *MNRAS*, **328**, 616–622 (2001).
6. A. A. Esin, et al., *Ap. J.*, **555**, 483–488 (2001).
7. R. Fender, "Jets from X-ray binaries", in *Compact Stellar X-ray Sources*, edited by W. H. G. Lewin and M. van der Klis, Cambridge University Press, Cambridge, 2005, in press (astro-ph/0303339).
8. R. Fender, et al., *Ap. J.*, **519**, L165–L168 (1999).
9. F. Frontera, et al., *Ap. J.*, **592**, 1110–1118 (2003).
10. M. Gierlinski, and C. Done, *MNRAS*, **347**, 885–894 (2004).
11. M. Gierlinski, and C. Done, *MNRAS*, **342**, 1083–1092 (2003).
12. J. E. Grove, W. N. Johnson, R. A. Kroeger, K. McNaron-Brown, and J. Skibo, *Ap. J.*, **500**, 899–908 (1998).
13. J. Homan, et al., *Ap. J. Suppl.*, **132**, 377–402 (2001).
14. J. Homan, et al., *Ap. J.*, **586**, 1262–1267 (2003).

15. J. Homan, et al., *Ap. J.*, in press (2005); (astro-ph/0406334).
16. S. Kato, *PASJ*, **53**, 1–24 (2001).
17. A. Kubota, and K. Makishima, *Ap. J.*, **601**, 428-438 (2004).
18. J. C. Lee, C. S. Reynolds, R. A. Remillard, N. S. Schulz, E. G. Blackman, and A. C. Fabian, *Ap. J.*, **567**, 1102–1111 (2002).
19. K. Makishima, et al., *Ap. J.*, **308**, 635–643 (1986).
20. S. Markoff, H. Falcke, and R. Fender, *A&A*, **372**, L25–L28 (2001).
21. D. Markovic, and F. K. Lamb, F.K., *Ap. J.*, **507**, 316–326 (1998).
22. J. E. McClintock, and R. A. Remillard, "Black Hole Binaries", in *Compact Stellar X-ray Sources*, edited by W. H. G. Lewin and M. van der Klis, Cambridge University Press, Cambridge, 2005, in press (astro-ph/0306213).
23. A. Merloni, A. C. Fabian, and R. R. Ross, *MNRAS*, **313**, 193–197 (2000).
24. A. Merloni, M. Vietri, L. Stella, and D. Bini, *MNRAS*, **304**, 155–159 (2001).
25. J. M. Miller, et al., *Ap. J.*, **577**, L15–L18 (2002).
26. S. Miyamoto, S. Iga, S. Kitamoto, and Y. Kamado, *Ap. J.*, **403**, L39–L42 1993.
27. S. Miyamoto, and S. Kitamoto, *Ap. J.*, **374**, 741–743 1991.
28. E. H. Morgan, R. A. Remillard, and J. Greiner, *Ap. J.*, **482**, 993–1010 (1997).
29. R. A. Remillard, E. H. Morgan, J. E. McClintock, C. D. Bailyn, and J. A. Orosz, *Ap. J.*, **522**, 397–412 (1999).
30. R. A. Remillard, J. E. McClintock, J. A. Orosz, and A. M. Levine, *Ap. J.*, submitted; (astro-ph/0407025).
31. R. A. Remillard, M. P. Muno, J. E. McClintock, and J. A. Orosz, *Ap. J.*, **580**, 1030–1042 (2002).
32. R. A. Remillard, G. J. Sobczak, M. P. Muno, and J. E. McClintock, *Ap. J.*, **564**, 962–973 (2002).
33. M. Revnivtsev, M. Gilfanov, and E. Churazov, *A&A*, **380**, 520–525 (2001).
34. C. S. Reynolds, and M. A. Nowak, *Phys.Rept.*, **377**, 389-466 (2003).
35. L. Rezzolla, S'i Yoshida, T. J. Maccarone, and O. Zanotti, *MNRAS*, **344**, L37–L41 (2003).
36. N. I. Shakura, and R. A. Sunyaev, *A&A*, **24**, 337–366 (1973).
37. J. D. Schnittman, and E. Bertschinger, *Ap. J.*, **606**, 1098–1111 (2004).
38. G. J. Sobczak, J. E. McClintock, R. A. Remillard, C. D. Bailyn, and J. A. Orosz, *Ap. J.*, **520**, 776–787 (1999).
39. G. J. Sobczak, et al., *Ap. J.*, **544**, 993–1015 (2000).
40. L. Stella, M. Vietri, and S. M. Morsink, *Ap. J.*, **524**, L63–L66 (1999).
41. T. E. Strohmayer, *Ap. J.*, **552**, L49–L53 (2001).
42. T. E. Strohmayer, *Ap. J.*, **554**, L169–L172 (2001).
43. Y. Tanaka, and W. H. G. Lewin, "Black-hole Binaries", in *X–ray Binaries*, edited by W. H. G. Lewin, J. van Paradijs, and E .P. J. van den Heuvel, Cambridge U. Press, Cambridge, 1995, pp. 126–174.
44. J. A. Tomsick, P. Kaaret, R. A. Kroeger, and R. A. Remillard, *Ap. J.*, **512**, 892–900 (1999).
45. M. van der Klis, "A Review of Rapid Variability in X-ray Binaries", , in *Compact Stellar X-ray Sources*, edited by W. H. G. Lewin and M. van der Klis, Cambridge University Press, Cambridge, 2005, in press (astro-ph/0410551).
46. R. V. Wagoner, *Phys. Rept.*, **311**, 259–269 (1999).
47. F. Yuan, W. Cui, and R. Narayan, *Ap. J.*, in press (2005); (astro-ph/0407612).
48. A. A. Zdziarski, and M. Gierlinski, *PThPS*, **155**, 99-119 (2004).
49. S. N. Zhang, W. Cui, and W. Chen, *Ap. J.*, **482**, L155–L158 (1997).
50. S. N. Zhang, et al., *Ap. J.*, **477**, L95–L98 (1997).

Black Hole Formation in X-Ray Binaries: The Case of GRO J1655-40

V. Kalogera, B. Willems, M. Henninger, T. Levin, N. Ivanova

Northwestern University, Department of Physics and Astronomy, 2145 Sheridan Road, Evanston, IL 60208, USA

Abstract. In recent years proper motion measurements have been added to the set of observational constraints on the current properties of Galactic X-ray binaries. We develop an analysis that allows us to consider all this available information and reconstruct the full evolutionary history of X-ray binaries back to the time of core collapse and compact object formation. Here we present our analysis and results for GRO J1655–40. We find that a symmetric black hole (BH) formation event cannot be formally excluded, but that the associated system parameters are only marginally consistent with the currently observed binary properties. BH formation mechanisms involving an asymmetric supernova explosion with associated BH kick velocities of a few tens of km s^{-1}, on the other hand, satisfy the constraints much more comfortably. We also derive an upper limit on the BH kick magnitude of $\simeq 210$ km s^{-1}.

Keywords: binaries, evolution, X-rays, black holes
PACS: 97.60.Lf, 97.80.Jp

INTRODUCTION

The current observed sample of Galactic black-hole (BH) X-ray binaries (XRBs) provides us with a unique opportunity for understanding the formation of black holes in binaries. For these systems there exists a wealth of observational information about their current physical state: BH and donor masses, orbital period, donor's position on the H-R diagram and surface chemical composition, transient or persistent and Roche-lobe overflow (RLO) or wind-driven character of the mass-transfer (MT) process, and distances. Even more recently proper motions have been measured for a handful of these systems (Mirabel et al. 2001, 2002; Mirabel & Rodrigues 2003a), complementing the earlier measurements of center-of-mass radial velocities and giving us information about the 3-dimensional kinematic properties of these binaries.

In what follows we summarize how the currently available observational constraints for XRBs can be used to uncover this past history, considering the specific case of GRO J1655-40. The ultimate goal of this project is to examine the systematics of the derived compact object progenitor masses and requirements for supernova (SN) kicks and their dependence on compact object masses and associated mass loss at the core collapse events. Such a comprehensive investigation will not just reveal the origin of these Galactic XRBs, but may also offer insight to the physics of supernovae and the possible association of natal kicks with BH formation.

Recently, Greene, Bailyn, & Orosz (2001) and Beer & Podsiadlowski (2002) have undertaken detailed studies of the ellipsoidal light variations when the system is in quiescence and derived rather different constraints for the two component masses and the position of the donor in the H-R diagram. In this study we therefore consider both sets of constraints and study them as two separate cases (GBO and BP). For more details of the analysis and results, see Willems et al. (2004).

OUTLINE OF ANALYSIS METHODOLOGY AND BASIC ASSUMPTIONS

According to our current understanding, the formation of a BH XRB with a low- to intermediate-mass donor star requires a primordial binary with an extreme mass ratio and a primary (the BH progenitor) mass in excess of $\simeq 20-25 M_\odot$. If the period of the primordial binary is in the range from $\simeq 1$ to $\simeq 10$ yr, the primary is expected to become larger than its critical Roche lobe and, due to the extreme mass ratio, lose most of its hydrogen-rich envelope in a dynamically unstable common-envelope phase. Provided that enough energy is available to completely expel the envelope and avoid a merger, the common-envelope phase results in the formation of a tight binary consisting of a helium star (the core of the Roche-lobe filling primary) and a relatively unevolved low- to intermediate-mass main-sequence (MS) companion. A BH XRB is then formed when the helium star collapses into a BH and stellar evolution and/or orbital angular momentum losses (or exchange through tides) cause the secondary in its turn to fill its Roche lobe and transfer mass to the BH. This formation channel is analogous to that usually considered for the formation of NS XRBs and has been considered previously by, e.g., Romani (1996), Portegies Zwart, Verbunt, & Ergma (1997), Ergma & van den Heuvel (1998), Kalogera (1999), and Podsiadlowski et al. (2003).

Our goal in this analysis is to trace back the evolutionary history of GRO J1655-40 and to constrain the properties of the system's progenitor just before and right after the SN explosion that formed the BH. The method adopted to derive the pre- and post-SN constraints incorporates the following set of calculations.

We first use a stellar evolution code and calculate a grid of evolutionary sequences for binaries in which a BH is accreting mass from a Roche-lobe filling companion. To consider the full range of possibilities, we include sequences for both conservative (for sub-Eddington rates) and fully non-conservative MT. For each sequence, we examine whether at any point in time the calculated binary properties, i.e., BH and donor masses, donor effective temperature and luminosity, and orbital period, are in agreement with the observational measurements or derivations of these quantities within their associated uncertainties. Among these quantities, the orbital period is measured with the highest accuracy and hence presents the most stringent constraint to be satisfied. We consider a large number of MT sequences and, although many of them can satisfy *some* of the constraints, the majority of sequences clearly fail to simultaneously satisfy *all* of the constraints at any given time. In the case of GRO J1655-40 the successful sequences furthermore have to satisfy the additional requirement that the system exhibits transient behavior. For the time interval during which all other constraints are satisfied, the long-term MT rate from the donor star must therefore be lower than than the critical rate separating transient from persistent behavior (Dubus et al. 1999 and references therein). This last requirement restricts the successful sequences even further. With the remaining fully successful sequences, we derive the properties of the binary at the onset of the RLO phase: initial BH and donor masses, orbital period, and age of the donor star. The time at which the fully successful sequences satisfy all observational constraints furthermore provides an estimate for the donor's current age.

Next, we consider the kinematic evolutionary history of the XRB in the Galactic potential. In particular, we use the current position and the measured 3D velocity with their associated uncertainties to trace the Galactic motion back in time. Combined with the tight constraints on the current age of the system given by the successful MT sequences this allows us to determine the position and velocity of the binary at the time of BH formation (we denote these as the "birth" location and velocity). By subtracting the local Galactic rotational velocity at this position from the system's total center-of-mass velocity, we then obtain an estimate for the *peculiar* velocity of the binary right after the formation of the BH.

The "birth" or post-SN peculiar velocity holds information about the mass loss and possibly the natal kick associated with the BH's formation. In order to extract this information we must however also constrain the orbital period and orbital eccentricity right after the formation of the BH. We derive these constraints in the third step of our analysis: we consider pairs of post-SN orbital periods and eccentricities and integrate the equations governing the evolution of the orbit under the influence of tides and general relativity (possibly important for the most highly eccentric orbits) forward in time. By using the age of the donor star at the onset of RLO for each of the successful MT sequences as an estimate for the time expired since the formation of the BH, we are able to map the post-SN orbital parameters to those at the onset of RLO. Comparison with the binary properties at the onset of RLO given by the sequences then allows us to select only those pairs of post-SN orbital period and eccentricity that can match these properties at the right time (i.e. at the right age or evolutionary stage of the donor star).

In the fourth step of our analysis we use the derived post-SN masses, orbital period, eccentricity, and peculiar velocity and examine the orbital dynamics of the compact object formation allowing for a natal kick. Based on angular momentum and energy conservation we derive constraints on the pre-SN binary properties (BH progenitor mass and orbital separation) and the natal kick (magnitude and direction) that may have been imparted to the BH.

PROGENITOR CONSTRAINTS

The elements presented above can be combined to establish a complete picture of the evolutionary history of GRO J1655-40, the pre- and post-SN binary properties, and the dynamics involved in the core-collapse event that formed the BH. In the following paragraphs, we first, as an example, discuss the progenitor constraints derived following this procedure for a single successful MT sequence, and next summarize the conclusions from all successful MT sequences.

To illustrate the derivation of the constraints, we consider the conservative MT sequence BP62 (see Table 3 in Willems et al. 2004) with initial BH mass $M_{BH} = 4.5 M_\odot$, initial donor mass $M_2 = 2.5 M_\odot$, and initial orbital period $P_{orb} = 1.4$ days. The constraints obtained for the mass M_{He} of the BH's helium star progenitor, the magnitude V_k of the kick velocity imparted to the BH at birth, the pre-SN orbital separation A_{preSN}, and

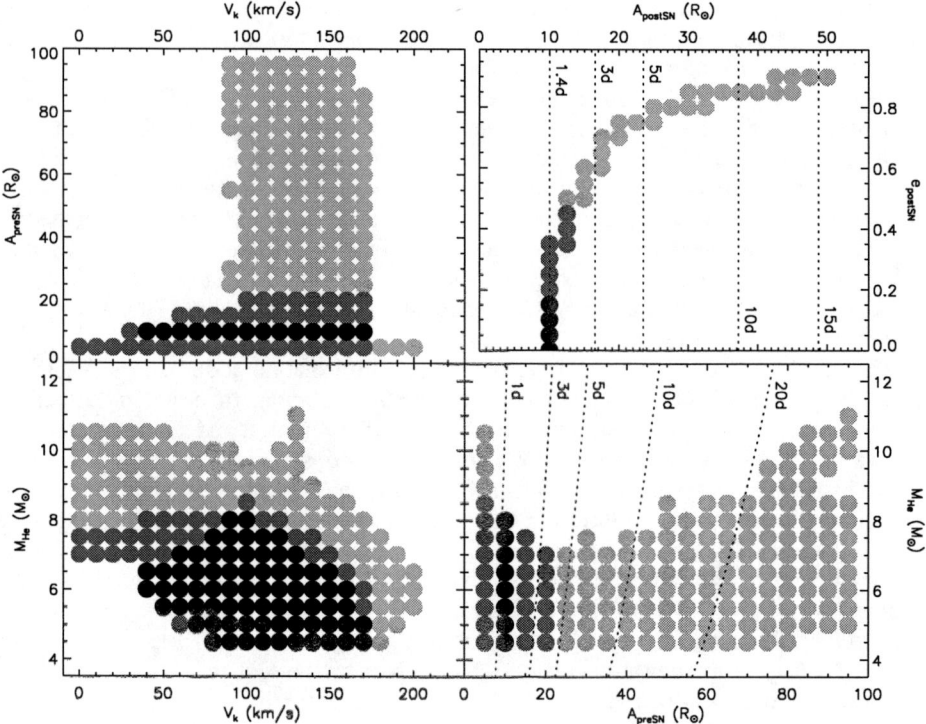

FIGURE 1. Constraints on the pre-SN helium star mass M_{He}, the magnitude of the kick velocity V_k, the pre-SN orbital separation A_{preSN}, and the post-SN orbital separation A_{postSN} and eccentricity e_{postSN} associated with the successful MT sequence BP62 (see Table 3 in Willems et al. 2004). Light grey circles correspond to progenitors for which RLO at periastron occurs more than 200 Myr earlier than dictated by the initial age of the donor star in the MT sequence, dark grey circles correspond to progenitors for which RLO at periastron occurs between 50 and 200 Myr earlier, and black circles correspond to progenitors for which RLO at periastron occurs within 50 Myr of the time dictated by the MT sequence. The black circles correspond to our preferred set of solutions. The dotted lines in the right-hand panels are lines of constant orbital period.

the post-SN orbital separation A_{postSN} and eccentricity e_{postSN} are presented in Fig. 1. The tight correlation between A_{postSN} and e_{postSN} (top right panel) arises from matching the post-SN orbital elements to those of the MT sequence at the onset of RLO. Pairs of A_{postSN} and e_{postSN} are considered to be acceptable if, after following the orbital evolution for a time equal to the age of the donor at RLO onset, the total angular momentum is equal to that given by the MT sequence within the uncertainties of the numerical calculations. This condition, however, does not guarantee that the eccentric orbit can accommodate the BH companion throughout the time interval between BH formation to RLO. In particular, for significantly eccentric orbits, RLO at periastron may occur at a time when the donor star has not yet reached the right evolutionary stage.

These pairs of $(A_{\text{postSN}}, e_{\text{postSN}})$ will therefore be valid post-SN orbital parameters only if RLO at periastron does not significantly affect the star and the orbit. In view of our limited knowledge of MT in eccentric binaries and its effects on the orbital elements, we include these solutions in the presentation of the progenitor constraints, although we consider them to be less compatible with the initial RLO conditions imposed by the MT sequence. In order to distinguish between solutions that lead to MT at periastron at times significantly close to RLO onset in the MT sequence and those that do not, we separate them into three groups in Fig. 1: solutions for which RLO at periastron occurs much too early (i.e. more than 200 Myr) are represented by light grey circles, solutions for which RLO at periastron occurs between 50 and 200 Myr too early are represented by dark grey circles, and solutions for which RLO at periastron occurs within 50 Myr of the proper time given by the MT sequence are represented by black circles. We consider the progenitor constraints associated with the latter solutions to be the most compatible with the MT sequence under consideration and therefore refer to them as our "preferred set" of solutions.

When RLO at periastron occurs within $\simeq 50$ Myr of the required donor age, $4.5 M_\odot \lesssim M_{\text{He}} \lesssim 8 M_\odot$ and $40 \,\text{km}\,\text{s}^{-1} \lesssim V_k \lesssim 170 \,\text{km}\,\text{s}^{-1}$. For our preferred set of solutions, the formation of the BH must therefore be accompanied by a natal kick. When solutions for which RLO at periastron occurs within $\simeq 50-200$ Myr of the required donor age are included, the ranges slightly broaden to $4.5 M_\odot \lesssim M_{\text{He}} \lesssim 8.5 M_\odot$ and $V_k \lesssim 170 \,\text{km}\,\text{s}^{-1}$. Hence, the derived constraints are fairly robust even when we allow RLO at periastron to occur up to $\simeq 200$ Myr too early. Somewhat larger variations are found when RLO at periastron occurs more than $\simeq 200$ Myr too early, but MT for these systems is likely to significantly affect both the star and the orbit so that the final parameters after $\simeq 480$ Myr (the required age of the donor at the start of RLO) of orbital evolution are probably incompatible with the initial conditions of the considered MT sequence and thus these solutions are not favored. Since highly eccentric post-SN orbital configurations always fall under the latter category, we do not consider post-SN orbital eccentricities larger than 0.9 in the derivation of any of the constraints.

To understand the core-collapse event leading to the formation of the BH, we are mainly interested in the constraints derived for the mass of the BH's helium star progenitor and the kick velocity that may have been imparted to the BH at birth. For the remainder of this section, we therefore restrict ourselves to presenting the constraints derived for these two quantities. Given that there is no way to distinguish which of the various successful MT sequences corresponds to the true progenitor of GRO J1655-40, we derive the constraints for each successful sequence and examine them collectively.

Since we expect the true MT sequence describing the evolution of GRO J1655-40 from the start of the X-ray phase to its present configuration to be encompassed by the successful MT sequences, we overlay the M_{He} and V_k constraints for all the successful conservative and non-conservative MT sequences found for GBO and BP system parameters in Fig. 2. It follows that even the combination of the extreme cases of fully conservative and fully non-conservative MT does not significantly relax the constraints on the progenitor masses and kick velocities with respect those obtained for individual sequences. For our preferred set of solutions, the helium star mass and kick velocity are constrained to $5.5 M_\odot \lesssim M_{\text{He}} \lesssim 10.5 M_\odot$ and $30 \,\text{km}\,\text{s}^{-1} \lesssim V_k \lesssim 160 \,\text{km}\,\text{s}^{-1}$ in the case

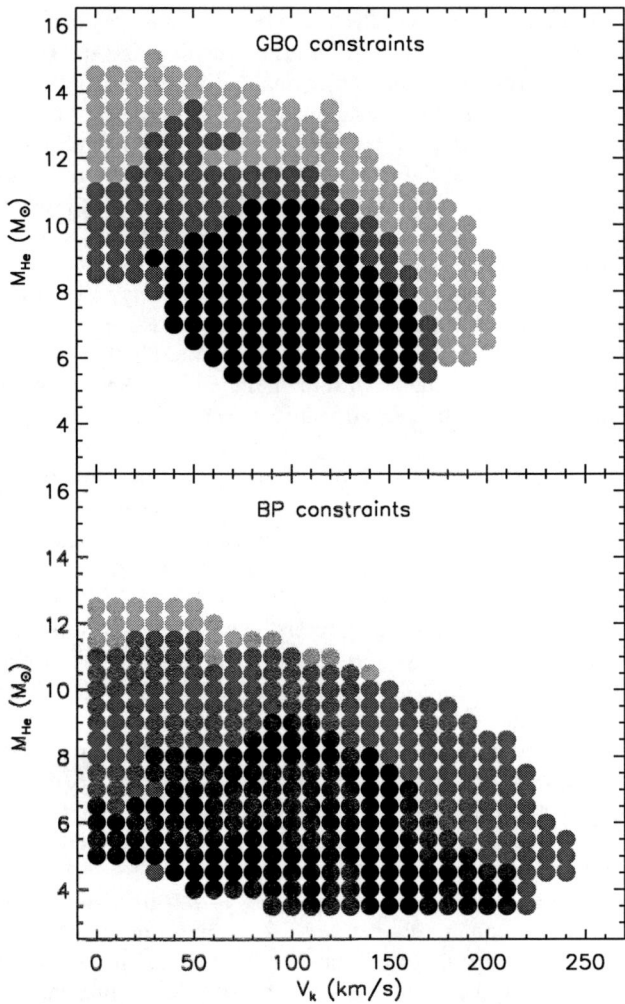

FIGURE 2. Constraints on the mass M_{He} of the BH's helium star progenitor and the magnitude V_k of the kick velocity that may have been imparted to the BH obtained by combining the results from all successful MT sequences.

of GBO parameters, and to $3.5 M_\odot \lesssim M_{He} \lesssim 9.0 M_\odot$ and $0\,\mathrm{km\,s^{-1}} \lesssim V_k \lesssim 210\,\mathrm{km\,s^{-1}}$ in the case of BP parameters. On average, higher progenitor masses furthermore tend to be associated with lower kick velocities since more mass loss requires less of kick to achieve the same result. A similar correlation was found by Willems et al. (2004) for the progenitor of PSR J0737-3039B.

We conclude that in the case of GBO parameters, it is likely that the BH in GRO J1655-40 received a kick at birth with a minimum kick velocity of $\simeq$ 30–

$40 \,\mathrm{km\,s^{-1}}$. In the case of BP parameters, a kick is much less of a requirement, although the zero-kick solutions are only marginally consistent with the observed binary properties. The main differences between the constraints derived for GBO and BP parameters are related to the higher initial BH masses and the correspondingly higher helium star masses which reduce the center-of-mass velocity imparted to the binary by a symmetric SN explosion in the case of GBO parameters.

DISCUSSION AND CONCLUSIONS

In this paper, we initiated a comprehensive study on the formation and evolution of Galactic XRBs. As a first application, we constrained the progenitor properties and the formation of the BH in the soft X-ray transient GRO J1655-40.

A crucial element emerging from the MT calculations performed to model the ongoing X-ray phase is the age of the donor star when all observational constraints are satisfied (i.e. the donor's *current* age). By using this age as an estimate for the time expired since the BH formation, we determined the possible birth sites and the post-SN peculiar velocity of the BH binary by following the motion of the system in the Galaxy backwards in time.

The MT sequences also yield the age of the donor at the onset of RLO. Knowledge of this age allows us to link the post-SN orbital parameters to those at the onset of RLO by numerically integrating the system of differential equations governing the orbital evolution due to tides and gravitational radiation. Post-SN orbital semi-major axes and eccentricities compatible with successful MT sequences range from $A_{\mathrm{postSN}} = 5 R_\odot$ to $A_{\mathrm{postSN}} = 15 R_\odot$ and from $e_{\mathrm{postSN}} = 0$ to $e_{\mathrm{postSN}} = 0.35$.

The solution of the equations describing the conservation of orbital energy and angular momentum during compact object formation allows us to derive constraints on the pre-SN orbital separation, on the mass of the BH's helium star progenitor, and on the magnitude of the kick velocity that may have been imparted to the BH at birth. Despite the large number of successful MT sequences found, the progenitor constraints turn out to be fairly robust to the initial parameters of the sequences at the onset of RLO. Combining the constraints obtained for all successful MT sequences for both conservative and non-conservative MT, yields $M_{\mathrm{He}} \simeq 5.5\text{--}10.5 M_\odot$ and $V_k \simeq 30\text{--}160 \,\mathrm{km\,s^{-1}}$ for GBO system parameters and $M_{\mathrm{He}} \simeq 3.5\text{--}9.0 M_\odot$ and $V_k \simeq 0\text{--}210 \,\mathrm{km\,s^{-1}}$ for BP system parameters (see Fig. 2). It is clear that symmetric SN explosions lie at the edge of our solution ranges and that BH kicks of at least a few tens of $\mathrm{km\,s^{-1}}$ are favored by the majority of the successful MT sequences.

The question of whether or not a natal kick was imparted to the BH in GRO J1655-40 was addressed previously by Brandt et al. (1995), Nelemans et al. (1999), and Fryer & Kalogera (2001). All three of these investigations used a lower limit on the present-day center-of-mass velocity given by the *current* radial velocity ($V_r = -114 \pm 19 \,\mathrm{km\,s^{-1}}$) instead of the actual post-SN peculiar velocity to constrain the mass loss and kick magnitude associated with the BH's formation. This neglects changes in the system's center-of-mass velocity resulting from its acceleration in the Galactic potential. We here find that, even though following the Galactic motion of the system backward in time yields post-SN peculiar velocities between $\simeq 45$ and $\simeq 115\,\mathrm{km\,s^{-1}}$, the additional constraints on the binary properties make the possibility of a symmetric SN explosion only marginally acceptable solutions.

ACKNOWLEDGMENTS

We are indebted to Laura Blecha for sharing the code used to follow the motion of GRO J1655-40 in the Galactic potential, and to Jon Miller, Jerome Orosz, Jeffrey McClintock, Klaus Schenker, and Gijs Nelemans for useful and stimulating discussions. This work is supported by a David and Lucile Packard Foundation Fellowship in Science and Engineering grant.

REFERENCES

1. Brandt, W.N., Podsiadlowski, P., & Sigurdsson, S. 1995, MNRAS, 277, L35
2. Beer, M. & Podsiadlowski, P. 2002, MNRAS 331, 351
3. Dubus, G., Lasota, J., Hameury, J., & Charles, P. 1999, MNRAS, 303, 139
4. Ergma, E. & van den Heuvel, E.P.J. 1998, A&A, 331, L29
5. Fryer, C. & Kalogera, V. 2001, ApJ, 554, 548
6. Greene, J., Bailyn, C.D., & Orosz, J.A. 2001, ApJ, 554, 1290
7. Kalogera, V. 1999, ApJ, 521, 723
8. Mirabel, I.F., Dhawan, V., Mignani, R.P., Rodrigues, I., & Guglielmetti, F. 2001, Nature, 413, 139
9. Mirabel, I.F., Mignani, R., Rodrigues, I., Combi, J.A., Rodríguez, L.F., & Guglielmetti, F. 2002, A&A, 395, 595
10. Mirabel, I.F. & Rodrigues, I. 2003a, Science, 300, 1119
11. Nelemans, G., Tauris, T.M., & van den Heuvel, E.P.J. 1999, A&A, 352, L87
12. Podsiadlowski, P., Rappaport, S., & Han, Z. 2003, MNRAS, 341, 385
13. Portegies Zwart, S.F., Verbunt, F., & Ergma, E. 1997, A&A, 321, 207
14. Romani, R.W. 1996, IAU Symp. 165: Compact Stars in Binaries, 165, 93
15. Willems, B., Henninger, M., Levin, T., Ivanova, N., Kalogera, V., McGhee, K., Timmes, T.X., Fryer, C.L. 2004, ApJ, submitted (astro-ph/0411423)
16. Willems, B., Kalogera, V., & Henninger, M. 2004, ApJ, 616, 414

The 2003 Outburst of the Galactic Microquasar V4641 Sgr (=SAX J1819.3-2525)

Dipankar Maitra* and Charles D. Bailyn*

Department of Astronomy, Yale University, New Haven, CT 06511, USA

Abstract. The black hole candidate V4641 Sgr (=SAX J1819.3-2525) went through a brief outburst during 2003 Aug 01 to Aug 08. During the outburst, activity was noted in optical, radio as well as X-rays. We report results of our X-ray spectral analysis and simultaneous X-ray and optical observations of the source during this outburst. Characteristics of the observed X-ray spectral energy distribution is discussed. The spectral results suggest occasional outflow/mass ejection and an obscured accretion disk. Simultaneous X-ray and optical R band lightcurves show strong correlation and the optical lags the X-rays, which can be explained if the origin of the optical emission is attributed to reprocessing of X-rays at the outer regions of the accretion disk.

Keywords: accretion, accretion disks — stars: black holes — X-rays: binaries — individual (V4641 Sgr)
PACS: 97.80.Jp, 98.70.Qy, 97.60.Gb, 97.60.Jd, 97.60.Lf

V4641 Sgr (=SAX J1819.3-2525) is presently known to be a compact binary system harboring an $9.6 M_\odot$ black hole and a $6.5 M_\odot$ B9 secondary[1]. X-ray activity was noted for the first time in 1999[3]. Rapidly changing radio structure was observed[4] during the preiod of strong X-ray and optical activity. The inferred superluminal proper motions together with the radio morphology suggested that V4641 Sgr is a relativistic jet source. This puts it in the category of galactic microquasars[5]. Activity was also observed in 2002 in X-ray, optical, and radio wavelenghts from the source [8, 9, 10].

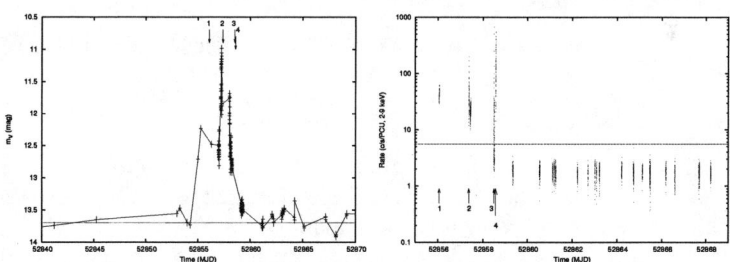

FIGURE 1. Left: Optical V band lightcurve showing the 2003 outburst of V4641 Sgr. Four vertical lines on the top labelled 1-4 are the times when significant X-ray activity was observed by RXTE. The horizontal dotted line represents the quiescent brightness of 13.7 magnitude. The data were taken by the CTIO 1.3m telescope operated by the Small and Moderate Aperture Research Telescope System (SMARTS) Consortium. Right: Background subtracted 2-9keV RXTE/PCA countrates during the 2003 outburst of V4641 Sgr. The dotted horizontal line is the background. Significant activity was seen only in dwells labelled 1-4 and are shown in detail in Fig. 2.

CP797, *Interacting Binaries: Accretion, Evolution, and Outcomes*,
edited by L. Burderi, L. A. Antonelli, F. D'Antona, T. Di Salvo,
G. Luca Israel, L. Piersanti, A. Tornambè, and O. Straniero
© 2005 American Institute of Physics 0-7354-0286-8/05/$22.50

V4641 Sgr went through a brief outburst during 2003 Aug 01 to Aug 08. Violent activity was noted in optical [11, 12, 13], radio [14, 15] as well as X-rays [12]. Optical V band observations from the Small and Moderate Aperture Research Telescope System (SMARTS[16]) 1.3m telescope during the outburst is shown Fig. 1. RXTE target of opportunity observations were triggered based on the optical observations. 22 pointed XTE observations of the source were carried out between MJD 52856 (2003 Aug 05) and MJD 52869 (2003 Aug 18, see Fig 1). However significant X-ray activity was noted in only the first 4 dwells. Background subtracted 2-60 keV PCA lightcurve during these four pointings are shown in Fig. 2.

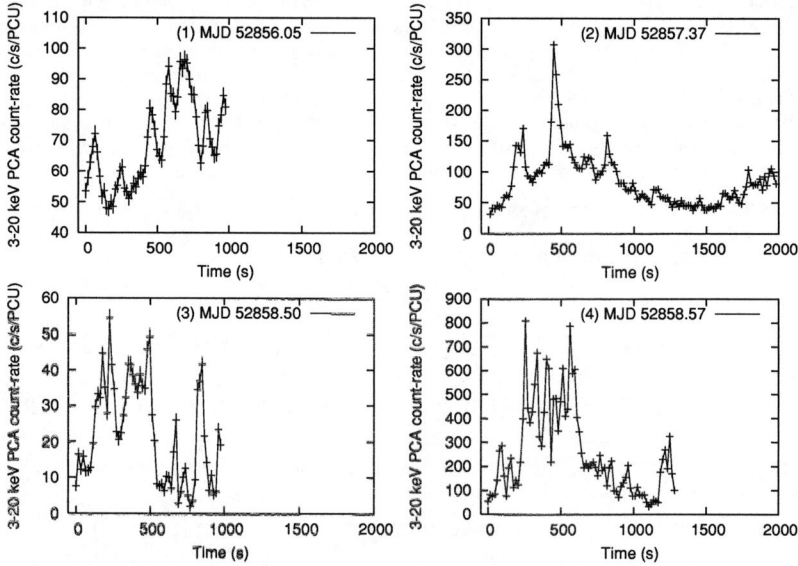

FIGURE 2. PCA background subtracted 3-20keV lightcurves with 16 second binning during four observing dwells. The starting time for the dwells are shown to the top-right of each dwell. Note the variability within the dwells and between the dwells, in particular dwells (3) and (4), which are only 1.7 hours apart whereas the average count-rate differs by a factor of ~ 20.

The energy spectra during the observations were very hard in nature, with almost no thermal blackbody component. Broad Fe emission line near 6.5 keV was observed in the spectra of all the dwells with very high equivalent widths ($\sim 0.3 - 2$ keV). In Fig. 3 we show the spectrum from the first 100 seconds of dwell (2). Except for the region around the iron line, a Comptonized powerlaw(*pexrav* model[17] in XSPEC) fits the continuum spectrum quite well ($\chi^2_\nu = 1.05$). Since the broad iron line feature makes it difficult to estimate the continuum parameters at low energies, we looked at the spectra from 10keV and above using PCA + HEXTE data.

It is assumed that the energy spectra in these regions is free of narrow features and lines. We expect that the radiation at these energies is created by non-thermal processes and therefore should have an intrinsic powerlaw spectral energy distribution. However, as seen in Fig. 4(a), a simple powerlaw fails to match the observed spectrum. The fits improved significantly ($\chi^2_\nu = 0.95$) upon inclusion of a Compton reflection

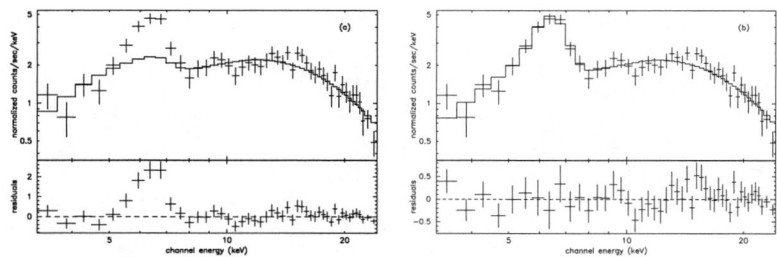

FIGURE 3. Spectral fits to the first 100 seconds of dwell (2). (a) Fitted 10-25keV with comptonized powerlaw only and extrapolated to lower energies. (b) Fitted 3-25keV, with a Gaussian line and compton reflected powerlaw.

component, as shown in Fig. 4(b). The strong iron emission line, Compton reflection feature and absence of blackbody in the spectra during the outburst suggest that the X-ray emitting disk is probably obscured from the observer, accompanied by the formation of an envelope/outflow enshrouding the central source as suggested by Revnivtsev et al. (2002)[18]. In fact, during 100s-200s of dwell (2) we found a flare which could not be modelled with the standard value of absorption column density of $2.3 \times 10^{21} cm^{-2}$ for this source[19]. Allowing n_H to vary resulted in an interesting plot shown in Fig. 5, which shows that the flare is accompanied by a sharp increase in absorption column. Although it is difficult to determine n_H accurately using the medium resolution of XTE, the selectively enhanced absorption of very soft photons during this flare is clearly real. The thick warm envelope/outflow scenario of Revnivtsev et al.(2002) can explain the observed/fitted variation of column density although other scenarios like variation of partial covering fractions cannot be ruled out.

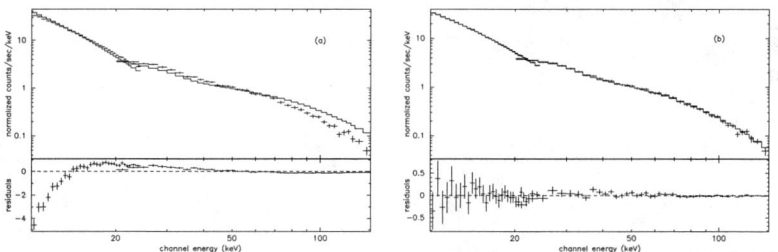

FIGURE 4. Spectral fits in 10-150keV energy range for dwell (4). (a) A simple powerlaw gives a very poor fit to the spectrum. (b) A Compton reflected powerlaw fit to the data. It is evident that a reflection component is required.

Simultaneous X-ray observations using RXTE and optical R band (from Lu-Lin Observatory, Taiwan; see Fig. 6) observations were obtained for about 22 minutes starting MJD 52858.57 (2003 Aug 07). The optical lightcurve showed strong correlation with X-ray data (Bailyn et al., in preparation). The optical light was lagging behind the X-rays by about 25-30s as suggested by the cross-correlation function between the X-rays and the optical (Fig. 6). From the orbital ephemeris[1], the binary phase was 0.06 during the time of simultaneous X-ray/optical observation. At the orbital phase during the simultaneous X-ray/optical observation of V4641 Sgr, the optical delay due to reprocessing at the surface of the secondary only, should be ~ 5 seconds. For this binary system, the distance from the black hole to the inner Lagrangian point is about 29 light seconds, therefore an average lag of 25-30s seems reasonable for reprocessing. The small binary phase further implies that a very small fraction of reprocessed light from the secondary surface reached the observer. Therefore, in the reprocessing scenario, most of the optical was created by reprocessing of X-rays from the central engine, near the outer edge of the accretion disk. From the lag timescale, a jet scenario seems possible but unlikely.

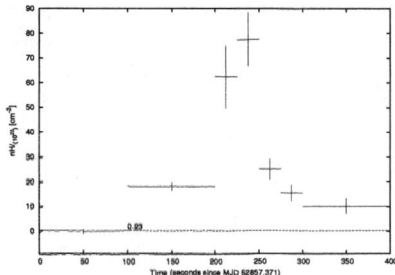

FIGURE 5. Variation of column density of the warm absorber during the hard flare seen in dwell (2).

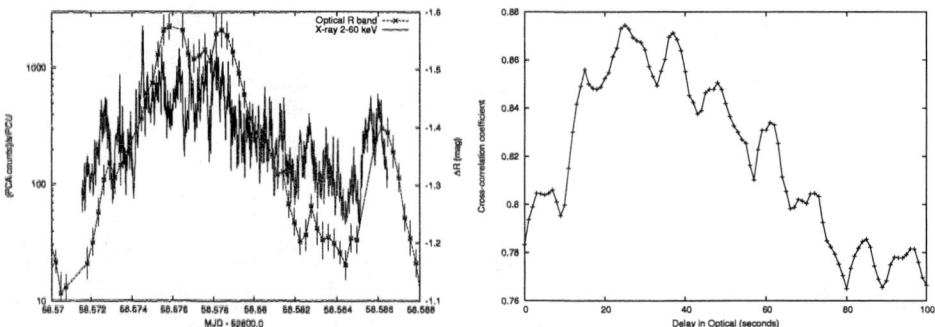

FIGURE 6. (Left) Simultaneous optical and X-ray lightcurve taken on 2003 Aug 7th (Dwell 4). Optical R band data taken by NCU Lu-Lin Observatory, Taiwan. X-ray data are one second binned PCA 2-60 keV background subtracted count-rates. (Right) The cross-correlation function between the optical and X-rays for the data in Fig.4. The peak(s) suggest a delay in the arrival of the optical by about 25-30s. If the optical is produced by reprocessing of the X-rays, then such a delay is expected.

The X-ray spectrum between 3-200 keV during the interval (MJD52858.571 - MJD52858.587) is mainly composed of a hard comptonized powerlaw of photon index 1.36 ± 0.02. A broad Fe emission line was noted near 6.2 ± 0.1 keV with an equivalent width of ~ 362 eV. No blackbody component was found in the spectrum. When the spectrum was fitted with the absorption column density fixed to the standard value of $2.3 \times 10^{21} cm^{-2}$, we found that the fitted countrates in the soft bands below 5 kev were overestimated. Allowing the n_H column density to vary with other parameters fixed results in a slightly better fit with fitted n_H about factor of 2 higher than the standard value. As discussed in the previous section, a similar drop in soft countrates is also observed in other observations of the source (Maitra & Bailyn, in preparation). The average 2-300 keV flux during this period was 1.77×10^{-8} $ergs/cm^2/s$ which corresponds to about 16% isotropic Eddington luminosity for a $10 M_\odot$ black hole at a distance of 10kpc.

Conclusions:

- V4641 Sgr went through a short lived, hard flaring outburst between 2003 Aug 01 through Aug 08, during which X-ray fluctuations by a factor of 10 on timescales of minutes were seen. Similar variability was also observed in optical bands.
- Compton reflected powerlaw photons dominated the spectra with little or no thermal blackbody detected and broad Fe Kα emission line near 6.5keV were seen with equivalent widths $\sim 0.3 - 2$keV.
- Simultaneous X-ray and optical data showed strong correlation between X-ray and optical light curves. The optical is delayed from the X-rays by about 25-30s.
- Large variation in soft (3-5keV) X-rays were seen in timescales of ~ 100s which could be due to an outflow from the black-hole.

ACKNOWLEDGMENTS

This work was supported by the National Science Foundation grant AST 0407063.

REFERENCES

1. J. A. Orosz, E. Kuulkers, M. van der Klis, J. E. McClintock, M. R. Garcia, P. J. Callanan, C. D. Bailyn, R. K. Jain, and R. A. Remillard, *ApJ*, **555**, 489–503 (2001).
2. V. P. Goranskij, *Astronomicheskij Tsirkulyar*, **1024**, 3–4 (1978).
3. J. in 't Zand, J. Heise, A. Bazzano, M. Cocchi, L. di Ciolo, and J. M. Muller, *IAU Circ.*, **7119** (1999).
4. R. M. Hjellming, M. P. Rupen, R. W. Hunstead, D. Campbell-Wilson, A. J. Mioduszewski, B. M. Gaensler, D. A. Smith, R. J. Sault, R. P. Fender, R. E. Spencer, C. J. de la Force, A. M. S. Richards, S. T. Garrington, S. A. Trushkin, F. D. Ghigo, E. B. Waltman, and M. McCollough, *ApJ*, **544**, 977–992 (2000).
5. I. F. Mirabel, and L. F. Rodriguez, *Nature*, **371**, 46 (1994).
6. S. J. Tingay, D. L. Jauncey, R. A. Preston, J. E. Reynolds, D. L. Meier, D. W. Murphy, A. K. Tzioumis, D. J. McKay, M. J. Kesteven, J. E. J. Lovell, D. Campbell-Wilson, S. P. Ellingsen, R. Gough, R. W. Hunstead, D. L. Jones, P. M. McCulloch, V. Migenes, J. Quick, M. W. Sinclair, and D. Smits, *Nature*, **374**, 141 (1995).
7. R. M. Hjellming, and M. P. Rupen, *Nature*, **375**, 464 (1995).
8. C. B. Markwardt, and J. H. Swank, "Sax J1819.3-2525," in *IAU Circ.*, 2002, p. 3.
9. M. Uemura, T. Kato, R. Ishioka, K. Tanabe, S. Kiyota, B. Monard, R. Stubbings, P. Nelson, T. Richards, C. Bailyn, and R. Santallo, *PASJ*, **54**, L79–L82 (2002).
10. M. P. Rupen, V. Dhawan, and A. J. Mioduszewski, "V4641 Sagittarii," in *IAU Circ.*, 2002, p. 2.
11. M. Buxton, D. Maitra, C. Bailyn, L. Jeanty, and D. Gonzalez, *The Astronomer's Telegram*, **170** (2003).

12. C. Bailyn, D. Maitra, M. Buxton, L. Jeanty, and D. Gonzalez, *The Astronomer's Telegram*, **171** (2003).
13. I. Khamitov, Z. Aslan, K. Yakut, I. Bikmaev, R. Gumerov, N. Sakhibullin, R. Sunyaev, M. Revnivtsev, and M. Pavlinsky, *The Astronomer's Telegram*, **174** (2003).
14. M. P. Rupen, A. J. Mioduszewski, and V. Dhawan, *The Astronomer's Telegram*, **172** (2003).
15. M. P. Rupen, V. Dhawan, and A. J. Mioduszewski, *The Astronomer's Telegram*, **175** (2003).
16. C. D. Bailyn, D. Depoy, R. Agostinho, R. Mendez, J. Espinoza, and D. Gonzalez, *Bulletin of the American Astronomical Society*, **31**, 1502 (1999).
17. P. Magdziarz, and A. A. Zdziarski, *MNRAS*, **273**, 837–848 (1995).
18. M. Revnivtsev, M. Gilfanov, E. Churazov, and R. Sunyaev, *A&A*, **391**, 1013–1022 (2002).
19. J. M. Dickey, and F. J. Lockman, *ARA&A*, **28**, 215–261 (1990).

SESSION 5

ACCRETION ON WHITE DWARFS AND NOVAE

Magnetic Accretion Onto White Dwarfs

Gaghik H. Tovmassian

Observatorio Astrónomico Nacional, Instituto de Astronomía, UNAM, México[1]

Abstract. The influence of the magnetic field on process of the accretion onto White Dwarfs in Cataclysmic Variables (CVs) is discussed. Except for the Polars or AM Her objects, the strength of magnetic field can not be measured directly in CVs by modern techniques. But there is growing evidence that most of the types of Cataclysmic Variables classified on the basis of their observational characteristics are behaving in one or the other way under the influence of the magnetic field of the accreting White Dwarf, among other things. Here, we discuss the bulk of CVs that are traditionally considered as *non magnetic* and review the properties that could be best explained by the magnetic governed accretion process.

Keywords: Cataclysmic Variables – Accretion – Magnetic Field
PACS: 97.80.Ũd, 97.10.Gz

INTRODUCTION

Cataclysmic Variables are at the low mass end of objects known as Interactive Binaries, where the primary, more massive and accreting component is a White Dwarf. The secondary is a pre-main sequence late type star that fills its Roche lobe and loses matter through the inner Lagrangian point L_1 to its compact companion. The transferred matter forms an accretion disc that usually is also the most luminous component of the binary system. Accretion is the cause of modulated brightness behavior with the Dwarf Novae (DN) class showing semi-periodic outbursts. According to the traditional models of Cataclysmic Variables, the accretion disc forms as a result of exchange of the angular momentum between the elements or particles comprising the disc, which otherwise move in Keplerian orbits in a ring with its radius being uniquely determined by the angular momentum. Thus, the ring (or torus) spreads out into a disk. It is obvious then that between the inner edge of the disc and the surface of the accreting star, which rotates with a different velocity, the excess mechanical energy of disk's element must be dissipated and its excess angular momentum transferred away, before that element can be accreted onto the stellar surface. This region is called the boundary layer. The processes occurring in the boundary layer and their observational fingerprints are not very well understood and remain a topic of controversy and debate in-spite of it relatively simple general picture [27]. Since the discovery of the magnetic CVs [28], it became obvious that at least some CVs possess a primary White Dwarf (WD) with a magnetic field strong enough to disrupt formation of the accretion discs and channel the transfered material through the magnetic lines directly onto the magnetic poles on the surface of the WD. It is also strong enough to overcome the spin-up torque of the accreting

---

[1] P.O. Box 439027, San Diego, CA, 92143-9027, USA

matter (see e.g., [12]) and synchronize rotation of the WD with orbital period. Magnetic field strength of these can be measured directly from observations and subsequently they were called Polars. Later these were joined by Intermediate Polars (IP) which are notorious for showing spin period of asynchronously rotating primary and often signs of presence of accretion disc. The magnetic field in IPs could not be measured directly, but it is unambiguously established that the variety of periods observed there in X-rays and optical are result of spinning WD which beams an intense X-ray emission modulated with P_{spin}. It is also universally accepted that inner parts of the accretion disk in IPs are truncated within corresponding Alfvén radius and the matter from there channeled onto the surface of the WD through the magnetic field lines, very much alike Polars. These two types of Cataclysmic Variables are defined as magnetic CVs and are the topic of another review talk at this conference [3]. Here, however, I will show that many observed features in different sub-classes of CVs usually considered as non-magnetic can be generally explained in terms of truncated accretion discs as in IPs and that the magnetic governed accretion plays significant role across the entire family of Cataclysmic Variables.

WHY ARE WHITE DWARFS MAGNETIC?

It is natural to compare properties of isolated White Dwarfs and primaries of CVs in order to find out how binary evolution and accretion processes influence their physics. And comparison of magnetic properties of these seemingly similar stars reveals significant differences. Wickramasinghe and Ferrario [37] published a large study of magnetic White Dwarfs where they show that isolated WDs are remnants of Ap and Bp stars with fossil magnetic fields of order of 0.1-1000 MG. They are thought to have significantly higher mean masses than their non-magnetic counterparts. They constitute about 5% of WD population. In contrast White Dwarfs in interacting binaries do not reach so high magnetic fields, and their masses are not much different from overall distribution of masses in CVs. And even taking into account only CVs recognized as magnetic (Polars and Intermediate Polars $\approx 10^5$ to 10^8 Gauss) with significant and often measurable strength of magnetic fields, the fraction of them easily reaches 25% of the total. While the lack of very high magnetic field CVs is a matter of ongoing discussions and speculations, the disparity of numbers can be attributed to processes taking place in close binary systems, rather than selection effects.

More recently Aznar Cuadrado et al.[2] discovered that probably up to 25% of White Dwarfs posses low magnetic fields of a few kG. They were previously considered as non-magnetic. Tout et al. [31] suggest that their magnetic field also has fossil origin of a cloud from which the stars emerged, and if so, all WDs born from stars over $2M_\odot$ might be a low field magnetic White Dwarfs. Regardless of the origins of magnetic field in WDs, it is important to stress for further consideration that the number of magnetic compact stars is much higher than previously thought.

CV TYPES THAT ARE SUSPECTED TO BE MAGNETIC

SW Sex stars

Similarly, the number of CVs in which magnetic driven accretion plays a significant role seems to be much higher than previously accepted. In most of the cases, it can not be measured directly, but there is observational evidence indicating influence of the magnetic field in the process of accretion on the WDs in a broad range of CV types. The most obvious is the case of much debated SW Sex stars. They were distinguished [29] for their peculiar emission and absorption line behavior. First thought to be eclipsing systems they were considered a rarity, but soon many other systems appeared showing one or other characteristics known as SW Sex phenomenon. According to Hellier [10], more than 20 systems show SW Sex phenomenon. Although he himself remains skeptical of the magnetic model for the explanation of SW Sex phenomenon, he admits that the evidence is mounting (see references therein). He assumes that periodic modulation of polarized emission is not enough evidence because the spin period of the WD does not come up in the photometric observations persistently, over a large time sets, as it easily happens in Intermediate Polars (IP). But it also can be argued that our inability to observe spin/beat modulations in SW Sex proves that there are many other CVs which do not show apparent IP photometric characteristics, but do accrete under magnetic field influence.

LS Peg and V795 Her, both members of SW Sex group are found to have circular polarization [23, 22]. This is direct evidence of magnetic field presence and its modulation with short periods most probably binds it with the spin period of WD. LS Peg shows circular polarization modulations with 0.3% amplitude and 29.6 min period. Simultaneously it shows emission line flaring with period corresponding to the beat period, if 29.6 min is considered as a spin period of WD. Circular polarization of 0.12% peak-to-peak amplitude was also detected in V795 Her with periods close to the optical quasi-periodic oscillations (QPO).

But except detection of circular polarization there are many indirect indications of magnetic driven accretion onto white dwarf in SW Sex systems. The same kind of emission line flaring with short periods as in V795 Her is detected in a number of other SW Sex objects: BT Mon and DW UMa. They are also typical to many Intermediate Polars (FO Aqr for example). A good example to demonstrate link between these seemingly different classes of CVs is V533 Her [21]. It erupted as Nova in 1963. In 1979, Patterson [17] reported rapid 63.5 sec variability classifying it as a DQ Her system (magnetic). But later this period disappeared and the system emerged as a 3.53 hour non-eclipsing SW Sex object showing among other SW Sex features flaring of emission lines.

Turning our attention to Nova remnants we find RR Cha, another Nova remnant that turned into an IP. Woudt & Warner [38] discovered a 1950 sec stable period and positive and negative superhumps in the system. Meanwhile Rodriguez-Gil and Potter [24] observed variable circular polarization and noted some distinct SW Sex features in its spectra. Here we touch upon another phenomenon (negative superhumps) that is common to number of systems. Patterson et al. [18] in a study of yet another two SW Sex objects find QPOs with periods around 1000 sec and negative superhumps. They believe

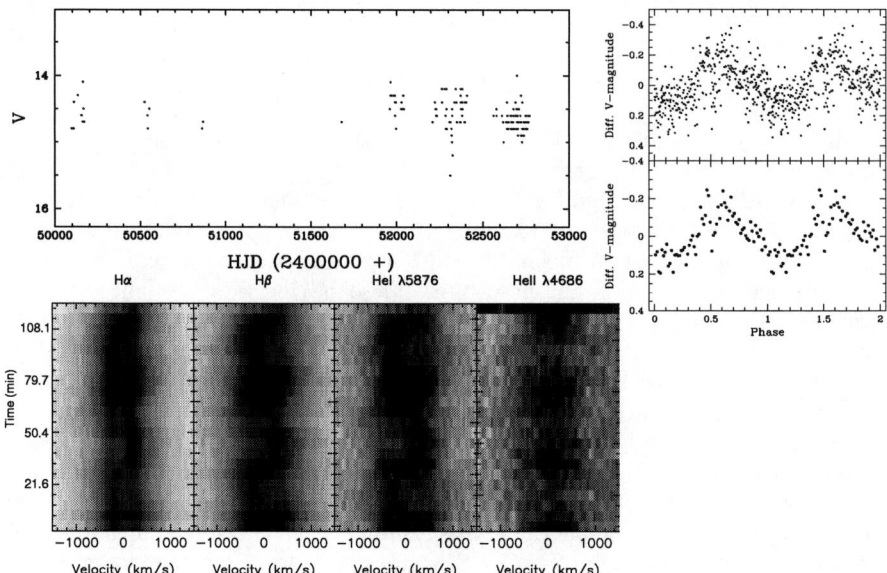

FIGURE 1. DW Cnc exhibiting features proper to a)VY Scl objects ($\approx$ 2 mag drop in brightness from quiescence levels); b) IPs (spin period of magnetic WD); c) SW SEX objects (emission line flaring). Adopted from Rodriguez-Gil et al. (2004)

the presence of negative superhumps can be best ascribed to the strong magnetism of the white dwarf. The warping of the disc is the most natural way of explanation of negative superhumps. It is widely agreed that the cause of warping is a magnetic field. But there are different approaches to whether the source of the magnetic field is the primary [13] or the secondary [16].

All these independent approaches to the SW Sex phenomenon show that the number of objects experiencing it are quite large and that the best explanation offered to explain variety of features is magnetic accretion onto a white dwarf.

VY Scl stars

Among other features shown by SW Sex objects are so called VY Scl characteristics. Or rather SW Sex sometimes considered to be part of larger VY Scl type of CVs. VY Scl are another growing group of CVs that are increasingly associated with magnetic CVs. Here, it is first worth mentioning DW Cnc. It was very recently studied by Rodriguez-Gil et al. [20]. They show that DW Cnc is a short period CV, $P_{orb} = 86$ min, which also shows 38.51 min photometric variability identified as a spin period of magnetic WD. Emission line flaring, another feature common to SW Sex objects, is present too. Most interestingly, this short period CV shows low states, down to 2 magnitude from its quiescence and no outbursts (see Fig. 1a-c borrowed from [20] exhibiting these features). The cyclical low states (anti dwarf nova behavior) is a main characteristic of VY Scl objects. They are believed to be concentrated in the 3-6 hours period range, same as SW Sex. A few years ago, we [6] demonstrated that the VY Scl star V751 Cyg shows a transient soft X-ray emission. The X-ray emission appears when the system is in low state. The interpretation was that V751 Cyg behaves very much alike super-soft X-ray binaries, e.g., RX J0513.9-6951. Later, however, Hameury & Lasota [7] suggested that VY Scl objects contain magnetic WDs (with magnetic field of order of 5×10^{30} G cm^3 corresponding to 0.4 mG for 0.7M$_\odot$ WD) based on their models and absence of outburst of VY Scl objects in low/intermediate states. This offers an alternative explanation of the soft X-ray emission in the low state. The spectrum of V751 Cyg in a low state, obtained almost simultaneously with the X-ray observations, shows a spectrum similar to a mCV. It has highly variable continuum, He II line becomes intense compared to the high state, lines are narrow, X-ray spectrum is soft. We considered the magnetic/polar scenario based on spectral appearance in the process of preparation of 1998 paper. The reason why we dismissed it, is still fundamental: how to switch off and on magnetic field or magnetic driven accretion between low and high luminosity states. Is it possible that with increased $\dot M$ the magnetic field can not cope anymore with the amount of incoming matter and the accretion geometry changes? Hameury & Lasota are primarily concerned with conditions of accretion disc and argue that the truncation of the disk is a key to the VY Scl phenomenon, but they do not reflect upon this question. But it needs to be answered. So far very little has been done to explore interaction and dependence of magnetic field to the mass transfer rate of accreted material. However there are hopeful signs that the problem can be tackled. Cumming [4] examined the problem to some depth. According to him compressional heating by accreting material can maintain interiors of WD in a liquid state. It allows to a decrease in the ohmic decay times to a few 10^9 years in contrast to isolated WDs, where ohmic decay time is always longer than the cooling time. He shows that as a consequence of accretion significant changes in surface magnetic field can occur. He also demonstrates that the higher is the magnetic field of the system, the lower the mass accretion rate. It is not immediately clear if the decrease of mass transfer/accretion rate in a VY Scl system provokes extension of magnetosphere and further truncation of the disc or just the decreased disc luminosity allow us to observe mCV features in the spectrum. Or what a Polar would look like if you increase the accretion rate an order or two from usual 10^{-11} M$_\odot$/year value. Certainly this problem needs additional research.

VY Scl and SW Sex objects comprise considerable part of CVs at the upper edge of Period Gap. But the problem is not confined only to VY Scl or SW Sex objects. There were reports of apparently ordinary Dwarf Novae displaying features like those of VY Scl. Recently explored DW Cnc is just one such case. Another one was reported in [30, 5]. SS Aur, a classical SS Cyg, system was caught in a low state for a short period of time. The object was down about a magnitude from its usual quiescence level. There were apparent changes in the spectrum of the object: instead of a power law corresponding to the accretion disc spectrum, two blackbody curves corresponding to the stellar components of this binary system nicely fit the observed flux distribution. The temperatures of components derived from this fitting were later confirmed from UV observations and parallax determinations by HST [8, 9, 26]. Most importantly it exhibited quasi-periodic photometric variations (Fig 2) with periods around 20 min. That is exactly what one would expect if the above described scenario is right: the diminished disc luminosity either from decreased mass transfer or truncation of the disc (or most probably from both at the same time) reveals periodic light variations best explainable in terms of IP model.

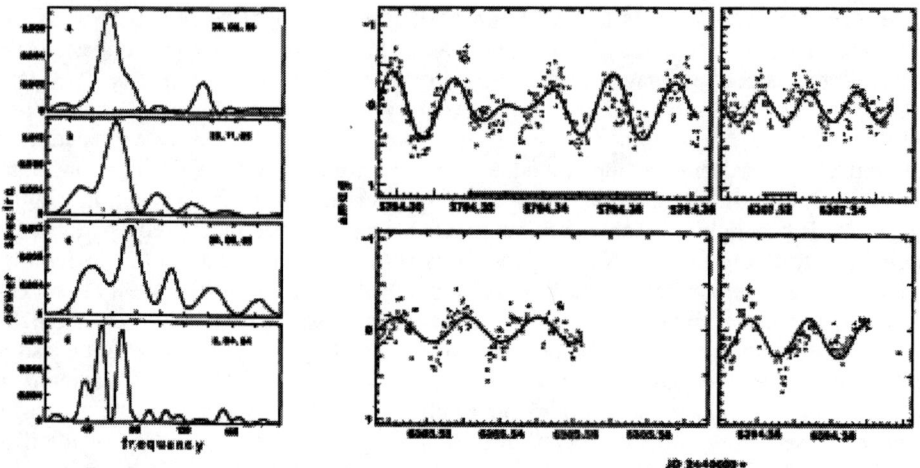

FIGURE 2. Quasiperiodic variations detected in the light curve of SS Aur during low luminosity states. On the left panel the power spectra of four different nights are presented, on the right corresponding light curves with *sin* fits. Adopted from Tovmassian (1986).

According to AAVSO light curves SS Aur was in a low state for very short time in contrast to usual VY Scl objects and our observation of quasi-periodic variations at that moment was completely accidental. We may assume that other Dwarf Novae also experiment episodes of low state with short duration and they are mostly unnoticed. It would be interesting to conduct a systematic search of such events and examine light curves for presence of periodic variations. Certainly it is necessary and would be easier to do that for members of VY Scl type of objects.

OTHER DWARF NOVAE

The DNe constitute one of the most numerous group of objects among CVs. The suggestion that many more DNe might experience short lasting low states (and their number is not limited to a few known cases) is completely speculative. However, another argument which favors the presence of magnetic driven accretion in DNe comes from observations of outbursts. Outbursts are the main feature that distinguish them from rest of CVs. Another well known fact is that during outbursts DNe show quasi-periodic oscillations of three distinct types [34]. These oscillations have long been associated with boundary layer based on observation of eclipsing systems, but their nature was not clearly understood and described. The study of quasi-periodic oscillations is complicated by the fact that they are not observed in every DNe, their magnitudes are small and highly variable and high time resolution, high precision photometry is required. One remarkable system that best suited for such study is VW Hydri. Warner et al. [32, 33], Woudt & Warner [38] in series of papers present results of long term study of QPOs mostly concentrated on this object, but not limited to it. They developed a Low Inertia

Magnetic Accretor (LIMA) model which allows to explain origin of QPOs and the existing relation between different types (different frequencies). The essence of the model is that the rapidly rotating equatorial belt, formed as a result of accretion of matter through disc on a surface of WD, enhances magnetic field of the primary. The magnetic field of a primary that is expected to be weaker than in regular Intermediate Polars, nevertheless reaches enough strength to channel accreting matter the way it does in IPs, but onto the equatorial belt instead of magnetic poles. The QPOs then arise due to a prograde traveling wave at the inner edge of the disc that reprocesses high energy radiation from accreting zones close to the primary. The frequency may be variable since the belt spins up during high accretion phase and decelerates after. The details of this model and certain relation existing between frequencies of QPOs common not only for CVs but also higher mass X-ray binaries are discussed in the Warner's [35] presentation included in this volume.

Interestingly, Huang et al. [11] detected inverse P Cyg profiles during superoutburst of VW Hyi and concluded that detached disc and structured gas flow is necessary for best-fitting model to describe their observation. Subsequently in [25] the same group demonstrate existence of equatorial belt around the WD after the outburst. On the other hand, X-ray observations of high inclination system OY Car [36] prove that X-rays come from an area much smaller than WD, probably upper polar region of the white dwarf, which testify that at least some DNe might have magnetic field strong enough to channel the accretion to the magnetic pole.

Some theoretical aspects of how the magnetic field can be induced/enhanced by a shear and influence the processes in boundary layer were considered by Armitage [1]. Completely different approach taken by Lasota [14] on the basis of Disc Instability Model (DIM) leads again to the idea that the internal parts of the accretion discs in most of DNe should be destroyed by the magnetic field and final stage of accretion occurs through magnetic lines. It is an extension of the idea first proposed for VY Scl objects, now applied to the OY Car, a classical Dwarf Nova for which the necessity of truncated disc was raised earlier but was attributed to the disc evaporation [15].

CONCLUSIONS

There is a growing observational evidence that the number of magnetic WDs is larger than was thought. Observations of limited sample of isolated WDs show that as many as 25% of WDs might have magnetic field strength of order of a few kG. Regardless of that fact, the number of systems considered as magnetic in Interactive Binaries with WD as a primary is unusually high compared to the distribution of isolated WDs. In addition to this, there are numerous groups of CVs traditionally not considered as magnetic, which increasingly require the presence and influence of the magnetic field on the accretion process in order to explain their observational characteristics. Probably the Intermediate Polar scenario of accretion on WDs in CVs, where the inner disc is truncated and matter channeled to the primary along magnetic lines is universal and accretion processes influence magnetic field strength in accreting compact objects.

Retter and Naylor [19] suggested that properties of CVs and thus their classification on both sides of Period Gap depends on their periods and mass transfer rate. Their scheme however does not include finer subdivision of CVs. If the hypothesis that the magnetic field plays important role in shaping properties of above-mentioned objects is correct, it could be stated that the classification of CVs is a function of their orbital period, mass transfer rate and magnetic field strength.

REFERENCES

1. Armitage, P. J. 2002, MNRAS, 330, 895
2. Aznar Cuadrado, R., Jordan, S., Napiwotzki, R., Schmid, H. M., Solanki, S. K., & Mathys, G. 2004, A&A, 423, 1081
3. Cropper, M., 2005, AIP Conf. this volume, Interacting Binaries, in preparation
4. Cumming, A. 2002, MNRAS, 333, 589
5. Efimov, Y. S., Tovmasyan, G. G., & Shakhovskoi, N. M. 1986, Astrophysics, 24, 131
6. Greiner, J., Tovmassian, G. H., Di Stefano, R., Prestwich, A., González-Riestra, R., Szentasko, L., & Chavarría, C. 1999, A&A, 343, 183
7. Hameury, J.-M. & Lasota, J.-P. 2002, A&A, 394, 231
8. Harrison, T. E., McNamara, B. J., Szkody, P., McArthur, B. E., Benedict, G. F., Klemola, A. R., & Gilliland, R. L. 1999, Ap.J.L., 515, L93
9. Harrison, T. E., McNamara, B. J., Szkody, P., & Gilliland, R. L. 2000, Ap.J., 120, 2649
10. Hellier, C. 2004, Revista Mexicana de Astronomia y Astrofisica Conference Series, 20, 148
11. Huang, M., Sion, E. M., Hubeny, I., Cheng, F. H., & Szkody, P. 1996, Ap.J., 458, 355
12. King, A. R. & Whitehurst, R. 1991, MNRAS, 250, 152
13. Lai, D. 1999, Ap.J., 524, 1030
14. Lasota, J.-P. 2004, Revista Mexicana de Astronomia y Astrofisica Conference Series, 20, 124
15. Meyer, F. & Meyer-Hofmeister, E. 1994, A&A, 288, 175
16. Murray, J. R., Chakrabarty, D., Wynn, G. A., & Kramer, L. 2002, MNRAS, 335, 247
17. Patterson, J. 1979, Ap.J.L., 233, L13
18. Patterson, J., et al. 2002, PASP, 114, 1364
19. Retter, A. & Naylor, T. 2000, MNRAS, 319, 510
20. Rodríguez-Gil, P., Gänsicke, B. T., Araujo-Betancor, S., & Casares, J. 2004, MNRAS, 349, 367
21. Rodríguez-Gil, P. & Martínez-Pais, I. G. 2002, MNRAS, 337, 209
22. Rodríguez-Gil, P., Casares, J., Martínez-Pais, I. G., & Hakala, P. J. 2002, ASP Conf. Ser. 261: The Physics of Cataclysmic Variables and Related Objects, 533
23. Rodríguez-Gil, P., Casares, J., Martínez-Pais, I. G., Hakala, P., & Steeghs, D. 2001, Ap.J.L., 548, L49
24. Rodríguez-Gil, P. & Potter, S. B. 2003, MNRAS, 342, L1
25. Sion, E. M., Cheng, F. H., Huang, M., Hubeny, I. & Szkody, P. 1996, Ap.J.L., 471, L41
26. Sion, E. M., Cheng, F., Godon, P., Urban, J. A., & Szkody, P. 2004, A.J., 128, 1834
27. Smak, J. 2001, Lecture Notes in Physics, Berlin Springer Verlag, 563, 110
28. Tapia, S. 1977, Ap.J.L., 212, L125
29. Thorstensen, J. R., Ringwald, F. A., Wade, R. A., Schmidt, G. D., & Norsworthy, J. E. 1991, A.J., 102, 272
30. Tovmasian, G. G. 1988, Advances in Space Research, 8, 329
31. Tout, C.A., Wickramasinghe, D. T., Ferrario, L., 2004, MNRAS, in press
32. Warner, B. & Woudt, P. A. 2002, MNRAS, 335, 84
33. Warner, B., Woudt, P. A., & Pretorius, M. L. 2003, MNRAS, 344, 1193
34. Warner, B. 2004, PASP, 116, 115
35. Warner, B. 2005, AIP Conf. this volume, Interacting Binaries, in preparation
36. Wheatley, P. J. & West, R. G. 2003, MNRAS, 345, 1009
37. Wickramasinghe, D. T. & Ferrario, L. 2000, PASP, 112, 873
38. Woudt, P. A. & Warner, B. 2002, MNRAS, 333, 411
39. Woudt, P. A. & Warner, B. 2002, MNRAS, 335, 44

High-angular Resolution Imaging of Interacting Binaries

Karovska Margarita

Harvard-Smithsonian Center for Astrophysics, 60 Garden Street, MA 02138, USA

Abstract.
I describe examples of long-term studies of interacting binary systems which have detected dramatic changes in the spatial and spectral distribution of the emission related to changes in accretion processes. These studies demonstrate that there are many aspects of accretion processes in general, and of wind accretion processes in particular, that are not yet understood. The key to further accretion studies is resolving a wide range of interacting binaries and studying their components and mass flows. A sub-milliarcsecond resolution is required to carry out detailed studies of nearby interacting binaries. I discuss future prospects for resolving accreting binary systems using the Stellar Imager - a many-element interferometer in space.

Keywords: interacting binaries, symbiotics, accretion
PACS: 97.80.-d, 97.80.Gm

INTRODUCTION

Accretion is a very important energy source for a wide variety of astronomical objects, including many interacting binaries. The nearby symbiotic binaries provide a unique laboratory for studying accretion processes because the systems can be resolved at many wavelengths and the components can be studied individually. This is not currently possible in most accreting systems.

Symbiotic systems are interacting binaries showing a composite spectrum with signatures of a late-type giant and a high-temperature component, often a compact object. They are some of the most fascinating interacting binary systems because of their dramatic transformations, and extremely complex circumbinary environment. Symbiotic systems are among the likely candidates for progenitors of bipolar planetary nebulae. They have been also invoked as potential progenitors of at least a fraction of type Ia Supernovae.

The interaction between the components in symbiotic systems is believed to occur via wind accretion and a Roche lobe overflow. The accretion processes are poorly understood partly because the wind accretion is much more complicated than tidal interaction. The amount of the accreted material depends on the wind properties (e.g. density and velocity), the binary parameters (e.g. orbital period and separation), and the dynamics of the flow. To determine these parameters it is necessary to resolve the binary system and study the individual components.

The circumbinary environment of several nearby symbiotic systems can be studied using multi-wavelength high-angular resolution techniques. However, so far the individual components have been resolved clearly in only one symbiotic system - Mira AB.

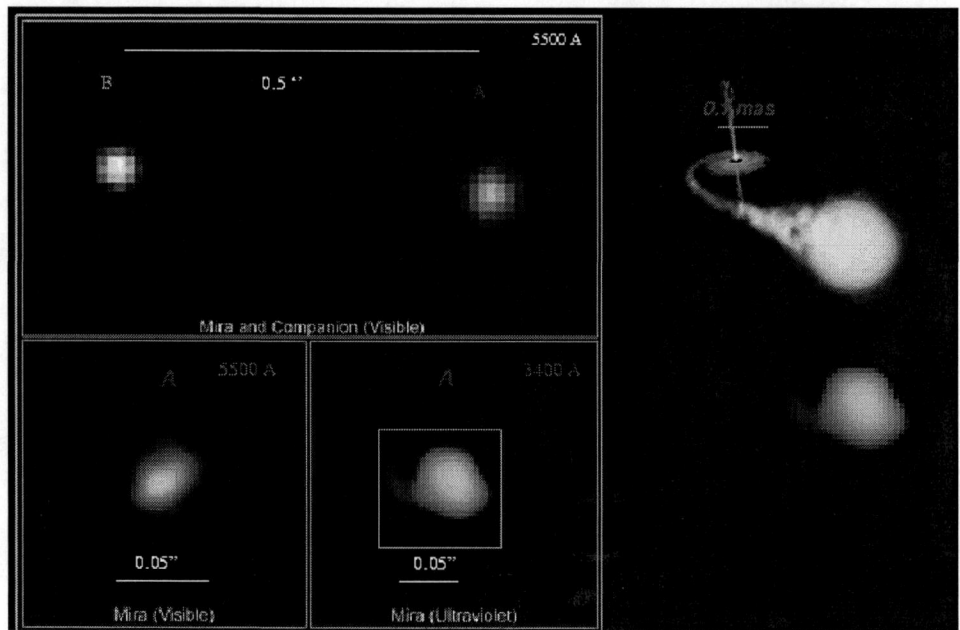

FIGURE 1. Upper panel - 1995 HST/FOC image of Mira A (right) and its nearby hot companion Mira B (left). Lower panel - Deconvolved optical image of Mira A showing a significant asymmetry in the atmosphere, and the first UV image of Mira A showing an outflow toward Mira B [3]. Image on the right shows the UV image of Mira A and an artistic conception of an interacting binary with a 0.1 milliarcsecond accretion disk.

LONG-TERM STUDIES OF NEARBY SYMBIOTIC BINARIES

Mira AB

Mira AB is an interacting binary system composed of an aging cool giant (Mira A) losing mass at a rate of $\sim 10^{-7}$ $M_\odot/yr$ [1], and an accreting companion (Mira B) about ~ 70 AU (~ 0.6") away. This nearby (~ 130 pc; [2]) detached binary is one of very few wind accreting systems that has been spatially resolved and for which the energy distribution of both components can be determined *unambiguously* [3]. Multi-wavelength studies of Mira AB wind accretion and mass transfer at wavelengths ranging from X-rays to radio provide a basis for understanding wind accretion processes in many other astronomical systems that currently cannot be resolved. Therefore this nearest symbiotic-like system offers a test-bed for detailed studies of wind accretion processes and accretion theory [4].

In 1995, using the HST FOC, we resolved the system for the first time spatially and spectrally at UV and optical wavelengths and studied the interacting components individually [3] (see also STScI-PRC1997-26). Figure 1 shows the HST images of the system and its individual components. The images of Mira A showed an outflow toward

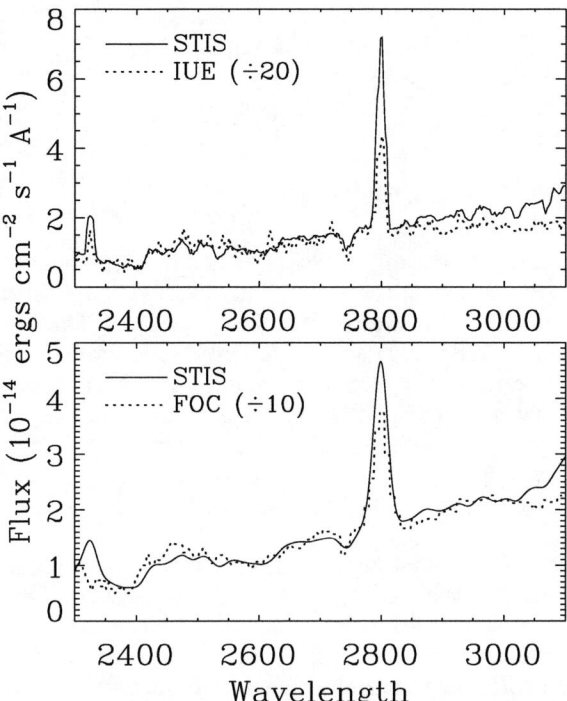

FIGURE 2. Comparison of the HST/STIS near-UV spectrum of Mira B with previous observations from IUE (top panel) and HST/FOC (bottom panel). In both panels, the STIS spectrum is rebinned and deresolved to match the resolution of the other observation. The peaks at 2325 Å and 2800 Å are C II] and Mg II lines, respectively. Note that the IUE and FOC fluxes had to be reduced by factors of 20 and 10, respectively, to match the STIS data [5].

Mira B. The accretion disk around Mira B was note resolved in these images. The actual size of the disk is not known, but it has been estimated $\sim$0.01 -0.1 AU, or $\sim$ 0.1 - 1 mas e.g. [6].

In the past few years we have been witnessing changes in the spectral energy distribution (SED) of Mira AB, especially at UV wavelengths [3][7][8]. In addition to the general fading of the accretion luminosity, another baffling development was the appearance of a forest of Lyα-fluoresced H_2 emission lines in, which dominated the HST spectra in 1999, despite not being seen at all in the 1995 observations or by IUE [7]. A similar drop in the accretion luminosity and appearance of a set of Lyα-fluoresced H_2 emission lines (Werner band lines) were also seen in the FUSE spectra obtained in 2001 observations [8].

During the low-state period observed by HST and FUSE in 1999–2001 the wind absorption in the Mg II h & k lines shows that the accretion-driven mass loss rate from Mira B was about 20 times lower than it was during the high-state observed a decade earlier (during the IUE era). Our long-term observations of the UV and optical variability in the Mira AB system indicate that the high-state is expected to occur in the 2004/2005 time frame (e.g. Fig 6 in [8]. Recently we resolved this interacting binary for the first time in X-rays using Chandra and detected an unprecedented soft X-ray outburst [9]. This recent activity detected at X-ray and UV wavelengths signals that the system is indeed undergoing dramatic changes.

CH Cyg

CH Cyg is another very interesting nearby (D~250 pc) symbiotic system composed of an evolved M6-7 III star and an accreting white dwarf. There is also evidence for a possible presence of an enigmatic third body in the system [10]. The components of this system have not been resolved using current telescopes and interferometers.

CH Cyg has undergone several outbursts in the past decades, each preceded by extended intervals of quiescence [11]. CH Cyg has shown dramatic jet activity observed at optical, UV, and radio wavelengths e.g. [12]. In fact, this system is one of the very few symbiotic systems showing jet activity, that are close enough so the jet itself and/or its interaction with the surrounding medium can be studied in detail using multi-wavelength observations.

Recently we detected a jet in this system using archival Chandra X-ray observations. This is only the second X-ray jet detected in a symbiotic system. Figure 3 shows the comparison between the HST 5007 Å image of the jet and the Chandra X-ray image of an extended source associated with the jet. The X-ray emission is likely originating in the region of interaction between the expanding jet material and the circumbinary envelope of this system [13]

Imaging with a sub-milliarcsecond resolution is required to separate the components of this system and of many other symbiotics, and to study the jet forming and collimation regions.

CONCLUSIONS AND FUTURE OBSERVATIONS

The long-term studies of these nearby interacting binaries demonstrate that there are many aspects of accretion processes that are not yet understood. The long-term temporal variability of these systems underlines the fact that it is crucial to monitor them especially at X-ray and UV wavelengths, where the most prominent signatures of accretion processes can be found. It is also clear the importance of resolving these binaries, their individual components, and the mass flows in these systems.

Increasing the angular resolution to sub-milliarcsecond level at UV and X-ray wavelengths will lead to unprecedented opportunities for detailed studies of accretion phenomena in symbiotics and in many other interacting binaries. Imaging with a sub-milliarcsecond resolution will result in an advance of at least two orders of magnitude compared to that provided by HST.

The Stellar Imager (SI) - a 'Vision' mission in the NASA's Sun-Earth Connection Roadmap - is a multi-aperture sub-milliarcsecond resolution interferometer in space that will provide the necessary tools for unprecedented studies of accretion processes in many interacting binaries [14](also see *http://hires.gsfc.nasa.gov./si/*).

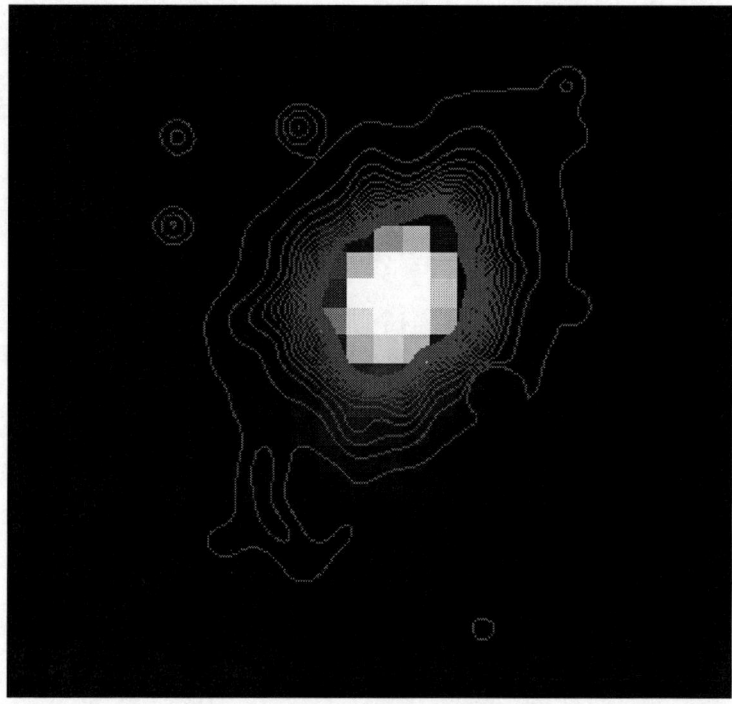

FIGURE 3. Jet activity in CH Cyg: Chandra X-ray image with overlayed contours of the HST 5007 Å image.

The SI main science driver is understanding the effects of stellar magnetic fields, the dynamos that generate them, and the internal structure and dynamics of the stars in which they exist. The ultimate goal is to achieve the best possible forecasting of solar/stellar magnetic activity and its impact on life in the Universe.

The science goals of SI require an ultra-high angular resolution, at ultraviolet wavelengths, on the order of 0.1 milliarcsecond and thus baselines on the order of 0.5 km. These requirements call for a large, multi-spacecraft (30 mirrors) imaging interferometer in orbit around the Sun-Earth L2 point.

The SI sub-milliarcsecond resolution in the UV-optical domain, combined with the unique spectral diagnostics available at these wavelengths, will be a unique resource for studies of interacting binaries in the 21st century. The ultra-high SI's resolution will make it also an invaluable resource for many other areas of astrophysics, including studies of AGN's, supernovae, young stellar objects, QSO's, and black hole environments.

ACKNOWLEDGMENTS

The results presented in this paper are based on the work carried out as part of a long term collaboration with several colleagues including Drs. B. Wood, M. Marengo, J.

Raymond, W. Hack, E. Guinan, P. Nisenson, and J. Mattei. Support for this work was provided by NASA through grants number GO-05822.01-94A and GO-08298.01-99A from the Space Telescope Science Institute, which is operated by AURA, Inc., under NASA contract NAS5-26555. M. K. is a member of the Chandra Science Center, which is operated under contract NAS-839073, and is partially supported by NASA, and of the SI vision team (PI. K. Carpenter).

REFERENCES

1. Knapp, G. R., Young, K., Lee, E., & Jorissen, A. 1998, ApJS, 117, 2.
2. Perryman, M. A. C., et al. 1997, A&A, 323, L49.
3. Karovska, M., Hack, W., Raymond, J., & Guinan, E. 1997, ApJ, 482, L175.
4. Livio, M. 1988, in Symbiotic Phenomena, Proccedings of IAU Coll. No . 103.
5. Wood, B. E., Karovska, M., & Hack, W. 2001, ApJ, 556, L51.
6. Reimers, D., & Cassatella, A. 1985, ApJ, 297, 275.
7. Wood, B. E., Karovska, M., & Raymond, J., W. 2002, ApJ,575, 1057.
8. Wood, B.E., and Karovska, M. 2004, ApJ, 601, 502
9. Karovska, M et al. 2005, ApJ, 623, L137.
10. Hinkle, K.H. et al 1993, AJ, 105, 1074.
11. Karovska, M., Carilli, C., and Mattei, J.,1998, JASVO, 26, 97.
12. Corradi R.L.M. et al., 2001, ApJ, 560, 912.
13. Karovska, M. et al. 2005, in prep.
14. Carpenter, K. et al. 2004, SPIE, 5491_28.

On the excitation of hydrodynamical turbulence in accretion discs

O.A. Kuznetsov

Keldysh Institute of Applied Mathematics, 4 Miusskaya sq., 125047 Moscow, Russia
Institute of Astronomy, 48 Pyatnitskaya str., 119017 Moscow, Russia

Abstract. Our 2D numerical simulations confirm a possibility of excitation of hydrodynamic turbulence in the accretion disk due to shear instability under finite-amplitude perturbations and at high Reynolds number. These results disproves the well-known claims about the impossibility to excite the turbulence in accretion disks by hydrodynamic shear instability. Our estimations have shown that the development of turbulence results in the value of accretion rate corresponding to the α_{SS}-coefficient in the range $0.08 \div 0.15$.

Keywords: Accretion Discs – Hydrodynamics: turbulence
PACS: 97.10.Gz

INTRODUCTION

The problem of the turbulence excitation in accretion discs is of great importance since the turbulent (or eddy) viscosity can provide an appropriate rate of the angular momentum expulsion and, consequently, appropriate value of accretion rate. The value of molecular viscosity defined by free path λ and thermal velocity c_s as $v_{mol} = c_s \cdot \lambda$ is too low, so the flow in the disk has extremely large Reynolds number $\Re = r\Omega_K H / v_{mol}$ (here H is a disc thickness, Ω_K is angular Keplerian velocity of the disc rotation). Accordingly, Lynden-Bell [1], Shakura [2], and Lynden-Bell & Pringle [4] suggested a possible origin of turbulence due to similar mechanism as those occurring in laboratory flows. Shakura [2] and Shakura & Sunyaev [3] introduced the so called turbulent viscosity and replaced viscosity coefficient as $v_{mol} = c_s \cdot \lambda \rightarrow v_{turb} = \alpha_{SS} \cdot c_s \cdot H$, where α_{SS} is an heuristic constant. Since $H \gg \lambda$ the turbulent viscosity coefficient is many order of magnitude larger than molecular one, and provides reasonable values for accretion rate.

How does the turbulence develop in accretion disc? It is known that flow in a pipe (Poiseuille flow) can show turbulence developing under some conditions. The main aim of this paper is to find the turbulence development in the accretion disc considering it as a (curved) pipe.

SYSTEM OF EQUATIONS AND RESULTS

We can describe the flow in the disc by 2D (i.e. z-integrated) Euler non-viscous equations in rotating coordinate frame

$$\partial_t(r\sigma) + \partial_r(r\sigma u_r) + \partial_\varphi(\sigma u_\varphi) = 0,$$

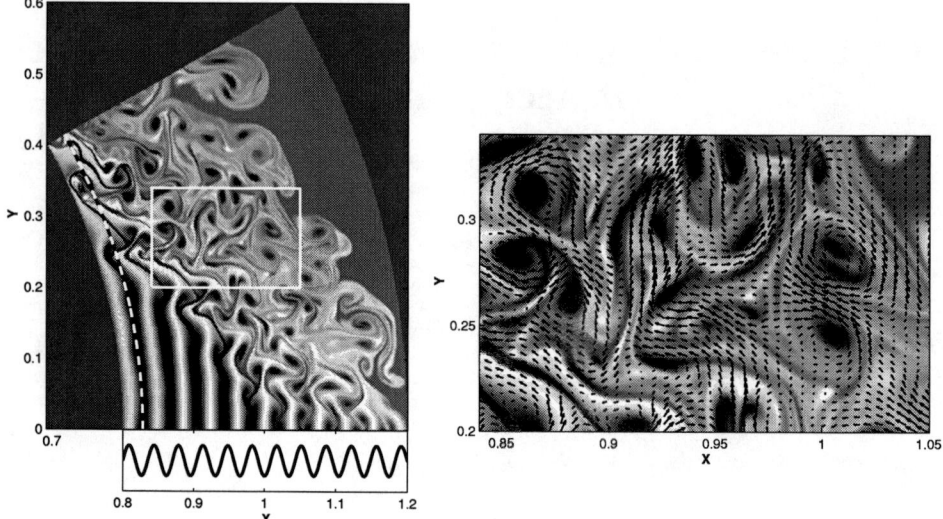

FIGURE 1. *Left:* Entropy distribution. Bold black-white line divides the supersonic and subsonic regions for unperturbed (initial) flow. A sinusoid on the bottom edge shows the perturbation of pressure/density. *Right:* Entropy distribution and perturbed velocity field for a part of the flow.

$$\partial_t(r\sigma u_r) + \partial_r(r\sigma u_r^2) + r\partial_r\Pi + \partial_\varphi(\sigma u_r u_\varphi) = -r\sigma\partial_r\Phi + \sigma u_\varphi^2 + 2\Omega_1 u_\varphi r\sigma,$$

$$\partial_t(r\sigma u_\varphi) + \partial_r(r\sigma u_r u_\varphi) + \partial_\varphi(\sigma u_\varphi^2 + \Pi) = -\sigma u_r u_\varphi - 2\Omega_1 u_r r\sigma,$$

$$\partial_t(r\Upsilon) + \partial_r(r(\Upsilon+\Pi)u_r) + \partial_\varphi((\Upsilon+\Pi)u_\varphi) = -r\sigma u_r \partial_r\Phi,$$

with perfect gas equation of state $\Pi = (\gamma-1)(\Upsilon - 1/2\sigma(u_r^2+u_\varphi^2))$. Here $\sigma = \int_{-\infty}^{\infty}\rho dz$, $\Pi = \int_{-\infty}^{\infty}Pdz$, $\Upsilon = \int_{-\infty}^{\infty}Edz$, $E = \rho(\varepsilon + 1/2(u_r^2+u_\varphi^2))$, $\Phi = -GM/r - 1/2\Omega_1 r^2$, other notations are standard. We exploit the non-dimensional variables using arbitrary values r_0 and ρ_0 as scales for distance and density, and $\sqrt{GM/r_0}$ and $\sqrt{r_0^3/GM}$ as scales for velocity and time. Dimensionless rotational period at $r=1$ is equal to $P_0 = 2\pi$.

We have considered a piece of the Keplerian disc $0 < \varphi < \pi/6$, $r_0 < r < r_1$, $r_0 = 0.8$, $r_1 = 1.2$ and choose the angular velocity of rotation of coordinate frame so that the right edge of the disc would be in rest $\Omega_1 = \Omega_K(r_1)$. The initial values were adopted as $u_{r0} = 0$, $u_{\varphi 0} = \Omega_K r - \Omega_1 r = r^{-1/2} - \Omega_1 r$, $\sigma_0 = 1$, $\Pi_0 = c_0^2\sigma_0/\gamma$, $c_0 = 0.5$. We have disturbed the disc on the lower edge $\varphi = 0$ as $\sigma = \sigma_0(1+A\sin(kr))$, $\Pi = \Pi_0(1+A\sin(kr))$ with $k = 160$ (it corresponds to the wavelength $\lambda = 2\pi/k = 0.04$) and amplitude of perturbations 50% (i.e. $A = 0.5$).

Figure 1 shows the entropy distribution for the total computation domain (left) and entropy distribution and perturbed velocity field for a part of the flow. It is seen that the flow contains both cyclonic and anticyclonic vortices. Figure 2 shows the visualization of perturbed velocity field using Line Integral Convolution method for two different moments of time. In Figs. 1–2 we can see that the flow is turbulent except left lower

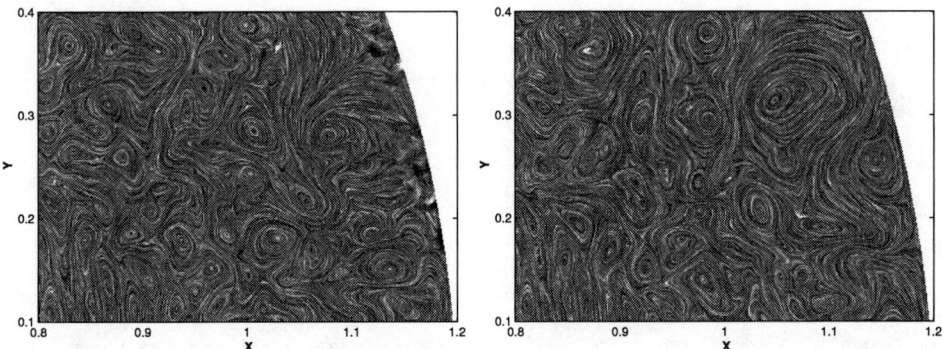

FIGURE 2. Visualization of perturbed velocity field using the Line Integral Convolution method, for two moments of time.

corner (supersonic and transonic zone) and right upper corner (zone of unperturbed flow).

BALBUS & HAWLEY ANALYSIS VS. NONLINEAR ANALYSIS IN LOCAL COORDINATE FRAME

Our results are partly in contradiction with analysis made by Balbus et al. [5], Hawley et al. [6]. Their analysis has shown that the radial and azimuthal components of perturbed kinetic energy are governed by

$$\frac{d}{dt}{}^1\!/_2\sigma\tilde{u}_r^2 = \left[2\Omega_K + \frac{\tilde{u}_\varphi}{r}\right]\sigma\tilde{u}_r\tilde{u}_\varphi - \tilde{u}_r\partial_r\Pi - {}^1\!/_2\sigma\tilde{u}_r^2 \cdot div\vec{\tilde{u}},$$

$$\frac{d}{dt}{}^1\!/_2\sigma\tilde{u}_\varphi^2 = \left[\frac{\varkappa^2}{2\Omega_K} + \frac{\tilde{u}_\varphi}{r}\right]\sigma\tilde{u}_r\tilde{u}_\varphi - \frac{\tilde{u}_\varphi\partial_\varphi\Pi}{r} - {}^1\!/_2\sigma\tilde{u}_\varphi^2 \cdot div\vec{\tilde{u}},$$

with $d/dt = \partial_t + \tilde{u}_r\partial_r + (\Omega_K - \Omega_1 + \tilde{u}_\varphi/r)\partial_\varphi$ – Lagrangian time derivative, $\varkappa^2 = 2\Omega_K(\Omega_K r^2)'/r$ – epicyclic frequency, $\vec{\tilde{u}} = \vec{u} - \vec{u}_0$. After linearizing, we get

$$\frac{d}{dt}{}^1\!/_2\sigma\tilde{u}_r^2 = \left[2\Omega_K\right]\sigma\tilde{u}_r\tilde{u}_\varphi + \dots, \quad (1)$$

$$\frac{d}{dt}{}^1\!/_2\sigma\tilde{u}_\varphi^2 = -\left[\frac{\varkappa^2}{2\Omega_K}\right]\sigma\tilde{u}_r\tilde{u}_\varphi + \dots, \quad (2)$$

with opposite signs in r.h.s. (for Rayleigh-stable flows) so both positive and negative values of $\sigma\tilde{u}_r\tilde{u}_\varphi$ produces the decreasing of either ${}^1\!/_2\sigma\tilde{u}_r^2$ or ${}^1\!/_2\sigma\tilde{u}_\varphi^2$ and perturbations decay.

Nevertheless, the linear approach appears to be not valid for large-amplitude perturbation so non-linear analysis would be desirable. Let us choose an arbitrary point

$(r_\star,\varphi_\star)$ and introduce a new local polar coordinate frame as $r\cos\varphi = r_\star\cos\varphi_\star + \varpi\cos\phi$, $r\sin\varphi = r_\star\sin\varphi_\star + \varpi\sin\phi$. We are going to bring a local non-linear analysis with $\varpi \ll r_\star$ so we can adopt $u_{\varphi 0}(r)$ and $\partial_r u_{\varphi 0}(r)$ in the form $u_{\varphi 0}(r_\star)$, $\partial_r u_{\varphi 0}(r_\star)$. Moreover, with no loss of generality, we can put $\varphi_\star = 0$. The equations for perturbed velocity in the new coordinate system read as

$$\frac{d}{dt}\tilde{u}_\varpi = \left(\frac{\tilde{u}_\phi + (\Omega_K(r_\star)-\Omega_1)r_\star\cos\phi}{\varpi} + B\right)\tilde{u}_\phi + A\tilde{u}_\varpi - \frac{\partial_\varpi \Pi}{\sigma},$$

$$\frac{d}{dt}\tilde{u}_\phi = -\left(\frac{\tilde{u}_\phi + (\Omega_K(r_\star)-\Omega_1)r_\star\cos\phi}{\varpi} + C\right)\tilde{u}_\varpi - A\tilde{u}_\phi - \frac{\partial_\phi \Pi}{\sigma\varpi},$$

with $d/dt = \partial_t + (\tilde{u}_\varpi + (\Omega_K(r_\star)-\Omega_1)r_\star\sin\phi)\partial_\varpi + \varpi^{-1}(\tilde{u}_\phi + (\Omega_K(r_\star)-\Omega_1)r_\star\cos\phi)\partial_\phi$, $A = {}^3\!/_2\Omega_K(r_\star)\cos\phi\sin\phi$, $B = \Omega_1 - \Omega_K(r_\star)({}^3\!/_2\sin^2\phi - 1)$, $C = 2\Omega_1 + {}^1\!/_2\Omega_K(r_\star) - B$. Multiplying these equations with $\tilde{u}_\varpi$ and $\tilde{u}_\phi$, and combining to continuity equation we obtain the non-linear version of equations (1,2):

$$\frac{d}{dt}{}^1\!/_2\sigma\tilde{u}_\varpi^2 = \left[\frac{\tilde{u}_\phi + (\Omega_K(r_\star)-\Omega_1)r_\star\cos\phi}{\varpi} + B\right]\sigma\tilde{u}_\varpi\tilde{u}_\phi + \left[A\right]\sigma\tilde{u}_\varpi^2 + \ldots,$$

$$\frac{d}{dt}{}^1\!/_2\sigma\tilde{u}_\phi^2 = -\left[\frac{\tilde{u}_\phi + (\Omega_K(r_\star)-\Omega_1)r_\star\cos\phi}{\varpi} + C\right]\sigma\tilde{u}_\varpi\tilde{u}_\phi - \left[A\right]\sigma\tilde{u}_\phi^2 + \ldots.$$

It is seen that production of ϖ- and ϕ-components of perturbed kinetic energy is not sign-definite mainly due to angular dependence. Moreover, the equations contain nonlinear terms $\sim \varpi^{-1}$ corresponding to strong production of turbulence at small scales as it usually occurs in 2D case (see [7]).

Our results also show that both cyclonic and anticyclonic vortices corresponds to *minimum* of pressure. This is in contradiction, in particular, to the standard rotational flow model (e.g., the Earth atmosphere) where cyclones produce low pressure (cloudy sky) and anticyclones produce high pressure (clear sky). To resolve let us come back to the non-linear analysis. Let as assume that the vortex is steady-state and axially symmetrical: $\partial_t = \partial_\phi = \tilde{u}_\varpi = 0$. Averaging and taking into account $\overline{\cos\phi} = \overline{\sin\phi} = \overline{\cos\phi\sin\phi} = 0$, $\overline{\sin^2\phi} = \overline{\cos^2\phi} = {}^1\!/_2$, we obtain

$$\frac{\partial_\varpi \Pi}{\sigma} = \frac{\tilde{u}_\phi^2}{\varpi} + B\tilde{u}_\phi, \qquad (3)$$

with $B \approx +1$ for our calculations. Introducing the Rossby number $\mathfrak{Ro} = |\vec{\tilde{u}}|/2\Omega_1\varpi$ we can rewrite (3) as

$$\frac{\partial_\varpi \Pi}{\sigma} = \mathfrak{Ro}\cdot 2\Omega_1|\vec{\tilde{u}}| + B\tilde{u}_\phi.$$

Low values of Rossby numbers corresponds to the so-called geostrophic approximation when non-linear term disappears. In this case $\partial_\varpi \Pi \propto \tilde{u}_\phi$ and $\tilde{u}_\phi > 0$ (cyclones) corresponds to $\partial_\varpi \Pi > 0$ (low pressure in the center), while $\tilde{u}_\phi < 0$ (anticyclones) corresponds to $\partial_\varpi \Pi > 0$ (high pressure in the center). Our results show $\mathfrak{Ro} \sim 2 \div 4$ so this case is far from geostrophic approximation due to non-linear interaction so all cyclones and almost all anticyclones produce the minimum of pressure.

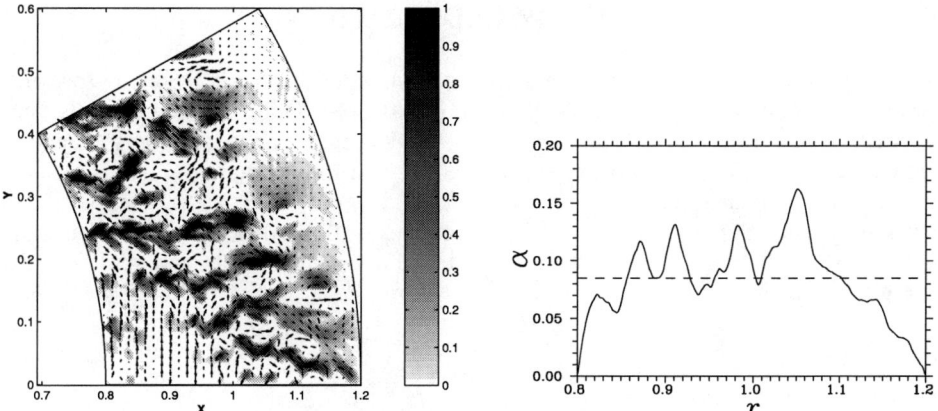

FIGURE 3. *Left:* Distribution of α_{SS} coefficient. White color corresponds to zones of decretion. *Right:* Azimuthally averaged distribution of α_{SS} coefficient. Dashed line shows the mean value $\bar{\alpha}_{SS} = 0.085$.

THE α_{SS} COEFFICIENT

Let us combine simple formulas

$$v_{turb} = \alpha_{SS} \cdot c_s^2 \cdot \Omega_K^{-1} \qquad u_r = -{}^3/_2 \frac{v_{turb}}{r}$$

to get

$$\alpha_{SS} = {}^2/_3 \frac{|u_r| \cdot \Omega_K \cdot r}{c_s^2}.$$

Figure 3 shows distribution of α_{SS} coefficient for zones where accretion takes place, i.e. where $u_r < 0$ (zones with $u_r > 0$ or decretion zones are shown by white color). Azimuthal averaging over all zones of accretion gives the distribution $\alpha_{SS}(r)$ with mean value $\bar{\alpha}_{SS} = 0.085$. This result is in the good agreement with theoretical and observational estimations of α_{SS} for accretion discs in dwarf novae during outburst (see [8], [9] and [10]).

CONCLUSIONS

Summarizing, we can assert that our results deals with non-linear finite-amplitude hydrodynamical shear instability which leads to the development of turbulence. We can draw the following conclusions from our study:

- Pure hydrodynamical turbulization of accretion disk is possible.
- The development of turbulence deals with non-linear finite-amplitude hydrodynamical shear instability.
- The value of α_{SS} coefficient is $\simeq 0.1$.

ACKNOWLEDGEMENTS

The work was partially supported by Russian Foundation for Basic Research (projects NN 02-02-16088, 02-02-17642, 03-01-00311, 03-02-16622), by Science Schools Support Program (project N 162.2003.2), by Federal Programme "Astronomy", by Presidium RAS Programs "Mathematical modelling and intellectual systems", "Nonstationary phenomena in astronomy", and by INTAS (grant N 00-491). Author thanks Russian Science Support Foundation for the financial support as well as D.V.Bisikalo, G.S.Bisnovatyi-Kogan, V.M.Chechetkin, A.M.Fridman, A.V.Koldoba and D.Molteni for useful discussions.

REFERENCES

1. D. Lynden-Bell, Nature **223**, 690 (1969).
2. N.I. Shakura, Astron. Zhourn. **49**, 921 (1972).
3. N.I. Shakura and R.A. Sunyaev, A&A **24**, 337 (1973).
4. D. Lynden-Bell and J.E. Pringle, MNRAS **168**, 603 (1974).
5. S.A. Balbus, J.F. Hawley and J.M. Stone, ApJ **467**, 76 (1996).
6. J.F. Hawley, S.A. Balbus and W.F. Winters, ApJ **518**, 394 (1999).
7. G.K. Batchelor, *The Theory of Homogeneous Turbulence* Cambridge Univ. Press, Cambridge (1992).
8. J. Smak, Acta Astron. **34**, 161 (1984);
9. J. Smak, Acta Astron. **48**, 677 (1998);
10. J. Smak, Acta Astron. **49**, 391 (1999).

Rapid Oscillations in Cataclysmic Variables, and a Comparison with X-Ray Binaries

Brian Warner and Patrick A. Woudt

Department of Astronomy, University of Cape Town, Rondebosch 7700, South Africa

Abstract. We compare some of the properties of rapid oscillations in cataclysmic variables and X-Ray binaries. In addition to the earlier recognition that both types possess the same correlation between high and low frequency quasi-periodic oscillations, we have now found that the dwarf nova VW Hyi in its late stages of outburst shows the 1:2:3 oscillation harmonics that are seen in some neutron star and black holes X-Ray binaries. We point out that the behaviour of the dwarf nova WZ Sge has some similarities to those of accreting millisecond pulsars.

Keywords: Binary Stars: Oscillations – X-Ray Binary Stars
PACS: 97.10.Sj, 97.80.Jp

INTRODUCTION

The rapid oscillations seen in cataclysmic variables (CVs) and X-Ray binaries (XRBs), although separated by orders of magnitude in time scales, show many common features in their phenomenology. The former have been recognised since the singular discovery of highly stable 71 sec modulations in the optical light curve of DQ Her, exactly 50 years ago, in July 1954 (Walker 1956), followed by the opening of a floodgate of phenomena associated with the discovery of similar but less stable oscillations in outbursting dwarf novae (Warner & Robinson 1972). The first observed XRB rapid modulations date to the days of EXOSAT and the quasi-periodic 5 – 50 Hz flux variations seen in the low mass XRB GX 5-1 (van der Klis et al. 1985). With the advent of RXTE in 1996 the time resolution was increased to where kilohertz oscillations could be detected, with the result that now a quarter of the neutron star XRBs (Swank 2004) and a sixth of the black hole XRBs (McClintock & Remillard 2004) have been observed with quasi-periodic oscillations (QPOs).

OSCILLATIONS IN CVS

The rich phenomenology of oscillations in CVs has been reviewed recently by Warner (2004). There are at least three distinct types of rapid oscillation in CVs, which can exist separately or simultaneously:

1. Dwarf Nova Oscillations (DNOs). These are oscillations typically in the range 3 – 40 s, usually appearing in high mass transfer ($\dot{M}$) discs, i.e. dwarf novae in outburst and nova-like variables. They have not been observed in all such CVs, and are of varying amplitude, often disappearing altogether. They show a period-$\dot{M}$ relationship, with shortest periods at highest $\dot{M}$, and are of moderate coherence with small

abrupt changes of period and almost continuous phase noise. 'Double DNOs' are occasionally seen, with frequency differences equal to the frequency of the QPO (see (3) below) present at the time. DNOs are sinusoidal modulations; the companion in a double DNO is at longer period and usually possesses a first harmonic (Warner & Woudt 2002, hereafter WW1; Woudt & Warner 2002, hereafter WW2).

2. Longer-period DNOs. These lpDNOs have periods ~ 4 times those of DNOs and show little if any variation with $\dot{M}$ (or optical luminosity). They are less commonly present than DNOs, are occasionally seen even in quiescence of dwarf novae, and can also appear doubled on rare occasions (Warner, Woudt & Pretorius 2003, hereafter WWP).

3. Quasi-Periodic Oscillations. The QPOs of interest here are those with periods $P_{QPO} \sim 15 P_{DNO}$ (there are others of longer period probably related to the rotation of the slower rotating primaries - Patterson et al. 2002). These QPOs have a short coherence length, typically growing and decaying in ~ 5 cycles, and changing phase or period on that time scale if they are longer lived (WW2).

It is useful to have a model in mind that ties these variations together. The following is the one advanced by WW1, which in its fundamentals was first proposed by Paczynski (1978). This is, in principle, an extension of the standard intermediate polar model (see Chapters 7 and 8 of Warner (1995)) to lower field strengths. The recent discovery of kilogauss magnetic fields in a number of DA stars (Aznar Cuadrado et al. 2004) demonstrates that such low fields are probably common in isolated white dwarfs and therefore also in CVs (though they are below the detection limit of direct measurement).

Provided that the primary has a sufficiently weak magnetic field ($\lesssim 5 \times 10^5$ G) accretion through a disc onto its surface generates a freely moving equatorial belt. There is spectroscopic evidence of such a belt after dwarf nova outbursts (e.g. Sion et al. 1996; Godon et al. 2004). If the primary's field is stronger then the system is an intermediate polar with very stable rotation. The rapidly rotating belt will enhance whatever field the primary has, resulting in magnetic channeling of accreting mass even for weak field primaries, but onto the equatorial belt. The low inertia of the belt is what allows it to be spun up and down by magnetic connection to the inner edge of the accretion disc – giving the $\dot{M}$-P_{DNO} relationship seen in outbursting dwarf novae. We refer to this as the Low Inertia Magnetic Accretor (LIMA) model (WW1).

There is a key observation that supports the idea that at least some of the white dwarfs in dwarf novae have magnetic fields strong enough to channel accretion, despite the absence yet of direct detection by spectroscopic or polarimetric means, and this comes from the X-Ray eclipse of OY Car observed shortly after outburst. Wheatley & West (2003) deduce that the X-Ray emitting region is considerably smaller than the white dwarf and is at high primocentric latitude.

The QPOs are thought (by WW1) to be due to a prograde traveling wave at the inner edge of the disc, probably excited by field winding arising from the non-synchronous rotation of the disc and the primary. The traveling wave can intercept and/or reprocess radiation from near the primary, thereby generating the optical QPOs. The revolving anisotropic radiation from the accretion zones on the belt (that causes the DNOs) can sweep across the traveling wave and be reprocessed at the beat period, producing the double DNOs.

The lpDNOs may be an additional channel of accretion from the disc, but along field lines that connect to the body of the primary, rather than to the belt.

Not all CVs in their high $\dot{M}$ phases show DNOs or QPOs. This shows that there is a parameter that determines the presence or absence of oscillations. In the LIMA model this is simply the strength of the magnetic field of the equatorial belt, which is determined by both the field of the primary and any shear enhancement that takes place.

THE EVOLUTION OF DNOS THROUGH OUTBURST IN VW HYDRI

At maximum luminosity of a VW Hyi outburst DNOs are seen at 14 s, both in soft X-Rays (van der Woerd et al. 1987) and the optical (WW2). Until almost the end of outburst DNOs are seen only intermittently, but when visible they are found to increase slowly and systematically in period to $\sim$ 20 s when the star has descended to $\sim$ 1.5 mag above minimum. At this phase of outburst DNOs are almost always present; as the star continues to fade WW2 discovered that there is a rapid increase in period, doubling to $\sim$ 40 s in about 5 h. This has been ascribed (WW1) to propellering, caused when the equatorial belt with its associated magnetic field is rotating at higher angular velocity than the inner edge of the accretion disc, and is consistent with the near cessation of EUV flux (which is a direct monitor of $\dot{M}$ onto the primary) precisely during the phase of rapid DNO deceleration (Mauche 2002). Following this deceleration phase there has been uncertainty in the evolution of the DNOs. Figure 8 of WW2 showed that the DNO periods at some point decrease by a factor of about two and the oscillations themselves become apparently less coherent. We have gathered more light curves of VW Hyi covering these final phases of outburst, the analysis of which has uncovered a remarkable behaviour.

Late stage DNOs in VW Hydri

When the DNOs have increased to $P_{DNO} \sim$ 39 s there is a sudden frequency doubling – seen in Figure1. We have not managed to be at the telescope, with a clear sky, when this happens, but our various observing runs suggest that it happens in less than $\sim$ 15 min, and (as seen by the later behaviour) may happen essentially instantaneously. As we do not have the crucial information we cannot yet claim that the period exactly halves, but again from subsequent behaviour we can be fairly certain that the fundamental has been replaced by its first harmonic. The period of the now dominant 1st harmonic continues to increase systematically and with little scatter as VW Hyi decreases in brightness until at a period of $\sim$ 28 s a new periodicity appears, which within error of measurement is at a period 2/3rds of the 1st harmonic, so here are sure that we are dealing with a 2nd harmonic.

The evolution of the DNOs from this point is complex in the sense that there is a gradual move from dominance by the 1st harmonic to dominance by the 2nd harmonic, with often both being simultaneously present – and, very importantly, there is an occasional appearance of the associated fundamental, showing that the description as fundamental,

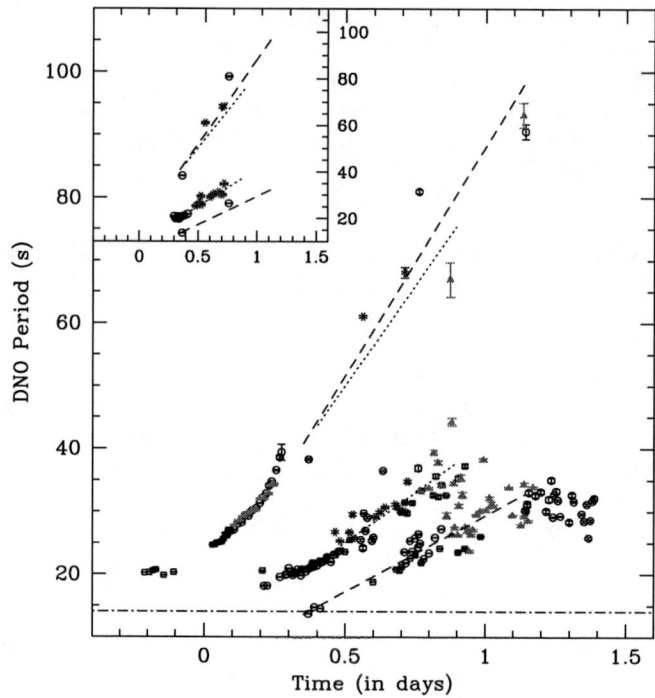

FIGURE 1. The evolution of DNO periods at the end of normal and super outbursts of the dwarf nova VW Hyi. Different symbols mark the various kind of outbursts, ranging from short (asterisks), normal (open circles), long (open squares) to super (filled triangles) outbursts. The dotted and dashed lines show the results of a least-squares fit to the first and second harmonic, respectively, and are multiplied by a factor of two and three to illustrate the inferred evolution of the fundamental DNO period. The dashed-dotted horizontal line at 14.1 s represents the minimum observed DNO period at maximum brightness. The inset highlights two observing runs in which the fundamental, the first and/or second harmonic of the DNO period were present simultaneously.

1st and 2nd harmonics is quantitatively justified. Figure 2 shows examples of Fourier transforms (FTs) with various combinations of the harmonics present.

In Figure 3 we show a modified form of Figure 1, in which the harmonics have been replaced by their implied fundamentals. From this we see that the fundamental continues to increase at a rapid rate, with increasing scatter, until it reaches ~ 105 s, after which VW Hyi has reached quiescence and we have not detected DNOs. The fundamental period increases almost linearly from 30 s to 105 s in ~ 27.5 h, i.e. $\dot{P} = 1.06 \times 10^{-3}$.

We have not detected such 1:2:3 ratios in the DNOs of any CV other than VW Hyi – but VW Hyi is also unique in having DNOs that are most prominent at the end of its outburst.

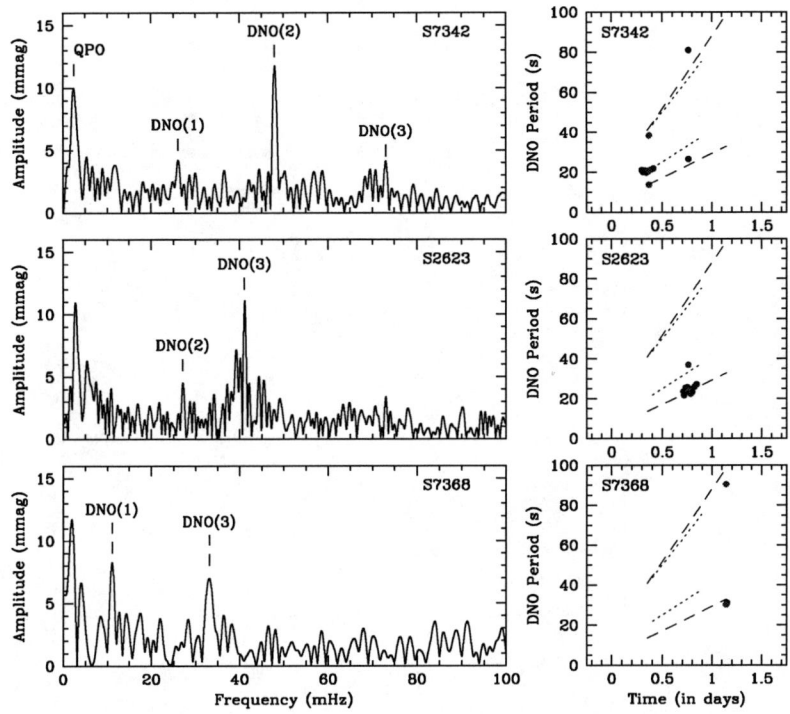

FIGURE 2. Examples of Fourier transforms of VW Hyi in which combinations of the fundamental and the first two harmonics of the DNO period are present.

Similarity to XRBs

As pointed out in WW1 and WWP, CVs and XRBs show the same correlation of their high frequency and low frequency oscillations, viz. the ratio of ~ 15. This relationship, which extends over 6 orders of magnitude in frequency, is shown in Figure 4.

The 3:2:1 period ratios seen in VW Hyi are similar to those already known for a few years in XRBs (McClintock & Remillard 2004). For example, for black hole binaries, XTE J1550-564 has strong X-Ray signals at 276 and 184 Hz and a weak signal at 92 Hz, which are in the ratio 3:2:1; GRO J1655-40 has 450 and 300 Hz oscillations (Remillard et al. 2002); and 240 and 160 Hz (ratio 3:2) are present in H1743-322 (Homan et al. 2004). For the neutron star XRB, Sco X-1, Abramowicz et al. (2003) claim a 3:2 period ratio.

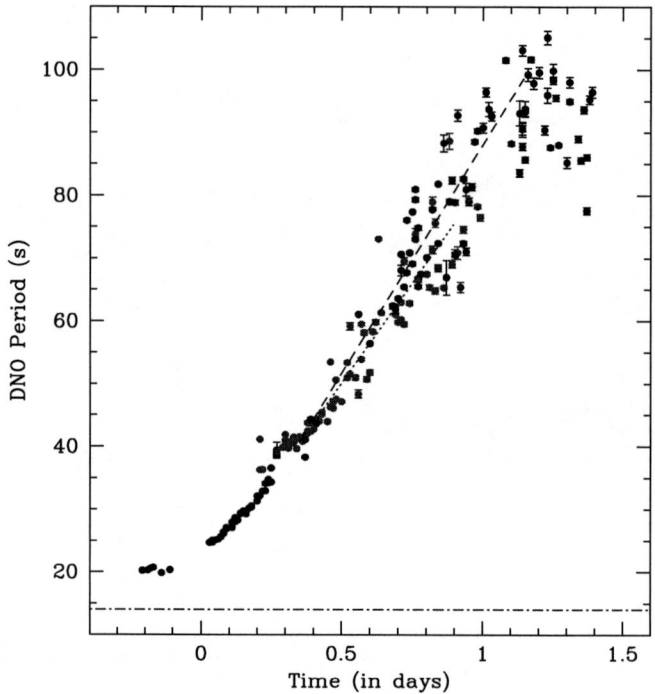

FIGURE 3. The evolution of the fundamental DNO period in VW Hyi at the end of normal and super outbursts. The dotted, dashed and dashed-dotted lines are as in Figure 1.

A MODEL FOR HARMONICS IN VW HYI

A possible explanation of the appearance of harmonics in the DNOs of VW Hyi is as follows. At high $\dot{M}$, i.e. maximum of outburst, the DNOs are rarely present, but when they are they are seen (in different outbursts) at 14.1 s. We identify this as the minimum period of the equatorial belt – i.e. the keplerian period at the surface of the primary, which leads to a mass of 0.70 $M_\odot$. At high $\dot{M}$ the inner edge of the disc is close to the primary and we can see only the upper accretion zone once per rotation. But, at an orbital inclination of $\sim 63°$ for VW Hyi, the inner edge of the disc eventually retreats (with lowering $\dot{M}$) sufficiently for us to see the zone on the far side (i.e., no longer hidden by the primary); when this happens the keplerian period at the inner edge (which will be similar to the rotation period of the magnetically coupled equatorial belt), with the above parameters, should be ~ 45 s. This accounts for the observed frequency doubling at $P_{DNO} \sim 40$ s.

As the inner parts of the disc are cleared out by the expanding magnetosphere, eventually the observed DNO should become the reprocessed beam – the traveling

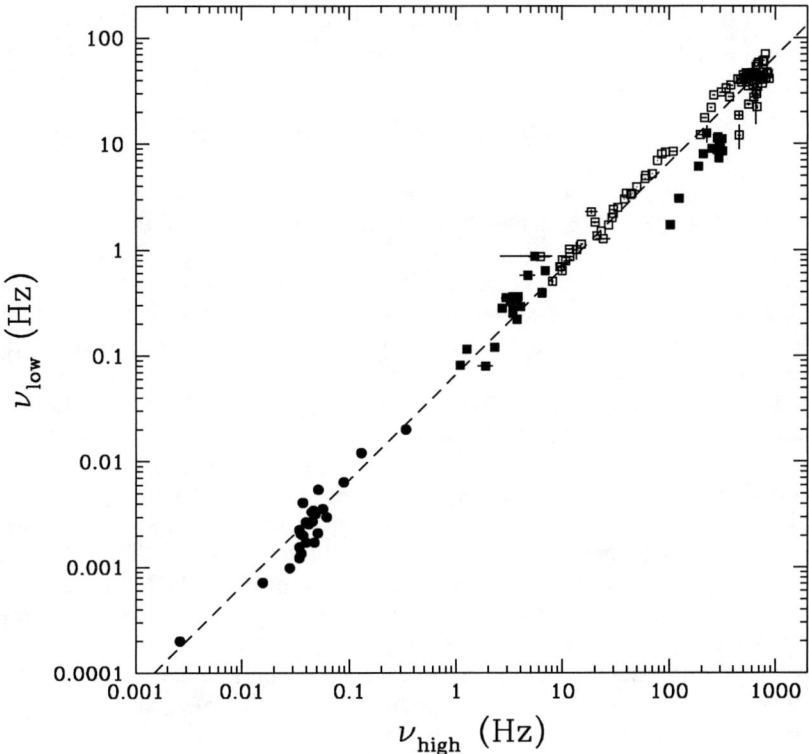

FIGURE 4. The Two-QPO diagram for X-ray binaries (filled squares: black hole binaries, open squares: neutron star binaries) and 26 CVs (filled circles). Each CV is plotted only once. The X-Ray binary data are from Belloni et al. (2002) and were kindly provided by T. Belloni. The dashed line marks $P_{QPO}/P_{DNO} = 15$ (From WWP).

(QPO generating) wave subtends the largest and nearest reprocessing site. The frequency of the observed DNO would then be the first harmonic of the difference between the frequencies of the equatorial belt and the traveling wave, which we can call $\tilde{\omega}$.

The traveling wave (according to WW1) is a region of deceleration and pile-up of disc material, and therefore constitutes a region of higher density at the inner edge of the disc. The magnetic field of the equatorial belt sweeps across this at frequency $\tilde{\omega}$, thereby modulating $\dot{M}$ onto the primary. This is an application of the beat-frequency model, first introduced for CVs (Warner 1983) and later independently for XRBs (Alpar & Shaham 1985; Lamb et al. 1985). The outcome is that the first harmonic, at frequency $2\tilde{\omega}$ will have sidebands at $\pm \tilde{\omega}$, i.e. the amplitude modulated frequency set $\tilde{\omega}$, $2\tilde{\omega}$ and $3\tilde{\omega}$. The modulations at $\tilde{\omega}$ and $2\tilde{\omega}$ will have additional components, of other amplitudes and phases, so their relative amplitudes will not be that of simple amplitude modulation, and may vary with time. Similar effects are seen in the orbital sidebands of intermediate polars (Warner 1986).

It should be noted that in this model there is no physical oscillation at the second harmonic, $3\tilde{\omega}$, it is purely a Fourier component in the light curve, as actually observed.

THE ROBERTSON-LEITER MODEL

As black holes are not supposed to have magnetic fields it is not automatically obvious that the same magnetically channeled accretion model for CVs and neutron stars is applicable to black hole XRBs. However, in a series of papers Robertson & Leiter have suggested that all of the central objects in galactic black hole candidate XRBs have magnetic moments. The evidence comes from (1) the conclusion that in neutron star binaries the power-law part of the X-Ray flux spectrum arises in the neutron star magnetosphere, where calculations and observations are in close agreement, and the recognition that the same model works for black hole XRBs (Robertson & Leiter 2002); (2) the realisation that a magnetic star collapsing towards its event horizon can be slowed by the radiation pressure ensuing from pair annihilation that itself is a result of pair production by the compressed magnetic field – such a "Magnetospheric, Eternally Collapsing Object" maintains an intrinsic magnetic moment outside its event horizon for orders of magnitude longer than a Hubble time (Robertson & Leiter 2003); (3) models of jets that arise from thin disc interaction with MECOs produce the Radio-IR correlation with Mass and X-Ray luminosity observed in neutron star and Black Hole binaries (and also in AGNs) (Robertson & Leiter 2004).

This recognition that black holes can have magnetospheres that interact with accretion discs permits a unified model that may explain the similarity of behaviour of DNOs and QPOs in CVs and X-Ray binaries seen in Fig. 4.

WZ SAGITTAE

In quiescence the dwarf nova WZ Sge has optical modulations at 27.87 and 28.95 s; their beat period at 744 s is commonly seen in the light curve and is responsible for recurring absorption dips. These have been interpreted in terms of the LIMA model (WW1), where the magnetic field of the primary is strong enough to prevent easy slippage of the equatorial belt in quiescence (Warner 2004), in which case the 27.87 s period should be thought of as an lpDNO.

During outburst weak oscillations near 6.5 s have been detected (Knigge et al. 2002), which may be the DNOs associated with the lpDNOs (i.e. with rotation of white dwarf itself, and also oscillations at 15 s that increased to 18 s (Welsh et al. 2003), which could be from a spun-up equatorial belt.

The reason for including WZ Sge here is because it appears to be an analogue of the accreting millisecond pulsars (MSPs) in X-Ray binaries, where $\dot{M}$ is very low, the spin rate is relatively high (a factor of only a few above rotational break-up period), and only a small amount of gas manages to leak onto the primary (Galloway et al. 2002; Rappaport, Fregeau & Spruit 2004). X-Ray transients containing MSPs have disc instabilities just as in dwarf novae. $\dot{M}$ during outburst is directly measured from the X-Ray flux; the millisecond flux pulsations are sinusoidal and are present during the whole factor of 50 –

100 of $\dot{M}$ variation during outburst. Optical and UV DNOs are present during superoutbursts of WZ Sge, and in quiescence, but their behaviour is complicated by reprocessing of the high energy rotating beam off various stationary and moving components in the binary system. The magnetic field of the primary in WZ Sge appears not to be strong enough to prevent a freely moving equatorial band from developing during the highest $\dot{M}$ delivered in outburst. In contrast, in the MSP XRBs the pulses are monoperiodic and have small but measurable period derivatives, but here there is no free equatorial belt and we observe only the X-Rays from the accretion region itself.

CONCLUSIONS

CVs show an increasing range of rapid modulation phenomena that have analogues among the XRBs. It appears increasingly likely that magnetically channeled accretion is the underlying cause for these properties, implying magnetic moments for some stellar mass black holes.

ACKNOWLEDGMENTS

BW is supported by research funds from the University of Cape Town; PAW is supported by research funds from the University, from the National Research Foundation, and by a strategic award given by the University to BW.

REFERENCES

1. Abramowicz, M.A., Bulik, T., Bursa, M. and Kluzniak, W., A&A, **404**, L21 (2003).
2. Alpar, M.A., and Shaham, J., Nature, **316**, 239 (1985).
3. Aznar Cuadrado, R., Jordan, S., Napiwotzki, R, Schmid, H.M., Solanki, S.K., and Mathys, G., A&A, in press (2004).
4. Belloni, T., Psaltis, D., and van der Klis, M., ApJ, **572**, 392 (2002).
5. Galloway, D.K., Chakrabarty, D., Morgan, E.H., and Remillard, R.A., ApJL, **576**, 137 (2002).
6. Godon, P., Sion, E.M., Cheng, F.H., Szkody, P., Long, K.S., and Froning, C.S., ApJ, **612**, 429 (2004).
7. Homan, J., et al., ApJ., in press (2004).
8. Knigge, C., Hynes, R.I., Steeghs, D., Long, K.S., Araujo-Betancor, S., and Marsh, T.R., ApJ Lett., **580**, L151 (2002).
9. Lamb, F.K., Shibazaki, N., Alpar, M.A., and Shaham, J., Nature, **317**, 681 (1985).
10. Mauche, C.W., ApJ, **580**, 423 (2002).
11. McClintock, J.E., and Remillard, R.A., in *Compact Stellar X-Ray Sources*, edited by W.H.G. Lewin & M. van der Klis, Cambridge University Press, in press (2004).
12. Paczynski, B., in *Nonstationary Evolution of Close Binaries*, edited by A. Zytkov (Polish Sci. Publ.: Warsaw), 89 (1978).
13. Patterson, J., et al., PASP, **114**, 1364 (2002).
14. Rappaport, S.A., Fregeau, J.M., and Spruit, H., ApJ, in press (2004).
15. Remillard, R.A., Muno, M.P., McClintock, J.E., and Orosz, J.A., ApJ, **580**, 1030 (2002).
16. Robertson, S.L., and Leiter, D.J., ApJ, **565**, 447 (2002).
17. Robertson, S.L., and Leiter, D.J., ApJL, **596**, 203 (2003).
18. Robertson, S.L., and Leiter, D.J., MNRAS, **350**, 1391 (2004).
19. Sion, E.M., Cheng, F.H., Huang, M., Hubeny, I., and Szkody, P., ApJL **471**, 41 (1996).
20. Swank, J., in *X-Ray Timing 2003: Rossi and Beyond*, AIP Conf. Ser., in press (2004).

21. van der Klis, M., et al., Nature, **316**, 225 (1985).
22. van der Woerd, H., et al., A&A, **182**, 219 (1987).
23. Walker, M.F., ApJ, **123**, 68 (1956).
24. Warner, B., in *Cataclysmic Variables and Related Objects*, edited by M. Livio & G. Shaviv (Reidel: Dordrecht), 269 (1983).
25. Warner, B.,MNRAS, **219**, 347 (1986).
26. Warner, B., *Cataclysmic Variable Stars*, Cambridge University Press (1995).
27. Warner, B., PASP **116**, 115 (2004).
28. Warner, B., and Robinson, E.L., Nature Phys. Sci., **239**, 2 (1972).
29. Warner, B., and Woudt, P.A., MNRAS, **335**, 84 (2002). (WW1)
30. Warner, B., Woudt, P.A., and Pretorius, M.L., MNRAS, **344**, 1193 (2003). (WWP)
31. Welsh, W.F., Sion, E.M., Godon, P., Gaensicke, B.T., Knigge, C., Long, K.S., and Szkody, P., ApJ, **599**, 509 (2003).
32. Wheatley, P.J., and West, R.G., MNRAS, **345**, 1009 (2003).
33. Woudt, P.A., and Warner, B., MNRAS, **333**, 411 (2002). (WW2)

High-Speed Photometry of AM CVn Stars

Patrick A. Woudt*, Brian Warner*, Joseph Patterson[†] and Catherine Espaillat**,[†]

*Department of Astronomy, University of Cape Town, Rondebosch 7700, South Africa
[†]Department of Astronomy, Columbia University, 550 West 120th Street, New York, NY 10027, USA
**Department of Astronomy, University of Michigan, 830 Dennison Building, Ann Arbor, MI 48109, USA

Abstract. We present high-speed photometry of three AM CVn stars discovered in the last few years (ES Cet, '2003aw' and SDSS J1240). ES Cet has a very stable photometric modulation with a single period of 620.2117 ($\pm$ 0.0002) s (Espaillat et al. 2005) which is identical to its spectroscopic period. The 2-sigma upper limit on the period derivative ($\dot{P}$) of ES Cet is $< 1.5 \times 10^{-11}$, based on a 3-year observational baseline. Since the discovery of '2003aw' in a supernova search, this AM CVn star has been observed during two outbursts and has a superhump modulation of $P_{sh} = 2041.5 \pm 0.3$ s. At quiescence, 2003aw is fairly faint ($V \sim 20.4$ mag) and no orbital period has yet been derived for this system. Two of the AM CVn stars (V803 Cen and CR Boo) show rapid oscillations similar to the (longer-period) dwarf nova oscillations and quasi-periodic oscillations seen in hydrogen-rich high-$\dot{M}$ CVs (Warner 2004).

Keywords: Accretion, Accretion disks – Binary Stars: Cataclysmic Variables
PACS: 97.10.Gz, 97.80.Ũd

INTRODUCTION

AM CVn stars mark a small subset of the family of cataclysmic variable stars (CVs); they are helium-transferring close interacting binaries, consisting of two helium white dwarfs. Some of the main characteristics of these systems are the short orbital periods in the range of several minutes up to ~ 70 minutes, the low mass-ratio (q) of secondary mass to primary mass (M_2/M_1), and high mass-transfer rates ($\dot{M} \sim 10^{-8}$ $M_\odot$ y^{-1}) for the shortest period ultracompact AM CVn stars. A comprehensive overview of the nature of AM CVn stars is given by Warner (1995) and Nelemans, Yungelson & Portegies Zwart (2004). There are currently only eleven certified and two candidate AM CVn stars known, with (orbital) periods ranging from 321 s to 3906 s; all thirteen objects are listed in Table 1.

The all-sky distribution of the known AM CVns (including the two candidates) is shown in Fig. 1; the clumpiness in the distribution is a clear indication that only the tip of the AM CVn iceberg has been recovered to date. Spectroscopic and photometric surveys to faint brightness limits such as the Sloan Digital Sky Survey (see Roelofs et al. 2004) and the Faint Sky Variability Survey (Groot et al. 2003) will probably result in the more systematic discovery of further AM CVns.

Among the AM CVn stars, there appears to be a parallel behaviour with respect to the hydrogen-rich CVs: there are systems with stable accretion discs and high rates of mass-transfer ($\dot{M}$) (the equivalents of nova-like variables), unstable high $\dot{M}$ systems (the

TABLE 1. Certified and candidate AM CVn stars

Star	V (mag)	P_{orb} (sec)	P_{sh} (sec)	References
RX J0806*	21.1	321.25		Israel et al. (2002, 2003); Ramsay, Hakala & Cropper (2002)
V407 Vul*	19.9	569.38		Cropper et al. (1998)
ES Cet	16.9	620.2117		Warner & Woudt (2002a); Espaillat et al. (2005)
AM CVn	14.1	1028.7	1051.2	Solheim et al. (1998); Skillman et al. (1999); Nelemans, Steeghs & Groot (2001)
HP Lib	13.7	1102.7	1119.0	O'Donoghue et al. (1994); Patterson et al. (2002)
CR Boo	13.0 – 18.0	1471.3	1487	Wood et al. (1987); Patterson et al. (1997)
KL Dra	16.8 – 20	1500	1530	Wood et al. (2002)
V803 Cen	13.2 – 17.4	1612.0	1618.3	Patterson et al. (2000)
CP Eri	16.5 – 19.7	1701.2	1715.9	Abbott et al. (1992)
'2003aw'	15.0 – 20.4		2041.5	Woudt & Warner (2003a); Nogami et al. (2004)
SDSS J1240	19.6	2238[†]		Roelofs et al. (2004); Woudt, Warner & Pretorius (2004)
GP Com	15.7 – 16.0	2974		Nather, Robinson & Stover (1981); Marsh, Horne & Rosen (1991)
V396 Hya	17.6	3906		Ruiz et al. (2001); Woudt & Warner (2001)

[*] Still to be confirmed as a bona fide AM CVn system
[†] Private communication (P.J. Groot)

nova-likes of VY Scl type), intermediate $\dot{M}$ dwarf novae, and low $\dot{M}$ systems possibly permanently in a low state. The stable high $\dot{M}$ systems are at shortest orbital periods ($P_{orb} \lesssim 1200$ s), for P_{orb} between ~ 1200 and ~ 2200 s all the AM CVns known so far show repeated high and low states, and for $P_{orb} \gtrsim 2200$ s the AM CVn systems appear in a permanent low state. The periods at which these transitions occur are still ill-determined given the small number of objects known to date (see Table 1).

RECENTLY DISCOVERED AM CVN STARS

ES Ceti

With a stable photometric period of 620.2117 s (Warner & Woudt 2002a; Espaillat et al. 2005) and a spectrum dominated by He I and He II emission lines and with no hydrogen present (see Woudt & Warner 2003b; Steeghs 2003), ES Cet's identification as an ultracompact AM CVn is unambiguous. In contrast, the AM CVn credentials of the two other candidate AM CVn systems with (orbital?) periods less than 10 minutes

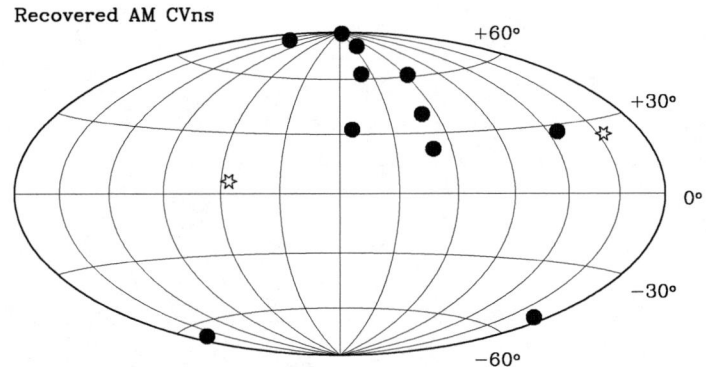

FIGURE 1. The distribution in Galactic coordinates of known AM CVn stars (filled circles) and two possible candidates (open stars). In this Aitoff projection, the Galactic centre is in the middle of the diagram with increasing Galactic longitudes towards the left.

– RX J0806.3+1527 (Israel et al. 2002; Ramsay, Hakala & Cropper 2002; Israel et al., these proceedings) and V407 Vul (Cropper et al. 1998) – remain the subject of some discussion (e.g. Woudt & Warner 2004; Norton, Haswell & Wynn 2004).

High-speed photometry of ES Cet was obtained over the last three years with the 1.9-m and 1.0-m telescopes of the South African Astronomical Observatory (SAAO) and the UCT CCD, and with the 1.3-m McGraw-Hill telescope at the MDM observatory on Kitt Peak. The result of the analysis of this data set was presented by Espaillat et al. (2005). Somewhat surprisingly (as the photometric signal of most AM CVns is dominated by a superhump signal), ES Cet's photometric modulation is very stable over this 3-year baseline, too stable to be associated with a superhump modulation. The O–C diagram of the UCT CCD photometry is shown in Fig. 2, and shows a signal consistent with a single constant period of 620.2117 s and with a limit on the period derivative of $\dot{P} < 1.5 \times 10^{-11}$ s/s. Support for the interpretation that the photometric modulation corresponds to the orbital period (and not a superhump period) comes from phase-resolved spectroscopy; the spectroscopic period is identical to the photometric period (Woudt & Warner 2003b; Steeghs 2003).

Despite the apparent lack of superhumps in the photometric data, the general conditions for superhump modulations appear to have been fulfilled (high $\dot{M}$ and low mass-ratio q) – the following parameters are derived for ES Cet: $\dot{M} \sim (2-4) \times 10^{-9}$ $M_\odot$ y^{-1} based on a distance estimate of 350 pc, and $q \sim 0.15$ (Espaillat et al. 2005). It is entirely possible, however, that in ES Cet the mass-transfer stream impacts directly on to the primary (as proposed by Marsh & Steeghs (2002) for the AM CVn candidate V407 Vul[1])

---

[1] Subsequent spectroscopy of V407 Vul (Steeghs 2003) has revealed a reddened K star at the position of V407 Vul. This has cast serious doubts on the AM CVn interpretation of this object (e.g., see Warner 2005). In the discussion here, the comparison between V407 Vul and ES Cet is based purely on their similar periods; spectroscopically, V407 Vul and ES Cet are completely different.

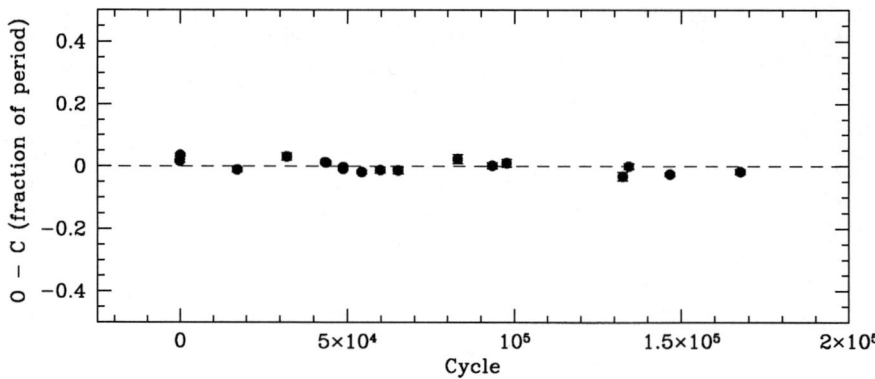

FIGURE 2. The O–C diagram of all the UCT CCD photometry obtained of ES Cet with respect to the ephemeris given by Espaillat et al. (2005) with $P_{orb} = 620.2117$ s, covering a 3-year baseline.

and an accretion disk has not been formed (hence the absence of superhumps). The orbital period of ES Cet is only slightly longer than the period in V407 Vul (569 s versus 620 s). Thus, given the small binary separation in ES Cet ($\sim 9.9 \times 10^9$ cm, Warner & Woudt 2002a), direct impact accretion is a distinct possibility (Marsh & Steeghs 2002; Espaillat et al. 2005). Espaillat et al. (2005) estimate the size of the emitting region in ES Cet to be $4.5 \times 10^8 d_{300}^{1.0}$ cm, too small to be an accretion disk but comparable to the size of the white dwarf, or a substantial spot on the white dwarf.

The AM CVn star in Hydra ('2003aw')

'2003aw' was discovered in the course of a supernova search in 2003 February and identified spectroscopically as an AM CVn system by Chornock & Filippenko (2003). It is the second AM CVn system found by means of supernova searches; KL Dra is the other example (Jha et al. 1998).

High-speed photometry of '2003aw' in outburst (2003 March/April) was presented by Woudt & Warner (2003a) and a superhump period of 2041.5 ± 0.3 s was clearly detected both in the high state (V $\sim$ 18.0–18.5 mag) and in the intermediate state (V $\sim$ 19.3–19.8 mag). During the latter phase, the brightness is modulated furthermore on a quasi-periodic period of ~ 16 h, with a range of > 0.4 mag (the true amplitude was undetermined as observations were made from a single observing site). Similar behaviour – a high amplitude (~ 1.1 mag), long-period (~ 20 h) quasi-periodic modulation – is also seen in two other AM CVn stars that show repeated high and low states, namely CR Boo and V803 Cen (Patterson et al. 1997, 2000).

During the high state, '2003aw' has short-lived 'outbursts' of $\Delta V \sim 2$ mag, similar to what is seen in some intermediate polars (Schwarz et al. 1988; van Amerongen & van Paradijs 1989). The origin of such a short-lived outburst is still unclear, but this behaviour in '2003aw' has been confirmed by observations during a later outburst (Nogami

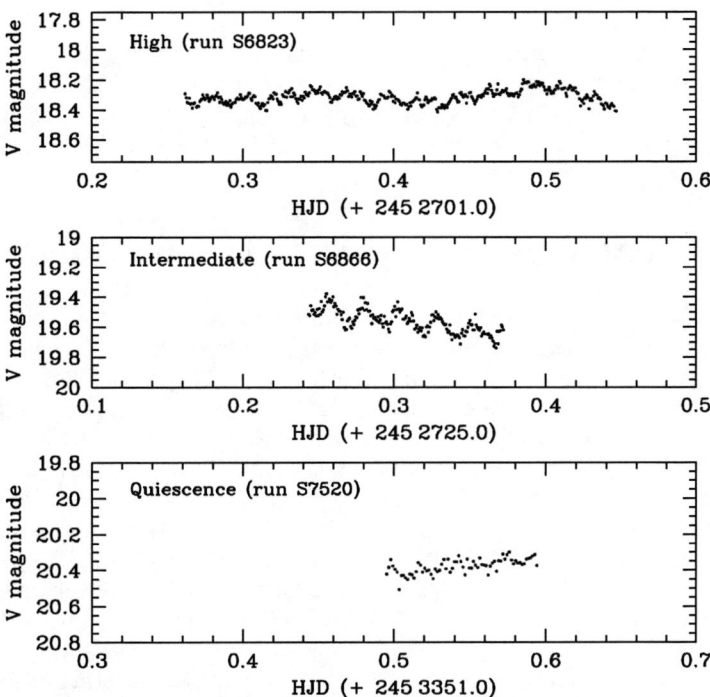

FIGURE 3. Light curves of '2003aw' in the high-state (upper panel), intermediate state (middle panel) and at quiescence (lower panel) obtained with the SAAO 1.9-m telescope and the UCT CCD.

et al. 2004) and is not seen in any of the other 'outbursting' AM CVn systems. Observations of '2003aw' by Nogami et al. (2004) during a second reported outburst (2004 May), confirm the superhump period measured earlier by Woudt & Warner (2003a). During this outburst '2003aw' was observed both during superoutburst (V $\sim$ 15.0 mag) – a phase presumably missed during the 2003 outburst – and during high state.

In quiescence (see the light curve of '2003aw' shown in the lower panel of Fig. 3), '2003aw' is fairly faint (V $\sim$ 20.4 mag) and does not show a clear orbital modulation. Phase-resolved spectroscopy of '2003aw' at quiescence with a 10-m class telescope will be the best way to determine its orbital period.

SDSS J1240

SDSS J124058.03–015919.2 (SDSS J1240 hereafter) is the latest addition to the family of AM CVn stars. It was discovered recently (Roelofs et al. 2004) as a candidate AM CVn star following a systematic search for such systems in public release data from the Sloan Digital Sky Survey. Our photometry of SDSS J1240 shows little or no flickering, and no orbital modulation can be detected from the photometry (Figure 17 of Woudt,

Warner & Pretorius 2004).

Phase-resolved spectroscopy using the VLT has resulted in the orbital period determination of SDSS J1240 (P.J. Groot, private communication). SDSS J1240 appears spectroscopically as a low-$\dot{M}$ AM CVn star with an orbital period of 2238 s (very close to that of '2003aw' which shows repeated high states), it is not unlikely that SDSS J1240 could have occasional outbursts.

RAPID OSCILLATIONS IN AM CVN STARS

There are about 100 CVs that show rapid oscillations with periods ranging from a few seconds to a few hundred seconds. A wide range of physical processes in CVs produce modulations on these time scales. In moderately magnetic systems (intermediate polars; see, e.g., Norton, Wynn & Somerscales 2004) the rapid oscillation, which is highly coherent, is related to the spin period of the white dwarf. In systems of high mass-transfer rate ($\dot{M}$) – dwarf novae in outburst and nova-likes – the rapid oscillations are associated with dwarf nova oscillations (DNOs) and quasi-periodic oscillations (QPOs).

Properties of the DNOs and QPOs in high $\dot{M}$ CVs have been discussed extensively in a recent review (Warner 2004). Warner (2004) lists DNO and/or QPO periods for 76 CVs, including four AM CVn systems – V803 Cen, AM CVn, HP Lib and CR Boo. In a recent series of papers on DNOs and QPOs in CVs (Woudt & Warner 2002b; Warner & Woudt 2002b; Warner, Woudt & Pretorius 2003), the Low Inertia Magnetic Accretor (LIMA) model has been put forward to explain the DNOs in terms of a temporarily enhanced (during periods of increased $\dot{M}$) weak magnetic field allowing magnetically channelled accretion onto a rapidly rotating equatorial belt. Evidence for such a belt in the dwarf nova VW Hyi was presented by Sion et al. (1996) based on HST spectroscopy shortly after an outburst of VW Hyi.

V803 Cen and CR Boo

V803 Cen is quite similar in its photometric behaviour to '2003aw' described earlier. It has repeated high states ($V \sim 13.2$ mag), exhibits the cycling behaviour during its intermediate brightness phase (see Fig. 4) and has low states around $V \sim 17.4$ mag.

It has, however, one feature not seen in either '2003aw' or CR Boo, and that is a sporadic 176-s brightness modulation (O'Donoghue, Menzies & Hill 1987; O'Donoghue & Kilkenny 1989). A preliminary analysis of the CBA (Center for Backyard Astrophysics) data archive of V803 Cen indicates that this signal mostly – though not exclusively – occurs around $V \sim 13.6 - 14.3$ mag when V803 Cen is in the cycling phase. This 176-s modulation has a fairly low coherence (the period can be anywhere in the range of $174.8 - 176.7$ s) and varies in amplitude ($\sim 3 - 20$ mmag) from night to night. The 176-s signal is consistent in behaviour with the longer-period DNOs (lpDNOs), and is interpreted as accretion onto the main body of the white dwarf primary through a weak magnetic field (Warner, Woudt & Pretorius 2003). The alternative explanation of a DB pulsator is less likely, as the 176-s signal is only present during periods of

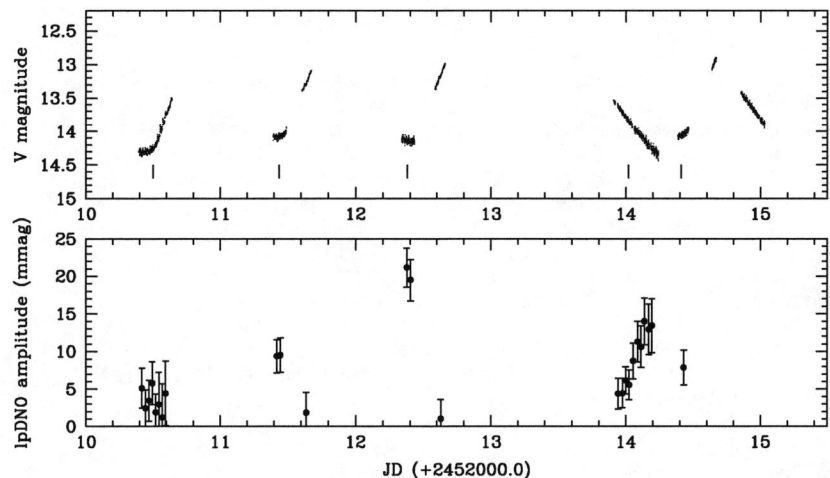

FIGURE 4. Data from the CBA data archive for V803 Cen in 2001 April. The cycling behaviour is apparent in the upper panel and the lower vertical marks below the light curve indicate when the 176-s brightness modulation was present. The lower panel shows the amplitude of the 176-s signal during these observations. Note the absence of the 176-s modulation during the brighter phases.

enhanced brightness when the accretion luminosity dominates and even then it is not always present.

High-speed photometry of CR Boo was obtained with the UCT CCD in 2003 March (Warner, Woudt & Pretorius 2003; Pretorius 2004) during which DNOs and QPOs were found to be simultaneously present. With a DNO period of 21.6 s and a QPO at 291 s – the ratio of QPO to DNO period is 13.5 – CR Boo falls on the extension of the two-QPO diagram for low-mass X-ray binaries together with 25 hydrogen-rich CVs (see Figure 4 of Warner & Woudt, these proceedings). It is interesting to note that this extention therefore is not only limited to hydrogen-rich systems.

In terms of the LIMA model, the (lp)DNOs and QPOs observed in V803 Cen and CR Boo are interpreted as signals of a weak magnetic field (temporarily enhanced during periods of high $\dot{M}$) channelling the accretion on to the primary. As such, it is the first (indirect) evidence that some AM CVn primaries are weakly magnetic.

ACKNOWLEDGMENTS

PAW is supported by research funds from the University of Cape Town and from the National Research Foundation; BW is supported entirely by the University of Cape Town. We thank Jennie McCormick, Fred Velthuis and Jonathan Kemp for the data shown in Figure 4.

REFERENCES

1. Abbott, T.M.C., Robinson, E.L., Hill, G.J., and Haswell, C.A., ApJ, **399**, 680 (1992).
2. Chornock, R., and Filippenko, A.V., IAU Circ., **8084** (2003).
3. Cropper, M., Harrop-Allin, M.K., Mason, K.O., et al., MNRAS, **293**, L57 (1998).
4. Espaillat, C., Patterson, J., Warner, B., and Woudt, P.A., PASP, **117**, in press (2005).
5. Groot, P.J., Vreeswijk, P.M., Huber, M.E., et al., MNRAS, **339**, 427 (2003).
6. Israel, G.L., Covino, S., Stella, L., et al., ApJ, **598**, 492 (2003).
7. Israel, G.L., Hummel, W., Covino, S., et al., A&A, **386**, L13 (2002).
8. Jha, S., Garnavich, P.M., Challis, P., and Kirshner, R., IAU Circ., **6983** (1998).
9. Marsh, T.R., and Steeghs, D., MNRAS, **331**, L7 (2002).
10. Marsh, T.R., Horne, K., and Rosen, S., ApJ, **366**, 535 (1991).
11. Nather, R.E., Robinson, E.L., and Stover, R.J., ApJ, **244**, 269 (1981).
12. Nelemans, G., Steeghs, D., and Groot, P.J., MNRAS, **326**, 621 (2001).
13. Nelemans, G., Yungelson, L.R., and Portegies Zwart, S.F., MNRAS, **349**, 181 (2004).
14. Nogami, D., Monard, B., Retter, A., et al., PASJ, **56**, L39 (2004).
15. Norton, A.J., Haswell, A.C., and Wynn, G.A., A&A, **419**, 1025 (2004).
16. Norton, A.J., Wynn, G.A., and Somerscales, R.V., ApJ, **614**, 349 (2004).
17. O'Donoghue, D., and Kilkenny, D., MNRAS, **236**, 319 (1989).
18. O'Donoghue, D., Menzies, J.W., and Hill, P.W., MNRAS, **227**, 347 (1987).
19. O'Donoghue, D., Kilkenny, D., Chen, A., et al., MNRAS, **271**, 910 (1994).
20. Patterson, J., Fried, R.E., Rea, R., et al., PASP, **114**, 65 (2002).
21. Patterson, J., Kemp, J., Shambrook, A., et al., PASP, **109**, 1100 (1997).
22. Patterson, J., Walker, S., Kemp, J., et al., PASP, **112**, 625 (2000).
23. Pretorius, M.L., MSc thesis, University of Cape Town (2004).
24. Ramsay, G., Hakala, P., and Cropper, M., MNRAS, **332**, L7 (2002).
25. Roelofs, G.H.A., Groot, P.J., Steeghs, D., and Nelemans, G., RMxAC, **20**, 254 (2004).
26. Ruiz, M.T., Rojo, P.M., Garay, G., and Maza, J., ApJ, **552**, 679 (2001).
27. Schwarz, H.E., van Amerongen, S., Heemskerk, M.H.M., and van Paradijs, J., A&A, **202**, L16 (1988).
28. Sion, E.M., Cheng, F.H., Huang, M., et al., ApJ, **471**, L41 (1996).
29. Skillman, D.R., Patterson, J., Kemp, J., et al., PASP, **111**, 1281 (1999).
30. Solheim, J.-E., Provencal, J.L., Bradley, P.A., et al., A&A, **332**, 939 (1998).
31. Steeghs, D., Workshop on Ultracompact Binaries, Santa Barbara (2003) See http://online.kitp.ucsb.edu/online/ultra_c03/steeghs/
32. van Amerongen, S., and van Paradijs, J., A&A, **219**, 195 (1989).
33. Warner, B., Ap&SS, **225**, 249 (1995).
34. Warner, B., PASP, **116**, 115 (2004).
35. Warner, B., Proc. IAU JD5 (Sydney GA), astro-ph/0310243 (2005).
36. Warner, B., and Woudt, P.A., PASP, **114**, 129 (2002a).
37. Warner, B., and Woudt, P.A., MNRAS, **335**, 84 (2002b).
38. Warner, B., Woudt, P.A., and Pretorius, M.L., MNRAS, **344**, 1193 (2003).
39. Wood, M.A., Casey, M.J., Garnavich, P.M., and Haag, B., MNRAS, **334**, 87 (2002).
40. Wood, M.A., Winget, D.E., Nather, R.E., et al., ApJ, **313**, 757 (1987).
41. Woudt, P.A., and Warner, B., MNRAS, **328**, 159 (2001).
42. Woudt, P.A., and Warner, B., MNRAS, **333**, 411 (2002b).
43. Woudt, P.A., and Warner, B., MNRAS, **345**, 1266 (2003a).
44. Woudt, P.A., and Warner, B., astro-ph/0310494 (2003b).
45. Woudt, P.A., and Warner, B., RMxAC, **20**, 120 (2004).
46. Woudt, P.A., Warner, B., and Pretorius, M.L., MNRAS, **351**, 1015 (2004).

Formation of the "precessional" spiral wave in the cool accretion disk in semidetached binaries

D.V. Bisikalo*, A.A. Boyarchuk*, P.V. Kaygorodov*, O.A. Kuznetsov*,[†] and T. Matsuda**

*Institute of Astronomy, 48 Pyatnitskaya str., 119017 Moscow, Russia
[†]Keldysh Institute of Applied Mathematics, 4 Miusskaya sq., 125047 Moscow, Russia
**Department of Earth and Planetary Sciences, Kobe University, Nada Kobe 657-8501, Japan

Abstract.
   We suggest that an additional spiral density wave can exist in the inner parts of the cool accretion disk, where gas-dynamical perturbations are negligible. This spiral wave is due to the retrograde precession of the streamlines in the binary system. The results of a three-dimensional gas-dynamical simulation have shown that a considerable increase in the accretion rate (by an order of magnitude) is associated with the formation of the "precessional" spiral wave. Basing on this fact we suggest a new mechanism for the superoutbursts and superhumps in binaries.

Keywords: Binary Stars – Accretion Disk – Gas-dynamical Modeling
PACS: 97.80.Űd, 97.10.Gz

The qualitative analysis and three-dimensional gasdynamical simulations of the flow structure in semi-detached binaries for low gas temperatures ($\sim 10^4$ K) enabled us to identify characteristic features of the structure of cool accretion disks [1, 2]. In general, the flow structure is qualitatively the same as for the case of high gas temperatures [3, 4, 5]. The gasdynamical structure of the flow is governed by the stream of matter from L_1, the accretion disk, the circumdisk halo, and the circumbinary envelope. The interaction between the stream and disk is shockless, and the interaction of the matter of the circumdisk halo and circumbinary envelope with the stream leads to the formation of a shock with the form of a "hot line" along the edge of the stream. However, reduction of gas temperature leads to several differences as well. A cool accretion disk becomes considerably denser (compared to the matter in the stream) and thinner, and its shape is nearly circular rather than quasielliptical. A second arm of the tidal spiral shock (see [6, 7]) is formed; neither arm of the shock approaches the accretor, and both are located in the outer parts of the disk. Considering the weak influence of the stream on the dense inner parts of the disk and the fact that all the shocks (the hot line and the two arms of the tidal shock) are located in the outer parts of the disk, we can distinguish an additional element of the flow structure in the cool case: the presence of an inner region of the accretion disk where the influence of the gas-dynamical perturbations noted above can be neglected.
   The formation of a region in the inner part of the disk that is free from gasdynamical perturbations enables us to treat this region simply as a slightly eccentric disk immersed in the gravitational field of the binary. It is known (see, e.g., [8, 9]) that particles revolving around one of the binary components will precess due to the influence of

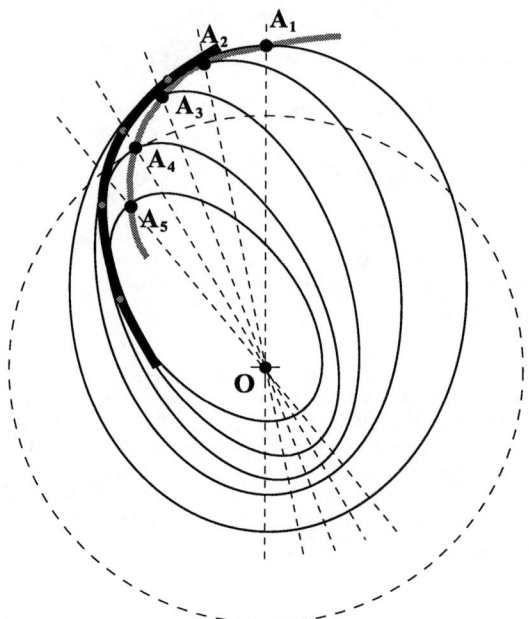

FIGURE 1. Schematic representation of the generation of spiral structures in the inner parts of a cool disk, where gas-dynamical perturbations are negligible.

the companion. This precession is retrograde, and the rate of the precession decreasing as the particle approaches the accretor according to the law

$$\frac{P_{prec}}{P_{orb}} = {}^4\!/_3 \frac{\sqrt{1+q}}{q}\left(\frac{r}{A}\right)^{-3/2}, \quad (1)$$

where P_{prec} is the period of the orbital precession, P_{orb} is the orbital period of the binary, q is the component mass ratio, r is the orbital radius, and A is the binary separation. For the accretion disk the orbits must be replaced by streamlines. Streamlines cannot intersect and can only touch by being tangent to each other. If the orbits precess so that the precession of distant streamlines tends to be faster, these distant streamlines will constantly overtake those with smaller semimajor axes. Since the streamlines in a gas-dynamical disk cannot intersect, an equilibrium solution is established with time and all the streamlines begin to precess with the same angular velocity, i.e., to display rigid-body rotation. Accordingly, distant orbits must turn through a larger angle opposite to the direction of the rotation of the disk material, since the precession is retrograde. The precession rate is confined within the range delimited by the precession rates of the outer ('fast') orbits and inner ('slow') orbits.

Let us consider a solution with the semimajor axes of the streamlines misaligned with respect to some chosen direction by an angle (turn angle) that is proportional to the semimajor axis of the orbit (Fig. 1). It is obvious that such a solution should contain spiral structures. In particular, due to the nonuniformity of the motion along the

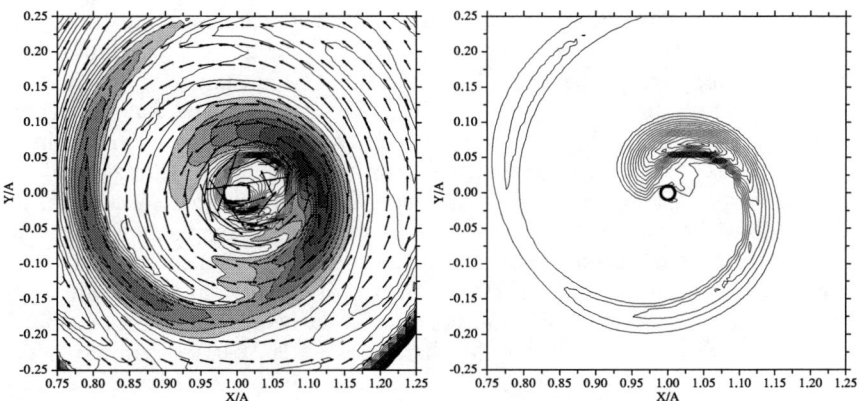

FIGURE 2. Left: contours of density and velocity vectors in the equatorial plane of the binary; right: contours of the radial flux of matter in the central parts of the disk.

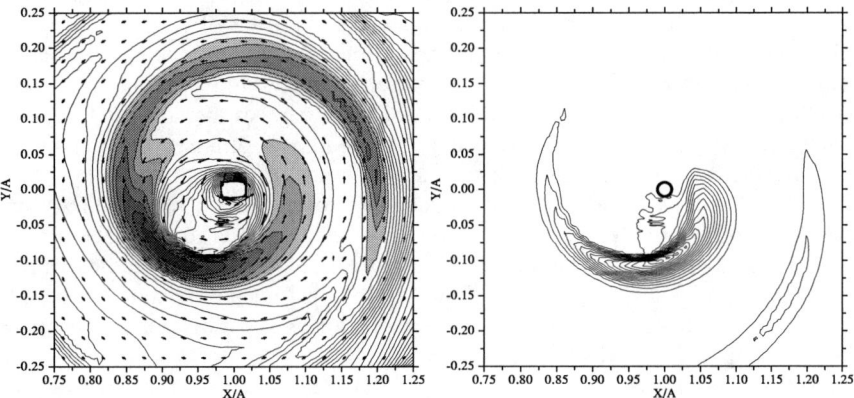

FIGURE 3. Same as in Fig. 2 for $t = t_0 + 3P_{orb}$.

streamline and the formation of a maximum density at apastron, the curve connecting the apastrons (the black circles labelled $A_1, A_2, \cdots, A_5$ in Fig. 1) will form a spiral density wave. The curve connecting the places of maximum approach of the streamlines (gray circles in Fig. 1) is likewise a spiral density wave. The rotational velocity of this wave is determined by the mean precession rate of the streamlines. The presence of the density wave, as well as the fact that the velocity of the particles increases after passing apastron leads to an increase in the radial component of the mass flux $F_{rad} \sim \rho u_r$ due to the increase of both the density ρ and the radial velocity component u_r directed toward the accretor. The increase in the radial component of the mass flux behind the wave increases the accretion rate in the region where the precessional wave approaches the accretor.

To investigate the possible existence of a spiral "precessional" wave in the inner regions of cool accretion disks, we performed a 3D gas-dynamical simulation of the disk structure for the case when radiative cooling reduces the gas temperature to $\sim 10^4$ K [2, 10]. The results are presented in Figs 2 and 3. The left panel of Fig. 2 shows the density distribution and velocity vectors in the equatorial plane for the steady-state solution. The right panel shows the distribution of the radial flux of matter in central regions of the disk for the same time. These distributions show the formation of a dense, circular accretion disk and a compact circumdisk halo in the binary. The interaction of the circumdisk halo and circumbinary envelope with the stream results in the formation of the hot line located outside the disk. A two-armed tidal spiral shock is formed in the disk, with both of its arms located in the outer part of the disk, so that they do not reach the accretor. Another spiral wave in the inner part of the disk is also clearly visible. These simulations were made for a binary with characteristics close to those of the dwarf nova IP Peg: $M_1 = 1.02 M_\odot, M_2 = 0.5 M_\odot$, and $A = 1.42 R_\odot$. The computation results show that the precession of the inner spiral wave is retrograde (see Figs. 2 and 3) and the velocity of its revolution in the inertial frame (i.e., the observers frame) is ~ 0.13 of a revolution per binary orbital period. The distribution of the radial flux of matter in the disk shows that, starting from the outer radius of the wave, this flux increases as it approaches the accretor, and reaches a maximum that is more than an order of magnitude higher than the flux in regions of the disk outside the wave. Thus, the accretion rate will be enhanced by more than a factor of ten due to the formation of this wave.

The "precessional" spiral density wave in the inner parts of the disk can be applied for explanation of superoutbursts and especially superhumps in binaries (the first application of this mechanism to explanation of the superotbursts in SU UMa stars was made by us in [10]) as follows: the beating between the orbital motion of the binary and the precessional motion of the density waves is displayed as a superhump on the orbital light curve. According to current theory, superhumps can only appear in binaries with tidally unstable accretion disks. Simulations show that the tidal instability can only occur if the disk radius exceeds a certain value, the 3:1 resonance radius. This implies that eccentric disks (which generate the superhumps according to current theory) can be present only in CVs with small mass ratios: $q = M_2/M_1 \leq 0.33$ [11]. There are binary systems with superhumps where the mass ratio is well above the criticall value. The estimates of the mass ratio for TV Col, for example, show that q can be as much as 0.92 [12, 13]. The "precessional" wave is formed even in systems with large mass ratio, and this mechanism can be useful for interpretation of the superhumps in systems with different mass ratio. Indeed, application of the "precessional" spiral wave to the explanation of the observational manifestations in OY Car (SU UMa type binary with $q = 0.147$) was successfull [10]. On the other hand the 3D gasdynamical simulations of the system with $q \simeq 0.5$ (see Figs 2 and 3) show that the formation of the "precessional" spiral density wave takes place in such a kind of systems as well.

The physical reasons of the formation of the "precessional"-type density wave is the retrograde precession of the elliptical orbit in binary system with period P_{prec} defined by Eq. (1). The beating between P_{prec} and P_{orb} results in modulation with period $P_{SH} = \dfrac{P_{prec} P_{orb}}{P_{prec} - P_{orb}}$, which is the superhump period ('$-$' sign in denominator deals with a retrogade direction of precession). As a rule, P_{SH} results few percents larger than P_{orb}:

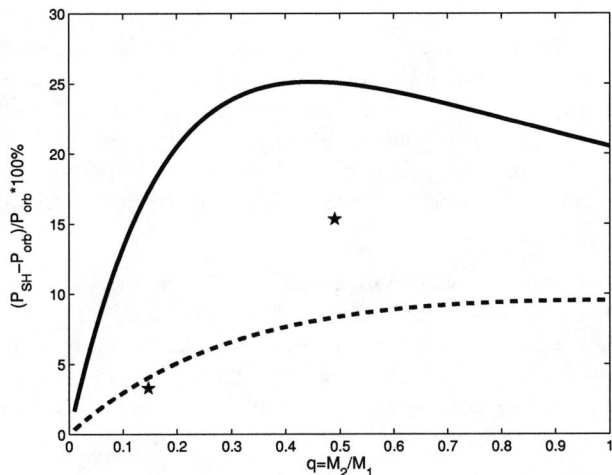

FIGURE 4. The deviation between P_{SH} and P_{orb} (in percents): dashed line – $f_0(q)$ in accordance to [8, 9]; solid line – $f(q)$ in accordance to [9, 15]; asterisks – our 3D gasdynamical calculations for a binary with $q \simeq 0.5$ and OY Car ($q = 0.147$) [2, 10].

$(P_{SH}/P_{orb} - 1) \cdot 100\% \simeq 10\%$.

To estimate the maximal value of P_{SH} as a function of q let us use the radius of large closed orbit calculated by Paczyński which can be considered as the maximal radius of the disk [14]. For $q < 1$ this expression can be approximated as $R = \dfrac{0.6}{1+q}$. So adopting $r \leq R$ we can estimate the deviation between P_{SH} and P_{orb} as

$$\frac{P_{SH}}{P_{orb}} - 1 \leq f_0(q). \tag{2}$$

In fact, formula (1) exploits only the first term of infinite serie for Laplace coefficient $1/2 b_{1/2}^{(0)}$. The full formula looks like [9, 15]

$$\frac{P_{prec}}{P_{orb}} = \frac{\sqrt{1+q}}{q} \left[1/2 \frac{1}{\sqrt{r}} \frac{d}{dr} \left(r^2 \frac{d(1/2 b_{1/2}^{(0)})}{dr} \right) \right]^{-1}, \tag{3}$$

$$1/2 b_{1/2}^{(0)}(r) = 1 + 1/4 r^2 + 9/64 r^4 + 25/256 r^6 + 1225/16384 r^8 + 3969/65536 r^{10} +$$
$$+ 53361/1048576 r^{12} + \cdots, \tag{4}$$

which gives more exact dependence

$$\frac{P_{SH}}{P_{orb}} - 1 \leq f(q). \tag{5}$$

Figure 4 depicts the dependencies $f_0(q)$ and $f(q)$. It is seen that our 3D gasdynamical calculations for binary systems with $q = 0.147$ and $q = 0.5$ are in a good agreement with theoretical estimates. The possibility of the formation of the "precessional" spiral wave in binaries with large mass ratios can be useful for interpretation of observations of these systems. There are no any reasonable explanation of the appearance of the superhumps at large mass ratios, so the application of the "precessional" spiral wave model to interpretation of binary systems with large q (like TV Col) looks very prominent.

Acknowledgements The work was partially supported by Russian Foundation for Basic Research (projects NN 02-02-16088, 02-02-17642, 03-01-00311, 03-02-16622), by Science Schools Support Program (project N 162.2003.2), by Federal Programme "Astronomy", by Presidium RAS Programs "Mathematical modelling and intellectual systems", "Nonstationary phenomena in astronomy". O.A.K. thanks Russian Science Support Foundation for the financial support.

REFERENCES

1. D.V. Bisikalo, A.A. Boyarchuk, P.V. Kaigorodov, and O.A. Kuznetsov, *Astron. Reports* **47**, 809–820 (2003).
2. D.V. Bisikalo, A.A. Boyarchuk, P.V. Kaigorodov, O.A. Kuznetsov, and T. Matsuda, *Astron. Reports* **48**, 449–456 (2004) [astro-ph/0403053].
3. D.V. Bisikalo, A.A. Boyarchuk, V.M. Chechetkin, O.A. Kuznetsov, and D. Molteni *MNRAS* **300**, 39–48 (1998).
4. D.V. Bisikalo, A.A. Boyarchuk, O.A. Kuznetsov, and V.M. Chechetkin *Astron. Reports* **44**, 26–35 (2000) [astro-ph/9907087].
5. A.A. Boyarchuk, D.V. Bisikalo, O.A. Kuznetsov, and V.M. Chechetkin, "Mass Transfer in Close Binary Stars", in *Advances in Astronomy and Astrophysics*, vol. 6, Taylor & Francis, London (2002).
6. K. Sawada, T. Matsuda, and I. Hachisu, *MNRAS* **221**, 679–686 (1986).
7. K. Sawada, T. Matsuda, M. Inoue, and I. Hachisu, *MNRAS* **224**, 307–322 (1987).
8. B. Warner, *Cataclysmic Variable Stars*, Cambridge University Press, Cambridge (1995).
9. M. Hirose and Y. Osaki, *PASJ* **42**, 135–163 (1990).
10. D.V. Bisikalo, A.A. Boyarchuk, P.V. Kaigorodov, O.A. Kuznetsov, and T. Matsuda, *Astron. Reports* **48**, 588–596 (2004) [astro-ph/0403057].
11. R. Whitehurst, *MNRAS* **232**, 35–51 (1988).
12. C. Hellier, *MNRAS* **264**, 132–144 (1993).
13. A. Retter, C. Hellier, T. Augusteijn, T. Naylor, T.R. Bedding, C. Bembrick, J. McCormick, and F. Velthuis, *MNRAS* **340**, 679–686 (2003).
14. B. Paczyński, *ApJ* **216**, 822–826 (1977).
15. K.J. Pearson, *MNRAS* **346**, L21–L25 (2003).

Mass and angular momentum loss during RLOF in Algols

W. Van Rensbergen, C. De Loore & D. Vanbeveren

Astrophysical Institute, Vrije Universiteit Brussel, Pleinlaan 2, B-1050 Brussels, Belgium

Abstract. We present a set of evolutionary computations for binaries with a B-type primary at birth. Some liberal computations including loss of mass and angular momentum during binary evolution are added to an extensive grid of conservative calculations. Our computations are compared statistically to the observed distributions of orbital periods and mass ratios of Algols. Conservative Roche Lobe Over Flow (RLOF) reproduces the observed distribution of orbital periods decently but fails to explain the observed mass ratios in the range $\in [0.4\text{-}1]$. In order to obtain a better fit the binaries have to lose a significant amount of matter, without transferring too much angular momentum.

Keywords: : Algols, binaries, mass loss, angular momentum loss
PACS: 45.20.Jj; 97.80-d: 95.10-Eg; 97.10.Nf

INTRODUCTION

Eggleton [3] introduced the denomination "liberal" to make a distinction between binary evolution with mass and angular momentum loss and the conservative case where no mass and consequently no angular momentum leave the system. Refsdal et al. [7] showed that the binary AS Eri is the result of liberal binary evolution; the amounts of mass and angular momentum lost by the system are however uncertain. Sarna [9] showed that only 60 % of the mass lost by the loser of β Per was captured by the gainer and that 30 % of the initial angular momentum was lost during RLOF. Hence it is clear that the liberal evolutionary scenario is important for binary evolution calculations. However the amount of mass and angular momentum that has to be removed from interacting systems is far from obvious. With the Brussels simultaneous evolution code (a description is given in "The Brightest Binaries", Vanbeveren et al. [11]) we calculated a representative grid of conservative evolution of binaries with a B type primary at birth. Application of the criterion of Peters [5] allows then to determine for each of the evolutionary sequences the beginning and ending of various Algol-stages. This criterion states that in the semi-detached system:

- The less massive star fills its Roche lobe (RL)
- The most massive star does not fill its RL and is still on the main sequence
- The less massive star is the coolest, the faintest and the largest

The grid allows to determine an expected distribution of orbital periods and mass ratios of Algols. These can be compared to the well established observed distributions (Budding et al. [1]).

Considering the examples mentioned above (Refsdal et al. [7], Sarna [9]) it may be expected that the match between the observations and the conservative results are far from satisfactory. Therefore we add liberal calculations to our conservative library. The detailed evolutionary tracks can be found in http://www.vub.ac.be/astrofys/

DETAILS OF THE CONSERVATIVE CALCULATION

Our grid contains binaries with sufficiently small initial periods to lead to Case A RLOF (RLOF A): i.e. during H core burning of the donor (initially primary star that becomes the less massive after Algol-ignition). The track across the HRD is traced for every system in our grid. A binary lives its era of "Algolism" (De Loore and Van Rensbergen [2]) when it obeys the criterion of Peters [5]. Every binary shows its Algol A (Algol during H core burning) aspect for some time during RLOF A. The drastical change of the mass ratio and the orbital period during this process can be followed in detail for every system in our grid. It happens frequently that RLOF A is succeeded by Case B RLOF (RLOF B): i.e. during H shell burning of the donor. Systems that have sufficiently large initial orbital periods so that RLOF A does not occur will also show a short living Algol B (Algol during H shell burning) appearance. These systems have been considered in a previous paper (Van Rensbergen [13]). Figure 1 shows a typical example of a case AB. The $(7+4.2)M_\odot$ binary with an intial period of 2.5 d starts as Algol A with q=1 and P=2.06 d. It ends this Algol A stage after some 20 million years with q=0.27 and P=6.87 d. From these inital values the system remains an Algol B for some 1.5 million years during H shell burning of the donor. The system eventually evolves into a long periodic (P>50 d) Algol B with $q \approx 0.1$. The grid contains some 240 more evolutionary tracks.

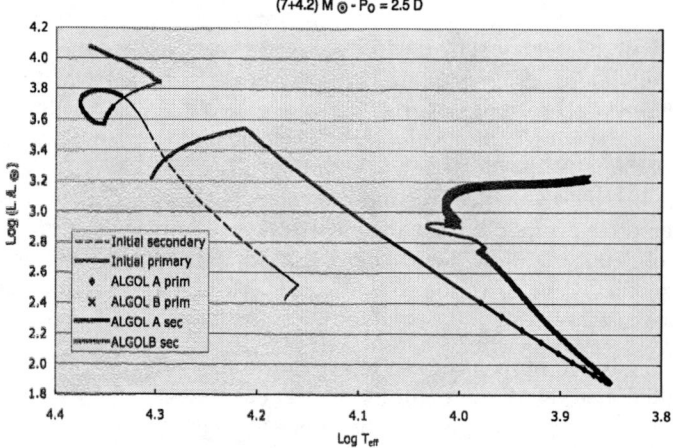

FIGURE 1. *Conservative evolution of a $(7+4.2)M_\odot$ binary with an intial period of 2.5 d. The system remains Algol A for $\approx$ 20 million years before it is Algol B for $\approx$ 1.5 million years.*

CONSERVATIVE SIMULULATION

From our grid of conservative calculations we made a simulation of the distribution of orbital periods and mass ratios of Algols using a Monte Carlo algorithm. As starting conditions we selected:

- The primaries IMF of Salpeter [8]: $\zeta(M) \div M^{-2.35}$
- an initial distrubution of orbital periods from Popova et al. [6]: $\Pi(P) \div \frac{1}{P}$
- an initial mass ratio distribution as derived by Van Rensbergen [12] from the non-evolved systems of the catalogue of Spectroscopic Binaries shown on http://sb9.astro.ulb.ac.be/

The distribution of the initial mass ratio $q = \frac{M_2^0}{M_1^0}$ obeys relation (1):

$$\Psi(q) \div (1+q)^{-\alpha}\ ;\ \alpha = 3.37\ for\ early\ B\ \&\ 1.47\ for\ late\ B\ primaries \qquad (1)$$

Figure 2 shows the observed distribution of orbital periods of Algols. Here we find more A than B-cases, because the Algol A-phase goes on for a fraction of the nuclear time scale, whereas the Algol B-phase lasts only for a fraction of the much shorter Kelvin-Helmholtz time scale. RLOF A produces Algols that follow the observed distribution rather well. Since the periods of the Algol B cases peak towards the long periods, a contribution of a few % of B cases to the Algol population will mimic the observed period distribution well.

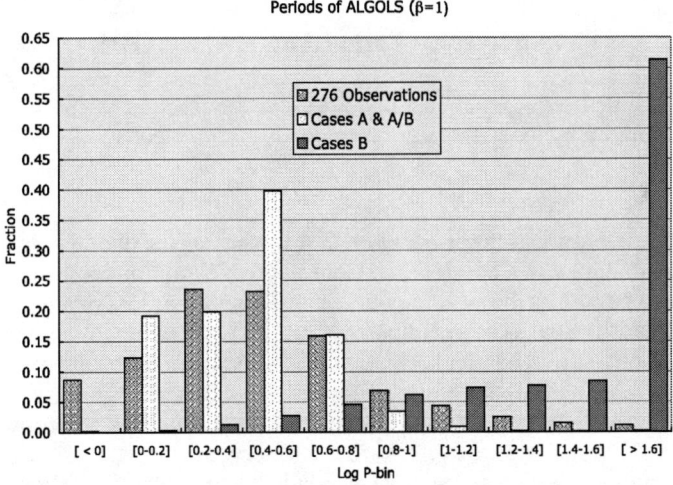

FIGURE 2. *Observed distribution of the orbital periods of Algols compared to conservative evolution. Cases B produce an excessive amount of long periods. Cases A follow the observed distribution better. There is no doubt that there are far more Algol A than Algol B cases.*

Figure 3 shows the observed mass ratio distribution of Algols. The mass ratio q is now defined as the observers do: $q = \frac{M_{donor}}{M_{gainer}}$.

The mass ratios of the Algol B cases peak towards the smallest mass ratios, whereas RLOF A produces a majority of Algols in the q-bin [0.2-0.4]. As a consequence an admixture of cases A and B will never reproduce well the observed mass ratio distribution in the q-bins [0.4-1]. We may conclude that liberal binary evolution is needed to describe the mass ratio distribution of Algols.

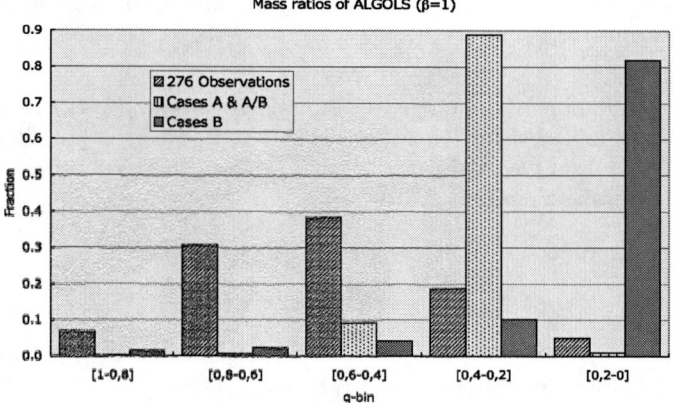

FIGURE 3. Observed distribution of mass ratios of Algols compared to conservative evolution. Cases B produce more than 80% Algols with the smallest mass ratio $q \in [0-0.2]$. Cases A produce most of their Algols with $q \in [0.2-0.4]$. The fact that almost 80% Algols are observed with $q \in [0.4-1]$ excludes conservative evolution as the major channel through which Algols can be formed.

CONFINING THE LIBERAL MODEL

Mass loss is defined by a parameter β describing the fraction of the mass lost by the loser that is accreted by the gainer:

$$\dot{M}_{gainer} = -\beta \, \dot{M}_{donor} \quad \text{with} \quad 0 \leq \beta \leq 1 \qquad (2)$$

Conservative evolution is described with $\beta = 1$, whereas the liberal case uses values of $\beta < 1$ which are not known beforehand. Conservative evolution implies that no angular momentum can be lost by the system, whereas the amount of loss of angular momentum in the liberal case is also a free parameter if no physics restricts the assumptions. It is clear that realistic hydrodynamical calculations should learn us the appropriate choice of the amounts of mass and angular momentum which are lost by a binary at any moment of the RLOF process. In the mean time we have performed a number of liberal evolutionary calculations and compared the results with the observations.

Our calculations reveal that the time dependent parameter $\beta(t)$ should sufficiently often differ drastically from 1.

For a given value of β, the amount of angular momentum lost by the system is defined by the position of the site where matter is recoiled from the system. Soberman et al. [10] have argued that matter can be trapped in a Keplerian ring after transit across the second Lagrangian point L_2. The radius of the ring is η times the semi major axis of the orbit. This Keplerian ring passes outside L_2 which is located at $\eta \approx 1.25$. This yields a minimum value of $\eta \approx 1.25$. Hydrodynamical calculations of Lubow and Shu [4] locate the ring at $\eta \approx 3$. This yields a maximum value of $\eta \approx 3$. A Keplerian ring located at $\eta \approx 2.25$ takes away as much angular momentum as the co-rotating point L_2 would do. A value of $\eta \approx 2.25$ is thus a fair value in the interval $[1.25-3]$ that can be used to calculate the change of the orbital period of a binary as a consequence of loss of mass and angular momentum through a ring which rotates with Keplerian velocity around the center of mass of the system (Soberman et al. [10]).

Our calculations reveal that if the system loses angular momentum across points located near $\eta \approx 2.25$ the orbital periods shrink drastically. Most binaries become mergers before they show Algolism. The obtained distribution of orbital periods of Algols is completely shifted towards the shortest periods. Since this conclusion conflicts with observations, the mass that leaves the system carries rather the angular momentum of points located near $\eta \approx 0$ than near $\eta \approx 2.25$.

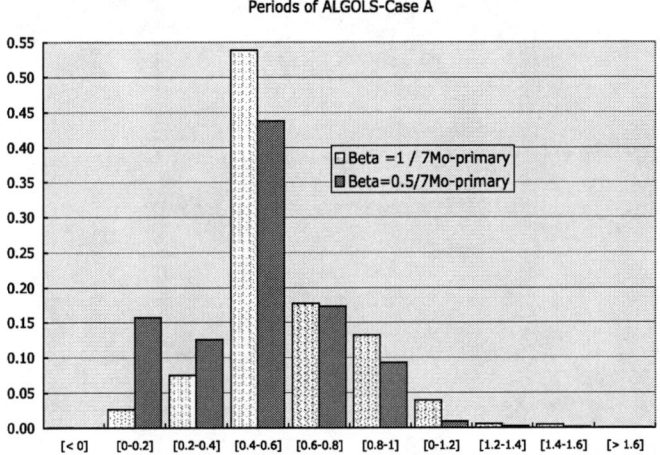

FIGURE 4. *Liberal evolution with a lot of mass loss ($\beta=0.5$) and a little loss of angular momentum leaves the distribution of orbital periods of Algols almost unaltered. This figure shows the result for Algols issued from RLOF A and a 7 $M_\odot$ primary at birth.*

For all these (statistically sound but physically not well understood) reasons we performed a number of calculations with constant $\beta = 0.5$ and angular momentum lost at the edge of the gainer. Until now only the representative cases with a $7M_\odot$ primary at birth and initial periods leading to RLOF A have been performed. Figure 4 shows that the assumption of mass loss through a point near $\eta = 0$ does not alter the conservative distribution of the orbital periods of Algols. A conclusion that meets the observations since the conservative orbital period distribution matched the observations fairly well. Figure 5 shows that our liberal assumption ($\beta=0.5$, $\eta \approx 0$) deviates radically from the conservative case. The q-bins $[0.4-1]$ which were not populated in the conservative scenario are now populated properly as required by the observations (see figure 3).

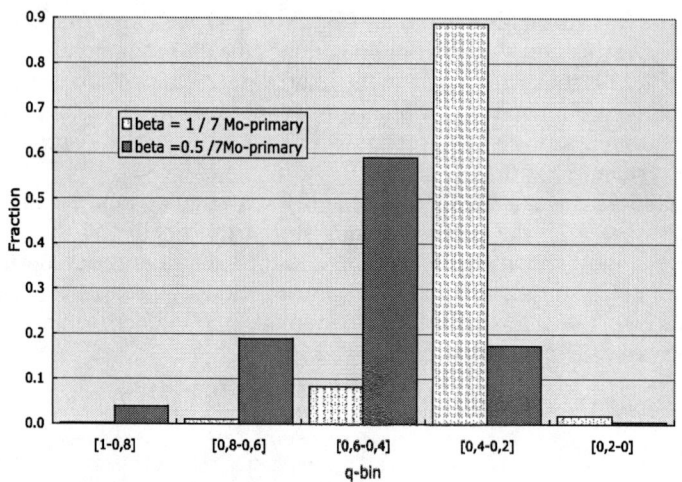

FIGURE 5. *Liberal evolution with a lot of mass loss ($\beta = 0.5$) and a little loss of angular momentum changes the mass ratio distribution of Algols drastically. The observed mass ratios in the q-interval [0.4-1] which were not produced by conservative evolution are created by this liberal model. This plot shows the result for Algols issued from RLOF A and a $7M_\odot$ primary at birth.*

REFERENCES

1. Budding, E., Erdem, A., Cicek, C., Bulut, I., Soydugan, E., Soydugan, V., Bakis, V., Demircan, O., *A&A*, **217**, 263-268, 2004.
2. De Loore, C., and Van Rensbergen, W., in *Zdenek Kopal's Binary Star Legacy*, ed. H. Drechsel, accepted.
3. Eggleton, P.P., *New Astron. Rev.*, **44**, 111-117, 2000.
4. Lubow, S., and Shu, F., *ApJ*, **198**, 383-405, 1975.
5. Peters, G., in *"The Influence of Binaries on Stellar population Studies"*, ed. D. Vanbeveren, *ASSL*, **264**, 79-94, 2001.
6. Popova, E., Tutukov., A, and Yungelson, L., *ApSS*, **88**, 55-80, 1982.
7. Refsdal, S., Roth, M., and Weigert, A., *A&A*, **36**, 113-122, 1974.
8. Salpeter, E.E., *ApJ*, **121**, 161-167, 1955.
9. Sarna, M.S., *MNRAS*, **262**, 534-542, 1993.
10. Soberman, G., Phinney, E., and Van den Heuvel, E., *A&A*, **327**, 620-635, 1997.
11. Vanbeveren,D., Van Rensbergen, W., and De Loore, C., *The Brightest Binaries*, *ASSL*, **232**, 1998.
12. Van Rensbergen, W., in *"The Influence of Binaries on Stellar population Studies"*, ed. D. Vanbeveren, *ASSL*, **264**, 21-35, 2001.
13. Van Rensbergen, W., in *"Stellar Astrophyics"*, eds. K. Cheng, K. Leung and T. Li, *ASSL*, **298**, 117-125, 2003.

RX J0806.3−1527: Ten Years of Phase Coherent Monitoring in the Optical and X-ray Bands

G.L. Israel*, S. Dall'Osso*, V. Mangano*, L. Stella*, S. Covino[†], D. Fugazza[†], S. Campana[†], G. Marconi[**], S. Mereghetti[‡] and U. Munari[§]

*INAF − Osservatorio Astronomico di Roma, Via Frascati 33, I−00040 Monteporzio Catone (Roma), Italy
[†]INAF − Osservatorio Astronomico di Brera, Via Bianchi 46, I−23807 Merate (Lc), Italy
[**]European Southern Observatory, Casilla 19001, Santiago, Chile
[‡]INAF − Istituto di Astrofisica Spaziale e Fisica Cosmica, Sezione di Milano "G.Occhialini" Via Bassini 15, I-20133 Milano, Italy
[§]INAF - Osservatorio Astronomico di Padova, I-36012 Asiago, Italy

Abstract. RX J0806.3−1527 is thought to be a 321s orbital period (the shortest known) double white dwarf binary system. According to the double degenerate binary (DDB) scenario this source is expected to be one of the strongest Gravitational Wave (GW) emitter candidates. In the last years RX J0806.3−1527 has been studied in great details, through multiwavelength observational campaigns. We present here the timing results obtained thanks to a 3.5-year long optical monitoring campaign carried out by the Very Large Telescope (VLT) and the Telescopio Nazionale Galileo (TNG) which allowed us to unambiguously detect and study the orbital period derivative (spin–up at a rate of $\sim 10^{-3}$s yr^{-1}) of the 321s modulation based on phase coherent techniques, to detect the linear polarisation (at a level of about 2%), and to study the broad band energy spectrum. The VLT/TNG observational strategy we used allowed us, for the first time, to infer a P-$\dot{P}$ coherent solution for the archival X-ray observations over a 9.5 years baseline.

Keywords: cataclysmic variables — white dwarfs — ultracompact binaries — X-rays binaries
PACS: 97.20.Rp, 97.80.Fk, 97.80.Gm, 97.80.Jp

INTRODUCTION

RX J0806.3−1527 was discovered in 1990 with the *ROSAT* satellite during the All-Sky Survey (RASS) [1]. However, it was only in 1999 that a periodic signal at 321s was detected in its soft X-ray flux with the *ROSAT* HRI [2, 3]. Subsequent deeper optical studies allowed to unambiguously identify the optical counterpart of RX J0806.3−1527, a blue $V = 21.1$ ($B = 20.7$) star [4, 5]. B, V and R time-resolved photometry revealed the presence of a $\sim 15\%$ modulation at the ~ 321 s X-ray period [5, 6]. The VLT spectral study revealed a blue continuum with no intrinsic absorption lines [5]. Broad (FWHM ~ 1500 km s^{-1}), low equivalent width ($EW \sim -2 \div -6$ Å) emission lines from the He II Pickering series (plus additional emission lines likely associated with He I, C III, N III, etc.; for a different interpretation see [7]) were instead detected [5]. These findings, together with the period stability and absence of any additional modulation in the 1 min–5 hr period range, are interpreted in terms of a double degenerate He-rich binary (a subset of the AM CVn class; see [8]) with an orbital period of 321 s, the shortest ever recorded. Moreover, RX J0806.3−1527 was noticed to have optical/X-ray properties similar to those of RX J1914.4+2456, a 569 s modulated soft X-ray source

CP797, *Interacting Binaries: Accretion, Evolution, and Outcomes*,
edited by L. Burderi, L. A. Antonelli, F. D'Antona, T. Di Salvo,
G. Luca Israel, L. Piersanti, A. Tornambè, and O. Straniero
© 2005 American Institute of Physics 0-7354-0286-8/05/$22.50

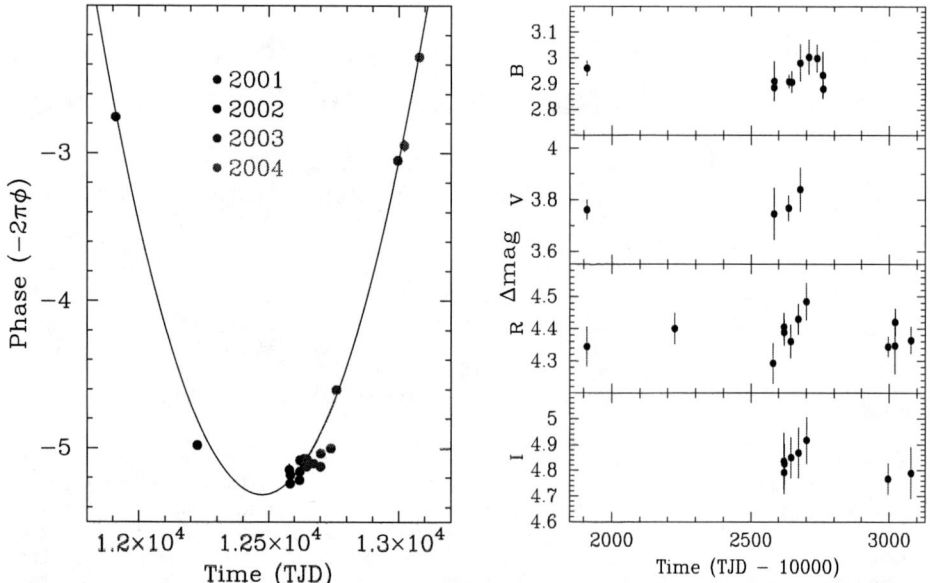

FIGURE 1. Left panel: Results of the phase fitting technique used to infer the P-$\dot{P}$ coherent solution for RX J0806.3−1527; the linear term (P component) has been corrected, while the quadratic term (the $\dot{P}$ component) has been kept for clarity. The best $\dot{P}$ solution inferred for the optical band is marked by the solid fit line. Right panel: 2001-2004 optical flux measurements at fdifferent wavelengths.

proposed as a double degenerate system [9, 10, 11].

More recently, the detection of spin–up was reported, at a rate of ∼6.2×10^{-11} s s^{-1}, for the 321 s orbital modulation, based on optical data taken from the Nordic Optical Telescope (NOT) and the VLT archive, and by using incoherent timing techniques [12, 13]. Similar results were reported also for the X-ray data (ROSAT and Chandra; [14]) of RX J0806.3−1527 spanning over 10 years of uncoherent observations and based on the NOT results [12]. Finally, a nearly-simultaneous Chandra/VLT observational campaign was carried out and the presence of nearly anti correlation between the optical and X–ray modulation was reported [15, 16]. The X-ray spectral study also allowed to characterise the emission mechanism; a black body with a temperature of ∼60 eV [16].

The study of RX J0806.3−1527 and RX J1914.4+2456 has posed, in the last years, serious questions about their possible origin (see [17] for a review of the proposed theoretical models). Moreover, there is a number of additional reasons for studying RX J0806.3−1527 and the related objects. Among other are the study of gravitational waves, and the possibility that RX J0806.3−1527 is a progenitor of the He-accreting DDBs AM CVns. For this reason we performed, during the last 3.5 years, several observations ranging from IR to X-ray band in order to unveil the nature of RX J0806.3−1527. Here we report on the latest results obtained from (a) the VLT/TNG photometric monitoring aimed at coherently measure the P and $\dot{P}$ of RX J0806.3−1527 (see also [18]), (b) VLT-FORS1 polarimetry observations, and (c) an *XMM–Newton* pointing.

OPTICAL OBSERVATIONS

After the successful first optical time-resolved observations of RX J0806.3−1527 during 2001 (January 1st at TNG, and November 11th), we started a relatively long-term project aimed at monitoring the source 321 s modulation. We obtained 21 pointings between November 2002 and May 2004 (11 at VLT and 10 at TNG) scheduled in a way such that it was possible to keep the phase coherence among observations (the first observations were obtained at 1st-2nd nights, 9-10th nights, 19-20th nights, 49-50th nights, etc.). Four different optical bands (B, V, R and I) have been used in order to study and monitor the pulse shape and pulsed fraction as a function of wavelength and time.

Such a strategy resulted quite efficient in reaching the purposes of the timing analysis, and allowed us also to extend the coherent solution backward to the 2001 optical observations. The best optical solution we found for P-$\dot{P}$ is for P=321.53040(4) s, $\dot{P}$=−3.6(1)×10^{-11} s s^{-1} (90% uncertainties are reported; for more details see [19]; see also Figure 1, left panel). Moreover, we found a slightly energy–dependent pulse shape with the pulsed fraction increasing toward longer wavelengths, from ∼12% in the B-band to nearly 14% in the I-band (see lower right panel of Figure 2) [19]. We also detected variability, at a level of 4% of the optical pulse shape as a function of time (see upper right panel of Figure 2 right).

The relatively high accuracy obtained for the optical phase coherent P-$\dot{P}$ solution (in the January 2001 - May 2004 interval) allowed us to extrapolate it backward to the ROSAT observations without loosing the phase coherency, i.e. only one possible period cycle consistent with our P-$\dot{P}$ solution. Given the wider time interval spanned by the X–ray observations (9.5 years) an even more accurate solution was possible. After taking into account the ROSAT photon arrival time spacecraft clock time - UTC correction (in order to compare the ROSAT data with the Chandra and *XMM–Newton* observations), our best X-ray phase coherent solution is P=321.53038(2) s, $\dot{P}$=−3.661(5)×10^{-11} s s^{-1} (for more details see [19]). Figure 2 (left panel) shows the optical (2001-2004) and X–ray (1994-2002) light curves folded by using the above reported P-$\dot{P}$ coherent solution, confirming the amazing stability of the X–ray/optical anti-correlation first noted by [16] (see inset of left panel of Figure 2).

In February 2003 we collected VLT FORS1 linear polarimetry data for the optical counterpart to RX J0806.3−1527 in the B-band. RX J0806.3−1527 was found to be linearly polarised at a level of ∼2.0±0.3% (after correcting the the field average polarisation, ∼0.7%). It is worth noting that the marginal detection of circular polarisation, at level of about 0.5%, has been recently reported, and ascribed to a ∼10^6 Gauss magnetic field [7].

THE *XMM–NEWTON* OBSERVATION

On 2002 November 1st we had the chance to observe RX J0806.3−1527 with the *XMM–Newton* instrumentations for about 26000 s. This ensured us a dataset of unpiled–up photons allowing us to increase the spectral analysis accuracy. Left panel of Figure 3 shows the results of the phase-resolved spectroscopic study we carried out (similar to the analysis reported in Figure 4 of [16] for the *Chandra* data). The *XMM–Newton*

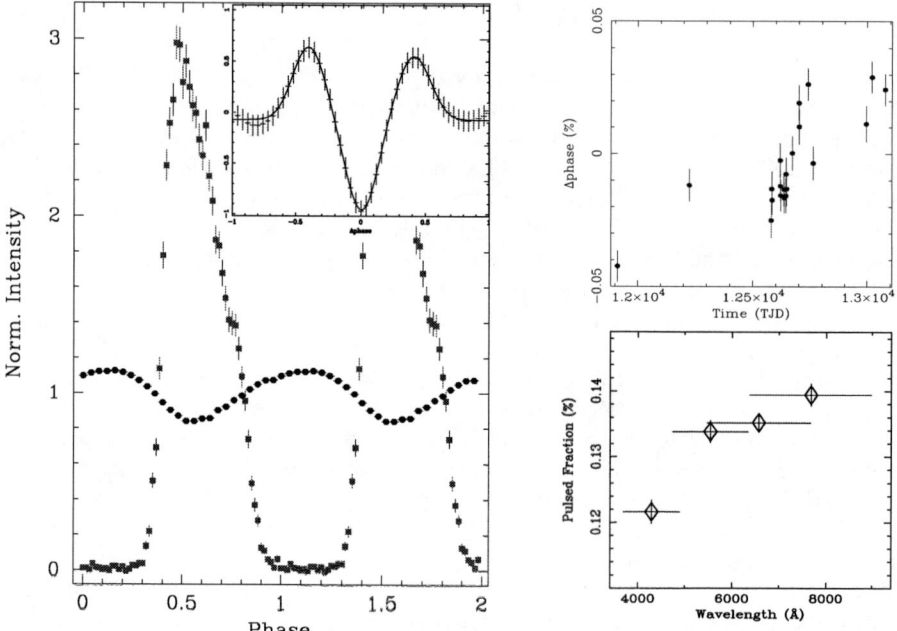

FIGURE 2. Left Panel: The 1994–2002 phase coherently connected X–ray folded light curves (filled squares; 100% pulsed fraction) of RX J0806.3−1527, together with the VLT-TNG 2001-2004 phase connected folded optical light curves (filled circles). Two orbital cycles are reported for clarity. The inset shows the cross correlation function (CCF) analysis performed on the X-ray/optical light curves. A nearly anti-correlation was found. Right panels: Analysis of the phase variations induced by pulse shape changes in the optical band (upper panel), and the pulsed fraction as a function of optical wavelengths (lower panel).

data show a lower value of the absorption column, a relatively constant black body temperature, a smaller black body size, and, correspondingly, a slightly lower flux. All these differences may be ascribed to the pile–up effect in the Chandra data, even though we can not completely rule out the presence of real spectral variations as a function of time. In any case we note that this result is in agreement with the idea of a self-eclipsing (due only to a geometrical effect) small, hot and X–ray emitting region on the primary star. Timing analysis did not show any additional significant signal at periods longer or shorter than 321.5 s, (in the 5hr-200ms interval). The *XMM–Newton* OM allowed us to obtain a first look at the source in the UV band (see right panel of Figure 3) confirming the presence of the blackbody component inferred from IR/optical bands.

DISCUSSION

In this contribution we briefly listed the main results obtained thanks to a four years optical and X-ray monitoring of RX J0806.3−1527, a DDB with an orbital period of

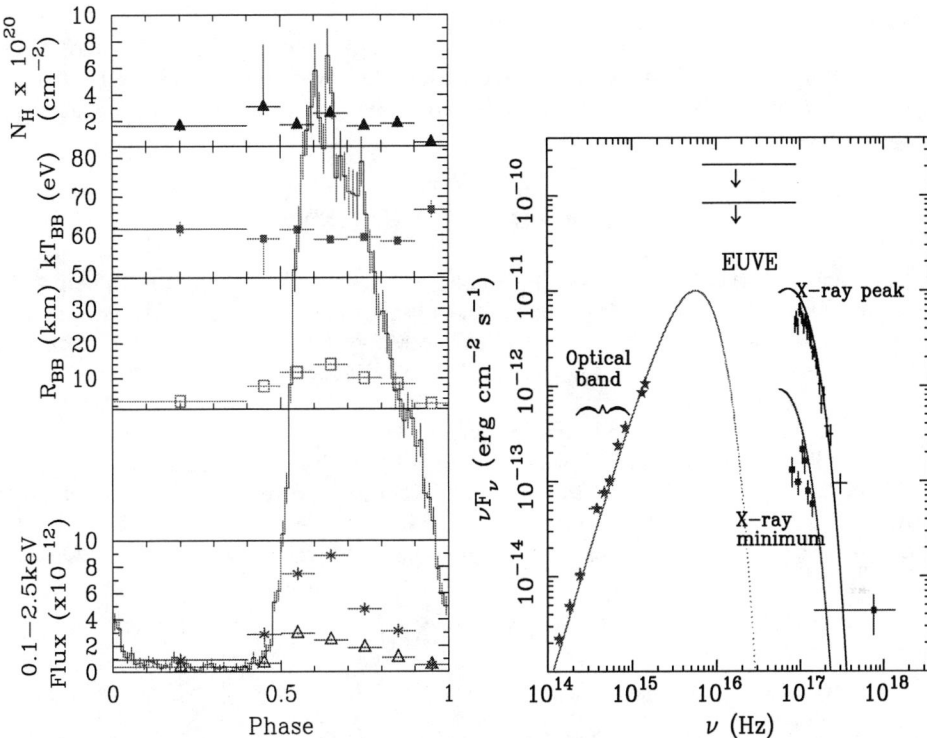

FIGURE 3. Left panel: The results of the *XMM–Newton* phase-resolved spectroscopy (PRS) analysis for the absorbed blackbody spectral parameters: absorption, blackbody temperature, blackbody radius (assuming a distance of 500 pc), and absorbed (triangles) and unabsorbed (asterisks) flux. Superposed is the folded X-ray light curve. Right panel: Broad-band energy spectrum of RX J0806.3−1527 as inferred from the *Chandra*, *XMM–Newton*, VLT and TNG measurements and *EUVE* upper limits. The dotted line represents one of the possible fitting blackbody models for the IR/optical/UV bands.

only 5.4min. Even though the optical monitoring is still active (in fact is quite important to continue the $\dot{P}$ study to look for variations or, more interestingly, a $d\dot{P}/dt$ component) a number of important implications can be considered.

It is now assessed that the orbital period is decaying at a rate that is nearly consistent with that predicted by the ultracompact binary model in which the loss or orbital angular momentum is dominated by gravitational radiation and in which there is no mass exchange between the two stars [14]. In this respect the unipolar inductor model is a possible scenario to account for the observed optical and X–ray properties of RX J0806.3−1527[20].

The X–ray and optical emissions are anti correlated at a level of stability (no variations on a baseline of at least 4 years) which is only observed in phase-locked binary systems. The anti correlation is by far and so far the strongest (indirect) suggestion of the binary nature of the 321 s modulation, where the X–rays illuminate the companion surface originating the phase–shifted optical modulation.

The detection of linear and circular polarisation might imply the presence of a relatively faint magnetic field, however we note that the low polarisation level is also consistent with other possibilities, such as the Thomson scattering (sometimes observed in eclipsing binaries).

ACKNOWLEDGMENTS

GianLuca Israel (GLI) thanks Rosario Gonzalez–Riestra (XMM team) for his help in setting the successful 2002 *XMM–Newton*/VLT observational campaign. GLI is also grateful to the VLT and TNG Team, in particular to Marco Pedani (TNG), for their effort in optimising and executing the FORS1 and DoLoRes observations of RX J0806.3−1527 during the last 4 years !

This work is Based on observations made with the Italian Telescopio Nazionale Galileo (TNG) operated on the island of La Palma by the Centro Galileo Galilei of the INAF (Istituto Nazionale di Astrofisica) at the Spanish Observatorio del Roque de los Muchachos of the Instituto de Astrofisica de Canarias. Based on observations collected at the European Southern Observatory, Chile (268.D-5737, 070.D-652 and 072.D-0717).

REFERENCES

1. K. Beuermann, H.-C. Thomas, K. Reinsch, et al., A&A, **347**, 47 (1999)
2. G.L. Israel, M.R. Panzera, S. Campana, et al., A&A, **349**, L1 (1999)
3. V. Burwitz. and K. Reinsch, *X-ray astronomy : stellar endpoints, AGN, and the diffuse X-ray background*, Bologna, Italy, eds White, N. E., Malaguti, G., Palumbo, G., AIP conference proceedings, 2001 **599**, pp. 522
4. G.L. Israel, L. Stella, W. Hummel, S. Covino and S. Campana, IAU Circ., **7835** (2002)
5. G.L. Israel et al., A&A, **386**, L13 (I02) (2002)
6. G. Ramsay, P. Hakala, M. Cropper, et al. MNRAS, **332**, L7 (2002)
7. K. Reinsch, V. Burwitz and R. Schwarz, Revista Mexicana de Astronomia y Astrofisica Conference Series, 2004, **20**, pp. 122, see astro–ph/0402458
8. B. Warner, Ap&SS, **225**, 249 (1995)
9. M. Cropper, M.K. Harrop-Allin, K.O. Mason, et al., MNRAS, **293**, L57 (1998)
10. G. Ramsay, M. Cropper, K. Wu, K.O. Mason, P. Hakala, MNRAS, **311**, 75 (2000)
11. G. Ramsay, K. Wu, M. Cropper, et al., MNRAS, **333**, 575 (2002)
12. P. Hakala, G. Ramsay, K. Wu, etal., 2003, MNRAS, **343**, L10 (2003)
13. P. Hakala, G. Ramsay, and K. Byckling, MNRAS, **353**, 453 (2004)
14. T. Strohmayer, ApJ, **593**, L39 (2003)
15. G.L. Israel, L. Stella, S. Covino, et al., ASP Conf. Ser. 315: IAU Colloq. 190: Magnetic Cataclysmic Variables, 2004, **315**, pp. 338, see astro-ph/0303124
16. G.L. Israel, S. Covino, L. Stella, et al., ApJ, **598**, 492 (2003)
17. M. Cropper, G. Ramsay, K. Wu, ASP Conf. Ser. 315: IAU Colloq. 190: Magnetic Cataclysmic Variables, 2004, **315**, pp. 324, see astro-ph/0302240
18. G.L. Israel, et al., Memorie della Societa Astronomica Italiana Supplement, 2004, **5**, 148
19. G.L. Israel, G.L. et al., A&A, submitted (2005)
20. K. Wu, M. Cropper, G. Ramsay, K. Sekiguchi, MNRAS, **331**, 221 (2002)

The Supersoft X-Ray Source CAL 83: A Massive White Dwarf

Thierry Lanz*,†, Marc Audard†, Frits Paerels† and Gisela A. Telis†

*Department of Astronomy, University of Maryland, College Park, MD 20742, USA
†Columbia Astrophysics Laboratory, Columbia University, New York, NY 10027, USA

Abstract. We have obtained *Chandra* HRC-S/LETG spectroscopy of the prototypical supersoft source CAL 83 in the Large Magellanic Cloud. The data reveal a very rich absorption line spectrum from the hot white dwarf photosphere. We have recently completed the analysis of the *Chandra* spectrum and of an earlier XMM-*Newton* RGS spectrum of CAL 83 with new non-LTE line-blanketed model atmospheres that explicitly include 74 ions of the 11 most abundant species. We have successfully matched the *Chandra* and XMM-*Newton* spectra, and have thus derived the basic stellar parameters of the hot white dwarf. In particular, we have obtained the first direct spectroscopic evidence that the white dwarf is massive ($M > 1M_\odot$). We also found no spectral signatures of a wind from the white dwarf. These results provide direct support for supersoft sources as likely progenitors of SN Ia.

Keywords: X-ray Binaries – White Dwarfs – Stellar Masses – Stellar Atmospheres
PACS: 97.10.Ex; 97.10.Nf; 97.20.Rp; 97.80.Jp

SN IA PROGENITORS

The identification of Type Ia supernova (SN Ia) progenitors remains a major open problem. Theoretical models suggest that SNe Ia arise from the thermonuclear explosion of a carbon-oxygen white dwarf (WD) that has grown to the Chandrasekhar mass M_{Ch} [1, 2]. When they form, CO WDs have masses between 0.6 and 1.2 $M_\odot$ depending on the star initial mass. SN Ia progenitors, therefore, must be in close binaries ($P > 8$ hours) where the WD can gain several tenths of solar mass donated by a companion to reach M_{Ch}. Because we lack a direct determination of the progenitor properties, we have to rely on indirect arguments to determine their nature.

Two main scenarios have been proposed, involving either the merger of two WDs (the double-degenerate [DD] scenario; [3]) or a single WD accreting from a normal companion (the single-degenerate [SD] scenario; [4]). The DD scenario provides a natural way to explain the absence of hydrogen lines in SN Ia spectra, but hardly explains why SN Ia could be "standard candles" since the explosion should depend on the respective properties of the two WDs. Moreover, the few double WD systems identified to date that could merge in a Hubble time have a total mass smaller than M_{Ch} [5], though a few promising systems have been recently discovered in new large-scale surveys [6].

The SD scenario is thus generally favored today. The most promising candidates to date are close binary supersoft X-ray sources (CBSS) that were revealed by ROSAT [7]. Their ultrasoft emission (no emission above 0.5 to 1 keV) is not consistent with accreting neutron stars or black holes, thus leading to the now classic model of an accreting WD that sustains steady nuclear burning of the hydrogen-rich accreted material [8]. Based on this model, Rappaport et al. [9] discussed the formation and evolution of CBSS, reproducing their typical luminosities, effective temperatures and orbital periods. They estimated that the rate of Galactic SN Ia associated with the evolution of CBSS could reach ≈ 0.006 yr^{-1}. Further observational support was recently brought to the SD scenario by the detection of a narrow Hα emission line in SN2002ic, providing the first evidence of hydrogen-rich circumstellar material associated with a SN Ia [10].

The total mass gained by the WD through accretion is the major issue of the SD scenario. The accretion rate needs to be finely tuned to result in a net mass gain. At low rates, the accreted material accumulates on the WD surface up to a critical point leading to a nova outburst. At high rates, hydrogen starts burning in a shell and results in a swollen atmosphere very much like the atmosphere of a red giant star. In both cases, the WD does not gain much mass since a large fraction of the accreted mass is lost during the nova outbursts or via a wind. Steady hydrogen burning without much photospheric swelling might occur at intermediate accretion rates (few $10^{-7} M_\odot$ yr^{-1}; [8]). If the WD can retain the gas accreted from a relatively-high mass donor, this may result in a sufficient increase of the WD mass to reach M_{Ch}. This issue has been hotly debated in the last decade, but the most recent theoretical works indicate that the WD may gain enough mass to undergo a SN Ia event. In particular, Yoon et al. [11] showed that fast WD rotation stabilizes the He-burning shell, and therefore makes mass gain more likely. Ivanova & Taam [12] revisited recently the evolution of binaries consisting of a WD and a main-sequence star, and characterized the different conditions leading to different outcomes, such as double WDs, SNe Ia, or accretion-induced collapse.

The exact nature of SN Ia progenitors still remains an open problem. Determining the properties of CBSS may, therefore, provide the strongest empirical case supporting the SD scenario.

X-RAY SPECTROSCOPY OF CAL 83

The *Einstein* observatory survey of the Large Magellanic Cloud (LMC) revealed two sources with an ultrasoft spectrum, CAL 83 and CAL 87 [13]. CAL 83 was later identified with a variable, blue, point-like source with an orbital period of 1.04 days [14, 15]. Soft X-ray spectra of CAL 83 have been subsequently obtained with *ROSAT* PSPC and *BeppoSAX* LECS, and have been modeled using blackbodies, LTE and NLTE model atmospheres of WDs [16]. Typical temperatures for CAL 83 and other CBSS are in the range, $kT \approx 30 - 80$ eV, after accounting for interstellar extinction [17]. Because of the limited spectral resolution of these observations, no spectral features were visible. Significant advances may now be expected from *Chandra* and *XMM-Newton* spectrometers due to their high spectral resolution.

XMM-Newton RGS spectroscopy of CAL 83 was first obtained in 2000 April aiming at deriving the fundamental stellar and binary parameters [18]. Because of the RGS cut-off at about 40Å, we have subsequently obtained in 2001 August an A02 *Chandra* LETGS spectrum covering the whole useful spectral range. The two spectra are very similar in both spectral energy distribution and line features. Details on the observations and data reduction are given in [19].

We have constructed a series of non-LTE (NLTE) line-blanketed model atmospheres of hot WDs with our model atmosphere program, TLUSTY [20]. Detailed emergent spectra are then calculated with our spectrum synthesis code, SYNSPEC[1]. TLUSTY allows for explicit departures from LTE for a large set of chemical species and arbitrarily complex model atoms. This feature enables us to account extensively for the line opacity from heavy elements, an essential feature as the observed X-ray spectrum of CAL 83 reveals strong line opacity. The NLTE model atmospheres explicitly include over 10^4 individual energy levels of 74 ions from the 11 cosmically most abundant species: H, He II-III, C V-VII, N V-VIII, O V-IX, Ne V-XI, Mg IX-XII, Si IX-XIV, S IX-XVI, Ar VIII-XVI, Ca IX-XIX, and Fe XIII-XXV, see [19]. About 200,000 bound-bound transitions are accounted for calculating the opacity and the transition rates for solving statistical equilibrium. To produce the final spectra with SYNSPEC, we built an extensive line list based on Peter van Hoof Atomic Line List extracted from http://www.pa.uky.edu/~peter/atomic/, complemented with a list of transitions from the Opacity Project [21]. Note that the latter list is built from theoretical energies (hence, with some uncertainties in wavelengths) and does not include fine structure.

The treatment of electron scattering in TLUSTY was upgraded to consider Compton scattering using a nonrelativistic diffusion approximation through a Kompaneets-like term in the radiative transfer equation [22]. We have compared model atmospheres and predicted spectra which were computed using Compton or Thomson scattering. At the considered temperatures ($T \approx 500,000$ K), the differences are very small and become visible only in the high-energy tail, above 2 keV. At these energies, the predicted flux is very low and, indeed, no flux has been observed in CBSS at these energies. Furthermore, the changes result in little feedback on the calculated atmospheric structure. We conclude that we can therefore safely use Thomson scattering in modeling the very hot WD atmospheres in CBSS.

To compare the model spectra to the *Chandra* LETG and *XMM-Newton* RGS data, we applied to the model a correction for the interstellar extinction [23], the appropriate instrumental response matrix, and a normalization factor $(R_{\mathrm{WD}}/d)^2$. The total interstellar extinction is proportional to the hydrogen column density, $N_{\mathrm{H}} = 6.5 \pm 1.0 \ 10^{20}$ cm^{-2} [24]. The model spectra were scaled to the *Chandra* LETG spectrum to match the observed flux level longward of 45 Å, yielding the normalization factor $(R_{\mathrm{WD}}/d)^2$. The WD radius immediately follows from the adopted distance to the LMC, $d = 50 \pm 3$ kpc, based on RR Lyrae and eclipsing binaries [25, 26].

We display in Fig. 1 our best model fit to the *Chandra* LETG spectrum along with the model sensitivity to $\log g$. The best model spectrum has been normalized to match the observed LETG flux between 45 and 50 Å, yielding a WD radius of $R_{\mathrm{WD}} = 7.3 \ 10^8$ cm $\approx 0.01 R_\odot$. The top panel show an excellent agreement between the model and the observed spectrum. The general energy distribution and most spectral features are well reproduced, for instance the features at 24, 29, 30, 32, and 36 Å. On the other hand, the observed absorption at ≈ 27 Å is not reproduced by our model. A cursory look might suggest that the model predicts an emission feature there but, overplotting the predicted continuum flux, we see that the model essentially misses line

---

[1] The two programs are available at http://tlusty.gsfc.nasa.gov

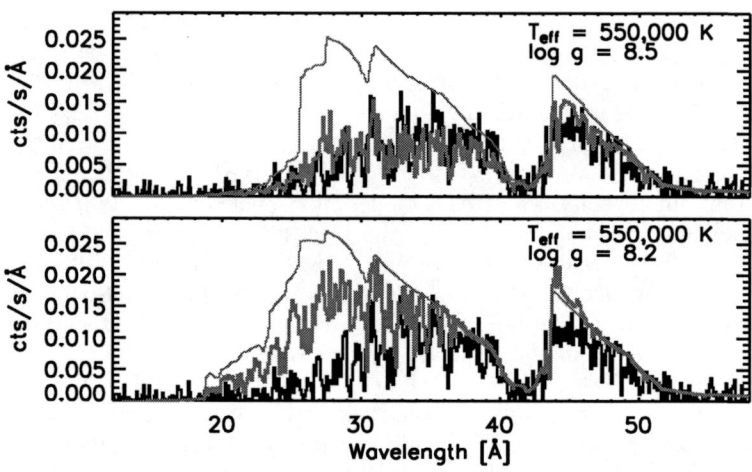

FIGURE 1. Best model fit to the *Chandra* LETG spectrum. The theoretical spectrum is normalized to the observations between 45 and 50 Å, yielding a WD radius of $0.01R_\odot$ and a WD mass of $1.3M_\odot$. The bottom panel shows the model sensitivity to surface gravity, hence to the WD mass (here, $0.65M_\odot$). The predicted continuum spectra are also displayed to illustrate the importance of line opacity.

opacity around 27-28 Å. In this respect, our analysis definitively demonstrates that we observe a photospheric absorption spectrum, with no obvious evidence of emission lines. Fig. 1 presents two models with a different surface gravity, clearly demonstrating that our data allow us to determine $\log g$. The higher flux in the low gravity model is the result of a higher ionization in the low gravity model atmosphere, decreasing the total opacity in the range between 20 and 30 Å. This effect is large enough to determine the surface gravity with a good accuracy, typically ± 0.1 dex on $\log g$. From a small grid of NLTE model atmospheres covering a range in temperature, the best match is obtained for $T_{\text{eff}} = 550,000 \pm 25,000\,\text{K}$ ($kT_{\text{eff}} = 46 \pm 2\,\text{eV}$). The WD mass, $M_{\text{WD}} = 1.3 \pm 0.3 M_\odot$, and WD luminosity, $L_{\text{WD}} = 9 \pm 3\,10^3 L_\odot$, straightforwardly follow, where uncertainties were propagated as dependent errors. Our results indicate that the WD surface composition is consistent with a LMC composition of the accreted material.

X-RAY VARIABILITY

Greiner & Di Stefano [27] discussed extensively the X-ray off-states and optical variability of CAL 83. After the original 1996 April off-state caught by Kahabka et al. [28], Greiner & Di Stefano reported a second off-state (1999 November). They revisited the two contrasting models put forward to explain the off-states, namely a drop in accretion resulting in a cessation of nuclear burning, or conversely a higher accretion rate inducing an increase of the photospheric radius and a shift of the emission to lower frequencies, but they raised a number of issues for the two models.

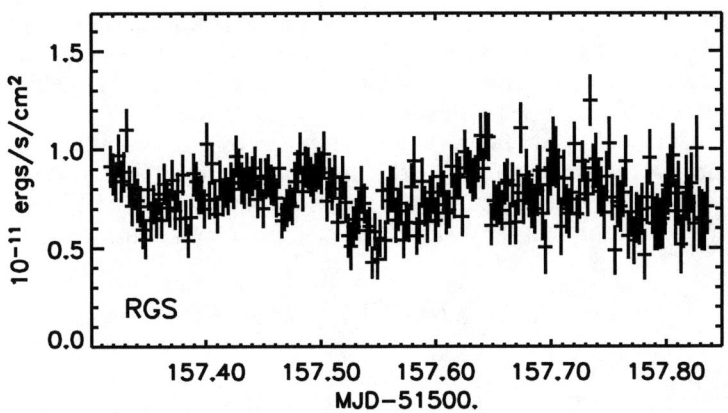

FIGURE 2. X-ray light curve of CAL 83 built from XMM-*Newton* RGS observations.

Lanz et al. [19] reported a third off-state (2001 October), thus demonstrating the recurrent nature of CAL 83 which cannot be considered as the prototypical CBSS undergoing steady nuclear burning anymore. While the overall X-ray flux level remained constant over the 16 months separating the 2000 April XMM-*Newton* observation from the 2001 August *Chandra* observation, they discovered substantial short timescale X-ray variations (see Fig. 2). Within a tenth of the orbital period ($P = 1.04$ d), the RGS flux (20 − 37 Å) and EPIC pn flux (0.2 − 0.8 keV) could vary by as much as a factor of two.

Calculations of WD envelopes [29] showed that the X-ray turnoff time after the cessation of nuclear burning could be very short for massive WDs. An observed turnoff time as short as 20 days implies $M_{WD} \geq 1.35 M_\odot$ [30]. Gänsicke et al. [24] noted that CAL 83 has a low luminosity for a CBSS and is found close to the stability limit of steady burning, which is thus consistent with the idea of unstable burning and might also be related to our discovery of short term variability of the X-ray flux. Although the actual process responsible of the off-states cannot be definitively established, the characteristic timescale further supports the idea of a massive WD in CAL 83.

CONCLUSIONS

Our NLTE model atmosphere analysis shows that CAL 83 contains a massive WD, $M_{WD} = 1.3 \pm 0.3 M_\odot$. Because low-mass models do not provide a match to the observed spectrum that is as good as the fits achieved with high-mass models, we have concluded that $M_{WD} > 1.0\ M_\odot$ is a robust lower limit for the WD mass. In addition, our model is consistent with the WD surface having a hydrogen-rich composition with LMC metallicity. Finally, our analysis does not reveal any evidence of an outflow from the WD.

X-ray off-states indicate that CAL 83 is a source undergoing unstable nuclear burning.

This is consistent with its low luminosity. The short timescale of the off-states (about 50 days) provides a supporting evidence that CAL 83 WD is massive ($M_{WD} \geq 1.35 M_\odot$). A better characterization of the off-states is required to definitively establish the mechanism(s) responsible of the off-states.

Within the model of SN Ia progenitors proposed by Hachisu et al. [31], our results would place CAL 83 in a late stage, after the strong accretion and wind phase when the accretion rate drops below the critical rate for sustaining steady nuclear burning. Our results therefore make CAL 83 a very likely candidate for a future SN Ia event.

ACKNOWLEDGMENTS

This work was supported by grants from NASA through the Astrophysics Theory Program (NRA 00-01-ATP-153) and to Columbia University for XMM-*Newton* mission support and data analysis.

REFERENCES

1. F. Hoyle, and W. A. Fowler, *ApJ*, **132**, 565–590 (1960).
2. D. W. Arnett, *Ap&SS*, **5**, 180–212 (1969).
3. I. Iben Jr., and A. V. Tutukov, *ApJS*, **54**, 335–372 (1984).
4. J. Whelan, and I. Iben Jr., *ApJ*, **186**, 1007–1014 (1973).
5. R. A. Saffer, M. Livio, and L. R. Yungelson, *ApJ*, **502**, 394–407 (1998).
6. R. Napiwotzki, N. Christlieb, H. Drechsel, et al., *ESO Messenger*, **112**, 25–30 (2003).
7. J. Trümper, G. Hasinger, B. Aschenbach, et al., *Nature*, **349**, 579–583 (1991).
8. E. P. J. van den Heuvel, D. Bhattacharya, K. Nomoto, et al., *A&A*, **262**, 97–105 (1992).
9. S. A. Rappaport, R. Di Stefano, and J. D. Smith, *ApJ*, **426**, 692–703 (1994).
10. M. Hamuy, M. M. Phillips, N. B. Suntzeff, et al., *Nature*, **424**, 651–654 (2003).
11. S.-C. Yoon, N. Langer, and S. Scheithauer, *A&A*, **425**, 217–228 (2004).
12. N. Ivanova, and R. E. Taam, *ApJ*, **601**, 1058–1066 (2004).
13. K. S. Long, D. J. Helfand, and D. A. Grabelsky, *ApJ*, **248**, 925–944 (1981).
14. A. P. Cowley, D. Crampton, J. B. Hutchings, et al., *ApJ*, **286**, 196–208 (1984).
15. A. P. Smale, R. H. D. Corbet, P. A. Charles, et al., *MNRAS*, **233**, 51–63 (1988).
16. A. N. Parmar, P. Kahabka, H. W. Hartmann, et al., *A&A*, **332**, 199–203 (1998).
17. P. Kahabka, and E. P. J. van den Heuvel, *ARA&A*, **35**, 69–100 (1997).
18. F. Paerels, A. P. Rasmussen, H. W. Hartmann, et al., *A&A*, **365**, L308–L311 (2001).
19. T. Lanz, G. A. Telis, M. Audard, et al., *ApJ*, **619**, in press, astro–ph/0410093 (2005).
20. I. Hubeny, and T. Lanz, *ApJ*, **439**, 875–904 (1995).
21. M. J. Seaton, et al., *The Opacity Project, Vol. 1*, Bristol: Intitute of Physics, 1995.
22. I. Hubeny, O. Blaes, J. H. Krolik, and E. Agol, *ApJ*, **559**, 680–702 (2001).
23. M. Balucinska-Church, and D. McCammon, *ApJ*, **400**, 699–700 (1992).
24. B. T. Gänsicke, A. van Teeseling, K. Beuermann, and D. de Martino, *A&A*, **333**, 163–171 (1998).
25. C. Alcock, D. R. Alves, T. S. Axelrod, et al., *AJ*, **127**, 334–354 (2004).
26. J. V. Clausen, J. Storm, S. S. Larsen, and A. Giménez, *A&A*, **402**, 509–530 (2003).
27. J. Greiner, and R. Di Stefano, *A&A*, **387**, 944–954 (2002).
28. P. Kahabka, F. Haberl, A. N. Parmar, and J. Greiner, *IAUC*, **6467**, 2 (1996).
29. M. Kato, *ApJS*, **113**, 121–129 (1997).
30. P. Kahabka, *A&A*, **331**, 328–334 (1998).
31. I. Hachisu, M. Kato, and K. Nomoto, *ApJ*, **470**, L97–L100 (1996).

Modelling the Evolution of Nova Outbursts

D. Prialnik* and A. Kovetz[†]

*Department of Geophysics and Planetary Sciences, Sackler Faculty of Exact Sciences, Tel Aviv University, Ramat Aviv 69978, Israel
[†]School of Physics and Astronomy, Sackler Faculty of Exact Sciences, Tel Aviv University, Ramat Aviv 69978, Israel

Abstract.
The theory of classical nova outbursts is reviewed. It shows that the different nova characteristics can be reproduced by varying the values of three basic and independent parameters: the white dwarf mass M_{WD}, the temperature of its isothermal core T_{WD} and the mass transfer rate $\dot{M}$. The parameter space is shown to be constrained by several analytical considerations. We present a grid of multicycle nova evolution models that spans the entire grid of parameter combinations for C-O white dwarfs. The full grid covers the entire range of observed nova characteristics, even those of peculiar objects, which have not been numerically reproduced until now. Most remarkably, runs for very low $\dot{M}$ lead to very high values for some characteristics, such as outburst amplitude $A \geq 20$, high super-Eddington luminosities at maximum, heavy element abundance of the ejecta $Z_{ej} \approx 0.63$ and high ejected masses $m_{ej} \approx 7 \times 10^{-4} \, M_\odot$. Some hitherto unpublished results of the long term evolution of a low mass accreting white dwarf are also presented.

Keywords: accretion, accretion disks — binaries: close — novae, cataclysmic variables — white dwarfs
PACS: 26.50.+x, 97.10.Tk, 97.30.Qt, 97.80.Gm

INTRODUCTION

It is a well accepted fact that Classical Novae (CN) are due to recurrent thermonuclear runaways (TNR) on the surfaces of white dwarfs (WD) in close, mass-transferring binary systems (Starrfield, Sparks & Truran [29]). The main characteristics of nova outbursts, as derived from observations (e.g. Duerbeck [3]), may be summarized as follows:

- Sudden ($\sim$ 1 day) brightening by 9^{mag} of a constant faint star;
- L $\sim$ L_{Edd}(1 M$_\odot$) — type A-F star — for a few days to several months;
- Mass ejection: $\sim(1\text{-}10)\times 10^{-5}M_\odot$ with velocities of 100-4000 km/sec;
- Composition of ejecta enriched in heavy elements: Z=0.04-0.60;
- Return to initial brightness within one month to several years.

The development of what has become known as *the CN theory* may be traced back to the 1950s, with the pioneering work of Mestel and Schatzman on degenerate stars. This was followed, starting more than 15 years later, by decades of nova outburst computer-simulations, which led to the recognition that three independent parameters control the behavior of a classical nova eruption: the white dwarf mass, the white dwarf core temperature (or luminosity, or age: they are all correlated) and the mass accretion rate from the companion Shara, Prialnik & Shaviv [26], Prialnik [15]. Increased computer power, and codes including better physics, led to an ever-increasingly sophisticated series of eruption simulations. Once multi-cycle simulations became possible, the arbitrariness of initial conditions could be eliminated, and systematic surveys throughout the three-parameter space became a real possibility. An extended grid of multicycle nova evolution models was computed by Prialnik & Kovetz [18], followed recently by an even more extended grid covering the entire parameter space (Yaron, Prialnik, Shara & Kovetz [32]). In summary, the cornerstones in the development of the CN theory are:

- 1950 (±2 yr) — Schatzman [21], Mestel [11]: first suggestion of explosive H-burning on the surface of accreting WDs.
- 1967 — Giannone & Weigert [6]: first numerical simulation of H-burning on the the surface of a WD.
- 1972 (±2 yr) — Starrfield, Sparks & Truran [29]: first detailed numerical calculations of CN evolution.
- 1976 — Bath & Shaviv [2]: first study of nova winds.
- 1978 — Kippenhahn & Thomas [9]: first investigation of shear mixing caused by accretion on a rotating WD.
- 1982 — Fujimoto [4]: first analytical study of the nova cycle instability.
- 1984 — Prialnik & Kovetz [16]: diffusion-convection mechanism for nova ejecta Z-enrichment.
- 1986 — Prialnik [13]: first calculation of a complete cycle of nova evolution.
- 1986 — Shara, Livio, Moffat & Orio [24]: the "hibernation" or "variable accretion rate" scenario for novae.
- 1986 — Starrfield, Sparks & Truran [31]: first study of outbursts on O-Ne-Mg WDs.
- 1995 — Prialnik & Kovetz [18]: first CN extensive parameter investigation (with diffusive mixing).
- 1999 — Kercek, Hillebrandt & Truran [8]: first 3-D CN calculation.
- 2004 — Yaron, Prialnik, Shara & Kovetz [32]: CN complete parameter study (with diffusive mixing).

The cyclic process that emerges from the numerous simulations of CN has several distinct phases, as listed below

- 1. Roche–lobe overflow $\Longrightarrow$ Mass accretion onto WD at high rate
- 2. Shrinking of secondary $\Longrightarrow$ Roche–lobe underflow
- 3. Hibernation $\Longrightarrow$ Diffusion
- 4. Angular momentum loss $\Longrightarrow$ Decrease of binary separation
- 5. Roche–lobe overflow $\Longrightarrow$ Mass accretion onto WD at high rate
- 6. Hydrogen ignition below WD surface $\Longrightarrow$ Thermonuclear runaway
- 7. Convection $\Longrightarrow$ Mixing of C-O into envelope, leading to high Z
- 8. Ejection of (CNO enriched) matter $\Longrightarrow$ Increase of binary separation
- 9. Illumination of secondary by L_{WD} $\Longrightarrow$ Bloating of secondary back to • 1.....

The meaning of "hibernation" is mass transfer at a lower rate than immediately prior to and following the outburst. During this time there is some elemental mixing due to molecular diffusion at the boundary between the accreted material and the WD core. In particular, traces of hydrogen penetrate—through a helium-rich buffer layer—into the C-O core. The explosive ignition of hydrogen thus occurs at some depth below the C-O boundary, and convection subsequently mixes the C-O above the front into the accreted envelope. Other mixing mechanisms between WD and accreted material at the boundary have been proposed (Kippenhahn & Thomas [9], Alexakis et al. [1]), but their effects have not been as thoroughly investigated as that of diffusion-convection.

NUMERICAL SIMULATIONS OF NOVA OUTBURSTS

Nova outburst simulations are usually carried out by Lagrangian stellar evolution codes, including nuclear reaction networks, equations of state and opacity coefficients in various degrees of sophistication. Some are quasi-static, others solve the momentum equation. The hydrodynamic Lagrangian stellar evolution code used by Prialnik & Kovetz [18] and by Yaron, Prialnik, Shara & Kovetz [32], for example, includes OPAL opacities, an extended nuclear reactions network comprising 40 heavy element isotopes and a mass-loss algorithm that applies a steady, optically thick supersonic wind solution (following the phase of rapid expansion). In addition, diffusion is computed for all elements, accretional heating is taken into account and convective fluxes are calculated according to the mixing length theory. This code was used for extensive parameter studies covering the $[M_{WD}, T_{WD}, \dot{M}]$ parameter space. With every extension of the parameter space, new features emerge for the resulting outburst characteristics. The question therefore arises: is the parameter space limited? And, if extended to its limits, is it possible to account for all the peculiar objects observed that may still be classed as novae?

Following Yaron, Prialnik, Shara & Kovetz [32], we may show that the various constraints imposed on parameters by nova and stellar evolution theory result in a confined space. The characteristic timescale for cooling of a WD, τ_{cool}, is basically a function only of the WD temperature (Mestel [11]). On the other hand, the accretion timescale τ_{acc} is directly determined by the accretion rate and indirectly affected by the other parameters through their influence on the mass required to ignite hydrogen, $\tau_{acc} = m_{acc}(M_{WD}, T_{WD}, \dot{M})/\dot{M}$. A nova outburst can take place only if $\tau_{cool} > \tau_{acc}$; otherwise the temperature cannot rise to the point of thermonuclear instability (Shara [23]). A limiting surface is thus obtained by equating these timescales:

$$\tau_{cool}(T_{WD}) - m_{acc}(M_{WD}, T_{WD}, \dot{M})/\dot{M} = 0. \tag{1}$$

The source of energy for classical nova outbursts is nuclear energy released during the TNR, by burning a fraction f of the hydrogen content of the accreted mass. Thus

$$E_{nuc} = fX m_{acc} Q, \tag{2}$$

where: $X \approx 0.7$ is the hydrogen mass fraction in the outer layers of the nova companion star, and $Q \approx 6 \times 10^{18}$ erg g^{-1} is the energy released per gram of burnt hydrogen. The greatest part of the energy released at outburst is used in lifting the ejected shell from the gravitational potential wall of the white dwarf. This gravitational energy may be roughly approximated by

$$E_{grav} = \frac{GM_{WD}m_{ej}}{R_{WD}}. \qquad (3)$$

Obviously, a mass ejecting outburst can take place only if $E_{nuc} > E_{grav}$. Therefore, a limiting surface may be defined by requiring $E_{nuc} = E_{grav}$,

$$\frac{m_{acc}(M_{WD},\dot{M},T_{WD})}{m_{ej}(M_{WD},\dot{M},T_{WD})} - \frac{G}{fXQ}\frac{M_{WD}}{R_{WD}(M_{WD})} = 0. \qquad (4)$$

For the nova progenitor to be able to accrete material during the quiescence phase, the accretion luminosity $L_{acc} = \alpha GM_{WD}\dot{M}/R_{WD}$ must be lower than the Eddington critical luminosity $L_{Edd} = 4\pi cGM_{WD}/\kappa_s$. Otherwise, radiation pressure would push away and dissipate the accreted material. In fact, the total (net) luminosity of the accreting star should be lower than the Eddington limit, but in most cases the WD intrinsic luminosity is negligible compared with L_{acc}. Thus a third limiting surface is obtained by equating L_{acc} and L_{Edd},

$$\dot{M} - \frac{4\pi c}{\alpha \kappa_s}R_{WD} = 0. \qquad (5)$$

We note that the result is independent of the WD temperature, although it might be indirectly affected by it to some extent through a more realistic opacity coefficient. Since R_{WD} is inversely correlated with M_{WD}, it is not surprising that the allowed accretion rates decrease with increasing M_{WD}.

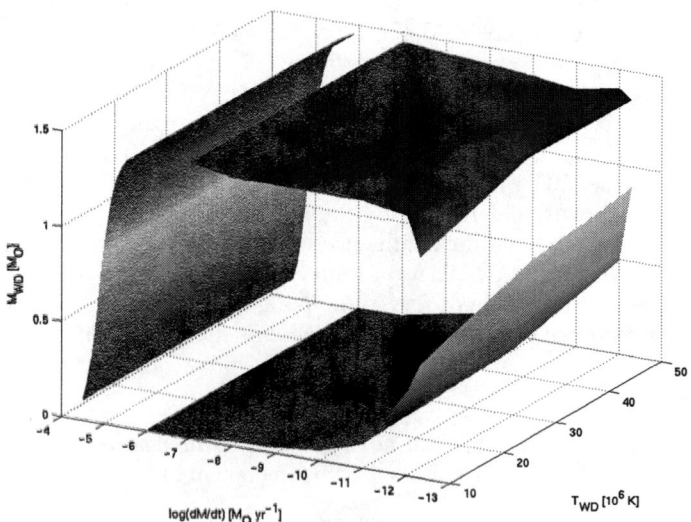

FIGURE 1. The combined restricting surfaces, confining a volume within the parameter space where conditions for nova outbursts are satisfied. The bottom surface corresponds to the heating versus cooling criterion; the top surface results from nuclear versus gravitational energy balance; the left surface relates to the accretion versus Eddington luminosity.

The three surfaces obtained above describe a restricted, tube-shaped region within the parameter space, as shown in Fig. 1, where conditions for nova outbursts are expected to be satisfied. Actually, there are two further limits on T_{WD}, so that only a *segment* of the tube is relevant. A rough estimate for the upper limit on T_{WD} may be obtained by setting the Fermi parameter $\varepsilon_F \propto (\ln P - 2.5 \ln T)$ to zero at the base of the accreted layer, where the pressure is of the order of 10^{19} dyn cm^{-2}.

A lower limit on T_{WD} results from the following considerations. When material starts accumulating on the WD surface, its temperature is lower than the core temperature and also lower than the ignition temperature (otherwise hydrogen would ignite immediately and quietly, rather than explosively under degenerate conditions). As the material becomes compressed, it releases gravitational energy, which is absorbed, in part, by the accreted layer, while in part it is conducted into the core. If the absorption of heat is sufficiently effective to raise the temperature of the hydrogen-rich material then, eventually, the temperature at the bottom of the hydrogen-rich layer will become high enough for hydrogen to ignite. The nuclear luminosity, low at first, will soon become the dominant energy source. Therefore, the restrictive condition for a nova outburst to occur is that compressional heating be sufficiently effective in order to raise the temperature to the ignition value of roughly 15×10^6 K required by the CNO cycle. Thus, a lower limit to the WD core temperature exists for any combination of WD mass and accretion rate, but for high accretion rates this limit is so low that it requires more than the age of the universe for a cooling WD to reach it. Hence the constraint on T_{WD} becomes significant only in the case of low accretion rates.

The additional constraints on T_{WD} determine the ends of the tube, and nova models indeed fall within this volume (Yaron, Prialnik, Shara & Kovetz [32]). It would have been very instructive if we were also able to place on this graph the positions of (parameter combinations of) observed novae. Unfortunately, the estimation of all three parameters for any observed eruption is still problematic and involves too many uncertainties. A unique estimation of the three basic parameters that lead to a set of observed characteristics, first attempted by Prialnik & Kovetz [18] is an ambitious aim yet to be achieved.

RESULTS OF NOVA EVOLUTION MODELS

The effect of each of the three input parameters on the resulting nova characteristics was discussed by Prialnik & Kovetz [18] and also by Schwartzman, Kovetz & Prialnik [22]. The outcome is not always self-evident and the trends are not necessarily monotonic. To illustrate the behavior of some of the more interesting characteristics, we show in Fig. 2 the model results over the $(M_{WD}, \dot{M})$ plane for a given value of T_{WD}.

Mass and composition of the ejecta

The masses of nova shells and the composition of nova ejecta have been the most controversial issues of nova theory, mainly because their ranges of variation are so wide.

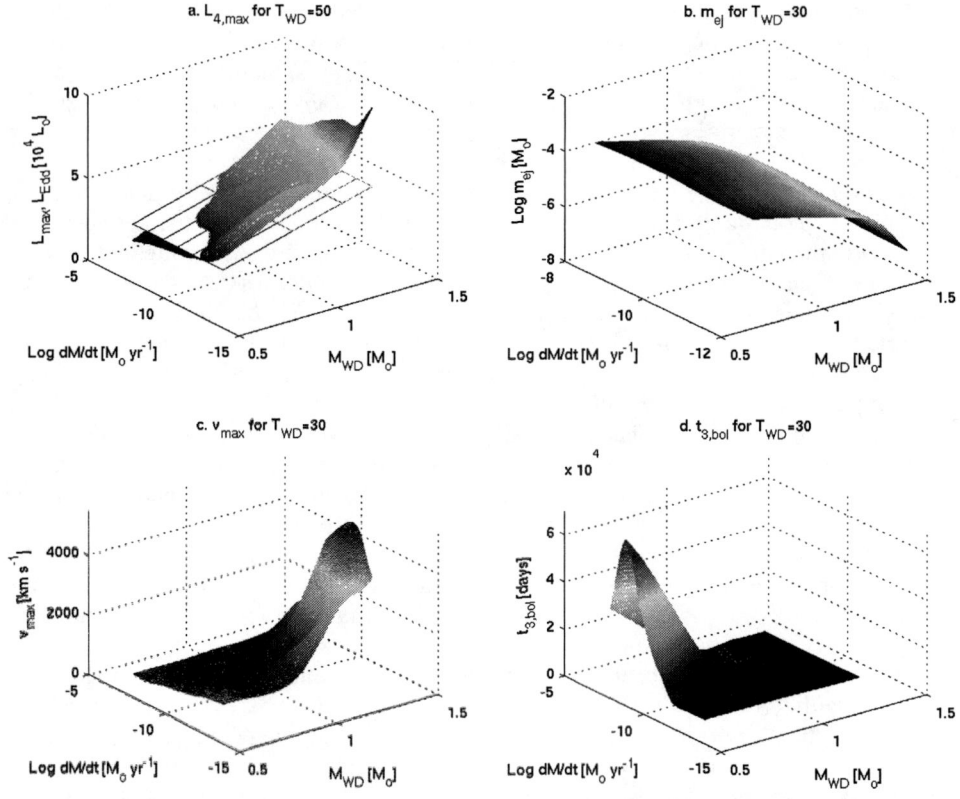

FIGURE 2. Plots of four CN properties, as labeled, over the $[M_{WD}, \dot{M}]$ plane, for a representative WD temperature (in 10^6 K). In panel a, the plane represents the critical luminosity L_{Edd}, marking the domain of super-Eddington luminosity.

The mass balance of a CN cycle may be summarized as follows

$$m_{env} = m_{acc} + m_{rem} + m_{core} \qquad (6)$$
$$m_{rem} = m_{env} - m_{ej}, \qquad (7)$$

where m_{env} is the mass of the envelope, between the burning shell and the surface, m_{acc} is the accreted mass, m_{rem} is part of the envelope of the previous cycle that was left on the WD at the end of mass ejection, m_{core} is the mass of the WD core that was mixed into the accreted material (by one or more mixing mechanisms) and m_{ej} is the ejected mass. An example of how these mass components vary with the WD mass—for given accretion rate and WD temperature—is given in Table 1. They span more than 3 orders of magnitude. As a rule, the highest ejected masses are obtained for the lowest WD

TABLE 1. Example: $T_{WD} = 10$; $\log \dot{M}_{acc} = -10$

Mass/$M_\odot$	0.65	1.00	1.25	1.40
Accreted	2.55	8.40	19.10	5.90
Core	0.21	1.32	2.70	1.00
Envelope	3.01	14.37	31.60	10.20
Remnant	0.25	4.65	9.80	3.30
Ejected	2.76	9.72	21.80	6.90
power	$\times 10^{-4}$	$\times 10^{-5}$	$\times 10^{-6}$	$\times 10^{-7}$

masses through the entire range of accretion rates. This is because the accreted masses necessary to trigger a TNR are higher for lower mass WDs whereas, at the same time, the weaker gravitational potential enables more massive ejecta, despite the typically lower outburst intensity.

Knowing the mass components, and assuming consecutive CN outbursts to be similar, one may easily deduce the composition of the ejecta

$$m_{env}X = m_{acc}X_\odot \qquad (8)$$
$$m_{env}Y = m_{acc}Y_\odot + m_{rem}(X+Y) \qquad (9)$$
$$m_{env}Z = m_{acc}Z_\odot + m_{rem}Z + m_{core} \qquad (10)$$

and by inverting these relations, one may obtain the various masses based from observed ejecta composition. This consistency test of the CN theory was discussed by Prialnik [14], Fujimoto & Iben [5], and Sparks [27].

Although the extended parameter studies cover the entire range of WD masses, practically up to the Chandrasekhar limit, they only consider C-O WDs (with C and O in equal mass fractions). This may appear unrealistic, since very massive WDs are believed to emerge after the carbon burning stage in the evolution of a star, and they should be composed of O, Ne and Mg (Gutierrez et al. [7]), although the mass fractions of these elements and their dependence on the WD mass are still uncertain. However, the grid of C-O models produces results that cover the *entire* range of observed characteristics (ignoring the breakdown of the total heavy element mass fraction Z_{ej}). This seems to indicate that they are more widely applicable than might have been expected by the restriction to C-O composition of the WD. In fact, the mechanism of nova eruptions revolves around thermonuclear instability under conditions of electron degeneracy. Electron degeneracy, in turn, is determined by $\mu_e = <A/Z>$, which has the same value for helium and carbon burning products. In addition, the CNO cycle is not sensitive to the initial CNO breakdown. Consequently, we should not expect significant differences in the outburst characteristics of WDs that differ only in composition, except for the *breakdown* of Z_{ej}. The results of calculations for two illustrative models after replacing the carbon in the WD core by neon are summarized in Table 2 and they indeed confirm the prediction. Thus the results obtained for the C-O models should be applicable to novae in general.

TABLE 2. Ejecta Composition - C-O vs. O-Ne Models

	1.25, 10, −11		1.40, 10, −9	
	C-O	O-Ne	C-O	O-Ne
$m_{ej}/M_\odot$	$3.61E-05$	$3.67E-05$	$4.74E-07$	$2.89E-07$
Y_{ej}	0.3155	0.3123	0.4732	0.5129
Z_{ej}	0.2092	0.2179	0.1521	0.1508
C^{12}	$3.353E-02$	$1.970E-02$	$2.430E-02$	$1.143E-02$
C^{13}	$3.136E-02$	$2.089E-02$	$2.485E-02$	$8.148E-03$
N^{14}	$6.545E-02$	$3.086E-02$	$9.392E-02$	$4.706E-02$
N^{15}	$6.660E-02$	$3.321E-02$	$4.191E-03$	$1.147E-03$
O^{16}	$1.165E-02$	$9.241E-05$	$2.092E-04$	$1.402E-04$
O^{17}	$6.157E-04$	$7.729E-05$	$1.342E-05$	$8.500E-06$
Ne	$3.384E-07$	$6.579E-02$	$1.372E-07$	$1.358E-03$
Na	$3.535E-09$	$6.541E-04$	$1.047E-10$	$1.141E-06$
Mg	$2.861E-08$	$9.611E-03$	$5.238E-09$	$3.640E-05$
Al^{26}	$2.098E-08$	$7.929E-03$	$6.355E-09$	$4.258E-05$
Al^{27}	$1.027E-08$	$3.137E-03$	$1.251E-09$	$7.662E-06$
$\geq$ Si	$1.046E-07$	$2.600E-02$	$6.087E-07$	$8.147E-02$

Long-term evolution of a low-mass WD accreting at high rate

An important question related to nova evolution through repeated cycles is whether the WD may grow in mass and eventually reach the Chandrasekhar limit to become a SNIa progenitor. The results of numerical simulations indicate that a growing WD mass may be obtained only for a narrow range of high accretion rates. Long-term, detailed evolution calculation spanning more than 1000 mild cycles of accretion, expansion, mass-loss, contraction and resumed accretion were carried out for a WD of 0.65 $M_\odot$ initial mass accreting at a rate of $10^{-6} M_\odot$ yr^{-1} (Prialnik & Kovetz, unpublished). Indeed it was found that the WD grows steadily in mass, although at a lower effective rate. This was due to some mass loss that followed each expansion episode. During the entire evolution time the nuclear luminosity remained almost constant at $1.3 \times 10^4 L_\odot$ and a helium layer built up on top of the WD C-O core. The bolometric luminosity fluctuated slightly around this value. Fluctuations in R, (and hence also in effective temperature) were caused by expansion of the WD envelope beyond the separation distance of the binary, at which point accretion was interrupted until the envelope shrank back within this distance. We found that varying the separation between $0.5 R_\odot$ and $1 R_\odot$ did not affect the results. The central density of the WD kept rising, as shown in Fig. 3, the central temperature also rose, but to a much lesser extent.

The significant continuous change, which eventually led to the end of this evolutionary episode, was the rise in the temperature of the burning shell, at the base of the envelope, from 7×10^7 K to over 10^8 K. At this point helium ignited in a strong flash that ejected a large fraction of the previously accreted mass. Thus the net mass gain was even lower than before. The structure of the star at the onset of the helium flash is shown in Fig. 3. In conclusion, it does not appear possible for the mass of a WD to grow significantly by accretion, simply because of the very low efficiency of the process, which would demand a high mass donor star.

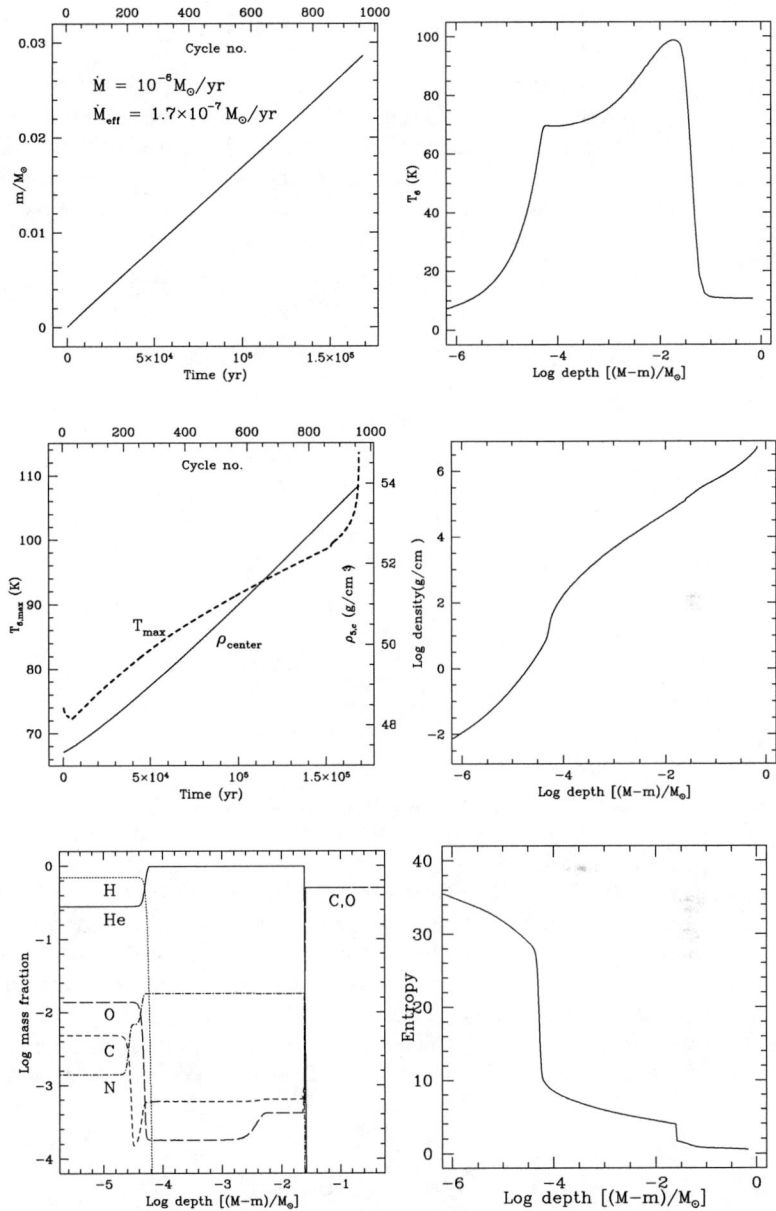

FIGURE 3. *Top-left:* Evolution of the total mass. *Middle-left:* Evolution of the central density and of the maximal temperature (in the burning shell). *Bottom-left:* Mass fraction profiles within the star as a function of depth in mass - on a logarithmic scale. *Top-right:* Temperature profile at the end of 1000 cycles, prior to helium ignition. *Middle-right:* Density profile. *Bottom-right:* Entropy profile (in arbitrary units).

Comparison with observations

Some of the successes of what became known as *the CN theory* include an explanation of the UV emission and high bolometric luminosities of novae after eruption (Starrfield, Sparks & Truran [30]); fast and moderately slow nova models (Prialnik, Shara & Shaviv [19]); natural enrichment of ejecta in heavy elements via mixing from the underlying white dwarf (Prialnik & Kovetz [17], Kutter & Sparks [10]); and novae recurring on timescales as short as a few years (Starrfield, Sparks & Shaviv [28]).

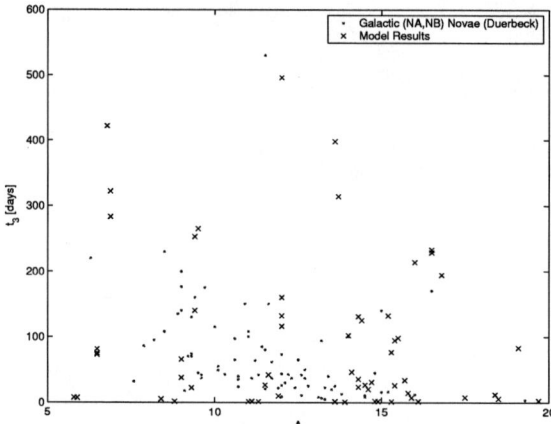

FIGURE 4. Time of decline t_3 versus outburst amplitude for observed galactic novae (from Duerbeck [3]) and model results (from Yaron, Prialnik, Shara & Kovetz [32]).

One of the most important characteristics of a classical nova outburst is the time of decline of the visual magnitude, which serves as a basis for the "speed class" classification of novae and on which the role of novae as standard candles is based. In Fig. 4 we show the distribution of the time of decline t_3 versus the outburst amplitude A for 79 observed galactic novae from the Duerbeck catalog (Duerbeck [3]), together with results of 75 nova outburst models obtained for various parameter combinations. First, we note the general tendency of decreasing t_3 for increasing A is clearly seen. Secondly, the model results provide full coverage of observations in the $[A,t_3]$ plane. In this respect, it should be noted that the density of points on the plane has no significance, since the calculated models represent parameter combinations (they are not based on a population synthesis model).

CONCLUSIONS

The entire range of observed nova characteristics is reproduced by models assuming extended ranges of the three basic parameters. Even exceptional observed values, such as an outburst amplitude of over 19 magnitudes, or very high Z, are covered by recent computational results. Interestingly enough, maximum values for the various characteristics are obtained for the lower-$\dot{M}$ values, down to the lowest accretion rate value of

TABLE 3. Comparison of model results with observations

Property	Observed	Exceptions	Calculated
M_{max}	−6 to −9	−10 (V1500 Cyg)	−5.73 to −10.2
A	7 − 16	19.3 (V1500 Cyg)	5.8 − 20.9
t_3	4 − 300 d	SymN:more	0.76 d−67 yr
m_{ej}	1 − 30	RN:less	0.005 − 66
$10^{-5} M_\odot$			
Z_{ej}	0.04 − 0.41	0.86 (V1370 Aql)	0.021 − 0.63
Y_{ej}	0.21 − 0.48	0.1 (V1370 Aql)	0.12 − 0.60
v_{exp}	350 − 2500	SymN:∼ 100	120 − 3900
km s^{-1}			

5×10^{-13} $M_\odot$ yr^{-1}. Low $\dot{M}$ values lead to maximum outburst luminosities surpassing the Eddington luminosities (calculated for electron scattering opacities) by factors of up to a few tens, for the whole range of WD masses. The highest super-Eddington luminosities (with correspondingly highest derived outburst amplitudes) are obtained for fast to very-fast novae.

There has been a long-standing problem deriving ejected mass values as high as those deduced observationally. It turns out, however, that high ejected masses *can* be obtained (e.g., 6.6×10^{-4} $M_\odot$) for a low M_{WD}, accreting at a low $\dot{M}$.

A puzzling question related to the investigation of novae is the ultimate fate of the WD. Will it be losing or gaining mass after multiple successive cycles? Will it be able to reach M_{Ch}, thus acting as a possible SNIa precursor, or not? The ratio m_{ej}/m_{acc} falls below unity only in a small region of the parameter space, thus strongly reducing the possibility of SNIa to result from accreting WDs. Nevertheless, this region does lead continuously from low mass to the Chandrasekhar limit provided the accretion rate remains very high all along. Therefore, it is possible, at least in principle, for a WD to grow by accretion up to M_{Ch}. We have shown, however, that the process would be extremely inefficient and would require a prohibitively massive binary companion.

REFERENCES

1. Alexakis, A., et al 2004, ApJ, 602, 931
2. Bath, G. & Shaviv, G. 1976, MNRAS, 175, 305
3. Duerbeck, H. W. 1987, Space Sci. Reviews, 45, 1
4. Fujimoto, M. Y. 1982, ApJ, 257, 752
5. Fujimoto, M. Y. & Iben, I. Jr. 1992, ApJ, 399, 646
6. Giannone, P. & Weigert, A. 1967, ZfAp, 67, 41
7. Gutierrez, J., Garcia-Berro, E., Iben, I. Jr., Isern, J., Labay, J., and Canal, R. 1996, ApJ , 459, 701
8. Kercek, A., Hillebrandt, W. & Truran, J. W. 1999, A&A, 345, 831
9. Kippenhahn, R. & Thomas, H.-C. 1978, A&A, 63, 265
10. Kutter, G. S. & Sparks, W. M. 1987, ApJ , 321, 386
11. Mestel, L. 1952, MNRAS , 112, 583

12. Payne-Gaposchkin, C. 1957, The Galactic Novae (Amsterdam:North Holland)
13. Prialnik, D. 1986, ApJ, 310, 222
14. Prialnik, D. 1990, in Physics of Classical Novae, A. Cassatella & I. Viotti, eds. (Berlin: Springer), p. 351
15. Prialnik, D. 1995, in Cataclysmic Variables, A. Bianchini, M. Della Valle, M. Orio, eds. (The Netherlands: Kluwer), p. 217.
16. Prialnik, D. and Kovetz, A. 1984, ApJ , 281, 367
17. Prialnik, D. and Kovetz, A. 1985, ApJ , 291, 812
18. Prialnik, D. and Kovetz, A. 1995, ApJ , 445, 789
19. Prialnik, D., Shara, M. & Shaviv, G. 1978, A&A , 62, 339
20. Regev, O. & Shara, M. M. 1989, ApJ , 340, 1006
21. Schatzman, E. 1950, AnAp, 13, 384
22. Schwartzman, E., Kovetz, A. & Prialnik, D. 1994, MNRAS , 269, 323
23. Shara, M.M. 1981, ApJ , 243, 926
24. Shara, M.M., Livio, M., Moffat, A. M. & Orio, M. 1986, ApJ , 311, 163
25. Shara, M. M., Prialnik, D. & Shaviv, G. 1977, A&A , 61, 363
26. Shara, M. M., Prialnik, D. & Shaviv, G. 1980, ApJ , 239, 586
27. Sparks, W. 1995, in Cataclysmic Variables, A. Bianchini, M. Della Valle, M. Orio, eds. (The Netherlands: Kluwer), p. 211.
28. Starrfield, S., Sparks, W. & Shaviv, G. 1988, ApJL , 325, L35
29. Starrfield, S., Sparks, W. & Truran, J. 1972, ApJ , 176, 169
30. Starrfield, S., Sparks, W. & Truran, J. 1976, ApJ , 208, 819
31. Starrfield, S., Sparks, W. & Truran, J. 1986, ApJL , 303, L5
32. Yaron, O., Prialnik, D., Shara, M. M. & Kovetz, A. 2005, ApJ, in press

The first three years of the outburst and light-echo evolution of V838 Mon and the nature of its progenitor

U. Munari* and A. Henden[†]

INAF National Institute of Astrophysics, I-36012 Asiago (VI), Italy
[†]*Univ. Space Research Assoc./U. S. Naval Obs., P. O. Box 1149, Flagstaff AZ 86002-1149, USA*

Abstract. V838 Mon has undergone one of the most mysterious stellar outbursts on record, with (*a*) a large amplitude ($\Delta B \sim 10$ mag) and multi-maxima photometric pattern, (*b*) a cool spectral type at maximum becoming cooler and cooler with time during the descent, until it reached the never-seen-before realm of L-type supergiants, never passing through optically thin or nebular stages, (*c*) the development of a spectacular, monotonically expanding light-echo in the circumstellar material, and (*d*) the identification of a massive and young B3V companion, unaffected by the outburst. In this talk we review the photometric and spectroscopic evolution during the first three full years of outburst, the light-echo development and infer the nature of the progenitor, which was brighter and hotter in quiescence than the B3V companion and with an inferred ZAMS mass of ~ 65 $M_\odot$.

Keywords: V838 Mon, novae, light-echo, massive star evolution, outburst, L-type supergiants
PACS: 43.35.Ei, 78.60.Mq

THE OUTBURST EVOLUTION

V838 Mon made headlines in early January 2002, when it was discovered in outburst by [1]. The unusually cool spectrum (completely unlike that of a classical nova) and the multi-maxima light-curve helped to keep attention focused on the object for the next three months, until the discovery in late March by [2] of a light-echo rapidly developing around V838 Mon. The presence of the first Galactic light echo in ~ 70 years fostered a massive, multi-wavelength observing campaign for V838 Mon. A high spatial resolution imaging series of the light-echo expansion and evolution was collected with HST by [3], which has been expanded by new images secured within the Hubble Heritage Program[1]. An account of the spectroscopic, photometric and polarimetric evolution of V838 Mon during the first season of visibility was presented by [4], [5] [6], [7], [8], [9], [10], [11]. A major observational constraint was the discovery by [12] and [13] that V838 Mon is a binary system containing a normal B3V star, implying that the outbursting component is young and massive. BaII, LiI and s−element lines were prominent in the outburst spectra, while Balmer lines emerged with modest emission only during the central phase of the outburst. The outburst never reached optically thin conditions and the spectra never went through a nebular stage.

---

[1] http://heritage.stsci.edu

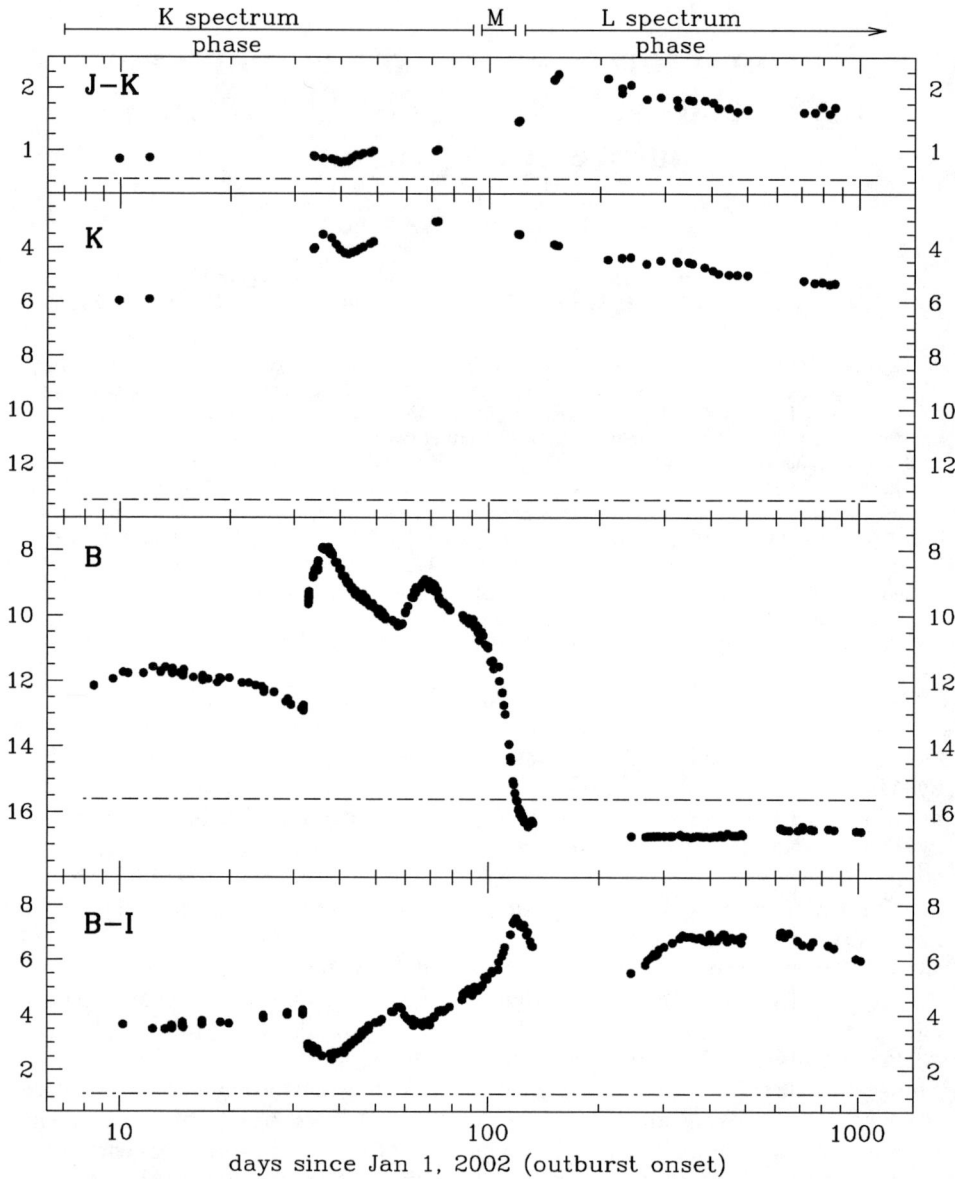

FIGURE 1. Optical and infrared lightcurves of V838 Mon from 2002-2004 observations with the USNO 1.0 and 1.55m telescopes.

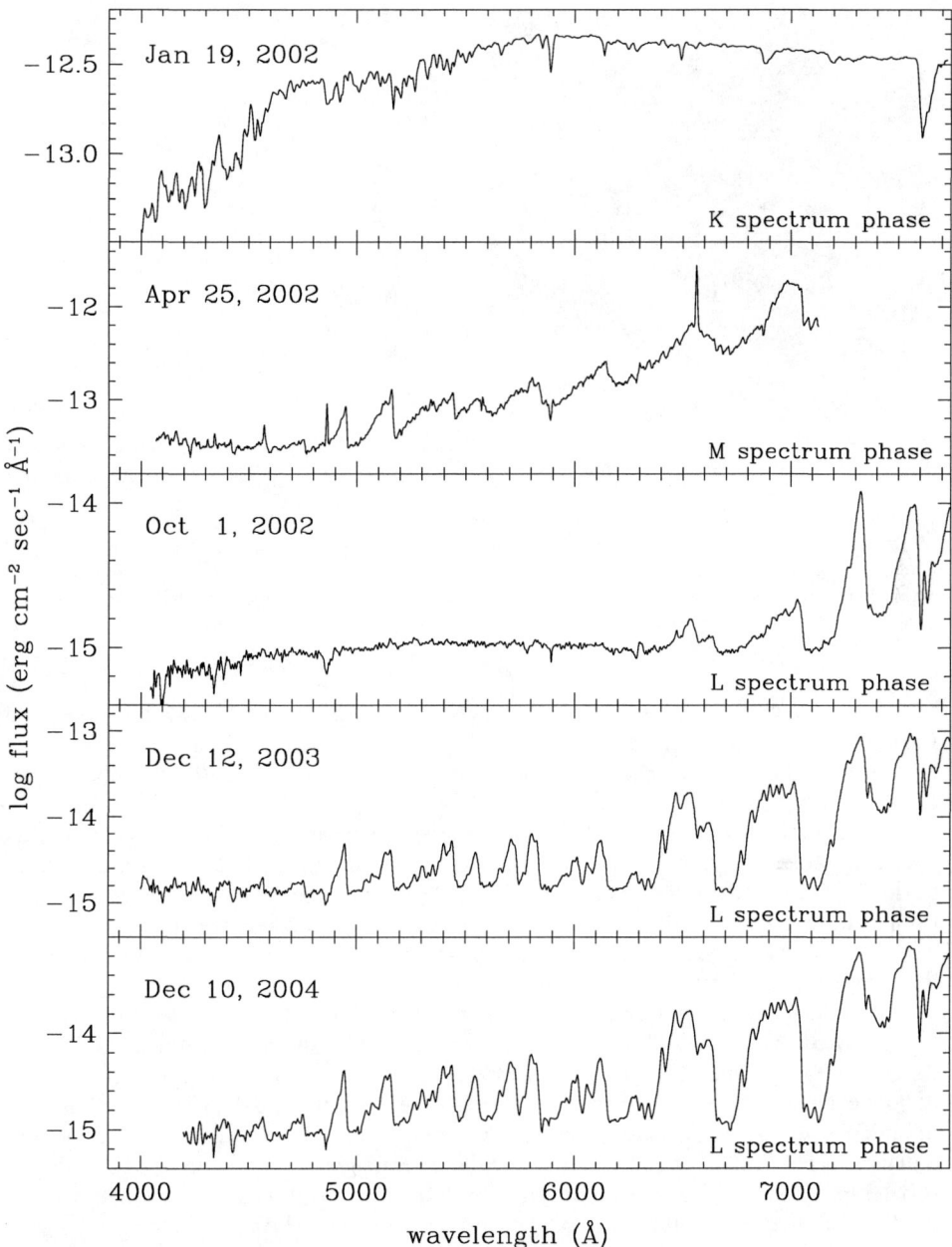

FIGURE 2. Spectral evolution of V838 Mon (sample spectra from an extensive monitoring of the whole event performed with the 1.82 m Asiago telescope, to be reported in detail by Munari et al. elsewhere).

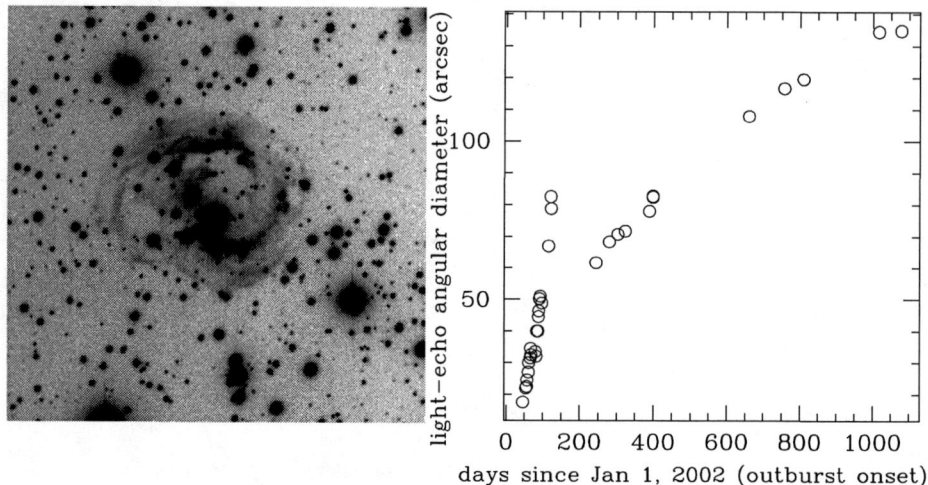

FIGURE 3. *Left panel.* The light echo imaged in R_C band with the 1.55 m USNO telescope on Oct 15, 2004 (48 min total exposure). The longest dimension of the light-echo occurs at position angle 100° where it extends 2.37 arcmin. Field is 4.5 arcmin square. North is up, East to the left. *Right panel.* The expansion rate of the outer edge of the light echo in the east-west direction. Based on NOFS ground-based imagery in multiple bandpasses.

An updated optical and infrared lightcurve of the eruption of V838 Mon is presented in Figure 1 and the general spectral evolution is highlighted in Figure 2. The outburst lightcurve and spectral evolution can be easily divided into three distinctive phases: the K, M and L supergiant spectrum phases.

During the first 90 days of the outburst, the star went through three maxima and its spectrum moved back and forth within the K supergiant types. The monotonic cooling during this phase (from $B-I_C \sim 3.5$ to ~ 5.0, cf. Figure 1) was interrupted by warming at each of three optical brightness maxima. This *K-supergiant* phase was characterized by a cool wind that gave rise to P-Cyg profiles with terminal velocities of ~ 500 km sec^{-1} that progressively diminished to ~ 250 km sec^{-1} by day +90.

Around day 90, the star entered an *M-supergiant* phase almost identical to that displayed by the M31-RV stellar eruption observed in 1988 in the Andromeda Galaxy as described by [14] and [15]. [16] investigated in detail the similarities, highlighting how in both events the transition to an M-supergiant spectrum was marked by a free-fall drop in optical brightness not accompanied by dust formation. The outbursting component in a matter of few weeks swept through the whole sequence of M spectral types while the B band brightness of V838 Mon dropped by 6 mag, stopped only by the emergence of the B=16.73 mag spectrum of the B3V companion by day ~ 120.

By day ~ 130, the photosphere of the outbursting component had cooled so much to enter the realm of *L supergiant* spectral types, a type of star never seen before anywhere in the Galaxy ([12], [17]). Very strong H_2O, CO and AlO absorption bands dominate the infrared, while huge VO and TiO shape the optical. Since the time of coolest temperature reached around day 180, the temperature of the L-supergiant has been smoothly and

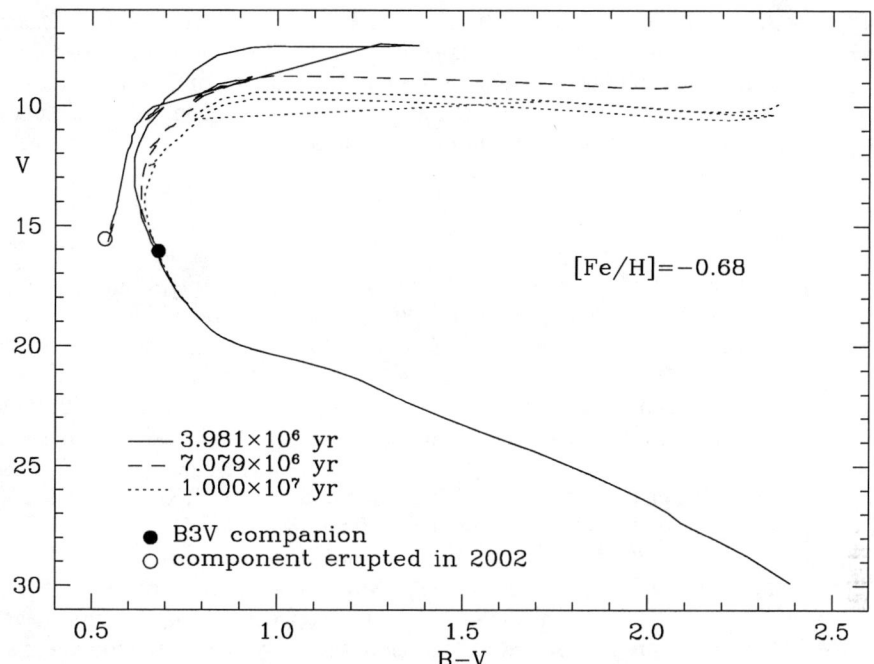

FIGURE 4. Comparison of the pre-outburst photometry of the two components of V838 Mon with Padova isochrones Z=0.004 and three ages scaled to the system distance (10 kpc) and reddening (E_{B-V}=0.87). The isochrones include the effect of mass loss and reddening-induced distortion.

constantly increasing even if by small amounts (cf. $J-K$ and $B-I$ in Figure 2). As a result, the optical spectra that in Oct 2002 were dominated by the B3V companion even in the V band, are now again showing the molecular absorptions down to ∼4800 Å.

THE LIGHT-ECHO EVOLUTION

The light-echo was first discovered by [2] on U-band images taken in February, 2002, about 10 days after the first large outburst (a hundred-fold increase in brightness with a relatively sharp peak). The initial shell was circularly symmetric, showing obvious structure that later HST images by [3] resolved into concentric rings, each giving a photometric snapshot of the light curve variations. These rings expanded, with the rate of expansion of the outermost ring shown in Figure 3. Note the dramatic break in the expansion rate, with the early expansion proceeding at about 800mas per day, while the later expansion proceeding at about 90mas per day. The early expansion is probably coming from a light sheet in front of the outbursting star, and is very nearly symmetric, with axial ratios at most 10% different. The later expansion may be evidence of the light echo proceeding through circumstellar material, and shows more than 20% asymmetry.

The light-echo has also faded, now difficult to image even at R-band, and has become more extended in the east-west direction than north-south. An inner void appeared about 70 days after the main outburst, matching the rapid decline from the last photometric peak. The bright cloud to the north has remained in all images. More detailed analysis of the ring expansion and structure evolution is in progress.

THE PROGENITOR

The energy distribution of V838 Mon in quiescence has been throughly investigated by [18], clearly showing how *the progenitor of the outbursting component was brighter and hotter than the B3 V companion*. The best fit to the V838 Mon quiescence data is the B3 V plus a 50 000 K star with $V=15.55$ and $B-V=+0.535$, both equally reddened by $E_{B-V}=0.87$. The location of the components of the binary is compared with Padova theoretical isochrones in Figure 4 (scaled to the system $E_{B-V}=0.87$ reddening and 10 kpc distance), for the [Fe/H]$=-0.7$ metallicity appropriate to the galacto-centric distance of V838 Mon. The Padova theoretical isochrones (that include the effect of mass loss) are corrected for reddening dependent deformation following [19].

The isochrone fitting the position of the two components indicates a system age of 4 million yr and a mass of ~ 65 $M_\odot$ on the ZAMS for the progenitor of the outbursting component. The outburst experienced in 2002 does not appear to be the terminal event in the life of the massive progenitor, but instead more probably was a thermonuclear shell flash in the outer layers of the star as could be expected in the case of He after most of the H-rich outer envelope has been blown away by the strong wind that characterize the Wolf-Rayet type of stars that occupy this part of the HR diagram.

REFERENCES

1. Brown N.J. 2002, IAU Circ 7785
2. Henden, A., Munari, U., Schwartz, M.B., 2002, IAUC 7859
3. Bond, H.E., Panagia, N., Sparks, W.B., Starrfield, S.G., Wagner, R.M., 2002, IAUC 7892
4. Munari, U., Henden, A., Kiyota, S., et al., 2002, A&ALett 389, L51
5. Banerjee, D.P.K., Ashok, N.M. 2002, A&A 395, 161
6. Goranskii, V.P., Kusakin, A.V., Metlova, N.V. et al. 2002, Astron.Lett. 28, 691
7. Kimeswenger, S., Lederle, C., Schmeja, S. et al. 2002, MNRAS 336, L43
8. Crause, L.A., Lawson, W.A., Kilkenny, D. et al. 2003, MNRAS 341, 785
9. Wisniewski, J.P., Morrison, N.D., Bjorkman, K.S. et al. 2003, ApJ 588, 486
10. Desidera, S., Giro, E., Munari, U. et al. 2004, A&A 414, 591
11. Kipper, T., Klochkova, V.G., Annuk, K. et al. 2004, A&A 416, 1107
12. Desidera, S., Munari, U. 2002, IAUC 7982
13. Munari, U., Desidera, S., Henden, A., 2002, IAUC 8005
14. Rich, R.M., Mould, J., Picard, A., et al., 1989, ApJ 341, L51
15. Mould, J., Cohen, J., Graham, J.R., et al., 1990, ApJ 353, L35
16. Boschi F., Munari U. 2004, A&A 418, 869
17. Evans A., Geballe T.R., Rushton M.T. et al. 2003, MNRAS 343 1054
18. Munari, U., Henden, A., Vallenari, A. et al. 2004, A&A, submitted
19. Fiorucci, M., & Munari, U. 2003, A&A, 401, 781

SESSION 6

RAPID VARIABILITY AND SECULAR VARIATION OF LMX-RAY BINARIES

The Accretion Disc Corona in LMXBs: size and temperature – fundamental consequences

M. Bałucińska-Church* and M. J. Church*

School of Physics and Astronomy, University of Birmingham, UK; Astronomical Observatory, Jagiellonian University, Poland

Abstract. The nature of the continuum X-ray emission in low-mass X-ray binaries has been controversial for many years, with workers polarised between the Eastern model with small central Comptonizing region and alternative models. We present measurements of accretion disc corona (ADC) radial extent by dip ingress timing which show unambiguously that the ADC is very extended, typically r_{ADC} ~50,000 km, ruling out the Eastern model. We secondly present measurements of ADC temperature which indicate that the ADC extends outwards to the position at which hydrostatic equilibrium fails. We also show that the ADC temperature decreases with increasing source luminosity, the high values in fainter sources of 25 keV or more requiring a presently unknown heating mechanism. These results suggest that ADC formation which is poorly understood takes place via illumination of the disc by the central source. Finally, we discuss the relevance of the extended ADC to the correct form for Comptonization models to be used in spectral fitting.

Keywords: accretion, accretion discs – binaries: close – stars: neutron – X-rays: binaries
PACS: 95.85.Nv, 97.10.Gz, 97.60.Jd, 97.80.Jp

INTRODUCTION

The nature and geometry of the X-ray emitting regions in Low Mass X-ray Binaries (LMXB) has been controversial for many years. Two very different approaches were developed in mid-1980s: the Western model and the Eastern model. In the Western model, dominance of the Comptonized emission in the LMXB spectra was recognized by modelling the spectra with the General Thermal model having the form of a power law with a high energy cut-off corresponding to the energy limit of Comptonizing electrons in the accretion disc corona [1]. Only very bright sources required an additional blackbody component thought to originate in the boundary layer between the accretion disc and the surface of a neutron star. A radically different approach was embodied in the Eastern model proposed by Mitsuda and his colleagues [2], but extensively used since then. In this case, seed photons from the neutron star are Comptonized in a small central region and thermal emission originates as multi-temperature disc blackbody.

Study of the dipping sources, i.e. LMXBs viewed at high inclination angle, which exhibit orbital-related X-ray dips due to absorption in the bulge in the outer disc [3], can resolve the problem. Spectral evolution during dipping strongly constrains emission models because the non-dip and several dip spectra must all be fitted by the same model, allowing only absorption terms to vary. This led to the proposal of the Birmingham model [4], closely related to, and developed from the Western model, having a neutron star blackbody in *all* sources plus Comptonized emission from an extended ADC. This model describes well not only all of the dipping sources but also the other classes of

LMXBs [5] (for a review see [6]).

In all LMXBs, the dominant emission component is the non-thermal Comptonized emission originating in the ADC; but the ADC has been very poorly understood. The dipping sources however, provide basic parameters of the corona. Firstly, they allow direct measurement of the radial extent of the ADC by the technique of "dip ingress timing" described in the next section. Study of the broadband spectra of these sources using, for example, *BeppoSAX* provides the high energy Comptonization cut-off energy, and thus the electron temperature (T_e) of the ADC. In many sources, T_e is high, i.e. in the range 10 – 100 keV [7], and this is not understood. Thus by making these measurements with the dipping sources we have obtained the most basic parameters: the radial size and the temperature of the ADC. Other basic aspects however, are not understood: notably, the process by which the ADC is formed, and the process resulting in the high measured electron temperatures.

In this paper we present our results on the radial extent and temperature of the ADC, and then discuss the implications these have for understanding formation and heating of the ADC. Finally, we discuss the consequences of the very extended ADC which apply in both neutron star systems and black hole systems for the correct form of the Comptonization term in spectral fitting, i.e. the specific form appropriate to an extended ADC fed by seed photons from an extended region of the accretion disc below the ADC.

SIZE AND TEMPERATURE OF THE ADC

Size of the ADC. Spectral evolution during dipping shows unambiguously that the Comptonized emission from the ADC is extended (e.g. [7]). Moreover, dipping shows that the ADC is flat ($h/r < 1$), like the accretion disc, since for a spherical ADC, 100% deep dipping would not be observed. Definite proof of the extended size follows from the technique of "dip ingress timing" since the ingress or egress time depends on the size of the ADC provided the absorber has larger angular extent than the ADC, and this has to be the case when dipping is 100% deep at any energy. The radial extent of the flat ADC, r_{ADC}, is related to the accretion disc radius, r_{AD} via the equation

$$\frac{2\pi r_{ADC}}{\Delta t} = \frac{2\pi r_{AD}}{P} \quad (1)$$

where P is the orbital period and Δt is the ingress time.

Fig. 1 shows the measured radial extent of the ADC in the sources XBT 0748-676, XB 1916-053, XB 1254-690 and X 1624-490, as a function of the luminosity in the band 1–30 keV [8]. The figure displays two significant results: firstly, it confirms that the ADC is very extended; secondly, there is strong correlation between the radius of the ADC and the total luminosity of the source. The radius of the corona varies between 20,000 km to 60,000 km, equivalent to $\sim 7 - 65\%$ of the accretion disc. In the case of XB 1916-053, our measured ingress times fall within the range of values obtained by Narita [9], who also found some evidence for a dependence of ADC size on mass accretion rate in data on this one source having limited variation in luminosity. Least-squares fitting provides the power law dependence $r_{ADC} = L_{Tot}^{0.88\pm0.16}$ at 99 per cent confidence, suggesting a simple proportionality. Dip ingress timing in the black hole binary Cygnus X-1 reveals

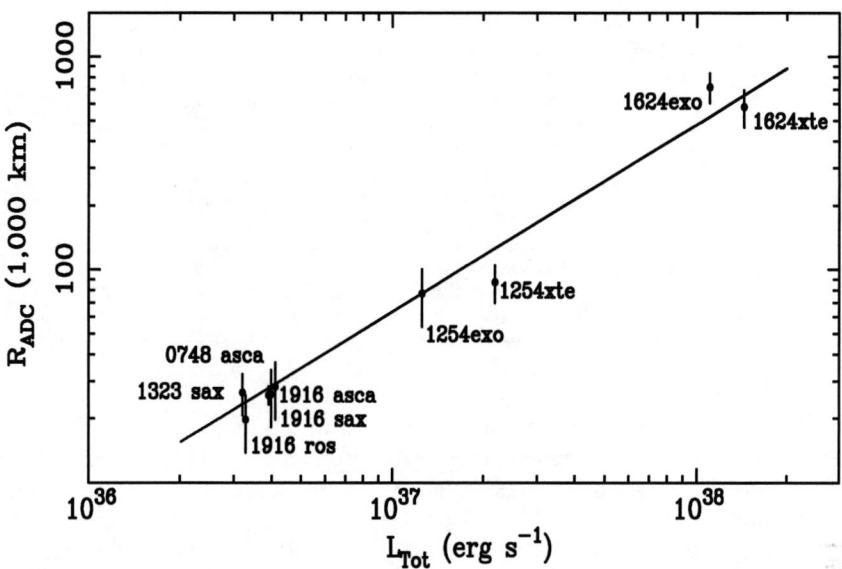

FIGURE 1. Measured values of ADC radius as a function of total source luminosity in the band 1 – 30 keV

that the ADC is not so large ($r_{ADC} \sim 5,000$ km) as in neutron star systems, but is still very extended [6].

Temperature of the ADC. We have also determined a mean electron temperature in the ADC plasma in several sources, as part of a continuing programme. Using broadband spectra from *BeppoSAX* extending to 100 keV, we have obtained the Comptonization cut-off energy and hence the mean electron temperature T_e of the Comptonizing ADC assuming either low optical depth τ or high τ. These temperatures suggest a probable mechanism that determines r_{ADC}. It is expected that the ADC will be in hydrostatic equilibrium providing kT is less than the gravitational binding energy. The critical radius at which equilibrium fails is the Compton radius $r_C \sim GMm_p/kT_e$, and for larger radii, the ADC will dissipate as a wind. In Table 1, we compare values of r_{ADC} and r_C. For the bright source X 1624-490 there is a good agreement for high optical depth, while for the other two, relatively faint sources XB 1916-053 and XB 1323-619, there is agreement for low optical depth. We thus have limited evidence that the size of the ADC is determined by hydrostatic equilibrium, i.e. the dependence on luminosity results primarily in change of kT_e which influences the size via the Compton radius.

FORMATION AND HEATING OF THE ADC

Formation of the ADC. The process by which the ADC is formed is not understood, and theoretical suggestions for formation mechanisms have been divided between in-

TABLE 1. Approximate values of the Compton radius or limiting radial size of an ADC in hydrostatic equilibrium for sources having well-determined E_{CO}. Ranges of electron temperature are given, with the lower limit for high optical depth in the corona, and the higher limit for low optical depth. Corresponding values of r_C are given in the next two columns and can be compared with approximate values of r_{ADC} (last column) from Fig. 1.

	E_{CO} keV	kT_e keV	r_C (High τ) 10^9 cm	r_C (Low τ) 10^9 cm	r_{ADC} 10^9 cm
XB 1916-053	80	26.7 - 80.0	7.2	2.4	3.0
XB 1323-619	44	14.7 - 44.0	13.1	4.4	2.7
X 1624-490	12	4.0 - 12.0	48.2	16.1	60

trinsic processes within the accretion disc, and extrinsic processes. Intrinsic processes, for example, involve mini magnetic loops [10] or instabilities [11] (see full discussion in [8]). External formation involves the central X-ray source (e.g. Fabian et al. [12], Jimenez-Garate et al. [13] and Różańska & Czerny [14]). Illumination leads to deposition of energy in the disc which is re-distributed vertically at all radii leading to a vertical structure of plasma density and temperature. The tenuous, hot upper layer may be identified with the ADC. Our results in Fig. 1 suggest that the illumination type of model is more likely by the very extended nature of the ADC.

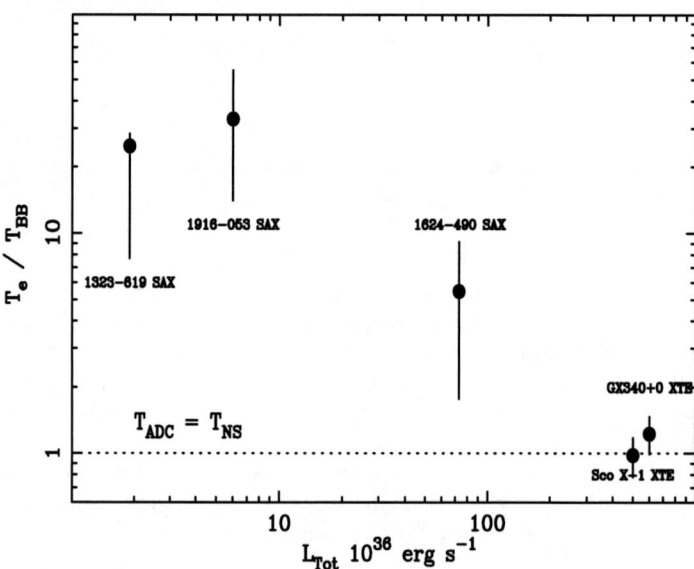

FIGURE 2. Temperature of the ADC relative to the blackbody temperature of the neutron star obtained from spectral fitting as a function of total source luminosity in the band 1 – 30 keV

Heating of the ADC. The high electron temperatures derived from broadband X-ray spectra conflict with simple theoretical expectations. In XB 1916-053 for example,

kT_e is at least 25 keV, and there is abundant evidence for similar high temperatures in other sources. If we assume that the neutron star provides the energy heating the ADC, the second law of thermodynamics would not allow kT_e to be larger than the neutron star temperature; i.e. about 1.5 keV (providing there is a direct transfer of energy between the two). If we argue that the heating originates in Compton processes, the temperature cannot be more than the inverse Compton temperature, i.e. about 3 keV. As kT_e is in fact substantially higher than these values, there must be a presently unknown heating mechanism. To establish this we have begun an observational program based on broadband spectra including data from *BeppoSAX* extending to $\sim$100 keV, and also from *Rossi-XTE* in which data from the PCA and HEXTE instruments can extend to $\sim$50 keV. Using *RXTE*, we can add measurements with the non-dipping sources GX 340+0 and Sco X-1 [15] to those shown in Table 1. With these broadband spectra we obtain both the ADC temperature and the neutron star blackbody temperature (T_{BB}), and then compare these, for a wide range of X-ray luminosities (L). Fig. 2 shows the results in the form of the ratio T_e/T_{BB} as a function of L. For brighter sources with L at about the Eddington limit, the cut-off energy of the Comptonized spectrum falls dramatically to $\sim$5 keV, implying an electron temperature $\sim$1.5 keV for an optically thick corona. Thus the temperature ratio is close to unity for the brightest sources, while in the fainter sources, the ADC is much hotter. It thus appears that some sort of equilibrium may be established in the more luminous sources; however, the elevated temperatures in fainter sources (i.e. Atoll sources) demand a heating mechanism that is presently unknown.

CONSEQUENCES OF THE LARGE ADC

The extended size of the ADC has several important consequences, relating to the controversy between emission models for LMXBs, the correct form of the Comptonization model in spectral fitting, and the formation of the ADC (already discussed). The large radial extent of the ADC rules out the Eastern model with its assumed small, central Comptonizing region associated with the atmosphere of the neutron star or with the inner disc. If we take the size of such a region to be less than 50 km, it can be seen that this is about 1000 times smaller than a typical measured value of r_{ADC}.

The extended size determines the nature of the seed photons that are Comptonized in the inner disc, since an extensive fraction of the inner accretion disc is covered by corona. Consequently, the seed photons consist predominantly of disc blackbody photons integrated from the inner disc edge to r_{ADC}, i.e. with a range of temperatures appropriate to the range of radial positions. At the outer radius, the blackbody temperature is quite low, the result being that the seed photon spectrum continues to rise with decreasing energy down to 0.01 keV or less. Because of the geometry of a flat ADC over a flat disc, there will be insignificant contribution of the neutron star to the seed photons.

When Comptonized, the spectrum is stretched between 0.01 – 100 keV having a form that is determined by the seed photon spectrum, i.e. a cut-off power law above 1 keV, the spectrum continuing to rise at energies below this (0.01 — 1.0 keV) (see the full discussion in [8]). The relevance of this is that it contrasts markedly with the form of the spectrum used in implementation of the Eastern model. In this case, it is assumed that the seed photons have simple blackbody form arising on the neutron star or inner disc with

$kT \sim 1$ keV. Thus the spectrum of both the seed photons and the Comptonized output *fall* sharply below ~ 3 keV. The extended ADC and the Eastern model have extremely different forms, and of course using an incorrect form in spectral fitting can lead to errors. The strongest differences between the extended ADC Comptonization model, and models (existing in XSPEC based on the Eastern model, will be in the energy band $0.1 - 3$ keV where the extended ADC model has very high flux but the Eastern model does not.

ACKNOWLEDGMENTS

This work was supported in part by the Polish KBN grant KBN-1528/P03/2003/25 and PBZ-KBN-054/P03/2001.

REFERENCES

1. N. E. White, L. Stella, and A. N. Parmar, *A&A*, **324**, 363–378 (1988).
2. K. Mitsuda, H. Inoue, N. Makamura, and Y. Tanaka Y., *PASJ*, **41**, 97–111 (1989).
3. N. E. White, and J. H. Swank, *ApJ*, **253**, L61–L66 (1982).
4. M. J. Church, and M. Bałucińska-Church, *A&A*, **300**, 441–445 (1995).
5. M. J. Church, and M. Bałucińska-Church, *A&A*, **369**, 915–924 (2001).
6. M. J. Church, *Adv. Space Res.*, **28**, 323–335 (2001).
7. M. J. Church, A. N. Parmar, M. Bałucińska-Church, T. Oosterbroek, D. Dal Fiume, and M. Orlandini, *A&A*, **338**, 556–562 (1998).
8. M. J. Church, and M. Bałucińska-Church, *Mon. Not. R. Astron. Soc.*, **348**, 955–963 (2004).
9. T. Narita, J. E. Grindlay, P. F. Bloser, and Chou Y., *ApJ*, **593**, 1007–1012 (2003).
10. A. A. Galeev, R. Rosner, and G. S. Vaiana, *ApJ*, **229**, 318–326 (1979).
11. Paczyński B., *Acta Astron.*, **28**, 241–251 (1978).
12. A. C. Fabian, P. W. Guilbert, R. R. Ross, *Mon. Not. R. Astron. Soc.*, **199**, 1045–1051 (1982).
13. M. A. Jimenez-Garate, J. C. Raymond, D. A. Liedahl, *ApJ*, **581**, 1297–1327 (2002).
14. A. Różańska, and B. Czerny, *Acta Astron.*, **46**, 223–252 (1996).
15. R. Barnard, M. J. Church, and M. Bałucińska-Church, *A&A*, **45**, 237–247 (2003).

Timing Neutron Stars

M. van der Klis

Astronomical Institute "Anton Pannekoek", University of Amsterdam

Abstract. This lecture deals with aspects of the rapid X-ray variability of low magnetic field neutron stars in low-mass X-ray binaries. The relations of the kilohertz quasi-periodic oscillations with orbital motion, neutron star spin, as well as with other, slower variability components are reviewed. Possible relations with the variability of black holes are discussed.

Keywords: X-rays, neutron stars, low-mass X-ray binaries, black holes, pulsars
PACS: 97.80.Jp, 98.70.Qy, 97.60.Gb, 97.60.Jd, 97.60.Lf

1. INTRODUCTION

Two prominent aims of X-ray binary studies are the direct determination of the properties of the strong gravitational fields near neutron stars and black holes, and of the supranuclear density matter in the interior of a neutron star. In the inner few kilometers of the accretion flow onto a low-magnetic-field neutron star or stellar-mass black hole of mass M the accreting matter is moving close to the Schwarzschild radius. This means we are looking at the motion of matter in strongly curved spacetime and can learn about gravitation in the strong-field regime, a regime where classical physics fails and general relativity has not yet been tested to any degree of confidence by means other than extrapolation from observations in weak gravity. In the case of a neutron star, the properties of the accretion flow constrain fundamental neutron star parameters such as mass and radius, and thereby (via stellar structure theory) the equation of state (EOS) of the ultradense matter at the star's core, whose properties are the subject of much speculation involving, among others, pion and kaon condensates, hyperons, and even strange matter. Measurements of the varying emission of hot spots on the surface of a spinning neutron star can likewise constrain both gravitation theory and neutron star parameters.

X-ray timing as well as spectroscopy can be used to study the dynamics of the inner accretion flows and the neutron star spins; for best results one wants to combine these techniques. Both X-ray binaries and active galactic nuclei (AGNs) are useful to study the accretion flow in the strong-field region; both types of system have advantages, and again for best results one wants to study both, and describe the results in one coherent picture. An important aspect of compact object studies that is unique to X-ray binaries is the possibility to compare systems that are similar except in the type of compact object they contain. A particularly interesting comparison is that between low-magnetic field neutron stars and black holes, which are both expected to allow gravity-dominated accretion flows down to well into the strong-field region. This comparison, at least in principle, allows to determine which of the observed properties are caused by the unique characteristics of a black hole, and which are in common with other accreting compact objects. Future additional techniques will allow to address these issues in new ways. One is direct imaging of the strong-field region, which becomes possible (in principle) initially in radio to IR/sub-mm wavelengths for the galactic center and AGNs, and only later in X-rays, as the latter requires the development of large baseline X-ray interferometers, which is very challenging indeed. Another is gravitational wave detection of accretion and merging events involving compact objects; ground based gravitational wave experiments such as LIGO and later space based ones such as LISA are directed towards this aim.

My charge for this review is to concentrate on neutron-star X-ray binary timing. I shall put particular emphasis on the millisecond time variability of the X-rays emitted by the accreting low magnetic field neutron stars in low mass X-ray binaries, as it is in these neutron stars that a plasma flow not dominated by the neutron star magnetic field extends down into the strong gravity region. The link of the timing phenomena observed in these neutron stars with those in stellar mass black holes, where the same basic physical situation applies, is also addressed. The emphasis on *millisecond* variability comes from the consideration that the dynamical time scales of stellar mass compact objects are of this order; hence the main characteristic time scales in the strong-field region, such as the orbital time scale, as well as the spins of weakly magnetic neutron stars spun up by accretion, tend to be of the order of milliseconds as well. I shall be very brief in this write-up on the topic of the data analysis of rapid X-ray variability and just mention that, certainly in the millisecond domain the analysis method of choice for detecting the variability and measuring its characteristic frequencies is the calculation of power spectra using Fourier technique (see, e.g., van der Klis 1989, van der Klis 2004) — examples of such power spectra are shown below. Narrow features in such power spectra are called quasi-periodic oscillations (QPOs) and broad ones are called noise components.

It is interesting to note that a number of millisecond phenomena that were predicted during the 1970's and 80's have all been discovered with the Rossi X-ray Timing Explorer (RXTE) satellite within a few years following its launch on December 30, 1995. Millisecond wavetrains due to clumps orbiting near the innermost stable circular orbit from general relativity were predicted by Sunyaev (1973) and it is likely that the kilohertz QPO signals discovered with RXTE in 1996 are caused by orbital motion effects of this kind. Short-lived periodic signals at the neutron star spin frequency during X-ray bursts were predicted by Livio & Bath (1982) and discovered with RXTE as burst oscillations also in 1996. Millisecond accreting pulsars were also predicted in the 1980's (Alpar at al. 1982, Radhakrishnan & Srinivasan 1984) shortly after the discovery of the millisecond radio pulsars and finally found with RXTE in 1998. The pulsations and most likely the burst oscillations are diagnostic of the millisecond neutron star spins; the various high frequency QPOs nearly certainly probe the accretion flow, likely, the inner disk. The RXTE Proportional Counter Array (PCA) remains the only instrument that can study these fundamental phenomena.

2. KILOHERTZ QUASI-PERIODIC OSCILLATIONS

The fastest variability components in X-ray binaries are the kilohertz quasi-periodic oscillations (kHz QPOs), which were discovered with RXTE in 1996 (van der Klis et al. 1996, Strohmayer et al. 1996) and occur in a wide variety of low magnetic-field neutron star systems. The phenomenon has been seen in more than 20 systems and does not occur in black-hole candidates — it is sufficiently characteristic to be considered a neutron star signature. Two QPO peaks (the 'twin peaks') occur in the power spectrum of the X-ray flux variations. They move up and down in frequency together in the 300–1200 Hz range in correlation with source state and often, luminosity. The higher-frequency one of these two peaks is called the 'upper kHz QPO', with a frequency ν_u, the lower-frequency one the 'lower kHz QPO' with frequency ν_ℓ; towards the edges of their observed frequency range, the peaks can also occur alone. The typically 300-Hz peak separation $\Delta \nu \equiv \nu_u - \nu_\ell$ usually decreases by a few tens of Hz when both peaks move up by hundreds of Hz (a few examples of a decrease in $\Delta \nu$ towards lower kHz QPO frequency have been reported as well). Weak sidebands to the lower kHz QPO have been reported in a number of sources (Jonker et al. 2000). See van der Klis (2000, 2004) for more extensive reviews of kHz QPOs.

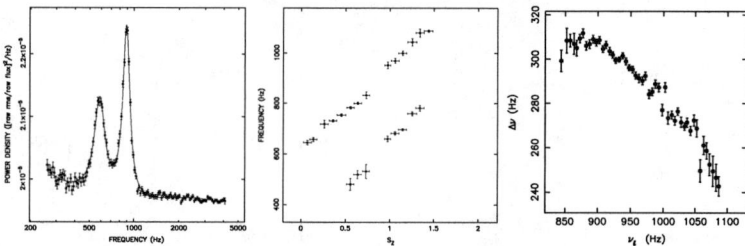

FIGURE 1. Left: Power spectrum showing twin kHz QPOs in Sco X-1. Middle: Twin kHz QPO frequency dependence on S_z, an $\dot{M}$ indicator, in GX 17+2. Right: The variation in kHz QPO peak separation as a function of the lower kHz QPO frequency in Sco X-1.

3. THE ORBITAL MOTION INTERPRETATION OF KHZ QPOS

kHz QPOs occur at the frequencies expected for orbits in the inner accretion disk and orbital motion at some preferred radius in the inner disk is an interpretation that underlies nearly all models for the phenomenon. If a kHz QPO peak at frequency ν corresponds to stable orbital motion around a neutron star, one can immediately set limits on neutron star mass M and radius R (Miller at al. 1998) from the constraints that the orbit must be outside both the star and the innermost stable circular orbit from general relativity (the ISCO, at $6GM/c^2$ in a Schwarzschild geometry). Fig. 2 shows these limits in the neutron star mass-radius diagram for several values of ν. The currently highest value of ν_u, identified in most models with the orbital frequency, is 1329±4 Hz (van Straaten at al. 2000), emperiling the hardest equations of state. For 1329 Hz these constraints imply $M < 1.65\,\mathrm{M}_\odot$ and $R < 12.4$ km; with corrections for frame dragging (shown to first order in j in Fig. 2) these numbers become $1.9\,\mathrm{M}_\odot$ and 15.2 km (van Straaten et al. 2000).

A specific model outlining how orbital motion at the inner edge of the disk could in fact modulate the X-rays was proposed by Miller et al. (1998). In their 'sonic point model' the inner edge of the Keplerian flow is at the sonic radius, r_{sonic}, where the radial inflow velocity becomes supersonic. This radius tends to be near the ISCO but radiative stresses change its location, as required by the observation that the kHz QPO frequencies vary. At r_{sonic} clumps form that orbit the star with orbital frequency v_{sonic}. Matter coming from these clumps accretes onto the neutron star following a fixed spiral-shaped trajectory in the frame corotating with the clumps' orbital motion (Fig. 3). At the 'footpoint' of a clump's spiral flow the matter hits the surface and the local accretion rate, and hence emission, is enhanced. The footpoint travels around the neutron star at the clump's orbital angular velocity, so irrespective of the star's spin rate the observer sees a hot spot move around the surface with frequency v_{sonic}. In the interpretation of Miller et al., this produces the upper kHz peak at v_u. The high Q of the QPO implies that all clumps are near one precise radius and live for several 0.01 to 0.1 s, and allows for relatively little fluctuations in the spiral flow.

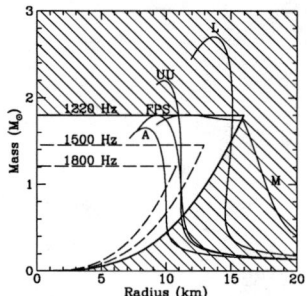 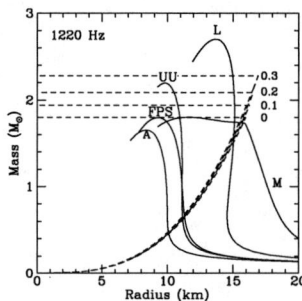

FIGURE 2. Constraints on neutron-star mass and radius from orbital motion. *Left:* for zero spin, orbital frequencies as indicated; the hatched area is excluded. *Right:* with first order corrections for frame dragging for the values of Kerr parameter j indicated. Mass-radius relations for representative EOS are shown. (from Miller et al. 1998)

In this model the frequency of the QPO is set by the inner radius of the disk, r_{sonic}. Miller at al. propose that this parameter in turn is determined by the balance between the gravitational and radiative stresses (mostly azimuthal drag and radial force) on the material in the disk. Normally, the kHz QPO frequency is seen to *in*crease when, presumably due to an increase in mass accretion rate $\dot{M}$, the luminosity L_x increases. In the model the increase in radiation reaching the inner disk edge from the stellar surface, which tends to make the inner radius larger and hence would make the frequency lower, is more than compensated by the scattering effects of the larger amount of accreting material between r_{sonic} and neutron star, so that the net effect is that when $\dot{M}$ rises the radius becomes smaller and the frequency increases in correlation with L_x, as observed. This implies that the model predicts that an increase in radiation *without* an increase in accretion rate should make the radius larger and hence the frequency lower. Work by Wenfei Yu (Yu & van der Klis 2002) provides some support for this interpretation: in 4U 1608–52 the QPO frequency was found to drop when the luminosity increased due to a process that was previously interpreted (Revnivtsev et al. 2001) as probably due to changes in the nuclear-burning rate on the neutron star surface.

One immediate prediction of the orbital interpretation is that, assuming general relativity, there should be an upper bound on the observed frequency set by the ISCO. In

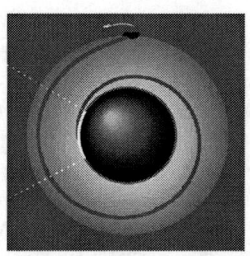 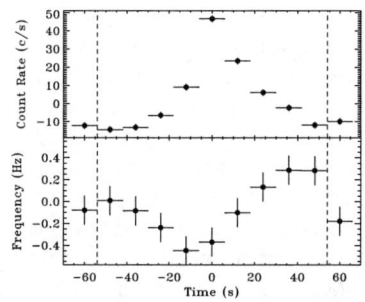

FIGURE 3. Left: The clump with its spiral flow and the emission from the flow's footpoint (dashed lines) in the Miller et al. (1998) model. Right: Light curve of milliHertz nuclear-burning oscillations (*top*) and associated kHz QPO frequency variations (*bottom*) in 4U 1608−52. The nuclear fueled X-ray emission from the neutron star surface increases the radius of the inner edge of the disk and thereby lowers the QPO frequency. This effect is *opposite* to that of variations in accretion-fueled emission observed simultaneously, where flux increases make the frequency go *up*. From Yu and van der Klis (2002).

this context, it is interesting to note that the maximum observed kHz QPO frequencies in well-studied sources are constrained to a relatively narrow range of $v_u = 1000$–1250 Hz, with only a few outlyers (the full range is 840 to 1329 Hz). If this is the ISCO frequency, then the neutron-star masses are near $2M_\odot$ (Zhang et al. 1997). If a magnetosphere limits the orbital frequencies, then to obtain similar maximum frequencies at very different accretion rates requires a rather tight correlation between magnetic field strength and accretion rate, for which there is no obvious reason (White and Zhang 1997).

4. RELATION OF KHZ QPOS WITH NEUTRON STAR SPIN — BURST OSCILLATIONS AND ACCRETING MILLISECOND PULSARS — BEAT FREQUENCY MODELS

Burst oscillations (see Chakrabarty, these proceedings) although exhibiting drifts of up to a few Hz, are generally interpreted as due to the neutron star spin (Strohmayer et al. 1996). The general picture for their formation mechanism is that some pattern of hot spots forms on the star's surface during the burst, which moves only slowly relative to the solid surface and hence spins around with the neutron star rotation. These burst oscillations have a frequency related to the kHz QPO peak separation Δv (they seem to occur near either Δv or $2\Delta v$). However, the correspondence with v_{spin} is not exact; discrepancies of several 10% have been reported (see van der Klis 2004 for an overview). Nevertheless, a beat frequency model for kHz QPOs, explaining the presence of the lower kHz QPO as due to a beat of the upper kHz QPO with the spin was explored in some detail (Miller et al. 1998). In this model the orbital frequency v_{orb} at the inner edge of the disk interacts with the neutron star spin v_{spin} to produce a third frequency, the beat frequency $v_{beat} = v_{orb} - v_{spin}$; v_{orb} is identified with v_u and v_{beat} with v_ℓ (see §2).

This model predicts $\Delta v = v_u - v_\ell$ to be constant at v_{spin}, contrary to observations.

However, as noted by Lamb and Miller (2001), if the clumps in their orbits at the inner edge of the disk gradually spiral down (slightly, on a trajectory not to be confused with the much faster downspiralling of the 'deorbited' matter in the spiral flows emanating from the clumps), then the observed beat frequency will be higher than the actual beat frequency at which beam and clumps interact, because then during the clumps' lifetime the travel time of matter from clump to surface gradually diminishes. Also the observed orbital frequency will be lower than the actual v_{orb} at the sonic radius, because the angle between clump and footpoint gradually becomes smaller. All this puts the lower kHz peak closer to the upper one, and thus decreases Δv, more so when at higher L_x due to stronger radiation drag the spiralling-down is faster, as observed (Lamb & Miller 2001). As the exact way in which this affects the relation between the frequencies is hard to predict, this complication makes testing the model more difficult.

However, the model still firmly predicts that $\Delta v \approx v_{spin}$. That Δv is sometimes near *half* the burst oscillation frequency v_{burst} then implies that in those cases $v_{burst} = 2v_{spin}$, which could result from the presence of two symmetric hot spots on the star, and predicts the presence of a subharmonic to v_{burst} at v_{spin}, i.e., in this interpretation, at $v_{burst}/2$. Such a subharmonic was reported on one occasion (Miller 1999), but not confirmed in further work (Strohmayer and Markwardt 2002).

The discovery of the first accreting millisecond pulsar SAX J1808.4–3658 (Wijnands and van der Klis 1998), followed by several more accreting millisecond pulsar discoveries (see Markwardt, these proceedings) finally provided the definitive confirmation of the long standing prediction that accretion induced spin up in a low-mass X-ray binary can produce a millisecond spin period neutron star (see §1). It facilitated two further breakthroughs when in 2002 twin kHz QPOs *and* burst oscillations were found in SAX J1808.4–3658. In this object $v_{pulse} = 401$ Hz, with an upper limit on the amplitude of the 200.5-Hz subharmonic, obtained from a coherent pulsation analysis, of <0.014% (see Wijnands et al. 2003). So very likely, $v_{spin} = 401$ Hz. This implies that the measurement of Δv at 196±4 Hz admits two important conclusions: (i) Δv *is* related to v_{spin}, but (ii) it *can* be $v_{spin}/2$. Conclusion (ii) is in direct conflict with the beat-frequency interpretation, which firmly predicts the commensurability should be with v_{spin}, not with $v_{spin}/2$. This conflict can not be resolved by assuming multiple hot spots on the star or multiple orbiting clumps, all of which only serves to make $v_{beat} = n(v_{orb} - v_{spin})$, where $n = 2, 3, \ldots$ (see Wijnands at al. 2003). The only way to save the model would be that $v_{spin} = v_{pulse}/2$, i.e., two hot spots exist on the star that are sufficiently symmetric for the amplitude at the pulse frequency subharmonic to be below the very low upper limit quoted above. Additionally one would have to assume that the clumps in the inner disk see only one of the two hot spots (then $v_{beat} = v_{orb} - v_{spin} \implies \Delta v = v_{spin} = 200.5$ Hz, and $v_{pulse} = 2v_{spin} = 401$ Hz.) All of this seems unlikely. This led Lamb & Miller (2003) to abandon their original beat-frequency model (where the beat interaction occurs at the inner disk edge with a radiation pattern rotating with the neutron star spin) and introduce a relativistic disk-spin resonance model instead, where a beat interaction also plays a role, but now it occurs at a resonant radius relatively far out in the disk, where a resonant wave traveling around the star at v_{spin} interacts with a radiation pattern rotating with the inner disk edge orbital frequency. In another millisecond pulsar, XTE J1807–294, twin kHz QPOs were found that have $\Delta v \approx v_{spin} = 191$ Hz (Markwardt 2004 priv. comm.);

clearly, just as is the case with the burst oscillation frequencies, the pulse frequencies, too, can be either $\Delta\nu$ or $2\Delta\nu$.

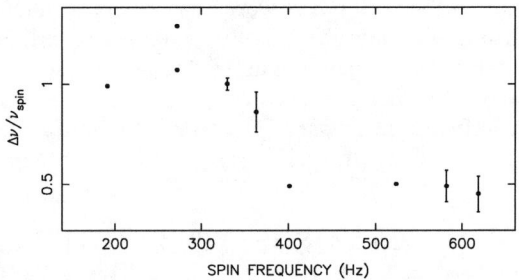

FIGURE 4. $\Delta\nu/\nu_{spin}$ vs. ν_{spin}, after van der Klis 2004.

Fig. 4 summarizes the current evidence for the spin – $\Delta\nu$ commensurability; the data available now seem to suggest that $\Delta\nu \approx \nu_{spin}$ for $\nu_{spin} \lesssim 400\,\text{Hz}$ and $\Delta\nu \approx \nu_{spin}/2$ for $\nu_{spin} \gtrsim 400\,\text{Hz}$. This might indicate that at high ν_{spin} some preferred radius in the disk required for generating a kHz QPO falls within the inner disk edge and hence is no longer active (Wijnands et al. 2003). Lamb & Miller (2003) in their spin resonance model provide another explanation, which makes use of the fact that at the 'spin-resonance' radius, where the beat interaction takes place, $\nu_{orb} \approx \nu_{spin}/2$. (This is because the resonance condition is $\nu_\phi + \nu_\theta = \nu_{spin}$, where $\nu_\phi \equiv \nu_{orb}$ is the azimuthal orbital frequency and ν_θ the vertical epicyclic frequency. At the radius where the resonance occurs $\nu_\phi \approx \nu_\theta$.) At this radius a wave pattern propagates around the star in step with ν_{spin}, but the matter itself orbits at $\sim \nu_{spin}/2$ and details in the clumpiness of the flow at the spin resonance radius determine which frequency dominates (at some level both are expected to occur).

The burst oscillation frequency in SAX J1808.4–3658 is $\sim 400\,\text{Hz}$, i.e., approximately equal to the spin frequency, demonstrating that $\Delta\nu \approx \nu_{burst}/2$ does not necessarily imply $\nu_{burst} = 2\nu_{spin}$. Extending the interpretation $\nu_{burst} = \nu_{spin}$ to all observed cases, as had been previously argued on other grounds (Strohmayer & Markwardt 2002), and including the other accreting millisecond pulsars as well, 16 millisecond spins have now been detected in low-mass X-ray binaries (a further 5 low-mass X-ray binary, LMXB, neutron stars have much slower measured spins and presumably stronger B fields: GRO J1744–28, 2A 1822–371, Her X-1, 4U 1626–67, GX 1+4). The frequencies are in the range 270–619 Hz, which has been interpreted as evidence for a cut-off around 760 Hz, well below the limit set by observational constraints and indicating that a braking mechanism limits ν_{spin} (see Chakrabarty et al. 2003). If the stars spin at the magnetospheric equilibrium spin rates corresponding to their current L_x, this predicts a tight correlation between L_x and magnetic-field strength B (White & Zhang 1997; a similar possibility came up to explain the similar maximum kHz QPO frequencies at very different L_x). Another possibility is that gravitational radiation limits ν_{spin} by transporting angular momentum out as fast as accretion is transporting it in; this predicts these sources to be the brightest gravitational-wave sources, with a known ν_{spin} facilitating their detection (Bildsten 1998). From the kHz QPO sources, 10 more spins are under the same assumptions known up to a factor of 2; these are in the range (1 or 2)×(220–410 Hz). See van der Klis (2004) for a list of the frequencies.

I finally note that some caution is still indicated in interpreting the SAX J1808.4–3658 kHz QPO result, as the twin peaks were observed only once in this pulsar, its $\Delta \nu$ (195 Hz) is lower than in non-pulsars, and other commensurabilities also exist between the observed frequencies (see Wijnands et al. 2003). The frequency correlations of some pulsars are a factor $\sim$1.45 off the usual ones (see §5). Clearly, further detections of twin kHz QPOs in millisecond pulsars would help to clarify the systematics in this phenomenology.

5. RELATION OF KHZ QPOS WITH LOWER FREQUENCY PHENOMENA — RELATIVISTIC PRECESSION MODELS

Low-frequency (<100 Hz) QPOs and noise components in the rapid X-ray variability of LMXBs form a complex with several characteristic frequencies in the range 0.1–100 Hz, the 'low-frequency complex'. The components in the low-frequency complex all vary in frequency together, and usually in correlation with the kHz QPO frequencies. These frequency correlations are quite systematic. Fig. 5a displays the frequency correlations of four well-studied intermediate luminosity (atoll) and four weak LMXBs. The frequencies of all components are plotted vs. ν_u. The figure includes data from sources covering an order of magnitude in luminosity when in the same state, yet they display very similar power spectra and essentially the same frequency correlations. Tracks corresponding to the various variability components are clearly recognizable, indicating that the scheme of frequency correlations is to some extent universal. One component seems to vary little in frequency and remains in the 100–200 Hz range, hence it is called the 'hectoHz QPO'. This plot can usefully serve as a template against which to match the variability of other objects; the brightest LMXBs (Z sources) approximately match (Fig. 5b).

An interesting discrepancy occurs in the frequencies of some millisecond pulsars: in SAX J1808.4–3658 a pattern of correlated frequencies occurs very similar to Fig. 5a, but with relations that are offset from the usual ones. At low frequencies where the relations are clearest, the match can be restored by multiplying the ν_u (and ν_ℓ) values with $\sim$1.45, i.e., close to 1.5, which suggests a link with the 2:3 frequency ratios in black holes (van Straaten et al. 2004, see §6). Of the other millisecond pulsars, XTE J0929–314 and XTE J1807–294 behave in a similar way, but XTE J1751–305 and XTE J1814–338 are like ordinary atoll sources in this respect. This suggests the kHz QPOs form one group of correlated frequencies and the low frequency complex another, independent one; the case of GX 17+2 above supports this.

The observations of these frequency correlations, and the difficulties encountered by the beat-frequency interpretation described in the previous section, led to the suggestion that the observed frequencies are directly related to general-relativistic orbital frequencies. The term relativistic precession model (Stella & Vietri 1998) is used for these models. In this interpretation the observed frequencies are directly identified with orbital, epicyclic, and precession frequencies. The models need additional physics to pick out one or more preferred radii in the disk whose frequencies correspond to those observed. Stella & Vietri (1998, 1999) identify the upper kHz QPO frequency ν_u with the azimuthal orbital frequency ν_ϕ at the inner edge of the disk, and relate ν_ℓ and ν_h with,

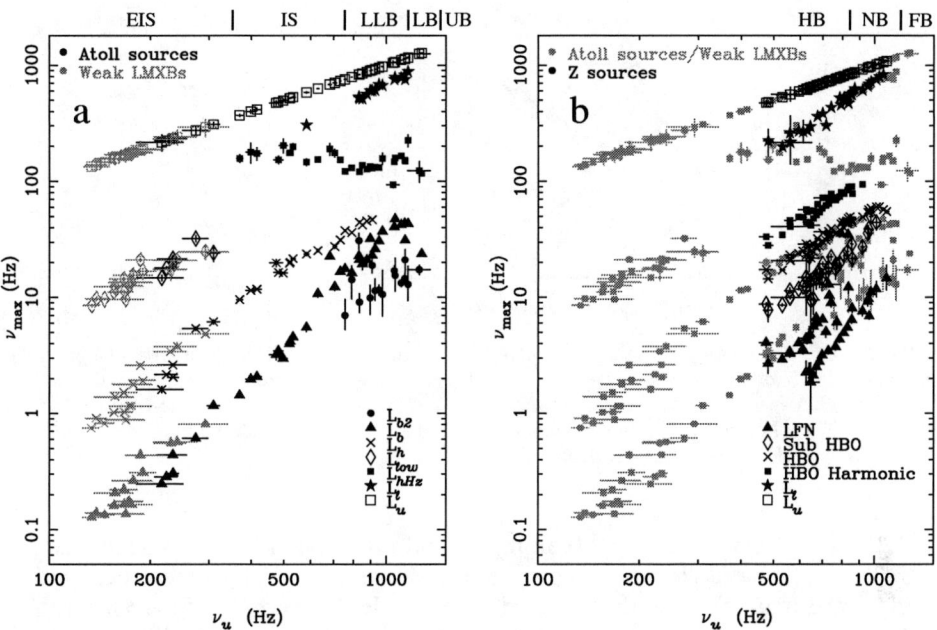

FIGURE 5. Frequency correlations. *(a)* Atoll sources and weak LMXBs, *(b)* Z sources compared with these objects. The characteristic frequencies of the components are plotted as indicated; approximate source state ranges are indicated at the top.

respectively, periastron precession (v_{peri}) and nodal precession (v_{nodal}) of this orbit.

So, as v_h in this interpretation is the Lense-Thirring precession frequency and v_u the corresponding orbital frequency, v_h is predicted to be proportional to v_u^2, which is indeed as observed (Fig. 6). However, the relation seems to be the same for neutron stars with very different spin frequency, whereas the relativistic precession model predicts them to all have different normalizations. The other prediction is that the kHz QPO peak separation $\Delta v \equiv v_r$, the radial epicyclic frequency. Stellar oblateness affects the precession rates and must be corrected for (Morsink & Stella 1999, Stella et al. 1999). This model does not explain why Δv is commensurate with the spin frequency, nor how neutron stars with different spins can have the same v_u vs. v_h relation (van Straaten et al. 2003). A clear prediction is that Δv should decrease not only when v_u increases (as observed) but also when it sufficiently decreases. There are observational indications for this, but the match of observed Δv vs. v_u relations to the predictions is not good. For acceptable neutron star parameters ($I_{45}/m = 0.5 - 2$), v_h is several times larger than the v_{nodal} predicted; perhaps this could be resolved by noting that in a warped disk geometry v_h could be $2v_{nodal}$ or $4v_{nodal}$ (Morsink & Stella 1999).

Relativistic precession models are very predictive, as the frequency relations are set by little more than compact-object parameters and general relativity, and in unmodified form most are contradicted by observations (e.g., Homan et al. 2002, van Straaten et

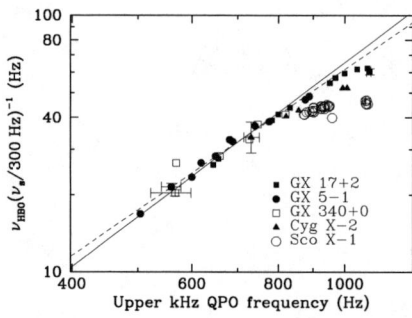
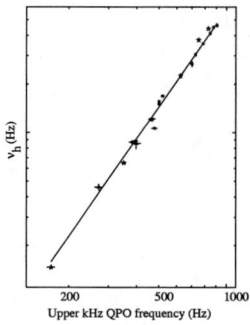

FIGURE 6. The relation between upper kHz QPO frequency and, *left*: horizontal-branch oscillation frequency in Z sources (scaled between sources by an inferred spin frequency), drawn line is for a quadratic relationship, dashed line is best power-law fit, *right*: ν_h in atoll sources, line is power law with index 2.01. In a Lense-Thirring interpretation a value of I_{45}/m of ~4 would be implied in both cases. From Psaltis et al. (1999b) and van Straaten et al. (2003).

al. 2004). A precise match between model and observations requires additional free parameters; Stella & Vietri (1999) propose that orbital eccentricity systematically varies with orbital frequency. Yet the observed quadratic dependencies between ν_u and ν_h are striking. Some disk oscillation models are able to produce some of the free particle frequencies used in relativistic precession models (Psaltis 2000, Psaltis & Norman 2000 and, e.g., Wagoner 1999). Non-relativistic disk oscillation models have been explored as well (e.g., Osherovich & Titarchuk 1999; see Miller 2003 for a critical discussion).

6. RELATION TO BLACK HOLE TIMING AND STATES – RELATIVISTIC RESONANCE MODELS

Just like low magnetic field neutron star low-mass X-ray binaries, black hole X-ray binaries show different source states (recurrent patterns of correlated X-ray spectral and timing properties). In black holes (like in neutron stars) source state is not well correlated to X-ray luminosity nor, presumably, mass accretion rate. The power spectra of black holes can be remarkably similar to those of neutron stars, particularly in hard spectral states (which despite what was just said about luminosity seem to be most common at low L_x and are known in black holes as the low hard state, LS, and in neutron stars as the 'extreme island state', EIS). Finally, just like in neutron stars, in black holes several different power spectral components are found whose frequencies vary in correlation. The frequency correlations found in black holes to a considerable extent coincide with those seen in neutron stars, which suggests similar timing phenomena are seen in these two types of systems.

The issue of whether we are seeing the same phenomena in neutron star and black hole systems is a very important one, as a phenomenon occurring in both types of system can not rely on any property unique to either type: the presence or absence of a solid surface, a horizon, a non-aligned magnetic field, spinning surface hot spots or frame dragging as strong as around near-extremal Kerr black holes can then all be excluded as ingredients for their formation. This leaves essentially only phenomena in the accretion flow (most likely the disk although there are other possibilities, e.g., a jet) for their explanation.

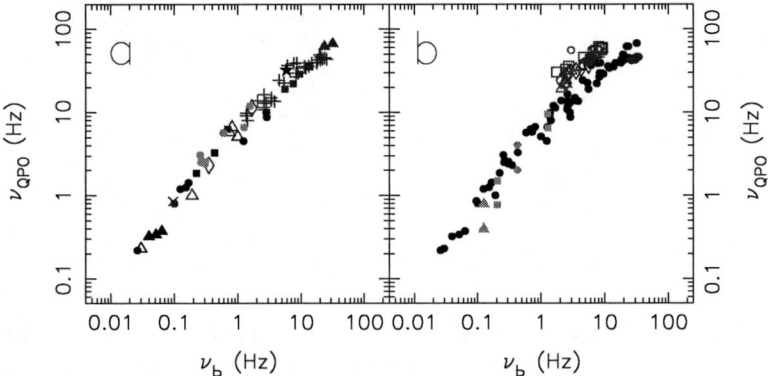

FIGURE 7. Frequencies of power spectral break and QPO are well-correlated across neutron stars and black holes. From Wijnands and van der Klis 1999.

As noted above, similar frequency correlations are seen between phenomena covering a wide range in coherence and frequency in both neutron stars and black holes. Wijnands & van der Klis (1999) noted that in atoll sources (including weak LMXBs) and black holes the characteristic frequency of the band-limited noise ν_b and of a hump or QPO often found above this break, ν_h, are correlated over 3 orders of magnitude (Fig. 7). This 'WK' relation is that between the two lower traces in Fig. 5. Psaltis et al. (1999a) were able to similarly select a set of variability components, which I shall refrain from fully describing here, from neutron stars and black holes that seem to define a common frequency correlation spanning nearly three decades in frequency, with the high and intermediate luminosity neutron stars populating the high-frequency range and the weak neutron stars and black holes in the low hard state the low-frequency one, and Cir X-1 filling in the gap in between (Figs. 8). This 'PBK' correlation combines features from different sources with very different Q values with relatively little overlap, and, as Psaltis et al (1999a) note, although the data are suggestive, they are not conclusive.

Further work produced many examples of power spectra confirming these correlations but as there is no direct observation of a gradual transition, the identification of the high-Q lower kHz QPO in the more luminous neutron stars with some of the low-Q components in the neutron-star and black-hole low states remains conjectural.

The relation of Figs. 7 and 8 suggest that physically similar phenomena cause the frequencies plotted there. If so, then these phenomena are extremely tunable, in some cases over nearly three orders of magnitude in frequency, and occur in neutron stars as well as black holes, which as mentioned above probably means they arise in the disk. The relativistic precession models already discussed above are one possible way to do this, although to actually match the observed frequencies for reasonable compact object parameters seems difficult with these models in their initial form.

A further problem to models of this type would seem to be the fact that these correlations may even extend to accreting white dwarfs (in cataclysmic variables), as proposed

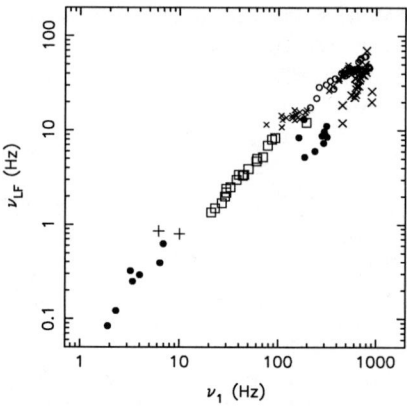

FIGURE 8. PBK relation (after Psaltis et al. 1999a) — see text. Filled circles represent black-hole candidates, open circles Z sources, crosses atoll sources, triangles the millisecond pulsar SAX J1808.4−3658, pluses faint burst sources and squares Cir X-1.

by Warner and Woudt (2002). If so, then by similar reasoning as above this would seem to exclude large amplitude general relativistic effects as a viable mechanism for the timing phenomena involved in the correlations (even then, orbital motion in the strong-field regime would still be implied in the neutron star and black hole cases). It is important to note, however, that even if the observation that the black holes, neutron stars, and white dwarfs all produce frequencies that follow the same correlations means that the same physics underlies the *correlations*, this does not strictly require that all *frequencies* participating in the correlations have similar physical origins in all three types of compact object. An alternative possibility is for example that it is a property of accretion disks to, given one oscillation frequency, produce a second one matching the correlation. A physical phenomenon that occurs in all compact object types (the common frequency-frequency correlation) would then derive from this accretion-disk property, which is not unique to any compact object type, while unique compact object properties (e.g., strong-field gravity effects in neutron stars and black holes) might well be involved in generating the original disk oscillation in the first place. This possibility was alluded to by Abramowicz et al. (2004) using the example of the 'ninth wave' phenomenon. It is good to keep in mind that these wide-ranging relations rely on identifying frequencies of different phenomena (QPOs as well as noise) in different sources with one another – eventually they need to be confirmed by detailed studies of the properties of these phenomena confirming that they *can* in fact be attributed to similar physical effects.

After the work by van Straaten on neutron stars (§5) first study of this kind including black holes was performed by Klein-Wolt (2004), who demonstrates the similarity between neutron stars and black holes in the low-hard/ extreme-island states can be used as a starting point to identify further correspondences between neutron stars and black holes in other states, and that further matches between neutron star and black hole frequency-frequency correlations can be found based on these correspondences.

Relativistic resonance models have been proposed for the black hole high-frequency

QPOs in which general relativity picks out particular radii in the disk as preferred radii, because only at these radii a pair of general relativistic orbital and epicyclic frequencies have a small integer ratio (e.g., $v_r/v_\theta = 2:3$) so that a resonant oscillation occurs in the flow at that radius. As noted above, the periodic forcing of the disk by the neutron star spin by means of magnetic or radiative stresses creates the potential for additional resonances in the flow, and several proposals have been made attempting to explain the neutron star kHz QPO frequency commensurabilities with neutron star spin in terms of such resonances (Wijnands et al. 2003, Kluzniak et al. 2004, Lamb and Miller 2003).

7. CONCLUSIONS

With RXTE we are observing X-ray variability on the dynamical time scale of the accretion flow in the inner, strong field gravity region. It seems likely that we are seeing the effect of orbital motion in that region, which implies we can put constraints on the EOS and can expect to see strong-field gravity effects in the properties of the variability that is detected. Most models do indeed involve strong field gravity as one of their ingredients. As noted in §6, the fact that frequency-frequency relations may exist covering black holes and neutron stars as well as white dwarfs is not in itself a sufficient argument to exclude strong field gravity effects for every phenomenon involved.

The expected millisecond spin frequencies of the low magnetic field neutron stars in low-mass X-ray binaries have been found, with a total of 26 neutron stars now determined to have millisecond spins from pulsations, burst oscillations and/or kHz QPOs. There are commensurabilities between the spin frequency and the QPO frequencies which are of such a nature that ordinary 'kinematic' beat frequency models can not explain them, and that instead require that matter in the disk is forced to move in step with the spin, presumably by either magnetic or radiative forcing of the disk flow by the star's spin. Most likely this involves some kind of resonant disk-star interaction, perhaps by the relativistic resonance mechanism where at a preferred radius in the disk one of the general relativistic epicyclic or precession frequencies resonates with the spin frequency.

The RXTE PCA remains the only instrument that can study the aperiodic millisecond phenomena discussed here. Much is still expected of this workhorse of X-ray astronomy. Hopefully, India's ASTROSAT can extend some of these capabilities into the future. Further breakthroughs can be predicted when larger area instruments become available. It will then become possible to directly relate the properties of the relativistically broadened Fe lines to those of the rapid variability and, with sufficient effective detector area, to see the Fe lines vary as a function of that variability. This will enormously enhance our understanding of the motions in the strong-field region as variability frequencies and line profile provide two complementary measurements of the same motions. In the simplest description these diagnose, respectively, the periods and the velocities of the same orbital motion: for example a luminous clump orbiting the compact object would be detected not just as a QPO signal but also as a feature periodically moving up and down the line profile. The ability to detect much weaker QPOs will allow to measure the weak sideband patterns to the main QPO peaks that are predicted to occur in, and are specific to, each model, discriminating between models. It will also allow to extend the frequency ranges over which the QPOs can be detected. Finally, it will become possible to take fundamentally superior approach to aperiodic millisecond timing by measure the QPO signals within their coherence time, which will essentially bring all the advantages of coherent analysis now only available for pulsations to QPOs. This will provide major additional constraints on the models.

REFERENCES

Abramowicz, M.A., Kluźniak, W., Stuchlik, Z., & Torok, G. 2004, astro-ph/0401464
Alpar, M. A., Cheng, A. F., Ruderman, M. A., & Shaham, J. 1982, *Nature*, 300, 728
Bildsten, L. 1998, *ApJ*, 501, L89
Chakrabarty, D., Morgan, E. H., Muno, M. P., Galloway, D. K., et al. 2003, *Nature*, 424, 42
Homan, J., van der Klis, M., Jonker, P. G., Wijnands, R., Kuulkers, E., et al. 2002, *ApJ*, 568, 878
Jonker, P. G., Méndez, M., & van der Klis, M. 2000, *ApJ*, 540, L29
Klein-Wolt, M., 2004, PhD thesis, Univ. of Amsterdam
Kluźniak, W., Abramowicz, M. A., Kato, S., Lee, W. H., & Stergioulas, N. 2004, *ApJ*, 603, L89
Lamb, F. K. & Miller, M. C. 2001, *ApJ*, 554, 1210
Lamb, F. K. & Miller, M. C. 2003, astro-ph/0308179
Livio, M. & Bath, G. T. 1982, *A&A*, 116, 286
Miller, M. C. 1999, *ApJ*, 515, L77
Miller, M. C. 2003, in Rossi and beyond, AIP Conf. Proc. 714, 365; astro-ph/0312449
Miller, M.C., Lamb, F.K., & Psaltis, D. 1998, *ApJ*, 508, 791
Morsink, S. M. & Stella, L. 1999, *ApJ*, 513, 827
Osherovich, V. & Titarchuk, L. 1999, *ApJ*, 522, L113
Psaltis, D. 2000, astro-ph/0010316
Psaltis, D., & Norman, C. 2000, astro-ph/0001391
Psaltis, D., Belloni, T., & van der Klis, M. 1999a, *ApJ*, 520, 262
Psaltis, D., Wijnands, R., Homan, J., Jonker, P.G., van der Klis, M., et al. 1999b, *ApJ* 520, 763
Radhakrishnan, V. & Srinivasan, G. 1984, Second Asian-Pacific Regional Meeting on Astronomy, 423
Revnivtsev, M., Churazov, E., Gilfanov, M., & Sunyaev, R. 2001, *A&A*, 372, 138
Stella, L. & Vietri, M. 1998, *ApJ*, 492, L59
Stella, L. & Vietri, M. 1999, Physical Review Letters, 82, 17
Stella, L., Vietri, M., & Morsink, S. M. 1999, *ApJ*, 524, L63
Strohmayer, T. E. & Markwardt, C. B. 2002, *ApJ*, 577, 337
Strohmayer, T.E., Zhang, W., Swank, J.H., Smale, A., Titarchuk, L., & Day, C. 1996, *ApJ*, 469, L9
Sunyaev, R. A. 1973, Soviet Astronomy, 16, 941
van der Klis, M. 1989, in Timing Neutron Stars, NATO ASI C262, p. 27
van der Klis, M. 2000, *ARA&A*, 38, 217
van der Klis, M. 2004, in Compact Stellar X-ray Sources, Lewin and van der Klis (eds.), CUP in press
van der Klis, M., Swank, J.H., Zhang, W., Jahoda, K., Morgan, E.H., et al. 1996, *ApJ*, 469, L1
van Straaten, S., Ford, E. C., van der Klis, M., Méndez, M., & Kaaret, P. 2000, *ApJ*, 540, 1049
van Straaten, S., van der Klis, M., & Méndez, M. 2003, *ApJ*, 596, 1155
van Straaten, S., van der Klis, M., & Wijnands, R. 2004, astro-ph/0410505
Warner, B. & Woudt, P. A. 2002, *MNRAS*, 335, 84
Wagoner, R. W. 1999, Phys. Rep., 311, 259
White, N.E., & Zhang, W. 1997, *ApJ*, 490, L87
Wijnands, R. & van der Klis, M. 1998, *Nature*, 394, 344
Wijnands, R. & van der Klis, M. 1999, *ApJ*, 514, 939
Wijnands, R., van der Klis, M., Homan, J., Chakrabarty, D., et al. 2003, *Nature*, 424, 44
Yu, W. & van der Klis, M. 2002, *ApJ*, 567, L67
Zhang, W., Strohmayer, T.E., & Swank, J.H. 1997, *ApJ*, 482, L167

X-ray studies of three binary millisecond pulsars

N.A. Webb*, J.-F. Olive* and D. Barret*

Centre d'Etude Spatiale des Rayonnements, 9 avenue du Colonel Roche, 31028 Toulouse, France

Abstract. It is thought that millisecond pulsars with white dwarf companions are born from X-ray binaries. The majority of known systems have been studied uniquely in the radio domain, which limits our understanding of such systems. We present here the X-ray observations of the millisecond pulsar PSR J0218+4232 and the two faint millisecond pulsars PSR J0751+1807 and PSR J1012+5307, which we discuss in conjunction with radio observations. We confirm the previously detected X-ray pulsations of PSR J0218+4232 and we show that its folded lightcurve is strongly dependent on energy. We present evidence to suggest that the broad band X-ray spectrum for this pulsar may not be a simple power law, but that there is some evidence for an excess of soft thermal emission over the power law spectrum, in particular from the strongest pulse, in support of a heated polar cap model for this pulsar. We also present the X-ray spectra of the two faint millisecond pulsars as well as some evidence to suggest that both of these millisecond pulsars show pulsations in the X-ray band. We then discuss the implied nature of the magnetic field configuration as a means of discriminating between competing magnetic field evolution theories in millisecond pulsars.

Keywords: X-rays: stars – pulsars: individual: PSR J0218+4232; PSR J0751+1807; PSR J1012+5307 – Radiation mechanisms: non-thermal; thermal – Magnetic fields
PACS: 97.60.Gb

INTRODUCTION

Binary millisecond pulsars (MSPs) found in the field are thought to be 'recycled' pulsars. Accretion onto the pulsar from a close companion is believed to transfer angular momentum to the neutron star, spinning it up to periods of milliseconds [1, 2]. These short period pulsars have long apparent ages and pulsed optical, X-ray, and γ-ray fluxes significantly below those expected for canonical pulsars with similar periods [1]. These millisecond pulsars also have very low spindown rates ($\dot{P}$) and consequently low surface magnetic fields (B_s), as the surface magnetic field is believed to be proportional to $(\dot{P}P)^{0.5}$. This can be seen clearly in, for example [3], their figure 1.

What is unclear, however, is how the surface magnetic field can evolve from high values ($\sim 10^{11}$-10^{13} G) to lower B_s values of only $\sim 10^8$-10^9 G [see 4]. Several mechanisms have been proposed. Renewed accretion, which is thought to spin up the pulsar, could also bury the magnetic field [5], reducing the strength of the observed field. This method would also systematically increase the neutron star mass. Alternatively during spin up, the surface magnetic field evolution could mirror changes in the core magnetic field configuration. The strong interaction between the core's superfluid neutrons, arranged into very dense almost parallel arrays of quantized vortex lines, and its magnetized superconducting protons provide the conditions necessary to produce the pulsar's high magnetic field. During spin up the vortex lines move inwards, 'dragging' the surface magnetic field to the pole of the hemisphere it started in, which would account for both the change in the magnetic field configuration and its strength ('spin-up squeezed' model) [6]. This

CP797, *Interacting Binaries: Accretion, Evolution, and Outcomes*,
edited by L. Burderi, L. A. Antonelli, F. D'Antona, T. Di Salvo,
G. Luca Israel, L. Piersanti, A. Tornambè, and O. Straniero
© 2005 American Institute of Physics 0-7354-0286-8/05/$22.50

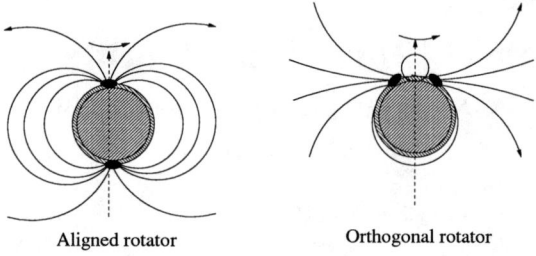

FIGURE 1. Magnetic configurations of 'spun-up squeezed' MSPs (adapted from [7]). The size of the crust has been exaggerated. **Left:** Aligned rotator. **Right:** Orthogonal rotator.

would imply that a strongly spun up MSP would have *either* an orthogonal *or* an aligned magnetic field configuration (see Fig. 1 for a schematic of these configurations).

Of the hundred or so MSPs known [see the latest version of the ATNF Pulsar Catalog, 4] the large majority have been detected in the radio domain only. Just a dozen of these rotationally powered MSPs have been detected in the X-ray domain. MSPs are interesting objects to study, especially in the X-ray domain, as the X-ray emission is the result of highly accelerated particles in a strongly magnetised environment. Thermal emission can be from a heated region, most likely the polar cap, which is bombarded by high energy particles. Non thermal emission, characterised by hard power law spectra, is thought to arise from the magnetosphere. Alternatively, the emission can be from a pulsar driven synchrotron nebula or interaction of relativistic pulsar winds with either a wind from a close companion star or the companion star itself. Thus we can use X-ray observations to understand both the surface and the environment of the pulsar and eventually to determine unknown quantities, such as the orientation of the magnetic field and the equation of state of the matter at the surface of the neutron star.

OBSERVATIONS AND DATA REDUCTION

We observed three MSPs, PSR J0218+4232, PSR J0751+1807 and PSR J1012+5307, with the X-ray observatory *XMM-Newton*. The observations were made using the MOS cameras in full frame mode and the pn in timing mode, all three configurations used the thin filter. The data were reduced in the standard manner, using Version (V.) 5.3.3 of the *XMM-Newton* SAS (Science Analysis Software) for the PSR J0218+4232 MOS data and V. 5.4.1 for the MOS data of the other two MSPs. For the pn data we took advantage of the development track version of the SAS, where improvements have been made to the *oal* ('Observation Data File' access layer) task (V. 3.106) [8], see [9, 10] for details.

RESULTS

In [9] we confirmed the previously detected X-ray pulsations of PSR J0218+4232 and showed that the Full Width at Half Maximum (FWHM) of the fairly broad pulses are $\delta\phi_1 = 0.112 \pm 0.038$ for the pulse centred at phase $\phi_1 = 0.242 \pm 0.008$ and

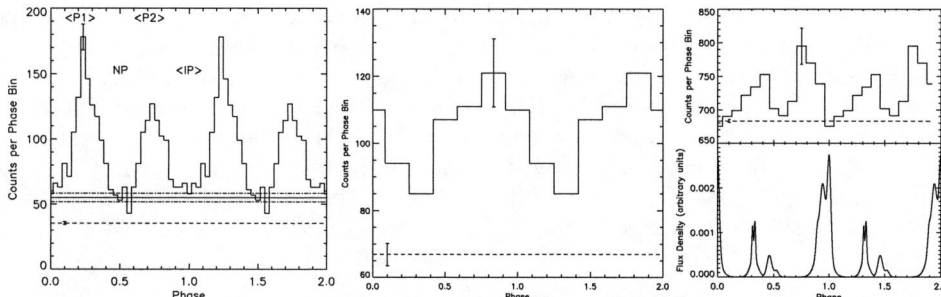

FIGURE 2. **Left:** PSR J0218+4232 X-ray lightcurve (0.6-12.0 keV). Pulses 1 (P1) and 2 (P2) are indicated. Between them is the non-pulsed region (NP). Between pulses 2 and 1 is the interpulse (IP). The solid line shows the DC level and the dashed-dotted lines give the $\pm 1\sigma$ error. (see [9] for more details). **Middle:** PSR J0751+1807 X-ray lightcurve (0.6-7.0 keV). **Right:** Upper panel: PSR J1012+5307 X-ray lightcurve (0.6-5.0 keV). Lower panel: PSR J1012+5307 radio profile (see [10] for more details). **In each panel** two cycles are plotted. The error bars are $\pm 1\sigma$. The dashed line shows the background level.

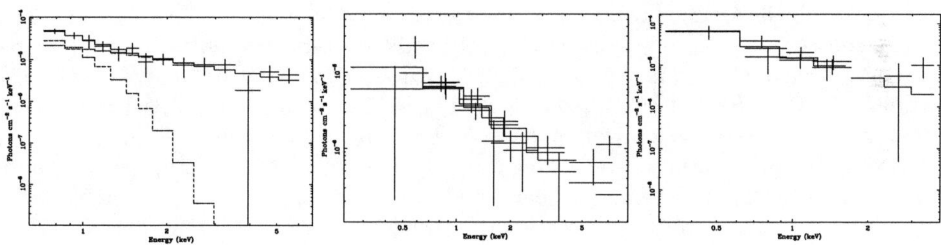

FIGURE 3. **Left:** X-ray spectrum for pulse 1 minus the DC and background contributions of PSR J0218+4232. The solid line shows the best blackbody plus power law fit and the two components are shown with dashed lines (see [9] for more details). **Middle:** X-ray spectrum for PSR J0751+1807 (see [10] for more details). **Right:** X-ray spectrum for PSR J1012+5307 (see [10] for more details).

$\delta\phi_2 = 0.121 \pm 0.056$ for the pulse centred at $\phi_2 = 0.733 \pm 0.014$ (see Fig. 2, left hand side). We also showed that the folded lightcurve is strongly dependent on energy. We presented evidence to suggest that the broad band X-ray spectrum for this MSP may not be a simple power law. We found that the XMM-Newton spectra were well fitted with a two component model (blackbody plus power law (see Fig. 3, left hand side), in the same way as for PSR J0437-4715 [11]). In doing this we also recover the same power law photon indices found by [12], when fitting BeppoSAX data between 2 and 10 keV, which supports the assertion that there is additional emission below 2 keV, not observable with BeppoSAX. Taking the blackbody temperature $(2.90 \pm 0.70 \times 10^6$ K$)$ from the fit to pulse 1, we find a radius of emission of 0.37 ± 0.33 km (90% confidence).

In [10] we showed for the first time the broad-band X-ray spectra of the two faint MSPs, PSR J0751+1807 and PSR J1012+5307. We found that a power law model best fits the PSR J0751+1807 spectrum, $\Gamma = 1.59 \pm 0.20$ (see Fig. 3, middle panel). A power law is also a good description of the spectrum of PSR J1012+5307, $\Gamma = 1.78 \pm 0.36$ (see

Fig. 3, right-hand side), however, a blackbody can not be excluded as the best fit to this data. We also provided some evidence that both of these MSPs show pulsations in the X-ray band. PSR J0751+1807 shows a single, broad pulse per period (see Fig. 2, middle panel). Fitting the lightcurve with a Lorentzian [as 13] we found that the FWHM of the pulse is $\delta\phi_1$=0.311±0.1, centred at phase ϕ_1=0.38±0.04 (errors are 90% confidence). PSR J1012+5307, however, shows two fairly narrow pulses per period (see Fig. 2, right-hand side). Fitting the lightcurve with two Lorentzians, the FWHM of the pulses is $\delta\phi_1$=0.10±0.07, centred at ϕ_1=0.36±0.04 and $\delta\phi_2$=0.09±0.03, at ϕ_2=0.80±0.03.

MAGNETIC FIELD CONFIGURATION

Recently, [14] examined the six MSPs for which both X-ray and radio lightcurves were known, to provide evidence to support the theory that recycled MSPs should have mainly orthogonal or aligned magnetic field configurations, which should result if the 'spin-up squeezed' theory for the magnetic field evolution proposed by [6] is correct. Of these six, he concluded that five (PSR 1821-24, PSR 1937+21, PSR J0030+0451, PSR J2124-3358 and PSR J0218+4232) showed evidence for being orthogonal rotators. The evidence for this assertion was that these pulsars showed two radio sub-pulses of comparable strength separated in time by about half a period. The X-ray lightcurves were also similar. However PSR 1821-24 and PSR J0218+4232 showed extra sub-peak structure in the radio lightcurves, which were not seen in the X-ray lightcurves. [14] concludes that this structure is the result of the double crossing of a hollow cone beam, rather than the emission from three truly separate beams and that this hypothesis is supported by the X-ray lightcurves which do not show this extra structure. The final MSP, PSR J0437-4715, is the only MSP of the six showing a single very broad X-ray lightcurve and is thought to be an almost aligned beam (and dipole moment). The X-ray spectrum is soft, indicative of a heated polar cap, which one would expect to see in an aligned rotator configuration.

The orthogonal rotator hypothesis for PSR J0218+4232 is in conflict with both [15] and [16] who state that this MSP is an aligned rotator using the fact that almost two thirds of the soft X-ray (0.1-2.4 keV) emission is non-pulsed [15] and using polarimetric radio observations showing magnetic inclinations consistent with 0° at both 410 and 610 MHz [16]. As stated above, we found that the XMM-Newton spectra were well fitted with a two component model (blackbody plus power law), in support of a heated polar cap model for this pulsar. The expected area of the surface polar cap with a conventional central dipole is $A_{pc} \sim \pi\Omega R^3/c$, where Ω is $2\pi/pulse\ period$, R is the typical radius of the pulsar, assumed to be 10^6cm as [17] and c is the speed of light in a vacuum. This gives a polar cap radius of 3.0 km, much greater than the radius found through fitting the X-ray spectrum. However, the predicted polar cap area for a 'spun-up squeezed' MSP in an aligned configuration is $A_{pc} \sim \pi\Omega R\Delta^2/c$, where Δ is the thickness of the crust, typically $0.1R$, following [17]. This gives a polar cap radius of 300 m, consistent with the observationally determined radius. This adds support to the aligned rotator hypothesis, as we do not expect to see a heated polar cap in the orthogonal rotator geometry and the measured size of the polar cap is consistent with the radius expected in the 'spun-up squeezed' aligned MSP configuration. [18] state that with their polar cap model, they can reproduce the wide double-peaked profile without assuming an aligned rotator, if the hollow-cone model is invoked as [14], but the wide double-peaked profile can also be produced in the aligned configuration. Taking all of these arguments into account, along with the fact that the radio profile for PSR J0218+4232 is fairly broad, it may be that this is an aligned MSP and not orthogonal as tentatively proposed by [14], although there is still a lot to be learned about expected observations of MSP in the aligned configuration.

With the X-ray observations of PSR J0751+1807 and PSR J1012+5307 we enlarge the original sample size of [14] by 33%, which is an important increase with which to test the magnetic field configuration of MSPs.

Considering PSR J0751+1807 initially, this pulsar shows a single broad X-ray pulse reminiscent of the MSP PSR J0437-4715 [19, 11] and a broad, double-peaked, radio pulse (1.4 GHz, 2.7 GHz) spanning almost one third of the orbit [20, 21]. These characteristics could indicate that this MSP is also an aligned rotator. This pulsar has an X-ray spectrum best fitted by a single hard power law, not typical of an aligned rotator. However, a blackbody model for this MSP can not be ruled out, as the spectrum presented in [10] is of poor quality [having only 14 degrees of freedom in the spectral fit, 10]. Taking the blackbody temperature of this MSP (0.32 ± 0.04 keV, error at the 90% confidence level [10]) we find a radius of emission of 380^{+116}_{-80} m. This is six times smaller than that predicted using the conventional central dipole model, but is almost consistent with that predicted by the 'spun-up squeezed' aligned configuration model, which gives a radius of 245m. If we take the 3σ error, rather than the 90% confidence limit, we find a radius of 380^{+196}_{-137}m, consistent with the radius predicted by the 'spun-up squeezed' aligned configuration model, in support of an aligned rotator for this MSP. Polarimetric radio observations should however help resolve the orientation of this MSP.

PSR J1012+5307 shows two fairly narrow X-ray pulses and two double peaked narrow radio pulses (100 GHz) as well as a sub-pulse [10]. The nature of the emission is unclear, as we cannot discriminate between a power law and a blackbody model fit to the MSP spectrum. The separation of the two X-ray pulses is consistent with 180°, a signature of an orthogonal rotator [7], although this does not rule out the aligned rotator hypothesis. From evolutionary arguments [e.g. 22], [16] state that the spin axis of PSR J1012+5307 is likely to be nearly aligned with the orbital angular momentum axis. They explain that the large width of the pulse profile and the overall similarity of the position-angle swing to that of PSR J0218+4232 lead them to believe that this could be a nearly aligned rotator with a wide beam. Their "rotating vector model" fits to the shallow position-angle swing indicate that the magnetic inclination is instead very close to 0°, supporting the classification as an aligned rotator.

SUMMARY

Using X-ray and radio observations we find that PSR J0218+4232 may be an aligned rotator, from its fairly broad X-ray and radio pulses, large non-pulsed fraction in X-rays, magnetic inclination consistent with 0° and evidence for a heated polar cap consistent with the expected size for a MSP in the 'spun-up squeezed' aligned configuration. It is possible that PSR J0751+1807 has the same configuration, as it shows a single, broad radio and X-ray pulse, which has a radius of emission consistent with that expected for such a configuration. It is unclear whether PSR J1012+5307 is an aligned or an orthogonal rotator. None the less, these observations lend support to the idea that 'recycled' MSPs have either orthogonal or aligned magnetic fields, in support of the 'spun-up squeezed' model invoked to explain the evolution of the magnetic field in MSPs.

ACKNOWLEDGMENTS

We are grateful to M. Ruderman for comments on this manuscript.

REFERENCES

1. M. A. Alpar, A. F. Cheng, M. A. Ruderman, and J. Shaham, *Nature*, **300**, 728–730 (1982).
2. G. Radhakrishnan, V. Srinivasan, *Current Science*, **51**, 1096–1099 (1982).
3. E. P. J. van den Heuvel, *Journal of Astrophysics and Astronomy*, **5**, 208–233 (1984).
4. G. Hobbs, R. Manchester, A. Teoh, and M. Hobbs, "The ATNF Pulsar Catalog," in *Young Neutron Stars and Their Environments*, edited by F. Camilo, and B. M. Gaensler, IAU Symposium 218, Astronomical Society of the Pacific, San Francisco, 2004, p. 139.
5. R. Romani, *Nature*, **347**, 741–743 (1990).
6. M. Ruderman, *Astrophysical Journal*, **366**, 261–269 (1991).
7. K. Chen, and M. Ruderman, *Astrophysical Journal*, **408**, 179–185 (1993).
8. M. G. F. Kirsch, W. Becker, S. Benlloch-Garcia, F. A. Jansen, E. Kendziorra, M. Kuster, U. Lammers, A. M. T. Pollock, F. Possanzini, E. Serpell, and A. Talavera, "Timing accuracy and capabilities of XMM-Newton," in *X-Ray and Gamma-Ray Instrumentation for Astronomy XIII*, edited by K. A. Flanagan, and O. H. W. Siegmund, 5165, SPIE, 2004, pp. 85–95.
9. N. Webb, J.-F. Olive, and D. Barret, *Astronomy and Astrophysics*, **417**, 181–188 (2004).
10. N. Webb, J.-F. Olive, D. Barret, M. Kramer, I. Cognard, and O. Löhmer, *Astronomy and Astrophysics*, **419**, 269–276 (2004).
11. V. E. Zavlin, G. G. Pavlov, D. Sanwal, R. Manchester, J. Trümper, P. Halpern, and W. Becker, *Astrophysical Journal*, **569**, 894–902 (2002).
12. T. Mineo, G. Cusumano, L. Kuiper, W. Hermsen, E. Massaro, W. Becker, L. Nicastro, B. Sacco, F. Verbunt, A. G. Lyne, I. H. Stairs, and S. Shibata, *Astronomy and Astrophysics*, **355**, 1053–1059 (2000).
13. L. Kuiper, W. Hermsen, F. Verbunt, S. Ord, I. Stairs, and A. Lyne, *Astrophysical Journal*, **577**, 917–922 (2002).
14. M. Ruderman, "Implications of Low Energy X-ray Emission from Millisecond Radio Pulsars," in *X-ray and γ-ray Astrophysics of Galactic Sources*, 2004.
15. L. Kuiper, W. Hermsen, F. Verbunt, and T. Belloni, *Astronomy and Astrophysics*, **336**, 545–552 (1998).
16. I. H. Stairs, S. E. Thorsett, and F. Camilo, *Astrophysical Journal Supplement Series*, **123**, 627–638 (1999).
17. M. Ruderman, "A Biography of the Magnetic Field of a Neutron Star," in *The Electromagnetic Spectrum of Neutron Stars*, edited by A. Baykal, S. Yerli, M. Gilfanov, and S. Grebenev, NATO-ASI, 2005.
18. Q. Luo, S. Shibata, and D. B. Melrose, *Monthly Notices of the Royal Astronomical Society*, **318**, 943–951 (2000).
19. V. E. Zavlin, and G. G. Pavlov, *Astronomy and Astrophysics*, **329**, 583–598 (1998).
20. M. Kramer, K. M. Xilouris, D. R. Lorimer, O. Doroshenko, A. Jessner, R. Wielebinski, A. Wolszczan, and F. Camilo, *Astrophysical Journal*, **501**, 270–285 (1998).
21. M. Kramer, C. Lange, D. R. Lorimer, D. C. Backer, K. M. Xilouris, A. Jessner, and R. Wielebinski, *Astrophysical Journal*, **526**, 957–975 (1999).
22. D. Bhattacharya, and E. P. J. van den Heuvel, *Physics Reports*, **203**, 1–124 (1991).

Echo Tomography of Sco X-1 using Bowen Fluorescence Lines

J. Casares*, T. Muñoz-Darias*, I.G. Martínez-Pais*, R. Cornelisse[†], P.A. Charles[†], T.R. Marsh**, V.S. Dhillon[‡] and D. Steeghs[§]

*Instituto de Astrofísica de Canarias, 38200 La Laguna, Tenerife, Spain
[†]School of Physics & Astronomy, Univ. of Southampton, Southampton SOB17 1BJ, UK
**Dept. of Physics, Univ. of Warwick, Coventry CV4 7AL, UK
[‡]Dept. of Physics & Astronomy, Univ. of Sheffield, Sheffield S3 7RH, UK
[§]Harvard-Smithsonian Center for Astrophysics, Cambridge MA 02138, USA

Abstract. We present preliminary results of a simultaneous X-ray/optical campaign of the prototypical LMXB Sco X-1 at 1-10 Hz time resolution. Lightcurves of the high excitation Bowen/HeII emission lines were obtained through narrow interference filters with ULTRACAM, and these were cross-correlated with X-ray lightcurves. We find evidence for correlated variability, in particular when Sco X-1 enters the Flaring Branch. The Bowen/HeII lightcurves lag the X-ray lightcurves with a light travel time which is consistent with reprocessing in the companion star.

Keywords: binaries: close – X-rays: binaries – stars: neutron – stars: individual: Sco X-1
PACS: 95.75.Wx, 95.85.Nv, 97.10.Gz, 97.60.Jd, 97.80.Jp

INTRODUCTION: IRRADIATION IN LMXBS

Optical emission in persistent low mass X-ray binaries (hereafter LMXBs) is triggered by reprocessing of the powerful, almost Eddington limited, X-ray luminosity ($L_x \simeq 10^{38}$ erg s^{-1}) in the gas around the compact object. This is supported by independent arguments such as (i) the statistical distribution of de-reddened $(U-B)$, $(B-V)$ colours (or $F_v \sim$ const., see [20]) which can be accounted for by redistribution of high energies into UV+optical through irradiation models (e.g. [21]); (ii) the detection of optical counterparts of Type I X-ray bursts, with delay times consistent with binary separations (e.g. [7]); (iii) the presence of the broad emission feature at $\lambda\lambda 4640$-50 associated with a blend of CIII/NIII/OII powered by X-ray photoionization and Bowen fluorescence emission (hence referred to as *the Bowen blend*; [12], [17]); (iv) suppression of outburst cycles caused by irradiation-induced heating of the outer disc [19], [11].

The accretion disc subtends the largest solid angle as viewed by the X-ray source, and is therefore responsible for the majority of the irradiation component. The spectroscopic features of the weak companion star, on the other hand, are completely swamped by the disc's reprocessed light, with the exception of a few long-period LMXBs with evolved companions such as Cyg X-2 ([1]). Therefore, dynamical studies have classically been restricted to the analysis of X-ray transients during quiescence (e.g. see [4]).

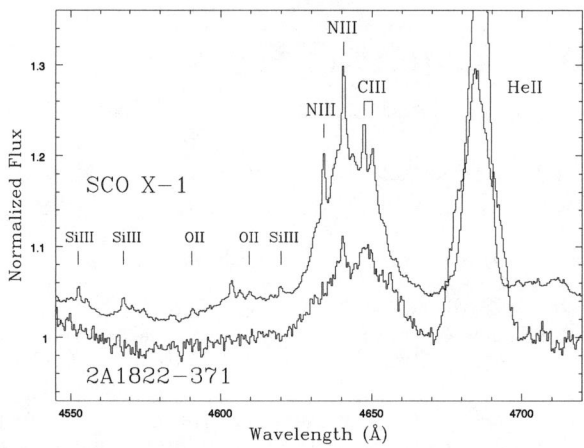

FIGURE 1. Summed spectra of Sco X-1 (top) and 2A1822-371 (bottom) in the rest frame of the companion star. Adapted from [18] and [2].

Fluorescence Emission from Donor Stars

However, this situation has changed recently thanks to the discovery of narrow emission components arising from the donor star in Sco X-1 [18]. High resolution spectroscopy, obtained with ISIS at the WHT, revealed many narrow high-excitation emission lines, the most prominent associated with NIII $\lambda\lambda 4634\text{-}41$ and CIII $\lambda\lambda 4647\text{-}50$ at the core of the broad Bowen blend (Fig. 1). The NIII lines are powered by fluorescence resonance through cascade recombination which initially requires seed photons of HeII Lyα. These narrow components are not resolved (i.e. their FWHM is the instrumental resolution) and they move in antiphase with respect to the wings of the HeII $\lambda 4686$ line, which approximately trace the motion of the compact star. Both properties (narrowness and phase offset) imply that these components originate in the irradiated face of the donor star. This work represents the first detection of the companion star in Sco X-1 and opens a new window for extracting dynamical information and thereby deriving mass functions in a population of ~ 20 LMXBs with established optical counterparts.

We now know that this property is not peculiar to Sco X-1 but is a feature of persistent LMXBs, as demonstrated by the following examples:

(i) Radial velocities of narrow Bowen lines in the black hole candidate GX 339-4, detected during the 2002 outburst, led to a mass function in excess of 5.8 $M_\odot$ and hence provided the first dynamical proof for a black hole [10].

(ii) Velocity information from the Bowen NIII $\lambda 4640$ line in the eclipsing ADC (*accretion disc corona*) pulsar 2A 1822-371, established a lower limit to the neutron star's mass of 1.14 $M_\odot$. Moreover, the radial velocity curve of the NIII emission is perfectly consistent with the donor's phase, as expected from the pulse time delay of the 0.59s spin period of the neutron star [2].

(iii) Sharp NIII $\lambda 4640$ Bowen emission has been detected in 4U 1636-536, 4U 1735-444 and the transient millisecond pulsar XTE J1814-338, which lead to donor velocity semi-amplitudes in the range 200-300 km s^{-1} [3].

Echo-Tomography

One of the most exciting prospects for this new technique is the possibility to perform echo-tomography using the Bowen lines. Echo-tomography is an indirect imaging technique which uses time delays between X-ray and UV/optical lightcurves as a function of orbital phase in order to map the reprocessing sites in a binary [16]. The optical lightcurve can be simulated by the convolution of the (source) X-ray lightcurve with a transfer function which encodes information about the geometry and visibility of the reprocessing regions. The transfer function quantifies the binary response to the irradiated flux as a function of the lag time and it has two main components, the accretion disc and the donor star. The latter is strongly dependent on the inclination angle, binary separation and mass ratio and, therefore, can be used to set tight constraints on these fundamental parameters. Successful echo-tomography experiments have been performed on several X-ray active LMXBs using X-ray and broad-band UV/optical lightcurves. The results indicate that the reprocessing flux is mostly dominated by the large contribution of the accretion disc (e.g. [9], [15], [8]).

Exploiting emission-line reprocessing rather than broad-band photometry has two potential benefits: a) it amplifies the response of the donor's contribution by suppressing most of the background continuum light (which is associated with the disc); b) since the reprocessing time in the lines is instantaneous, the response is sharper (i.e. only smeared by geometry) and also the transfer function is easier to compute (see [13]). Therefore, we decided to undertake a simultaneous X-ray/optical campaign on the prototypical LMXB Sco X-1 with the aim of performing echo-tomography so as to search for the reprocessed signatures of the donor using Bowen/HeII lines. As a first step, here we present our preliminary cross-correlation analysis which provides evidence for delayed echoes consistent with reprocessing in the companion star.

OBSERVATIONS

Simultaneous X-ray and optical data of Sco X-1 were obtained on the nights of 17-19 May 2004. The full 18.9 hr orbital period was covered in 12 snapshots, yielding 20.1 ks of X-ray data with the RXTE PCA. Only 2 PCA detectors (2 and 5) were used and the pointing offset was set to 0°.71 due to the brightness of Sco X-1. The data were analysed using the FTOOLS software and the times corrected to the solar barycenter. The STANDARD-2 mode data, with a time resolution of 16s, were used to produce a colour-colour diagram which showed that Sco X-1 was in the Normal Branch on 17 and 19 May and in the Flaring Branch on 18 May. The STANDARD-1 mode, with a time resolution of 0.125s, was used for the variability analysis.

The optical data were obtained with ULTRACAM on the 4.2m WHT at La Palma. ULTRACAM is a triple-beam CCD camera which uses two dichroics to split the light

TABLE 1. Observing log

Date	Exp. time (secs)	Seeing	Orbital Phases*	Number of RXTE windows	X-ray State
17 May 2004	0.1	< 1"	0.07-0.35	4	Normal Branch
18 May 2004	0.25-1	1"-5"	0.34-0.73	5	Flaring Branch
19 May 2004	0.3	1"-2"	0.55-0.95	5	Normal Branch

* Computed using ephemeris from [18]

into 3 spectral ranges: Blue (<λ3900), Green($\lambda\lambda$3900-5400) and Red (>λ5400). It uses frame transfer 1024x1024 Marconi CCDs which are continuously read out, and are capable of time resolution down to 500 Hz by reading only small selected windows (see [5] for details). ULTRACAM is equipped with a standard set of *ugriz* Sloan filters. However, since we want to amplify the reprocessed signal from the companion, we decided to use two narrow (FWHM =100 Å) interference filters in the Green and Red channels, centered at λ_{eff}=4660Å and λ_{eff}=6000Å. These will block out most of the continuum light and allow us to integrate two selected spectral regions: the Bowen/HeII blend and a featureless continuum, from which continuum-subtracted lightcurves of the high excitation lines can be derived. The images were reduced in the standard way with bias subtraction and flatfielding. Star counts were extracted using an optimal extraction algorithm [14] and lightcurves were obtained relative to a comparison star which is 96 arcsecs NW of Sco X-1. Lightcurves of the Bowen/HeII lines were computed by subtracting the Red (continuum) and Green channel lightcurves. The seeing was 1-1.5 arcsecs most of the time, except for the first two RXTE windows of the second night when it rose to over 3 arcsecs. Optical observations during the first 3 RXTE visits on 19 May were not possible because of clouds. The exposure time was initially set to 0.1s but was increased to 0.25s, because of weather conditions. Integrations of 1s were used for the first window on 18 May, when the seeing was worst. An observing log is presented in Table 1.

RESULTS

Figure 2 presents the X-ray and Bowen/HeII lightcurves corresponding to the nights of 17 and 18 May. Sco X-1 was at the bottom of the Normal Branch during most of the first night but moved towards the Flaring Branch in the last RXTE visit, showing a 50 percent increase in flux. The amplitude of the X-ray variability is less than 1% for the first three windows and no clear correlation with Bowen/HeII is evident on long timescales. On 18 May Sco X-1 was in the Flaring Branch and exhibited large amplitude variability, with large flares similar to that seen during the third RXTE visit. The left panel in Fig. 3 presents a 10 min segment of the third RXTE window, with each tickmark corresponding to 43s. We note significant X-ray variability, at the 10 percent level, and some correlated structures can be identified in the Bowen/HeII lightcurve towards the end of this interval. This window is centered at orbital phase 0.53 i.e. near the superior conjunction of the donor star, when the irradiated face of the donor presents the largest visibility and the

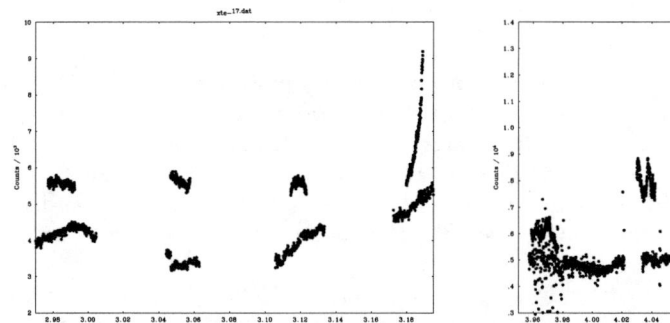

FIGURE 2. X-ray and Bowen/HeII lightcurves for the nights of 17 May (left panel) and 18 May (right panel). The top lightcurves correspond to the RXTE data and have been averaged in 4s bins. The bottom lightcurves represent the (continuum-subtracted) Bowen/HeII emission lightcurves and have been averaged in 8s bins. The Bowen/HeII counts are relative to the comparison star and have been multiplied by a factor 23500 for display clarity.

light-travel delay is expected to be at a maximum (around 12s for the binary parameters in [18]). We have calculated cross-correlation functions [6] for several time intervals within this window, after subtracting a low-order polynomial fit to the lightcurves in order to remove low-frequency variations. The right panel in Fig. 3 presents one of the cross-correlation functions, and shows a clear peak centered at a lag of $\sim$ 10-15s. This is in good agreement with the expected delay time for reprocessing in the companion star at this particular orbital phase. We have also detected correlated variability in other windows and we expect to be able to combine this information in order to set constraints on the binary parameters of Sco X-1.

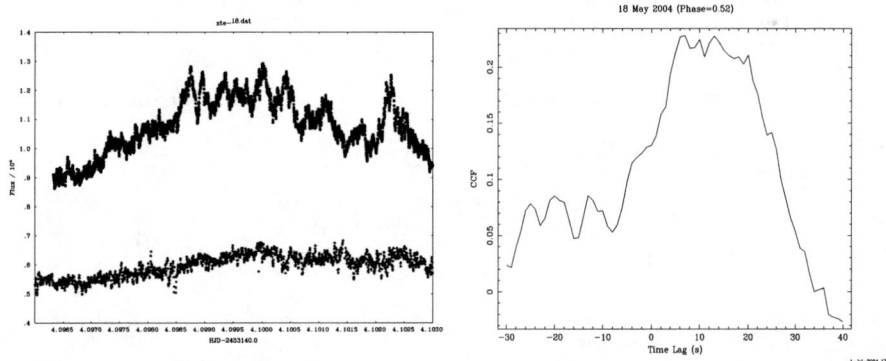

FIGURE 3. Left panel: 10 min detail from the third RXTE visit during 18 May. Top lightcurve represents the RXTE data whereas the bottom lightcurve is the Bowen/HeII data. Right panel: Cross-correlation function of the two lightcurves for the last 3 min interval.

SUMMARY

We have obtained simultaneous X-ray/optical photometry of Sco X-1 at 1-10 Hz time resolution using RXTE and WHT+ULTRACAM. The use of narrow interference filters in ULTRACAM has provided us with lightcurves of the Bowen blend + HeII λ4686 emission lines. Our preliminary analysis shows evidence for correlated variability, with the Bowen/HeII lightcurves lagging the X-rays. The observed time delays are consistent with reprocessing in the companion star and the correlations are most evident when Sco X-1 is in the Flaring Branch.

Future work requires a systematic search for correlated variability at different orbital phases and as a function of X-ray energy, and to use the information in order to constrain the binary parameters (mainly inclination, binary separation and mass ratio) using appropriate synthetic transfer functions [13].

ACKNOWLEDGMENTS

The author acknowledges support by the Spanish MCYT grant AYA2002-0036 and the programme Ramón y Cajal.

REFERENCES

1. Casares J., Charles P.A. & Kuulkers E. 1998, *ApJ*, 493, L39.
2. Casares J., Steeghs D., Hynes R.I., Charles P.A. & O'Brien K. 2003, *ApJ*, 590, 1041.
3. Casares J., Steeghs D., Hynes R.I., Charles P.A., Cornelisse R. & O'Brien K. 2004, *RevMexAA*, 20, 21.
4. Charles P.A. & Coe M.J. 2004, in *Compact Stellar X-ray Sources*, eds. W.H.G. Lewin & M. van der Klis, CUP (astro-ph/0308020).
5. Dhillon V.S. & Marsh T.R. 2001, *NewA Rev.*, 45(1-2), 91.
6. Gaskell C.M. & Peterson B.M. 1987, *ApJ*, 65, 1.
7. Grindlay J.E., McClintock J.E., Canizares C.R., Cominsky L., Li F.K., Lewin W.H.G. & van Paradijs J. 1987, *Nature*, 274, 567.
8. Hynes R.I. 2005, in *The astrophysics of cataclysmic variables and related object*, eds. J.M. Hameury and J.P. Lasota (astro-ph/0410218).
9. Hynes R.I., O'Brien K., Horne K., Chen W. & Haswell C.A. 1998, *MNRAS*, 299, L37.
10. Hynes R.I., Steeghs D., Casares J., Charles P.A.& O'Brien K. 2003, *ApJ*, 583, L95.
11. King A.R., Kolb U. & Burderi L. 1996, *ApJ*, 464, L127.
12. McClintock J.E., Canizares C.R. & Tarter C.B. 1975, *ApJ*, 198, 641.
13. Muñoz-Darias T., Martínez-Pais I.G. & Casares J. 2005, in *INTERACTING BINARIES: Accretion, Evolution and Outcomes*, eds. L.A. Antonelli et al., ASP Conf. Ser., in press.
14. Naylor T. 1998, *MNRAS*, 296, 339.
15. O'Brien K 2001, in *The proceedings of the first Galway workshop on high time resolution astrophysics*, ASP Conf. Ser. (astro-ph/0110267).
16. O'Brien K., Horne K., Hynes R.I., Chen W., Haswell C.A. & Still M.D. 2002, *MNRAS*, 334, 426.
17. Schachter J., Filippenko A.V. & Kahn S.M. 1989, *ApJ*, 340, 1049.
18. Steeghs D. & Casares J. 2002, *ApJ*, 568, 273.
19. van Paradijs J. 1996, *ApJ*, 464, L139.
20. van Paradijs J. & McClintock J.E. 1995, in *X-Ray Binaries*, eds. W.H.G. Lewin, J. van Paradijs and E.P.J. van den Heuvel (CUP 26, Cambridge), p58.
21. Vrtilek S.D. et al. 1990, *A&A*, 235, 162.

Spectral changes during six years of Scorpius X-1 monitoring with BeppoSAX Wide Field Cameras

Patrizia Santolamazza*, Fabrizio Fiore[†], Luciano Burderi[†] and Tiziana Di Salvo**

*ASI Science Data Center, c/o ESA-ESRIN, via Galileo Galilei, I-00044 Frascati, Italy
[†]INAF- Osservatorio Astronomico di Roma, via di Frascati 33, I-00040 Monteporzio, Italy
** Dipartimento di Scienze Fisiche ed Astronomiche, Universita' di Palermo, via Archirafi 36, 90123 Palermo, Italy

Abstract. We analyse a sample of fifty-five observations of Scorpius X-1 available in the *Beppo*SAX Wide Field Camera public archive and spanning over the six years of *Beppo*SAX mission life. Spectral changes are initially analysed by inspection of colour-colour and colour-intensity diagrams, we also discuss the shift of the Z tracks in these diagrams. Then we select two long observations for spectral fitting analysis, a secular shift is evident between the tracks in these observations. We finally extract spectra along the tracks and discuss the best fit model, the parameter variations along the track and between tracks, and their link to the accretion rate.

Keywords: accretion, accretion discs – binaries: close – X-rays: binaries – stars: neutron – stars: individual: Sco X-1
PACS: 95.75.Wx, 95.85.Nv, 97.10.Gz, 97.60.Jd, 97.80.Jp

INTRODUCTION

Neutron star low mass X-ray binaries (LMXB) are divided into two classes based upon the different morphology of their tracks in colour-colour and colour-intensity diagrams and their correlated timing behaviour. Six of the brightest LMXBs are classified as Z-type since they trace a Z-shaped pattern on such diagrams. The three branches of the Z are named Horizontal (HB), Normal (NB) and Flaring (FB) branches respectively, while atoll-type sources are characterized by an "island" and a "banana" state [1]. The spectral state of LMXBs is most easily determined by using X-ray colour-colour and colour-intensity diagrams, in fact track morphology is due to spectral variations on timescales of weeks, days or hours. Tracks are commonly interpreted as an accretion sequence [2], since it is thought that the mass accretion rate $\dot{M}$ is the parameter governing spectral variations along the track (see however [3]). Secular shifts and shape changes of the Z track were reported for several Z sources such as Cyg X-2 [4], whose shifts were interpreted in terms of occultation of the emitting region by a precessing accretion disk, or very recently for LMC X-2 [5], in which secular shifts of 2.5-10% are seen. Conversely significant secular variations of the track of Scorpius X-1 were never observed [6]. Scorpius X-1 is the brightest extra-solar X-ray source and is a LMXB of the Z-type showing a high level of activity in the X-ray, optical and radio bands, where radio jets were recently observed. In the case of Scorpius X-1, the complete Z track is generally traced out in a few hours to a day. Its timing and spectral properties were studied by using data from many observatories such as EXOSAT [6] and RXTE [3, 7]. This is the first systematic study of Scorpius X-1 using BeppoSAX Wide Field Cameras (WFC). WFCs are two coded mask instruments (WFC1 & WFC2) with a wide field of view of 40x40 deg, pointing away from each other and perpendicular to the Narrow Field Instruments. In the large majority of cases WFCs observed random sky positions during primary narrow field instruments' observations, giving the opportunity to monitor a large number of sources over the full six year satellite lifetime. Here we select fifty-five observations for a total monitoring duration of more than 600 hr and a total net exposure of the source of about 200 hr.

COLOUR-COLOUR AND COLOUR-INTENSITY DIAGRAMS

In order to obtain colour-colour and colour-intensity diagrams we define the total intensity as the count rate in the 1.7-19.1 keV band, the soft colour as the ratio [3.5-6.4keV / 2.0-3.5keV] and the hard colour as the ratio [9.5-16.4keV / 6.4-9.5keV].

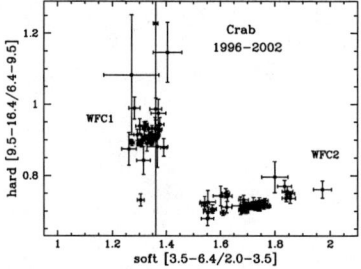
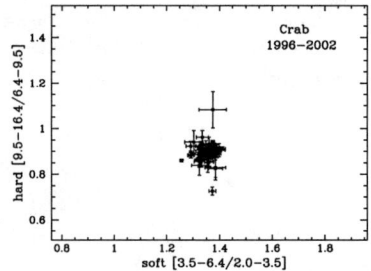

FIGURE 1. Crab colour-colour diagram over six years, each point represents an observation. Left panel shows raw data, the separation between WFC1 (upper left clump of points) and WFC2 (lower right clump) observations is apparent. Right panel shows offset-corrected scaled data, the diagram dimensions are on the same scale of Scorpius X-1 colour-colour diagram in right panel of Figure 2 for comparison.

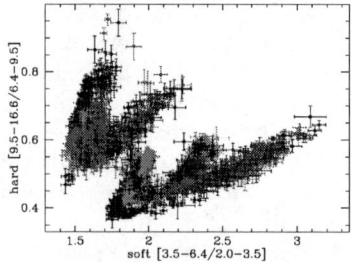
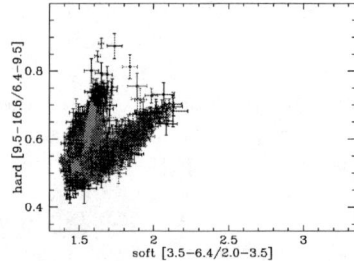

FIGURE 2. Scorpius X-1 colour-colour diagram over six years. Left panel shows raw data, the separation between WFC1 (upper left clump of tracks) and WFC2 (lower right clump) observations is apparent. Right panel shows offset-corrected scaled data.

Since we want to compare, on these diagrams, observations pointed off the source by different offset angles and at different epochs, the detector aging and the spatial response variations must be taken into account by scaling data to a common reference condition, for instance to the center of the detector at a certain epoch. Scaling factors were calculated taking into account the response of each camera at the source position in the field of view (FOV) for each epoch, then count rates and hardness ratios were scaled to the central quadrant of WFC1 at the date January 2002. These scaling factors change the intensity by 1% at most, while for the soft colour and the hard colour the maximum change is 15% and 8% respectively. Systematic residuals were checked using observations of the Crab Nebula and an empirical correction to systematic residual effects was estimated [8] as a function of the off-axis angle. The maximum empirical corrections found are 13%, 7%, 2% for the intensity, hard colour and soft colour respectively, over the 0 to 20 deg range. Therefore corrections are applied to intensity and hard colour only. Figure 1 shows the Crab Nebula colour-colour diagram. A clear separation between WFC1 and WFC2 data and a noticeable spread within each camera data set is seen in the uncorrected diagram (left panel), while points from both WFCs overlap in the corrected diagram (right panel). Average Crab colours are: <soft>=1.34, <hard>=0.89 with a spread of $3\sigma/$<soft>=5% $3\sigma/$<hard>=7%.

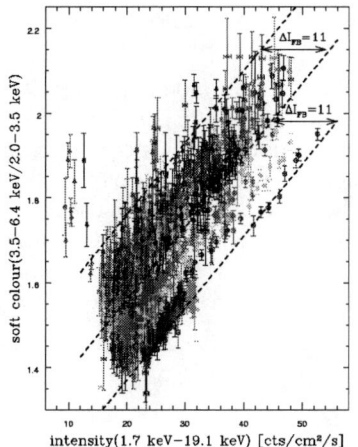 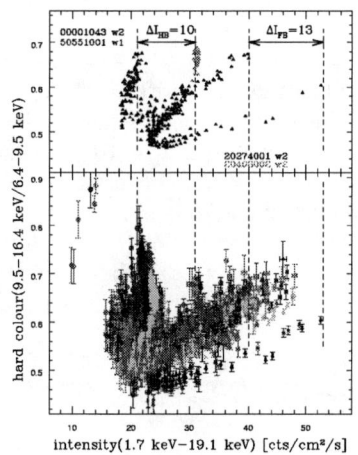

FIGURE 3. Scorpius X-1 corrected colour-intensity diagrams over six years. Parallel tracks appear due to large intensity variations. Arrows indicate line spacing of $\Delta I = 10\text{-}13$ cts/cm^2/s. Left panel shows the soft colour versus intensity, the dashed lines mark, from left to right, the upper envelope to all data, the lower envelope to the bulk of "normal" observations and the lower envelope to all data. The right panels show the hard colour versus intensity, the lower panel includes all observations while the upper panel shows only two normal and two shifted observations. The vertical dashed lines mark, from left to right, the two most widely intensity-shifted (almost vertical) horizontal branches and the tip of the the two most widely intensity-shifted flaring branches.

Figure 2 shows Scorpius X-1 colour-colour diagrams over six years. In the left panel a clear separation appears between the Z-tracks measured with WFC1 and located in the upper left part of the diagram and those measured with WFC2 and located in the lower right part. The right panel shows scaled and offset-corrected data, a spread is still present in this diagram, vertex points can differ by as much as Δsoft=6% and Δhard=14%. Thus observations suggest that secular variations are present. We checked that about 10% of tracks in the sample is shifted with respect to the remainder of "normal" observations. Secular shifts are more evident in the hardness-intensity diagrams in Figure 3, where some tracks are characterized by softer hardness ratios and higher total intensity. Large intensity variations produce parallel tracks in both diagrams. The soft colour is well correlated with intensity (see left panel in Figure 3) but tracks, which look like slanting lines, may differ by up to 30% in intensity at equal soft colour. Also in the right panels of Figure 3 the tips of the flaring branches may differ by 30% and horizontal branches by 50% in intensity, while in the Crab hardness-intensity diagram (not shown) the largest percent variation for intensities is 16%. It is known that on long timescales the total luminosity of LMXBs may vary secularly and that temporal variability indices like the characteristic frequencies of quasi-periodic oscillations and spectral parameters form a set of parallel lines when plotted versus the total luminosity, each line reflecting the short-term correlation with luminosity and the offset reflecting the average luminosity difference in different epochs [see 9, 10]. A possible explanation for the "parallel tracks" phenomenon was proposed by [10].

SPECTRA

Spectral analysis is performed in the energy range 2.0-22 keV along the Z-track of two observations, the "normal" observation 20143001 of January 25 1997 and the "most shifted" observation 20274001 of March 13 1997. Figure 4 shows corrected colour-colour diagrams of both observations with boxes delimiting the data used to make spectra. Our best fit model is made up of a Comptonization component, a Gaussian component with a fixed width σ=0.5 keV to model an iron line component and an absorbing column density set to the Galactic value $N_H = 0.3 \times 10^{22}$ cm^{-2}. A 2% systematic error was assumed for all spectra. The general behaviour of spectral changes is the same

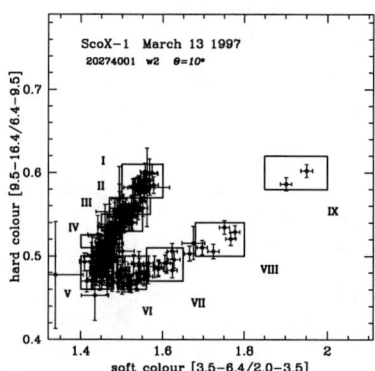
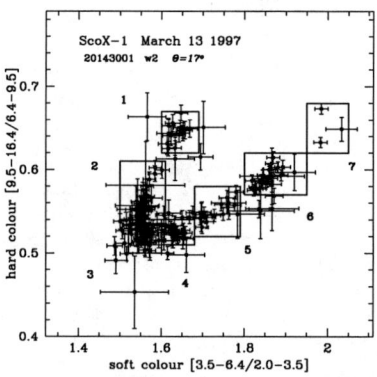

FIGURE 4. Scorpius X-1 corrected colour-colour diagrams of observations 20274001 (left panel) and 20143001 (right panel). The overlaid boxes delimit the data used to make spectra and are numbered from HB-NB to FB.

along both tracks. As an example spectra of observation 20274001 and best fit models are shown in the upper panel of Figure 5 while lower panel shows residuals in terms of sigmas with error bars of size one. Residuals are close to zero in the low energy range of either normal or "shifted" observations, consequently our choice of the Galactic value for N_H seems not to affect results. Residuals at energy higher than 18 keV suggest that a further hard tail component is needed. This is also supported by the comparison with

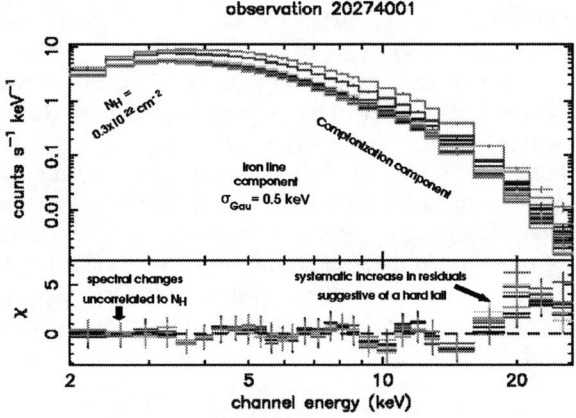

FIGURE 5. Spectral fits to spectra along the track of observation 20274001. Upper panel shows data and fitted models, labels indicating model components are positioned in the corresponding component energy ranges. Lower panel shows residuals in terms of sigmas, arrows indicate where expected (or unexpected!) variations occur.

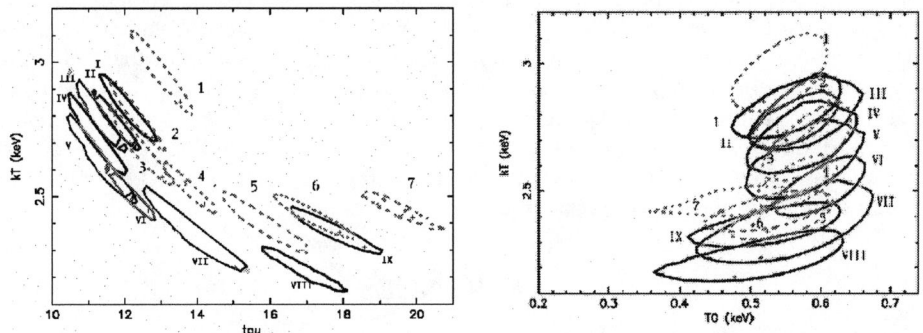

FIGURE 6. 90% χ^2 confidence level contour plot for the Comptonization component parameters. The hot plasma temperature (kT) is reported as a function of the optical depth (tau) in left panel and of the seed photon temperature (T0) in right panel. Contour numbers correspond to spectrum boxes in Figure 4, solid lines belong to observation 20274001 while gray dotted lines belong to observation 20143001.

Crab Nebula spectra (not shown), which show no such systematic distortion and the ratios of data over model show random distribution around unity.

In Figure 6 we report, for both observations, the 90% χ^2 confidence level contour plots for two Comptonization component parameters. The hot plasma temperature is plotted as a function of the optical depth in left panel and of the seed photon temperature in right panel. There is a strong correlation between the plasma temperature and the optical depth, the temperature decreases by about 20% from the HB-NB to the FB while the optical depth increases by 50% from the vertex (V), namely the transition point between NB and FB, to the FB. We interpret the plasma temperature decrease from HB-NB to FB in terms of Compton cooling due to increasing mass accretion rate, and therefore to increasing seed photon number. The $\dot{M}$ increase is accompanied by a growing number of plasma particles in the corona which contributes to the optical depth. Luminosity variations are also interpreted as an effect of the $\dot{M}$ increase, however we note that the total luminosity follows the intensity behaviour seen in colour-intensity plots, where the total count rate increases on the FB but slightly decreases from the HB-NB to the V. The optical depth behaves like the total luminosity.

The luminosity along the track of the shifted observation is systematically higher by 10-20% but also the ranges of variation of the optical depth and the plasma temperature seem to be affected. Comparison between these two parameters on corresponding branches (cfr Figure 6) shows that their ranges move to lower values by as large as 10%. This decrease matches with the softer hardness ratios found in the previous paragraph. So the two parameters seem to correlate to luminosity changes as average luminosity varies in different epochs.

CONCLUSIONS

Comparison of Scorpius X-1 colour-colour and hardness-intensity diagrams of fifty-five observations shows secular shifts of the tracks in 10% of the sample. Spectra from two observations are fitted with a constant column density, a Comptonization component and an iron line component. The spectral analysis shows great spectral changes from the HB-NB to the FB. The plasma temperature decreases along the track while the optical depth increases together with the total luminosity. The interpretation of these changes in terms of inverse Compton cooling confirms the widespread idea that it is the mass accretion rate $\dot{M}$ that varies along the track. The analysed observations show a systematic luminosity difference and shifted tracks. Comparison between spectral parameters on corresponding branches suggests that the secular shift is also accompanied by correlated secular changes in the optical depth and plasma temperature.

REFERENCES

1. G. Hasinger, and M. van der Klis, *A&A*, **225**, 79 (1989).
2. G. Hasinger, W. C. Priehorsky, and J. Middleditch, *ApJ*, **337**, 843 (1989).
3. R. Barnard, M. J. Church, and M. Balucinska-Church, *A&A*, **405**, 237 (2003).
4. E. Kuulkers, M. van der Klis, and B. A. Vaughan, *A&A*, **311**, 197 (1996).
5. A. P. Smale, J. Homan, and E. Kuulkers, *ApJ*, **590**, 1035 (2003).
6. S. W. Dieters, and M. van der Klis, *MNRAS*, **311**, 201 (2000).
7. C. Bradshaw, B. J. Geldzahler, and E. B. Fomalont, *ApJ*, **592**, 486 (2003).
8. P. Santolamazza, F. Fiore, L. Burderi, and T. Di Salvo, *Proc. of the BeppoSAX Symposium: "The Restless High-Energy Universe"*, p. 644 (2003).
9. M. Méndez, M. van der Klis, and E. C. Ford, *ApJ*, **561**, 1016 (2001).
10. M. van der Klis, *ApJ*, **561**, 943 (2001).

The evolution of low–mass X–ray binaries

H. Ritter

Max–Planck–Institut für Astrophysik, Karl–Schwarzschild–Str. 1, D–85741 Garching, Germany

Abstract. This brief review addresses our current understanding (or lack thereof) of the long-term evolution of low-mass X-ray binaries. Similarities and differences between the evolution of low-mass X-ray binaries and cataclysmic variables will be emphasized. Particular attention is paid to the following points: a) the evolutionary history and the evolutionary state of the donor star, b) systemic loss of orbital angular momentum, c) other modes of mass and angular momentum loss from the binary system, d) the possible role of irradiation of either the donor star or the accretion disc by accretion luminosity from the vicinity of the compact star, and e) low-mass X-ray binaries as progenitors of binary millisecond pulsars.

Keywords: stars: evolution – binaries: close – X–rays: binaries
PACS: 97.30.Qt; 97.80.Gm; 97.80.Jp

INTRODUCTION

In the context of this brief review, low-mass X-ray binaries (LMXBs) are semi-detached binaries in which either a neutron star (NS) or a stellar mass black hole (BH) primary accretes from a low-mass secondary ($M_2 \lesssim 1.5 M_\odot$). In cases where the primary is a BH stable mass transfer is also possible if the donor star is of intermediate mass with $M_2 \lesssim M_{BH}$. Such systems are usually referred to as intermediate mass X-ray binaries (IMXBs). This article mainly addresses aspects of the evolution of individual systems whereas collective properties and population synthesis of LMXBs and IMXBs are addressed in [1].

The principles of the long-term (secular) evolution of semi-detached binaries have been outlined in detail in [2], and will not be repeated here for reasons of space limitation. Based on time scale arguments it is obvious that mass transfer in the vast majority (if not all) of the observed LMXBs and IMXBs is thermally and adiabatically stable. Therefore, Eqs. (1.8) and (1.12) in [2] apply for describing the secular evolution of these systems. Although these equations are of little practical use for actually computing the mass transfer rate in a semi-detached binary, they illustrate clearly which properties of a binary system have to be known to be actually able to compute the mass transfer rate and thus the secular evolution numerically. First the precise evolutionary state of the donor, i.e. its mass, chemical composition and thermal structure, and thus the full evolutionary history before the onset of mass transfer need to be known. Second, the rate of systemic loss of orbital angular momentum and thus the mechanisms of angular momentum loss at work must also be known. Third, one has to model (or specify) the loss of mass from the binary and the orbital angular momentum which is carried away with it. In the next two sections these aspects are addressed separately and in more detail.

The structural similarity of LMXBs to catalysmic variables (CVs), where in the latter the compact accretor is a white dwarf (WD) rather than a NS or a BH, suggests that also the long-term evolution of LMXBs is similar to that of CVs. However, there are also a number of important differences between LMXBs and CVs which can lead to quite different evolutionary behaviour. This will be discussed in more detail in the penultimate section. Finally, it is widely accepted that LMXBs are progenitors of binary millisecond pulsars (msPSRs). Therefore, in the final section we examine what can be learned about the evolution and population of LMXBs from comparing the properties of known LMXBs with those of binary msPSRs.

THE DONOR STARS OF LMXBS

Observations

Observationally very little is known about the evolutionary state of the donor stars in LMXBs. In the majority of cases not even the donor's spectral type is known. The main reason for this is that during the X-ray active phases of an LMXB the light of the relatively faint secondary is swamped by accretion luminosity, and therefore remains undetectable. Spectral information about the donor and with it stellar parameters become available only if either the system is transient, in which case the secondary can be seen during quiescence, or if the donor itself is relatively luminous, i.e. if it is either an intermediate mass star as in the IMXBs V1033 Sco (GRO J1655–40), IL Lup (4U 1543–37) or V4641 Sgr (SAX J1819.3–2525), or a giant as in V404 Cyg (GS 2023+338), V395 Car (2S 0921–63) or V1487 Aql (GRS 1915+105) (for a compilation of binary and stellar parameters of LMXBs and IMXBs as well as the relevant references see [3]). Because the property of being a transient X-ray binary is strongly correlated with a combination of long orbital period ($P \gtrsim$ few days) and large mass ratio M_1/M_2 (see [4]) it is not surprising that most of the known donor stars are found in BH-LMXBs and show signs of advanced nuclear evolution. Cases in point are V1487 Aql, V395 Car, and V404 Cyg, where the donor is obviously a low-mass giant, and V1033 Sco and V4641 Sgr where the donor is either already in the Hertzsprung gap or near the terminal age main sequence. In addition, there are the ultra-short period LMXBs with orbital periods 11 min $\lesssim P_{orb} \lesssim$ 60 min [3], where the donor is a very low-mass degenerate dwarf. Finally there are also systems like V1341 Cyg (Cyg X-2) where the donor is a low-mass He-star in a phase of very late Case B mass transfer to its NS primary [5, 6].

Theoretical considerations

Stellar evolution, single or binary, is an initial value problem. As has been pointed out in the introduction, computing the secular evolution of a LMXB requires knowledge of the complete evolutionary history of the donor star up to the onset of the mass transfer phase one wishes to model. Whereas the evolution of single stars is comparatively well understood, binary evolution is not. The main reason for this is that from the beginning of its evolution (both components on the zero age main sequence) a binary system is subject to loss of mass and angular momentum from the system via various mechanisms (e.g. common envelope evolution [7, 8], ejection of mass via the Jeans mode, see e.g.

[5], mass loss via stellar winds, systemic angular momentum loss, e.g. via magnetic braking [9, 10, 11], and, last but not least, explosive mass loss in the supernova explosion with which the compact component is formed and the associated supernova kicks), all of which are not sufficiently well understood to allow for a parameter-free description of the resulting mass and angular momentum loss. Thus the evolutionary state of the donor star at the onset of the X-ray phase and the subsequent evolution of a LMXB, depend both on numerous more or less free parameters. As a result, an unambiguous forward integration in time of the evolution of a binary star to the LMXB state is not possible at present. It is in this context that binary population synthesis techniques (for a review see [1]), by comparing theoretical predictions with the observed properties of a whole population, can help constrain some of the free parameters or isolate the crucial mechanisms involved in the formation of LMXBs.

MASS AND ANGULAR MOMENTUM LOSS PROCESSES

In the following the main mass and angular momentum loss processes which are potentially important for the formation and the evolution of LMXBs are briefly discussed.

Mass loss via stellar winds

The progenitors of BHs (and some NSs) start their evolution as very massive stars on the main sequence. A massive star loses a substantial fraction of its initial mass already during its main sequence evolution via a strong stellar wind. Later during its evolution towards the compact component of a LMXB the massive primary loses its remaining hydrogen-rich envelope in a common envelope phase (discussed below). The resulting He-burning core is itself a very luminous (Wolf-Rayet) star which loses mass via a stellar wind at a high rate until core collapse ends this phase. Whereas the wind mass loss rates of massive stars on or near the main sequence can now be modelled with some accuracy [12], the same is hitherto not true for the post common-envelope He-stars. Depending on the rather uncertain mass loss rates of He-stars, the remnant core mass at the time of the supernova explosion does not or does allow for the formation of a massive stellar mass BH (with a mass of $10 M_\odot \lesssim M_{BH} \lesssim 15 M_\odot$, see e.g. [13]). When computing the angular momentum loss which goes with the wind mass loss it is usually *assumed* that the matter carries away the specific orbital angular momentum of the source (so-called Jeans mode). Thereby the interaction of the stellar wind with the companion's orbital motion with respect to the wind, i.e. the possible transfer of orbital angular momentum of the companion to the wind (see e.g. [14] for the same effect in a different context) is neglected.

Consequential angular momentum loss (CAML)

CAML is loss of orbital angular momentum as a direct consequence of mass transfer in a binary system. CAML is one of the most troublesome aspects of binary evolution to deal with because its computation is tantamount to modelling the (magneto-) hydrodynamical interaction of the outflowing gas with the binary's orbital motion and to modelling the mass loss process itself. This is not what is done in usual binary evolution models. Rather CAML is described by two (almost) free parameters, one specifying the fraction of the transferred mass that is lost from the system, the other the specific angular momentum carried away. Only in exceptional cases can one argue for simple approximations such as CAML in the Jeans mode. A case in point is the ejection of transferred matter from the compact star via jets which do not interact strongly with the components' orbital motion. A possible example for such a situation is discussed in [5].

Common envelope (CE) evolution

CE evolution is probably the single most important process involved in the formation of very close compact binaries [7, 8]. Because to this date it has not been possible to compute the evolution of a binary through the CE phase self-consistently from the onset of mass transfer to the end of envelope ejection, this process is usually dealt with in a parametrized form [7]. Two parameters enter the problem in the simplest approximation, one an efficiency parameter (usually denoted as α_{CE}) and a parameter λ which is a measure for the binding energy of the envelope to be unbound by the CE process. Because α_{CE} and λ appear only as a product $\alpha_{CE}\lambda$, CE is essentially parametrized by this one quantity. Recent attempts to compute λ (e.g. [15, 16, 17]) which have yielded ambiguous results because of the freedom in the choice of the core-envelope boundary, do not help much as long as we do not also have a better idea of how large α_{CE} is. Therefore the treatment of CE evolution, i.e. the choice of $\alpha_{CE}\lambda$, is the biggest uncertainty in modelling the formation of compact binaries (see also [1]).

Systemic loss of orbital angular momentum

This mode of angular momentum loss which operates also in the absence of mass transfer is not only important for driving mass transfer in semi-detached compact binaries but also for contracting detached post CE binaries to the onset of mass transfer. In this context two mechanisms need to be mentioned:

a) Gravitational radiation [18]
   This is the only systemic angular momentum loss mechanism for which the rate of angular momentum loss is computable from first principles.
b) Magnetic braking (MB)
   This process, which has been invoked for more than 20 years in the evolution of close binaries [19, 20] using various semi-empirical prescriptions [9, 10], has recently come under attack: the authors of [11] argue that MB in close binaries is much less efficient than according to [9, 10]. Yet estimates of the surface temperatures of accreting WDs in CVs [21] suggest that the mass transfer rates in long-period CVs ($P \gtrsim 3^h$) is more in line with what [9, 10] yield rather than with predictions of [11]. Nevertheless, there is currently much uncertainty about the efficiency

of MB in close binaries. And even if [9, 10] yield the correct order of magnitude of braking, one has to keep in mind that the braking rates derived by [9, 10] are semi-empirical estimates, not calculations from first principles, and contain at least one or even several free parameter(s), the value(s) of which is (are) not precisely known.

Supernova explosions and supernova kicks

The mass loss associated with a supernova explosion, if it occurs in a binary system, either makes the post-explosion orbit more eccentric than the pre-explosion orbit was, or even leads to the break-up of the binary [22, 23]. However, this is not the only effect a supernova explosion has on the orbital elements of a binary.

The fact that many of the observed (isolated) radio pulsars and even a few BH-LMXBs have very high space velocities with respect to the galactic rotation [24, 25, 26] (among the pulsars up to $\sim 10^3 \text{kms}^{-1}$) clearly indicates that in a core collapse supernova and the subsequent formation of a NS or a BH an additional huge momentum can be imparted to the compact star. Because neither the modulus nor the direction of this momentum can be predicted by theory, the effect of such a supernova kick is to essentially randomize the orbital elements of a binary in which such a supernova occurs. Thus the outcome of a binary evolution from a well-defined initial state beyond the first supernova explosion can only be determined in a statistical sense. What has to be done in practice is to assume a particular momentum distribution, e.g. chosen such as to best reproduce the observed distribution of the space velocities of radio pulsars [24], and a random orientation of the momentum vector.

DIFFERENCES BETWEEN LMXBS AND CVS

Although LMXBs and CVs are superficially similar, LMXBs differ in a number of ways from CVs and so does probably also their evolution. One of these differences we have just noted: supernova explosions and supernova kicks which are unavoidable in the formation of LMXBs but do not occur at all in the formation of CVs. Among the other important differences we note the following:

The mass of the accretor

The average mass of a NS ($\sim 1.4 M_\odot$) is about 2.5 times larger than that of an average WD ($\sim 0.6 M_\odot$) and the mass of a typical stellar mass BH ($\sim 5 - 15 M_\odot$) is even bigger. Therefore, stable mass transfer in LMXBs is possible with correspondingly more massive donor stars. In BH-LMXBs even intermediate mass donor stars are possible as e.g. in V4641 Sgr, V1033 Sco and IL Lup [3]. This in turn means that nuclear evolution of the donor star as a driving mechanism of mass transfer is much more important than in CVs. Furthermore, in BH-LMXBs with an intermediate mass donor even stable thermal time scale mass transfer is possible if the donor has evolved beyond the terminal age main sequence, as is the case in V1033 Sco (see e.g. [27]).

On the other hand, for given donor mass M_2 the mass ratio M_1/M_2 is systematically larger in a LMXB than in a CV. Therefore mass transfer is more stable which means that for a given angular momentum loss rate $\dot{J}$ or rate of expansion of the donor $\dot{R}_2$ due to nuclear evolution the resulting mass transfer rate is lower. On top of that we note that with increasing mass ratio gravitational radiation [18] as well as magnetic braking according to [9, 10] become less efficient.

Furthermore, for a given donor radius R_2 the corresponding Roche radius of the primary and with it the outer radius of the accretion disc grow with the mass ratio, i.e. are systematically larger in LMXBs than in CVs. This, together with the tendency to reduced mass transfer leads to an increased tendency for the occurrence of disc instabilities, i.e. for LMXBs to be transients [4].

The gravitational binding energy at the "surface" of the accretor

A direct consequence of the fact that the gravitational binding energy at the surface of a NS or at the innermost stable circular orbit of a BH is of the order of 100 MeV/nucleon, i.e. ~ 1000 times larger than the corresponding value for a WD is that the accretion luminosity per unit mass accreted is higher by a factor ~ 1000 in LMXBs than in CVs, and that in LMXBs the bulk of the energy emerges as (hard) X-rays whereas in CVs most of the energy emerges as (far) UV and optical radiation. As a consequence of the much higher accretion luminosities generated in LMXBs and of the fact that the emitted radiation is much harder, i.e. more penetrating, irradiation by accretion luminosity of the other components of a LMXB, i.e. of the donor star and the accretion disc becomes an important issue.

Irradiation of the accretion disc

Irradiation of the outer parts of the accretion disc by accretion luminosity emerging from the vicinity of the compact star plays an important role in LMXBs. On the one hand, as was first noted by [28], irradiation of the disc makes LMXBs visually much brighter than what one would expect based on viscous heating of the disc alone. On the other hand, irradiation of the disc allows for stationary accretion in systems which would otherwise be transient and undergo dwarf nova-like outbursts [29]. Even more important are the consequences which irradiation has for the properties of transient systems [29, 30]. Irradiation results in much longer outbursts, with higher accretion rates that can easily exceed the Eddington limit of the compact star. During an outburst the irradiated parts of the accretion disc are essentially emptied. As a consequence, the duration of the quiescence during which the disc is refilled also lasts much longer. Thus it is irradiation of the accretion disc which is to a large extent responsible for the characteristics of transient LMXBs. We note however, that other factors such as disc evaporation [30, 31] during quiescence also have an important influence on the outburst behaviour of X-ray transients.

Irradiation of the donor star

That irradiation of the donor star by accretion luminosity might drastically change the long-term evolution of LMXBs was first suggested in [32]. A detailed yet still simplified analysis [33, 34, 35, 36] shows that in semi-detached compact binaries (i.e. LMXBs and CVs), where the donor is irradiated by accretion luminosity, mass transfer could proceed through a limit cycle in which phases of irradiation-driven mass transfer alternate with phases during which mass transfer virtually stops, and in some cases the system even becomes detached. Whether or not this irradiation-driven feedback loop really works is currently unclear. Its occurrence depends on a free parameter which essentially measures the fraction of the accretion luminosity that is absorbed in the donor's photosphere. And the precise value of that parameter is strongly model-dependent and difficult to determine.

LESSONS FROM MS-PULSAR BINARIES

It is now widely accepted that NS-LMXBs are immediate progenitors of msPSR binaries (for a review see e.g. [37]). msPSR binaries are characterized by spin periods of the NS in the range $1.6 \text{ ms} \lesssim P_{spin} \lesssim 60 \text{ ms}$, and orbital periods in the range $0.1^d \lesssim P_{orb} \lesssim 700^d$ [38]. When compared with the observed ms-PSR binary "phase space" (P_{orb}, P_{spin}), we find that LMXBs with known (P_{orb}, P_{spin}) [3] cover only a very small part of the pulsar "phase space", i.e. the LMXBs in question are all characterized by short orbital periods $P_{orb} \lesssim 1^d$ and the majority of them (11 out 12 systems) have very short spin periods $P_{spin} \lesssim 5$ ms. Therefore, if LMXBs are progenitors of ms-PSR binaries with long orbital periods $(P_{orb} \gtrsim 1^d)$ this means that *we do not see the progenitors of the msPSR binaries with long orbital periods $(P_{orb} \gtrsim 1^d)$*. One possbile reason for why this is so is that long-period LMXBs have very extended accretion discs and, therefore, are all transient X-ray sources [39]. Because of the large disc size and the effect of irradiation of the disc during an outburst, the outbursts during which the irradiated parts are practically emptied last very long. As a consequence of the almost complete depletion of the irradiated parts of the disc, the subsequent quiescence, during which the disc is refilled and the X-ray luminosity is very low, lasts very long too. Because of the small duty cycle (ratio of the duration of an outburst relative to the duration of a full outburst cycle) which gets smaller the longer the orbital period [39], and because of their very low X-ray luminosities during quiescence, such long-period transient X-ray binaries stand a very low chance of being detected either in outburst or during low state. Therefore, if this is the reason for why we do not see the progenitors of long-period msPSR binaries, this implies that *there exists a large hidden population of transient, long-period NS-LMXBs in quiescence* [39, 40]. Finally we note that an analogous argument applies also to long-period BH-LMXBs like V1487 Aql.

ACKNOWLEDGMENTS

The author thanks Ron Webbink for carefully reading the manuscript and for helpful suggestions.

REFERENCES

1. Podsiadlowski, Ph., *this volume*.
2. Ritter, H., in *Evolutionary Processes in Binary Stars*, edited by R. A. M. J. Wijers, M. B. Davies, and C. A. Tout, NATO ASI Ser. C, Vol. 477, Kluwer, Dordrecht, 1996, pp. 223–248.
3. Ritter, H., and Kolb, U., *A&A*, **404**, 301–303 (2003), with regularly updated material available at the following web sites: http://www.mpa-garching.mpg.de/RKcat, or at http://www.physics.open.ac.uk/RKcat/.
4. King, A. R., Kolb, U., and Burderi, L., *ApJ*, **464**, L127–L130 (1996).
5. King, A. R., and Ritter, H., *MNRAS*, **309**, 253–260 (1999).
6. Kolb, U., Davies, M. B., King, A. R., and Ritter, H., *MNRAS*, **317**, 438–446 (2000).
7. Webbink, R. F., *ApJ*, **277**, 355–360 (1984).
8. Taam, R. E., and Sandquist, E. L., *ARA&A*, **38**, 113–141 (2000).
9. Verbunt, F., and Zwaan, C., *A&A*, **100**, L7–L9 (1981).
10. Mestel, L., and Spruit, H. C., *MNRAS*, **226**, 57–66 (1987).
11. Andronov, N., Pinsonneault, M., and Sills, A., *ApJ*, **582**, 358–368 (2003).
12. Kudritzki, R.-P., and Puls, J., *ARA&A*, **38**, 613–666 (2000).
13. Woosley, S. E., Langer, N., and Weaver, T. A., *ApJ*, **411**, 823–839 (1993).
14. Livio, M., Govarie, A., and Ritter, H., *A&A*, **246**, 84–90 (1991).
15. Dewi, J. D. M., and Tauris, T. M., *A&A*, **360**, 1043–1051 (2000).
16. Tauris, T. M., Dewi, J. D. M., *A&A*, **369**, 170–173 (2001).
17. Podsiadlowski, Ph., Rappaport, S., Han, Z., *MNRAS*, **341**, 385–404 (2003).
18. Landau, L. D., and Lifshitz, E. M., in *The Classical Theory of Fields*, 4th edition, Pergamon, New York (1975).
19. Spruit, H. C., and Ritter, H., *A&A*, **124**, 267–272 (1983).
20. Rappaport, S., Verbunt, F., and Joss, P. C., *ApJ*, **275**, 713–731 (1983).
21. Townsley, D. M., Bildsten, L., *ApJ*, **596**, L227–L230 (2003).
22. Verbunt, F., in *Evolutionary Processes in Binary Stars*, edited by R. A. M. J. Wijers, M. B. Davies, and C. A. Tout, NATO ASI Ser. C, Vol. 477, Kluwer, Dordrecht, 1996, pp. 201–222.
23. Kalogera, V., *ApJ*, **471**, 352–365 (1996).
24. Arzoumanian, Z., Chernoff, D. F., and Cordes, J. M., *ApJ*, **568**, 289–301 (2002).
25. Mirabel, I. F., Migniani, R., Rodrigues, I., Combi, J. A., Rodriguez, L. F., and Guglielmetti, F., *A&A*, **395**, 595–599 (2002).
26. Mirabel, I. F., Dhawan, V., Migniani, R. P., Rodrigues, I., and Guglielmetti, F., *Nat*, **413**, 139–141 (2001).
27. Kolb, U., King, A. R., Ritter, H., and Frank, J., *ApJ*, **485**, L33–L36 (1997).
28. van Paradijs, J., and McClintock, J. E., *A&A*, **290**, 133–136 (1994).
29. King, A. R, and Ritter, H., *MNRAS*, **293**, L42–L48 (1998).
30. Lasota, J.-P., *New Astron. Rev.*, **45**, 449–508 (2001).
31. Meyer-Hofmeister, E., and Meyer, F., *A&A*, **372**, 508–515 (2001).
32. Podsiadlowski, Ph., *Nat*, **350**, 136–138 (1991).
33. King, A. R., Frank, J., Kolb, U., and Ritter, H., *ApJ*, **467**, 761–772 (1996).
34. King, A. R., Frank, J., Kolb, U., and Ritter, H., *ApJ*, **482**, 919–928 (1997).
35. Ritter, H., Zhang, Z.-Y., and Kolb, U., *A&A*, **360**, 969–990 (2000).
36. Büning, A., and Ritter, H., *A&A*, **423**, 281–299 (2004).
37. Bhattacharya, D., and van den Heuvel, E. P. J, *Phys. Rep.*, **203**, 1–124 (1991).
38. Manchester, R. N., Hobbs, G. B., Teoh, A., and Hobbs, M., *AJ*, in press (2005) = astro-ph/0412641.
39. Ritter, H., and King, A. R., in *Evolution of Binary and Multiple Star Systems*, edited by Ph. Podsiadlowski, S. Rappaport, A. R. King, F. D'Antona, and L. Burderi, ASP Conf. Ser. Vol. **229**, pp. 423–432

(2001).
40. Ritter, H., and King, A. R., in *The Physics of Cataclysmic Variables and Related Objects*, edited by B. T. Gänsicke, K. Beuermann, and K. Reinsch, ASP Conf. Ser. Vol. **261**, pp. 531–432 (2002).

Binary Population Synthesis: Theory and Applications

Philipp Podsiadlowski*, Saul Rappaport[†], Eric Pfahl**, Zhanwen Han[‡] and Martin E. Beer[§]

*Dept. of Astrophysics, University of Oxford, OX1 3RH, UK
[†]Dept. of Physics, M.I.T., 77 Massachusetts Ave., Cambridge, MA 02139
**Chandra Fellow: Dept. of Astronomy, University of Virginia, 530 McCormick Road, Charlottesville, VA 22903
[‡]National Astronomy Observatory/Yunnan Observatory, The Chinese Academy of Sciences, P.O. Box 110, Kunming, 650011, China
[§]Theoretical Astrophysics Group, University of Leicester, Leicester, LE1 7RH, UK

Abstract. Binary population synthesis can be an important tool to link binary evolution theory with observed classes of binaries. Here we review some of its limitations and illustrate how it can be used effectively to study well-defined problems, using low-/intermediate-mass X-ray binaries, cataclysmic variables and Type Ia supernova progenitors as case studies.

Keywords: binaries: cataclysmic variables – binaries: close – stars: Neutron Stars – stars: X-ray binaries – supernovae: Type Ia
PACS: 97.80.Fk, 97.80.Gm, 97.80.Jp, 97.60.Bw, 97.60.Gb, 97.60.Jd, 96.60.Lf, 97.10.Cv, 97.10.Zr, 97.30.Qt

1. INTRODUCTION

The simulation of binary populations (binary population synthesis [BPS]) provides an important link between stellar evolution theory and observed populations of binary systems; it can be used to constrain theoretical uncertainties and to help uncover gaps in our physical understanding. However, as a tool BPS also has its limitations. The purpose of this review is to provide an overview of the technique and outline some of its limitations (§ 2) and then apply it to individual classes of binary systems, in particular low-/intermediate-mass X-ray binaries (§ 3), cataclysmic variables/AM CVn systems (§ 4) and the progenitors of Type Ia supernovae (§ 5), demonstrating successes and failures of the method in each case and their implications.

2. BINARY POPULATION SYNTHESIS (BPS)

Binary population synthesis involves the simultaneous statistical simulation of large populations of single and binary stars, starting with a distribution of unevolved systems and following them through various phases of stellar evolution and phases of mass transfer, including the effects of supernovae, dynamical interactions (e.g. in a cluster environment), etc. (for an extensive list of references see, e.g., Han et al. 2001). At present, many of these complex phases are often treated using simple analytic or semi-quantitative prescriptions that give a reasonable description of the individual phases, but are not always fully reliable at the detailed quantitative level. However, in many cases it is possible to improve the reliability of the simulations by including realistic binary evolution calculations (see § 3 to 5), at least for the phases of mass transfer that are directly amenable to observations (e.g. the X-ray phase in accreting compact systems).

The method itself is inherently limited by our incomplete knowledge of the primordial properties of binaries and our relatively poor understanding of many physical phases involving binary interactions and advanced phases of stellar evolution.

The Initial Properties of Binaries

All BPS simulations require as input the distributions of the initial properties of binaries (i.e. the orbital period distribution, the distribution of the primary mass and the mass ratio distribution). In particular, the mass-ratio distribution is poorly constrained and at some level controversial. These uncertainties are generally not a major qualitative problem and generally do not affect the overall viability of particular channels; of course, they affect the frequencies of particular channels (typically by not more than a factor of a few, though in some extreme cases it may be as much as a factor of 10).

The Common-Envelope (CE) Phase

Common-envelope (CE) evolution is undoubtedly the least understood binary interaction (see, e.g., Iben & Livio 1993; Taam & Sandquist 2000; Podsiadlowski 2001). It typically involves the spiral-in of a companion star inside the envelope of a (super-)giant donor star and in many case the ejection of the envelope, transforming an initially wide binary into a very close binary (Paczyński 1976). Most typically, it occurs when the radius of the mass-losing star expands more rapidly than the radius of its Roche lobe, leading to *dynamical mass transfer*. The conditions for the occurrence of dynamical mass transfer are not very well determined. In BPS simulations it is still occasionally assumed that mass transfer from a star with a convective envelope is dynamically unstable if the mass ratio q of the mass donor to the mass accretor is larger than a critical value $\simeq 0.7$ (this is the appropriate value for a fully convective polytropic star). However, this does not take into account the stabilizing effect of the compact core of the giant (e.g. Hjellming & Webbink 1987), and indeed full binary evolution calculations show that a much more typical critical mass ratio is 1.2 (1.1 – 1.3), 70 % larger than the commonly used value (see, e.g., Han et al. 2002). Indeed, there is some observational evidence that a common-envelope phase may not always lead to a dramatic spiral-in phase (e.g. Webbink 1986; Eggleton & Tout 1989; Podsiadlowski, Joss & Hsu 1992; Nelemans et al. 2001).

One of the biggest uncertainties in modelling CE evolution is the condition that leads to CE ejection. The most commonly used criterion is that the CE is ejected when the orbital energy times some efficiency factor α_{CE} exceeds the binding energy of the envelope, but this simple formula involves numerous uncertainties, in particular whether the binding energy is estimated from a simple analytic expression or realistic calculated envelope structures (e.g. Dewi & Tauris 2000) and whether the ionization energy should be included in the energy balance (see Han et al. 2002 for discussions). The simplistic application of such a criterion can also lead to the violation of energy conservation (by up to a factor of 10 in some published studies). Moreover, in cases where the spiral-in becomes self-regulated and where all the energy released in the spiral-in can be radiated away at the surface of the CE (Meyer & Meyer-Hofmeister 1979; Podsiadlowski 2001), an energy criterion is no longer appropriate. Finally, it is also not clear whether this treatment is applicable to CE phases where the donor star has a radiative envelope (as may happen when stars start to fill their Roche lobes in the Hertzsprung gap). Indeed, it seems more likely that this leads to a frictionally driven wind, at least initially, rather than a classical CE phase (Podsiadlowski 2001).

Non-Conservative Mass Transfer

Another major uncertainty in modelling binary evolution is the treatment of non-conservative mass transfer, in particular the amount of specific angular momentum that is lost from the system. Different reasonable prescriptions can give very different evolutionary paths. In this context, an important case involves the treatment of contact phases during early case B mass transfer (when a massive donor star crosses the Hertzsprung gap). If all of the systems that binary evolution models predict to experience a contact phase (Pols 1994; Wellstein, Langer & Braun 2001) were to merge, this would probably overpredict binary mergers and their resulting supernovae (Podsiadlowski et al. 1992).

Mass Transfer in Eccentric Binaries

One aspect that has received very little attention to-date are the consequences of mass transfer in very eccentric binaries. While it is often a good assumption that a system will have tidally circularized before the onset of Roche-lobe overflow, there are distinct situations where this is not the case. In very wide binaries (e.g. VV Cephei binaries), the evolutionary timescale may be shorter than the circularization timescale (Podsiadlowski et al. 1992), but the more important cases involve tidal/collisional capture systems and post-supernova binaries where tidally driven mass transfer may start immediately after the capture/supernova event (Podsiadlowski et al. 2005).

BPS Health Warning

As these very selected examples demonstrate, there are numerous quite fundamental uncertainties in BPS simulations. This does not mean that BPS is not a useful tool, but a tool that should be used with some care. It is most useful for answering specific questions with well-defined physical uncertainties rather than when it is used like a black box. Moreover, it is possible – at least in principle – to observationally calibrate some of the BPS uncertainties using well-defined evolutionary channels (e.g. short-period sdB binaries to constrain the CE parameters; Han et al. 2002, 2003), although it is often not clear whether these calibrations can be generalized to other binary classes. In many cases (but not all cases) it is important to use fully realistic binary calculations instead of simplified analytic models. All of this requires improvements of the binary and stellar physics input, coupled to stringent observational tests.

3. LOW- AND INTERMEDIATE-MASS X-RAY BINARIES

It has been established for more than 15 years that the standard theory for low-mass X-ray binaries (LMXBs) is not consistent with the observed characteristics of the observed population (see Ruderman, Shaham & Tavani 1989), a fact that is still being widely

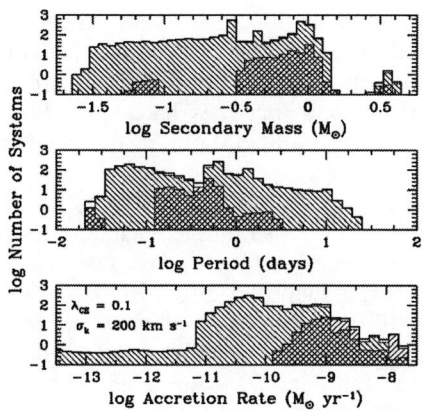
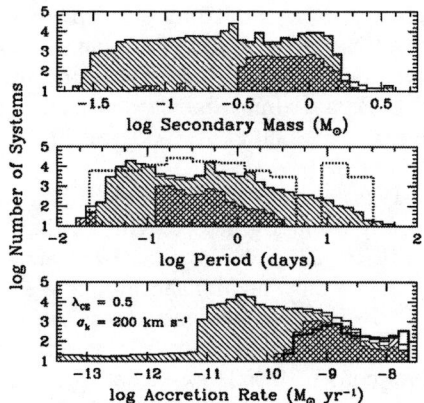

FIGURE 1. Distributions at the current epoch of the orbital period, donor mass and mass accretion rate onto the neutron star for various assumptions about the CE ejection parameter (*left:* inefficient; *right:* efficient). The hatched regions indicate persistent (+45°) and transient (−45°) X-ray sources, and the enclosing solid histogram gives the sum of these two populations. Overlayed (dotted histogram) on the theoretical period distribution in the figure on the right is the distribution of 37 measured orbital periods. The thick, solid distribution on the bottom panel of the right figure illustrates how the inclusion of X-ray irradiation effects might change the theoretical $\dot{M}$ distribution. (From Pfahl et al. 2003.)

ignored in the X-ray binary community. One of the reasons for this is that the assessment of the importance of the discrepancies requires detailed and in particular *realistic* BPS simulations which have only been performed in the last few years.

The Importance of Intermediate-Mass X-Ray Binaries: the Case of Cyg X-2

One of the major recent developments in the field of X-ray binary research has been the realization that X-ray binaries with intermediate-mass companion stars (IMXBs) are much more important than believed previously. Of particular importance is the case of Cyg X-2. The *observations* by Casares et al. (1998) showed that the companion of the X-ray binary Cygnus X-2, formerly classified as a low-mass X-ray binary, was far too luminous and far too hot to be consistent with a sub-giant in a 10-d orbit. The *theoretical* resolution of this surprising observation (King & Ritter 1999; Podsiadlowski & Rappaport 2000) was that the system must have originated from an IMXB rather than an LMXB, where the mass of the companion star must originally have been around $3.5\,M_\odot$. However, this implies that the system must have survived as a binary despite an extremely high mass-transfer rate ($\sim 10^{-5}\,M_\odot\,\mathrm{yr}^{-1}$) – several orders of magnitude above the Eddington accretion rate – ejecting most of the transferred mass from the system, perhaps in the form of an equatorial outflow, as seen from the relativistic jet system SS 433 (Blundell et al. 2001).

Problems with the Standard Model and Irradiation-Driven Evolution

Podsiadlowski, Rappaport & Pfahl (2002) and Pfahl, Rappaport & Podsiadlowski (2003) systematically investigated *theoretically* the role of IMXBs and found, not surprisingly, that IMXBs are much easier to form than traditional LMXBs, since these systems can more easily survive as binaries, both the common-envelope phase and the supernova in which the neutron star is formed. After the initial high mass-transfer phase, IMXBs are almost indistinguishable from LMXBs; but since they have much higher birthrates, Pfahl et al. (2003) estimate that 80–95 % of all L/IMXBs in fact originate from IMXBs.

Pfahl et al. (2003) were the first to study the population of L/IMXBs using realistic binary evolution models. One of their main conclusions was that the standard model for L/IMXBs failed to reproduce some of the main features of the observed population. The two most significant failures are: (1) the overproduction of L/IMXBs by a factor of 10–100 (though consistent with the birthrate of binary millisecond pulsars), and (2) the luminosity distribution, where the theoretical distribution neither produces enough luminous L/IMXBs (with $L_X > 10^{37}\,\mathrm{ergs\,s^{-1}}$) nor reproduces the observed correlation between X-ray luminosity and orbital period (Podsiadlowski et al. 2002).

One major omission in the standard model is that it does not take into account the strong X-ray irradiation of the secondary which can fundamentally change the evolution of the system by either driving a wind from the secondary (Ruderman et al. 1989) or by driving expansion of the secondary (Podsiadlowski 1991). Even a modest expansion of the secondary ($\sim 10\%$) can drive mass-transfer cycles (Hameury et al. 1993) where the mass-transfer rate ($\dot{M}$) is larger than the rate without irradiation effects by a factor > 10, which at the same time shortens the X-ray active lifetime by a proportionate amount. Pfahl et al. (2003) demonstrated that the inclusion of such mass-transfer cycles could potentially solve both of the major problems mentioned above, by increasing the typical observed X-ray luminosity by a factor of 10 or more and at the same time eliminating the L/IMXB overproduction problem, but still producing enough binary millisecond pulsars.

At the present time, the effects of irradiation on the secondary are still very poorly understood. Phillips & Podsiadlowski (2002) have shown that the external irradiation can dramatically distort the shape of the companion which has important implications for modelling ellipsoidal lightcurves and determining radial-velocity curves of the secondary. One of the key uncertainties is how much energy is transported from the irradiated side to the back side by irradiation-driven circulation. Even the transport of only 1 % of the intercepted irradiation energy can have drastic effects on the appearance and the further evolution of the secondary. To help answer these questions, Beer has developed a custom-designed 3-d stellar hydrodynamics code to study the irradiation-induced circulation (initially using a polytropic equation of state, which is now being extended to include a thermodynamic equation; Beer & Podsiadlowski 2002a,b). Some of his preliminary results show that the circulation velocities are a significant fraction of the sound speed and that a substantial amount of energy is transported to the backside in the form of kinetic energy (rather than thermal energy) where it is thermalized and raises the temperature by more than 1000 K in the case of an LMXB companion.

Observations will play an essential role in constraining the theoretical models (in particular the turbulent viscosity in the outer shear layer). These constraints may involve ellipsoidal light curves, phase-dependent spectral variations and distortions of radial-velocity curves. Indeed many of these effects have already been observed in a number of systems (e.g. HZ Her/Her X-1, Cyg X-2, Nova Sco, AA Dor).

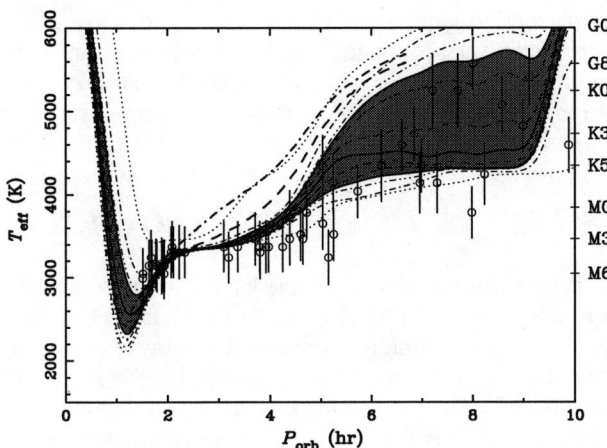

FIGURE 2. Comparison of the theoretical distribution of effective temperatures/spectral types for cataclysmic variables as a function of orbital period with the observational data of Beuermann et al. (1998). The various curves show the intrinsic distribution of temperatures where the shaded region includes 50 per cent of all systems around the median (only for systems before they reached the period minimum). The circles with error bars give the temperatures of systems with spectral types taken from Beuermann et al. (1998). (From Podsiadlowski et al. 2003.)

4. CATACLYSMIC VARIABLES AND AM CVN BINARIES

One of the most striking properties of the population of cataclysmic variables (CVs) is the period gap between 2 and 3 hr in their period distribution. The standard explanation for this gap (the disrupted magnetic braking model; Rappaport, Verbunt & Joss 1983; Spruit & Ritter 1983) is that magnetic braking becomes ineffective when the donor star becomes fully convective (at a mass around $0.3 M_\odot$). This allows the donor star to relax to thermal equilibrium, i.e. shrink, making the system detached until gravitational radiation can bring the donor star back into contact (at a shorter orbital period). While it is relatively straightforward to model this gap with unevolved main-sequence stars (Kolb 1993; Howell, Nelson & Rappaport 2001), it is not *a priori* clear that the period gap feature survives if a distribution of white dwarf masses and different evolutionary states of the secondary are considered, in particular since observations clearly show that some CVs must have at least somewhat evolved secondaries (essentially all systems with periods longer than about 7 hr; see Beuermann et al. 1998 and Fig. 2). Such systems evolve rather differently from unevolved stars (Podsiadlowski, Han & Rappaport 2003): (1) they experience a much shorter period gap, if at all, and the period range is shifted towards lower periods; (2) they can evolve towards ultra-short periods (as short as ~ 7 min), contributing at least in part to the population of AM CVn binaries.

This implies that the realistic modelling of the CV population requires full binary evolution calculations. Podsiadlowski et al. (2003) have performed such a study where they integrated a grid of binary evolutionary models into a BPS simulation of short-period white-dwarf binaries. They found: (1) CVs with evolved secondaries tend to somewhat fill in the period gap (10–20%), but do not destroy it; (2) there is a large variation of the predicted mass-transfer rate $\dot{M}$ for orbital periods longer than $\sim 5\,\mathrm{hr}$, probably consistent with observations, but not for shorter periods. Since the latter is not consistent with the observed spread in $\dot{M}$, other mechanisms, e.g. nova-induced mass-transfer variations (Kolb et al. 2001), may have to be invoked.

5. PROGENITORS OF TYPE IA SUPERNOVAE AND SN 2002IC

In recent years, Type Ia supernovae (SNe Ia) have been used successfully as cosmological probes of the Universe (Riess et al. 1998; Perlmutter et al. 1999). However, the nature of their progenitors has remained somewhat of a mystery. There is almost universal agreement that they represent the disruption of a degenerate object; but this is also where the agreement ends. There are numerous progenitor models (for detailed reviews see, e.g., Branch et al. 1995; Ruiz-Lapuente, Canal & Isern 1997), but most of these have serious theoretical/observational problems or do not appear to occur in sufficient numbers to explain the observed frequency of SNe Ia in our Galaxy ($\sim 3 \times 10^{-3}\,\mathrm{yr}^{-1}$; Cappellaro & Turatto 1997).

The two main classes of progenitor models are the double-degenerate (DD) model and single-degenerate Chandrasekhar models. The double-degenerate model (Iben & Tutukov 1984; Webbink 1984) involves the merger of two CO white dwarfs with a combined mass in excess of the Chandrasekhar mass ($\sim 1.4 M_\odot$). While this model has the advantage of being quite common (see, e.g., Iben & Tutukov 1986; Yungelson et al. 1994; Han et al. 1995; Iben, Tutukov & Yungelson 1997; Han 1998; Nelemans et al. 2001), it seems more likely that the disruption of the lighter white dwarf and the accretion of its debris onto the more massive one leads to the transformation of the surviving CO white dwarf into an ONeMg white dwarf which subsequently collapses to form a neutron star (i.e. undergoes accretion-induced collapse; AIC) rather than experience a thermonuclear explosion (e.g. Nomoto & Iben 1985). There may be a small parameter range where AIC can be avoided, but it is unlikely to account for more than a small number of SNe Ia.

The arguably most favoured class of models at the present time involves single-degenerate scenarios, where the white dwarf accretes from a non-degenerate companion star (Whelan & Iben 1973; Nomoto 1982). In these models, the companion star can be in different evolutionary phases and may either be a hydrogen-rich star or a helium star. One of the major problems with these models is that it is generally difficult to increase the mass of a white dwarf by accretion due to the occurrence of nova explosions and/or helium flashes (Nomoto 1982) which may eject most of the accreted mass. There is a narrow parameter range where a white dwarf can accrete hydrogen-rich material and burn it in a stable manner, but this requires rather special circumstances.

One promising channel that has been identified in recent years relates them to super-soft X-ray sources (van den Heuvel et al. 1992; Rappaport, Di Stefano & Smith 1994; Li & van den Heuvel 1997). In this channel, the companion star is a somewhat evolved main-sequence star or subgiant of $2-3 M_\odot$, transferring mass on a thermal timescale to a white dwarf. The assessment of the importance of this channel again requires BPS simulations that use fully realistic binary evolution calculations. Han & Podsiadlowski

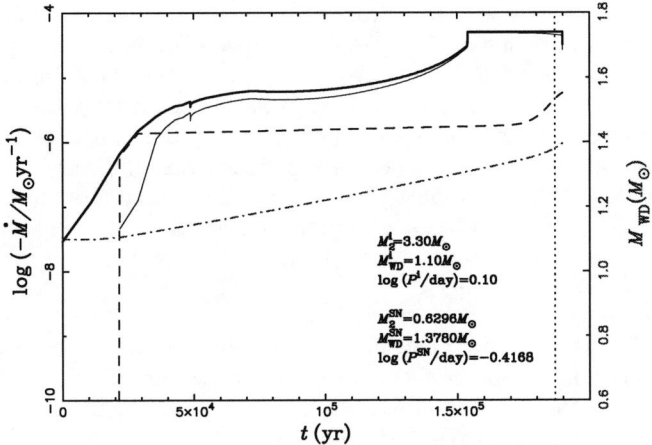

FIGURE 3. The evolution of the mass transfer rate (thick solid curve), mass loss rate from the system (thin solid curve) and WD accretion rate (dashed curve) as a function of time illustrating the case of a delayed dynamical instability, as may be applicable to SN 2002ic. Note the high mass-loss rate in the last 3×10^4 yr before the explosion (which has been limited to $0.5 \times 10^{-4} M_\odot\,\mathrm{yr}^{-1}$ in the calculations). The dot-dashed curve shows the evolution of WD mass (right axis). The vertical dotted curve indicates the time of the explosion. (From Han & Podsiadlowski 2005.)

(2004) performed such a simulation, calculating a grid of 2300 binary sequences using the accretion formalism of Hachisu et al. (1999). They found that the predicted frequency was close to the observed frequency, but perhaps too low by a factor of at most a few. This is in contrast to the claim by Fedorova, Yungelson & Tutukov (2004) that the frequency is at least a factor of 10 lower than the observed rate. This difference is a good illustration for the range of uncertainty associated with BPS simulations. Both studies use only slightly different input assumptions, perfectly reasonable in either case. The key question that remains is whether this difference between the observed and the theoretical estimates (a factor of 2 to 10) is a serious discrepancy or acceptable within the limitations of our understanding of the physical model? In particular, the efficiency at which a white dwarf can accrete, one of the major physical inputs into the simulations, is very poorly determined. Recently, Yoon & Langer (2004) showed that the inclusion of rotation dramatically changes the accretion behaviour, perhaps completely avoiding helium flashes. While it is clear that the inclusion of rotation should increase the regime in which a white dwarf can accrete, it also increases the critical mass at which the white dwarf explodes (e.g. the Chandrasekhar mass), i.e. require more mass to be accreted. How this will affect the population of SN Ia progenitors will be one of the challenges for the near future.

The case of SN 2002ic

A few years ago the first SN Ia, SN 2002ic, was discovered for which circumstellar hydrogen has been unambiguously detected (Hamuy et al. 2003). The detection of hydrogen in a SN Ia has long been considered one of the cornerstone observations required to help distinguish between different progenitor models, in particular between single-degenerate models with a hydrogen-rich donor star and the double-degenerate model. However, the amount of circumstellar hydrogen inferred in the case of SN 2002ic is much larger than one would naively expect if the companion star were a slightly evolved star of a few solar masses as, e.g., in the supersoft scenario: estimates range from a minimum of $\sim 0.5 M_\odot$ up to $6 M_\odot$ (Wang et al. 2004; Chugai & Yungelson 2004; Deng et al. 2004). Moreover, from the observed interaction of the supernova ejecta with the circumstellar medium and the lightcurve one can deduce that this matter must be located within $10^{17} - 10^{18}$ cm of the supernova and hence must have been ejected from the pre-supernova system within the last $\sim 10^4$ yr (this, however, is strongly dependent on the assumed ejection velocity).

It has been suggested that this may imply a very different explosion type for SN 2002ic (e.g. a SN 1 1/2; Hamuy et al. 2003). Indeed, the amount of circumstellar material is not consistent with a *typical* supersoft model. However, the detailed mass-transfer history depends strongly on the initial system parameters (the initial WD mass, $M_{\rm WD}^0$, the initial secondary mass, M_2^0, and the initial orbital period, $P_{\rm orb}^0$). To explain systems like SN 2002ic requires that the mass-transfer rate increases dramatically to values well in excess of $10^{-4} M_\odot$ yr^{-1} when the white dwarf approaches the Chandrasekhar mass. At such rates, very little mass can be accreted by the white dwarf and most of it must be lost from the system. This implies that the WD must have accreted most of its mass before this phase. This type of evolution can indeed be realized for binary systems near the upper edge of the allowed parameter space in the $P_{\rm orb} - M_2$ plane (see, e.g., Fig. 5 of Han & Podsiadlowski 2004), where systems are close to experiencing a delayed dynamical instability. An example for this evolution is shown in Figure 3. However, in order to be able to accrete enough mass in the limited amount of time before the delayed dynamical instability sets in, Han & Podsiadlowski (2005) had to adopt an accretion efficiency that was a factor of 2.5 higher than the value they used in their standard parametrization (based on the Hachisu et al. 1999 formalism). A higher accretion efficiency would also increase estimates for the overall frequency of SNe Ia. Thus, despite its unusual properties, SN 2002ic may still provide an important constraint for SNe Ia and may help to constrain some of the most uncertain BPS parameters in this progenitor channel.

REFERENCES

. Beer, M. E., & Podsiadlowski, Ph. 2002a, MNRAS, 335, 358
. Beer, M. E., & Podsiadlowski, Ph. 2002b, in Tout, C. A., & Van Hamme, W., eds, Exotic Stars as Challenges to Evolution, ASP Conf. Proc., Vol. 279 (ASP, San Francisco), p. 253
. Beuermann, K., Baraffe, I., Kolb, U., & Weichold, M. 1998, A&A, 339, 518
. Blundell, K. M., Mioduszewski, A. J., Muxlow, T. W., Podsiadlowski, Ph., & Rupen, M. P. 2001, ApJL, 562, L79
. Branch, D., Livio, M., Yungelson, L. R., Boffi, F. R., & Baron, E. 1995, PASP, 107, 1019
. Cappellaro, E., & Turatto, M. 1997, in Ruiz-Lapuente, P., Canal, R., & Isern, J.,eds, Thermonuclear Supernovae (Kluwer, Dordrecht), P. 77
. Casares, J., Charles, P., & Kuulkers, E. 1998, ApJ, 493, L39
. Chugai, N. N., & Yungelson, L. R., 2004, Astronomy Letters 30, 65
. Deng, J. et al. 2004, ApJ, 605, 37
. Dewi, J., & Tauris, T. 2000, A&A, 360, 1043
. Eggleton, P. P., & Tout, C. A. 1989, in Batten, A. H., ed., Algols, Kluwer, Dordrecht, P. 164
. Fedorova, A. V., Tutukov, A. V., & Yungelson, L. R. 2004, Astronomy Letters, 30, 73

- Hachisu, I., Kato, M., Nomoto, K., & Umeda, H. 1999, ApJ, 519, 314
- Hameury, J. M., King, A. R., Lasota, J. P., & Raison, F. 1993, A&A, 277, 81
- Hamuy M. et al. 2003, Nature 424, 651
- Han, Z. 1998, MNRAS, 296, 1019
- Han, Z., Eggleton, P. P., Podsiadlowski, Ph., Tout, C. A., & Webbink, R. F. 2001, in Podsiadlowski Ph. et al., eds, Evolution of Binary and Multiple Star Systems, ASP Conf. Ser., Vol. 229, P. 205
- Han, Z., & Podsiadlowski, Ph. 2004, MNRAS, 350, 1301
- Han, Z., & Podsiadlowski, Ph. 2005, MNRAS, submitted
- Han, Z., Podsiadlowski, Ph., & Eggleton, P. P. 1995, MNRAS, 272, 800
- Han, Z., Podsiadlowski, Ph., Maxted, P. F. L., Marsh, T. R., & Ivanova, N. 2002, MNRAS, 336, 449
- Han, Z., Podsiadlowski, Ph., Maxted, P. F. L., & Marsh, T. R. 2003, MNRAS, 341, 669
- Hjellming M. S., & Webbink R. F. 1987, ApJ, 318, 794
- Howell S. B., Nelson, L. A., & Rappaport S., 2001, ApJ, 550, 897
- Iben, I., Jr., & Livio M., 1993, PASP, 105, 1373
- Iben, I., Jr., & Tutukov, A. V. 1984, ApJS, 54, 335
- Iben, I., Jr., & Tutukov, A. V. 1986, ApJ, 311, 753
- Iben, I., Jr., & Tutukov, A. V., & Yungelson, L. R. 1997, ApJ, 475, 291
- King, A. R., & Ritter, H. 1999, MNRAS, 309, 253
- Kolb U., 1993, A&A, 271, 149
- Kolb U., Rappaport S., Schenker K., & Howell S., 2001, ApJ, 563, 958
- Li, X.-D, & van den Heuvel, E. P. J. 1997, A&A, 322, L9
- Meyer, F., & Meyer-Hofmeister, E. 1979, A&A, 78, 167
- Nelemans, G., Yungelson, L. R., Portegies Zwart, S. F., & Verbunt, F. 2001, A&A, 365, 491
- Nomoto, K. 1982, ApJ, 253, 798
- Nomoto, K., & Iben, I., Jr. 1985, ApJ, 297, 531
- Paczyński B., 1976, in Eggleton P.P., Mitton S., Whelan J., eds, Structure and Evolution of Close Binaries, Kluwer, Dordrecht, P. 75
- Perlmutter, S., et al. 1999, ApJ, 517, 565
- Pfahl, E., Rappaport, S., & Podsiadlowski, Ph. 2003, ApJ, 597, 1036
- Phillips, S. N., & Podsiadlowski, Ph. 2002, MNRAS, 337, 431
- Podsiadlowski, Ph. 1991, Nat, 350, 136
- Podsiadlowski Ph., 2001, in Podsiadlowski Ph. et al., eds, Evolution of Binary and Multiple Star Systems, ASP Conf. Ser., Vol. 229, P. 239
- Podsiadlowski, Ph., Han, Z., & Rappaport, S. 2003, MNRAS, 340, 1214
- Podsiadlowski, Ph., Joss, P. C., & Hsu, J. J. L. 1992, ApJ, 391, 246
- Podsiadlowski, Ph., Mardling, R. A, & Rappaport, S. 2005, in preparation
- Podsiadlowski, Ph., & Rappaport 2000, ApJ, 529, 946
- Podsiadlowski, Ph., Rappaport, S., & Pfahl. E. 2002, ApJ, 565, 1107
- Pols, O. R. 1994, A&A, 290, 119
- Rappaport, S., Di Stefano, R., & Smith, J. D. 1994, ApJ, 426, 692
- Rappaport S., Verbunt F., & Joss P. C., 1983, ApJ, 275, 713
- Riess, A., et al. 1998, AJ, 116, 1009
- Ruderman, M., Shaham, J., & Tavani, M. 1989, ApJ, 336, 507
- Ruiz-Lapuente, P., Canal, R., & Isern, J. 1997, Thermonuclear Supernovae (Kluwer, Dordrecht)
- Spruit H. C., & Ritter H., 1983, A&A, 124, 267
- Taam R. E., & Sandquist E. L., 2000, ARA&A, 38, 113
- van den Heuvel, E. P. J., Bhattacharya, D., Nomoto, K., & Rappaport, S. 1992, A&A, 262, 97
- Wang L. et al. 2004, ApJ 604, L53
- Webbink, R. F. 1984, ApJ, 277, 355
- Webbink, R. F. 1986, in Leung K.-C. & Zhai D. S., eds, Critical Observations versus Physical Models for Close Binary Systems, Gordon & Breach, New York, P. 403
- Wellstein, S., Langer, N., & Braun, H. 2001, A&A, 369, 939
- Whelan, J., & Iben, I., Jr. 1973, ApJ, 186, 1007
- Yoon, S.-C., & Langer, N. 2004, A&A, 419, 623
- Yungelson, L. R., Livio, M., Tutukov, A. V., & Saffer, R. 1994, ApJ, 420, 336

Optical spectroscopy of (candidate) ultra-compact X-ray binaries

Gijs Nelemans* and Peter Jonker[†]

*Institute of Astronomy, Cambridge, UK
and
Department of Astrophysics, Radboud University Nijmegen, The Netherlands
[†]Harvard-Smithsonian Center for Astrophysics, Cambridge, MA, USA

Abstract. We present (preliminary) results of our systematic spectroscopic study of (candidate) ultra-compact X-ray binaries. Most candidates are confirmed and we found the first optical spectra of (pure) carbon-oxygen accretion discs.

Keywords: X-ray binaries – optical observations
PACS: 95.75.Fg, 97.10.Cv, 97.10.Gz, 97.80.Jp

INTRODUCTION

Ultra-compact X-ray binaries (UCXBs) are close binaries with periods less than about one hour, in which a neutron star (or possibly a black hole) accretes material from a companion star. Their short periods rule out ordinary hydrogen-rich companion stars, since these stars are too big; they do not fit in the Roche lobe [e.g. 1]

In recent years a renewed interest has developed in these systems for a number of reasons. Improved observing facilities (e.g. large optical telescopes and sensitive X-ray satellites) make it possible to study them in more detail. But more importantly the number of known systems has increased. In particular the discovery of three transient UCXBs [e.g. 2, 3, 4] in which the millisecond pulsations of the accreting neutron star were seen, has been an exciting development.

We have started a systematic spectroscopic study of known and candidate UCXBs (excluding systems in globular clusters) in order to confirm/reject the candidates, to constrain possible formation scenarios and to open the way to study the chemically peculiar accretion process that is expected to operate in these binaries.

CURRENT OPEN QUESTIONS

The formation of UCXBs has been proposed through three channels, in which the donor stars are either white dwarfs, helium stars, or the remnants of evolved main sequence stars [see 5, and references therein]. However, there are still many open questions about UCXBs. One of the first priorities is to increase the number of known systems and to find their orbital periods. Besides the orbital period there are a number of properties that indicate that a particular system could be an UCXB. Van Paradijs and McClintock [6] studied the absolute magnitudes of X-ray binaries and derived a relation between the

TABLE 1. Overview of the UCXB candidates.

Name	Period (min)	Ne	M_V	confirmed?
4U 1543-624	18[a]	y		✓
4U 0614+09	?	y	5.4	✓
2S 0918-549	?	y	6.9	(✓)
4U 1822-00	?		5.5	(✓)
4U 1556-605	?	y		×
XB 1905+000	?		4.9	??

[a] Wang and Chakrabarty [7]

absolute magnitude, the orbital period and the X-ray luminosity, based on the assumption that the absolute magnitude is dominated by the irradiated accretion disc, whose surface is determined by the size of the binary, i.e. the orbital period. They empirically gauged this relation. Systems with absolute magnitudes fainter than about $M_V = 4$ have a good chance to be UCXBs.

Juett et al. [8] found features in the X-ray spectrum of the 20 min UCXB 4U 1850-087, which they attributed to enhanced Ne in the system and found similar features in 3 other systems, making them good UCXB candidates. In Table 1 we list our selection of candidates, based either on their absolute magnitudes or the the precence of the "Ne feature".

Optical spectroscopy of UCXBs might also be a good way to study the formation of UCXBs, as the different formation scenarios will lead to different chemical composition of the transferred material and thus of the accretion disc. The first constraints on the chemical composition have come from the properties of the type I X-ray bursts observed from the 11 min globular cluster sources 4U 1820-30, suggesting helium rich material [9]. In the following section we will describe the results of our observations, which are already summarized in the last column of the table.

OPTICAL SPECTROSCOPY: RESULTS

The VLT and FORS2

We used the FORS2 spectrograph on the 8.2m Very Large Telescope (VLT) of the European Southern Observatory at Paranal to obtain optical spectra of our candidate UCXBs. The observations were taken in the spring of 2003 and 2004. In 2003 we used the 1400V and 600RI holographic grisms, with a 1" slit, using 2x2 on-chip binning. This setup resulted in coverage of 4620 – 5930 Å with mean dispersion of 0.64 Å/pix for the 1400V spectra and 5290 – 8620 Å with mean dispersion of 1.63 Å/pix for the 600RI spectra. The 2004 spectra are taken with the 600B and 600RI grisms. The 600B grism covers the range 3325 – 6367Å with dispersion of 1.48 Å/pix. All spectra were reduced using standard IRAF tasks.

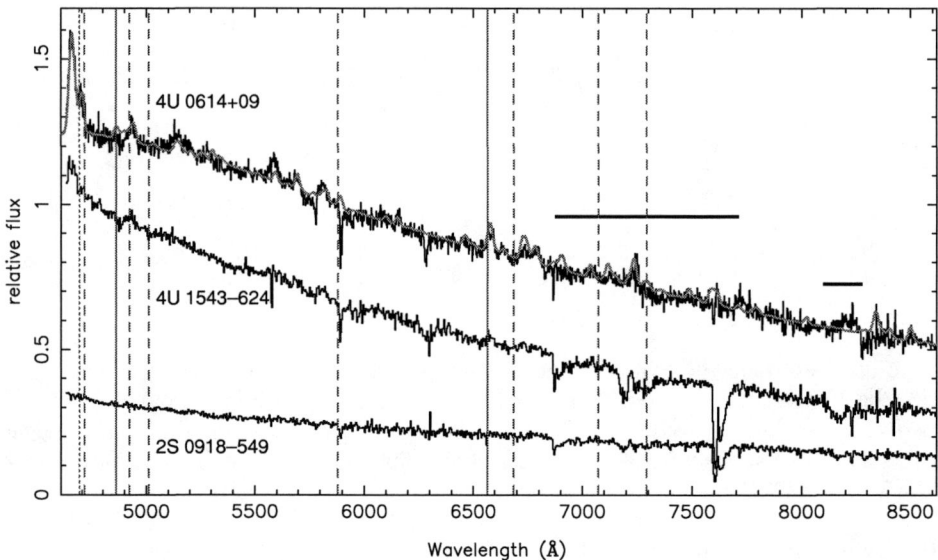

FIGURE 1. VLT spectra of 4U 0614+09, 4U 1543-624 and 2S 0918-549, showing lines from a carbon-oxygen accretion disc. From Nelemans et al. [10]

4U 0614+09, 4U 1543-624 and 2S 0918-549

The results of our 2003 programme are published in Nelemans et al. [10] and are summarized in Fig. 1. We identified the features in the spectrum of 4U 0614+09 as relatively low ionization states of carbon and oxygen. This clearly identifies this system as an UCXB and suggests the donor in this system is a carbon-oxygen white dwarf. The similarity of the spectrum of 4U 1543-624 suggests it is a similar system, while for 2S 0918-549 the spectrum didn't have a high enough S/N ratio to draw firm conclusions, but is also is consistent with being a similar system (and clearly does not show the characteristic strong hydrogen emission lines of low-mass X-ray binaries). We therefore concluded that all these systems are UCXBs.

The known UCXB 4U 1626-67 compared to 4U 0614-09

In Fig. 2 we compare the blue part of our 4U 0614+09 spectrum with the VLT spectrum of the 42 min. binary 4U 1626-67, which harbours a 7 sec X-ray pulsar [11]. The similarities are remarkable. This is interesting, as strong line emission in the X-ray spectrum of 4U 1626-67 has been identified with O and Ne lines [12].

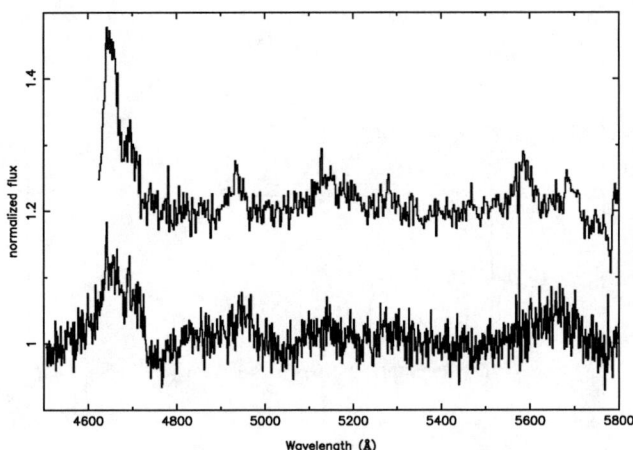

FIGURE 2. VLT spectra of Comparison of the carbon-oxygen spectrum of 4U 0614+09 with the spectrum of the 42 min system 4U 1626-67.

The preliminary 2004 results

The results of the 2004 programme will be published in a forthcoming paper, but we will give some preliminary results here.

4U 1822-00

Due to the faintness of 4U 1822-00 its spectrum is of low quality. However, just as with 2S 0918-549, the spectrum does not show hydrogen or helium lines, making it clearly different from the spectra of hydrogen rich systems. Provisionally we classify this system as an UCXB.

4U 1556-60

Based on its "Ne feature" 4U 1556-60 is a good UCXB candidate. However, its optical spectrum (Fig. 3) shows a classical low-mass X-ray binary spectrum with strong Balmer lines (4101, 4340, 4961 and 6563 Å) and lines from HeII (4686 (very strong), 5411 and 6678 Å). There is also strong emission at the Bowen blend, a C and N complex around 4640Å that is driven by He fluorescence. This system thus probably is not an UCXB, suggesting that the "Ne feature" is not a unique property of UCXBs.

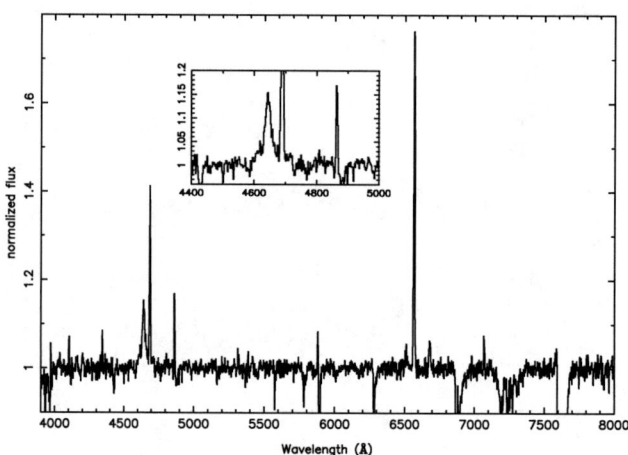

FIGURE 3. Normalized spectrum of 4U 1556-60, showing strong H and HeII emission, plus the Bowen blend at 4640Å.

XB 1905+00

The spectrum of XB 1905+00 shows the standard features of an early G star. This puzzling result is possibly due to a chance alignment. The acquisition image of the object obtained with a seeing of 0.6 arcsec suggest the source actually is a blend of two objects. In that case the optical counterpart of XB 1905+00 would be the fainter of the two stars. Note that this system was in quiescence at the time of our observation, making the counterpart much fainter than when it was found by Chevalier and Ilovaisky [13].

The known UCXB XB 1916-05

In Fig. 4 we show the spectrum of the 50 min UCXB XB 1916-05. Again it shows some features that are very similar to the spectra of 4U 0614+09 and 4U 1626-67, but in addition broad emission around 4540Å. A possible origin could be He, as there is a HeII line at 4541Å. Some of the other features also coincide roughly with positions of HeII lines, but these are very close (within a few Å) of the Balmer lines, making identification very difficult.

CONCLUSIONS

We have started a systematic study of the optical spectra of known and candidate UCXBs. The first results are both interesting and promising: we have confirmed the

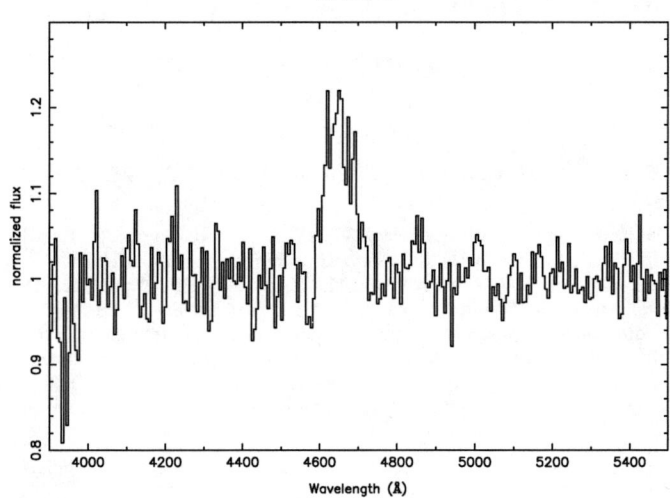

FIGURE 4. Normalized (and rebinned) spectrum of the known UCXB XB 1916-05, showing broad features, like 4U 0614, but possibly also hints of He and H emission.

ultra-compact nature of many of the candidates and have uncovered a variety of optical spectra, from pure carbon-oxygen spectra to spectra probably still showing signs of helium (and possibly hydrogen). One candidate (4U 1556-60) shows a classical low-mass X-ray binary spectrum, making it unlikely that this system is an UCXB. The faintness of these systems makes spectroscopic period determination almost impossible, so that periods will have to be found photometrically, as was recently done with 4U 1543-624 [7].

REFERENCES

1. L. A. Nelson, S. A. Rappaport, and P. C. Joss, *ApJ*, **304**, 231–240 (1986).
2. C. B. Markwardt, J. H. Swank, T. E. Strohmayer, J. J. M. i. Zand, and F. E. Marshall, *ApJ*, **575**, L21–L24 (2002).
3. R. A. Remillard, J. Swank, and T. Strohmayer, *IAU Circ.*, **7893**, 1 (2002).
4. C. B. Markwardt, M. Juda, and J. H. Swank, *IAU Circ.*, **8095**, 2 (2003).
5. P. Podsiadlowski, S. Rappaport, and E. D. Pfahl, *ApJ*, **565**, 1107–1133 (2002).
6. J. van Paradijs, and J. E. McClintock, *A&A*, **290**, 133–136 (1994).
7. Z. Wang, and D. Chakrabarty, *ApJ*, **616**, L139–L142 (2004).
8. A. M. Juett, D. Psaltis, and D. Chakrabarty, *ApJ*, **560**, L59–L63 (2001).
9. L. Bildsten, *ApJ*, **438**, 852–875 (1995).
10. G. Nelemans, P. G. Jonker, T. R. Marsh, and M. van der Klis, *MNRAS*, **348**, L7 (2004).
11. J. Middleditch, K. O. Mason, J. E. Nelson, and N. E. White, *ApJ*, **244**, 1001–1021 (1981).
12. N. S. Schulz, D. Chakrabarty, H. L. Marshall, C. R. Canizares, L. C. Lee, and H. J., *ApJ*, **563**, 941 (2001).
13. C. Chevalier, and S. A. Ilovaisky, *Space Science Reviews*, **40**, 443 (1985).

An absorbed view of a new class of INTEGRAL sources

E. Kuulkers

ISOC, ESA/ESAC, Urb. Villafranca del Castillo, P.O. Box 50727, 28080 Madrid, Spain

Abstract. The European γ-ray observatory *INTEGRAL* has found a group of hard X-ray sources which are highly absorbed, i.e., with column densities higher than about 10^{23} cm^{-2}. Here I give an overview of this class of *INTEGRAL* sources. The X-ray, as well as the optical/IR, properties of these sources and their location in the sky suggest that they belong to the class of high-mass X-ray binaries, some of them possibly long-period X-ray pulsars. The donors in these binaries are most probably giant or supergiant stars. I suggest that the soft X-ray spectrum below $\sim$5 keV of IGR J16318−4848, as well as in several other X-ray binaries (e.g., XTE J0421+56), can be described by emission from a compact object which is strongly absorbed by a partionally ionised dense envelope.

Keywords: X-ray – Accretion and accretion disks – Mass loss and stellar winds – Emission-line stars – X-ray binaries
PACS: 95.85.Nv, 97.10.Gz, 97.10.Me, 97.30.Eh, 97.80.Jp

INTRODUCTION

Hard X-rays (typically $\gtrsim$20 keV) and γ-rays are not easily absorbed by matter and thus are highly penetrating. Such radiation is, therefore, ideal to probe high-energy emitting sources in dense regions. Since its launch in October 2002 *INTEGRAL* (*Inter*national *G*amma-*R*ay *L*aboratory; Winkler et al. 2003) is revealing hard X-ray/soft γ-ray sources which were not easily spotted in earlier soft X-ray (typically $\lesssim$10 keV) observations. This invited review deals with this apparent new class of X-ray sources which show intrinsically high absorption along the line of sight, i.e., orders of magnitude higher than the usual interstellar absorption. I will describe their properties, and discuss their nature.

INTEGRAL carries on-board four instruments[1]: SPI, a hard X-ray/γ-ray spectrometer; IBIS, two coded mask hard X-ray/γ-ray imagers, ISGRI and PICsIT; Jem-X, two identical coded mask X-ray imagers; OMC, an optical monitor. The sources described in this paper are mainly found by IBIS/ISGRI (Ubertini et al. 2003, Lebrun et al. 2003). It has a 8.3°x8° fully-coded field-of-view, while it is partially coded out to 29°x29°; the angular resolution is 12'. This makes it ideal for observations in crowded regions.

The large field-of-view of IBIS makes it ideal to map the hard X-ray/γ-ray sky. During the first year of Galactic Plane observations about 120 point sources were detected down to $\simeq$1 mCrab (30–100 keV; Bird et al. 2004). Among them there are previously unknown sources, such as the well-known example IGR J16318−4848 (see below). About 86% of

---

[1] For a full description of the instruments, as well as an account of the first results, I refer to the special A&A Letters *INTEGRAL* issue 411 (2003).

the Galactic hard X-ray emission up to ~ 100 keV can be attributed to these high-energy point sources (Lebrun et al. 2004).

NEW, HIGHLY ABSORBED, INTEGRAL SOURCES

As an illustration of the usefulness of the combination of a large field-of-view and high sensitivity at hard X-ray/soft γ-rays, *INTEGRAL* discovered its first source, IGR J16318−4848, soon after nominal operation started, on January 29, 2003 during a routine Galactic Plane Scan (Courvoisier et al. 2003). Re-analysis of archival *ASCA* data revealed that its position coincides with a highly absorbed ($N_H \sim 10^{24}$ cm^{-2}) source with some hint of an Fe emission line (Murukami et al. 2003, Revnivtsev et al. 2003). Two weeks after the *INTEGRAL* detection, an *XMM-Newton* TOO observation indeed unveiled a variable and heavily absorbed source (2×10^{24} cm^{-2}), which emitted strong emission lines (Schartel et al. 2003). The emission complex could be resolved into three components, with centroid energies of 6.4 keV, 7.1 keV and 7.5 keV. They are most naturally interpreted as low ionised emission from Fe Kα, Kβ and Ni Kα (de Plaa et al. 2003, Walter et al. 2003, Matt & Guainazzi 2003).

After IGR J16318−4848, *INTEGRAL* found many more new sources (hereafter called IGR sources). Up to April 2005 more than 50 of these IGR sources have been reported.[2] Some were identified with already known sources (e.g., IGR J17464−3213 = H1743−322), but most of them are new ones.[3] About one third of the IGR sources can be classified. Most of them are either persistent or transient low-mass X-ray binaries (LMXBs) or high-mass X-ray binaries (HMXBs). Some of them have been classified as either being cataclysmic variables (e.g., IGR J17303−0601), accreting millisecond X-ray pulsar (IGR J00291+5934), AGN (e.g., IGR J18027−1455), or the central source of our Galaxy, Sgr A* (IGR J17456−2901). Still, about two third of them are unclassified, and some work lies ahead of us. The distribution of these sources is shown in Fig. 1. It seems that they are all distributed along the galactic plane, with concentrations in the direction of the Galactic Center and Galactic arms (see, e.g., Lutovinov et al. 2005b). One must note, however, that a lot of the *INTEGRAL* observations are concentrated on regions around the Galactic plane and the detection of new (especially transient) sources may, therefore, be biased towards these regions.

Of the (up to now well-studied) IGR sources, ten of them show very strong absorption ($N_H \gtrsim 10^{23}$ cm^{-2}), i.e., one to two orders of magnitude higher than the Galactic value of around 10^{22} cm^{-2}. It is this class of sources which I concentrate on for the rest of this paper, and I will refer to them as highly absorbed IGR sources. They are listed in Table 1; one thing which immediately catches the eye is that all but one (IGR J19140+0951) are in the direction of the Norma-arm tangent region. I will come back to this later.

Some of the highly absorbed IGR sources seem to be more or less persistent (such as IGR J16318−4848; see, e.g., Matt et al. 2005); some of them are clearly transient

---

[2] See http://isdc.unige.ch/~rodrigue/html/igrsources.html for an up-to-date list and further information.
[3] Note, however, that many of the new sources do have catalogued *ROSAT*, *ASCA*, and/or *BeppoSAX* soft-energy counterparts.

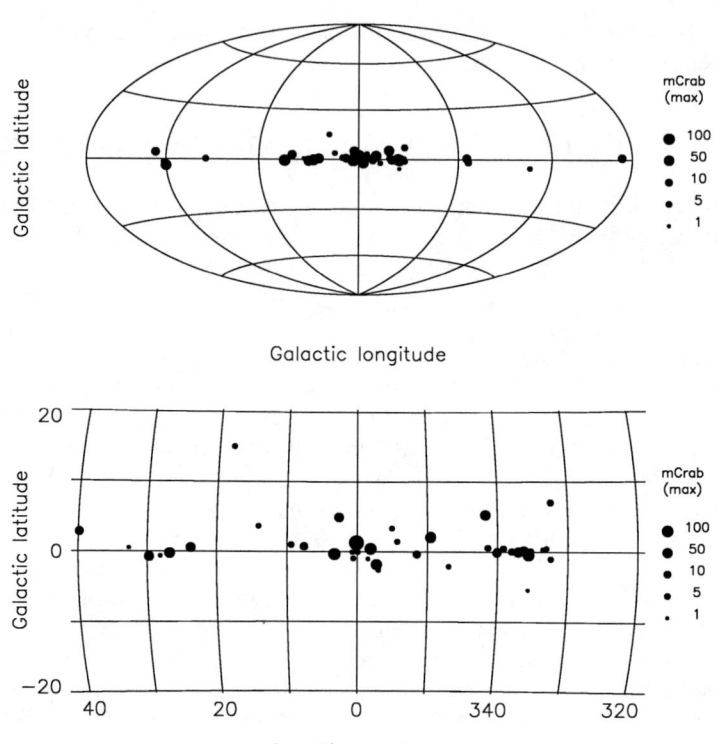

FIGURE 1. Galactic distribution of IGR sources. The size of the symbol • indicates the maximum observed X-ray intensity in units of mCrab in the 20-40 keV or 20-60 keV band. Flux values were mostly taken from those reported in the Astronomer's Telegrams and IAU Circulars.

TABLE 1. Properties of highly absorbed ($N_H \gtrsim 10^{23}$ cm^{-2}) *INTEGRAL* sources

IGR source	N_H (10^{23} cm^{-2})	Γ	E_{cut} (keV)	P_{pulse} (min)	references[a]
IGR J16195−4945	≃1	∼0.6			[1]
IGR J16318−4848	≃20	1.7–2.1	≃15		[2],[3]
IGR J16320−4751	≃2	0.5–1.1	≃12	≃22	[4],[5],[6]
IGR J16358−4726	≃4	∼0.7	≃16	≃100	[7],[8]
IGR J16393−4643	≃6	∼1.3	≃11	≃15	[8],[9],[10]
IGR J16418−4532	≳1				[11]
IGR J16493−4348	≃1	∼1.4			[12]
IGR J16465−4507	≃7	∼1	≃32	≃4	[8]
IGR J16479−4514	≃1	∼1.4			[8]
IGR J19140+0951	≃1[b]	≃1.6			[13]

[a] References: [1] Sidoli et al. (2005), [2] Walter et al. (2003), [3] Matt & Guainazzi (2003), [4] Rodriguez et al. (2003), [5] Foschini et al. (2004), [6] Lutovinov et al. (2005a), [7] Patel et al. (2004), [8] Lutovinov et al. (2005b), [9] Combi et al. (2004), [10] Walter (2005), [11] Walter et al. (2004), [12] Markwardt et al. (2005), [13] Rodriguez et al. (2005a)
[b] This is an observed maximum value which is only reached occasionally.

(e.g., IGR J16358−4726: Patel et al. 2004; IGR J16465−4507: Lutovinov et al. 2005b). The highly absorbed IGR sources vary in brightness on time scales of minutes to hours, as well as from observation to observation, both at soft and hard X-ray energies (e.g., IGR J16318−4848: Walter et al. 2003, Matt & Guainazzi 2003, Matt et al. 2005; IGR J16320−4751: Rodriguez et al. 2003, Foschini et al. 2004). In IGR J16318−4848 the line emission also varies on time scales of $\sim$15 min and longer (Matt & Guainazzi 2003). Up to now, four of the (well-studied) highly absorbed IGR sources have been seen to also vary on a regular time scale (between $\simeq$4 to $\simeq$100 min, see Table 1). IGR J16358−4726, for example, displayed a strong periodic flux modulation with a peak-to-peak pulse fraction of $\sim$70%. These have been interpreted as neutron star pulse periods. The absorption column is also seen to vary, from observation to observation (e.g., IGR J16318−4848: Revnivtsev 2003; IGR J16320−4751: Rodriguez et al. 2003; IGR J19140+0951: Rodriguez et al. 2005a).

In Table 1, I give the parameters for the (cut-off) power-law model[4], which is usually used to fit the soft X-ray (*ASCA*, *XMM-Newton*, *Chandra*) spectral data, most of the time, in combination with *INTEGRAL* spectral data. The spectra are hard ($\Gamma \lesssim 2$), and show evidence for high-energy cut-off values. However, one must note here that the soft-energy spectra were not taken simultaneously with the hard X-ray spectra. Since the sources are (highly) variable, both at soft and hard X-ray energies, one must be cautious with the spectral fitting results (see, e.g., Walter et al. 2003). Of the highly absorbed IGR sources, only IGR J16138−4848 shows very strong emission lines (see above). The others only show weak (or undetectable) line emission.

X-ray and optical/IR emission

As shown in the previous Section, the highly absorbed IGR sources have similar X-ray properties: hard X-ray (cut-off; $E_{cut} \gtrsim 10$ keV) power-law ($\Gamma \lesssim 2$) emission and some, or most of the time, strong ($N_H \gtrsim 10^{23}$) and variable absorption. Four of them have been found to show (long-period; $\gtrsim$4 min) 'pulse' periods. These properties are rather typical for accreting X-ray pulsars in HMXBs (see, e.g., White et al. 1983), suggesting that the highly absorbed IGR sources are HMXBs. The highly absorbed IGR sources are located in the direction of the Norma spiral-arm tangent, except for IGR J19140+0951. The latter is located in the direction of the Sagittarius spiral-arm tangent; various HMXB X-ray pulsars are located in that direction, as well as other IGR sources (see, e.g., Molkov et al. 2004). These regions have an enhanced concentration of young massive stars (e.g., Grimm et al. 2002), and supports the HMXB hypothesis for the highly absorbed IGR sources. If they lie indeed in the above-mentioned spiral arms, their unabsorbed $\sim$2–100 keV X-ray luminosity would be around 10^{35} to a few times 10^{36} erg s^{-1}, also very typical for HMXBs. Note that various other of the less absorbed IGR sources are classified as bonafide HMXB/Be-X-ray transients (see, e.g., Negueruela 2005).

---

[4] The cut-off power-law model is given by $e^{-N_H \sigma(E)} A_{pl} E^{\Gamma} e^{E/E_{cut}}$, where N_H is the absorption column density, $\sigma(E)$ the absorption cross-section, A_{pl} the power-law normalization, Γ the power-law index, and E_{cut} the cut-off energy.

There are a few well-known HMXBs which show strong ($N_H \gtrsim 10^{23}$ cm^{-2}) and variable absorption, as well as strong Fe Kα emission and X-ray pulsations: GX 301−2 (Swank et al. 1976, Endo et al. 2002), Vela X-1 (Haberl & White 1990), XTE J0421+56 (CI Cam; Boirin et al. 2002). The latter is an atypical Be/X-ray binary (see below). Note that GX 1+4, not an HMXB but a rare type of symbiotic LMXB (see below), showed also similar spectra during an extended low state (Naik et al. 2005). Other known HMXBs exist which show similarly hard and strongly absorbed X-ray spectra (e.g., EXO 1722−363: Tawara et al. 1989, Takeuchi et al. 1990, see also Walter 2005; 4U 1909+07: Levine et al. 2004). The fact that the absorption is seen to vary in the HMXBs and the highly absorbed IGR sources, suggests it is intrinsic. For example, in GX 302−1 the absorption varies between 3×10^{23} and 2×10^{24} cm^{-2}, and is indeed connected to its 41 day orbit (e.g., White & Swank 1984, Endo et al. 2002).

Long X-ray 'pulse' periods have been found previously in HMXBs (but in those which show only moderate X-ray absorption), such as SAX J2239.3+6116 (21 min: in 't Zand et al. 2001), 4U 2206+54 ($\simeq$60 min; Masetti et al. 2004) and 4U 0114+650 (2.7 hr: Finley et al. 1992). 4U 2206+54 and 4U 0114+650 show noticeably similar pulse profiles as those seen in the highly absorbed IGR sources. Most of the X-ray pulsar HMXBs with pulse period longer than typically a few minutes are considered to be pulsars fed by a stellar wind (e.g., Nagase 1989). In the classical Corbet (1986) diagram the highly absorbed IGR sources either fall in the group of supergiant systems or the long orbital period ($\gtrsim$30 days) Be/X-ray transients. So far, only IGR J19140+0951 has a reported orbital period ($P_{orb} \simeq 13.55$ days, Corbet et al. 2004).[5] If the highly absorbed IGR sources indeed contain (slow) pulsars, their compact object is evidently a neutron star. However, if this interpretation is not correct, a black hole can not be excluded either (this is especially true for those IGR sources with no identified pulsations).

The presence of strong absorption in the X-ray domain shows that the compact object must be embedded in a dense circumstellar envelope, originating from a dense stellar wind from the donor. This (relatively cold) envelope also serves as the source of the fluorescent emission, especially in IGR J16318−4848 (e.g., Walter et al. 2003; Matt & Guainazzi 2003; Revnivtsev et al. 2003).

About 70% of the mass donors in HMXB are classical Be stars, the rest are blue supergiants. A big fraction of the accretion powered X-ray pulsars in the Be-systems are transients. Be stars show rich emission lines their spectra. There is, however, a subclass of objects which also shows forbidden lines, as well as a near-IR excess. These are the B[e] stars; they include many objects of different types and evolutionary status (e.g., Lamers et al. 1998). The typical mass-loss rates in these stars are $\dot{M} \gtrsim 10^{-6}$ M$_\odot$ yr^{-1}.

---

[5] Of course, it is assumed here that the highly absorbed IGR sources are binaries; this, however, remains to be verified. In this respect it is interesting to note that no convincing orbital period has been reported for XTE J0421+56 either (see, e.g., Hynes et al. 2002, for a discussion). Note also, that the HMXB interpretation was questioned by Patel et al. (2004), when discussing the periodic X-ray variations in IGR J16358−4726. They argued that even extremely small amounts of accretion can spin up the star to shorter periods than those observed. The only way out might be a pulsar in a Be/X-ray binary, which is able to spin down due to the propeller effect during the long quiescence period in between outbursts. Interestingly, the pulsations seen in 4U 0114+650 have been interpreted as pulsations from its early B star donor (Finley et al. 1992).

A few of the highly absorbed IGR sources have been identified in the optical and/or IR. The IR spectra of IGR J16318−4848 are rich in emission lines, i.e., various order H-lines, He I and II, low excitation permitted lines, as well as forbidden iron lines; some lines show P-Cygni profiles (Filliatre & Chaty 2004). Many of these IR spectral lines can be identified in XTE J0421+56 too (see Clark et al. 1999). For both XTE J0421+56 (e.g., Clark et al. 1999, Hynes et al. 2002) and IGR J16318−4848 (Filliatre & Chaty 2004) it has been suggested that they have a supergiant B[e] donor present in a dense and absorbing circumstellar environment. Comparable IR spectra are also seen in GX 1+4. Its donor is, however, a cool giant star, and the IR spectra show in addition late-type features, such as CO-bands (Clark et al. 1999). These features are not seen in either XTE J0421+56 or IGR J16318−4848. The donor in IGR J16465−4507 has also been suggested to be an (early) supergiant (Smith 2004), whereas in IGR J16320−4751 it is either a cool giant or supergiant (Rodriguez et al. 2003). Early-type stars are also found in the error circles for some of the other highly IR absorbed sources.

The near-IR excess in B[e] stars points to the presence of hot circumstellar dust. Both in IGR J16320−4751 (Rodriguez et al. 2003) and IGR J16465−4507 (Smith 2004) there is evidence for such a near-IR excess. The IR spectra of IGR J16318−4848 suggest a similar configuration (Filliatre & Chaty 2004).

The column density derived from the optical extinction is found to be one to two orders of magnitude less than that derived from the X-ray measurements (IGR J16318−4848: Walter et al. 2003, Filliatre & Chaty 2004; IGR J16320−4751: Rodriguez et al. 2003; IGR J16465−4507: Smith 2004). This suggests that the dense circumstellar envelope must be rather compact and concentrated towards the compact object (e.g., Revnivtsev et al. 2003).

Soft X-ray excess

Soft X-ray excess emission between 0.3 and 5 keV has been reported for IGR J16318−4848 (when fitting the observed X-ray spectrum with a power-law spectrum to standard absorption column, see footnote on page 4). It seems to be consisting of two parts, one above and one below ∼2 keV. Partial covering could account for the excess between ∼2−5 keV. If the covering fraction is less than 1, part of the X-ray illuminated surface should be directly visible, producing a Compton reflection component (Matt & Guainazzi 2003, Matt et al. 2005, and references therein). It is not clear whether such a component is present or not (Walter et al. 2003, Matt & Guainazzi 2003). The excess between 0.3−2 keV could not be easily explained (Matt & Guainazzi 2003). Hints of a soft X-ray excess are seen as well in IGR J16320−4751 (Rodriguez et al. 2005b).

There is evidence for an (independent) soft component in the X-ray spectrum of XTE J0421+56 in outburst, subsequent decay, and quiescence, below a few keV (Boirin et al. 2002, Ishida et al. 2004, and references therein). A similar feature is seen in the soft X-ray spectrum of CH Cyg, a symbiotic star containing a white dwarf (CH Cyg; Ezuka et al. 1998). Modeling the parts below and above ∼2 keV as separate emission components (Ezuka et al. 1998; Ishida et al. 2004), or modeling the emission with a partial covering absorption model (e.g., Boirin et al. 2002) seemed to work fine. Because of the similarity with CH Cyg, it was proposed that XTE J0421+56 contains a white dwarf (Ishida et al. 2004); this is, however, hard to reconcile with the observed X-ray properties of XTE J0421+56 (e.g., Hynes et al. 2002, and references therein) and IGR J16318−4848. It is interesting to note that soft X-ray spectra during X-ray dips in the light curves of LMXBs and HMXBs also show an excess in emission below typically 4 keV (see, e.g., Kuulkers et al. 1998, and references therein). Such dips are thought to be due to strong absorption (up to a few 10^{23} cm^{-2}) of emission from the inner parts of the accretion disk and compact object by the cooler outer parts of the accretion disk.

Wheatley (2001) showed that the soft X-ray spectrum of CH Cyg can be solely described by emission from the white dwarf which is strongly absorbed by a partionally ionised wind from the red giant. Recently, a partionally ionised absorber has been proposed as well to explain the soft X-ray spectra during the dips in LMXBs (Boirin et al. 2005). Since the soft spectral properties of XTE J0421+56 and IGR 16318−4848 are very similar to CH Cyg, it is logical to suggest such a model for these sources as well. The strong emission from the compact object is able to ionize the immediate environment. Such a ionised region can be responsible for, e.g., the observed IR He II emission in XTE J0421+56 and IGR 16318−4848. If this works, than there may not be a need for a Compton reflection component and one can conclude that the whole X-ray emitting region is covered by the circumstellar material.

SUMMARY

The above described X-ray and optical/IR properties suggests that the highly absorbed IGR sources are HMXBs containing either a neutron star or black hole in orbit around a (super)giant donor. The stellar wind accreting onto the compact object could form a dense envelope in which absorption, fluorescence and ionization takes place. This circumstellar envelope does not seem to cover much of the (super)giant donor. Because of the wavelength window *INTEGRAL* is able to observe, we are now starting to find more of this previously poorly known class of sources. Indeed, thanks to INTEGRAL "we can see clearly now ..." (White 2004).

ACKNOWLEDGMENTS

I am indebted to the conference organizers and editors, who allowed me to review these intriguing class of sources driven by results from *INTEGRAL*. I thank Deepto Chakrabarty for drawing my attention to the IR observations of GX 1+4, and Jérôme Rodriguez for discussions on an earlier draft of this paper.

REFERENCES

1. A.J. Bird, et al., *ApJ*, 607, L33 (2004).
2. L. Boirin, et al., *A&A*, 394, 205 (2002).
3. L. Boirin, et al., *A&A*, in press [astro-ph/0410385] (2005).

4. J.S. Clark, et al., *A&A*, 348, 888 (1999).
5. J.A. Combi, et al., *A&A*, 422, 1031 (2004).
6. R.H.D. Corbet, *MNRAS*, 220, 1047 (1986).
7. R.H.D. Corbet, et al., *ATel*, 269 (2004).
8. T.-J. Courvoisier, et al., *IAUC*, 8063 (2003).
9. J. de Plaa, et al., *IAUC*, 8076 (2003).
10. T. Endo, et al., *ApJ*, 574, 897 (2002).
11. H. Ezuka, et al., *ApJ*, 499, 388 (1998).
12. P. Filliatre, and S. Chaty, *ApJ*, 616, 469 (2004).
13. J.P. Finley, et al., *A&A*, 262, L25 (1992).
14. L. Foschini, et al., in *The INTEGRAL Universe*, ESA SP-552, 247 (2004).
15. H.-J. Grimm, et al., *A&A*, 391, 923 (2002).
16. F. Haberl, and N.E. White, *ApJ*, 361, 225 (1990).
17. R.I. Hynes, et al., *A&A*, 392, 991 (2002).
18. J.J.M. in 't Zand, et al., *A&A*, 380, L26 (2001).
19. M. Ishida, et al., *ApJ*, 601, 1088 (2004).
20. E. Kuulkers, et al., *ApJ*, 494, 753 (1998).
21. H.J.G.L.M. Lamers, et al., *A&A*, 340, 117 (1998).
22. F. Lebrun, et al., *A&A*, 411, L141 (2003).
23. F. Lebrun, et al., *Nat*, 428, 293 (2004).
24. A.M. Levine, et al., *ApJ*, 617, 1284 (2004).
25. A. Lutovinov, et al., *A&A*, 433, L41 (2005a).
26. A. Lutovinov, et al., *A&A*, submitted [astro-ph/0411550] (2005b).
27. C.B. Markwardt, et al., *ATel*, 465 (2005).
28. N. Masetti, et al., *A&A*, 423, 311 (2004).
29. G. Matt, and M. Guainazzi, *MNRAS*, 341, L13 (2003).
30. G. Matt, et al., these Proceedings (2005).
31. S.V. Molkov, et al., *AstL*, 30, 534 [astro-ph/0402416] (2004).
32. H. Murukami, et al., *IAUC*, 8070 (2003).
33. I. Negueruela, in *The Many Scales of the Universe - JENAM 2004 Astrophysics Reviews*, KAP, in press [astro-ph/0411759] (2004).
34. F. Nagase, *PASJ*, 41, 1 (1989).
35. S. Naik, et al., *ApJ*, 618, 866 (2005).
36. S.K. Patel, et al., *ApJ*, 602, L45 (2004).
37. M. Revnivtsev, *AstL*, 29, 719 [astro-ph/0304353] (2003).
38. M.G. Revnivtsev, et al., *AstL*, 29, 587 [astro-ph/0303274] (2003).
39. J. Rodriguez, et al., *A&A*, 407, L41 (2003).
40. J. Rodriguez, et al., *A&A*, 432, 235 (2005a).
41. J. Rodriguez, et al., in preparation (2005b).
42. N. Schartel, et al., *IAUC*, 8072 (2003).
43. L. Sidoli, et al., *A&A*, 429, L47 (2005).
44. D.M. Smith, *ATel*, 338 (2004).
45. J. Swank, et al., *ApJ*, 209, L57 (1976).
46. Y. Takeuchi, et al., *PASJ*, 42, 287 (1990).
47. Y. Tawara, et al., *PASJ*, 41, 473 (1989).
48. P. Ubertini, et al., *A&A*, 411, L131 (2003).
49. R. Walter, talk presented at the *Internal INTEGRAL Science Workshop*, ESA/ESTEC, Noordwijk, The Netherlands [see http://www.rssd.esa.int/INTEGRAL/workshops/Jan2005/] (2005).
50. R. Walter, et al., *A&A*, 411, L427 (2003).
51. R. Walter, et al., in *The INTEGRAL Universe*, ESA SP-552, 417 (2004).
52. P.J. Wheatley, *AIP Conf. Ser.*, 599, 1007 (2001).
53. N. White, *Nat*, News & Views, 428, 264 (2004).
54. N.E. White, and J. Swank, *ApJ*, 287, 856 (1984).
55. N.E. White, et al., *ApJ*, 270, 711 (1983).
56. C. Winkler, et al., *A&A*, 411, L1 (2003).

Exploring the Nature of Weak Chandra Sources near the Galactic Centre

R. M. Bandyopadhyay*, J. C. A. Miller-Jones*,†, K. M. Blundell*, F. E. Bauer**,‡, Ph. Podsiadlowski*, Q. D. Wang§, S. Rappaport¶ and E. Pfahl||

*Dept. of Astrophysics, Oxford University, Oxford OX1 3RH, U.K.
†University of Amsterdam, Kruislaan 403, Amsterdam, The Netherlands
**Institute of Astronomy, Cambridge University, Cambridge CB3 0HA, U.K.
‡Columbia University, New York, NY 10027, USA
§Dept. of Astronomy, University of Massachusetts, Amherst, MA 01003, USA
¶Center for Space Research, MIT, Cambridge, MA 02139, USA
||Dept. of Astronomy, University of Virginia, Charlottesville, VA 22903, USA

Abstract. We present results from the first near-IR imaging of the weak X-ray sources discovered in the *Chandra*/ACIS-I survey (Wang et al. 2002) towards the Galactic Centre (GC). These ~800 discrete sources, which contribute significantly to the GC X-ray emission, represent an important and previously unknown population within the Galaxy. From our VLT observations we will identify likely IR counterparts to a sample of the hardest sources, which are most likely X-ray binaries. With these data we can place constraints on the nature of the discrete weak X-ray source population of the GC.

Keywords: Accretion - Stars: binaries - Stars: X-rays - Stars: infrared
PACS: 97.80.Jp

THE *CHANDRA* GALACTIC CENTRE SURVEY

In July 2001 Wang et al. (2002) performed an imaging survey with *Chandra*/ACIS-I of the central $0.8 \times 2°$ of the Galactic Centre (GC), revealing a large population of previously undiscovered discrete weak sources with X-ray luminosities of $10^{32} - 10^{35}\,\mathrm{erg\,s^{-1}}$. The nature of these ~800 newly detected sources, which may contribute ~10% of the total X-ray emission of the GC, is as yet unknown. In contrast to the populations of faint active Galactic nuclei (AGN) discovered from recent deep X-ray imaging out of the Galactic plane, our calculations suggest that the extragalactic contribution to the hard point source population over the entire Wang et al. survey is $\leq 10\%$, consistent with the log(N)-log(S) function derived from the *Chandra* Deep Field data (e.g. Brandt et al. 2001). The harder (≥ 3 keV) X-ray sources (for which the softer X-rays have been absorbed by the interstellar medium) are likely to be at the distance of the GC, while the softer sources are likely to be foreground X-ray active stars or cataclysmic variables (CVs) within a few kpc of the Sun. The distribution of X-ray colours (Figure 1) suggests that only a small fraction of the *Chandra* sources are foreground objects. The combined spectrum of the discrete sources shows emission lines characteristic of accreting systems such as CVs and X-ray binaries (XRBs). These hard, weak X-ray sources in the GC are therefore most likely a population of XRBs; candidate classes include quiescent black hole binaries or quiescent low-mass XRBs, CVs, and high-mass wind-accreting neutron

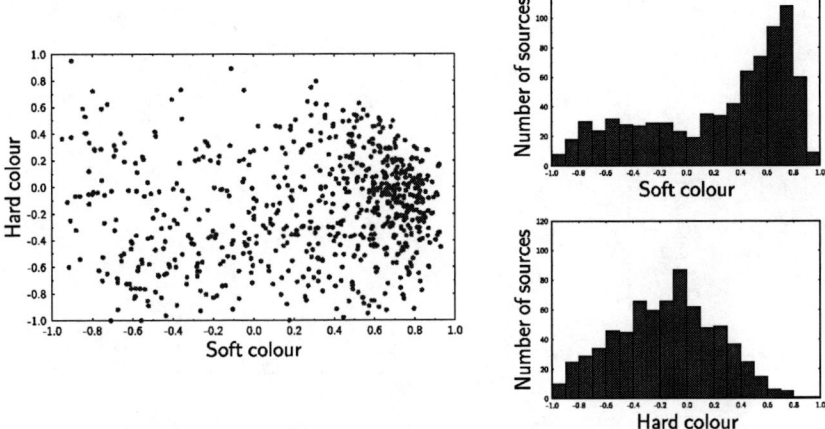

FIGURE 1. Characteristics of the X-ray source population detected in the *Chandra* mosaic. Left panel: colour-colour diagram; right panel: histogram of the number of soft and hard X-ray sources in the GC field. Soft colour is defined using the 1-3 and 3-5 keV bands, and hard colour using the 3-5 and 5-8 keV bands. The distribution of X-ray colours suggests that only a small fraction of the *Chandra* sources are foreground objects.

star binaries (WNSs).

WHAT ARE THESE POINT SOURCES?

Pfahl *et al.* (2002) have considered in detail the likely nature of these *Chandra* sources and concluded on the basis of binary population synthesis (BPS) models that many, if not the majority, of these systems are WNSs. Depending on the mass of the companions, the WNSs may belong to the "missing" population of wind-accreting Be/X-ray transients in quiescence or the progenitors of intermediate-mass X-ray binaries (IMXBs; $3 \lesssim M/M_\odot \lesssim 7$). The existence of tens of thousands of quiescent Be/XRBs in the Galaxy has been predicted since the early 1980s (Rappaport & van den Heuvel 1982; Meurs & van den Heuvel 1989), while it has only recently been recognized that IMXBs may constitute a very important class of XRBs that had not been considered before (King & Ritter 1999; Podsiadlowski & Rappaport 2000). The Wang *et al. Chandra* survey may contain as many as 10% of the entire Galactic population of WNSs. In addition to the WNSs, Pfahl *et al.* estimate that a small fraction of the *Chandra* sources could be CVs or transient low-mass XRBs/black-hole binaries.

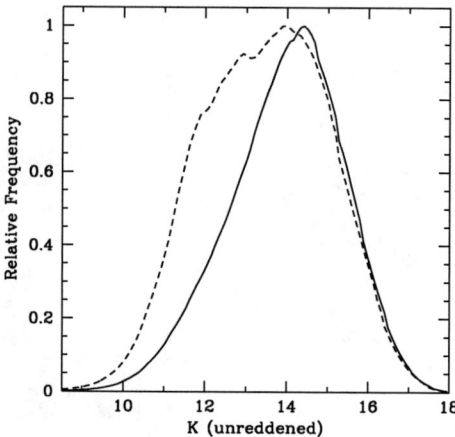

FIGURE 2. Distributions of intrinsic K magnitudes for the mass donors in WNS XRBs for two WNS formation models (Pfahl et al. 2002). The main difference between the two models is the relative proportion of binaries with intermediate- (solid line) and high-mass (dashed line) companions.

OUR VLT IMAGING PROGRAM

The first step in determining the nature of this population is to identify counterparts to the X-ray sources. The successful achievement of our goals requires astrometric accuracy and high angular resolution to overcome the confusion limit of the crowded GC. The 2MASS survey is severely confusion limited in the GC and is of insufficient depth (K=14.3) to detect the majority of the expected counterparts. We therefore constructed a survey program using the ISAAC IR camera on the VLT to obtain high-resolution JHK images in order to identify a statistically significant number of counterparts to the X-ray sources on the basis of the *Chandra* astrometry. We imaged 26 fields within the *Chandra* survey region, containing a total of 85 X-ray sources. In constructing our VLT program, we preferentially selected for hard X-ray sources from the *Chandra* survey, as the soft sources are most likely to be foreground. Of the hardest sources we selected those detected with an S/N$\geq$3 and which were imaged in the central area of the ACIS-I field (due to the off-axis characteristics of the *Chandra* point spread function, greater astrometric accuracy can be obtained for sources in the central regions of the field).

For the early-type donors of the WNSs, we would expect intrinsic magnitudes of K=11-16, with the peak of the magnitude distribution at $K \sim 14$ (Figure 2); these are therefore readily distinguishable from the majority of late-type donors expected for black hole X-ray transients which generally have $K \geq 16$ in quiescence. The average extinction towards the GC is $K \sim 3$; therefore with our images, which have a magnitude limit of K=20, we should detect most of the WNSs.

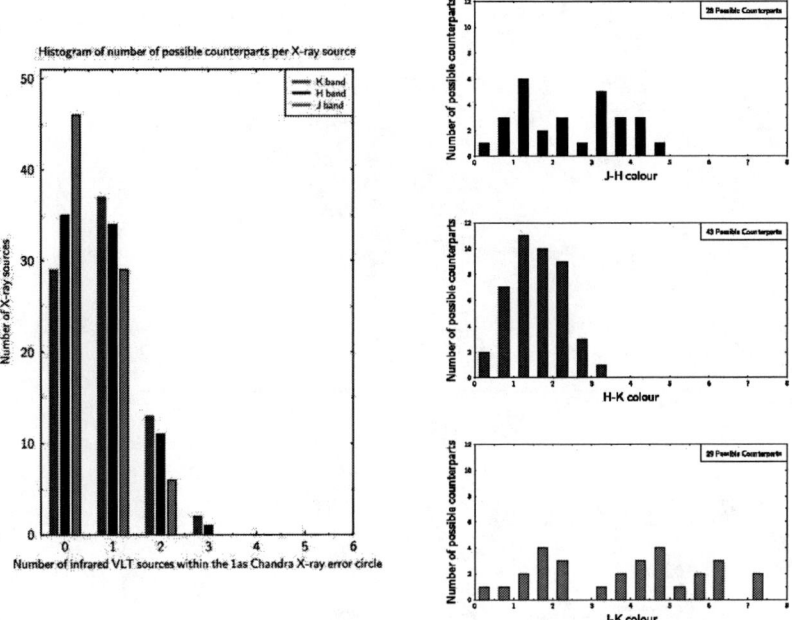

FIGURE 3. Left panel: histogram of the number of candidate IR counterparts per *Chandra* source within a 1" X-ray error circle. Right panel: distribution of IR colours of potential counterparts.

RESULTS: IR MAGNITUDES AND COLOURS

For 65% of the X-ray sources in our VLT fields, there are one or two resolved K-band sources within the 1" *Chandra* error circle; only a small number of X-ray sources have more than two potential counterparts (Figure 3). Over 50% of the *Chandra* sources have no potential J-band counterparts, and only a few of the potential IR counterparts have colours consistent with unreddened foreground stars (Figure 4). This is consistent with the expectation that the majority of the detected X-ray sources are heavily absorbed and thus are at or beyond the GC.

The magnitude and colour distribution of the identified candidate counterparts is redder than expected for WNS systems (see Figures 3 and 4). For an average GC extinction of A_K ~2-3, the peak of the expected reddened K magnitude distribution for the WNSs is ~16-17. The peak of the observed reddened K magnitudes for the potential counterparts is ~14-15, with an *(H-K)* colour of ~1-2, as expected for later-type stars. However, some potential counterparts do have colours consistent with early-type stars.

There are no K-band counterparts for ~35% of the *Chandra* sources. This is larger than the expected fraction of background AGN from the Chandra Deep Field estimate, though other groups have predicted larger fractions (up to 50%). However, we note that

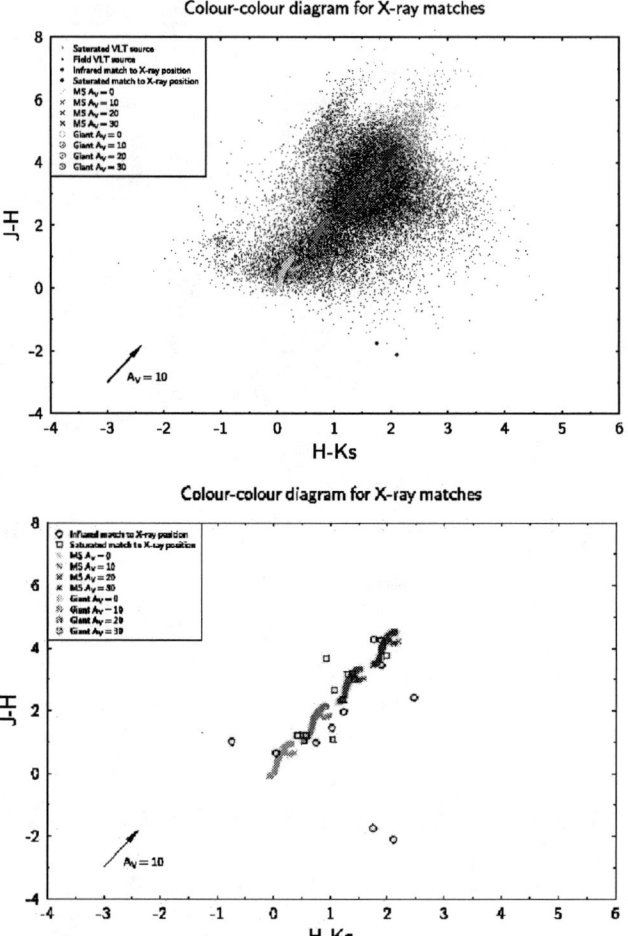

FIGURE 4. Top panel: colour-colour diagram showing all stars in our VLT fields. Bottom panel: colour-colour diagram of all potential IR counterparts to the *Chandra* sources for which we have full three-colour information. The theoretical main sequence and giant branch are indicated at visual extinctions of 0, 10, 20, 30. The IR colour-colour diagram illustrates that vast majority of field stars are consistent with highly reddened main sequence stars; thus most of the stars (including potential X-ray counterparts) are at the distance of the GC (or beyond).

the extinction in the GC is extremely variable, even in the K-band. This is evident from visual inspection of our VLT data, which clearly show areas of heavier than average extinction in the form of dust patches and lanes. Therefore we need to carefully determine which X-ray sources actually have no IR counterpart down to the $K=20$ magnitude limit, and which are located in areas of locally heavy extinction.

THE PROGRAM CONTINUES: IR SPECTROSCOPY

We have selected 36 of the best candidate counterparts for follow-up IR spectroscopy (candidate magnitudes $K \sim 12$-17); the goal of these observations will be to identify the X-ray source counterparts via detection of accretion signatures. The primary accretion signature in the K-band which distinguishes a true X-ray counterpart from a field star is strong Brackett γ emission; this technique of identifying XRB counterparts has been verified with observations of several well-studied GC XRBs (see *e.g.* Bandyopadhyay *et al.* 1999). As these *Chandra* sources are weaker in X-rays than the previously known population of Galactic XRBs, and thus have lower accretion rates, the emission signature will likely be somewhat weaker than in the more luminous XRB population. However, the Brackett γ accretion signature is clearly detected in the IR spectra of CVs, which are only weak X-ray emitters with a similar X-ray luminosity range to the *Chandra* sources (see *e.g.* Dhillon *et al.* 1997 for IR emission signatures in CVs). Therefore we expect the spectroscopic identification to be definitive even for these low-luminosity X-ray sources. For our brighter targets ($K=12$-14) the spectra we obtain will allow us not only to identify the counterpart via its emission signature but also to spectrally classify the mass donors if absorption features are detected, a crucial step in determining the nature of this new accreting binary population.

Identifying IR counterparts to these newly discovered X-ray sources provides a unique opportunity to obtain a census of the various populations of accreting binaries in the GC and may ultimately allow a determination of each system's physical properties. As this *Chandra* survey may contain 1% of the entire population of accreting binary systems in the Milky Way, our results will have important implications for our understanding of XRBs in the Galaxy, including their formation, evolutionary history, and physical characteristics. The results of this observational program will represent an important "calibration" point for BPS codes so that they can be more reliably applied to the study of other types of XRBs that have evolved from massive stars, including ultraluminous X-ray sources (ULXs) in external galaxies. Finally, the combination of the imaging data and our spectroscopic follow-up will allow us to identify the nature of an entirely new population of X-ray emitters within our Galaxy.

REFERENCES

1. Wang, Q.D., Gotthelf, E.V., & Lang, C.C., 2002, Nature, 415, 148.
2. Brandt, W.N., et al., 2001, AJ, 122, 2810.
3. Pfahl, E., Rappaport, S., & Podsiadlowski, Ph., 2002, ApJL, 571, L37.
4. Rappaport, S. & van den Heuvel, E.P.J., 1982, Proc. IAU Symposium 98, p. 327.
5. Meurs, E.J.A. & van den Heuvel, E.P.J., 1989, A&A, 226, 88.
6. King, A.R. & Ritter, H., 1999, MNRAS, 309, 253.
7. Podsiadlowski, Ph. & Rappaport, S., 2000, ApJ, 529, 946.
8. Bandyopadhyay, R.M., Shahbaz, T., Charles, P.A., & Naylor, T., 1999, MNRAS, 306, 417.
9. Dhillon, V.S., et al., 1997, MNRAS, 285, 95.

Chemical composition of secondary stars in LMXBs: implications on the progenitors of black holes and neutron stars

Jonay I. González Hernández

Instituto de Astrofísica de Canarias, E-38205 La Laguna, Tenerife, SPAIN: jonay@ll.iac.es

Abstract.
Recent studies of chemical abundances of secondary stars in low mass X-ray binaries have opened a new window to obtain information on the progenitors of black holes and neutron stars. Secondary stars could have captured a significant amount of the ejected matter in the supernova explosions that originated compact objects in these systems. Thus, anomalous atmospheric abundances of these stars may be a signature of nucleosynthetic products in supernovae. The detailed chemical analysis of secondary stars may provide constraints on many parameters involved in supernova explosion models like the mass cut, the amount of fallback matter, possible mixing processes, and explosion energies and geometries. Such chemical analysis has been made in two black hole binaries: GRO J1655−40 (Israelian et al. 1999) and A0620−00 (González Hernández et al. 2004) where we have found several elements enhanced in the secondary stars. These element abundances compared with element yields in supernova explosion models suggest that the progenitors of compact objects were massive stars in the mass range 25–40 $M_\odot$. We have also studied the neutron star binary Cen X-4 (González Hernández et al. 2005, submitted) where such strong anomalous abundances, with respect to element abundances in stars with similar Fe content, have not been found. However, element abundances appear to be super solar which might be explained if the secondary could be polluted by the ejected matter (also containing Fe) in a SN explosion of a progenitor with a He core of roughly 4 $M_\odot$.

Keywords: black hole physics—stars:abundances—stars:individual (Nova Sco 94, A0620−00, Cen X-4)—stars:neutron—stars:X-rays:low-mass—binaries
PACS: 97.60.Lf, 97.60.Jd, 97.80.Jp

INTRODUCTION

It is believed that massive stars with masses greater than roughly 8 $M_\odot$ are the progenitors of compact objects, either neutron stars or black holes. Hydrostatic nucleosynthesis produces heavy nuclei in the short-lived evolution of these massive stars which ends as supernova (SN hereafter) where explosive nucleosynthesis modifies appreciably the chemical composition of the ejecta.

Secondary stars in primordial low mass X-ray Binaries (LMXBs) might have captured a significant amount of the ejected matter, and therefore the study of the chemical composition of the secondary star can provide information about nucleosynthetic products in SNe and consequently, constrain the mass of the progenitor of the compact object in these systems.

Convection (Langer 1991) and rotation (Maeder & Meynet 2000; Heger et al., 2000) influence the structure and evolution of massive stars and subsequently the uncertainties in the treatment of these parameters limit our understanding of the evolution of the progenitors of compact objects. In addition, uncertainties in various aspects of the supernova explosion models (as e.g. the mass cut, the amount of fallback, mixing processes, and the energy and the symmetry of the SN explosion) affect the predictions of the final remnant mass and the chemical composition of any ejecta captured by the companion.

With the aim of obtaining information on the link between compact objects and their progenitor stars, Israelian et al. (1999) measured element abundances in the secondary star of the black hole binary Nova Scorpii 1994. They found α-elements enhanced by a factor of 6–10. Since these elements cannot be produced in a low mass secondary star, this was interpreted as evidence of a SN event that originated the compact object whose progenitor was in the mass range 25–40 $M_\odot$. Afterwards, these over-abundances were compared with a variety of supernova models, including standard as well as hypernova models (for various helium star masses, explosion energies, and explosion geometries) and a simple model of the evolution of the binary and the pollution of the secondary (Brown et al., 2000; Podsiadlowski et al., 2002). Later on, Orosz et al. (2001) also found overabundances (by a factor 2–10 with respect to solar) of some of these elements in the secondary star of the LMXB system J1819.3−2525 (V4641 Sgr). In addition, another evidence of anomalous abundances has also been found in the ultraviolet spectra of XTE J1118+48 suggesting that the accreting material has been CNO processed which may give us information on the evolutionary status of the secondary star in this system (Haswell et al. 2002).

In this paper I will sumarize the results we have obtained for the black hole binary A0620–00 and the neutron star binary Cen X-4.

CHEMICAL ANALYSIS

The chemical analysis of secondary stars in A0620–00 and Cen X-4 were carried out using optical spectra obtained with the UV-visual Echelle Spectrograph (UVES) at the European Southern Observatory (ESO), with the 8.2m *Very Large Telescope* (VLT). Twenty short exposures were obtained covering the spectral region between 4800 and 6800 Å with a resolving power $\lambda/\delta\lambda \sim 40000$. Invidual spectra were corrected from radial velocities and combined to improve the signal-to-noise ratio.

In our analysis we tried to obtain the stellar parameters of the secondary star by using synthetic spectral fits to the average spectrum of the secondary star, taking into account the effect of the veiling from the accretion disk on the stellar features. We selected several spectral features of FeI and using a grid of LTE models of atmospheres provided by Kurucz and the LTE code MOOG from Sneden (1973), we generated a grid of synthetic spectra for these features in terms of five free parameters, three to characterize the star atmospheric model (effective temperature, $T_{\rm eff}$, surface gravity, $\log g$, and metallicity, [Fe/H]) and two further parameters to take into account the effect of the accretion disk emission in the stellar spectrum. This was assumed to be a linear function of wavelength and thus characterized by two parameters: veiling at 4500 Å, $f_{4500} = F_{\rm disc}^{4500}/F_{\rm cont,star}^{4500}$, and the slope, m_0.

The average observed spectrum was compared with each of the 1,500,000 synthetic spectra in the grid via a χ^2 minimization procedure and using a bootstrap Monte Carlo

TABLE 1. Stellar and veiling parameters of these LMXBs

Parameter	A0620–00	Cen X-4
$T_{\rm eff}$	4900 ± 100 K	4500 ± 100 K
$\log g$	4.2 ± 0.3	3.9 ± 0.3
[Fe/H]	0.15 ± 0.20	0.25 ± 0.10
f_{4500}	0.30 ± 0.05	1.85 ± 0.10
m_0	-0.00014 ± 0.00002	-0.00071 ± 0.00003

method, we defined the confidence regions for the five free parameters whose most likely values are given in Table 1. Veiling was found to be less than 15 per cent in A0620–00 and 60 per cent in Cen X-4 at 5000 Å and decreasing toward longer wavelengths. Average surface gravities were found to be lower than typical values in late-type main sequence stars which indicates that these stars are expanded, filling their Roche lobes.

Using the derived five parameters we analyzed several spectral features of Fe, Ca, Al, Ti and Ni. In Table 2 I show these element abundances referred to the solar values adopted from Anders & Grevesse (1989) [1] with their associated total errors (from González Hernández et al. 2004; González Hernández et al. 2005, submitted).

TABLE 2. Chemical abundances

Element	$[E/H]_{\rm LTE}$	
	A0620–00	Cen X-4
Fe	0.14 ± 0.20	0.23 ± 0.10
Ca	0.10 ± 0.20	0.21 ± 0.17
Ti	0.37 ± 0.23	0.40 ± 0.17
Ni	0.27 ± 0.10	0.35 ± 0.10
Al	0.40 ± 0.12	0.30 ± 0.17

DISCUSSION

Heavy elements

We are searching for anomalous abundances as an evidence that the secondary star could be polluted by the ejecta in the supernova explosion that originated the compact object in these systems. In Fig. 1 I show the observed element abundances relative to iron in the secondary star of A0620–00 (solid-lined crosses) and Cen X-4 (dashed-lined crosses) in comparison with those in stars of the solar neighbourhood with similar Fe content. In both secondary stars, the [Ca/Fe] ratio is consistent with abundances of stars with similar iron content, while Ni and Ti appear to be moderately enhanced. Al is clearly overabundant in the secondary star of A0620–00 whereas in the secondary star of Cen X-4 is consistent with the Galactic abundance trend of this element in the relevant range of metallicities.

These results must be understood in the framework of possible evolutionary scenarios of these systems.

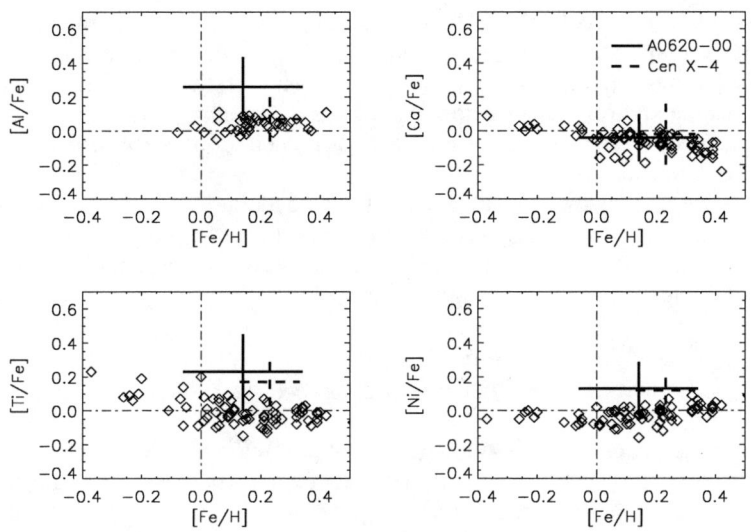

FIGURE 1. Abundances of secondary stars in A0620–00 (solid-lined wide crosses) and in Cen X-4 (dashed-lined wide crosses) in comparison with these element abundances in G and K metal-rich dwarf stars. Galactic trends of Ca, Ni, and Ti were taken from Bodaghee et al. (2003) while Al from Feltzing & Gustafsson (1998). The size of the cross indicates the error. The dashed-dotted lines indicate solar abundance values.

The secondary star of A0620–00

The evolutionary scenario proposed by de Kool et al. (1987) for the A0620–00 system begins with a massive star of roughly 40 $M_\odot$ and a companion of 1 solar mass. The short-lived massive star would rapidly develop a helium core of roughly 14 $M_\odot$ which would come to an end in SN explosion. Considering a spherically symmetric SN explosion we can calculate the pre-SN system parameters and therefore the amount of the ejected matter which is eventually captured by the companion star. We assume that the mass captured by the companion star is a well-mixed with the mass of its convective envelope at the time of the SN explosion. We used 40 $M_\odot$ spherically symmetric core-collapse explosion models from Umeda & Nomoto (2002 and 2003, private communication) for two different explosion energies, different mass cuts, fallback masses, capture efficiencies and secondary masses. The high mass of the compact object in this system ($11 \pm 1.9\ M_\odot$, Gelino et al. 2001 [3]) makes it quite difficult that a significant mass fraction of iron could have escaped from the collapsing matter in the supernova event because iron is formed in the inner layers of the star. Then assuming an initial chemical composition for the secondary star similar to the element abundances of field stars with similar iron content we can estimate the expected abundances of the newly chemical-enriched secondary star. By comparing these expected abundances with the observed abundances (see González Hernández et al. 2004) we found that only models with mass cuts as high as 11–12.5 $M_\odot$, depending on the capture efficiency assumed, can fit the observed abundances of the secondary star. In models with lower mass cuts, the amount of Al in the ejecta is too high in comparison with the other heavy elements analyzed, even under strong assumptions such as large amounts of fall-back matter and/or mixing efficiency fixed at 1. This result depends on neither the explosion energy nor the mass of the secondary.

In summary, the element abundances of the secondary star of A0620–00 can be explained if a progenitor with a $\sim 14\ M_\odot$ helium core explode with a mass cut in the range 11–12.5 $M_\odot$, such that no significant amount of iron could escape from the collapse of the inner layers.

The secondary star of Cen X-4

The compact object in Cen X-4 is a neutron star of mass in the range $M_{NS} = 0.5 - 2.5\ M_\odot$ (Torres et al. 2002). The mass cut in the SN explosion, that divides the ejecta from the collapsing core, is low enough that a significant fraction of iron could also escape from collapsing matter. In this case we have assumed that the initial chemical composition of the secondary star was solar and we have computed the expected abundances after contamination from the ejecta in the supernova. We used spherically symmetric core-collapse explosion models from Maeda et al. (2002, private communication) of a He core of mass $M_{He} \sim 4\ M_\odot$ for an explosion energy of 10^{51} ergs. The expected abundances presented in González Hernández et al. (2005, submitted) can fit very well the observed abundances in the secondary star for lower mass cuts. However, these element abundances are strongly sensitive to the mass cut, and therefore for a slighlty higher mass cut the expected abundances cannot reproduce the observed abundances.

On the other hand, the high center-of-mass velocity of the system suggests that the compact object received a *natal* kick in the supernova explosion due to an asymmetric mass ejection and/or an asymmetry in the neutrino emission (see Lai et al. 2001 for a review, and references therein). Thus, we have also inspected the pollution of the secondary star from the ejecta of a non-spherically symmetric SN explosion using models from Maeda et al. (2002, private communication). On the contrary to the spherical case, Al is dramatically enhanced in all the model computations and this effect is more relevant for higher mass cuts. Thus, none of the model computations produce acceptable fits, even for lower mass cuts.

To summarize, the element abundances of the secondary star of Cen X-4 are super solar and could be the primordial abundances of the system. However, if the secondary star captured a significant amount of the ejecta in the SN that gave rise to the neutron star, the observed abundances can be explained with a spherically symmetric SN explosion of a progenitor of with a $\sim 4\ M_\odot$ helium core.

Futher chemical analysis of secondary stars will be carried out in order to extract more information about the connection between compact objects and their progenitors.

ACKNOWLEDGMENTS

I would like to thank Rafael Rebolo, Garik Israelian and Jorge Casares for they collaboration in this project. I am also grateful to Keiichi Maeda, Hideyuki Umeda and Ken'ichi Nomoto for their contribution with their Supernova explosion models and for helpful discussions. This work has been partially financed by the Spanish Ministry project AYA2001-1657.

REFERENCES

1. Anders, E.,& Grevesse, N. 1989, Geochim. Cosmochim. Acta, 53, 197
2. Brown, G. E., Lee, C.-H., Wijers, R. A. M. J., Lee, H. K., Israelian, G., & Bethe, H. A. 2000, NewA, 5, 191
3. Gelino, D. M., Harrison, T. E., & Orosz, J. A. 2001, ApJ, 122, 2668
4. González Hernández, J. I., Rebolo, R., Israelian, G., Casares, J., Maeder, A., & Meynet, G. 2004, ApJ, 609, 988
5. González Hernández, J. I., Rebolo, R., Israelian, G., Casares, J., Keiichi, M., Bonifacio, P., & Molaro, P. 2005, submitted.
6. Haswell, C. A., Hynes, R. I., King, A. R., & Schenker, K. 2002, MNRAS, 332, 928
7. Heger, A., Langer, N., & Woosley, S. E. 2000, ApJ, 528, 368
8. Israelian, G., Rebolo, R., Basri, G., Casares, J., & Martín, E. L. 1999, Nature, 401, 142
9. de Kool, M., van den Heuvel, E. P. J., & Pylyser, E. 1987, A&A, 183, 47
10. Lai, D., Chernoff, D. F., & Cordes, J. M. 2001, ApJ, 549, 1111
11. Langer, N. 1991, A&A, 248, 531
12. Maeda, K., Nakamura, T., Nomoto, K., Mazzali, P. A., Patat, F., & Hachisu, I. 2002, ApJ, 565, 405
13. Maeder A.,& Meynet, G. 2000, A&ARA, 38, 143
14. Orosz, J. A., Kuulkers, E., van der klis, M., McClintock, J. E., Garcia, M. R., Callanan, P. J., Baylin, C. D., Jain, R. K., & Remillard, R. A. 2001, ApJ, 555, 489
15. Podsiadlowski, P., Nomoto, K., Maeda, K., Nakamura, T., Mazzali, P., & Schmidt, B. 2002, ApJ, 567, 491
16. Sneden, C. 1973, PhD Dissertation (Univ. of Texas at Austin)
17. Torres, M. A. P., Casares, J., Martínez-Pais, & Charles, P. A. 2002, MNRAS, 334, 233
18. Umeda, H.,& Nomoto, K. 2002, ApJ, 565, 385
19. Umeda, H.,& Nomoto, K. 2003, Nature, 422, 871

Stellar-Mass Black Hole Binaries as ULXs

S. Rappaport*, P. Podsiadlowski† and E. Pfahl**

*Department of Physics, 37-602B, MIT, Cambridge, MA 02139
†Department of Astrophysics, Oxford University, Oxford OX1 3RH
**Department of Astronomy, University of Virginia, Charlottesville, VA 22903-0818

Abstract. Ultraluminous X-ray sources (ULXs) with $L_x > 10^{39}$ ergs s^{-1} have been discovered in great numbers in external galaxies with *ROSAT*, *Chandra*, and *XMM-Newton*. The central question regarding this important class of sources is whether they represent an extension in the luminosity function of binary X-ray sources containing neutron stars and stellar-mass black holes (BHs), or a new class of objects, e.g., systems containing intermediate-mass black holes (100 – 1000 $M_\odot$). We have carried out a theoretical study to test whether a large fraction of the ULXs, especially those in galaxies with recent star formation activity, can be explained with binary systems containing stellar-mass black holes. To this end, we have applied a unique set of binary evolution models for black-hole X-ray binaries, coupled to a binary population synthesis code, to model the ULXs observed in external galaxies. We find that for donor stars with initial masses > 10 $M_\odot$ the mass transfer driven by the normal nuclear evolution of the donor star is sufficient to potentially power most ULXs. This is the case during core hydrogen burning and, to an even more pronounced degree, while the donor star ascends the giant branch, though the latter phases last only ∼5% of the main sequence phase. We show that with only a modest violation of the Eddington limit, e.g., a factor of ∼10, both the numbers and properties of the majority of the ULXs can be reproduced.

INTRODUCTION

The *Chandra X-ray Observatory* and the *XMM-Newton* mission have been used to study entire populations of accretion powered binary X-ray sources in external galaxies to distances exceeding ∼100 Mpc. The observed sources with X-ray luminosities, L_x, up to a few $\times 10^{38}$ ergs s^{-1} are very likely closely related to the high- and low-mass X-ray binaries that have been well studied for the past four decades in our own Galaxy and its neighbors. The discovery of ultraluminous X-ray sources (ULXs; defined as having $L_x > 10^{39}$ ergs s^{-1} in the 2-10 keV band) with *Einstein* (Fabbiano 1989), *ROSAT* (Roberts & Warwick 2000; Colbert & Ptak 2002; Ptak & Colbert 2004), and *ASCA* (Makashima et al. 2000) has been greatly extended by *Chandra* and *XMM-Newton* with their far superior sensitivity (see, e.g., reviews by Fabbiano & White 2004 [18]; Colbert & Miller 2004). These sources have been observed to L_x values as high as ∼ 5 × 10^{40} ergs s^{-1}, and are difficult to explain with conventional X-ray binaries. A key question which observations of these sources seek to answer is whether the compact object is (1) a neutron star of mass ∼ 1.4 $M_\odot$ or black hole of up to ∼ 15 $M_\odot$ (see, e.g., Tanaka & Lewin 1995; Greiner et al. 2001; Lee et al. 2002; McClintock & Remillard 2004), or (2) a black hole of "intermediate mass", e.g., 100 − 1000 $M_\odot$ (e.g., Colbert & Mushotzky 1999). This is the question we address in this article.

ULXs appear in different types of galaxies, including ellipticals (Angelini et al. 2001;

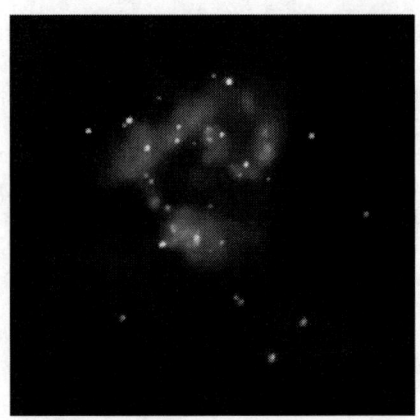
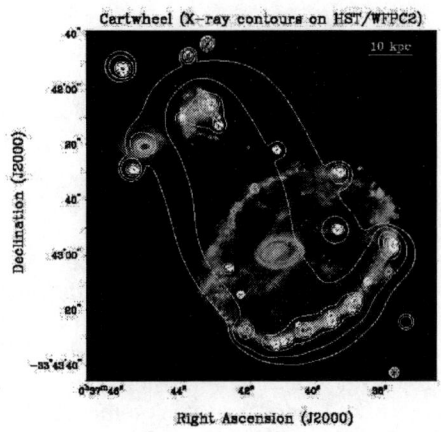

FIGURE 1. 72 ksec Chandra image of the central region ($4' \times 4'$) of the Antennae galaxies, revealing 49 bright X-ray sources, including 18 ULXs, in the starburst region produced by the galaxy collision (Fabbiano et al. 2001).

FIGURE 2. The Cartwheel galaxy as recorded in X-rays with *Chandra* (Gao et al. 2003, based on the data of Wolter & Trinchieri 2002; 2004). The hard X-ray contours are shown superposed on the HST/WFPC2 image ringed with bright HII regions.

Jeltema et al. 2003) where their luminosities are generally confined to $L_x < 2 \times 10^{39}$ ergs s^{-1} (Irwin et al. 2004). ULXs, a few with luminosities as high as $\sim 5 \times 10^{40}$ ergs s^{-1}, are especially prevalent in galaxies with starburst activity, including ones that have likely undergone a recent dynamical encounter (e.g., Fabbiano et al. 2001; Wolter & Trinchieri 2003; Belczynski et al. 2004; Fabbiano & White 2004; Colbert & Miller 2004). Two highly photogenic examples are the Antennae and Cartwheel galaxies. The Antennae galaxies (see Fig. 1; Fabbiano et al. 2001) include 49 very luminous X-ray sources, 18 of which are classified as ULXs (Zezas et al. 2002a,b), and are likely to be by-products of the star formation triggered by the collision of these galaxies (e.g., Hernquist & Weil 1993). Approximately half of the ULXs in the Antennae are identified with young star-forming regions while the other half are not associated with any star cluster (see, e.g., Fabbiano & White 2004). The Cartwheel galaxy reveals a substantial number of highly luminous point sources in a ring coinciding with starburst activity and punctuated by numerous HII regions (see Fig. 2; Wolter & Trinchieri 2003, 2004; Gao et al. 2003). This ring of star formation is apparently propagating outward from the original point of disturbance which was presumably triggered by the penetration of a smaller galaxy some 500 Myr ago. The *Chandra* sensitivity limit at the distance of the Cartwheel ($\sim$ 120 Mpc) is $L_x \simeq 5 \times 10^{38}$ ergs s^{-1}.

The ULXs found in these galaxies and numerous others have been suggested to harbor "intermediate mass black holes" (IMBHs, e.g., Colbert & Mushotzky 1999). The motivation for this is clear. The Eddington limit for spherically symmetric accretion is given by $L_{\text{Edd}} \simeq 2.5 \times 10^{38}(M/M_\odot)(1+X)^{-1}$ ergs s^{-1}, where M is the mass of the accretor and X is the hydrogen mass fraction in the accreted material. The Eddington limit for neutron stars is only $\sim 2 \times 10^{38}$ ergs s^{-1}, though a few accretion-powered

X-ray sources have typical persistent luminosities of $\sim 5 - 8 \times 10^{38}$ ergs s^{-1} (e.g., Levine et al. 1991, 1993)—not quite in the ULX range. IMBHs, by contrast, would have Eddington luminosities (L_{Edd}) of $\sim 10^{40} - 10^{41}$ ergs s^{-1} which nicely span the ULX range. Moreover, the expected spectra from IMBHs accreting substantially below their Eddington limit would have low inner-disk temperatures, as is inferred for some of the ULXs (Miller et al. 2003; Miller et al. 2004a,b; Cropper et al. 2004). However, before invoking a new type of hitherto unobserved object, it is sensible to ascertain whether stellar-mass black holes (hereafter, "BH") of mass $5 - 15\ M_\odot$ could explain many or most of the ULXs. The corresponding value of L_{Edd} for these sources is as high as $\sim 2 \times 10^{39}$ ergs s^{-1}, i.e., extends into the low end of the ULX range. Several Galactic BH transient X-ray sources have been suspected of exceeding their respective values of L_{Edd}, but the best case may be GRS 1915+105 where the observed L_x has been above 10^{39} ergs s^{-1} about $\sim$30% of the time in daily *RXTE—ASM* averages over the past 8 years (A. Levine, private communication). This source is reported to contain a $\sim 14\ M_\odot$ BH (Greiner et al. 2001). If taken at face value, however, sources with Eddington limited stellar-mass BHs would still fall short in explaining a significant fraction of the observed ULXs.

A number of ideas have been put forth for ways to circumvent the problem of how $\sim 10\ M_\odot$ BHs could have apparent L_x values as high as a few $\times\ 10^{40}$ ergs s^{-1}. King et al. (2001) suggested that the radiation may be geometrically beamed by a thick accretion disk so that the true value of L_x does not, in fact, exceed the Eddington limit. Körding et al. (2002) proposed that the apparently super-Eddington ULXs are actually emission from microblazer jets that are relativistically beamed along our line of sight. However, studies of the giant ionization nebulae surrounding a number of the ULXs (Pakull & Mirioni 2003) seem to confirm the full luminosity inferred from the X-ray measurements. Begelman (2002) [6] and Ruszkowski & Begelman (2003) found that in radiation pressure dominated accretion disks super-Eddington accretion rates of a factor of $\sim$10 can be achieved due to the existence of a photon-bubble instability in magnetically constrained plasmas. They propose that this instability results in a large fraction of the disk *volume* being composed of tenuous plasma, while the bulk of the *mass* is contained in high-density regions. The photons then diffuse out of the disk mostly through the tenuous regions, thereby effectively increasing L_{Edd}. This effect is shown to grow as $M^{1/5}$ and may yield a super-Eddington factor (hereafter "Begelman factor") of $\sim$10 in disks around stellar-mass BH systems.

Finally, we point out that the issue of whether the compact objects in ULXs are stellar-mass black holes (BH), intermediate-mass black holes (IMBH), or a mix, has profound implications beyond those associated with this particular class of X-ray source. If IMBHs are ultimately demonstrated to be present in at least some ULXs, they might well contribute to the elements involved in the formation of massive black holes (e.g., $> 10^6\ M_\odot$) via mergers that follow galaxy mergers (Haehnelt & Kauffmann 2000; Ebisuzaki et al. 2001; Milosavljevic & Merritt 2001).

STELLAR-MASS BLACK HOLE BINARY MODELS

Binary Population Synthesis

In two earlier studies, we have explored the formation and evolution of binary systems containing a stellar-mass BH and normal star (Podsiadlowski, Rappaport, & Han 2003, hereafter PRH; Rappaport, Podsiadlowski, & Pfahl, hereafter RPP). For these studies we developed two sets of tools which we herein combine to generate a simulated population of black-hole binaries. The first is a binary population synthesis code (BPS) where we start with a very large set of massive primordial binaries and generate a much smaller subset of these that evolve to contain a BH and relatively unevolved companion star. The product is a set of "incipient" BH X-ray binaries with a particular distribution of orbital periods, P_{orb}, donor masses, M_2, and BH masses, M_{BH}, for each of a number of different sets of input assumptions. For this part of the calculation, we employed various "prescriptions", based on single star evolution models for the primary, simple orbital dynamics associated with wind mass loss and transfer, assumptions about the magnitude of the wind mass loss from the primary as well as from the core of the primary after the common envelope, and natal kicks during the core collapse and formation of the BH. Simple energetic arguments were used to yield the final-to-initial orbital separation during the common-envelope phase wherein the envelope of the primary is ejected. (For detailed discussions and extensive references regarding these topics see PRH and RPP.)

The second important resource needed for a population study of ULXs (with stellar-mass BHs) is a grid of binary evolution models carried out with a Henyey-type stellar evolution code. These calculations are done with a full Henyey code so that at least the behavior of the donor star throughout its mass-transfer phase should be quite accurate and realistic. To our knowledge the PRH study was the first to carry out binary stellar evolution calculations through the X-ray phase of stellar-mass BH binaries (see also RPP; Kalogera et al. 2004; Portegies Zwart et al. 2004). Examples from our current grid of 52 binary evolution models are shown in Figs. 3 and 4. In Fig. 3 we show plots of the "potential" X-ray luminosity (L_x^{pot}) for 14 binary evolution sequences with initial donor masses of 2–17 $M_\odot$ from the original models presented in PRH (for the case of initially *unevolved* donors). Here, L_x^{pot} refers to the energy output expected in the absence of the Eddington limit; i.e., if the X-ray luminosity, L_x, were limited only by the mass transfer rate, $\dot{M}$, and the energy conversion efficiency which is dictated by the instantaneous spin of the BH (Bardeen 1970). We assumed that each BH starts with a mass equal to 10 $M_\odot$ and with zero spin (i.e., $j = 0$). We note that our binary models include, in addition to $\dot{M}$ and L_x, the evolution of P_{orb}, M_2, M_{BH}, and j (see Fig. 4 of PRH for the evolution of these other parameters). Fig. 4 shows the evolution of L_x for a sample of initially more evolved donors.

Experience has shown that population synthesis calculations for systems containing collapsed stars typically require $> 10^4$ active systems in order to achieve results of high statistical quality (see, e.g., Howell, Nelson, & Rappaport 2001). Since each of the 52 models we have computed required some half hour of cpu time it is therefore impractical at this time to run many thousands of BH binary models with a Henyey code to describe the population. However, we have found from our models that the evolutions

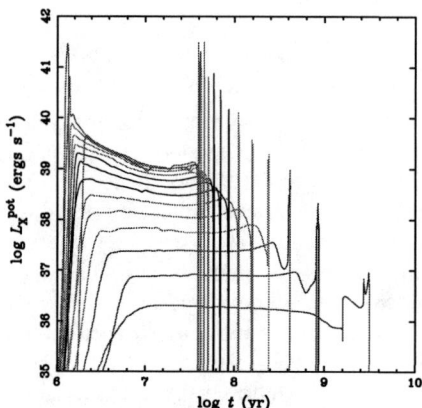
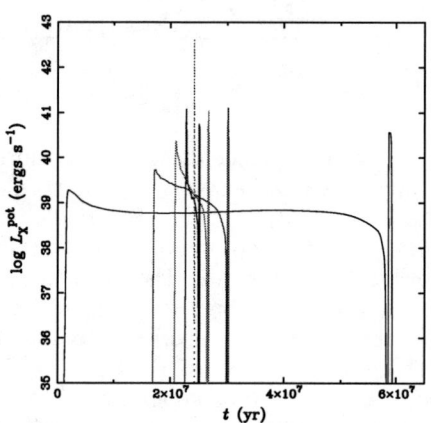

FIGURE 3. Potential X-ray luminosities as a function of time for 14 black-hole X-ray binary evolution sequences with masses ranging from 2 to 17 $M_\odot$ (bottom to top; from PRH and RPP). In all cases the donor star is unevolved at the start of mass transfer, and the black-hole mass is $10 M_\odot$. The spiky feature at the end of each evolution corresponds to the donor star ascending the giant branch.

FIGURE 4. Potential X-ray luminosities as a function of time for 5 BH X-ray binary evolution sequences; each curve corresponds to a different evolutionary state of the donor when mass transfer commences. In all cases the initial masses of both the donor and BH are $10 M_\odot$. The four solid curves correspond to the central H abundance of the donor star at the onset of mass transfer of 0.7, 0.35, 0.2, and 0.1, respectively. The dotted curve is for a donor that has already evolved off the MS (so-called early case B mass transfer).

in L_x form a nearly self-similar set. That is, if we take any two of the evolutionary models, but especially ones that are close in initial M_2 and $P_{\rm orb}$, then the plots of L_x vs. evolution time can be scaled in time and in L_x so they are nearly the same. This is quite different from the case of binary evolution for systems containing neutron stars (see, e.g., Podsiadlowski, Rappaport, & Pfahl 2002) where the evolution tracks for initially similar systems can diverge dramatically. Therefore, we have made use of this self-similar behavior to develop an interpolation scheme to produce an effectively much larger set of models.

We have used this binary evolution interpolation scheme to generate a population of 10^5 evolution tracks for BH binaries. For this study we used somewhat more general distributions of incipient systems in the $P_{\rm orb} - M_2$ plane that are inspired by the BPS results we found in PRH, rather than taken directly from the specific output for any particular model. We utilized three generic types of distributions for the incipient BH binaries. In the first, $P_{\rm orb}$ and M_2 were uniformly distributed between $\sim 0.6 - 4$ days and $2 - 17$ $M_\odot$, respectively (Models E and B), thereby allowing for the broadest contributions from all the binary evolution models that we have calculated. Second, we considered $P_{\rm orb}$ distributed uniformly as above, but M_2 is distributed in the same way as the stellar IMF as deduced by, e.g., Salpeter (1955), Miller & Scalo (1979), and Kroupa et al. (1993) (Models S and TS). Third, only the initially more massive secondary stars were assumed to be successful in ejecting the envelope of the primary, and the incipient

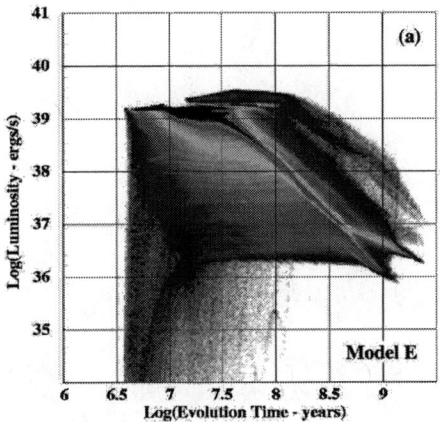 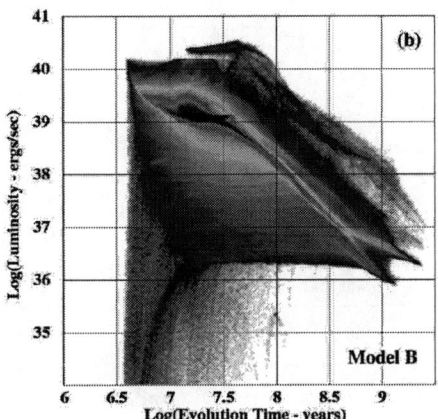

FIGURE 5. Simulated evolution of the bolometric X-ray luminosity function (LF) of BH binaries with time since an impulsive star formation event. 10^5 evolution tracks are represented in the image. The gray scale crudely represents the logarithm of the relative populations. The mass of the BH at the start of mass transfer is taken to be 10 $M_\odot$. The appropriate value of $L_{\rm Edd}$ for each system has been applied. The upper bar-like feature represents systems with donor stars on the giant branch with higher mass transfer rates. (From RPP.)

FIGURE 6. Simulated evolution of the X-ray luminosity function of BH binaries with time since an impulsive star formation event. The gray scale and other specifications are the same as for Fig. 5, except that the Eddington limit is allowed to be violated by up to a factor of 10 (see text for details), and each system radiates the luminosity that corresponds to the mass transfer rate and the energy conversion efficiency according to the instantaneous spin of the black hole (Bardeen 1970). (From RPP.)

population of BH binaries had secondary masses confined to $M_2 > 6\,M_\odot$ (Model 6), and were uniformly distributed above this value.

In addition to the three different distributions of incipient BH binaries in the $P_{\rm orb} - M_2$ plane described above, there are two different assumptions we made regarding the maximum L_x that can be radiated by a BH binary. First, the maximum L_x is taken to be just that given by $L_{\rm Edd}$ (Model E) which, in turn, is governed by the mass of the BH and the H/He composition of the accreted material. Second, we allow for the possibility that L_x up to 10 $L_{\rm Edd}$ can be attained (Model B; Begelman 2002; Ruszkowski & Begelman 2003). Finally, we consider that L_x is, under certain physical conditions, transient in its behavior (Model TS). The transient behavior is thought to arise from the well-known thermal-ionization disk instability (Cannizzo et al. 1982; van Paradijs 1996; King et al. 1996; Dubus et al. 2001; Lasota 2001). The specific prescription we used to determine whether a particular model at a given time would exhibit transient behavior is described in detail in PRH. When one of our sources is in an evolutionary phase where it would be a transient we simply allow, in a somewhat *ad hoc* manner, the mass transfer rate onto the BH to increase by a factor of 30, but then give that source a probability weighting of only 1/30 in the population, indicating low duty cycle "on states".

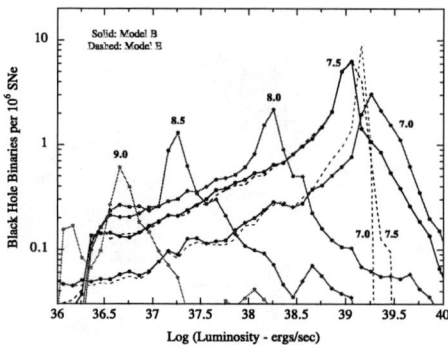
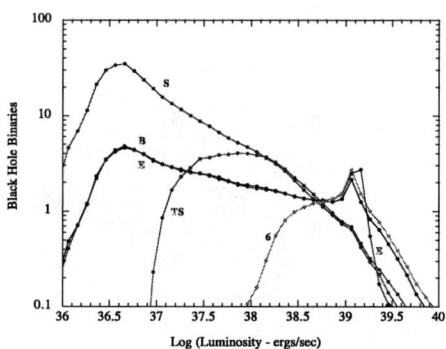

FIGURE 7. Calculated X-ray luminosity functions of BH binaries at five different epochs after an *impulsive* star formation event (units: BH-binaries per L_x bin). The time label on each curve indicates $\log(t_{ev})$ in years. Solid and dashed curves are for Model B (Fig. 6) and Model E (i.e., L_x limited to L_{Edd}; see Fig. 5), respectively. The distributions are normalized to a star formation event yielding 10^6 core-collapse SNe. (From RPP.)

FIGURE 8. Luminosity functions for a galaxy undergoing *continuous* star formation at a fixed age $\gg 10^8$ yr. The core-collapse SN rate is taken to be 0.01 yr^{-1}. The results are derived from the types of luminosity function "images" shown in Figs. 5 and 6 by weighting each column of pixels by the time interval represented by that pixel, and then summing each row of pixels over all columns. Models are defined in the text. (From RPP.)

Results of the BPS

Illustrative luminosity functions (LF) from the RPP study are shown in Figs. 5 and 6 as color images. These are the LFs expected for Models E and B, respectively (defined above) as a function of time since an impulsive star formation event. They result from 10^5 BH binary evolution tracks generated as described in detail in RPP. The gray-scale crudely represent the logarithm of the numbers of sources with a particular value of L_x at t_{ev}. The arrays of values from which the plots are generated (a 700×700 pixel array) can be used to produce quantitative LFs vs. time after a star formation event, or LFs for any arbitrary star formation rate history. Sample results are shown in Figs. 7 and 8. In the first of these, the LFs at a set of logarithmically spaced times are shown for Model B following a star formation event in which 10^6 core-collapse SNe were produced. In the second, the Model B results were used to generate the expected LF for the case of uniform, continuous star formation (see Fig. 8; normalized to 0.01 SNe yr^{-1}; Cappellaro et al. 1999).

We can also illustrate how, for external galaxies, our results can be used to relate the spatial offsets between ULXs and current star forming regions. For this we use the evolving luminosity functions after an impulsive star formation event (e.g., Figs. 5 and 6). In this exercise we make a crude model for the annular concentration of luminous X-ray sources in the Cartwheel galaxy. We assume that $\sim 5 \times 10^8$ years ago a smaller galaxy passed through its center and triggered a wave of star formation (Hernquist & Weil 1993) which has just now reached a radius marked by numerous HII regions and luminous X-ray sources. We choose random locations over the surface of the disk; the radial distance of each location then specifies the time since the star formation wave

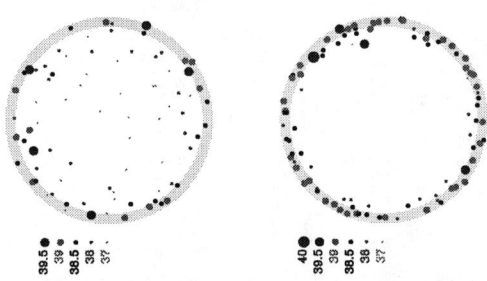
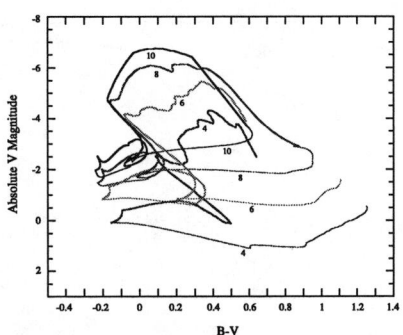

FIGURE 9. Simulated representation of the luminous BH X-ray binaries in a galaxy where a star formation wave was triggered by a catastrophic event at the center some 5×10^8 years ago. The left and right panels were produced for Models B and 6, respectively. The star formation wave is assumed to propagate outward at a constant speed. The X-ray luminosities were chosen via a Monte Carlo technique from the distribution for Model B (left panel) and Model 6 (right panel). This spatial distribution of sources is reminiscent of Chandra images of the Cartwheel galaxy. (From RPP.)

FIGURE 10. Evolution tracks in the HR diagram are shown for 4 of our models; with donor stars that had initial masses of 4, 6, 8, and 10 $M_\odot$ and were on the ZAMS at the start of mass transfer. Each binary system is represented by two different evolution tracks; the thin curve is the contribution from the donor star alone, while the thicker curve describes the track for the total system light–including the accretion disk.

has passed. The results shown in Fig. 6 are then used as a probability distribution for choosing L_x; this is repeated until 100 X-ray sources have been chosen (see left panel of Fig. 9). The shaded ring marks the outer 10% in radius. Note that while the ULXs are indeed concentrated in the outer ring, there are still a number of luminous sources ($> 10^{38}$ ergs s^{-1}) well inside the ring. We then repeated this exercise for Model 6 (for donors with $M_{2,i} > 6\,M_\odot$), and the sources are even more concentrated in the outer 10% of the simulated galaxy (right panel, Fig. 9).

Our models also have some predictive power concerning the optical appearance of BH–ULX binaries. We have computed tracks for our binary evolution models in the HR diagram, taking into account both the contribution from the donor star and the accretion disk. Light from the disk, in turn, results from reprocessing of X-radiation and from viscous heating. We computed the effective temperature of the disk as a function of radius for representative values of disk angular thickness and effective X-ray albedo. The contributions to the B and V bands from the donor stars were estimated from their $M_{\rm bol}$ and $T_{\rm eff}$ using conversion factors taken from Reed (1998). Illustrative results for 4 evolution tracks in the HR diagram are shown in Fig. 10 for initially unevolved donor stars of four different masses. It is clear that during the early part of the binary evolution, the contribution from the heated accretion disk leads to a brighter optical counterpart in the V band by about 1 magnitude. However, during the short-lived giant branch phase, when L_x is very much larger, the V-band magnitudes are enhanced by ~ 4 mag and the system colors are considerably bluer. Thus, we conclude that there is a substantial contribution to both the color and magnitude of the optical counterparts of the BH–ULXs from the accretion disk. These results need to be taken into account by observers trying to characterize the "donor star" from its location in an HR diagram.

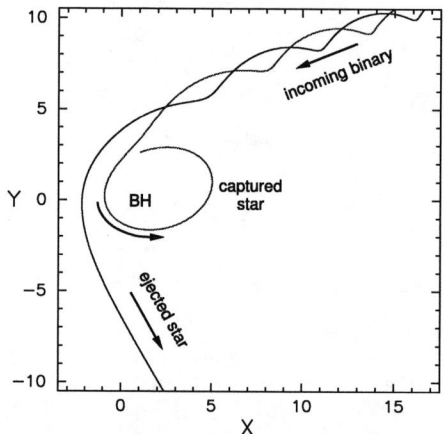

FIGURE 11. Exchange interaction between an incident binary consisting of two 30 $M_\odot$ stars (entering from the top right) and an IMBH (filled circle) – displayed in the rest frame of the IMBH. The IMBH disrupts the binary, captures one of the 30 $M_\odot$ stars, and the other is ejected. The motions depicted are projected onto an arbitrary $x-y$ plane from the original 3-D trajectory.

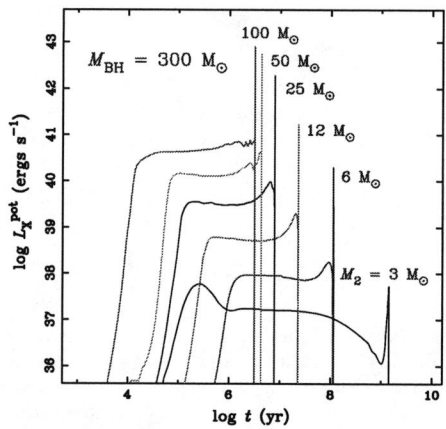

FIGURE 12. Potential luminosity for binary systems consisting of a 300 $M_\odot$ black hole and an initially unevolved donor star. Systems with donor stars having initial masses $> 16\,M_\odot$ will be ULXs for up to 10 Myr, including intervals while the donor is still on the main sequence. Donor stars with $M_{2,i}$ as low as 6 $M_\odot$ can appear as ULX sources, but only while the donor star is ascending the giant branch.

DISCUSSION AND CONCLUSIONS

Our models of stellar-mass BH binaries can easily account for the mass-transfer rates required to yield bolometric X-ray luminosities ($L_{x,\mathrm{bol}}$) of $\sim 2 \times 10^{39}$ ergs s^{-1} and the overall system formation rates without invoking any special parameters or circumstances. If we allow for enhancements in L_{Edd} by only a factor of 10, then the same holds for X-ray sources with $L_{x,\mathrm{bol}}$ up to $\sim 2 \times 10^{40}$ ergs s^{-1}. Greater loosening of the restrictions on L_{Edd} does not help the problem since $L_{x,\mathrm{bol}}$ is then limited by $\dot{M}$. After allowing for the fact that only about 1/3 of $L_{x,\mathrm{bol}}$ is expected to be emitted in the 2-10 keV band (this value is sensitive to the exact spectral shape) it is not entirely clear whether our models (binaries with stellar-mass BHs) can account for the ULXs that have L_x in the range $5 \times 10^{39} - 5 \times 10^{40}$ ergs s^{-1} unless significant X-ray beaming is invoked.

Because of the limitations that may be ultimately be imposed by the Eddington limit, intermediate-mass black hole (IMBH) accretors are also an attractive model to consider—at least for the most luminous of the ULXs. If such systems do indeed harbor IMBHs, then one proposed scenario for their formation and evolution could be as follows. During the formation of star clusters, there may be a runaway stellar collision process leading to the production of a very massive star which somehow evolves to core collapse before much of its envelope has been lost. While the dynamics of the runaway collision may be understood quantitatively (e.g., Portegies Zwart & McMillan 2002; Portegies Zwart et al. 2004) there is a great deal of uncertainty over how, and whether, such a massive star evolves to core collapse and the formation of an IMBH (but, see also

Miller & Hamilton 2002 for another formation scenario). If we allow for the presence of an IMBH at the center of a star cluster shortly after the formation of the cluster itself, then we have to consider how the IMBH acquires a mass-donating companion star. The dynamics of stars in the sphere of influence around the IMBH have been worked out both statistically (e.g., Bahcall & Wolf 1976) and with N-body simulations (e.g., Lin & Tremaine 1980; Baumgardt et al. 2004). Presumably the IMBH could capture a companion from the stars contained within this sphere of influence, either via tidal capture, or a mechanism that has been *heretofore unexplored* in this context, namely exchange encounters with primordial binaries in the cluster (Pfahl 2004; see discussion below and Fig. 11 for an example). After the IMBH has acquired a companion (potential donor star), but before steady mass transfer is possible, the star must survive tidal circularization and, more specifically, the tidal heating that accompanies the process. There has been much discussion in the literature about this issue (e.g., Ray et al. 1987; Podsiadlowski 1996; Alexander & Morris 2003), but few quantitative calculations have been carried out using a proper stellar evolution code (but, see Podsiadlowski 1996). Finally, the binary system has to survive encounters with other passing stars—either single or binary stars.

Once the companion to the IMBH has evolved to the point where mass transfer commences, then the issues come down to (i) whether $\dot{M}$ is adequate to power the most luminous ULXs with $L_x > 10^{40}$ ergs s^{-1}, and (ii) how long such systems would persist. Fig. 12 shows some illustrative examples of binary evolution involving an IMBH and a normal donor star. In particular the evolution of L_x^{pot} is shown for donor stars of 6 different initial masses, all commencing mass transfer while on the ZAMS. Note that systems with donor stars having initial masses $> 16\,M_\odot$ will be ULXs for up to 10 Myr.

ACKNOWLEDGMENTS

We thank Ed Colbert, Vicky Kalogera, Miriam Krauss, Ron Remillard, and Tim Roberts for helpful discussions. One of us (SR) acknowledges support from NASA RXTE Grant NAS5-30612. Ph.P. thanks the *Chandra* subcontract to MIT for partial support during his visit to M.I.T. EP was supported by NASA and Chandra Postdoctoral Fellowship Program through grant number PF2-30024.

REFERENCES

1. Alexander, T., & Morris, M. 2003, ApJ, 590, 25.
2. Angelini, L., Loewenstein, M., & Mushotzky, R.F. 2001, ApJ, 557, L35.
3. Bahcall, J.N., & Wolf, R.A. 1976, ApJ, 209, 214.
4. Bardeen, J.M. 1970, Nature, 226, 64.
5. Baumgardt, H., Makino, J., & Ebisuzaki, T. 2004, ApJ, in press [astro-ph/0406231].
6. Begelman, M. 2002, ApJ, 568, 97.
7. Belczynski, K., Kalogera, V., Zezas, A., & Fabbiano, G. 2004, ApJ, 601, L147.
8. Cannizzo, J.K., Ghosh, P., & Wheeler, J.C. 1982, ApJ, 260, L83.
9. Cappellaro, E., Evans, R., & Turatto, M. 1999, A&A, 351, 459.
10. Colbert, E., & Mushotzky, R. 1999, ApJ, 519, 89.

11. Colbert, E., & Ptak, A.F. 2002, ApJS, 143, 25.
12. Colbert, E.J.M., & Miller, M.C. 2004, talk at the Tenth Marcel Grossmann Meeting on General Relativity, Rio de Janeiro, July 20-26, 2003. Proceedings edited by M. Novello, S. Perez-Bergliaffa and R. Ruffini, World Scientific, Singapore, 2004 [astro-ph/0402677].
13. Cropper, M. Soria, R., Mushotzky, R.F., Wu, K., Markwardt, C.B., & Pakull, M. 2004, MNRAS, 349, 39.
14. Dubus, G., Hameury, J.-M., & Lasota, J.-P. 2001, A&A, 373, 251.
15. Ebisuzaki, T., et al. 2001, ApJ, 562, L19.
16. Fabbiano, G. 1989, ARA&A, 27, 87.
17. Fabbiano, G., Zezas, A., & Murray, S.S. 2001, ApJ, 554, 1035.
18. Fabbiano, G., & White, N.E. 2004, to appear in "Compact Stellar X-Ray Sources", eds. W.H.G. Lewin and M. van der Klis, (Cambridge Univ. Press: Cambridge) [astro-ph/0307077].
19. Gao, Y., Wang, D.Q., Appleton, P.N., & Lucas, R.A. 2003, ApJ, 596, L171.
20. Greiner, J., Cuby, J.G., & McCaughrean, M.J. 2001, Nature, 414, 522.
21. Haehnelt, M.G., & Kauffmann, G. 2000, MNRAS, 311, 576.
22. Hernquist, L., & Weil, M. 1993, MNRAS, 261, 804.
23. Hopman, C., Portegies Zwart, S.F., & Alexander, T. 2004, ApJ, 604, L101.
24. Howell, S.B., Nelson, L.A., & Rappaport, S. 2001, ApJ, 550, 897.
25. Irwin, J.A., , Bregman, J.N., & Athey, A.E. 2004, ApJ, 601, L143.
26. Jeltema, T.E., Canizares, C.R., Buote, D.A., & Garmire, G.P. 2003, ApJ, 585, 756.
27. Kalogera, V., Henninger, M., Ivanova, N., & King, A.R. 2004, ApJ, 603, L41.
28. King, A.R., Kolb, U., & Burderi, L. 1996, ApJ, 464, L127.
29. King, A.R., Davies, M.B., Ward, M.J., Fabbiano, G., & Elvis, M. 2001, ApJ, 552, L109.
30. Körding, E., Falcke, H., & Markoff, S. 2002, A&A, 382, L13.
31. Kroupa, P., Tout, C.A., & Gilmore, G. 1993, MNRAS, 262, 545.
32. Lasota, J.-P., 2001, New AR, 45, 449.
33. Lee, C.-H., et al. 2002, ApJ, 575, 996.
34. Levine, A., Rappaport, S., Putney, A., Corbet, R. & Nagase, F. 1991, ApJ, 381, 101.
35. Levine, A., Rappaport, S., Deeter, J.E., Boynton, P.E., Nagase, F. 1993, ApJ, 410, 328.
36. Lin, D.N.C., & Tremaine, S. 1980, ApJ, 242, 789.
37. Makashima, K., et al. 2000, ApJ, 535, 632.
38. McClintock, J.E., & Remillard, R. 2004, to appear in "Compact Stellar X-Ray Sources", eds. W.H.G. Lewin and M. van der Klis, (Cambridge Univ. Press: Cambridge) [astro-ph/0306213].
39. Miller, G.E., & Scalo, J.M. 1979, ApJS, 41, 513.
40. Miller, J.M., Fabbiano, G., Miller, M.C., & Fabian, A.C. 2003, ApJ, 585, L37.
41. Miller, J.M., Fabian, A.C., & Miller, M.C. 2004a, ApJ, 607, 931.
42. Miller, J.M., Fabian, A.C., & Miller, M.C. 2004b, ApJ, in press [astro-ph/0406656].
43. Miller, M.C., & Hamilton, D.P. 2002, MNRAS, 330, 232.
44. Milosavljevic, M. & Merritt, D. 2001, ApJ, 563, 34.
45. Pakull, M., & Mirioni, L. 2003, RMxAC, 15, 197.
46. Pfahl, E. 2004, in preparation for ApJ. ApJ, 597, 1036.
47. Podsiadlowski, Ph. 1996, MNRAS, 279, 1104.
48. Podsiadlowski, Ph., Rappaport, S., & Pfahl, E. 2002, ApJ, 565, 1107.
49. Podsiadlowski, Ph., Rappaport, S., & Han, Z. 2003, MNRAS, 341, 385 [PRH].
50. Portegies Zwart, S., & McMillan, S.L.W. 2002, ApJ, 576, 899.
51. Portegies Zwart S.F., Dewi, J., & Maccarone, T. 2004, MNRAS, 355, 413.
52. Ptak, A., & Colbert, E. 2004, ApJ, in press [astro-ph/0401525].
53. Reed, C. 1998, J. R.A.S. Canada, 92, 36.
54. Rappaport, S., Podsiadlowski, Ph., & Pfahl, E. 2004, MNRAS, in press [RPP].
55. Ray, A., Kembhavi, A. K., & Antia, H. M. 1987, A&A,184, 164.
56. Roberts, T., & Warwick, R. 2000, MNRAS, 315, 98.
57. Ruszkowski, M., & Begelman, M.C. 2003, ApJ, 586, 384.
58. Salpeter, E. E. 1955, ApJ, 121, 161.
59. Tanaka, Y., & Lewin, W.H.G. 1995, X-Ray Binaries, eds. W.H.G. Lewin, J. van Paradijs, & E.P.J. van den Heuvel (Cambridge: Cambridge University Press), p. 126.
60. van Paradijs, J. 1996, ApJ, 464, L139.

61. Wolter, A., & Trinchieri, G. 2003, MSAI, 73, 23.
62. Wolter, A., & Trinchieri, G. 2004, submitted to A&A [astro-ph/0407446].
63. Zezas, A., & Fabbiano, G. 2002a, ApJ, 577, 726.
64. Zezas, A., Fabbiano, G., Rots, A.H., & Murray, S.S. 2002b, ApJ, 577, 710.

Young Rotation-Powered Pulsars as Ultraluminous X-ray Sources in Star-Forming Galaxies

L. Stella* and R. Perna[†]

*Osservatorio Astronomico di Roma,
Via Frascati 33, 00040 Monteporzio Catone, Italy
[†]JILA and Dept of Astrophysical and Planetary Sciences, University of Colorado at Boulder, 440 UCB, Boulder, CO, 80309, USA

Abstract.
Nearby, (mostly) star-forming galaxies have revealed a number of ultraluminous X-ray sources (ULXs) with super-Eddington luminosities. This contribution discusses the possibility that a fraction of them could be young, Crab-like pulsars, the X-ray luminosity of which is powered by rotation. The pulsar birth parameters estimated from radio pulsar data are used to compute the steady-state pulsar X-ray luminosity distribution as a function of the star formation rate (SFR) in the galaxy. The fraction of sources with $L_x \gtrsim 10^{39}$ erg/s correlates with the SFR, ranging from $\sim 10\%$ in optically dull galaxies, to a few in starbursts galaxies. Furthermore, the X-ray luminosity of a few percents of galaxies is found to be dominated by a single bright pulsar with $L_x \gtrsim 10^{39}$ erg/s, roughly independently of its SFR. Observational diagnostics that can help distinguish a young pulsar population in ULXs are discussed.

Keywords: neutron stars — pulsars — cosmic X-ray sources — starburst galaxies
PACS: 97.60.Jd — 97.60.Gb — 98.70.Qy — 98.54.Ep

INTRODUCTION

It has been known since the ROSAT (and Einstein) surveys that the X-ray luminosity function of normal galaxies extends up to luminosities $\sim 10^{42}$ erg/s (e.g. Fabbiano 1988) for a review). However, the sensitivity and spatial resolution of the ROSAT and Einstein telescopes proved insufficient to resolve the X-ray emission into its main constituents. Recently, the superior spatial resolution and sensitivity afforded by the X-ray telescope on board *Chandra* demonstrated that an important fraction (up to a half) of the X-ray luminosity in starburst galaxies derives from a few very bright individual X-ray sources, the luminosity of which is often above the Eddington limit for a stellar mass black hole. Furthermore, optical follow-up studies of sources in deep *Chandra* images revealed a few early type galaxies that show no sign of nuclear activity and yet emit X-rays, likely from an individual source, at the $\sim 10^{41} - 10^{42}$ erg/s level (e.g. Fabbiano & White 2003 for a review).

A variety of models has been proposed to explain the origin of these ultraluminous X-ray sources. These include, among others, accreting intermediate mass black holes (IMBH) possibly in binaries, microquasars, transient super-Eddington accretors, background BL Lac objects, young supernova remnants (SNRs) in extremely dense environments, beamed X-ray binaries (e.g. Colbert & Mushotzky 1999; Makishima et al. 2000;

King et al. 2001; Begelman 2002; Miller et al. 2003). All the observational evidence gathered so far does not appear to point uniquely to any of the above scenarios. ULXs might actually be a heterogeneous population.

In this contribution, we examine the possibily that a sizeable fraction of these ultra-luminous X-ray sources are likely to be young, Crab-like pulsars, the X-ray emission of which is powered by their rotational energy rather than by accretion. A number of these sources is indeed expected in any star-forming region galaxy; these sources would not require any "special", unusual condition to output a large X-ray luminosity when they are very young. Here we compute their X-ray luminosity distribution, based on the birth parameters determined from pulsar radio studies; we then discuss observational diagnostics that can help identify them among the population of ULXs. More details can be found in Perna & Stella (2004).

PULSAR X-RAY LUMINOSITY

Most isolated neutron stars are X-ray bright throughout all their lifetime: at early times, their X-ray luminosity is powered by rotation (e.g. Michel 1991; Becker & Trumper 1997) after an age of $\sim 10^3 - 10^4$ yr, when the star has slowed down sufficiently, the dominant X-ray source becomes the internal energy of the star, and finally, when their internal heat is exhausted, accretion by the ISM continues to make them shine again, although to a low luminosity level (e.g. Blaes & Madau 1993; Popov et al. 2000; Perna et al. 2003) What is of interest for this work, is the origin of the X-ray luminosity at early times, when rotation is the main source of energy. During a Crab-like phase, relativistic particles accelerated in the pulsar magnetosphere are fed to a synchrotron emitting nebula, the emission of which is characterized by a powerlaw spectrum. Another important contribution is the pulsed X-ray luminosity (about 10 % of the total in the case of the Crab) originating directly from the pulsar magnetosphere.

Observationally there appears to be a correlation between the rotational energy loss of the star, $\dot{E}_{\rm rot} = I\Omega\dot{\Omega}$, and its X-ray luminosity, L_x. This correlation (with L_X in the 0.1-2.4 keV range) was first noticed in a small sample of radio pulsars by Seward & Wang (1988), and later examined in more detail by Becker & Trumper (1997). In the 2-10 keV band, the $L_x - \dot{E}_{\rm rot}$ correlation was first examined by Saito (1988) for a small sample of pulsars. The most comprehensive investigation up to date has been carried out by Possenti et al. (2002). They found, for a sample of 39 pulsars, that the $L_x - \dot{E}_{\rm rot}$ correlation is best fit by the relation

$$L_x = 10^{-15.3} (\dot{E}_{\rm rot}/{\rm erg~s}^{-1})^{1.34} {\rm ~erg~s}^{-1}. \qquad (1)$$

Although empirically established, there is no physical reason for which Eq.(1) should break down for values of $\dot{E}_{\rm rot}$ larger than the currently observed range. We therefore assume that it holds also for larger values of $\dot{E}_{\rm rot}$. Given the scatter in the data, we allow for a scatter in relation (1) by assigning, to a pulsar of rotational energy $\dot{E}_{\rm rot}$, an X-ray luminosity drawn from a log-Gaussian distribution of mean given by Eq.(1) and standard deviation (in log) $\sigma = 0.5$. We further impose the condition that the efficiency $\eta_x \equiv L_x/\dot{E}_{\rm rot}$ be always ≤ 1.

The rotational energy loss of the star, under the assumption that it is dominated by magnetic dipole losses, is given by

$$\dot{E}_{\rm rot} \simeq \frac{B^2 \sin^2\theta\, \Omega^4 R^6}{6c^3}, \qquad (2)$$

where R is the radius of the neutron star, which we assume fixed at a value of 10 km, B its magnetic field, $\Omega = 2\pi/P$ its angular velocity, and θ the angle between the magnetic and spin axes.

In order to compute the X-ray luminosity distribution of a population of young pulsars, the magnetic fields and the initial periods of the pulsars need to be known. Modeling the intrinsic properties of the Galactic population of pulsars, with the purpose of inferring their intrinsic properties, has been the subject of extensive investigation. The most recent, comprehensive analysis, based on large-scale 0.4 GHz pulsar surveys, has been carried out by Arzoumanian, Cordes & Chernoff (2002, ACC in the following). Here we use their inferred parameters under the assumption that spin down is only caused by dipole radiation losses and there is not a significant magnetic field decay. They find that the initial magnetic field strength (taken as Gaussian in log) has a mean $\langle \log B_0[G] \rangle = 12.35$ and a standard deviation of 0.4, while the initial birth period distribution (also taken as a log-Gaussian), is found to have a mean $\langle \log P_0(s) \rangle = -2.3$ with a standard deviation greater than 0.2. Here we adopt the value 0.3 while discussing the dependence of our results on σ (see §3.1). In agreement with ACC, we take $\sin\theta = 1$ in Eq.(2).

The spin evolution of the pulsars is then simply given by:

$$P(t) = \left[P_0^2 + \left(\frac{16\pi^2 R^6 B^2}{3Ic^3} \right) t \right]^{1/2}, \qquad (3)$$

where I is the moment of inertia of the star.

ACC derive that the pulsar birth rate $\dot{N}$ consistent with their initial parameters is of one pulsar every 760 yr. This value, which corresponds to a star formation rate (SFR) of $\sim 0.2 M_\odot$ yr^{-1} for a Salpeter IMF, is a factor of a few smaller than most other estimates in the literature. Here, we will show our results for this value of the SFR as well as for larger values that are appropriate for starburst galaxies.

It should be noted that, within the context of starbursts, Van Bever & Vanbeveren (2000) simulated the emitted hard X-ray luminosity from HMXBs and young SNRs, including their embedded pulsars. By adopting the Saito (1997) relation and a fixed (fiducial) value for the pulsar birth periods, they also found that some young pulsars can have X-ray luminosities exceeding the Eddington limit.

ULXS FROM A POPULATION OF YOUNG PULSARS

Montecarlo simulation of the luminosity function

The question we address here is the following: given a galaxy where the pulsar birth rate is $\dot{N}$, what is the number of sources with luminosity larger than L_X at any given time? Firstly, we wish to determine the probability distribution of a pulsar to have a given luminosity *at birth*. We determine this distribution by performing a Montecarlo simulation where, for any given pulsar, the initial period and magnetic field are randomly generated from the corresponding distributions described in §2. For that given set of parameters, Eqs. (2) and (1) are then used to compute the rotational energy loss and the corresponding X-ray luminosity. Figure 1 (left) shows the resulting probability distribution, i.e. the fraction of pulsars with an X-ray luminosity at birth larger than L_X, for three different values (0.3, 0.4, 0.5) of the width σ_{P_0} of the initial period distribution (for which ACC were able to derive only the lower limit $\sigma_{P_0} > 0.2$. While ACC cut their lower period distribution at 0.1 ms, we cut it at 0.5 ms, corresponding to a maximum velocity $\Omega_{max} \sim 12500$ rad/s, consistent with the break up speed (Cook et al. 1994) for typical NS equations of state (Wiringa 1988). The curves in Figure 1 (left) are obtained by randomly generating 20000 pulsars. As it can be seen from the figure, a substantial fraction of the pulsar population has an X-ray luminosity at birth which is super-Eddington. This fraction clearly increases with the width σ_{P_0} of the spin period distribution. Here we adopt the value $\sigma_{P_0} = 0.3$, which is consistent with the lower limit derived by ACC while being on the more conservative side.

In order to make a comparison with current observations, we need to answer the question posed at the beginning of the section, that is finding the number of sources with a given luminosity in any galaxy at any given time. Besides depending on the initial birth parameters P_0, B (and their time evolution), this number also depends on the birth rate of pulsars. As discussed above, we will use here both the pulsar birth rate derived by ACC, as well as higher rates as expected in starburst galaxies. We perform a Montecarlo simulation of the young pulsar population as follows: for a given birth rate $\dot{N}$, we take the total simulation time to be $T \gg 1/\dot{N}$, so that steady state is reached (typically $T \sim 10^4 1/\dot{N}$). The number of pulsars generated during this time is $N_{puls} = T/(1/\dot{N})$, with ages randomly drawn from a flat distribution between 0 and T. The birth parameters P_0, B are drawn from the distribution by ACC described in §2; the periods of the pulsars are evolved, for the pulsar ages, according to Eq.(3), and the luminosity as a function of time of each pulsar is estimated according to Eq.(1). For every choice of $\dot{N}$, the results are the average over 2000 different montecarlo realizations.

The four curves in Figure 1 (right) show the resulting X-ray luminosity distribution in a galaxy with a pulsar birth rate of $1/10, 1/50, 1/200, 1/760$ yr^{-1}. The luminosity function is described by a roughly "universal" function, i.e. a power law with index $\alpha \approx -0.4$, the normalization of which is proportional to the pulsar birth rate (or, equivalently, the SFR). An interesting comparison is the one between the X-ray luminosity function of the pulsars and that of the HMXBs. The latter, as shown by Grimm et al. (2003), also has an almost "universal' shape, which, in cumulative form, can be described by the function $N(>L_{38}) \approx 5.4$ SFR/(M$_\odot$yr^{-1})$[L_{38}^{-0.61} - 210^{-0.61}]$. The normalization depends on the SFR and for the Galaxy they quote a value of ~ 0.25M$_\odot$ yr^{-1} as found from a combination of different SFR indicators.

Figure 2 (left) shows a comparison between the integrated HMXB and pulsar counts for a galaxy with a SFR$\sim$ 1M$_\odot$/yr. The relative normalization between the two populations is calibrated to be the same as that for the Galaxy, i.e. so that a pulsar rate of 1/760 yr^{-1} corresponds to a SFR rate of 0.25 M$_\odot$/yr. The HMXB population generally

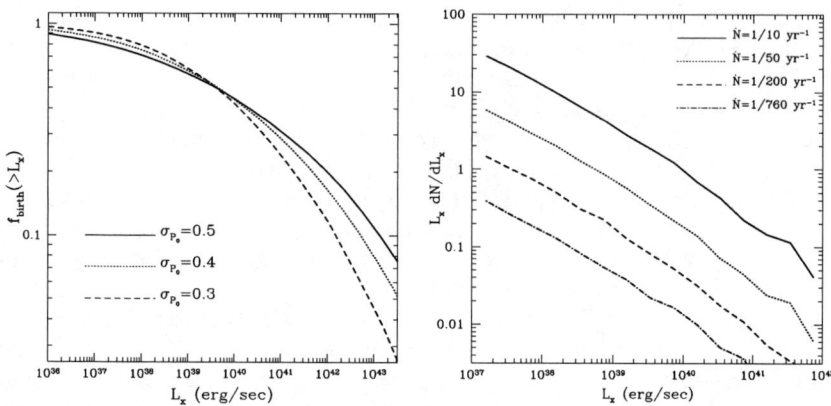

FIGURE 1. *Left*: Probability that a rotation-powered pulsar has an X-ray luminosity at birth larger than L_x for the pulsar birth parameters of the Arzoumanian et al. (2002) distribution, and for various values of the spread σ_{P_0} of the initial period distribution. *Right*: X-ray luminosity function (2–10 keV range) of the young pulsar population for different values of the pulsar birth rate.

dominates the X-ray luminosity at sub-Eddington luminosities; however at higher luminosities, where HMXBs drop out, the luminosity function of pulsars takes over. Our results show that, for a pulsar birth rate typical of our Galaxy, $\sim 7\%$ of galaxies are expected to have at least one source with luminosity $\gtrsim 10^{39}$ erg/s, and $\sim 0.3\%$ with luminosity $\gtrsim 10^{40}$ erg/s. Starburst galaxies, with SFR $\sim 10-20 M_\odot$/yr, are expected to each have at least one source with $L_x \gtrsim 10^{40}$ erg/s. Note that our simulation for the ACC pulsar rate (about 1/4 of the value shown in Fig.2 left), predicts that there should be ~ 1 Crab-like pulsar ($L_X \sim 10^{36} - 10^{37}$ erg/s) in our Galaxy, consistent with observations.

Another question of interest is the fraction of galaxies in which the *total* X-ray luminosity is dominated by that of a single young pulsar source. To address this issue, we ran 5000 random realizations of the steady-state pulsar population in a galaxy with a SFR similar to that of the Galaxy. We kept track of all the cases where the luminosity of a single source was $\geq 90\%$ of the total X-ray luminosity of the galaxy due to all the pulsar sources together. The fraction of galaxies the luminosity of which is dominated by a that of a single source according to the criterion above is shown in Figure 2 (right) as a function of the luminosity of the source. If all the X-ray sources in a galaxy were pulsars, then $\sim 10\%$ of galaxies would be dominated in X-rays by a single bright source. However, for a given SFR, one expects a number of HMXBs to be present. We therefore addressed the same question by including in the total luminosity of the galaxy also the contribution from HMXBs. To this aim, for each random realization of the pulsar population in the galaxy, we also randomly generated N_{HMXB} from their own luminosity distribution, where $N_{HMXB} \sim 80$ is the number of HMXBs corresponding to the SFR of the Galaxy (Grimm et al. 2003). Figure 2 (right) shows that, although the luminosity function of the HMXBs is a few times higher than that of the young pulsar population at luminosities $\lesssim 10^{40}$ erg/s, $\sim 2\%$ of galaxies will be dominated in X-rays by a single young pulsar. While the number of sources per galaxy with luminosity above a certain

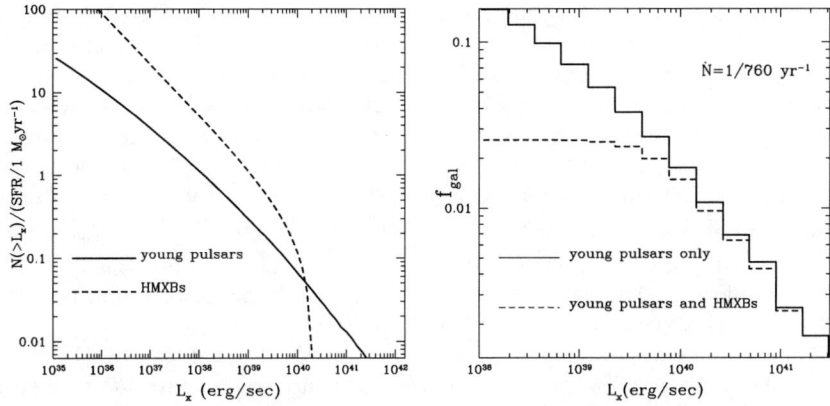

FIGURE 2. *Left*: Comparison between the pulsar and HMXB number counts, normalized to a SFR of $1 M_\odot/\text{yr}$. *Right*: Fraction of galaxies the X-ray luminosity of which (in the 2-10 keV range) is dominated ($> 90\%$) by a single young pulsar source with luminosity larger than L_x.

value roughly scales with the SFR (Fig. 2, left), the number of sources that dominate the total luminosity of the galaxy is roughly independent of the SFR, given the almost universal shape of the luminosity function of both pulsars and HMXBs.

It should also be noted that a young pulsar is probably surrounded by a bright SNR, and that the X-ray emission from a young SNR can be substantial, sometimes comparable to or larger than that of the associated pulsar (Immler & Lewin 2002). If the luminosity of the SNR is super-Eddington and larger than that of the pulsar, then the previously proposed SNR-ULX connection would hold. However, a recent study of young SNR luminosites by Bregman et al. (2003) showed that it is unlikely for young SNRs to make up a substantial fraction of the ULX population (see also Zesas (2002) for spectral studies). At any event, separating a possible contribution to the luminosity from the SNR could only be done spectroscopically, since the angular size of the X-ray emitting shell is expected to be well in subarcsec range. On the contrary a thermal-like spectral component could testify to the presence of the expanding SN shell. For more details on the effects of a young SN shell around a young pulsar see Perna & Stella (2004).

Observational diagnostics of the young pulsar population

We summarize below some observational diagnostics that can help identifying young rotation powered pulsars among the population of ULXs.

(a) X-ray spectra, variability and polarization. Young, rotation-powered pulsars display power law energy spectra. In a sample of pulsars studied with *Chandra*, Gotthelf (2003) found their photon indices to be in the range $0.6 < \gamma_{PSR} < 2.1$ (while their associated pulsar wind nebulae had $1.3 < \gamma_{PSR} < 2.3$). Pulsations might be detected with

high time resolution, high throughput, X-ray light curves of ULXs. The super-Eddington range of the distribution is produced by the subset of sources with smaller periods and larger magnetic fields. No appreciable intrinsic variability is expected on timescale of days to months. Aperiodix flux variability, if observed in these sources, would likely be due to external causes, such as the passage of a cloud along the line of sight to the source. A small, constant decline in luminosity is however predicted due to the pulsar spin down. For a spin down by pure dipole, the X-ray luminosity declines with time as $L_x = L_{x,0}(1+t/t_0)^{-2}$, where $t_0 \equiv 3Ic^3 P_i^2/B^2 R^6 (2\pi)^2 \sim 65 \, \text{yr} I_{45} B_{12}^{-2} R_{10}^{-6} P_{i,\text{ms}}^2$, having defined $I_{45} = I/(10^{45} \, \text{g cm}^2)$, $R_{10} = R/(10 \, \text{km})$, $P_{i,\text{ms}} = P_i/(1 \, \text{ms})$. For $t \lesssim t_0$ the flux does not vary significantly. Finally, based on the analogy with the Crab, a high polarization degree ($\sim 20\%$) is expected (Weisskopf et al. 1978).

(b) *Association with star forming regions and young SN remnants*. Newly born pulsars, even if endowed with large kick velocities, have not had the time to travel far from their birth sites, and therefore are expected to be mostly found in star forming regions. Hence, we expect that active, starbursts galaxies will contain a much larger fraction of ULXs than optically dull galaxies. For an optically dull galaxy with a SFR $\sim 0.2 M_\odot/\text{yr}$, we predict that there is a $\sim 7\%$ probability of finding a source with $L_x \gtrsim 10^{39}$ erg/s and a $\sim 0.3\%$ probability of finding a source with $L_x \gtrsim 10^{40}$ erg/s. Correspondingly, for an active, starburst galaxy with a SFR $\sim 10 M_\odot/\text{yr}$, we expect to find $\sim$ a few sources with $L_x \gtrsim 10^{39}$ erg/s and ~ 1 source with $L_x \gtrsim 10^{40}$ erg/s. Finally, the presence of a young SNR ($T_{SNR} \lesssim$ several tens of yr) surrounding the source is a fundamental ingredient of the young pulsar scenario. There is at least one case where the association between a ULX and a SNR has been established (Roberts et al. 2004)

(c) *Binarity*. A large number of stars are in binary sistems. Whether a binary survives the first SN explosion will mostly depend on the kick velocity of the NS and the mass of the companion; for high mass companions up to several tens of percent of binaries are expected to survive the first SN explosion (Brandt & Podsiadlowski 1995; Kalogera 1998). Therefore, a sizeable fraction of young pulsars is expected to be in a binary system. On the other hand, while binarity is required in most models of ULXs, for the young pulsar scenario it is not a requirement as the source of the X-ray luminosity is the rotational energy of the star.

(d) *Optical and radio emission*. Young pulsars are also detected in radio and often in optical (e.g. Becker & Trumper 1997). For the Crab, the ratio L_{opt}/L_x is $\sim 10^{-2}$, while for Vela it is $\sim 10^{-3}$. The statistics of optical detections in young pulsars is too small to yield predictions about the intensity of the optical radiation as a function of the X-ray one. However, if the ratio remains of the same order as for the Crab and Vela, then significant optical emission is expected, although prospects for detection appear promising only for sources in relatively closeby galaxies. The greatest majority of young X-ray pulsars are also radio emitters. Although the radio luminosity generally anti-correlates with age (Camilo 2002), the radio luminosity for the youngest pulsars is still many orders of magnitude below the X-ray luminosity ($L_{\text{radio}}/L_x \sim 10^{-5} - 10^{-10}$ at 1GHz), and radio detection in external galaxies is extremely difficult. The radio flux from a source associated with a young pulsar would instead be dominated by the radio emission from its young SN remnant.

Finally note that a contribution to the optical and longer wavelength emission can derive from a disk of fallback material from the SN explosion (e.g. Perna et al. 2000).

SUMMARY

The nature of ultraluminous X-ray sources is a current issue of debate. In this contribution we have discussed the possibility that a fraction of them could be associated with rotation-powered pulsars. We have performed a Montecarlo simulation which uses the pulsar birth parameters inferred from radio data analysis, together with an emphirical correlation between X-ray and spindown luminosity, to predict the X-ray luminosity function of this population. We have found that a few percent of optically dull galaxies are expected to possess young pulsars with highly super-Eddington luminosity, while starbursts galaxies can each have several of them. We have discussed possible diagnostics of these sources in the observed population of ULXs.

REFERENCES

1. Arzoumanian, Z, Cordes, J. M., Chernoff, D. F. 2002, ApJ, 568, 289 (ACC)
2. Becker, W., Trumper, J. 1997, A&A, 326, 682
3. Begelman, M. C. 2002, ApJ, 568, L97
4. Blaes, O., Madau, P. 1993, ApJ, 403, 690
5. Brandt, N., Podsiadlowski, P. 1995, MNRAS, 274, 461
6. Bregman, J. N., Houck, J. C., Chevalier, R. A., Roberts, M. S. 2003, ApJ, 596, 323
7. Camilo, F., Manchester, R. N., Gaensler, B. M.; Lorimer, D. R., Sarkissian, J. 2002, ApJ, 567L, 71
8. Colbert, E. J. M., Mushotzky, R. F. 1999, ApJ, 519, 89
9. Cook, G. B., Shapiro, S. L., Teukolsky, S. A. 1994, ApJ, 424, 823
10. Eracleous, M., Shields, J. C., Chartas, G., Moran, E. C. 2002, ApJ, 565, 108
11. Fabbiano, G. 1989, AR&A, 27, 87
12. Fabbiano, G., White, N. 2003, in "Compact Stellar X-ray Sources", Cambridge University Press (eds. W. Lewin, M. van der Klis)
13. Grimm, H.-J., Gilfanov, M., Sunyaev, R. 2003, MNRAS, 339, 793
14. Immler, S., Lewin, W. H. G. 2002, to appear in "Supernovae and Gamma-Ray Bursts," edited by K. W. Weiler (Springer-Verlag)
15. Kalogera, V. 1998, ApJ, 493, 368
16. King, A. R., Davies, M. B., Ward, M. J., Fabbiano, G., Elvis, M. 2001, ApJ, 552, L109
17. Makishima, K. et al. 2000, ApJ, 535, 632 Forman, W., Jones, C., Flanagan, K. 1995, ApJ, 483, 663
18. Michel, F. C. 1991, in "Theory of Neutron Star Magnetosphere", University of Chicago Press, Chicago, IL
19. Miller, J. M., Fabbiano, G., Miller, M. C., Fabian, A. C. 2003a, ApJ, 585, L37
20. Perna, R., Hernquist, L., Narayan, R. 2000, ApJ, 541, 344
21. Perna R., Narayan R., Rybicki G., Stella L., Treves, A., 2003, ApJ, 594, 936
22. Perna R., Stella, L., 2004, ApJ, 615, 222
23. Popov, S.B., Colpi, M., Treves, A., Lipunov, V. M., Prokhrov, M .E. 2000, ApJ, 530, 896
24. Possenti, A., Cerutti, R., Colpi, M., Mereghetti, S. 2002, A&A, 387, 993
25. Roberts, T. P., Warwick, R. S., Ward, M. J., Goad, M. R. 2004, preprint astro-ph/0401306
26. Saito, Y. 1998, Ph.D. Thesis, Univ. of Tokyo
27. Seward, F. D., Wang, Z. 1988, ApJ, 332, 1999
28. Van Bever, J., Vanbeveren, D. 2000, A&A, 358, 462
29. Weisskopf, M. C.; Silver, E. H.; Kestenbaum, H. L.; Long, K. S.; Novick, R. 1978, ApJ, 220, L117
30. Wiringa, R. B., Fiks, V., Fabrocini, A. 1988, Phys. Rev. C., 38, 1010
31. Zesas, A., Fabbiano, G., Rots, A. H., Murray, S. S. 2002, ApJ, 577, 710

SESSION 7

SUPERNOVAE TYPE Ia

Binaries, cluster dynamics and population studies of stars and stellar phenomena

Dany Vanbeveren

Astrophysical Institute, Vrije Universiteit Brussel, Pleinlaan 2, 1050 Brussels, Belgium.

Abstract. The effects of binaries on population studies of stars and stellar phenomena have been investigated over the past 3 decades by many research groups. Here we will focus mainly on the work that has been done recently in Brussels and we will consider the following topics: the effect of binaries on overall galactic chemical evolutionary models and on the rates of different types of supernova, the population of point-like X-ray sources where we distinguish the standard high mass X-ray binaries and the ULXs, a UFO-scenario for the formation of WR+OB binaries in dense star systems. Finally we critically discuss the possible effect of rotation on population studies.

1. INTRODUCTION

A population synthesis code calculates the temporal evolution of a population of stars in regions where star formation is continuous in time or in starbursts. Population number synthesis (PNS) predicts the number of stars of a certain type whereas population spectral synthesis (PSS) computes the effects that they have on the integrated spectrum.

Close binary evolution has been a main research topic in Brussels for about 30 years. The last decade, we investigated the effects of binaries on various aspects of population studies and we recently started to implement stellar dynamics (using direct N-body integration techniques) in our PNS and PSS codes. In the present paper we will summarize some of our recent results.

2. THE EFFECTS OF BINARIES ON GALACTIC EVOLUTION

Galactic chemical evolution (GCE). Although most of the existing GCE codes account in some parametrized way for the effects of supernova (SN) explosions of type Ia (SN Ia), they do not account in a consistent way for the evolution of the whole population of binaries. I sometimes get the impression that it is not always realized that the observed SN Ia rate indicates indirectly that a significant fraction of the intermediate mass stars are interacting binary members, or that the observations of OB populations including X-ray binaries, double neutron star binaries or SN type Ib indicate that also the massive close binary frequency is large.

In order to study the effects of binaries on GCE, one has to combine PNS (which includes massive and intermediate mass single stars and binaries) and a model that describes the formation and evolution of galaxies. Our group in Brussels is among the first ones who studied in a consistent way the effects of binaries on GCE. An extended review was published by De Donder and Vanbeveren (2004). In this paper we first present massive single star yields deduced from evolutionary calculations performed in Brussels which account for moderate convective core overshooting and for the effects of recent stellar wind mass loss rate formalisms. For the intermediate mass stars we use synthetic yields which consider all the dredge up phases during the AGB-phase (note that especially for the estimation of the carbon enrichment of a galaxy a correct implementation of the latter yields is essential). Secondly, we spent quite some time to tabulate binary yields as function of primary mass, binary mass ratio and period, using our extended library of binary tracks and we explain how to implement binaries in a GCE model. To decide whether or not an intermediate mass binary produces a SN Ia, we separately applied the single-degenerate (SD) model of Hashisu et al. (1999, and references therein) and the double-degenerate (DD) model. Next, since stellar evolution depends on the initial chemical composition of the gas out of which stars form, we linked our PNS code to a GCE model.

Obviously the inclusion of binaries in a GCE model increases significantly the number of parameters. We made a large number of simulations and when conclusions are formulated, we only restrained those that are largely independent from the parameter uncertainties. Summarizing:

- low and intermediate mass binaries enrich less (up to a factor 5) in carbon than single stars do. This is due to the fact that the dredge-up phases during the TP-AGB are suppressed by Roche lobe overflow in most binary systems.

- of course intermediate mass close binaries are essential in order to understand the SN Ia rate, thus to understand the variation of the galactic iron abundance.

- the inclusion of massive close binaries alters the galactic variation of α-elements by at most a factor 2-3. We leave it to the chemical evolutionary community to decide whether or not this difference is important enough in order to implement binaries in the codes. Notice however that such an implementation is NOT something simple like changing the yields in existing codes which do not account for the detailed evolution of a population of binaries.

- our models predict the appearance of merging degenerate binaries (black hole + neutron star or double neutron star binaries) a few million years after the formation of the galactic halo, early enough in order to explain the observed galactic variation of r-process elements. This conclusion depends essentially on the physics of case BB evolution after the spiral-in phase of OB+neutron star binaries.

- a chemical evolutionary model of the Galaxy must reproduce the G-dwarf distribution in the disk (the number of G-dwarfs as function of iron-abundance). The predicted distribution obviously depends on the variation of the iron-abundance and therefore depends on whether or not binaries are included, and on whether the SD and/or DD scenario is adopted in order to calculate the SN Ia rate. Our prediction fits the observations only when both the SD and DD model produces an SN Ia.

Note that Langer et al. (1998) investigated the galactic ^{26}Al and concluded that massive close binaries may be the dominant source for this radionuclide.

The rates of different types of SN. De Donder and Vanbeveren (2004) also investigated the effects of binaries on the temporal evolution of the galactic SN rates. SN Ia and most of the SN Ib result from binaries and therefore binaries are essential in order to calculate SN rates. Summarizing:

- In order to reproduce the present day SN Ia rate of our Galaxy, the progenitor intermediate mass interacting binary frequency must have been very high, at least 50%.
- the number of SN II relative to SN Ibc depends significantly on the properties of the massive binary population (binary frequency and mass ratio distribution). However, the ratio hardly depends on the metallicity and therefore the ratio in one particular galaxy varies very slowly as a function of time.
- while the observed number ratio SN Ia/SN Ibc is nearly equal in early and late type spirals ($\sim$1.6), the observed SN II/SN Ibc ratio is significantly different (about a factor of 2) in both type of spirals. Since most of the SN Ibc progenitors are binary components, these observed number ratios may suggest that the massive binary population relative to the intermediate mass binary population is similar in early and late type spirals but that the overall binary frequency (or the overall binary population) in both types of spiral galaxies is significantly different.
- the observed high SN Ia/SN Ibc ratio of $\sim$1.6 is difficult to reproduce with either the SD scenario or the DD scenario separately. The observed ratio is much better approached if both SN Ia scenarios act together and produce SN Ia. Notice that this conclusion was also reached when the observed and theoretically predicted G-dwarf distributions are compared (previous subsection).
- for all galaxies together in the observed sample of Cappellaro et al. (1999) the ratio SN II/SN Ibc is $\sim$5. Since in the sample all the main morphological types of galaxies are included, this value could be considered as some cosmological average. It follows from our simulations that to recover this average, we need a massive binary fraction (on the zero age main sequence) between 40% and 70% which could imply that the cosmological massive binary formation frequency may be of the order of 50%.

3. THE POPULATION OF POINT-LIKE X-RAY SOURCES

The standard high mass X-ray binaries. The scenario for the formation and evolution of the standard high mass X-ray binaries (HMXBs) proposed by Van den Heuvel and Heise (1972) has been confirmed frequently by detailed binary evolutionary calculations. We distinguish three X-ray phases:

- the OB star is well inside its critical Roche lobe and loses mass by stellar wind. The X-rays are formed when the compact star accretes mass from the wind (wind fed systems).
- The OB star is at the beginning of its RLOF phase and mass transfer towards the compact star starts gently (RLOF fed systems).

- The optical star is a Be star and X-rays are emitted when the compact star orbits inside the disk of the Be star (disk fed systems).

In the massive binary evolutionary simulations performed with the Brussels code, we detected a possible fourth phase: when the OB+compact companion binary survives the RLOF-spiral-in-common envelope phase and the optical star is at the end of its RLOF, burning helium in its core, it transfers mass at a very moderate rate similar as the rate at the beginning of RLOF. The star is overluminous with respect to its mass, the surface layers are nitrogen rich and have a reduced surface hydrogen abundance (X $\leq$ 0.4). Possible candidates with an overluminous optical companion are Cen X-3 and SMC X-1. The question here is how a binary can survive the spiral-in phase? Obviously, some binaries have to survive because we observe double neutron stars. Theoretically the survival probability becomes larger if one accounts in detail for the combined action of stellar winds and spiral-in. To illustrate, when after the formation of the compact star, the binary period is large enough so that an LBV-type or RSG-type stellar wind mass loss can start before the onset of the spiral-in, the importance of the latter process can be reduced significantly. Our simulations (with the RSG or LBV wind rates discussed by Vanbeveren et al., 1998a, see also DV) allow to conclude that it cannot be excluded that some HMXBs are RLOF-fed systems where the optical star is a core helium burning star at the end of the RLOF.

Most of the supernova type Ib/c happen in binaries and all HMXBs with a neutron star companion are expected to have experienced (and survived) such a supernova. Since a WC star is expected to be a type Ic progenitor, evolutionary calculations predict that the SN shell may contain lots of carbon and oxygen. When this WC star was a binary component and when the SN shell hits the OB companion star, quite some C and O may be accreted by the latter and abundance anomalies may be expected. Performing a detailed analysis of the CO abundances in the optical star of Cyg X-1 may be interesting. An observed overabundance may be an indication that the black hole progenitor experienced a supernova explosion, may be even a hypernova.

Ultra luminous X-ray sources (ULXs). ULXs are point-like X-ray sources with X-ray luminosities in excess of 10^{39} ergs^{-1} (Fabbiano, 1986).

Young supernova remnants (YSNRs) lose rotational energy and part of this is radiated as X-rays. Van Bever and Vanbeveren (2000) investigated the effects of YSNRs on the X-ray emission of starbursts and concluded that they could dominate the X-radiation (and identified as a ULX) if the initial rotation period of the neutron star at birth is very small, smaller than 0.01 sec. Whether or not such small initial periods are real is still a matter of debate.

Most of the ULXs can be explained by accretion of mass on a 10-20 $M_\odot$ stellar mass black hole (BH), but to explain the most massive ones in the same way, BH masses of 100 to several 100 $M_\odot$ are required (Colbert and Miller, 2004). The latter objects are generally referred to as intermediate mass BHs (IMBHs).

The BH masses predicted by stellar evolutionary computations depend on the effect of stellar wind mass loss on massive star evolution, and more specifically on the mass loss during the CHeB-WR-phase. Before 1998, most of the massive star evolutionary calculations used a WR-mass loss rate formalism which was based on theoretical interpretation of WR spectra with atmosphere models that assume homogeneity of the stellar wind (Hamann, 1994). However, already in 1996, at the meeting *Wolf-Rayet Stars in the Framework of Stellar Evolution* (eds. J.M. Vreux, A. Detal, D. Fraipont-Caro, E. Gosset, G. Rauw, Universite de Liege) Tony Moffat and John Hillier presented evidence that WR winds are inhomogeneous implying that the real WR mass loss rates were smaller by at least a factor 2-3. In 1998, we were among the first to perform and publish evolutionary computations of massive stars with such reduced WR mass loss rates (Vanbeveren et al., 1998a, b). At that time, the evolutionary-referees were not always in favor, to express it mildly. After, 1998, observational evidence was growing that indeed WR winds are inhomogeneous and that the rates are lower. Since 2000, everybody is using reduced WR-rates in their evolutionary code. A major consequence of lower WR mass loss rates is of course the final stellar mass before core collapse. In our 1998 calculations, stars with a metallicty $Z = 0.02$ and with an initial mass $\leq 120\ M_\odot$ end their life with a mass $\leq 20\ M_\odot$. When the WR stellar wind mass loss rate is metallicty dependent (as predicted by the radiation driven wind theory, Pauldrach et al., 1994 among many others), the pre-core collapse mass may be as large as 40-50 $M_\odot$ in small Z environments (like the SMC for example).

What about stars with initial mass larger than 120 $M_\odot$? First, let us recall that the observational evidence for the existence of stars with initial mass $> 120\ M_\odot$ is very poor (a candidate may be the Pistol Star, Figer et al., 1998). However, an important question is the following: if someway or another a star forms with a mass $> 120\ M_\odot$, how does it evolve? Hydrodynamic simulations (within its limitations) let us suspect that stars with a luminosity close enough to the Eddington value, will lose mass at very high rate. Luminous blue variables (LBVs) are stars with $\gamma = L/L_{edd}$ close to 1 (Aerts et al., 2004). One of the most famous LBVs is η Car, a star with $\gamma > 0.7$ and an observed average mass loss rate $\sim 10^{-3}\ M_\odot$/yr (high + low state). Evolutionary calculations reveal that already on the zero age main sequence, a 150 $M_\odot$ has a $\gamma = 0.9$, a 200 $M_\odot$ even has $\gamma = 0.96$. This let us suspect that stars with an initial mass $> 120\ M_\odot$ will suffer from a very high stellar wind mass loss rate already at zero age. The consequences are obvious: when a star is formed with a mass $> 120\ M_\odot$, it will very soon evolve into a state where it is almost undistinguishable from a star whose mass was $\sim 120\ M_\odot$ on the zero age main sequence. One may therefore be inclined to conclude that the maximum stellar BH masses quoted above may be real maxima.

The foregoing LBV-stellar wind argument and its effect on stellar evolution affect in a critical way the outcome of N-body dynamical computations of young dense stellar systems, and in particular the formation of IMBHs by runaway collision (Portegies Zwart et al., 2004). In order to investigate the LBV-effect on the formation of IMBHs, we recently decided to implement stellar dynamics in our PNS and PSS codes (Belkus et al., 2005). We use a standard N-body integration technique in order to follow the motion of N objects, an object can either be a single star or a binary, the interaction of two objects is treated with the chain regularization method as explained by Mikkola and Aarseth (1993, and references therein) and everything is combined with the PNS and PSS of starbursts discussed in Van Bever and Vanbeveren (2000, 2003). As a first order simulation, a cluster is generated with 3000 massive single stars (mass between 10 and 120 $M_\odot$) and a King (1966) distribution with parameters so that the simulation may be appropriate for MGG-11, one of the brightest star clusters in the central region of M82 (notice that the cluster probably contains a ULX, see Portegies Zwart et al., 2004). When due to real collisions a star is formed with a mass > 120 $M_\odot$, it evolves with a stellar wind mass loss rate $= 10^{-3}$ $M_\odot$/yr. When the mass drops below 120 $M_\odot$, we switch back to normal stellar evolution as it is implemented in our PNS/PSS code. Collision products are mixed instantaneously and since we follow the pre-collision stars in detail, we calculate the resulting chemical abundances of the mixed star from first principles. The further evolution of this merger is calculated with our stellar evolutionary code with the appropriate abundances. Figure 1 shows a typical simulation. A runaway collision starts after a few 10^5 yrs. However, after the major part of the runaway process there is enough time left for the merger to lose sufficient mass so that it becomes a *normal* 120 $M_\odot$ star. Notice that the term *normal* may be misleading. Due to the fact that we mix merger products, mergers (and the resulting stars after thermal relaxation) have nitrogen rich surface layers and their convective core may be significantly larger than the one of a real normal star (thus also the resulting BH mass at the end of its evolution; to illustrate, the BH mass of the merger star shown in figure 1 equals 40 $M_\odot$, which is a factor 2 larger than the BH mass of a normal 120 $M_\odot$ star).

The main conclusion of our calculations is that in order to study the possibility to form IMBHs in young dense stellar systems, a good knowledge of the LBV-type instability in very massive stars and the resulting mass loss rate is essential. In any case, due to the action of an LBV type instability in stars with a mass > 120 $M_\odot$, the formation of an IMBH by runaway collision of normal hydrogen burning stars in young dense stellar systems becomes much more unlikely.

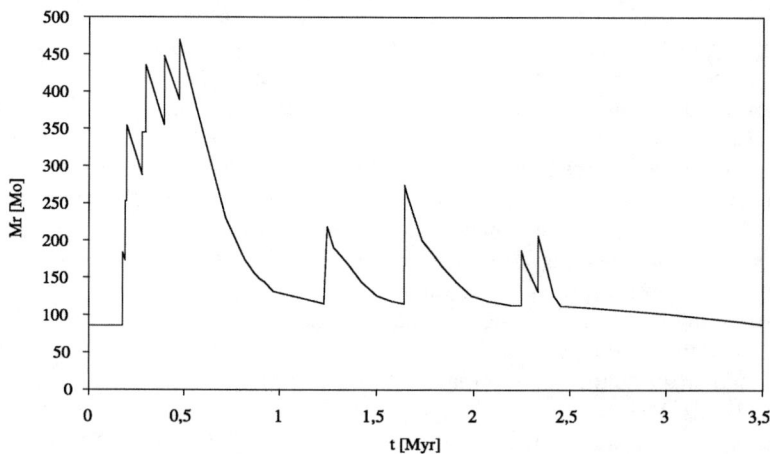

FIGURE 1. The variation of the mass Mr of the most massive star in the cluster.

4. A UFO-SCENARIO FOR WR+OB BINARIES

The formation of WR+OB binaries in young dense stellar systems may be quite different from the conventional binary evolutionary scenario as it was proposed by Van den

Heuvel and Heise (1972). Mass segregation in dense clusters happens on a timescale of a few million years which is comparable to the evolutionary timescale of a massive star. Within the lifetime of a massive star, close encounters may therefore happen very frequently. When we observe a WR+OB binary in a dense cluster of stars, its progenitor evolution may be very hard to predict. Our cluster simulations including the effects of dynamics discussed in the previous section predict the following unconventionally forming object-scenario (*UFO-scenario*) of WR+OB binaries. After 4 million years the first WR stars are formed, either single or binary. Due to mass segregation, this happens most likely when the star is in the starburst core. Dynamical interaction with another object becomes probable, especially when the other object is a binary. In our simulations, we encountered a situation where the WR star (a single WC-type with a mass = 10 $M_\odot$) encounters a 16 $M_\odot$ + 14 $M_\odot$ circularized binary with a period P = 6 days. The result of the encounter is the following: the two binary components merge and the 30 $M_\odot$ merger (which is nitrogen enhanced) forms a binary with the WC star with a period of ~80 days and an eccentricity e = 0.3. This binary resembles very well the WR+OB binary γ^2Velorum but it is clear that conventional binary evolution has not played any role in its formation.

5. THE EFFECT OF ROTATION ON POPULATION SYNTHESIS

PNS and PSS depend on stellar evolution and we need to know how evolution is affected by rotation. Rotation implies rotational mixing in stellar interiors and it can enhance the stellar wind mass loss compared to non-rotating stars. This enhancement may be important for stars that are close to the Eddington limit (LBVs and very massive stars) and therefore rotation may affect indirectly their evolution.

The observed distribution of rotational velocities has been investigated by Vanbeveren et al. (1998b) and we illustrated that the majority of the early B-type stars and of the O-type stars are slow rotators, slow enough to conclude that rotational mixing only plays a moderate role during their evolution (the effect is similar to the effect of moderate convective core overshooting). In the latter paper we argued that due to the process of synchronization in binaries, accounting for the observed binary period distribution, a majority of primaries in massive interacting binaries is expected to rotate slow enough so that the effect of rotation on their overall evolution is moderate as well.

The distribution has an extended tail towards very large rotational velocities, i.e. the distribution is highly asymmetrical which means that in order to study the effect of stellar rotation on population synthesis (the WR and the O type star population for example), it is NOT correct to use a set of evolutionary tracks calculated with an average rotational velocity corresponding to the observed average. This tail obviously demonstrates that there are stars which are rapid rotators. Binary mass gainers, binary mergers and stellar collision products in young dense stellar environments are expected to be rapid rotators and thus are expected to belong to the tail. The question however is whether or not one can approximate their evolution with rotating single star models.

Due to the dynamo effect, rotation generates magnetic fields (Spruit, 2002) which means that the evolutionary effect of rotation cannot be studied separately from the effects of magnetic fields. This was done only since recently (Maeder and Meynet, 2004;

see also Norbert Langer in the present proceedings) and (as could be expected) several of the stellar properties (size of the core, main sequence lifetime, tracks in the HR diagram, surface abundances etc.) are closer to those of models without rotation than with rotation only. Maeder and Meynet notice that single star evolution with rotation only explains the surface chemistry of the observed massive supergiant population, whereas single star evolution with rotation and magnetic fields does not. They use this argument to conclude that magnetic fields must be unimportant. However this argumentation is based on the assumption that most of the massive stars evolve as single stars do. I can think of at least three other processes which can make stars with non-solar CNO abundances: the RLOF process in interacting binaries where the surface layers of the mass loser but also of the mass gainer may becomes N-enriched, the merger of two binary components due to a highly non-conservative RLOF (common envelope phase) and last not least, the collision and merger process due to N-body dynamics in young dense stellar systems. Therefore before an argumentation as the one above has any meaning, one has to consider all these processes.

All in all, it is my personal opinion that, for the overall synthesis of pre-SN stellar populations, the effect of rotation is moderate.

REFERENCES

1. Aerts, C, Lamers, H.J.G.L.M., Molenberghs, G.: 2004, A&A 418, 639.
2. Belkus, H., Van Bever, J., Vanbeveren, D.: 2005, submitted.
3. Cappellaro, E., Evans, R., Turatto, M.: 1999 A&A 351, 459.
4. Colbert, E.J.H., Mushotzky, R.F.: 1999, Ap.J. 519, 89.
5. De Donder, E., Vanbeveren, D.: 2004, New Astron. Reviews 48, 861.
6. Fabbiano, G.: 1989, ARA&A 27, 87.
7. Figer, D. F., Najarro, F., Morris, M., et al.: 1998, ApJ. 506, 384.
8. Hachisu, I., Kato, M., Nomoto, K.: 1999, ApJ 522, 487.
9. Hamann, W.-R.: 1993, Space Sci. Rev. 66, 237.
10. Langer, N., Braun, H., Wellstein, S.: 1998, Proceedings of the 9th workshop on Nuclear Astrophysics. ed. W. Hillebrandt and E. Muller.: p.18
11. Maeder, A., Meynet, G.: 2004, A&A 422, 225.
12. Mikkola, S., Aarseth, S.J.: 1993, Celestial Mechanics and Dynamical Astronomy 57, 439.
13. Pauldrach, A.W., Kudritzki, R.P., Puls, J., et al.: 1994, A&A 283, 525.
14. Portegies Zwart, S.F., Baumgardt, H., Hut, P., et al.: 2004, Nature 428, 724.
15. Spruit, H.C.: 2002, A&A 381, 923.
16. Van Bever, J., Vanbeveren, D.: 2000, A&A 358, 462.
17. Van Bever, J., Vanbeveren, D.: 2003, A&A 400, 63.
18. Vanbeveren D., De Loore, C., Van Rensbergen, W.,: 1998a, A&A Rev. 9, 63.
19. Vanbeveren, D., De Donder, E., Van Bever, et al.: 1998b, NewA 3, 443.
20. Van den Heuvel, E.P.J., Heise, J.G.: 1972, Nature Phys. Sci. 239, 67.

Thermonuclear supernova models, and observations of Type Ia supernovae

E. Bravo*,†, C. Badenes** and D. García-Senz*,†

*Dept. Física i Enginyeria Nuclear, UPC, Av. Diagonal 647, 08028 Barcelona
†Institut d'Estudis Espacials de Catalunya, Barcelona
**Dept. Physics and Astronomy, Rutgers Univ., 136 Frelinghuysen Rd., Piscataway NJ 08854-8019

Abstract. In this paper, we review the present state of theoretical models of thermonuclear supernovae, and compare their predicitions with the constraints derived from observations of Type Ia supernovae. The diversity of explosion mechanisms usually found in one-dimensional simulations is a direct consequence of the impossibility to resolve the flame structure under the assumption of spherical symmetry. Spherically symmetric models have been successful in explaining many of the observational features of Type Ia supernovae, but they rely on two kinds of empirical models: one that describes the behaviour of the flame on the scales unresolved by the code, and another that takes account of the evolution of the flame shape. In contrast, three-dimensional simulations are able to compute the flame shape in a self-consistent way, but they still need a model for the propagation of the flame in the scales unresolved by the code. Furthermore, in three dimensions the number of degrees of freedom of the initial configuration of the white dwarf at runaway is much larger than in one dimension. Recent simulations have shown that the sensitivity of the explosion output to the initial conditions can be extremely large. New paradigms of thermonuclear supernovae have emerged from this situation, as the Pulsating Reverse Detonation. The resolution of all these issues must rely on the predictions of observational properties of the models, and their comparison with current Type Ia supernova data, including X-ray spectra of Type Ia supernova remnants.

Keywords: Hydrodynamic models – Supernovae: modelling, nucleosynthesis and evolution
PACS: 24.10.Nz, 26.30.+k, 26.50.+x, 95.85.Nv

INTRODUCTION

The huge increase in number, quality and diversity of observational data related to Type Ia Supernovae (SNIa) in recent years, combined with the advance in computer technology, have persuaded modellers to leave the phenomenological calculations that rely on spherical symmetry, and attempt more physically meaningful three-dimensional (3D) simulations. Although the more plausible models of the explosion always involve the thermonuclear disruption of a white dwarf [1], the current zoo of explosion mechanisms is still too large to be useful in cosmological applications of Type Ia supernovae or to make it possible to understand the details of the chemical evolution of the Galaxy. Nowadays, the favoured SNIa model is the explosion of a white dwarf that approaches the Chandrasekhar-mass limit owing to accretion from a companion star at the appropiate rate to avoid the nova instability [2, 3]. Going beyond this general picture into the details of the supernova explosion is not easy, especially with respect to the multidimensional models that are just beginning to appear in the literature. In particular, the prediction of the optical light curve or spectra of a 3D model is still out of reach, and therefore it is necessary to rely on other gross features of the observations in order to estimate the viability of a given model.

The most relevant property of SNIa is the homogeneity of their light curve and spectral evolution. The light curve is powered by the radioactive decay of ^{56}Ni and ^{56}Co [4], but the range of nickel masses allowed by the observations varies by about a factor five from low-luminosity SNIa up to normal events. Although the shape of the light curves can be described by a one-parameter relationship between brightness and width of the curve [5, 6, 7, 8], due to the dependence of opacity on temperature, there still remains a residual scatter of ~ 0.2 mag around the template curves. The main spectral features of normal (bright) SNIa at early photospheric phase include the absence of conspicuous lines of H and the presence of strong SiII absorption lines together with absorption lines of other intermediate mass elements (CaII, SII, OI) spanning a range of velocities from 8,000 up to 30,000 km s^{-1}. The nebular phase is dominated by Fe lines. Usually, the spectral evolution is attributed to the recession (in terms of lagrangian mass) of the photosphere through a layered chemical structure. Recent spectroscopic observations of a dozen *Branch-normal* Type Ia supernovae in the near infrared [9] suggest that the unburnt matter ejected has to be less than 10% of the mass of the progenitor. According to these results, the presence of a substantial amount of unburnt low-velocity carbon near the center of the star is rather improbable.

A relevant question for multidimensional simulations, is whether there is any significant observational evidence of departure from spherical symmetry in the SNIa sample. In this regard, there are several signs that the departure from spherical symmetry is not large: the low level of polarization of most SNIa, although there are exceptions [see, for instance, 10], the homogeneity of the profile of the absorption line of SiII [11], and the fact that galactic and extra-galactic young Type Ia supernova remnants (SNR) do not show large departures from spherical symmetry.

The spectral homogeneity of normal SNIa near maximum brightness is particularly relevant for the discussion below. By comparing the spectra of four normal SNIa (SN 1989B, SN 1990N, SN 1994D, and SN 1998bu) Thomas et al. [11] have shown that the absorption features of SiII displayed quite homogeneous profiles from event to event. Such homogeneity can be used to constrain the presence of chemically inhomogeneous clumps at the photosphere, through the effect they would have on the line profiles. Specifically, Thomas et al. [11] limited the size of the clumps to be less than $\sim 30\%$ of the radius of the photosphere.

To summarize the gross constraints imposed by SNIa observations, the *overall* shape of the supernova has to be spherical (low polarization), there are not large chemically inhomogeneous blobs at the photosphere at maximum light (homogeneous SiII line profiles), and the chemical composition of the ejecta has to retain a high degree of stratification. One has to keep in mind that subluminous as well as superluminous events display a peculiar spectral evolution. For the time being, however, as long as multidimensional simulations are concerned, the main objective is to explain the gross properties of normal SNIa. In the first part of this paper, we will review the results of recent 3D simulations of thermonuclear supernovae (TSN), and compare their predictions with observational data. In the second part, we will discuss the prospects for the use of X-ray spectra of supernova remnants to discriminate between the different explosion mechanisms or progenitor scenarios that are currently advocated to explain SNIa.

3D MODELS OF THERMONUCLEAR SUPERNOVAE

From a theoretical point of view, there remain several fundamental issues to be solved:

- What is the progenitor system and how does the white dwarf manage to reach the Chandrasekhar mass?
- What is the evolution of the white dwarf short before carbon runaway? and how, when, and where does the ignition process begin?
- How does the flame propagate through the white dwarf once ignited?
- Is a deflagration-to-detonation transition possible, and under which conditions?
- What is the role played by the rotation of the white dwarf?

Here, we will not discuss the presupernova evolution, which is addressed at length in other contributions to this volume. However, it is worth to remark that the output of the explosion in terms of its kinetic energy, density profile, and chemical composition is determined in the first few seconds after runaway, and is not much sensitive to the details of the presupernova evolution. The only features of the progenitor system that influence the explosion are the C/O ratio and metallicity of the white dwarf, its central density (determined by the accretion rate), and rotation (see the contributions by Piersanti et al. and Domínguez et al. in this volume). However, the supernova properties can be influenced on longer timescales by several characteristics of the progenitor system, owing to the interaction of the ejecta with the secondary star (this is probably the case for the peculiar supernovae SN2000cx and SN2002cx, see e.g. Thomas et al. [12], Li et al. [13]), with a circumstellar medium (normal Type Ia SN2003du and SN2002ic, see Gerardy et al. [14], Hamuy et al. [15]), or during the formation of the supernova remnant [16].

Spherically symmetric models and early two and three-dimensional calculations of TSN assumed that the ignition started in a central volume. This view was challenged by Garcia-Senz and Woosley [17], who showed that burning blobs formed during the convective preignition phase would be able to float and accelerate up to ~ 100 km s^{-1}, with the result that the flame would be rapidly scattered in a region $100-250$ km away from the center of the star. In this case, the flame would not begin just in a central volume but distributed in an exponentially increasing [18, 19] number of hot spots, whose velocity could reach $\sim 1\%$ of the sound velocity.

In order to simulate TSN it is always necessary to approximate the behavior of the flame below the scales not resolved by the hydrocode with a suitable model. This is not an easy task, due to the quite different regimes of thermonuclear flames at high densities, $\rho \sim 10^9$ g cm^{-3}, (thin flame of width < 1 cm, propagating with a low velocity of order $1-3\%$ of the sound velocity, and with a surface progressively corrugated by hydrodynamic instabilities on timescales of a few tenths of a second) and at the end of the explosion, $\rho <$ a few $\times 10^7$ g cm^{-3}, (thick flame of width similar to the white dwarf radius, subject to mixing between ashes and fuel *before* completion of the nuclear reactions, which favours the production of intermediate mass elements). The range of involved lengthscales spans ~ 9 orders of magnitude, which on one hand discards its direct resolution with any hydrodynamical code but, on the other hand, allows to use a statistical description of the flame. Up to now, there is no convergence between the different approximations adopted by different 3D hydrocodes.

Deflagrations and delayed detonations

The multidimensional calculations of deflagrations carried out so far [see 20, 21, 19, for the most recent results] have shown interesting deviations from what is predicted in spherically symmetric models:

1. The geometry of the burning front is no longer spherical owing to the important role played by buoyancy and hydrodynamic instabilities,
2. the chemical stratification of the ejecta is lost,
3. the amount of ^{56}Ni is sufficient to power the light curve, but it is localized in clumps distributed all along the radius of the white dwarf, and
4. an uncomfortably large amount of carbon and oxygen remains unburnt at the center of the white dwarf.

Three-dimensional simulations have also demonstrated that the flame evolves in quite a different way when calculated in 2D or 3D, due to the different degrees of freedom of the flow. Thus, earlier results of 2D models of thermonuclear supernovae have to be regarded with caution.

The results of the most up-to-date 3D simulations of deflagration supernovae (kinetic energy, K, and masses of ^{56}Ni, M_{56}, intermediate mass elements, M_{ime}, and unburned C+O, M_{CO}) are shown in Table 1. The results obtained under quite different initial conditions and using very different numerical techniques (PPM vs SPH, with degrees of spatial resolution varying by a factor ten, with different subgrid-scale models of the flame, etc) are remarkably homogeneous. The kinetic energy and the amount of ^{56}Ni are compatible with SNIa, but the amount of intermediate mass elements is rather low and the mass of C+O ejected in the explosion is too large, probably by a factor five or more. The convergence of the results obtained with different codes reflects the fact that the main trends of deflagration supernovae are well understood and incorporated into the calculations. There is little hope that further refinements in the methods used to simulate them will substantially change the outcome of current 3D deflagration models.

In the left panel of Fig. 1 it is shown the distribution of ^{56}Ni at the photosphere for the model starting from 30 bubbles of the same size [19]. The size of the clumps is too large to be compatible with the observational limits posed by the homogeneity of the spectral features of SNIa [11]. It is important to keep in mind that the main properties of 3D deflagrations are model independent, as they result from first principles. In particular, the deformation of the flame front due to hydrodynamical instabilities is unavoidable, because the timescale for developing the Rayleigh-Taylor instability is only a few tenths of a second, i.e. about a factor five lower than the time needed by the deflagration to reach the white dwarf surface. Once the spherical symmetry of the flame is lost it is quite difficult to restore it unless a very energetic and impulsive phenomenon, like a detonation, is invoked.

In spite of the success achieved by one-dimensional delayed detonation TSN models, the physical mechanism responsible for the transition from the deflagration to a detona-

TABLE 1. Results of 3D simulations of thermonuclear supernovae

Model	K (10^{51} erg)	M_{56} ($M_\odot$)	M_{ime} ($M_\odot$)	M_{CO} ($M_\odot$)
Deflagrations				
Central ignition with strong turbulence[21]	0.6	~ 0.5	~ 0.1	~ 0.7
Central ignition with realistic turbulence[21]	0.37			
Central ignition[20]	0.48	0.30	0.10	0.75
Ignition in 40 bubbles, large resolution[20]	0.45	0.33	0.23	0.64
Ignition in 30 bubbles of the same size[19]	0.43	0.43	0.07	0.67
Ignition in 90 bubbles of different sizes[19]	0.45	0.44	0.08	0.66
Ignition in 240 bubbles, very large resolution[22]	0.6	0.42	0.10	0.62
Delayed detonations				
Macroscopic transition to detonation[23]	0.75	0.54	0.16	0.34
Local transition in the central region[23]	0.48	0.43	0.10	0.48
Local transition at intermediate radius[23]	0.51	0.42	0.14	0.45
Local transition in the outer layers[23]	0.33	0.34	0.09	0.57
Transition in a central volume[24]	0.8	0.78*		
Off-center transition[24]	0.8	0.73		
Transition in a central volume at high density[24]	1.1	0.94		

* In this and the following models M_{56} represents the approximate yield of Fe-group nuclei

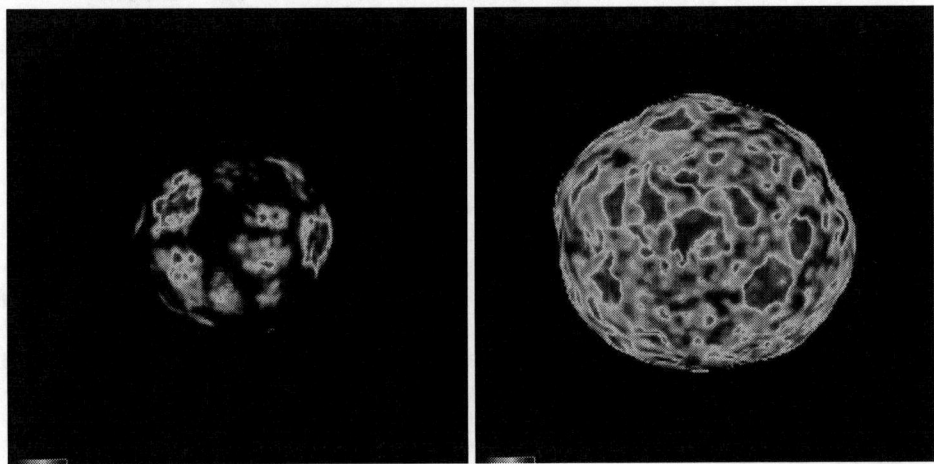

FIGURE 1. Mass fraction of ^{56}Ni at the photosphere 15 days after the explosion. *Left*: Deflagration model. *Right*: Delayed detonation model

tion at the convenient densities ($\sim 10^7$ g cm^{-3}) is still unknown. Up to now two different scenarios have been proposed:

- A local transition, induced when a fluid element burns with a supersonic phase velocity when the flame changes from the laminar to the distributed regime [25].

- A macroscopic transition, triggered by a complex topology of the flame that results in a fuel consumption rate larger than that obtained in a supersonic spherical front [26].

The first mechanism has been studied by Lisewski et al. [27] who found that the required mass of the detonator was too large, precluding the formation of a detonation at the densities of interest. The viability of the second mechanism has not been demonstrated so far. Therefore, delayed detonation calculations in 3D are constrained by the uncertainty about the transition density, assuming there is any transition at all. In Table 1 we show the results of the few 3D models of delayed detonation that have been computed up to now. The results obtained by different groups show a larger discrepancy than those derived from pure 3D deflagration simulations. The kinetic energy and the masses of ^{56}Ni and intermediate mass elements are, in general, larger than in 3D deflagration models. Nevertheless, the amount of unburned C-O ejected in the models computed by Garcia-Senz and Bravo [23] is still quite large. In contrast to this situation, the delayed detonation models computed by Gamezo et al. [24] showed that there was no fuel left at the center after the passage of the detonation front. The reason for this apparent discrepancy is the large density of the transition ($> 10^8$ g cm^{-3}) adopted by these authors. Both groups also obtained different results with respect to the stratification of the chemical composition in the ejecta.

The clumps formed during the deflagration phase are destroyed by the detonation waves. The distribution of ^{56}Ni resulting for the macroscopic transition delayed detonation model of Garcia-Senz and Bravo [23] (see Table 1) is shown in the right panel of Fig. 1. There, it can be seen how the size of the individual clumps is smaller than in the deflagration case, making 3D delayed detonations compatible with the spectral homogeneity of SNIa.

New explosion paradigms

Although most of the *mildly successful* 3D deflagration models calculated so far start from a large number of bubbles scattered through the central region of a white dwarf, nowadays it is not clear how many hot spots can be present at runaway. Thus, it is interesting to ask what would be the outcome of the explosion if the initial number of bubbles were small? At first sight, one can expect that the energetics of the explosion would be smaller than in the many bubbles models, probably giving rise to a failed explosion and a pulsation of the white dwarf. In this way, the uncertainty about the initial configuration of the flame has allowed the introduction of two new paradigms of TSN, the so-called Pulsating Reverse Detonation (PRD) and the Gravitationally Confined Detonation (GCD).

The PRD mechanism of explosion is a byproduct of the simulations of deflagrations carried out by Garcia-Senz and Bravo [19] starting from 6-7 bubbles. In these simulations, the nuclear energy generated during the deflagration phase was insufficient to unbind the star, as expected. Due to the ability of hot bubbles to float to large radii in 3D models, most of the thermal and kinetic energy resided in the outer parts of the structure, which resulted in an early stabilization of the central region (mostly made of cold C and

O, i.e. fuel) while the outer layers were still in expansion. A few seconds later, an accretion shock formed at the border of the central nearly hydrostatic core (whose mass was about $0.9\,M_\odot$). As a consequence, the temperature at the border of the core increased to more than 2×10^9 K, on a material composed mainly by fuel but with a non-negligible amount of hot ashes, thus giving rise to a highly explosive scenario. If a detonation were ignited at this point, it would probably propagate all the way inwards through the core, burning most of it and producing an energetic explosion, with a stratified composition. A one-dimensional follow-up calculation of the detonation stage produced a kinetic energy of 0.89×10^{51} erg and $0.35\,M_\odot$ of ^{56}Ni. At the same time, the amount of unburned C-O was reduced to $0.22\,M_\odot$.

The GCD mechanism of explosion of Plewa et al. [28] keeps some similarity with the PRD, but this time the runaway starts in a single hot bubble located close to the center of the white dwarf. Once again, the evolution is dominated by the bubble motion towards the surface, which hinders a substantial propagation of the deflagration front and determines the explosion failure. However, it is just following this failure of the deflagration, and the subsequent breakout of the bubble at the surface of the white dwarf, when the most interesting events take place. The bubble material is then spreaded around the surface, where it experiences a strong lateral acceleration while remaining gravitationally confined to the white dwarf. Finally, the material focuses at the pole opposite to the point of breakout, providing a high compression and attaining a high temperature ($> 2.2 \times 10^9$ K). The calculations of Plewa et al. [28] end at this point, but they claim that a detonation will probably form at the point of maximum temperature, propagating through the remaining of the white dwarf and burning it to Fe-group and intermediate mass elements. It has to be noted that this model was the result of a 2D calculation, and its results have still to be confirmed by full 3D simulations.

NEW WINDOWS TO SNIA: THE SN-SNR CONNECTION

The X-ray spectra of supernova remnants originated by SNIa contain important information regarding the physical mechanism behind the explosions. In the process of formation of the remnant, the supernova ejecta interact with the ambient medium surrounding the supernova progenitor, transfer mechanical energy to it, and are heated through shock waves to a state in which both the ambient medium and the ejecta emit X-rays. During the initial phase of the remnant evolution, when the SNR is still young, the emission in the high energy band is determined mainly by the properties of the ejecta. In general, the X-ray emission can have several components:

- A non-thermal continuum, related to the ambient magnetic field, non-maxwellian populations, etc.,
- a thermal continuum (bremsstrahlung), sensitive to the local state of the plasma, and
- thermal line emission, sensitive to the chemical abundances, ionization state, and thermal state of the plasma.

The X-ray line emission from young SNRs provides a convenient way to constrain the nucleosynthetic production and energetic properties of the explosion.

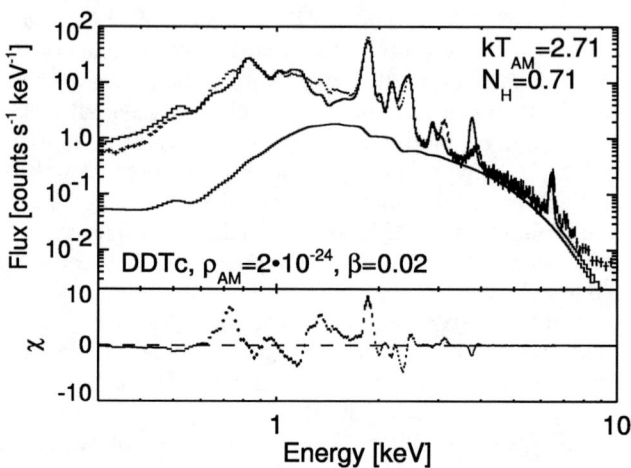

FIGURE 2. The best fit to the *XMM-Newton* spectrum of the Tycho SNR (solid line) is compared to the data (points). The spectrum emitted by the shocked interstellar medium is also shown (featureless solid curve). See Badenes [16] for details

In a recent work, Badenes et al. [29] showed that the differences in chemical composition and density profile of SNIa ejecta have indeed a deep impact on the thermal X-ray spectra emitted by young SNRs. Thus, it is possible to use the excellent X-ray spectra of Type Ia SNRs obtained by X-ray observatories, like *XMM-Newton* and *Chandra*, to constrain Type Ia SN explosion models. Similar approaches were taken in the past by Dwarkadas and Chevalier [30] and Itoh et al. [31], although these works either did not include spectral calculations or were limited to the study of a particular model (the W7 model). Kosenko et al. [32] have undertaken a similar enterprise, whose results are just beginning to appear in the literature. More recently, Badenes [16] has compared the spectra predicted by more than 400 supernova remnant models (generated combining 26 SNIa explosion models in 1D and 3D, including all the explosion mechanisms currently under debate, with different assumptions about the physical state of the ambient medium) with the spectra of well-known remnants, like the Tycho SNR.

The Tycho SNR, which is the best candidate for a Type Ia remnant, has been extensively studied with both *Chandra* and *XMM-Newton*, providing high-quality spectra with precise determinations of the flux and energy centroid of the spectral features produced by each element. One of the main properties of the X-ray spectra of the Tycho remnant is that the emission lines due to Fe and Si (and other intermediate mass elements, like S and Ca) are produced under quite different thermal conditions. Through a careful analysis, Badenes [16] has been able to prove that most of the explosion paradigms are incompatible with the spectra of Tycho. The best approximation to the X-ray spectrum of the Tycho SNR is obtained with a mildly-energetic ($K = 1.2 \times 10^{51}$ erg) 1D (i.e. chemically stratified) delayed detonation model, which synthesizes 0.74 $M_\odot$ of ^{56}Ni (see Fig. 2). It was shown in this work that, due to the absence of chemical stratification, the SNR models produced by 3D deflagrations and 3D delayed detonations were characterized by the homogeneity of the ionization and thermal properties of all the chemical elements. As we have explained before, such homogeneity is incompatible with the physical properties derived from the X-ray spectra of the Tycho SNR and other candidate Type Ia SNRs. It will be interesting to see whether the spectra predicted by the new 3D explosion paradigms of SNIa will retain these characteristics or, on the contrary, will become more similar to the ones corresponding to 1D explosion models.

SUMMARY

Since Hoyle and Fowler [1] proposed the white dwarf scenario for Type Ia SNe, the ideas about the way the star explodes have evolved. The 70's were the epoch of pure detonations. The 80's witnessed a flourishing of the deflagrations, mainly thanks to the popular W7 model. The 90's knew about delayed detonations in its various flavors. Nowadays, at the beginning of the 21st century, the future of TSN modelling resides most probably on new paradigms, like PRD and GCD.

Although, in the near future, the analysis of Type Ia supernovae will continue being based predominantly on optical observations, the realm of high energies is going to play an increasingly important role in the understanding of these objects. We have discussed some recent progress on the applications of X-ray spectra from young supernova remnants to the determination of the explosion mechanism. Future observations of gamma-rays from SNIa will allow a much more in depth knowledge of the amount of radioactive nuclei produced in the explosion, its distribution throughout the ejecta, and the eventual influence of the interaction with a secondary star in a binary system [see, e.g. 33].

ACKNOWLEDGMENTS

This research has been partialy supported by the CIRIT, the MEC programs AYA2002-04094-C03-01/02 and AYA2004-06290-c02-01/02, and by the EU FEDER funds.

REFERENCES

1. F. Hoyle, and W. Fowler, *ApJ*, **132**, 565 (1960).
2. P. Nugent, E. Baron, D. Branch, A. Fischer, and P. Hauschildt, *ApJ*, **485**, 812 (1997).
3. W. Hillebrandt, and J. Niemeyer, *ARA&A*, **38**, 191 (2000).
4. S. Colgate, and C. McKee, *ApJ*, **157**, 623 (1969).
5. A. Riess, W. Press, and R. Kirshner, *ApJ*, **473**, 88 (1996).
6. M. Hamuy, M. Phillips, N. Suntzeff, R. Schommer, J. Maza, and R. Aviles, *AJ*, **112**, 2391 (1996).
7. S. Perlmutter, et al., *ApJ*, **483**, 565 (1997).
8. M. M. Phillips, P. Lira, N. B. Suntzeff, R. A. Schommer, M. Hamuy, and J. Maza, *AJ*, **118**, 1766 (1999).
9. G. H. Marion, P. Höflich, W. Vacca, and J. Wheeler, *ApJ*, **591**, 316 (2003).
10. D. Kasen, P. Nugent, L. Wang, D. Howell, J. Wheeler, P. Hofflich, D. Baade, E. Baron, and P. Hauschildt, *ApJ*, **593**, 788 (2003).
11. R. Thomas, D. Kasen, D. Branch, and E. Baron, *ApJ*, **567**, 1037 (2002).
12. R. Thomas, D. Branch, E. Baron, K. Nomoto, W. Li, and A. Filippenko, *ApJ*, **601**, 1019 (2004).

13. W. Li, et al., *PASP*, **115**, 453 (2003).
14. C. Gerardy, P. Höflich, R. Fesen, G. Marion, K. Nomoto, R. Quimby, B. Schaefer, L. Wang, and J. Wheeler, *ApJ*, **607**, 391 (2004).
15. M. Hamuy, et al., *Nature*, **424**, 651 (2003).
16. C. Badenes, *Thermal X-Ray Emission from Young Type Ia Supernova Remnants*, Ph.D. thesis, Polytechnic University of Catalonia (UPC), Barcelona, Spain (2004).
17. D. Garcia-Senz, and S. Woosley, *ApJ*, **454**, 895 (1995).
18. S. E. Woosley, S. Wunsch, and M. Kuhlen, *ApJ*, **607**, 921 (2004).
19. D. Garcia-Senz, and E. Bravo, *A&A*, **in press** (2004).
20. M. Reinecke, W. Hillebrandt, and J. Niemeyer, *A&A*, **391**, 1167 (2002).
21. V. Gamezo, A. Khokhlov, E. Oran, A. Chtchelkanova, and R. Rosenberg, *Science*, **299**, 77 (2003).
22. J. Niemeyer, M. Reinecke, C. Travaglio, and W. Hillebrandt, "Small Steps Toward Realistic Explosion Models of Type Ia Supernovae," in [34], p. 151.
23. D. Garcia-Senz, and E. Bravo, "Influence of Geometry on the Delayed Detonation Model of SNIa," in [34], pp. 158–164.
24. V. Gamezo, A. Khokhlov, and E. Oran, *PhRvL*, **92**, 1102 (2004).
25. A. Khokhlov, *A&A*, **245**, 114 (1991).
26. S. Woosley, and T. Weaver, "Massive Stars, Supernovae, and Nucleosynthesis," in *Supernovae*, edited by S. Bludman, R. Mochkovitch, and J. Zinn-Justin, Les Houches Session LIV, Elsevier, Amsterdam, 1994, p. 63.
27. A. Lisewski, W. Hillebrandt, and S. Woosley, *ApJ*, **538**, 831 (2000).
28. T. Plewa, A. Calder, and D. Lamb, *ApJ*, **612**, L37 (2004).
29. C. Badenes, E. Bravo, K. Borkowski, and I. Domínguez, *ApJ*, **593**, 358 (2003).
30. V. Dwarkadas, and R. Chevalier, *ApJ*, **497**, 807 (1998).
31. H. Itoh, K. Masai, and K. Nomoto, *ApJ*, **334**, 279 (1988).
32. D. Kosenko, E. Sorokina, S. Blinnikov, and P. Lundqvist, *Advances in Space Research*, **33**, 392–397 (2004).
33. J. Isern, E. Bravo, and A. Hirschmann, *Advances in Space Research*, **in press** (2004).
34. W. Hillebrandt, and B. Leibundgut, editors, *From Twilight to Highlight: The Physics of Supernovae*, Springer, Berlin, 2003.

White dwarf merging and the emission of gravitational waves

J. Isern[*,†], E. García–Berro[*,**], J. Guerrero[*], P. Lorén–Aguilar[*,†] and J.A. Lobo[*,‡]

[*]*Institut d'Estudis Espacials de Catalunya*
Ed. NEXUS, C/ Gran Capità 2, 08034 Barcelona (Spain)
[†]*Institut de Ciències de l'Espai (CSIC)*
Campus UAB, Facultat de Ciències, Torre C-5, 08193 Bellaterra (Spain)
[**]*Departament de Física Aplicada (UPC),*
Av Canal Olímpic sn, 08860 Castelldefels (Spain)
[‡]*Departament de Física Fonamental (UB)*
c/Martí i Franquès 1, 08028 Barcelona (Spain)

Abstract. We have computed, using an SPH code, the process of merging of two white dwarfs with different masses and chemical compositions. We have found that, although relatively high temperatures are attained during the most violent phase of the merging process, the thermonuclear flame is rapidly quenched. Special attention has been paid to the emission of gravitational radiation. We have found that the signature of the merging is relatively weak.

Keywords: White dwarf stars, merging stars, gravitational radiation
PACS: 95.30.Lz, 93.30.S1, 95.55.Ym

INTRODUCTION

As a consequence of the emission of gravitational waves, merging is the final destiny of a good fraction (those which are separated by less than $\sim 3\,R_\odot$) of binary systems containing two white dwarfs. This property led to Iben & Tutukov [1] and Webbink [2] to propose the merging of two white dwarfs with a total mass larger than Chandrasekhar's mass as one of the possible scenarios for Type Ia supernova explosions. Despite the initial popularity of such a scenario, two arguments have been advanced against its viability: the negative result of the surveys aimed to identify possible progenitors and the prediction that the merging should produce, as a first step, an ONeMg white dwarf followed by its collapse to a neutron star [3].

Concerning the first argument, it is necessary to realize that, since white dwarfs are intrinsically dim objects, the search for binary systems composed by two of these objects is specially difficult. Several strategies have been advanced to detect them: 1) radial motion in spectroscopic binaries [4], 2) white dwarfs of very low mass [5], 3) white dwarfs showing composite spectra [6], and 4) common proper motion binaries [7]. Although several double degenerate systems have been found up to now, no one can be the progenitor of a Type Ia supernovae. Despite this negative result, it has been shown [8] that there is not a contradiction between the number of close binaries necessary to sustain the galactic supernova rate and the observed density of double degenerates. The reason is that white dwarfs are intrinsically dim and their properties are only known in the solar vecinity ($d \sim 100$ pc), while the supernova rate is related to the average properties of the Galaxy. In any case, the uncertainties in the scale height of the white dwarf population, necessary to connect the local and the averge properties, as well as in the treatment of the common envelope evolution prevent to reach any definite conclusion, for which reason it would be extremely convenient to determine the distribution of the double degenerates in the whole Galaxy.

Concerning the second argument, it is necessary to realize that the critical points that have to be elucidated are, first, to determine if the primary ignites during the process of merging and, second, which is the effective rate at which mass is transfered from the disk to the primary.

The purpose of this paper is to simulate the process of merging, in order to determine if the primary ignites, and to compute the emission of gravitational waves during the last phases of the collision in order to see if the future gravitational telescopes are appropriate for establishing the properties of the galactic population of binaries.

INPUT PHYSICS AND METHOD OF CALCULATION

The hydrodynamic evolution of the binary system has been computed using a lagrangian particle numerical code, the so–called Smoothed Particle Hydrodynamics (SPH). We will not describe here in detail the most basic equations of our numerical code since it has already been described in a previous paper [9]. Similar calculations were performed by other authors [10, 11, 12] but either with a reduced resolution or with less accurate physical inputs. The most important ingredients of the code are: i) the kernel, for which we have adopted the standard polynomic one [13]; ii) the gravitational forces which are evaluated using a binary tree [14]; iii) the integration scheme, for which we have chosen a predictor–corrector scheme with a variable time step [15]; iv) the equation of state, which is that of a Coulomb plasma plus an ideal electron gas; and v) the nuclear network, which is composed by 14 nuclei linked by α–capture reactions plus the $^{12}C+^{12}C$ and the $^{16}O+^{16}O$ reactions. Special care has been placed in the treatment of the pseudoviscosity. Two set of calculations have been performed, the first one using the algorithm of Balsara [16] and the second one using a Riemann solver [17] in order to minimize as much as possible the shear between the disk and the primary. The typical number of particles for each simulation is 20,000.

In order to obtain the initial equilibrium configurations, each individual model star was separately relaxed at a distance that was larger than the radius of the Roche lobe. A small negative acceleration was added to this system in order to bring both stars together. When the secondary started to fill the Roche lobe, the acceleration term was suppressed and the system was allowed to freely evolve.

CO+CO mergers

Table 1 displays the three cases considered here. All of them have a similar behavior that can be summarized as it follows. Initially, both stars are almost completely spherically symmetric; after some time, the secondary overflows its Roche lobe and an accre-

TABLE 1. CO+CO mergers

Case	0.6+0.8	0.6+1.0	0.8+1.0
Ratio	0.75	0.60	0.80
T_{max}(K)	1.4×10^9	1.6×10^9	2.0×10^9
Mass of the particles ($M_\odot$)	3×10^{-5}	3×10^{-5}	4×10^{-5}
Expelled mass ($M_\odot$)	2.82×10^{-3}	6.48×10^{-3}	6.52×10^{-3}
Duration (s)	50	65	20

tion stream, directed from the secondary to the primary, forms; as time goes on, the arm twists more and more until it becomes an accretion disk with cylindrical symmetry. The whole process lasts for about a minute.

It is interesting to note that the particles that are closer to the primary have smaller rotation velocities as a consequence of the dissipation introduced by the artificial viscosity, while those located in the outer part of the accretion stream have larger velocities and, ultimately, they are ejected from the system. In all the cases considered here, however, the total amount of mass that is expelled is very small, less then the 0.2% of the total mass. Another interesting feature is the rotation velocity around the center of masses of the system since, at the end of the process, the primary rotates as a rigid body while the secondary forms a keplerian disk without any shear in the region of contact. This behavior is a consequence of the transfer of orbital angular momentum to rotational angular momentum by the accreted matter and by the artificial viscosity.

SPH numerical codes are known to be affected by a large shear viscosity which results in an artificial kinetic energy dissipation in shear flows. In these calculations we have used the Balsara [16] artificial viscosity, which does not produce an excessive shear, and Riemann capture front techniques which, in principle do not introduce any shear. The results have been qualitatively the same for both cases. In principle, the numerical viscosity can also be reduced by increasing the resolution, which scales as $N^{1/3}$, but we have not found any substantial difference when the number of particles was increased by an order of magnitude.

The initial temperature of both stars was set to be 10^7 K. As a consequence of the impact of the accretion arm and tidal effects, the temperatures of the primary and the secondary increase noticeably and are larger in the region of impact. As table 1 shows, the maximum temperature that it is reached increases with the mass of the primary an the secondary, and it is high enough to ignite nuclear reactions. However, since the density is low, matter rapidly expands and the thermonuclear flash is rapidly quenched, thus avoiding a prompt thermonuclear explosion.

The accretion disk that it is formed is very thick and, in principle, unstable. However, because of the explicit nature of our integration scheme, we have not been able to follow its long term evolution. In fact, when the calculations were stopped, the disk was still showing some of the inhomogeneities caused by the reflection of the different shocks that it had experienced.

TABLE 2. He+He, ONe+He, CO+He mergers

Case	0.4+0.4	1.2+0.4	0.6+0.4
Ratio	1	1/3	2/3
T_{max}(K)	2.2×10^8	3.1×10^9	7.0×10^9
Mass of the particles ($M_\odot$)	4×10^{-5}	2×10^{-5}	2×10^{-5}
Expelled mass ($M_\odot$)	0	3.54×10^{-2}	3.32×10^{-3}
Duration (s)	600	180	569

He+He, ONe+He and CO+He mergers

Table 2 displays the characteristics of three simulations in which the primary has different masses and chemical compositions (0.4, 1.2 and 0.6 $M_\odot$ and He, ONe and CO, respectively), while the secondary is always a helium white dwarf of 0.4 $M_\odot$. The main differences between the collisions, in this early epoch, are essentially due to the depth of the gravitational well and to the mass ratio between the primary and the secondary.

In the case of the merging of a 0.4+0.4 $M_\odot$ system, the most important feature is the formation of two spiral accretion streams that collide in the center of mass of the system. At the end of the simulation, the final configuration is formed by a rotating object with cylindrical symmetry formed by material of both stars. The final distribution of temperatures is also axially symmetric and the maximum temperature never exceeds 2×10^8 K in the central region, the hottest point. This temperature is not enough to drive a thermonuclear flash, but just to drive the expansion and the late cooling of the outer layers. No mass is ejected from the system.

In the case 1.2+0.4 $M_\odot$, the coalescence is rather violent and the disruption of the binary is very rapid. In fact, most of the particles of the secondary form an accretion stream that collides with the compact surface of the primary, bounces back and eventually forms a toroidal structure around the ONe primary. This structure, however, lasts only for a few moments and finally another, more extended, accretion stream is formed, which again hits the surface of the primary leading to the formation of the definitive accretion disk.

The merging of a 0.6+0.4 $M_\odot$ system is very similar to the 0.8+0.6 $M_\odot$ case, since a narrow accertion stream forms as a consequence of the tidal interaction between both components of the binary system. As expected, because of the differences in the degree of compactness, the primary is practically not afected during the collision while the secondary is considerabily deformed until it is finally swallowed by the most massive star, leading to the formation of an accretion disk. Again, a region of high temperatures is formed around the equatorial plane of the primary. These regions reach a considerably high temperature and some material is burnt but, as before, matter expands and the thermonuclear runaway is rapidly quenched.

EMISSION OF GRAVITATIONAL WAVES

The emission of gravitational energy has been computed in the slow motion, weak–field quadrupole approximation [18]. The amplitude of the strain that characterizes the gravitational emission is given by:

$$h_{ij}^{TT}(\vec{r},t) = \frac{2G}{c^4 R} P_{ijkl}(\vec{n}) \int [2v^k v^l - x^k \partial_l \Phi - x^l \partial_k \Phi] \rho d^3x$$

where

$$P_{ijkl}(\vec{n}) \equiv (\delta_{ij} - N_i N_k)(\delta_{jl} - N_j N_l) - 1/2(\delta_{ij} - N_i N_j)(\delta_{kl} - N_k N_l)$$

is the transverse–traceless projection operator onto the plane orthogonal to the outgoing wave direction, $\vec{n}$, and Φ is the gravitational potential. The other symbols have their usual meaning and the strain is related with the plus and cross gravity wave amplitudes and the polarization tensors by:

$$h_{ij}^{TT}(\vec{r},t) = h_+ e_+ + h_\times e_\times$$

with $h_{+,\times} = A_{+,\times}/R$ and

$$\begin{aligned} A_+ &= \ddot{I}_{xx} - \ddot{I}_{yy} \\ A_\times &= 2\ddot{I}_{xy} \end{aligned}$$

for $i = 0$, and

$$\begin{aligned} A_+ &= \ddot{I}_{zz} - \ddot{I}_{yy} \\ A_\times &= -2\ddot{I}_{yz} \end{aligned}$$

for $i = \pi/2$, being

$$\ddot{I}_{kl} = \frac{G}{c^4} \int [2v^k v^l - x^k \partial_l \Phi - x^l \partial_k \Phi] \rho d^3x$$

In the case of an SPH code, this quantity can be easily obtained by converting the integral into a sum over all the particles. As a proof of consistency it is possible to perform two tests: to compute the gravitational emission of a single, isolated spherically symmetric star and the gravitational emission from a close pair of white dwarfs in a circular orbit.

In the first case, the gravitational emission should be zero. However, in SPH simulations, particles have relatively large masses and are allowed to freely move under the action of the gravitational potential and the pressure gradients, for which reason it is not obvious, at least a priori, if it is possible to obtain a stable, relaxed configuration that does not radiate gravitationally. To test such a behavior we have followed the evolution of an isolated white dwarf of 1 $M_\odot$, composed by 20,000 particles. The gravitational luminosity was found to be 10^{-6} $L_\odot$, which is small enough to consider that the method of calculation is reliable and well calibrated.

With regard to the second test it is convenient to recall that in the case of point–like mass sources, the gravitational emission is given by:

TABLE 3. Distances at which DD systems can be detected.

Case	v_0(mHz)	R_{max} (kpc)	ρ (10 kpc)
0.4+0.4	21	10	5.0
0.4+1.2	32	21	10.5
0.6+0.8	40	31	15.6
0.6+1.0	34	29	14.5
0.8+1.0	58	33	16.6

$$h_+ = \sqrt{2}\frac{\mu}{R}\frac{G^{5/3}}{c^4}(\omega M)^{2/3}(1+\cos^2 i)\cos 2\omega t$$

$$h_\times = 2\sqrt{2}\frac{\mu}{R}\frac{G^{5/3}}{c^4}(\omega M)^{2/3}\cos i \sin 2\omega t$$

where ω is the angular velocity of the system, μ is the reduced mass, M is the total mass, i is the observation angle with respect to the orbital plane and R is the distance. For our test we have assumed two white dwarfs of 1 $M_\odot$ represented by 20,000 particles each one. The initial separation was set to be 0.05 $R_\odot$ or, equivalently, $\omega = 7.94 \times 10^{-2}$ s^{-1}, that is far enough to preserve the initial spherical symmetry of both stars. Our results show that both the theoretical and the simulated amplitude and the frequency, $v \cong 0.025$ Hz, are in excellent agreement.

GRAVITATIONAL EMISSION DURING THE MERGING PROCESS

Table 3 displays the characteristic parameters of the simulations of mergers for which we have computed the emission of gravitational waves. As explained before, in all the cases the initial separation was larger than the corresponding Roche lobe radius of the less massive component and both stars were brought together by introducing a small acceleration. Consequently, the chirping phase was not computed self–consistently, although the agreement between the computed and expected amplitudes of the gravitational waves during this initial epoch of the coalescence was in excellent agreement.

An example of our results is displayed in figure 1 where the dimensionless strains h_+ and $h_\times$ are shown as a function of time for different inclinations of the binary system. The beginning and the final time of the merging itself are shown in both figures as thin dotted lines. This figure shows that before the coalescence proceeds, the emission of gravitational waves still has a sinusoidal pattern, but with an increasing frequency. That is, the close white dwarf chirps as a consequence of the spiral trajectory of both stars towards the center of mass. When the two white dwarfs start to coalesce, the amplitude of the dimensionless strains somehow increases first, but only during the first and most violent stage of the merger. At this point the detailed behavior depends on the mass and mass ratio of both stars. If one of the stars is very compact, the coalescence is very violent and strong accelerations appear which translate into several maxima in the

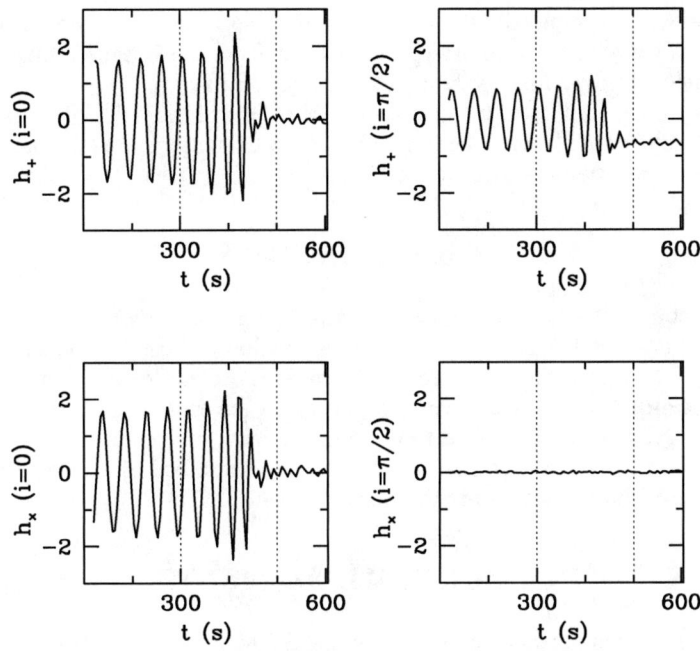

FIGURE 1. Gravitational emission from the merger of a 0.4+0.4 close white dwarf binary system. Strains are measured in units of 10^{-22}. The source is assumed to be at 10 kpc

emission rate. If both stars have similar mass, as it is the case of figure 1, the gravitational signal suddenly disappears in a short time scale, comparable to the orbital period, as a consequence of the fact that the two components of the binary system are disrupted around the center of masses, where a shocked region forms as a consequence of the impact between the two streams.

In order to check whether or not LISA would be able to detect the coalescence of a close white dwarf binary system we have to realize that the most prominent feature of the emitted signal is its sudden disappearance in a couple of orbital periods and that the gravitational wave emission during the coalescence does not increases noticeably, which means that the gravitational emission is dominated by the chirping phase. Hence, we have adopted the orbital separation of the two white dwarfs as exactly equal to that of the binary system when the mass transfer starts and we have assumed that the integration time of LISA is one year. Table 3 displays the frequency of these systems as well as the maximum distance where these systems have to be placed in order to get a signal to noise ratio of 5 or, equivalently, the signal to noise ratio that it would be obtained if these sources were placed at 10 kpc. It is clear from these figures that LISA will be able to detect almost all the systems present in the Galaxy ready to merge, although the probability to detect the sudden disappearance of one of them during the lifetime of the detector is rather small, since the rate of white dwarf mergers is of the order of 10^{-2} per year, with an uncertainty of a factor of 5.

CONCLUSIONS

We have computed the characteristics of merging white dwarf binaries for a wide range of masses and realistic chemical compositions of the components. Our results have shown that there is not a prompt thermonuclear explosion as a consequence of the coalescence process. We have also computed the emission of gravitational waves from these systems and we have found that the most noticeable feature is the sudden disappearance of the gravitational strains. By contrast, the chirping phase will be easily detectable by future space–borne interferometers like LISA.

ACKNOWLEDGMENTS

This work has been partially supported by the Spanish MCYT grants AYA2002–4094–C03–01/02 and AE–ESP2002–10451–E, by the European Union FEDER funds and by the CIRIT.

REFERENCES

1. I.Iben and A.V. Tutukov, *ApJS*, **54**, 335 (1984)
2. R.F. Webbink, *ApJ* **277**, 355 (1984)
3. H. Saio and K.Nomoto, *A&A*, **150**, L21 (1985)
4. E.L. Robinson and A.W. Shafter, *ApJ*, **322**, 296 (1987)
5. P. Bergeron, R.A. Saffer and J. Liebert, *ApJ*, **394**, 228 (1992)
6. P. Bergeron, J.L. Greenstein, and J. Liebert, *ApJ*, **361**, 190 (1990)
7. E.M. Sion, T.D. Oswalt, J. Liebert, and P. Hintzen, in *White Dwarfs*, Eds. G. Vauclair and E. Sion, Dordrecht: Kluwer Academic Publishers, 395 (1991)
8. J. Isern, M. Hernanz, M. Salaris, E. Bravo, E. García–Berro and A. Tornambé, in *Thermonuclear supernovae*, Eds. P. Ruiz-Lapuente, R. Canal and J. Isern, Dordrecht: Kluwer Academic Publishers, 127 (1997)
9. J. Guerrero, E. García–Berro, and J. Isern, *A&A*, **413** 257 (2004)
10. W. Benz, A.G.W. Cameron, W.H. Press and R.L. Bowers, *ApJ*, **348** 647 (1990)
11. F.A. Rasio and S. L. Shapiro, *ApJ*, **438** 887 (1995)
12. L. Segretain, G. Chabrier, and R. Mochkovitch, *ApJ*, **481**, 355 (1997)
13. J.J. Monaghan and J.C. Lattanzio, *A&A* **149** 135 (1985)
14. J. Barnes and P. Hut,*Nature*, **324** 446 (1986)
15. A. Serna, J.M. Alimi, and J.P. Chieze, *ApJ*, **461** 884 (1996)
16. D. Balsara, *J. Comp. Phys.*, **121** 357 (1995)
17. J.J. Monaghan, *J. Comp. Phys.*, **136**, 298 (1997)
18. C.W. Misner, K.S. Thorne, and J.A. Wheeler, *Gravitation*, New York: W.H. Freeman (1973)

White Dwarfs Undergoing Hydrogen Shell Burning in Single Degenerate Binary Systems

M. Orio[*,†], T. Rauch[**,‡], E. Leibowitz[§] and E. Tepedelenlioglu[¶]

[*]*INAF–Osservatorio Astronomico di Torino, Strada Osservatorio 20,*
I-10025 Pino Torinese (TO), Italy
[†]*Department of Astronomy, 475 N. Charter Str., University of Wisconsin, Madison WI 53706, USA*
[**]*Dr.-Remeis-Sternwarte, Sternwartstrasse 7, 96049 Bamberg, Germany*
[‡]*Institut für Astronomie und Astrophysik, Sand 1, 72076 Tübingen, Germany*
[§]*School of Physics and Astronomy and the Wise Observatory,*
Tel Aviv University, Tel Aviv 69978, Israel
[¶]*Physics Department, University of Wisconsin, 3500 University Ave., Madison WI 53706, USA*

Abstract. We review recent X-ray observations of potential type Ia supernova progenitors: white dwarfs in "single degenerate systems" that are undergoing hydrogen shell burning. Grating spectra prove to be very different from each other, and basically fall into two classes: in some systems we observe X-ray emission in a wind, or diffuse circumstellar material, while in others we detect the white dwarf atmosphere. When the white dwarf atmosphere can be observed atmospheric models show promising preliminary results. The light curves offer also important information. Short term light curves of two post-outburst novae have revealed non-radial oscillations of the white dwarf. Long term light curves, possibly coordinated with optical observations, are very important to determine when the transient behaviour of supersoft X-ray sources is due to an interplay of mass accretion rate variations, atmospheric expansion and irradiation of the disk, and when we are instead observing hydrogen shell flashes. Another important open question concerns the X-ray light curve of post-outburst recurrent novae. Recent observations of IM Nor seem to rule out RN as statistically significant SNe Ia progenitors.

INTRODUCTION

X-ray observations of supersoft X-ray sources offer three very useful ways to learn how hydrogen burning in a shell occurs on a hot white dwarf (hereafter WD). For bright sources, grating spectra have finally given us the possibility to analyse the physics of systems that with CCD-type detectors were classified as "supersoft" but were produced in reality through very different mechanisms.

The light curves are also very important, both on short and long time scales. Among classical novae, we are especially interested in the lightcurves of *Recurrent Novae*, most interesting as possible type Ia Supernovae (SNe Ia) progenitors. In the past, attempts were made to derive properties of hydrogen shell flashes by monitoring non-nova recurrent sources move in and out of the "supersoft X-ray window". Later, it became apparent that a source can move out of this window without undergoing a flash, just because the accretion rate varies and there is a complex interplay of atmospheric expansion and irradiation of the disk. We will also talk about a new phenomenon recently observed in two post-outburst classical novae, namely oscillations in the light curve that are attributed to non-radial g-mode pulsations of the WD.

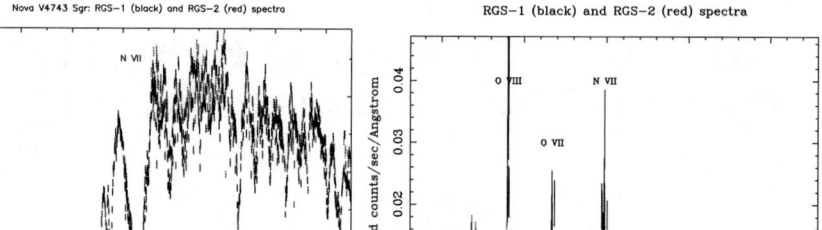

FIGURE 1. XMM-Newton RGS high S/N spectra of V4743 Sgr (left) and of Cal 87 (right). The RGS-1 spectrum is plotted in black and the RGS-2 one is plotted in red.

X-RAY GRATING OBSERVATIONS OF SUPERSOFT X-RAY SOURCES

X-ray spectra of completely different types were classified "supersoft" using CCD-type instruments, even where there was no clear evidence of a white dwarf as compact object (e.g. Bearda et al. 2002). The right panel in Fig. 1 shows the *XMM-Newton RGS* spectra, 6.5 months after the optical maximum, of a very hot WD burning hydrogen in a shell, the Galactic post nova V4743 Sgr, observed repeatedly with *Chandra LETG* and with *XMM-Newton* since it was extremely bright in X-rays. The atmospheric continuum of a very hot WD ($T_{eff} \approx 6-7 \times 10^5$ K) is detected, with deep absorption lines and absorption edges. Model atmospheres show that carbon is greatly depleted, an expected consequence of CNO processing (Rauch et al., 2004, in preparation). This is extremely important because it means that the spectrum originates in pre-outburst left-over material, that is the WD in this case is increasing its mass during its secular evolution. Some classical novae may thus be type Ia SN progenitors. In N V4743 Sgr the absorption lines are blue-shifted by $\simeq 2000$ km s^{-1}, and the blue-shift seems to have remained approximately constant for several months, a fact that is probably understood in terms of a source at super-Eddington luminosity. Work is in progress to fit the grating spectra of V4743 Sgr, and of other supersoft X-ray sources, with the models of the Tübingen group (e.g. Rauch 2003). We find that the spectrum of Nova V1494 Aql (right panel of Fig. 2) has many analogies with the one of N V4743 Sgr, although it has lower S/N. This was not the case of another nova, V382 Vel, which was observed with the *LETG* when only nebular emission seemed to be present, although the spectrum could still be classified as "supersoft" (Burwitz et al. 2002). The left panel in Fig. 1 shows instead the *emission* spectrum of the LMC source Cal 87 (Orio et al. 2004). Independent evidence indicates that the compact object in Cal 87 is indeed a very hot WD, but it is blocked by an accretion disk corona and a wind, in which red-shifted emission lines are produced. Past attempts to derive the WD parameters from the spectrum of Cal 87 observed with CCD-type instruments were not realistic. The EPIC-pn spectra of V4743 Sgr and Cal

87 appear relatively similar and quite featureless, because the forest of lines (either in emission or in absorption) is smoothed and "washed out" by the low spectral resolution. Other supersoft X-ray sources (e.g. Cal 83) seem to display a superimposed wind and WD spectrum, and they can be difficult to disentangle. A comparison of the spectrum of N V1494 Aql and of Cal 83, both observed with the *Chandra LETG*, is shown in Fig. 2. Despite the lower count rate and S/N, it is evident how much more complex the spectrum of Cal 83 is. The low S/N grating spectra for QR And and RX J0439.8-6809 (see Fig. 3) are not well understood yet, but they seem to be originated in the atmosphere of a WD with a very peculiar chemical composition.

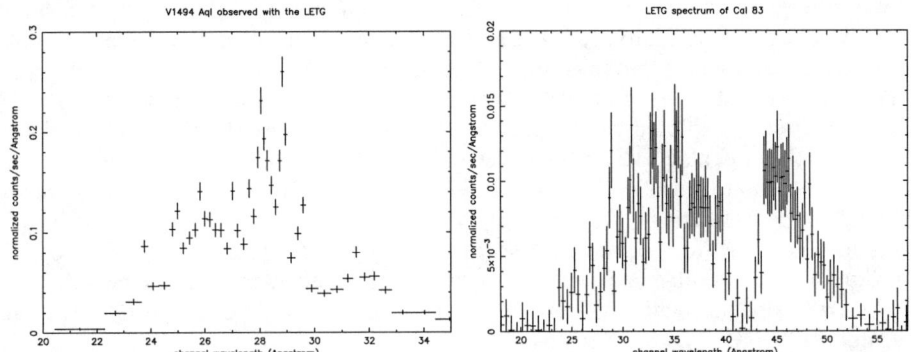

FIGURE 2. Chandra LETG low S/N spectra of N V1494 Aql (left, plotted binning with S/N=20) and of Cal 83 (right, plotted binning with S/N=20). Despite the lower S/N, it is evident that the spectrum of Cal 83 is much more complex and covers a wider energy band.

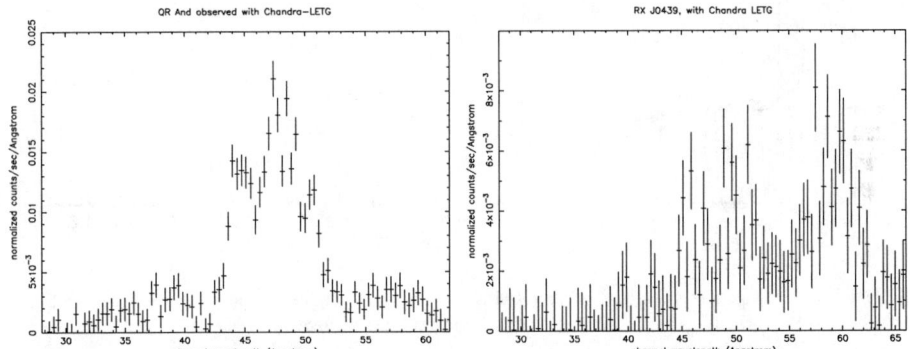

FIGURE 3. Chandra LETG low S/N spectra of QR And (left) and of RX J0439.8-6809 (right).

LONG AND SHORT TERM LIGHT CURVES

More than half of supersoft X-ray sources are transient or recurrent. The long term light curves, correlated with the optical ones, teach about the secular evolution of these systems. Are we witnessing the H shell flashes predicted by the theory, or do flux and spectral distribution vary because the mass accretion rate varies and the disk is irradiated? Recent *XMM* observations of RX J0533.7-7034 and RXJ0527.8-6954 (Orio & Lipkin 2004) indicate recurrence periods in supersoft X-rays >10 years, which seem to imply shell flashes, unlike Cal 83 which instead has "on" and "off" states on time scales of months (Greiner & Di Stefano 2002) due to variation in mass transfer rate and irradiation of the disk, or some M31 sources that seem to also vary on months time scales. One other important open question concerns Recurrent Novae and their statistical significance as type Ia SN progenitors. Recent observations of IM Nor (Orio et al., 2004) indicate that this recurrent nova, thought to be a most likely SN Ia progenitor because it accretes and ejects more mass than most RN, burned hydrogen for only less than 6 months after the outburst. Therefore, it does not accumulate sufficient mass to be a type Ia SN candidate, even if initially it was considered one of the most promising candidates because it accumulates more mass than other RN.

Non-radial oscillations of the WD, in an "instability strip" for pulsations that has not been theoretically explored yet, were discovered for V1474 Aql (Drake et al. 2003) and V4743 Sgr (Ness et al. 2004, Leibowitz et al. 2004). For V4743 Sgr, the most prominent peaks correspond to periods of 1308 and 1374 s. Periods of 665, 686 and 3270 s are also detected in the light curve. We are trying to assess the reality of these periodicities by studying a new method for the analysis of the X-ray light curves (Leibowitz 2005, in preparation). We are also analysing the light curves of other, non-nova supersoft X-ray sources. We also temptatively found a 10 s period in the light curve of QR And, that resembles a "DNO" (Dwarf Nova oscillation).

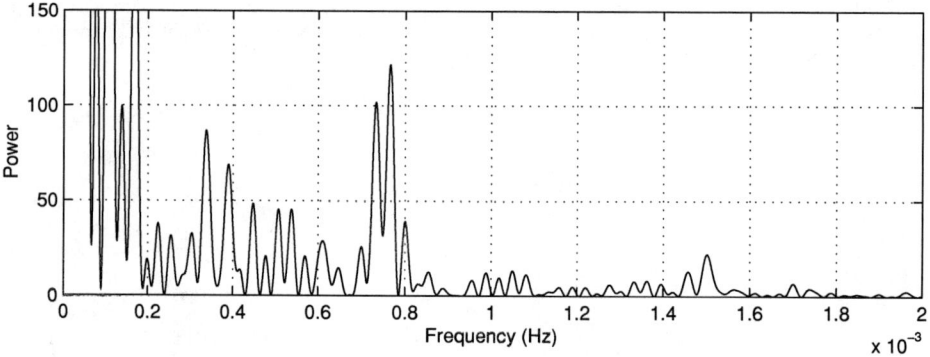

FIGURE 4. The power spectrum of the XMM-Newton light curve of the post nova V4743 Sgr, observed in 2004 April. A new method of analysis should help to determine the reality of these apparent periodicities.

CONCLUSIONS

With the new generation of X-ray telescopes, a new era has opened for the study of supersoft X-ray sources. The results are complex and at times difficult to interpret theoretically, but we have a glimpse into previously unknown physics. Ultimately, we want to determine basic physical facts and parameters that determine the final fate of these systems, including the possibility that they end with an accretion induced collapse or a SN Ia explosion.

REFERENCES

- Bearda, H., et al. 2002, A&A, 385, 511
- Burwitz, V., Starrfield, S., Krautter, J., & Ness, J-U. 2002, in Classical Nova Explosions, ed. M. Hernanz and J. Jose', AIP Conf. Proc. 637, 377
- Drake, J.J., et al. 2003, ApJ, 584, 448
- Greiner, J., & Di Stefano R. 2002, A&A, 387, 944
- Greiner, J., Di Stefano, R., Kong, A., & Primini, F. 2004 ApJ 610, 261
- Leibowitz, E., Orio, M., et al. 2004, preprint
- Ness, J-U., et al. 2003, ApJ, 594, L27
- Orio, M., Ebisawa, K., Heise, J., & Hartmann, W. 2004, Compact Binaries in the Galaxy and beyond, IAU Coll. 194, E. Sion & G. Tovmassian editors, RevMexAA(SC), 20, 210
- Orio, M., & Lipkin, Y. 2004, preprint
- Orio, M., Tepedelenlioglu, E., Starrfield, S., Woodward, C.E., & Della Valle, M. 2004, preprint
- Rauch, T. 2003, A&A, 403, 709
- Rauch, T., Orio, M., Gonzalez-Riestra, R., & Still, M. 2004, preprint

Galactic disk abundance ratios: constraining SNIa stellar yields

Cristina Chiappini

Osservatorio Astronomico di Trieste - OAT/INAF - Via G. B. Tiepolo 11, Trieste 34131, Italy

Abstract. Stellar abundance ratios of very good quality are now available for a large number of stars in the solar vicinity. Moreover, for an increasing number of stars informations on kinematics is also available. The combined information on abundance and kinematics enables one to select objects belonging to the different components of our Galaxy (thin disk, thick disk and halo). In this work we show that a careful comparison of our chemical evolution model for the Milky Way with the available abundance ratio measurements for stars in the solar neighborhood can be used to constrain the stellar yields. In particular, yields of Type Ia SNe are constrained by the abundance pattern of thin disk stars. Our results suggest that the 3-D models for SNIa explosion studied here lead to discrepancies once their predicted stellar yields are used as input in our chemical evolution model. These models produce flat Si/O and Mg/O ratios in disagreement with what is observed in thin disk stars. Moreover, our results indicate that larger quantities of Mg (at least a factor of 10 more than current theoretical predictions of either 1-D or multi-D models) need to be produced in SNIa.

Keywords: Chemical composition and chemical evolution; Abundances
PACS: 98.35.Bd; 97.10.Tk

INTRODUCTION

Many are the groups involved in the computation of SNIa progenitor and explosion models (see several contributions in this conference proceedings). The SNIa progenitor model is still an open issue, with some groups favoring the single degenerate scenario (where a C-O white dwarf (WD) accretes mass from an evolved companion star until it reaches the Chandrasekhar limit leading to a thermonuclear runaway), while others prefer the double-degenerate models (where two WDs in a close binary system merge, due to the loss of angular momentum, again reaching the Chandrasehkar mass and exploding by C-deflagration). Published SNIa yields assume the single degenerate scenario, and in the present work we will verify the effect of these different stellar yields on chemical evolution models for the Milky Way (MW). Our goal is to show that chemical evolution models for the MW offer an important extra-tool to constrain the stellar yields coming from different models of SNIa which can be then linked to some of the still uncertain parameters in these latter models. This is now possible thanks to the large and precise amount of abundance measurements available for stars in the solar vicinity. While very metal poor stars enable us to constrain the "zero point" of our models by constraining the stellar yields of core collapse SNII, thin disk stars abundance ratios (which clearly show the imprints of SNIa enrichment) can be used to constrain the SNIa nucleosynthetic output. In this work we suggest some of the "key" abundance ratios to be used in order to constrain the SNIa models.

SNIA MODELS AND THEIR PREDICTED YIELDS

Many are still the uncertainties involved in the computation of the single degenerate models. Different models might differ in the following points:

- the mass accretion rate (which is linked to the C-ignition central density)
- the flame speed after ignition (fast models vs. slow models)
- the explosion mechanism: pure deflagration models vs. delayed detonation models
- the deflagration/detonation transition densities in delayed detonation models
- 1-D vs. Multi-D models (with central vs. multi location ignition)

Models computed with the different parameters outlined above predict different nucleosynthetic outputs. By instance, in delayed-detonation models, lower transition densities favor larger amounts of matter experiencing incomplete Si-burning. As a consequence, these models produce less Fe and more intermediate mass nuclei (from Si to Ca). Delayed detonation models also produce more O-burning in the outer layers [1]. Multi-dimensional models (the ones studied here assume pure deflagration) tend to produce less Fe and more unburned material, producing larger quantities of carbon. Among these latter models, only 3-D models are "potentially" able to produce enough quantities of Fe [2]. However, current available calculations still tend to produce low quantities of this latter element.

In the next sections we show results for the chemical evolution model of the MW computed with the so-called "Two-Infall Model" [3, 4] once different stellar yields for SNIa are adopted. We show our predictions for "key" abundance ratios, namely: Fe/O, Mg/O, Si/O and S/O. The variation of these abundance ratios with metallicity depends strongly on the SNIa nucleosynthesis outputs as it will be shown in the next section. These predictions are then compared with abundance measurements in stars of the solar vicinity. To constrain the SNIa stellar yields we focus on the abundance data for thin disk stars.

Here we are going to compare our chemical evolution model predictions computed with 4 different sets of stellar yields for SNIa: two of them come from 1-D models, whereas the other two are from more recent 3-D models. For the 1-D models we adopted the yields computed by Iwamoto et al. [1] (their models W7 and WDD1). Their W7 model is a pure deflagration model, where the flame speed reaches 30% of the sound speed. WDD1 is a model where a deflagration induces a detonation when reaching low density layers. In this case the authors took their slow deflagration models (with flame speeds of 1.5% of sound speed) and transformed them into detonations when the density ahead of the flamed decreased to 3.0, 2.2 and $1.7 \, 10^7$ gcm^{-3}. The WDD1 model stands for the lowest transition density case.

For the 3-D models we adopt the stellar yields computed by Travaglio et al. [2]. The latter authors computed multidimensional (2-D and 3-D) hydrodynamical simulations of the thermonuclear burning phase in SNIa, in the single degenerate scenario. These are pure deflagration models. However, in contrast to 1-D models which adjust the burning speed to reproduce light curves and spectra, the multi-D model of Travaglio et al. does not contain adjustable parameters and variations in the yields are dependent on changes of initial conditions only. Here we consider two different cases for the 3-D models,

TABLE 1. SNIa stellar yields predicted by different SNIa models

	C	O	Fe	Mg	Si	S	Si/O	S/O
W7	0.048	0.143	0.626	0.0085	0.154	0.0846	1.1	0.6
WDD1	0.005	0.088	0.587	0.00755	0.272	0.160	3.1	1.8
Tb30-3d	0.278	0.339	0.439	0.00753	0.054	0.0262	0.16	0.08
Tc3-3d	0.337	0.417	0.336	0.0126	0.054	0.0257	0.13	0.06

namely: a) their best model, which is a pure deflagration one with multi off-center ignition (Tb30-3d) and b) a model with central ignition (Tc3-3d) (see [2] for details). Table 1 shows the stellar yields computed by the different SNIa models considered here, focusing on the abundance ratios that most differ among them. The variations go from a factor of ∼2 for Fe (and Mg) up to a factor of ∼60 for C, ∼30 for S/O and ∼20 for Si/O. Given these huge differences, we expect to see an impact on our chemical evolution model predictions. This is shown in the next section[1].

THE IMPACT OF THE DIFFERENT SNIA YIELDS ON CHEMICAL EVOLUTION MODEL PREDICTIONS

In this section we focus on the following abundance ratios: Fe/O, Si/O, Mg/O and S/O. The figures shown here cover a metallicity range from solar down to the end of thick disk metallicities. Solar abundances are taken from Asplund et al. [12]. The effect of varying the SNIa yields are seen for oxygen abundances larger than [O/H] ≥ -1, i.e., already in the halo/thick disk phase. However, the main differences are seen for the metallicity range that corresponds to the thin disk ([O/H] ≥ -0.5). In Fig. 1 it is shown our model prediction for [Fe/O] vs. [O/H] in the solar vicinity compared with data for halo stars [7] and thin disk stars [8, 9]. The effect of the temporal gap in the star formation assumed in our model between the formation of the halo and the thin disk [3] is clearly seen. During the halt in the star formation, oxygen (mainly produced in SNII) stops being ejected into the ISM whereas iron keep increasing due to the contribution of SNeIa born before the gap. This figure also shows the large differences in the predicted Fe/O abundance ratios depending on the adopted SNIa yields. We remind the reader that the models plotted here are in good agreement with the Fe/O abundance ratios down to the very low metallicities (see [10]). However, it is known that the uncertainties in the Fe yields (not only in the case of SNIa, but mainly in the case of SNII) are quite large.

In Fig. 2 we compare the predicted [Mg/O] and [Si/O] ratios as a function of [O/H]. In this case we are free from the uncertainties still affecting the iron yields. Oxygen seems to be a better understood element in the case of core collapse supernovae. In this way we can be more confident on our "zero point" which is given by the fit of the very metal poor stars as discussed above. In Fig. 2 we compare our model predictions with

---

[1] Here we will not discuss results related to carbon since in this case low and intermediate mass stars are also supposed to contribute and many are the still open problems on the C produced in such stars [5, 6].

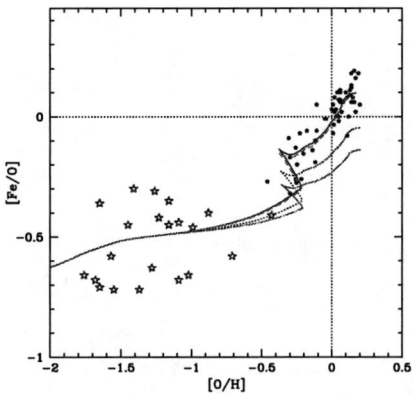

FIGURE 1. [Fe/O] vs. [O/H] diagram. Data are: halo stars (stars [7]), thin disk (dots [8, 9]). The curves represent our "two-infall" model predictions for different assumptions of stellar yields in SNIa. The different sets of adopted stellar yields are shown in Table 1 and are labelled here as: W7 (solid curve), WDD1 (dashed curve), Tb30-3d (dotted curve), Tc3-3d (dot-dashed curve). In this figure the effect of the "gap" in the star formation between the halo and thin disk formation is clearly seen (see [3] for details).

the stars of Bensby et al. [8, 9] belonging to the thin disk (the thick disk will be explicitly considered in a forthcoming paper). Two important things can be seen in Fig. 2: a) the multi-D models predict a flat Si/O ratio for thin disk stars which is not supported by the data. The 1-D models of Iwamoto et al. [1] predict instead an increasing Si/O abundance ratio as a function of metallicity in the thin disk stars metallicity range. The absolute value of Si/O is larger in the case of WDD1 model which seems to better agree with the data, b) all the models shown here predict a flat Mg/O ratio (or even slightly decreasing) as a function of oxygen. The thick-dashed line shows a model similar to the solid line (which was computed with SNIa yields from the W7 models of [1]), but for which we increased the Mg yield by a factor of 10. This model predicts a slightly increase of the Mg/O ratio at larger metallicities as it is suggested by Bensby et al. [8, 9] data. The predicted "solar abundance" in this case is 7.55, whereas that of a pure W7 model is 7.43. The observed value is 7.53 ± 0.09 [12].

François et al. [10] also suggested, based on a [Mg/Fe] vs. [Fe/H] plot that Mg should be increased in SNIa. However, as discussed before it is interesting to confirm the same finding now from a diagram not including Fe and hence less prone to uncertainties in the stellar yields of both SNII and SNIa.

Finally, in Fig. 3 we plot S/O vs. $\log(O/H) + 12$ (this time we show the absolute values, not normalized to solar). Here we show data from [11] and the solar value of [12]. The predicted trend again depends strongly on the adopted yields for SNIa. While 1-D models predict an increasing S/O ratio as the metallicity increases, 3-D models predict the opposite. Unfortunately there is very few data and it is not possible to constrain these models at the present. In fact, there is a lack of abundance measurements of both S and O in the same stars.

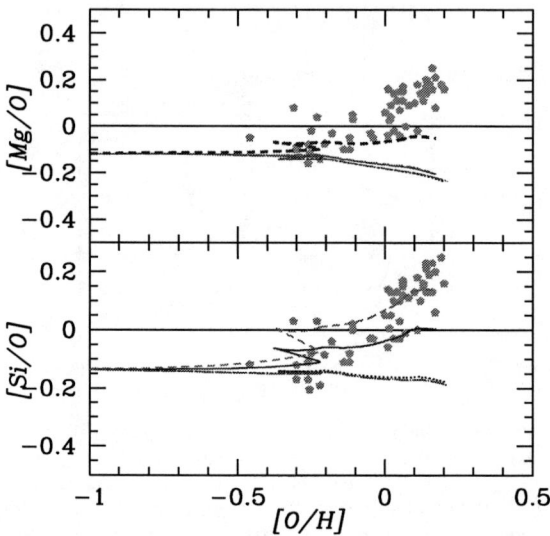

FIGURE 2. Data are thin disk stars from [8, 9]. The curves are labelled as in Fig. 1. In the upper panel, the thick-dashed line shows a model similar to the one computed with W7 yields, but where the stellar yield of Mg was increased by a factor of 10.

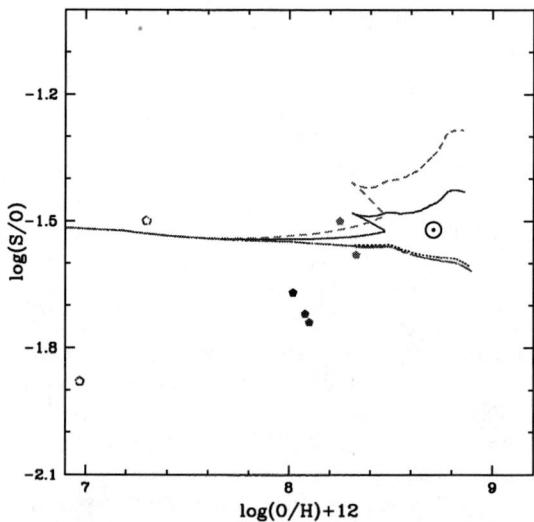

FIGURE 3. Very few data are avaible for both S and O in the same stars. The data shown here are from [11]. The indicated solar value is from [12]. Here we show the absolute values of S/O vs. log(O/H)+12 (while in the figures presented before the plots were normalized to the solar values). Notice the large differences predicted for S/O around solar metallicity once different sets of SNIa yields are adopted.

CONCLUSIONS

- With better data for both the disk and halo stars we are now in a position to start "testing" SNIa yields and the timescales for the bulk of SNIa enrichment of the interstellar medium in the solar vicinity.
- Our results suggest that W7 and WDD1 models lead to a better agreement of our model predictions with the abundance data in the solar vicinity. Still, these models seem to produce a little too much iron (W7) and a too large Si/S ratio (WDD1). The 3-D models studied here produce too much oxygen and too little iron, sulphur and silicon. When tested in our chemical evolution models these latter SNIa models produce a flat [S/O] ratio in the metallicity range corresponding to thin disk stars which is not supported by recent observations.
- More Mg from SNIa is needed in order to explain the abundance trend recently reported by Bensby et al. [8, 9] namely, a decreasing O/Mg ratio with metallicity in thin disk stars and to better reproduce the solar abundance of Mg.

Whether an increase of a factor of 10 in the Mg ejected by SNIa is physically plausible is still to be assessed by the experts in the field. Interestingly, after the conference, a paper by Stehle et al. [13] which applies a new technique called "Abundance Tomography" to derive abundance distributions in SN spectra found, in the case of the SN2002bo, a Mg abundance which is a factor of ~ 7 above the predicted by the SNIa models discussed here. It would be interesting to verify if this is common to other SNIa spectra.

ACKNOWLEDGMENTS

I am pleased to acknowledge partial financial support from MIUR/COFIN 2003028039 and INAF/COFIN. I thank the organizers for what was a very pleasant and interesting conference and L. Yungelson for the many interesting discussions. Finally I thank S. Feltzing for having send me her recent abundance measurements in electronic form.

REFERENCES

1. K. Iwamoto, F., Brachwitz, K., Nomoto et al., ApJS **125**, 439-462 (1999)
2. C. Travaglio, W. Hillebrandt, M. Reinecke and F. K. Thielemann, A&A **425**, 1029-1040 (2004)
3. C. Chiappini, F. Matteucci and R. Gratton, ApJ **477**, 765-780 (1997)
4. C. Chiappini, F. Matteucci and D. Romano, ApJ **554**, 1044-1058 (2001)
5. C. Chiappini, D. Romano and F. Matteucci, MNRAS **339**, 63-81 (2003a)
6. C. Chiappini, F. Matteucci and G. Meynet, A&A **410**, 257-267 (2003b)
7. J. Meléndez and B. Barbuy, ApJ **575**, 474-483 (2002)
8. T. Bensby, S. Feltzing and I. Lundström, A&A **415**, 155-170 (2004)
9. T. Bensby, S. Feltzing and I. Lundström, A&A **421**, 969-976 (2004)
10. P. François, F. Matteucci, R. Cayrel, M. Spite, F. Spite, C. Chiappini, A&A, **421**, 613-621 (2004)
11. P. E. Nissen, Y. Q. Chen, M. Asplund and M. Pettini, A&A **415**, 993-1007 (2004)
12. M. Asplund, N. Grevesse, J. Sauvaul, in "Cosmic abundances as records of stellar evolution and nucleosynthesis", eds. F. N. Bash and T. G. Barnes, (in press), astroph/0410214
13. M. Stehle, P. A. Mazzali, S. Benetti and W. Hillebrandt, MNRAS (in press), astroph/0409342

Clues on Type Ia Supernovae Progenitors

Luciano Piersanti and Amedeo Tornambé

INAF - Osservatorio Astronomico di Teramo, via M. Maggini, SNC, 64100, Teramo, Italy

Abstract. We show that in the framework of canonical stellar evolution it is hard, if not impossible, to determine the growth in mass of a CO White Dwarf, up to the Chandrasekhar limit by means of mass transfer from its companion in a binary system. This is the case either if matter is accreted from a normal companion with an H-rich envelope or if direct CO accretion occurs from a CO WD companion. At variance, we show that if the effects of rotation are taken into account in modeling the accretion process, a CO WD can increase its mass at the expenses of the degenerate CO companion up and beyond 1.4 $M_\odot$, so that an explosive event of the type Ia class is naturally produced. This theoretical finding revives the Double Degenerate scenario for type Ia SNe progenitors. In such a case the internal spread in the observational properties of type Ia SNe may be interpreted as a consequence of different total masses; hence differences between SNe Ia in nearby elliptical galaxies and the majority of those in spirals should be expected and the current use of type Ia SNe as cosmological distance indicators should be justified.

INTRODUCTION

It is widely accepted that type Ia Supernovae (SNe Ia) are produced by the thermonuclear disruption of Carbon Oxygen (CO) White Dwarfs (WDs) which has accreted their mass up to the Chandrasekhar limit via matter deposition from their companions in binary systems. Although this scenario well explains several observational properties of SNe Ia (see *e.g.* [1]), no clear consensus has been attained so far on the typology of the companion and on the evolutionary path followed by the system up to the explosion. The two most longliving evolutionary scenarios are the so called **Single Degenerate (SD)** and **Double Degenerate (DD)** scenarios. In the former it is claimed that the donor is a normal star with an H-rich envelope and that the accreted Hydrogen undergoes nuclear burning on the surface of the White Dwarf so that it is first converted into Helium and then into a CO mixture. In the latter, the binary system is composed by two CO WDs with total mass of the system of the order of or greater than M_{Ch}. In this case CO rich matter is directly accreted onto the WD as a consequence of a merging process of the two degenerate components (see below). In the following we comment on how for both cases there is no way *in the framework of canonical stellar evolution* to attain the physical conditions for central explosive C-ignition, independently of the initial mass of the WD and the chemical composition of the accreted matter, unless unrealistic assumptions are done. Moreover we show that, by including the lifting effects of rotation in the evolution of accreting WD, DD systems can increase in mass up and beyond 1.4 $M_\odot$. In fact in this case the accretion process becomes self-regulated on values for which central carbon burning is ignited in highly degenerate physical condition thus producing a type Ia event.

THE SINGLE DEGENERATE SCENARIO AND ITS FAILURE

A binary system composed by a CO WD and a normal star with an H-rich envelope can be produced as a consequence of one Common Envelope (CE) episode (for more details see [2]). The transfer of H-rich matter onto the WD occurs when the secondary component eventually overfills its own Roche Lobe. As suggested by Kipphenhan and Weigert [3] such a situation can occur in three different evolutionary phases: at the end of the Central Hydrogen Burning phase (*Case A*), during the RGB phase (*Case B*) or during the AGB phase (*Case C*). In Case B and C, since the donor has a deep convective envelope, the mass transfer occurs on the thermal time scale of the envelope so that the rate at which matter is accreted to the WD is very high (of the order of or greater than $\dot{M}_{Edd}$ - [4]). In this case the accretor expands to giant dimensions due to energy delivered by the accreted matter so that a second CE episode occurs and matter is not effectively deposited onto the degenerate component. On the other hand, in Case A the donor star has a radiative envelope so that the mass transfer is steady and occurs on the nuclear time scale. In this case the rate at which matter is deposited onto the WD is low ($\dot{M} \sim 10^{-8} - 10^{-9}$ M$_\odot$ yr^{-1}). Another interesting scenario is represented by a CO WD + a RGB star which does not overfill its own Roche Lobe. In this case the accretion occurs through the wind with a rate which is in general smaller than in the previous case. The occurrence of the second CE is thus avoided and matter can be steadily deposited onto the companion.

The final outcome of the accretion process is determined univocally by the thermal response of the WD to mass deposition. This problem has been investigated by several authors during the last two decades (for a complete list see [5]). The state of art of H-accreting CO WDs is summarized in Fig. 1, where the possible final outcomes are reported as a function of the WD total mass and the adopted accretion rate.

For high values of the accretion rate (say $\dot{M} > 1 \div 3 \times 10^{-7}M_\odot$ yr^{-1}), slightly depending on M$_{WD}$) the accreting star expands almost immediately, thus overfilling its own Roche Lobe and interacting with the already expanded companion. An additional CE episode occurs and no matter is accreted onto the WD. Note that in this case the accretion rate is smaller than $\dot{M}_{Edd}$ so that the expansion is due to the fact that the matter is deposited more rapidly than the rate at which the H-burning shells moves outward. Hence an extended H-rich envelope forms and the accreting stars resembles at all an AGB stars, thus moving toward the Hayashi track.

Lowering the accretion rate the region labeled *Steady Accretion* is encountered. In this case the accreted Hydrogen is steadily converted into Helium at the same rate at which matter is transferred to the WD. As a consequence of H-burning, an He-buffer is piled up until the onset of He-burning. 3α-reactions are ignited via a flash which, for the adopted values of $\dot{M}$, is not dynamical. In fact the He-buffer is accreted by the overlying H-burning shell very rapidly so that the compressional heating is very high and in addition thermal energy flows from the H-shell inward, thus preventing the degeneration of the He-shell. However, it has to be remarked that the energy delivered during the He-flash is so high to determine the expansion of the He-buffer and of the overlying H-shell to giant dimensions. According to the extant theoretical models, it can be argued that the donor and the accretor stars interact and determine the occurrence of a CE during which a great part (if not all) of the matter accreted in the previous evolution is lost by the star.

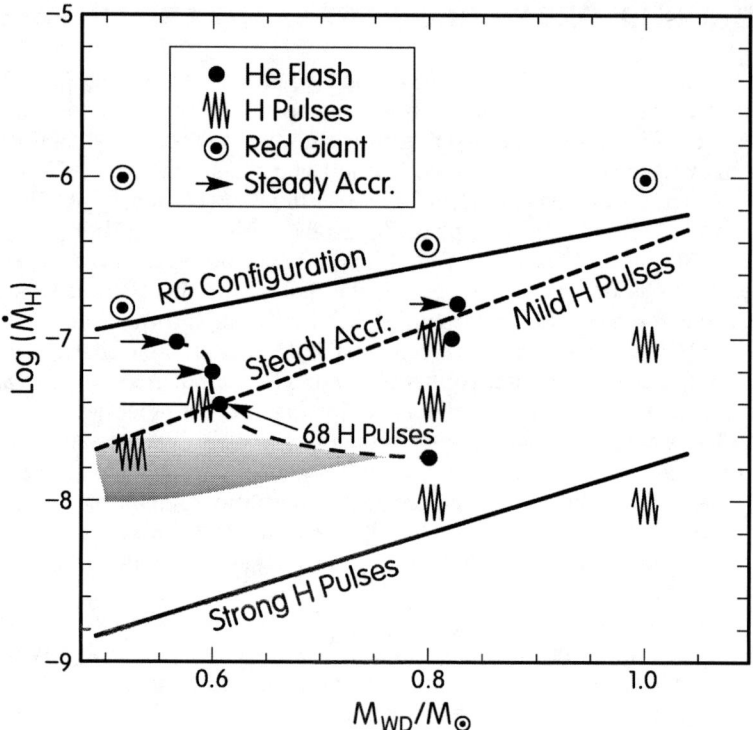

FIGURE 1. The parameter space M_{WD}-$\dot{M}$. Different symbols represent different final outcomes, as labeled inside the figure (see text).

Even for accretion rates in this range it seems impossible to produce the effective growth in mass of the WD.

For lower values of $\dot{M}$, the accreting WD experiences recurrent mild H-flashes which pile up an He-buffer up to the onset of the He-flash. In this case the ignition of He-burning can occur either in mildly degenerate physical conditions or with a dynamical explosive flash. The different behavior of the accreting model is determined by the accretion rate; in fact, for $\dot{M}$ of the order of $3-4 \times 10^{-8} M_\odot$ yr^{-1} the He- and H-shell are thermally coupled, so that thermal energy flows from the H-shell inward, keeping hot the underlying He-shell. In this case the base of the He-shell does not fully degenerate, 3α-reactions are ignited via a non-dynamical flash and, hence, as in the previous case a CE episode occurs. On the other hand, for lower accretion rates (the shaded region in Fig. 1) the He-buffer becomes more massive than ~ 0.1 M$_\odot$ so that the two shells thermally decouple. In this case, the accreting WD evolves as if He-rich matter was directly accreted at the same rate and, hence, the evolution is driven only by the compressional heating produced by the piled-up He-rich layers [6]. These models experience a dynamical He-flash due to the high degeneracy at the base of the He-shell. In this case the final outcome is an explosion of SNe Ia proportion but the produced nucleosynthesis does not fit the observed spectra (see [7]). Moreover the presence of ^{56}Ni in the external layers determines that the color at the epoch of maximum magnitude in the light curve is too blue with respect to observed SNe Ia.

Finally, for very low values of the accretion rate (say $\dot{M} < 10^{-9}$ $M_\odot$ yr^{-1} - region labeled *Strong H-pulses* in Fig. 1), the accreting WD experiences strong dynamical H-flashes, thus producing Nova-like events. In this case the matter accreted during the quiescent phase is ejected during the flash due to a combination of dynamical acceleration, wind mass loss and binary interactions with the companion. According to the extant numerical models, during the Nova outburst the CO core or the He layers underlying the H-flashing shell are secularly reduced in mass so that for these values of $\dot{M}$ there is no possibility for the WD to grow in mass at all [8].

According to the discussed scenario *it is very difficult to see how the mass of a CO WD can be increased to the Chandrasekhar mass by accretion of Hydrogen at any realistic rate* [9].

To end this section we must recall that various attempts have been made in the past to find mechanisms able to tune the accretion rate to values suitable for a stable accretion. This is the case of the "wind solution" by Hachisu, Kato & Nomoto [10]: according to this authors for high values of the accretion rate (the region labeled "RG Configuration" in Fig. 1) the expansion of the external layers of the accreting WD is prevented due to the effect of a strong wind which reduces the amount of matter effectively deposited. Such a solution has been criticized by several authors (*e.g.* [11]) since the efficiency of mass extraction from the Giant companion by the WD may be overestimated. The analysis of such mechanism is well beyond the scope of this review, so we simply note that, even if it was at work, He-burning in the He-buffer is ignited via a non-dynamical flash which, in any case determines the occurrence of a CE episode. This implies that any possible real path of the H-accretion process may be reconnected to the general scenario exposed here, thus making unreliable an effective growth in mass of the CO WD.

THE DOUBLE DEGENERATE SCENARIO AND THE QUEST FOR FINE TUNED ACCRETION RATE

Double Degenerate systems are composed by two CO WDs and they can form as a consequence of two CE episodes, either directly or indirectly as the final outcome of the evolution of He-stars [2]. A DD system can represent the progenitor of a SN Ia event if its total mass is of the order of or greater than M_{Ch} and the orbital separation is small enough (A$\sim$ $R_\odot$), so that the merging of the two components via Gravitational Wave Radiation (GWR) emission occurs in a time smaller than the Hubble time. In this case the less massive WD, which first overfills its own Roche Lobe, undergoes a dynamical mass transfer and, hence, it completely disrupts, thus forming a thick accretion disk around the more massive companion. This scenario, first proposed by [12], is very interesting on the theoretical point of view since there are no light elements at all in the progenitor system, so that the total lack of Balmer's lines in the optical spectrum can be easily explained; in addition, since CO-rich matter is directly accreted, the growth in mass of the WD up to M_{Ch} is not be affected by flash-driven instabilities as it occurs in the SD scenario.

In the past this scenario was considered not promising as candidate for SN Ia progenitor since the observational searches of DD systems with the right separation and total mass have provided negative results (see [13], [14] and [15]). However the very recent discovery by the SPY group [16] of two DD systems more massive than M_{Ch} and close enough has raised again the attention to this scenario.

We want to remark that according to the results of stellar population synthesis the right number of close DD systems suitable as SN Ia progenitors has to form. This evidence suggests that up to now the observational searches have been unsuccessful not because the right DD systems do not exist at all but because these objects are intrinsically very faint so they can not be detected using the current generation of telescopes.

While the observational problem seems to be solved, it has to be remarked that in the framework of classical stellar evolution DD systems can not be regarded as good candidate for SNe Ia progenitors. In fact, if a merging occurs, then the rate at which matter is effectively transferred from the accretion disk to the WD is very high ($\dot{M} \sim \dot{M}_{Edd}$ - see [17] and [18]). In this case off-center C-ignition occurs well before the accreting WD could attain M_{Ch} and the final outcome is an O-Ne-Mg WD which can eventually explode as Supernova but by core collapse [19]. Such an occurrence is due to the fact that for high values of $\dot{M}$, the energy delivered by the accreted matter onto the WD surface can not be removed via inward thermal diffusion. As a consequence a thermal energy excess remains localized in the external layers and, due to the continuous deposition of matter it increases up to the off-center ignition of C-burning (for a detailed discussion see [20]). Hence, in the framework of standard stellar evolution also DD systems are not promising as SNe Ia progenitors.

THE LIGHT AT THE END OF THE TUNNEL

The results previously summarized have been obtained in the framework of classical stellar evolution, *i.e.* by assuming that rotation is negligible in driving the evolution of an accreting WD. Such an assumption is based on the observational evidence that single WDs rotate at a very slow rate. But at least for DD systems such an assumption is not correct. In fact in this case the angular velocity of the two WDs is not a reminiscence of the previous evolution during the AGB phase but it is determined by the synchronization of the orbits in the binary system which occurs on a very short time scale [21]. Once synchronization has been achieved, it is maintained up to the merging time, so that the two component becomes fast rotators ($\omega_{Merg} \sim 0.1 - 0.2$ rad s^{-1}). In addition, due to tidal forces, part of the rotational energy acquired by the two components is converted into thermal energy so that they result hotter and more expanded. In this way the physical properties of the two WDs are dramatically different with respect to standard non-rotating object ([21] and [22]).

The inclusion of the lifting effect of rotation in modeling the evolution of WD accreting CO-rich matter completely changes the evolutionary scenario for DD systems. In particular rotation acts as the tuning mechanism of the accretion process. According to [22], after the merging, three different phases in the accretion process can be distin-

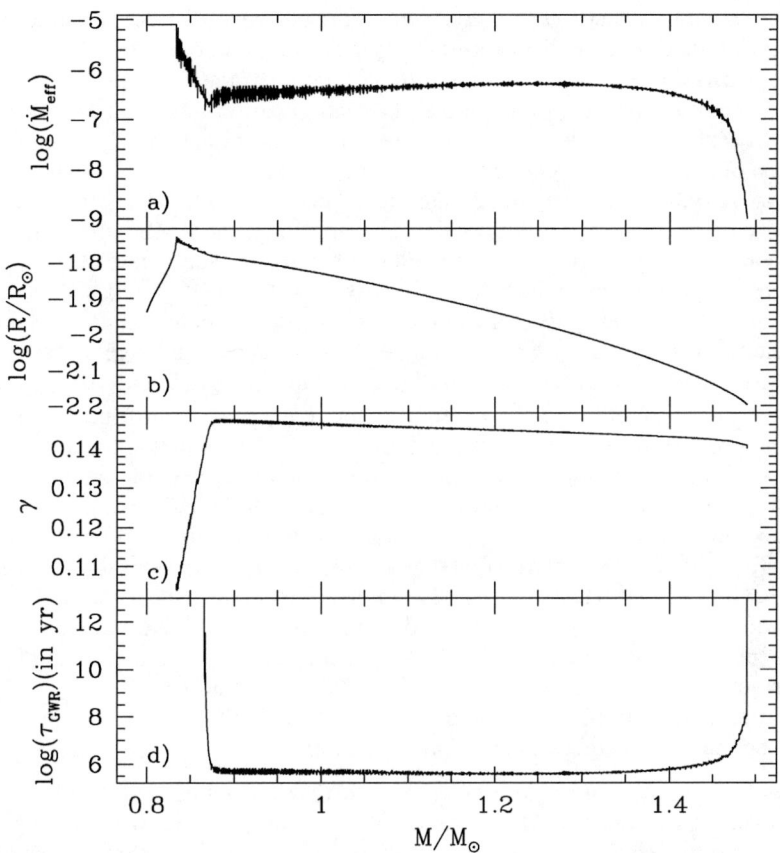

FIGURE 2. Evolution of some relevant quantities for a rotating WD the accretion rate for a rotating WD accreting CO-rich matter. We report the evolution of the accretion rate (panel a)), of the surface radius (panel b)), of the γ-parameter (panel c)) and of the time scale for the emission of GWR (see text).

guished

1. due to the fact that $\dot{M}$ is very high, a huge amount of gravitational energy is delivered on the WD surface and, since the inward thermal diffusion occurs on a time scale longer than the accretion process, a thermal energy excess forms in the external layers. As a consequence the WD expands and the surface gravity decreases up to the moment when the Roche instability occurs, *i.e.* the angular velocity of the accreting star exceeds the critical value ω_{cr} (see Fig. 2);

2. due to the occurrence of the gravitational instability, the accretion process becomes self-regulated and the value of $\dot{M}$ decreases, the exact value being determined

by the condition that $\omega_{WD} \leq \omega_{cr}$. Such an occurrence can be explained just by observing that after the Roche instability the accretion comes to a halt for while. As a consequence, inward thermal diffusion removes part of the energy excess in the external layers thus producing a contraction. In this way the critical angular velocity increases whilst the actual ω remains almost unchanged (the moment of inertia of the accreting WD remains constant). Hence ω_{WD} results smaller than ω_{cr} and accretion resumes so that, due to the deposition of angular momentum by the accreted matter, the WD continues to spin up. When the Roche instability occurs once again the cycle repeats. As a matter of fact, the effective rate of the accretion process is determined by the balance of these two processes. As it is depicted in Fig. 2 (panel a), $\dot{M}$ steadily decreases during this phase since, as the time elapses, the efficiency at which the thermal energy excess in the external layers is transferred inward decreases due the reduction of the local thermal gradient.

3. due to the contraction of the external layers (see panel b in Fig. 2) and to the continuous deposition of angular momentum the WD becomes a fast rotator. When the rotational energy becomes a relevant fraction of the gravitational energy ($\gamma_{cr} \sim 0.14$ - [23]) the accreting star becomes secularly unstable and it adopts an elliptical shape, acquiring a quadrupole momentum. In this condition gravitational wave radiation (GWR) is emitted so that the rotational energy and hence the total angular momentum of the WD decreases; this implies that GWR acts as a real brake and it becomes the leading mechanism determining the effective accretion rate. In particular, as it is shown in Fig. 2, the braking efficiency remains more or less constant for a long time (panel d in Fig. 2) so that $\dot{M}$ adopts a plateau value. As a matter of fact, in this way the total mass of the WD increases steadily up and beyond the non-rotating Chandrasekhar mass limit.

The results previously summarized clearly show that the inclusion of the lifting effects of rotation in the evolution of DD systems solve the longstanding theoretical problem of the effective increase of the accreting WD up M_{Ch}. In fact rotation acts as the real fine-tuning mechanism of the effective accretion rate onto the WD, thus preventing the off-center ignition of C-burning and stabilizing $\dot{M}$.

According to [22], the physical conditions suitable for an explosion of SN Ia proportion are attained for WD in the mass range 1.4 -1.5 $M_\odot$ (see also the contribution of Dominguez et al., this volume), the upper value corresponding to the Chandrasekhar limit for rigidly rotating stars. Note that this range of masses results even larger when differential rotation is accounted for (see the contribution by Yoon & Langer, this volume), the exact value being determined only by the total mass of the initial DD system.

Such a spread in the final value of M_{WD} could be responsible of the diversity in the observational properties of type Ia SNe. If this simple scenario is really at work some predictions can be done. First of all GWR emitted during the accretion process has an high frequency ($f_{GWR} \sim 0.1 - 0.5 Hz$ - see Fig. 3) so that they can be detected with the new generation of gravitational detectors (LISA). Moreover, by means of population synthesis calculations for the evolution of binary systems it is possible to derive the evidence that the majority of Supernovae Ia (say, 3 over 4) in nearby and distant spiral galaxies arise from short living systems and the progenitor DD systems are as massive as 1.6 - 1.8 $M_\odot$, whilst those in nearby ellipticals are all in the low mass range (1.4 $M_\odot$).

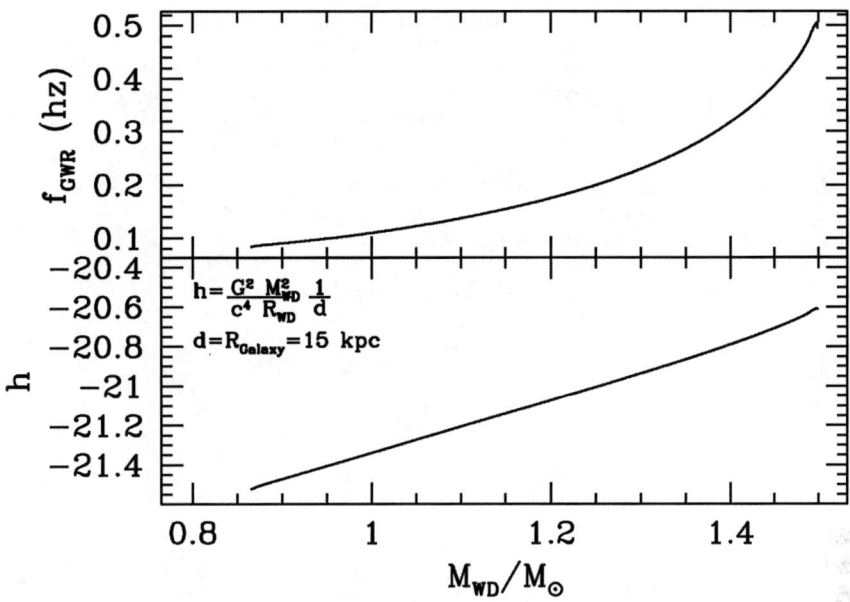

FIGURE 3. Evolution of the frequency (upper panel) and of the amplitude (lower panel) of GWR emitted during the self-regulated accretion phase (see text).

Concerning the high red-shift ellipticals, the mass of observed Type Ia SNe increases up to 1.6 - 1.8 $M_\odot$ for those inside the first billion years after the galaxy formation. All in all the observed type Ia supernovae could indeed resemble a single parameter family, even if we are inclined to believe that the variety could better be mimicked by an additional parameter which is the degree of rotation at the instant of central carbon ignition which, in turn, determines the density profile of the exploding objects.

REFERENCES

1. B. Leibundgut, *A&A Rv*, **10**, 179 (2000).
2. A. Tornambé, *MNRAS*, **239**, 771 (1989).
3. R. Kipphenhan, and A. Weigert, *Zeit. Astr.*, **65**, 221 (1967).
4. A. Tornambé, "Supernovae in Binary Systems," in *Menorca School of Astrophysics*, edited by E. Bravo, R. Canal, J. Ibanez, and J.Isern, 1992, p. 59.
5. L. Piersanti, S. Cassisi, I. Iben, and A. Tornambé, *ApJ*, **535**, 932 (2000).
6. L. Piersanti, S. Cassisi, I. Iben, and A. Tornambé, *ApJ Lett.*, **521**, L59 (1999).
7. S. Woosley, and T. Weaver, *ApJ*, **423**, 371 (1994).
8. D. Prialnik, and A. Kovetz, *ApJ*, **445**, 789 (1995).
9. S. Cassisi, I. Iben, and A. Tornambé, *ApJ*, **496**, 376 (1998).
10. I. Hachisu, M. Kato, and K. Nomoto, *ApJ Lett.*, **470**, L97 (1996).
11. M. Livio, "The Progenitors of Type Ia Supernovae," in *Type Ia Supernovae, Theory and Cosmology*, edited by J. Niemeyer, and J. Truran, Cambridge University Press, 2000, vol. 20, pp. 33–35.
12. I. Iben, and A. Tutukov, *ApJ Supp.*, **54**, 335 (1984).
13. E. Robinson, and A. Shafter, *ApJ*, **332**, 296 (1987).

14. A. Bragaglia, L. Greggio, A. Renzini, and S. D'Odorico, *ApJ Lett.*, **365**, L13 (1990).
15. R. Saffer, M. Livio, and L. Yungelson, *ApJ*, **502**, 394 (1998).
16. R. Napiwotzki, and e. al., "Close binary white dwarfs and supernovae Ia," in *Compact Binaries in the Galaxy and Beyond. IAU Colloquium 194*, edited by G. Tovmassian, and E. Sion, Revista Mexicana de Astronomía y Astrofísica (Serie de Conferencias), 2004, vol. 20, pp. 113–116.
17. W. Benz, A. Cameron, W. Press, and R. Bowers, *ApJ*, **348**, 647 (1990).
18. F. Rasio, and S. Shapiro, *ApJ*, **438**, 887 (1995).
19. J. José, M. Hernanz, and J. Isern, *A&A*, **269**, 291 (1993).
20. L. Piersanti, S. Gagliardi, I. Iben, and A. Tornambé, *ApJ*, **583**, 885 (2003).
21. I. Iben, A. Tutukov, and A. Fedorova, *ApJ*, **503**, 344 (1998).
22. L. Piersanti, S. Gagliardi, I. Iben, and A. Tornambé, *ApJ*, **598**, 1229 (2003).
23. J. Friedman, and B. Schutz, *ApJ Lett.*, **199**, L157 (1975).

Supernovae Type Ia: an Observational Perspective

John Danziger

Osservatorio Astronomico di Trieste - OAT/INAF - Via G. B. Tiepolo 11, Trieste 34131, Italy

Abstract. A review is presented of various observational features of the family of Type Ia (thermonuclear) supernovae. There are properties common to all SN Ia and there are differences within the Type Ia classification. One seeks to understand all of these properties within a physical framework providing a way of discriminating among the various types of models that have emerged over the past couple of decades.

Keywords: Supernova; spectra; light curves; rates; galaxies
PACS: 98.35.Bd; 97.10.Tk

INTRODUCTION

Here we briefly list some general facts that are commonly agreed to by virtually all workers in the field.

- SN Ia occur in galaxies of all Hubble types.
- Their presence in E, S0 galaxies suggests that their immediate progenitors are low mass stars.
- The progenitors must be highly evolved because of lack of hydrogen (and helium) lines in their spectra. (A recent variation of this situation will be discussed).
- The consensus (?) is that the explosion results from mass accretion onto a C-O degenerate white dwarf (WD). There is no agreement on whether the accretion process involves only one degenerate WD with a red giant companion, or coalescence of two WDs.
- Differences among SN Ia are observed to exist, but are generally smaller than among core-collapse SNe.
- Radioactive decay of ^{56}Ni - ^{56}Co - ^{56}Fe powers the light curve.
- Radio emission at neither early nor late times seems to occur.

LIGHT CURVES

Radioactive decay of ^{56}Co powering the light curve from 50 to several hundred days was first established by Colgate and McKee [1] following earlier suggestions that a source extended in time such as radioactive decay was necessary. This has led more recently to relatively simple models of γ-ray and positron deposition leading to light curves to match observations [2]. In fact since observations of light curves of a growing sample show a range in luminosity of an order of magnitude, this has led to estimates of a mass range of a similar order of magnitude of Fe produced in SN Ia. That range is 0.1 - 1.0 $M_\odot$ with a mean about 0.7 $M_\odot$. That range in luminosity is relevant to the discussion of SN Ia as standard candles.

Phillips [3] placed on a quantitative basis the systematics of the post-maximum decay rate and its relation to the luminosity at maximum, showing that intrinsically fainter objects decayed faster, the now well known Mv(max) vs. Δm_{15} correlation. This relation has been considerably improved in the past 10 years with larger better observed samples and observations in different colours [4]. Important for cosmological studies is that all SN Ia of any intrinsic luminosity can have their intrinsic luminosity known by measuring Δm_{15} and using this relation. The physical reasons for these differences remain to be demonstrated.

Other luminosity effects and their relationship to environment will be discussed later.

SPECTRAL CHARACTERISTICS

The differences that are seen in intrinsic luminosities of SN Ia have their counterpart in the spectra. Spectra both in the photospheric absorption line phase and in the nebular emission line phase show systematic differences that correlate with intrinsic luminosity. With some few exceptions however the same lines of the same ions are represented in all SN Ia. In fact these atomic species are qualitatively consistent with what theory of thermonuclear production would predict.

From an observational point of view quantitative testing is not straight-forward because spectrum synthesis in the early phases only describes what is visible above the photosphere, while in the nebular phase although one sees through the optically thin bubble, there are only a restricted number of emission lines that are visible and amenable to analysis. Nevertheless it is the analysis in the nebular phase that gives values of the mass of Fe consistent with those mentioned above from the modelling of the light curves with a range of an order of magnitude [5]. An analysis of emission line profiles also allows one to conclude something about distribution of matter in what could be a stratified or condensed envelope. For example, the analysis of the [CoIII]5908Å profile observed at day 143 in SN1991bg allowed the conclusion that there was no 'hole' in the ^{56}Ni distribution.

Other spectral characteristics pointing to possible real abundance differences are apparent at either limit of the range of intrinsic luminosity. For example SN1991T the intrinsically brightest known SN Ia had the SiII 6355Å line much weaker in absorption 15 days after outburst than that shown by the normal SN1994D at similar phase. This was ascribed [6] to a higher temperature in the former SN with much more SiIII present rather than an abundance effect. If this is correct it simply moves the question to why the temperature was higher.

Another example comes from the spectra at maximum light of SN1991bg the intrinsicaly faintest SN known. In contrast to spectra of intrinsically brighter SNe such as SN1991T and SN1994D there are very prominent absorption lines of the OI 7773Å triplet [5]. It is of interest to understand if in SN1991bg oxygen was less consumed in a weaker explosion or whether this again results from temperature differences near the photosphere. More detailed modelling would be necessary to resolve this question.

Large differences in photospheric velocities are evident by simple visual inspection. It is important that comparisons within a sample be made at similar phases because photospheric velocities evolve rapidly particularly in the the first 25 days after outburst. Quantitative measures of the absorption minimum of the SiII 6355Å line when plotted against time after explosion show a wide range of velocities for a sample at the same phase, with a positive correlation between velocity and intrinsic luminosity. At maximum light that range is at least 10000-16000 km/s. It is worth noting that even for relatively normal SN Ia such as SN1994D, SN1992A, SN1981B and others there are significant differences in velocity. Equally intriguing is the fact that these differences generally become much less and are sometimes negligible reaching values of the order of 9000km/s after 60 days.

The results from the emission line spectra in the nebular phase at 200-275 days are qualitatively similar in the sense of an inverse correlation of the FWHM with Δm_{15} or intrinsic luminosity. Quantitatively the velocities of the emission lines are rather similar to those measured in absorption 60 days after maximum. Although a detailed comparison has not been made a quick inspection reveals that the differences between individual SNe measured at the photospheric phase are preserved into the nebular phase - somewhat reassuring about the meaning of such measurements. Just as for the the photospheric velocities, a range of velocities 12000-15000 km/s is apparent for a given value of Δm_{15} [7].

ENVIRONMENTAL CHARACTERISTICS

Since SN Ia occur in all types of galaxies it has now become possible with reasonable samples to examine what bearing, if any, the parent galaxy might have on the characteristics of the SN. A plot of the photospheric velocity curves extending to 30 days past maximum shows that SNe in E (elliptical) galaxies tend to have the lowest velocities. This same trend can be shown by plotting the SiII 6355A velocity 10 days after maximum versus galaxy type. SNe in E galaxies have the lowest velocities (8500 km/s) with an increase progressing to later galaxy types as well as a greater scatter. This already suggests that SN Ia on average are drawn from a different population of progenitors in different galaxies. The scatter is also consistent with later galaxies having a wider range of population types as is of course recognised through stellar population studies.

Early work also indicated that intrinsically brighter SN Ia tended to occur in later type galaxies, a result that became apparent when careful corrections were made for reddening of greater importance in spirals [9]. The use of Δm_{15} (decay rate) for a larger sample tends to confirm this as might be expected [10]. There is also a tendency for faster decay to correlate with galaxy luminosity, a result not unrelated to the fact that in the samples the E galaxies tend to be the intrinsically brightest.

The attempts to establish age and metallicity effects have not been so convincing partly because both these parameters are not easy to define in a uniform way for different mixtures of stellar populations. Within the family of spiral galaxies it has been possible to examine the spatial distribution [11]. A comparative study of SN types has shown that SN Ia are more centrally concentrated than core-collapse SNe. This probably reflects the concentration of an older population in the bulges of spirals. A reported tendency of SN Ia to concentrate in spiral arms has been disputed.

The most recent results for SN rates shows that in the local universe SN Ia rates are always less than the SN II+Ib (core-collapse) rates except in E and S0 galaxies [12].

Polarization studies can give information about the degree of asymmetry of an expanding envelope of a SN, but they can also give evidence of circumstellar material close to the SN or interstellar matter along the line-of-sight. Whereas linear polariztion has been reported for an increasing number of core-collapse objects the results for SN Ia are limited to 2 cases at present. The sub-luminous SN1999by showed polarization varying with wavelength, namely ∼0 at wavelengths <5000Å up to 0.8% at 7000Å. Variation across the SiII 6355Å line shows that this polariztion arises in the envelope [13]. In the case of SN2001el 0.2-0.3% polarization before maximum changed to 0% one week after maximum, also showing that it probably arose in the envelope [14]. This could result from a slightly oblate electron scattering envelope viewed almost equator on. There was also reported a kinematically distinct CaII IR triplet feature showing 0.7% polarization which might be a clue to a possible binary nature and/or accretion disk.

SUMMARY OF DIFFERENCES

- Faintest are ∼ 2.5 magnitudes less luminous than brightest.
- Fainter are redder at maximum reflecting a lower temperature.
- Fainter have lower expansion velocities both for the photosphere and the inner Fe core.
- Fainter fade more quickly after maximum.
- Fainter are statistically more prevalent in early-type galaxies.
- Fainter produce less ^{56}Co where the total range is ∼ 10.
- Faintest do not have a secondary IR maximum.

PROGENITORS

Recently SN2002ic was observed with a 2-component Hα emission line normally ascribed to a core-collapse SN [15]. The light curve resembled that typical of an ejecta-wind interaction. Nevertheless the subtraction of a smooth continuum and Hα emission produced a spectrum similar to that of a normal SN Ia. This has been suggested as evidence for the progenitors of SN Ia coming from a WD accreting matter from a red giant companion with an associated wind of ∼ 3 $M_\odot$. This large mass was inferred because a spectral similarity to that of SN1997cy where modelling of the CSM had already been made, suggested this SN may have had a similar history [16]. More detailed modelling of the light curve of these two SNe [17] and spectra of SN2002ic [18] suggest two plausible models. The model of a WD accreting from a red giant companion is less favoured because of the similarity of all three events which include SN1999E [19] and the lack of a physical reason for rapid large mass loss immediately prior to the explosion. A Type 1.5 SN resulting from the explosion of the C-O core of a single star is preferred. Previous theoretical objections based on supposed slow growth rates of the core because of loss of the hydrogen envelope may not apply if the mass loss rate is lower owing to lower metallicity. In reality all three SNe occurred in intrinsically faint dwarf galaxies where the metallicity is probably low.

Whatever the most plausible model for progenitor evolution of these three SNe all of them seem to imply associated masses that would be difficult to reconcile with their occurrence in E galaxies. In fact so far no such SNe have been reported in E galaxies.

Added to this upsurge in possible identification of a progenitor is the recent report of the discovery of a DD (double degenerate) WD system that has orbital characteristics that could mean eventual coalescence [20].

CONCLUSIONS

In this short account it is difficult to summarize what has been accomplished to relate observations to the theoretical types of explosion that might occur. Unique conclusions are still lacking. In fact as mentioned above we still lack clear unequivocal evidence on the nature of the progenitor(s) and the types of explosion that occur, be they detonation, delayed detonation or deflagration, all of which have accomodated some but not all the observational constraints. There exists even the possibility of sub-Chandrasekhar mass explosions for which the faint SN Ia may be testimony. The possible reasons for the dispersion in luminosity can only be sorted out by understanding the role of metallicity, mass and rotation to name just three relevant parameters. It hardly needs saying that all of these parameters vary amongst normal stars, and unfortunately observations of SNe so far do not offer an easy way to discriminate such effects. Further more detailed observations of SN Ia may add insight into the question of metallicity. With luck one might eventually identify progenitors and glean something concerning mass. From an observational point of view the question of rotational effects is likely to remain elusive so there we will have to depend on theoretical modelling alone. Hopefully Tornambe, Piersanti and colleagues will provide this.

REFERENCES

1. S. Colgate, C. McKee, ApJ **157**, 623-643 (1969)
2. E. Cappellaro, P. Mazzali, S. Benetti, I. J. Danziger, et al., A&A **328**, 203-210 (1997)
3. M. M. Phillips, ApJ **413**, L105-L108 (1993)
4. M. M. Phillips, P. Lira, N. B. Suntzeff, R. A. Schommer, et al., AJ **118**, 1766-1776 (1999)
5. P. Mazzali, N. N. Chugai, M. Turatto, L. Lucy, et al., MNRAS **284**, 151-171 (1997)
6. P. Mazzali, I. J. Danziger, M. Turatto, A&A **297**, 509-534 (1995)
7. P. Mazzali, E. Cappellaro, I. J. Danziger, M. Turatto, et al., ApJ **499**, L49-L52 (1998)
8. D. Branch, S. van den Bergh, AJ **105**, 2231-2235 (1993)
9. S. van den Bergh, J. Pazder, ApJ **390**, 34-38 (1992)
10. M. Hamuy, S. C. Trager, P. A. Pinto, M. M. Phillips, et al., AJ **120**, 1479-1486 (2000)
11. S. van den Bergh, AJ **113**, 197-200 (1997)
12. E. Cappellaro, R. Evans, M. Turatto, A&A **351**, 459-466 (1999)
13. E. Howell, P. Hoflich, L. Wang, J. C. Wheeler, ApJ **556**, 302-321 (2001)
14. D. Kasen, P. Nugent, L. Wang, D. A.Howell, et al. ApJ **593**, 788-808 (2003)
15. M. Hamuy, M. M. Phillips, N. B. Suntzeff, J. Maza, et al., Nature **424**, 651-654 (2003)

16. M. Turatto, T. Suzuki, P. Mazzali, S. Benetti, et al., ApJ **534**, L57-L61 (2000)
17. N. N. Chugai, L. R. Yungelson, Ast.Lett. **30**, 65-72 (2004)
18. N. N. Chugai, R. A. Chevalier, P. Lundqvist, MNRAS **355**, 627-637 (2004)
19. L. Rigon, M. Turatto, S. Benetti, A. Pastorello, et al., MNRAS **340**, 191-196 (2003)
20. R. Napiwotzki, et al., IAU Colloquium 194, Eds. G. Tovmassian and E. Sion, Revista Mexicana de Astronomía y Astrofísica (Serie de Conferencias), **20**, 113-116 (2004)

Rotating Type Ia SN progenitors: explosion and light curves

I. Domínguez*, L. Piersanti†, E. Bravo**, S. Gagliardi†, O. Straniero† and A. Tornambé†

*Dpto. Física Teórica y del Cosmos, Universidad de Granada, 18071 Granada, Spain
†INAF-Osservatorio Astronomico di Collurania, 64100 Teramo, Italia
**Dep. Física i Enginyeria Nuclear, Universitat Politecnica de Catalunya, 08028 Barcelona, Spain

Abstract. High redshift SNe Ia have been recently used to calibrate the cosmological distance scale and to infer the existence of the dark energy. The reliability of such a method depends on the effective knowledge of the absolute brightness of this class of supernovae. This would require a complete understanding of the physics of SNeIa.

Starting from an accreting rotating white dwarf, the only progenitor that we found to be able to grow till the Chandrasekhar mass and undergo a thermonuclear explosion, we simulate the explosion, deriving the nucleosynthesis and the light curve. We explore the final outcome in the framework of a 1D delayed detonation model, where the characteristic density for which the transition from deflagration to detonation takes place is a free parameter.

Although preliminary, our results imply that rotating white dwarfs produce a range of explosive conditions, characterized by different ignition densities and total masses. Maximum luminosities of successfully explosive models differ up to 0.11 mag. In a few cases, the formation of a small highly neutronised remnant is found.

Keywords: Cosmological parameters – Stellar evolution – White Dwarfs – Binary systems – Supernovae
PACS: 97.60.Bw,26.30.+k,26.50.+k,97.80.6m

INTRODUCTION

Observations of high redshift SNe Ia (Schmidt, 1998; Perlmutter, 1999) have changed our view of the Universe. At present, these observations provides precise distances for galaxies up to $z \sim 1$ and allow us to extend the Hubble diagram (distance vs. redshift) toward the past. This study indicates that the universal expansion rate is increasing with time and, in turn, that an unknown positive energy contribution, capable to overcome the gravitational deceleration, is at work. This result coupled to the complementary detection of temperature fluctuations in the CMB (De Bernardis et al. 1999, Spergel et al. 2003), which basically indicates that the overal geometry should be flat, imply a revision of the standard cosmological model. In the framework of the ΛCDM model, a 30% only of the total energy would be provided by matter (mostly - 96% - Cold Dark Matter) and the remaining 70% would be dark (that means unknown) energy, which is formally expressed by the Eistein cosmological constant (Λ). Alternative models, where $\Lambda = 0$, necessarily require that the SNe Ia distance scale is wrong (Blanchard et al. 2003).

The absolute brightness at maximum of nearby SNe Ia may vary up of 2 and 3 mag in V and B, respectively. A clear correlation of this variations with the shape of the light curve (LC) was found (Phillips et al., 1987, 1999; Riess et al., 1997). Brighter SNe present a decline after maximum slower compared to dimmer SNe. This correlation is used to derive the absolute luminosity of cosmological SNe Ia and, in turn, the distance of their parent galaxy. Then, the use of SNe Ia as standard candles implicitly requires that the nearby and the distant supernovae follow the same maximum-decline rate relation.

Some observational evidences indicating a dependence of the SN Ia LC with the host galaxy type has been reported. The slower LCs (brighter SNe) are rare in ellipticals (Branch et al. 1996; Hamuy et al. 2000 and references therein; Ivanov et al. 2000). In addition, the number of Type Ia SNe, per unit mass, is 3 times larger in spirals as compared to ellipticals (Cappellaro et al. 1997). It seems, therefore, that the frequency and the LCs of SNe Ia may depend on the parent stellar population.

Type Ia SNe are considered to be the thermonuclear explosion of a carbon-oxygen (CO) white dwarf (WD) with a mass close to the Chandrasekhar limit (Hoyle & Fowler, 1960). The evolution of single stars with mass lower than about 7-8 $M_\odot$ is expected to produce CO WDs with a maximum mass of about 1 $M_\odot$. In more massive stars, C is ignited before the onset of the AGB phase (Becker & Iben, 1980, Dominguez et al. 1999). In which way a typical WD having mass in the range $0.6 - 1 M_\odot$ could reach the Chandrasekhar limit by accretion is still largely unexplained. Mass exchange in close binary systems containing at least one WD could provide the accretion mechanism capable to produce CO WDs massive enough. The hypothesis that SNe Ia progenitors are low or intermediate mass stars, rather than massive stars, may explain the occurrence of these supernovae in old stellar populations (like those of elliptical galaxies), where star formation halted several Gyrs ago.

Two unsatisfactory scenarios have been proposed. The first is generally addressed as the Single Degenerate scenario (SD), namely one WD accreted by a red giant companion filling its Roche lobe. In this case, a slow accretion of H onto the WD would give rise to a nova. Extant theoretical models (see the contribution of Prialnik to this meeting) show that during the nova explosion the previously accreted material is expelled in the interstellar medium, a fact that prevents the formation of a WD with mass approaching the Chandrasekhar limit. On the other hand, high accretion rates cause the expansion of the envelope, the primary component leaves the WD cooling sequence and comes back to the red giant branch. In this case, a common envelope configuration is attained with high mass loss from the external lagrangian points (Iben & Tutukov 1984). The second scenario, the Double Degenerate scenario(DD), refers to the coalescence or merging of two WDs. The total mass of the binary system should be equal to or larger than the Chandrasekhar mass. In this case, the less massive component forms a thick accretion disk around the more massive one. The consequent C and O accretion is expected to occur at a very high rate, $10^{-5} M_\odot yr^{-1}$ (Saio & Nomoto, 1985), until C ignites in an external layers and the burning is not explosive. Note that this prediction is based on SNe models that do not include the effects of rotation even if the synchronization of the orbital period with the spin may induce high angular velocities of the binary components. For this reason, several groups have recently considered, in their numerical simulations, rotating CO white dwarfs (Piersanti et al. 2003a, 2003b; Saio & Nomoto, 2004; Uenishi, Nomoto & Hachisu, 2003; Yoon & Langer, 2004). So far, Piersanti et al. (2003b; see also this volume) found that the angular momentum lost via gravitational wave radiation (GWR) counterbalances the angular momentum gained from the accreted matter and the accretion proceeds at a stable, moderate rate. With such a rate, the primary WD can accrete enough mass. On the other hand, rotation has a lifting effect over the structures allowing WD masses to grow above the critical Chandrasekhar mass. Later on, as a consequence of the disk frictional braking, the spin velocity slows down and C ignition takes place.

Starting from the progenitor models computed by Piersanti et al. (2003b), we have followed their thermonuclear explosions. Then we have calculated nucleosynthesis and light curves. In the following sections, we describe the results of these calculations.

MODELS

The progenitor evolutions, including the accretion phase, have been simulated by means of a 1D hydrostatic code, the FRANEC code (Chieffi et al. 1998; Straniero et al. 1997), properly modified to take into account accretion and rotation (see Piersanti et al. 2003b). Explosions and later phases have been computed by means of a 1D hydrodynamic codes (Bravo et al. 1993, 1996).

Initial Models

All the models are super-Chandrasekhar WDs that start to compress when the rotational velocity decreases. We follow the evolution of WDs with final total masses equal to 1.40 $M_\odot$, 1.43 $M_\odot$, 1.46 $M_\odot$ and 1.49 $M_\odot$. To slow down the WD a braking efficiency is assumed, physically motivated by the frictional viscosity due to the interaction of the WD with the accretion disk. We have assumed for the time scale of viscous dissipation $\tau_{vis} = 10^4 yr$ (see Piersanti et al. 2003b).

The models have been followed with the FRANEC hydrostatic code up to the moment when the ignition conditions are reached. All models ignite carbon at the center, except the most massive one that ignites at $10^{-4} M_\odot$ far from the center.

Our main results are summarized in Table 1, in which we show: (1) total mass, (2) viscous dissipation time, (3) ignition density, (4) ignition temperature, (5) total binding energy (1 foe=$10^{51} erg$), (6) total kinetic energy, (7) mass of ^{56}Ni produced in the explosion and (8) bolometric magnitude at the maximum time.

The most relevant properties for the following evolution are: ignition density, total mass and binding energy. Note that gravitational energy and thermal energy compensate each other; for a higher density (column 3) a higher temperature (column 4) is required to ignite.

Ignition densities (3rd column, Table 1) are similar for the models with $M_{tot} \leq 1.46$ $M_\odot$, being greater (3.3×10^9 g/cm^3, for the more massive one. The biding energy increases with mass, ranging from 0.50 foe to 0.67 foe. A greater binding energy implies that a greater part of the delivered nuclear energy should be employed to unbind the WD.

TABLE 1. Main properties of the models

M_{tot} ($M_\odot$)	τ_{vis} (yr)	ρ_{ig} (g/cm^3)	T_{ig} (K)	U_{bin} (foe)*	E_k (foe)	^{56}Ni ($M_\odot$)	M_{bol}
1.40	10^4	2.2×10^9	4.96×10^8	0.50	0.985	0.759	-19.36
1.43	10^4	2.1×10^9	4.88×10^8	0.55	0.995	0.752	-19.34
1.46	10^4	2.1×10^9	4.92×10^8	0.60	0.996	0.784	-19.38
1.49	10^4	3.3×10^9	5.22×10^8	0.67	0.925	0.819	-19.42
1.49	10^5	3.0×10^9	4.91×10^8	0.67	0.954	0.831	-19.45

* 1 foe=10^{51} erg

Explosions

In type Ia Sne, once C is ignited the burning front does not propagate by the pressure waves (detonation). If so, all the WD would be transformed into Fe peak elements and no intermediate mass elements would be produced. This is in contrast with the observations; in fact, a month after the explosion, the spectra is still dominated by intermediate mass elements. On the other hand, if subsequent zones are heated by conduction, the velocity of the flame would be too low and the shock wave would expand the WD till low densities, quenching the flame. However, turbulent mixing of the unburnt cold matter with the hot ashes likely occurs and instabilities, like the Rayleigh-Taylor instability, develop. In this case the velocity will be determined by the growth rate of the instabilities, being subsonic but faster than the conductive velocity.

In 1D models we rely on parameterization of this 3D turbulent mixing. The simple way is to assume a flame velocity equal to a few percent of the sound velocity. In this computations we have assumed a turbulent deflagration velocity that is a 3% of the sound velocity. More realistic prescriptions for the deflagration velocity could be adopted, but as shown by Domínguez and Höflich (2001) the final outcome is not very sensitive, within reasonable limits, to the flame velocity at this phase.

However observations require that all the WD is burnt and so the flame velocity has to be increased. The delayed detonation model proposed by Khokhlov (1991, 1995) fulfills this requirement. In this model when the burning front arrives to a certain density, called transition density ρ_{tr}, the deflagration velocity is accelerated till the sound velocity and a detonation occurs. For higher transition densities this point is reached earlier and the subsequent burning occurs at higher densities because the WD has less time to expand, and consequently more ^{56}Ni is produced.

Delayed detonation explosions of WDs with masses close to the Chadrasekhar mass, well explain all the observed properties, like the evolution of the light curves and spectra in the optical and IR and also the observed correlations, including the crucial maximum decline relation (Höflich & Khokhlov 1996, Höflich et al. 2002, 2003 and references therein). We decide to adopt the delayed detonation mechanism for all the simulations and fix the transition density, $\rho_{tr}=10^7$ g/cm^3 (being the density ahead the flame). Our hydrodynamic code has been modified to include the centrifugal force in the equation of motion; as we assume spherical symmetry, an average of this force is employed (see Domínguez et al. 1996 and Piersanti et al. 2003b). Around 700 isotopes are included in the code. In general, the explosive nucleosynthesis is similar in all models; the only remarkable difference occurs in the amount of neutronised elements produced; in the most massive model, with the higher ignition density, ^{54}Cr is overproduced by a factor of thousands.

Assuming a fixed transition density homogenizes our results (see Table 1) and similar amounts of the crucial ^{56}Ni are produced in all models (7th column).

Light Curves

Following the same prescription adopted in the explosion code, we include the centrifugal force in the hydrodynamic equation of motion in the light curve code.

In the most massive model, 1.49 $M_\odot$, the inner zone (0.163 $M_\odot$) collapses several seconds after the beginning of the explosion. We do not follow the evolution of this zone, that would left a remnant composed of neutronised elements, mainly ^{54}Cr. This is an interesting result that has to be checked in future simulations. The fact that this kind of remnant is left has two important implications: ignition at higher density does not imply a problematic overproduction of neutronised elements as compared with solar system abundances, and, the ejected envelopes present a range in total masses greater than the one obtained just considering rotation. The initially most massive model ejects finally the least massive envelope (1.33 $M_\odot$).

At the time of maximum, the difference in luminosity among these models is of 0.08 mag (Table 1, 8th column) in agreement with the small difference in the amount of ^{56}Ni produced (column 7th). For these models, the velocity of the photosphere increases from 9670 km/s in the more luminous (more massive model with the smaller kinetic energy) to 9950 km/s in the dimmer one. These results may be understood in terms of the binding energy and the ignition density. More massive models have a greater central density and a greater ignition density; more energy has to be converted to unbind the WD, implying less kinetic energy and a greater completely incinerated zone. However, electron captures are also favored in the inner high density region. As a consequence the final production of ^{56}Ni depends on both these factors.

CONCLUSIONS AND FINAL REMARKS

We have derived the WD properties, ignition conditions included, directly from previous evolution. We obtain that the expected rotation of the accreting WD gives a range of properties at the explosion time. We stress that all groups working in 1D, 2D and 3D explosion models assume the physical and chemical structure of the WD, its mass and the density at ignition. The obtained differences at maximum light are within the dispersion of the maximum-decline relation. This is the empirical relation used to derive the existence of the dark energy (or cosmological constant). In this context, the rotation of the progenitor does not question the Standard Model of Cosmology with the cosmological constant. Rotation, as well as evolutionary effect (Domínguez et al. 2001), should be taken into account to reduce the actual dispersion of the maximum-decline rate relation (0.17 mag, see Perlmutter et al. 1999). Note that to identify, in a

Hubble diagram, the nature of the dark energy a precision of 0.05 to 0.1 mag is needed (Albrecht & Weller, 2001).

Analyzing our results, we realize that we have to increase the spatial resolution in our explosion models (in this calculations 300 shells have been employed). This is needed for a better simulation of the transition to a detonation and of the formation of a remnant. Moreover, we have to check carefully if the formation of the remnant could be a numerical artifact or otherwise, establish the physical reasons.

We plan to compute the explosion and evolution of all the models derived from lower braking efficiencies (higher τ_{vis} - Piersanti et al. 2003b). Up to now, we have done the numerical simulations corresponding to the 1.49 $M_\odot$ model, with a characteristic viscous dissipation of 10^5 yrs (instead of the one used in all the models presented here, 10^4 yrs). The characteristics of this model are shown in the last row of Table 1. This is the most luminous model, increasing the difference at maximum light up to 0.11 mag.

In view of our results it is expected that rotation and braking efficiency, which are connected with properties of the binary system, like total mass, play a fundamental role in determining the properties at ignition and so, the explosion and light curves.

ACKNOWLEDGMENTS

This work has been partially supported by the MCyT grant AYA2002-04094-C03-02 and AYA2002-04094-C03-03, and by the Spanish-Italian cooperation INFN-CICYT.

REFERENCES

1. A. Albrecht, and J. Weller, *A&A Suppl.*, **197**, 6106 (2000).
2. S.A. Becker, and I.Jr. Iben, *Astrophys. J.*, 237, 111 (1980).
3. A. Blanchard, M. Douspis, M. Rowan-Robinson, and S. Sarkar, *Astronomy and Astrophysics*, **412**, 35 (2003).
4. D. Branch, W. Romanishin and E. Baron, *Astrophys J.*, **465**, 73 (1996)
5. E. Bravo, I. Domínguez, J. Isern, R. Canal, P. Höflich, J. Labay, *Astronomy and Astrophysics*, **269**, 187 (1993).
6. E. Bravo, E., A. Tornambe, A., I. Domínguez, J. Isern, *Astronomy and Astrophysics*, **306**, 811 (1996).
7. E. Cappellaro et al., *Astron. Astrophys.*, **322**, 431 (1997)
8. A. Chieffi, M. Limongi, O. Straniero *Astrophys. J.*, **502**, 737 (1998)
9. I. Domínguez, O. Straniero, A. Tornambe, J. Isern, *Astrophys. J.*, **472**, 783 (1996)
10. I. Domínguez, P. Höflich, *Astrophys. J.*, **528**, 854 (2000)
11. I. Domínguez, P. Höflich, O. Straniero, *Astrophys. J.* **557**, 279 (2001)
12. M. Hamuy et al., *Astron. J.* **120**, 1479 (2000).
13. I. Jr. Iben, A. V. Tutukov, *Astrophys. J. SS*, **54**, 335 (1984).
14. P. Höflich, A. Khokhlov, *Astrophys. J.*, **457**, 500 (1996)
15. P. Höflich, C. Gerardy, E. Linder et al. *Lecture Notes in Physics (review)*, **635**, 203 (2003)
16. P. Höflich, C. Gerardy, R.A. Fesen, S. Sakai, *Astrophys. J.*, **568**, 791 (2002).
17. P. Hoyle, W.A. Fowler, *Astrophys. J.*, **132**, 565 (1960).
18. V.D. Ivanov, M. Hamuy, P.A. Pinto, *Astrophys. J.*, **542**, 588 (2000).
19. A. Khokhlov, *Astrophys. J.*, **245**, 114 (1991).
20. A. Khokhlov, *Astrophys. J.*, **449**, 695 (1995).
21. S. Perlmutter et al., *Astrophys. J.*, **517**, 565, (1999).
22. L. Piersanti, S. Gagliardi, I. Jr. Iben, A. Tornambe, *Astrophys. J.*, **583**, 885 (2003).
23. L. Piersanti, S. Gagliardi, I. Jr. Iben, A. Tornambe, *Astrophys. J.*, **598**, 1229 (2003).

24. M.M. Phillips et al., *Pubbl. Astron. Soc. Pacific* **99**, 592 (1987)
25. M.M. Phillips, P. Lira, N.B. Suntzeff, R.A. Schommer, M. Hamuy, J. Maza, *Astron. J.*, **118**, 1766 (1999).
26. H. Saio, K. Nomoto, *Astronomy and Astrophysics*, **150**, L21 (1985).
27. H. Saio, K. Nomoto, *Astrophys. J.*, **615**, 444 (2004).
28. B.P. Schmidt et al., *Astrophys. J.*, **507**, 46 (1998).
29. D.N. Spergel et al., *Astrophys. J. Suppl.*, **148**, 175 (2003).
30. O. Straniero, A. Chieffi, M. Limongi, *Astrophys. J.*, **490**, 425 (1997).
31. T. Uenishi, K. Nomoto, I. Hachisu, *Astrophys. J.*, **595**, 1094 (2003)
32. S.C. Yoon, N. Langer, *Astronomy and Astrophysics*, **419**, 623 (2004).

SESSION 8

SECULAR EVOLUTION OF HIGH-MASS X-RAY BINARIES

XMM–Newton EPIC & OM Observations of Her X-1 over the 35 d Beat Period and an Anomalous Low State

S. Zane[*], G. Ramsay[†], Mario A. Jimenez-Garate[**], Jan Willem den Herder[‡], Martin Still[§], Patricia T. Boyd[§] and Charles J. Hailey[¶]

[*]*Mullard Space Science Laboratory, University College of London, Holmbury St Mary, Dorking, Surrey, RH5 6NT, UK*
[†] *Mullard Space Science Laboratory, University College of London, Holmbury St Mary, Dorking, Surrey, RH5 6NT, UK*
[**]*MIT Center for Space Research, 77 Massachusetts Avenue, Cambridge, MA 02139, USA*
[‡]*SRON, the National Institute for Space Research, Sorbonnelaan 2, 3584 CA Utrecht, The Netherlands*
[§]*NASA/Goddard Space Flight Center, Code 662, Greenbelt, MD 20771*
[¶]*Columbia Astrophysics Laboratory, Columbia University, New York, NY 10027, USA*

Abstract. We present the results of a series of *XMM–Newton* EPIC and OM observations of Her X-1, spread over a wide range of the 35 d precession period. We confirm that the spin modulation of the neutron star is weak or absent in the low state - in marked contrast to the main or short-on states. The strong fluorescence emission line at ~ 6.4 keV is detected in all observations (apart from one taken in the middle of eclipse), with higher line energy, width and normalisation during the main-on state. In addition, we report the detection of a second line near 7 keV in 10 of the 15 observations taken during the low-intensity states of the system. We discuss these observations in the context of previous observations, investigate the origin of the soft and hard X-rays and consider the emission site of the 6.4keV and 7keV emission lines.

Keywords: Accretion and accretion disks — stars : neutron — Pulsars : individual (Her X-1) — X-rays : stars
PACS: 01.30.Cc, 95.85.Nv, 97.10.Gz, 97.60.Gb

INTRODUCTION

Her X-1 is one of the best studied X-ray binaries in the sky. The binary system consists of a neutron star and an A/F secondary star. It has an orbital period of 1.7 d and the neutron star spin period is ~ 1.24 s. It is one of only a few systems which shows a regular variation in X-rays, over a "beat" period of 35 d, which is generally interpreted as the precession of an accretion disk that periodically obscures the neutron star beam. The cycle comprises: i) a 10 d duration main-on state, ii) a fainter 5 d duration short-on state, and iii) a period of lower emission in between. Exceptions to the normal 35 d cycle has been observed only in four occasions: in 1983, 1993, 1999 and in January 2004 the turn-on of the source has not been observed. During these "anomalous low states" (ALS; the period of which ranges from several months to 1.5 years) Her X-1 appears as a relatively faint X-ray source, with a strength comparable to that of the standard low state. The event registered last year was only the fourth one that has been seen since

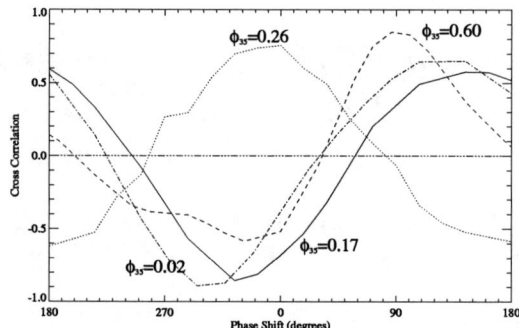

FIGURE 1. Cross correlation of the (0.3-0.7 keV) and (2-10 keV) light curves at four different Φ_{35}.

the discovery of Her X-1 in 1972, and the first one that could be observed with a high capability satellite as *XMM–Newton*.

Her X-1 has been observed by *XMM–Newton* on 15 separate occasions outside the ALS, giving good coverage over the beat period. Moreover, it has been observed 10 times during the ALS. The analysis of datasets taken before the ALS has been presented by Ref. [1], [2] and [3]. Here we report our main findings, referring the interested reader to the above papers for more details, and we present a preliminary analysis of the ALS.

TIMING ANALYSIS

We first focussed on datasets taken before the source entered the anomalous low state. We performed a search for pulsations in all datasets, confirming that, in marked contrast to the main or short-on states, the spin modulation of the neutron star is weak or absent in the low state. During the states of higher intensity, we observe a substructure in the broad soft X-ray modulation below ~ 1 keV, revealing the presence of separate peaks which reflect the structure seen at higher energies (see Ref. [3]).

The soft and hard X-ray lightcurves of Her X-1 are known to be shifted in phase (see Fig.2, central panel). Under the assumption that soft X-rays are due to the reprocessing of the pulsar beam by the inner edge of the disk, this is usually interpreted as evidence for a tilt angle in the disk[4],[5]. In our *XMM–Newton* data, we find the first evidence for a substantial and systematic change in the phase difference along the beat cycle, which is predicted by precessing disk models[6] (see Fig. 1).

THE FE Kα LINE

The strong emission line at ~ 6.4 keV is detected in all our *XMM–Newton* observations, with larger broadening and normalization during the main-on (see Fig. 2, left panel). The line centroids observed using the EPIC PN deviate by 4σ from the 6.40 keV neutral value: the Fe line emission originates in near neutral Fe (Fe XIV or colder) in the low and short-on state observations, whereas in the main-on the observed Fe Kα centroid

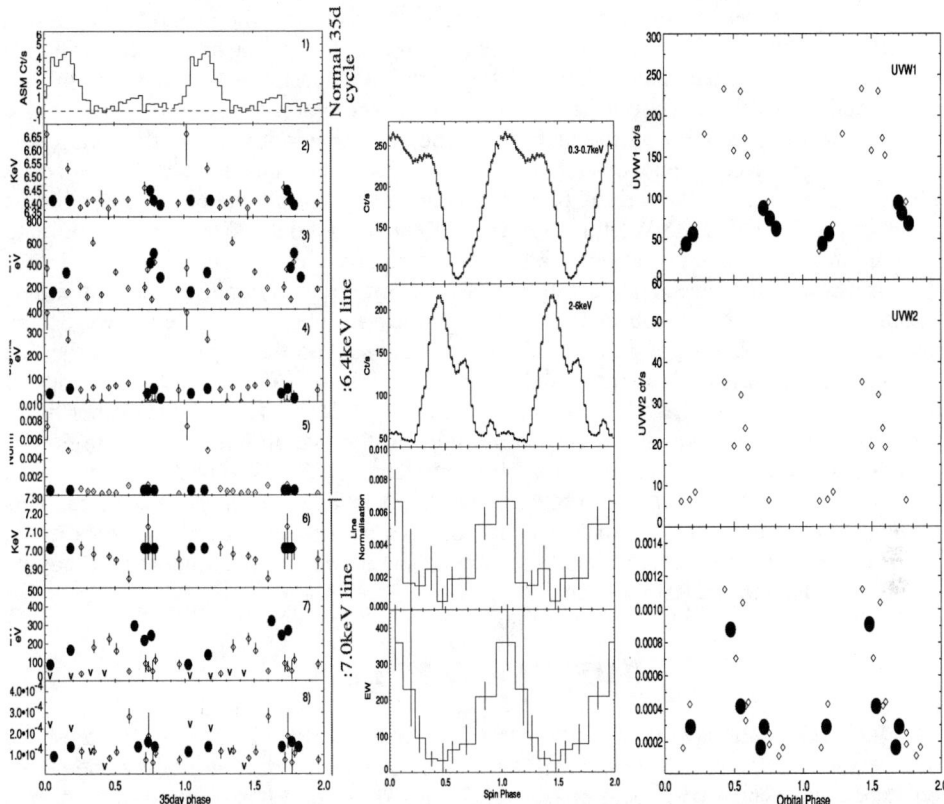

FIGURE 2. Left panel: Variation of the Kα and Fe XXVI line parameters along the beat cycle. From the top: 1) mean ASM light curve; 2)-5) central energy, equivalent width (EW), width and normalization of the prominent Kα Fe emission line; 6)-8) central energy, EW and normalization of the Fe line at ~ 7 keV. When the second feature is undetected, we show the 90% confidence level upper limits to the EW and normalization ("v" symbols). Central panel: The observation made at $\Phi_{35}=0.02$. From the top panel: the light curve in the 0.3-0.7 keV and 2-6 keV band (note the shift in phase between hard and soft X-ray emission); the Fe Kα line normalisation and EW as a function of the spin phase. Right: The UVW1 (top panel), UVW2 (central panel) data and the Fe 6.4 keV normalisation folded on the orbital period. The blue circles superimposed in the left and right panels are data taken during the 2004 ALS.

energies (6.65 ± 0.1 keV and 6.50 ± 0.02 keV at $\Phi_{35} = 0.02$ and 0.17) correspond to Fe XX-Fe XXI [7].

Possible reasons for this behaviour may be: 1) an array of Fe Kα fluorescence lines exists for a variety of charge states of Fe (anything up to Fe XXIII); 2) Comptonization from a hot corona for a narrower range of charge states centered around Fe XX; 3) Keplerian motion. The Keplerian velocity measured at $\Phi_{35} = 0.02$ and 0.17 is ~ 15500 and ~ 13000 km/s, respectively. This gives a radial distance of $\sim 2 - 3 \times 10^8$ cm, which is close to the magnetospheric radius for a magnetic field of $\sim 10^{12}$ G.

However, another possibility is that the region responsible for the Fe Kα line emission is different for lines observed at different beat phases. In fact, while data taken during the main-on clearly indicate a correlation between the fluorescent Fe Kα line and the soft X-ray emission (Fig. 2, central panel), suggesting a common origin in the illuminated hot spot at the inner edge of the disk, the same is not explicitly evident in data taken during the low state. Instead, at such phases the Fe Kα line is a factor > 5 weaker and is clearly modulated with the orbital period (see again Fig. 2, right panel). The correlation between the fast rising UVW1 flux and the Fe Kα detected outside the main-on point to a common origin, possibly in the disk and/or illuminated companion.

A fraction of the Fe Kα emission may arise from relatively cold material in a disk wind, such as commonly observed in cataclysmic variables [8]. However, we do not detect the Fe Kα line during the middle of the eclipse, and the upper limit on the line flux is ~ 10 or less of that measured outside the eclipse. Also, there is no Doppler signature of a wind in the HETG spectrum of the Fe Kα line [10]. On the other hand, the data reported here suggest a complex origin for the overall emission of the Fe Kα line. To our knowledge, a complex of lines which include all ionization states from Fe XVIII to XXIV Fe Kα has not been observed in any astrophysical source. This may still indicate an outflow of relatively cold gas or some complex dynamics in the disk/magnetosphere interface. Such phenomena should be time-dependent and may be monitored in the future using Astro-E2.

THE ~ 7 KEV FE LINE

XMM–Newton data have revealed, for the very first time for this source, the presence of a second Fe line at ~ 7 keV. The feature is only detected during the low and short-on states, and over several beat phases (see Fig. 3). Also, it has been confirmed by a *Chandra* HETGS observation of the source (the only one made during the low state) taken at $\Phi_{35} = 0.44 - 0.46$ [10].

The feature cannot be produced by fluorescence, and it is more likely to be a Fe XXVI line originating in widely extended photo-ionized plasma. This is consistent with the fact that also RGS data taken during the low and short-on states show the presence of photo-ionized gas [2]. Grating spectra exhibit several narrow recombination emission lines, the most prominent being C VI, N VI, N VII, O VII, O VIII and Ne IX. The line ratio $G = (f+i)/r$, as computed for all the helium-like ion complexes, is $G \simeq 4$, which indicates that photoionization is the dominant mechanism. Moreover, RGS spectra shows two radiative recombination continua of O VII and N VII, consistent with a low temperature of the emitting plasma (30000 K $< T <$ 60000 K) [2]. None of these features is detected during the main-on state.

The recombination X-ray line emission are not likely to originate in HZ Her, due to the absence of UV induced photoexcitation signatures in the He-like triplets (observed with HETGS) [10]. Instead, we propose that an extended, photoionized accretion disk atmosphere may be responsable for such features [2]. The evidence for the disk identification relies on the modeled structure and spectra from a photoionized disk, which is in agreement with the limit set spectroscopically on the density of the low-energy lines emitting region. This theoretical model has been developed by Ref. [9], who computed

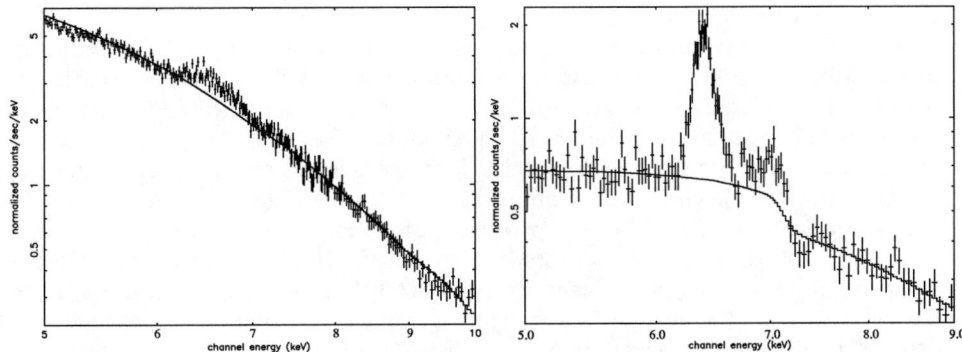

FIGURE 3. The spectral region around 6.6keV from $\Phi_{35}=0.02$ (top) and $\Phi_{35}=0.79$ (bottom). In both panel a solid line shows the best fit after the normalisation of the one or two Gaussian components have been set to zero.

the spectra of the atmospheric layers of a Shakura-Sunyaev accretion disk, illuminated by a central X-ray continuum. They found that, under these conditions, the disk develops both an extended corona which is kept hot at the Compton temperature, and a more compact, colder, X-ray recombination-emitting atmospheric layer (see Figure 9 in Ref. [2]).

Interestingly, we find that the Fe XXVI line detected by *XMM–Newton* may be a signature of the *hottest* external layers of the disk corona, which are located above the recombination-emitting layers. Again, the computed values of the density agree with the constraints inferred from the 7 keV line parameters [3].

In summary, most of the spectral lines discovered with EPIC and RGS can be associated with the illuminated atmosphere/corona of the accretion disk, which explains why they are more prominent during the low states when the direct X-ray beam from the pulsar is obscured by the accretion disk. Therefore, the variability of the Her X-1 spectrum lends support to the precession of the accretion disk, strengthen the interpretation of the low state emission in term of an extended source and open the exciting possibility to monitor spectroscopically the different atmospheric components of the disk during the transition from the low to the high state.

THE ANOMALOUS LOW STATE

Anomalous low states are rare and peculiar events, during which the source resides in a deep low state. While the mechanism that forces state changes is almost certainly variations in accretion disk structure, the engine ultimately driving structural evolution remains unknown. Each past anomalous low state, including the most recent one of January 2004[11], has been preceded by a period of enhanced spin-down, that has been interpreted in terms of an increasing torque leading to a reversal in the rotation of the inner disk. This also implies that the onset of an anomalous low state is accompanied by a large variation in the structure of the inner region of the accretion disk, that becomes increasingly warped.

In order to search for residual evidence of a 35 d cycle, we compared the line emission detected during the ALS with that observed in the several *XMM–Newton* datasets we have accumulated during the standard 35 d cycle. As we can see from Fig. 2, as far as the Fe complex is concerned, there is no an obvious difference in the spectral properties of the ALS and of the standard low states. The line features are consistent with being the same in these epochs and the orbital variation of the Kα Fe line shows the same correlation with the UVW1. This supports our scenario in which the Fe line emission of the low state originates in an extended component (disk atmosphere/corona and/or companion) instead that in the inner region of the disk. Moreover, it is consistent with the fact that at higher energies a significant Compton reflection component has been detected by RXTE in the spectrum of the ALS ([12]).

A detailed comparison of the recombination emission lines measured by RGS can shed more light on this issue. If our interpretation is correct, by measuring the line ratios during the anomalous low state allows to infer the ionization state of the plasma, column density and optical depth in the visible portion of the disk atmosphere/corona, ultimately constraining the accretion geometry.

REFERENCES

1. Ramsay, G., et al., *MNRAS*, **337**, 1185, (2002).
2. Jimenez-Garate, M.A., et al., *ApJ*, **578**, 391 (2002)
3. Zane, S., et al., *MNRAS*, **350**, 506 (2004)
4. Oosterbroek, T., et al., *A&A*, **327**,, 215 (1997)
5. Oosterbroek, T., et al., *A&A*, **353**,, 5755 (2000)
6. Gerend, B., & Boynton, P. *ApJ*, **209**,, 562 (1976)
7. Palmeri, P., et al., *A&A*, **403**, 1175 (2003)
8. Drew, J. *IAU colloquium 197, Accretion Phenomena and Related Outflows*, editeb by D.T. Wickramasinghe, L. Ferrario, & G.V. Bicknell (ASP Conf. Ser., 121; San Francisco:ASP), 1997, pp.465
9. Jimenez-Garate, M.A., et al., *ApJ*, **558**, 448 (2001)
10. Jimenez-Garate, M.A., et al., *ApJ* accepted (2005), astro-ph/0411780
11. Still, M. & Boyd, P., *ApJ*, **606**, L135 (2004)
12. Still, M. & Boyd, P., American Astronomical Society, HEAD meeting #8, #33.02, (2004)

The Properties of the Absorbing and Line Emitting Matter in IGR J16318-4848

G. Matt*, M. Guainazzi†, A. Ibarra† and E. Jimenez-Bailon†

*Dipartimento di Fisica, Universitá degli Studi "Roma Tre", Via della Vasca Navale 84, I–00146 Roma, Italy
†XMM-Newton Science Operation Center, RSSD of ESA, VILSPA, Apartado 50727, E-28080 Madrid, Spain

Abstract. We analyzed the 2003 Target of Opportunity XMM–Newton observation of IGR J16318-4848, to derive the properties of the matter responsible for the obscuration and for the emission of Fe and Ni lines. The line of sight material has a column density of about 2×10^{24} cm^{-2} but, from the Fe Kα line EW and Compton Shoulder, we argue that the average column density is a few $\times 10^{23}$ cm^{-2}, while the covering factor is about 0.1–0.2. The iron Kα line varies on time scales as short as 1000 s, implying a size of the emitting region less than 3×10^{13} cm. An ongoing XMM–Newton/INTEGRAL monitoring campaign is confirming the non–transient nature of the source.

Keywords: Accretion and accretion disks — X-ray binaries
PACS: 97.10.Gz; 97.80.Jp

INTRODUCTION

IGR J16318-4818 was discovered by the ISGRI detector of the IBIS instrument onboard the INTEGRAL satellite on January 29, 2003, with a 15–40 keV flux of 50–100 mCrab [3]. It was initially interpreted as a new "transient" X-ray source. However, a reanalysis of archival ASCA data revealed the presence of a source whose position was coincident with that of IGR J16318-4818 [14], and with a 2–10 keV observed flux of about 4×10^{-11} erg cm^{-2} s^{-1}. IGR J16318-4818 was the first discovered and most extreme object of a new class of highly absorbed X–ray binaries discovered by INTEGRAL (see [7] for a review).

The INTEGRAL discovery prompted a Target of Opportunity observation with XMM-Newton [6] on February 10, 2003. The EPIC spectra unveiled a variable and heavily absorbed source, with evidence for strong emission lines [16]. This is in agreement with the ASCA source, in which Revnivtsev et al. ([15]) suggested a column density $> 4 \times 10^{23}$ cm^{-2}. Preliminary spectral fits [4] indeed indicated that during the XMM-Newton observation IGR J16318-4818 was obscured by a Compton-thick absorber $N_H = (1.66 \pm 0.16) \times 10^{24}$ cm^{-2}. The emission complex could be resolved in three lines, with centroid energy of 6.410 ± 0.003 keV, 7.09 ± 0.02 keV and 7.47 ± 0.02 keV.

An optical counterpart has been found and studied in detail by Chaty & Filliatre ([2]). The companion is very likely a sgB[e] star; the distance is constrained to be in between about 1 and 6 kpc.

We have reanalyze the EPIC/pn spectrum with the aim to characterize the physical properties of the matter responsible for the obscuration of the X-ray source and the associated reprocessing features. This has been done by comparing the spectral fit results with Monte Carlo simulations (see [9, 10]). The detailed analysis can be found in [11]. A combined XMM–*Newton* and INTEGRAL analysis has been presented by [17].

DATA ANALYSIS AND RESULTS: TIME AVERAGED SPECTRUM

Details on the data reduction can be found in [11]. Here, suffice it to say that, for simplicity, we analyzed pn data only, extracting the spectrum in a 52" region centred on the source and using single and double events. After screening for high background periods, the net exposure time is 22.3 ks.

The XMM-Newton spectrum is characterized by a heavily absorbed continuum and 3 emission lines ([4]). The lines are most naturally interpreted as the Fe Kα and Kβ and the Ni Kα. We therefore fitted the spectrum (in the 5-13 keV energy band) with the simplest possible model: an absorbed power law plus three (unabsorbed) narrow (*i.e.* intrinsic width, σ, fixed to 1 eV) Gaussian lines. In the absorption model we left the iron abundance free to vary independently of the other elements. As the absorbing matter results to be Compton–thick (see below), we included also Thompson absorption (model CABS). It is worth noting that this model, ignoring the scattering of photons, is, strictly speaking, valid only for absorbing matter along the line of sight with a negligible covering factor (see below for a discussion).

The fit is reasonably good (χ^2=99.1/64 d.o.f.), but residuals around the iron Kα line are visible, most likely due to the Compton Shoulder (CS), as already observed in the reflection spectrum of the AGN in the Circinus Galaxy ([1, 13]), and expected on theoretical ground (see [10] and references therein). Modeling for simplicity the Fe Kα Compton Shoulder with a Gaussian with centroid energy fixed to 6.3 keV, and σ fixed to 50 eV, a significant improvement is found (χ^2=80.9/63 d.o.f., corresponding to 99.96% confidence level). The best fit results for this baseline model are summarized in Table 1. The observed 2-10 keV flux is 6.7×10^{-12} erg cm^{-1} s^{-1}. The flux corrected for absorption is instead 1.1×10^{-9} erg cm^{-1} s^{-1}, corresponding to a luminosity of $1.3\times10^{35}d_1^2$ erg s^{-1}, where d_1 is the distance to the source in units of 1 kpc. The luminosity suggests a neutron star as the compact object, but of course further confirmations are needed.

Given the large column density of the line–of–sight absorber, if the covering factor of the absorbing matter is large, a significant contribution from photons scattered towards the line of sight is expected, as discussed in [9]. We therefore fitted the spectrum with the Monte-Carlo model described in that paper. The fit is completely unacceptable. It must be noted that the model in [9] assumes spherical geometry and homogeneous matter, while the covering factor may be significantly smaller than one and the average column density smaller than that on the line–of–sight (see below).

The energies of the Fe and Ni Kα lines correspond to neutral or low ionized atoms ([5]). On the contrary, the Fe Kβ centroid energy is significantly larger than expected. This cannot be due to high ionization, not only because it does not agree with the Kα energy, but also because for significantly ionized matter the Kβ line becomes much fainter, to disappear completely for Fe XVII or more, when no M electrons remain. Instead the observed Kβ/Kα ratio ($0.20\pm^{0.02}_{0.03}$) is slightly larger than expected for neutral iron (see the discussion in [13]). It should however be noted that, given the proximity to

TABLE 1. Best fit results for the baseline model. Equivalent widths are calculated against the unabsorbed continuum.

Γ	$1.60^{+0.07}_{-0.11}$
N_H (10^{24} cm^{-2})	$1.91^{+0.03}_{-0.04}$
A_{Fe}	$0.89^{+0.04}_{-0.03}$
E (Fe Kα) [keV]	$6.401^{+0.001}_{-0.001}$
F (Fe Kα) [10^{-5} ph cm^{-2} s^{-1}]	$14.8^{+0.4}_{-0.6}$
EW (Fe Kα) [eV]	13
E (Fe Kβ) [keV]	$7.099^{+0.001}_{-0.006}$
F (Fe Kβ) [10^{-5} ph cm^{-2} s^{-1}]	$3.05^{+0.33}_{-0.38}$
EW (FeK β) [eV]	3
E (Ni Kα) [keV]	$7.45^{+0.05}_{-0.02}$
F (Ni Kα) [10^{-5} ph cm^{-2} s^{-1}]	$0.85^{+0.19}_{-0.21}$
EW (Ni Kα) [eV]	1
F (Fe Kα CS) [10^{-5} ph cm^{-2} s^{-1}]	$1.88^{+0.59}_{-0.68}$

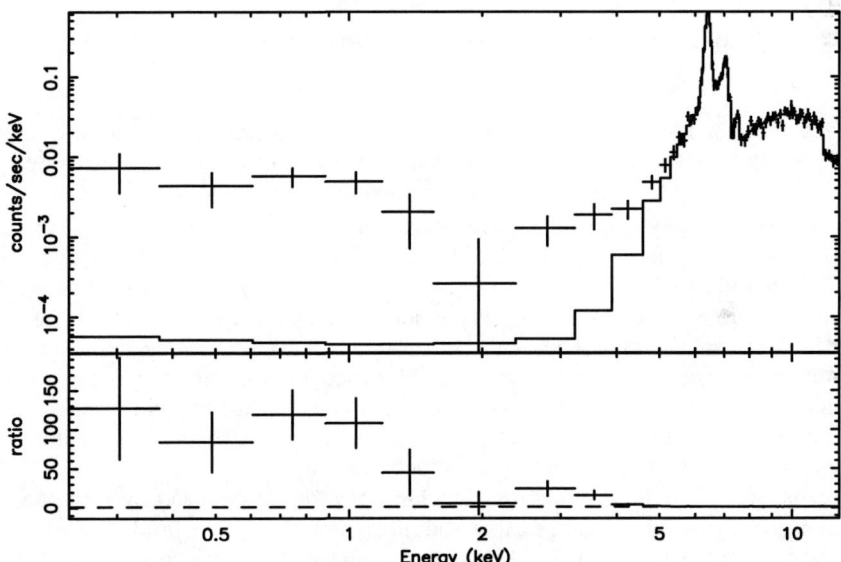

FIGURE 1. The overall 0.3-13 keV time integrated spectrum fitted with the baseline model

the iron edge, the parameters of the Kβ line are necessarily difficult to estimate, a task to be deferred to high resolution observations as those provided in the near future by ASTRO-E2.

The Ni to Fe Kα line ratio is about 6%, suggesting a possible Ni overabundance (see again the discussion in [13]). In Table 1, the EW of the lines with respect to the unabsorbed continuum (to make easier the comparison with the expected value for the

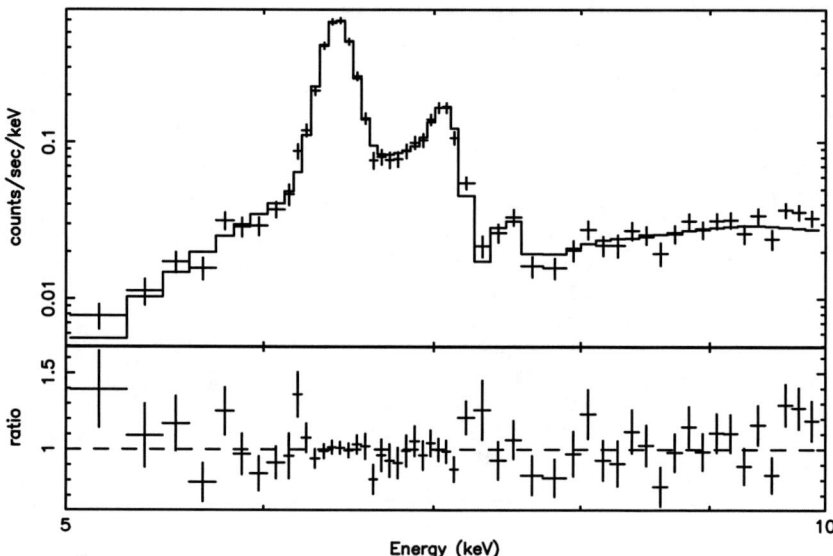

FIGURE 2. The same as the previous figure, but restricted to the 5-10 keV range.

iron Kα presented in [10]), are also given. The expected value of the ratio, f, between the Compton Shoulder and the line core is 0.44, to be compared with a measured value of $0.12^{+0.04}_{-0.05}$. f does not depend much on the geometry, but rather on the column density ([10]). The observed value would correspond to a column density of a few $\times 10^{23}$ cm^{-2}, for which values around 100 eV of the EW of the line core are expected (a value of about 20 eV is instead expected for 1.9×10^{24} cm^{-2}). The observed EW is instead about 13 eV. It is then possible that the matter is very inhomogeneous, with a denser blob just on the line of sight (which is what the fit can measure) but an average optical depth an order of magnitude less, and a covering factor, taking into account the uncertainties on the power law index, of about 0.1–0.2. The lower (with respect to the line of sight) average column density, along with the relative small covering factor, would explain the failure of the spherical transmission model in fitting the data.

Because there is evidence that the absorbing material has a covering factor less than 1, part of the X–ray illuminated surface should be directly visible, producing a Compton reflection component (e.g. [8]). As discussed above, the average column density is possibly as low as a few $\times 10^{23}$ cm^{-2}; however, below the iron line energy the reflection component for this column density is very similar to that for Compton–thick matter ([12]). This component could therefore account for the excess emission below 5 keV (see Fig. 1, where the whole 0.3-13 keV spectrum is shown, after being fitted with the baseline model), and down to about 2 keV (the further excess at lower energies should have a different origin, maybe a confusing source or a dust scattering halo). Fitting the 2-13 keV spectrum with the baseline model gives $\chi^2=104.9/68$ d.o.f., and a very flat ($\Gamma \approx 0.6$) power law. The addition of a pure Compton reflection component (with the photon index linked to that of the absorbed power law, and fixed to 1.6) improves the fit significantly, giving χ^2=79.9/68 d.o.f.. The value of R, 0.003, implies that the visible part of the illuminated matter is very small (R is equal to 1 for 2π visible solid angle, i.e. a covering factor of 0.5). The other parameters are similar to those listed in Table 1. The iron line EW with respect to the reflection component is very large, $\sim$28 keV, implying that almost all the line is related to the transmitted component.

The small value of R may, at the first glance, appears rather surprising, given the value of the covering factor deduced from the iron line EW and the CS (about 0.1). It may be explained if, e.g., the absorber has a flat configuration and is seen at high inclination.

DATA ANALYSIS AND RESULTS: TEMPORAL BEHAVIOUR

Due to lack of space, we cannot discuss in detail the temporal behaviour of the source, for which we defer the reader to [11]. We just recall here that IGR J16318-4848 exhibits a complex variability pattern during the XMM-Newton observation, with large flux variations. The iron Kα line varies on time scales as short as 1000 s, implying a size of the emitting region not exceeding $\sim 3 \times 10^{13}$ cm. A time–resolved spectral analysis shows that the variations of the line–of–sight column density (if any) cannot explain the observed flux variability, which therefore must be intrinsic.

WORK IN PROGRESS

A simultaneous XMM–*Newton* and INTEGRAL monitoring campaign is ongoing, with the aim of studying the source behaviour on different time scales. The campaign consists of three observations: the first two have been performed about one month apart (XMM–*Newton* observing dates: 2004-02-18 and 2004-03-20), the third about five months later the second one (2004-08-20). The spectral and temporal analysis of these observations is in progress, and is deferred to a future paper. Here, suffice is to note that the new XMM–*Newton* observations show that the source is still there, confirming that IGR J16318-4848 is not a transient source, even if a highly variable one. In fact, while the time–average flux of the first 2004 observation was about 20% higher than that of the 2003 TOO observation, during the second observation the flux dropped by about a factor 3, to partly recover in the third observation. Despite the large flux variability, the spectrum in these observations is very similar to that of the 2003 observation (see Fig. 3). Even if a $\sim$30% variation of the line-of-sight column density is apparent, this can explain only part of the observed variability, most of it being therefore intrinsic.

ACKNOWLEDGMENTS

We thank E. Kuulkers for useful discussions.

REFERENCES

1. Bianchi, S., et al., A&A, **396**, 793 (2002)

IGR16318−4848 XMM−Newton

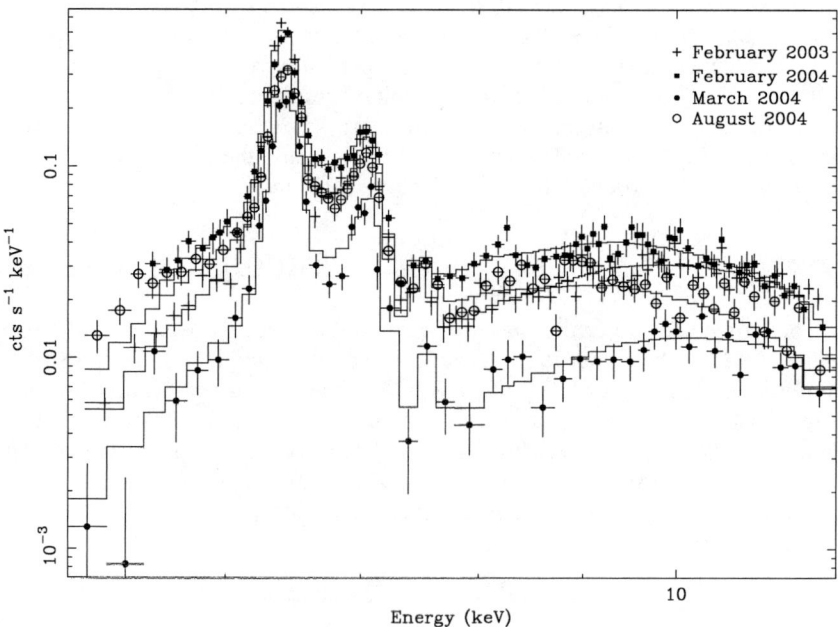

FIGURE 3. The spectra of all four XMM–*Newton* observations. Despite the large flux variations between different observations, the spectral shapes are very similar one another.

2. Chaty, S., Filliatre, P., 2004, Rev. Mex. AA, **20**, 65 (2004)
3. Courvoisier, T.-J., Walter ,R., Rodriguez, J., Bouchet, L., Loutovinon, A.A., IAU Circ. **8063** (2003)
4. de Plaa, J., et al., ATEL **#119** (2003)
5. House, L.L., ApJS, **18**, 21 (1969)
6. Jansen, F., et al., A&A, **361**, L1 (2001)
7. Kuulkers, E., this volume
8. Matt, G., Perola, G. C., and Piro, L., A&A, **247**, 25 (1991)
9. Matt, G., Pompilio, F., and La Franca, F., New As., **4/3**, 191 (1999)
10. Matt, G., MNRAS, **337**, 147 (2002)
11. Matt, G., and Guainazzi, M., MNRAS, **341**, L13 (2003)
12. Matt, G., Guainazzi, M., and Maiolino, R., MNRAS, **342**, 422 (2003)
13. Molendi, S., Bianchi, S., and Matt, G., MNRAS, **343**, L1 (2003)
14. Murakami, H., Dotani, T., and Wijnands, R., IAU Circ. **8070** (2003)
15. Revnivtsev, M., Sazonov, S., Gilfanov, M., and Sunyaev, R., Ast. Letters, **29**, 587 (2003)
16. Schartel, N., et al., IAU Circ. **8072** (2003)
17. Walter, R., et al., A&A, **411**, L427 (2003)

RXTE Observation of the Low-Mass X-ray Binary Pulsar GX1+4

Takayoshi Kohmura* and Shunji Kitamoto[†]

Physics Department, Kogakuin University, 2665–1, Nakano-cho, Hachioji, Tokyo, Japan, 192–0015
[†]*Physics Department, Rikkyo University, 3–34–1, Nishi-Ikebukuro, Toshima-ku, Tokyo, Japan, 171–8501*

Abstract. We present the study of the aperiodic time variation of the low mass X-ray pulsar GX 1+4 observed with the *Rossi X-Ray Timing Explorer* (RXTE) satellite. At the bright state in 2002, we discovered, for the first time, the delayed iron line as compared to the continuum X-rays in GX 1+4. Assuming that the iron emission lines are fluorescent lines in origin and are emitted by circumstellar matter, we derived 0.24±0.08 s for the time delay of the temporal variation of the iron line from the continuum X-rays. This result leads to a determination of $(7.2\pm2.4)\times10^9$ cm for the distance between the original X-ray source and reprocessor of the iron lines. If the iron lines is the fluorescent iron line from the cold matter at the Alfven shell, we can derive the magnetic field strength of GX 1+4 to be an 2.5×10^{13} G.

Keywords: stars : neutron — Pulsars : individual (GX 1+4) — X-rays : stars
PACS: 01.30.Cc, 95.85.Nv, 97.60.Gb

INTRODUCTION

The accreting X-ray pulsars are thought to be magnetized neutron stars with a normal star companion. X-ray emission of such a system is powered by the accretion of matter onto the neutron star from the companion star via an accretion disk or a stellar wind.

The accretion disk is believe to be formed around the neutron star in an accretion pulsar. The disk is disrupted by the strong magnetic field of the neutron star at the magnetospheric radius, where accretion matter is channeled to the polar regions of the star along the field lines (e.g., [4]). In the conventional picture of accretion onto a neutron star via an accretion disk, the magnetospheric radius, r_m, as the radius at which the magnetic pressure balances the ram pressure of a spherical accretion flow. Assuming a dipole field at a large distance from the neutron star, we have ([8])

$$r_m = 5.2 \times 10^8 L_{36}^{-2/7} (\frac{M}{1.4M_\odot})^{1/7}, B_{12}^{4/7}(\frac{R}{10^6})\text{cm}. \qquad (1)$$

Here L_{36} is the X-ray luminosity in unit of 10^{36} ergs s^{-1}, M and $M_\odot$ are the mass of the neutron star and the solar mass, and B_{12} is the surface polar magnetic field strength in unit of 10^{12} G ,respectively.

The low mass X-ray binary pulsar GX 1+4 is the most enigmatic of all X-ray binary pulsars. It is a slow pulsar with a spin period of about 2 minutes and its companion star is identified with the M6 III giant. The other basic parameters, such as the orbital period and the urface magnetic field of the neutron star of GX 1+4, are not known. Some report an unusually strong magnetic field of a few $\times 10^{13}$ G. According to the accretion torque model ([4]), the observed spin down rate require a very large magnetic field on the neutron star surface of $\sim 10^{14}$ G ([9]). The propeller effects observed in GX 1+4 also requires a surface magnetic field of a 3×10^{13} G ([1]; [2]). The X-ray spectrum of GX 1+4 shows the neutral or weakly ionized iron Kα lines at $\sim$6.4 keV ([3]; [7]), and it may come from cold gas around the neutron star.

If this binary has such a strong magnetic field, the size of Alfven shell is to be an order of 10^9 cm from equation (1). We can expect the delay time, $\sim$100 ms, between the time variation of iron lines and the continuum X-ray by the difference of the light path if the iron line is the fluorescent iron lines which is reprocessed by the irradiation of the continuum X-ray from the neutron star. In this paper, we report the discovery of the delayed iron line in GX 1+4 using *RXTE* data.

DELAYED IRON LINES IN GX 1+4

RXTE carried out weakly monitoring campaign on GX 1+4 from 2001 to 2002 for over a 2 yr. In 2002, GX 1+4 showed a transition from very bright state to the faint state ([2]). In the bright state, the pulsed fraction was strong. On the other hand, in the faint state, no X-ray pulsation was detected. In the faint state, it was interpreted that the propeller effect might manifest itself directly in the cessation of pulsation in GX 1+4, when the mass accretion rate was significantly low.

To detect a small delay time between the time variation of iron photons and the continuum X-ray, *RXTE* is the best experiment, which provides us high time resolution data with excellent statistics and reasonable energy resolution. We concentrate not on the low state but on the bright, and are attempting to analyze the aperiodic time variation rather than the coherent pulsations. The best method to use is the cross spectral analysis ([10]), since the cross spectra make it possible to distinguish the pulse component from the interesting aperiodic component. Using cross spectral analysis, we were succeeded in obtaining the delay time of iron lines as compared to the continuum X-ray in the high mass X-ray binary pulsar Centaurus X-3 (Cen X-3) by 6.5$\pm$1.6 ms (Figure 1; [5]; [6]). In Cen X-3 we determined the distance between the neutron star and the reprocessor of iron lines to be $(2.0\pm 0.5)\times 10^8$ cm.

To derive the total X-ray flux in the bright state, we modeled the observed spectra with *RXTE* PCA and HEXTE using the power-law with an exponential high-energy cutoff including absorption by cold gas and Gaussian emission line around 6.4 keV. A power-law index (1.17$\pm$0.03) is and the absorption, N_H is $(4.9\pm 0.2)\times 10^{22}$ cm^{-2}. Emission line centered at (6.46$\pm$0.01) keV is consistent with the iron line of neutral or weakly ionized iron with the equivalent width is to be $\sim$334 eV. The observed X-ray flux is 7.61×10^{36} ergs s^{-1} for the distance of 10 kps.

To apply the cross spectral analysis, we first made light curves of some energy bands. One energy band contained the iron lines and the other did not contained (e.g. 0.71–5.82, 6.04–7.33, 7.33–12.08,12.08–21.29, and 30.11–39.38 keV). We refer to the light curve which contain the iron lines as the "iron-band"(6.04–7.33 keV). We calculate the cross spectra between the X-ray light curve in the iron-band and the between them. Figure 1 shows the result of our cross spectral analysis as a function of the energy. We

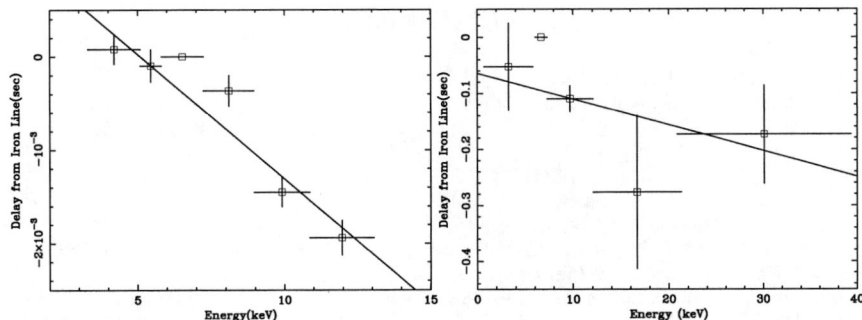

FIGURE 1. The time delay in the time variation from that of the iron-band as a function of the X-ray energy (Left: Centaurus X-3, Right: GX 1+4). Errors are one sigma confidence levels. The positive direction of the vertical axis represents the delay from the variation of the iron-band. The best-fit linear function is also displayed.

apply the simple linear function to this data, and the result is $-(4.4\pm4.0)\times10^{-3}$ E (keV) $+ (9.5\pm2.6)\times10^{-2}$ s. The deviation of the iron-band data at 6.45 keV from this best-fit model is $(9.0\pm3.0)\times10^{-2}$ s, where error is the one sigma level. Therefore, we detected the delayed iron-band intensity variation with almost the four sigma confidence level.

DISCUSSION

We derived a $(9.0\pm3.0)\times10^{-2}$ s time delay of the iron-band relative to that of the general trend of the other energy bands. However, this value may not necessarily represent an actual delay of the reprocessed X-rays due to the mixture effect of continuum X-rays in the iron band. The fraction of iron line photons in the iron-band is $\sim$30 % and the most of X-ray photons in iron band is the continuum X-rays. Therefore we have to consider the effect of the mixture of the delayed and the non-delayed components. As a result, we derived 0.24 ± 0.08 s for the time delay of the temporal variation of the iron line from the continuum X-rays. This result leads to a determination of $(7.2\pm2.4)\times10^{9}$ cm for the distance between the original X-ray source and reprocessor, which emits the iron lines. If the radius derived from above discussion is the Alfven radius derived from equation (1), we can obtain the magnetic field strength to be 2.5×10^{13} G. The advance of the hard X-ray variation was also discovered in GX 1+4. We think this advanced trend of the continuum X-ray is related to the X-ray emission mechanism on the neutron star.

ACKNOWLEDGMENTS

This research has made use of data obtained through the High Energy Astrophysics Science Archive Research Center Online Service, provided by the NASA/Goddard Space Flight Center.

REFERENCES

1. Cui, W., ApJ, **482**, L63 (1997)
2. Cui, W., & Smith, B., ApJ, **602**, 320 (2004)
3. Dotani, T., Kii, T., Nagase, F., Makishima, K., Ohashi, T., Sakao, T., Koyama, K., & Tuohy, I.R., PASJ, bf 41, 427 (1989)
4. Ghosh, P., & Lamb, F.K., ApJ, **234**, 296 (1979)
5. Kohmura, T., Kitamoto, S., & Torii, K., ApJ, **562**, 943 (2001)
6. Kohmura, T., PhD thesis, Osaka University (2002)
7. Kotani, T., Dotani, T., Nagase, F., Greenhill, J.G., Pravdo, S.H., & Angelini, L., ApJ, **510**, 369 (1999)
8. Lamb, F.K., Pethick, C.J., & Pines, D., ApJ, **184**, 271 (1973)
9. Makishima, K., et al., Nature, **333**, 746 (1988)
10. van der Klis, M., Hasinger, G., Stella L., Langmeier, A., van Paradijs, J., & Lewin, W.H.G., ApJ, **319**, L13 (1987)

The Double Pulsar System J0737−3039: News and Views

Marta Burgay*, Nichi D'Amico[†,*], Andrea Possenti*, Andrew Lyne**, Michael Kramer**, Maura McLaughlin**, Duncan Lorimer**, Dick Manchester[‡], Fernando Camilo[§], John Sarkissian[¶], Paulo Freire[||] and Bhal Chandra Joshi[††]

*INAF - Osservatorio Astronomico di Cagliari, Loc. Poggio dei Pini, Strada 54, 09012 Capoterra (CA), Italy
[†]Università degli studi di Cagliari, Dipartimento di Fisica, SP Monserrato - Sestu km 0.7, 09042 Monserrato (CA), Italy
**University of Manchester, Jodrell Bank Observatory, Macclesfield, Cheshire, SK11 9DL, UK
[‡]Australia Telescope National Facility, CSIRO, P.O. Box 76, Epping, New South Wales 2121, Australia
[§]Columbia Astrophysics Laboratory, Columbia University, 550 West 120 th Street, New York 10027, USA
[¶]Australia Telescope National Facility, CSIRO, Parkes Observatory, P.O. Box 276, Parkes, New South Wales 2870, Australia
[||]NAIC, Arecibo Observatory, Puerto Rico, USA
[††]National Center for Radio Astrophysics, P.O. Bag 3, Ganeshkhind, Pune 411007, India

Abstract. The double-pulsar system J0737−3039 is the most intriguing pulsar discovery of last decade. This binary system, with an orbital period of only 2.4-hr, provides a truly unique laboratory for relativistic gravitational physics. Its discovery enhances of about an order of magnitude the estimate of the merger rate of double neutron stars systems, opening new possibilities for the current generation of gravitational wave detectors. The high orbital inclination, moreover, offers the opportunity of using the radio beams from one pulsar as a probe for studying the magnetosphere of the other. In this contribution we summarize the present results and look at the prospects of future observations.

Keywords: pulsar: general – pulsar: individual (J0737−3039A, J0737−3039B)
PACS: 97.60.Gb; 97.60.Jd; 95.30.Sf; 95.10.Eg; 95.85.Bh

THE DISCOVERY OF THE FIRTST DOUBLE PULSAR

PSR J0737−3039A (hereafter simply 'A'; [1]), a 22-ms pulsar, was discovered in a 4' pointing of the Parkes High-Latitude Pulsar Survey (Burgay et al., in preparation). From the pronounced curvature of the pulsar's signal in the pulse phase vs time panel in the original detection plot (central panel in Fig. 1), denoting an high Doppler variation of the pulse period with time, it was immediately clear that this object was included in a short period binary. Follow-up observations confirmed that the system is indeed very tight with an orbital period of only 2.4 hr and a projected semimajor axis of about 1.4 light-seconds.

These orbital parameter, obtained in few days of observation shortly after the discovery, give a minimum mass for the companion of $\sim 1.24\,M_\odot$ immediately suggesting that

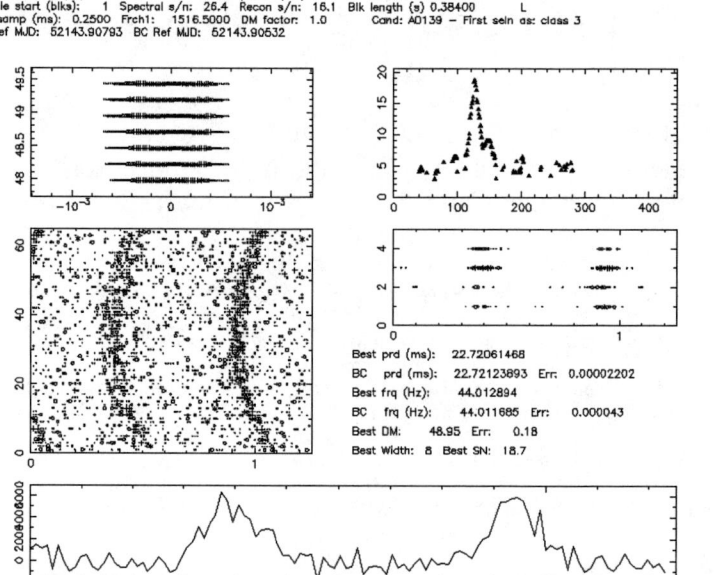

FIGURE 1. Original detection plot for PSR J0737−3039A. Clockwise from the top left the sub-plots show a gray-scale of the dependence of signal-to-noise ratio S/N on dispersion measure DM and offset (in ms) from the nominal period, the dependence of S/N on DM trial number, a grey-scale plot of S/N vs pulse phase in four sub-bands, the integrated pulse profile and a grey-scale plot of S/N vs pulse phase for 64 sub-integrations.

J0737−3039 is a double neutron star system. Moreover, after only 5 days of follow-up observations, an incredibly high value of the advance of the periastron $\dot{\omega} \sim 16.8$ deg/yr (the previously highest value measured being 5.33 deg/yr for PSR J1141-6545; [2]) was measured implying, if interpreted in the framework of general relativity, a total mass for the system of $\sim 2.56\ M_\odot$, again supporting the hypothesis of a double neutron star binary.

The above picture was confirmed few months later when, in a follow-up observation of pulsar A, a signal with a period of 2.7 s was detected ([3]). The newly discovered pulsar J0737-3039B (or simply 'B') has the same dispersion measure of A and shows Doppler variations in the pulse period that identify it as the companion. The first ever Double Pulsar System was discovered.

In the following some of the implications of this discovery are discussed.

TEST OF GENERAL RELATIVITY

When a new pulsar is discovered the parameters immediately available are its approximate period, dispersion measure and position within about half a beam-width of the radio-telescope. Precise parameters of the source can be obtained with regular monitoring of the new pulsar according to a procedure known as *timing*. For most binary systems including a radiopulsar, the determination of the position in the sky, of the rotational parameters and of the five keplerian parameters allows to satisfactorily reproduce the orbital motion and to predict the times of arrival of the pulse from the given source within the measurement errors.

That is not the case for close double neutron star binaries. In fact, due to their strong gravitational fields and rapid motions, these systems exhibit large relativistic effects [4]. When they are large enough, their measurement can be used for testing the predictions of General relativity and of other theories of gravity. The tests can be performed when a number of relativistic corrections to the Keplerian description of the orbit (the so-called post-Keplerian, hereafter PK, parameters) are measured: the PK parameters in each theory, infact, are only functions of the star masses and of the measured Keplerian parameters. With the two masses as the only unknowns, the measurement of three or more PK parameters over-constrains the system, hence providing a test-ground for theories of gravity [5]. In a theory that describes the binary system correctly, the PK parameters define lines in a mass-mass diagram (Fig. 2) that all intersect in a single point.

Such tests have been possible to date in only two double neutron star systems, PSR B1913+16 [6] and PSR B1534+12 [7]. For PSR B1913+16, the relativistic periastron advance, $\dot{\omega}$, the orbital decay due to gravitational wave damping, $\dot{P}_b$, and the gravitational red-shift/time dilatation parameter, γ, have been measured, providing a total of three PK parameters. For PSR B1534+12, Shapiro delay, caused by passage of the pulses through the gravitational potential of the companion, is also visible, since the orbit is seen nearly edge-on. This results in two further PK parameters, r (range) and s (shape) of the Shapiro delay. However, the observed value of $\dot{P}_b$ requires correction for kinematic effects, so that PSR B1534+12 provides four PK parameters usable for precise tests [7].

With a intense campaign of regular timing observations started immediately after the discovery, we measured in only 6 months A's $\dot{\omega}$ and γ and have also detected the Shapiro delay in the pulse arrival times of A due to the gravitational field of B [3]. After adding further 6 months of observation, we have now got also the first meaningful determination of the orbital period decay $\dot{P}_b = -1.2 \pm 0.3 \times 10^{-12}$ s/s. This provides five measured PK parameters[1], resulting in a M_A-M_B plot (Fig. 2) through which we can test the predictions of general relativity. We can see that a region satisfying all the constraints exists. In particular the current data, spanning only 12 months of observations, indicate an agreement of the observed with the expected Shapiro parameter s of $s_{obs}/s_{exp}=$ 1.00007±0.00220, where the uncertainties are likely to decrease quickly. The position of the allowed region in Fig. 2 also determines the inclination of the orbit to the line-of-sight. It turns out that the system is observed nearly edge-on with an inclination angle of about 88°.

The detection of B as a pulsar opens up opportunities to go beyond what is possible with previously known double neutron star binary systems. Firstly, we can exclude all

---

[1] Assuming that correction to $\dot{P}_b$ due to the motion of the binary system in the galactic gravitational potential are negligible or can be isolated by proper motion and distance measurements (shortly available).

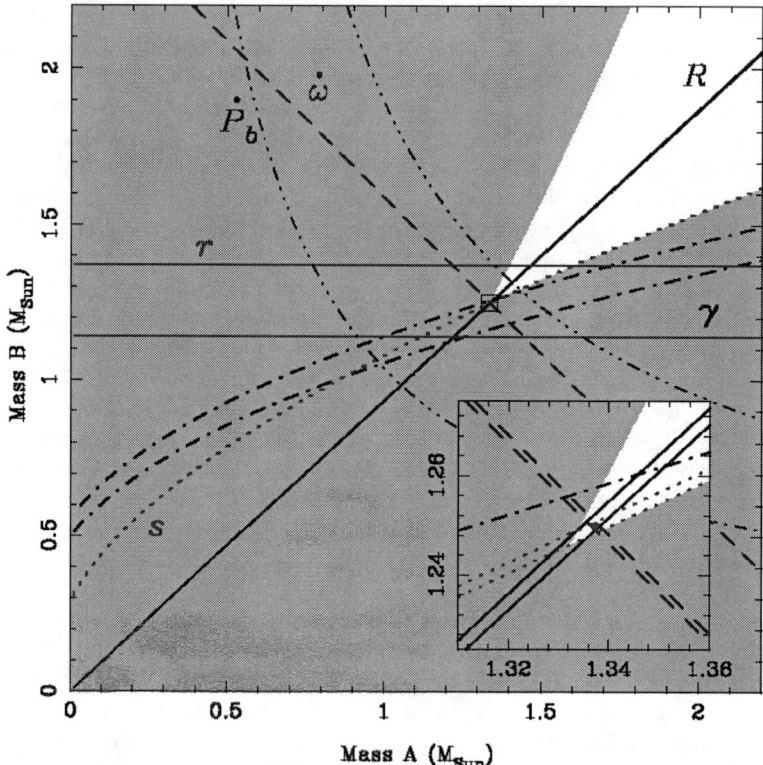

FIGURE 2. The observational constraints upon the masses M_A and M_B. The grey regions are those which are excluded by the Keplerian mass functions of the two pulsars. Further constraints are shown as pairs of lines enclosing permitted regions as predicted by general relativity: (a) the measurement of the advance of periastron $\dot{\omega}$, giving the total mass $M_A+M_B=2.588\pm0.003$ $M_\odot$ (dashed lines); (b) the measurement of $R = M_A/M_B = x_B/x_A = 1.069\pm0.006$ (solid lines); (c) the measurement of the gravitational red-shift/time dilation parameter γ (dot-dash lines); (d) the measurement of Shapiro parameter r (solid horizontal lines) and Shapiro parameter s (dotted lines); (e) the measurement of the orbital decay (triple dot-dash lines). Inset is an enlarged view of the small square which encompasses the intersection of the three tightest constraints, with the scales increased by a factor of 16. The permitted regions are those between the pairs of parallel lines and we see that an area exists which is compatible with all constraints.

regions in the M_A-M_B plot plane that are forbidden by the individual mass functions of A and B due to the requirement $\sin i \leq 1$ (where i is the inclination of the orbital plane). Secondly, with a measurement of the projected semi-major axes of the orbits of A and B, we obtain a precise measurement of the mass ratio, $R = M_A/M_B = x_B/x_A$, providing a further constraint in the M_A-M_B plot (Fig. 1). This relation is valid for most theory of gravity; in particular, the R-line is independent of strong-field (self-field) effects, providing a new constraint for tests of gravitational theories [4]. With five PK parameters already available for tests, this additional constraint makes this system the most over-determined double neutron star binary.

A REVISED DOUBLE NEUTRON STAR COALESCENCE RATE

The merging of a double-neutron-star system should produce a burst of emission of gravitational waves. Due to the energy budget and to the expected typical frequency of these events, they are among the primary targets for the current generation of ground-based gravitational waves detectors, which should be able to detect them up to a distance of about 20 Mpc. Hence, a key question is the occurrence rate of these double-neutron-star coalescences in a volume of universe of that radius. This rate can in turn be estimated on the basis of the rate of events in the Galaxy.

Among the double-neutron-star systems previously known, only three had tight enough orbits so that the two neutron stars will merge within a Hubble time. Two of them (PSR B1913+16 and PSR B1534+12) are located in the Galactic field, while the third (PSR B2127+11C) is found on the outskirts of a globular cluster. The contribution of globular cluster systems to the Galactic merger rate is estimated to be negligible [8]. Also, recent studies [9] have demonstrated that the current estimate of the Galactic merger rate $\mathscr{R}$ relies mostly on PSR B1913+16 characteristics. One can hence start by comparing the observed properties of B1913+16 and J0737−3039 systems. The latter will merge due to the emission of gravitational waves in ~ 85 Myr, a time-scale that is a factor 3.5 shorter than that for PSR B1913+16. In addition, the estimated distance of PSR J0737−3039A/B (500 - 600 pc, based on the observed dispersion measure and a model for the distribution of ionized gas in the interstellar medium [10]) is an order of magnitude less than that of PSR B1913+16. These properties have a substantial effect on the prediction of the rate of merging events in the Galaxy.

For a given class k of binary pulsars in the Galaxy, apart from a beaming correction factor, the merger rate $\mathscr{R}_k$ is calculated as $\mathscr{R}_k = N_k/\tau_k$ [9]. Here τ_k is the binary pulsar lifetime and N_k is the scaling factor defined as the number of binaries in the Galaxy belonging to the given class. The shorter lifetime of J0737−3039 system [τ_{1913}/τ_{0737} = (365 Myr)/(185 Myr)] implies a doubling of the ratio $\mathscr{R}_{0737}/\mathscr{R}_{1913}$. A much more substantial increase results from the computation of the ratio of the scaling factors N_{0737}/N_{1913}. The luminosity L_{400}=30 mJy kpc^2 of PSR J0737−3039A is much lower than that of PSR B1913+16. For a planar homogeneous distribution of pulsars in the Galaxy, the ratio $N_{0737}/N_{1913} \sim L_{1913}/L_{0737} \sim 6$. Hence we obtain $\mathscr{R}_{0737}/\mathscr{R}_{1913} \sim 12$. Including the moderate contribution of the longer-lived PSR B1534+12 system to the total rate, we obtain an increase factor for the total merger rate of about an order of magnitude.

Extensive simulations [11] give results consistent with this simple estimate and show that the peak of the merger rate increase factor resulting from the discovery of J0737−3039 system lies in the range 5-7 and is largely independent of the adopted pulsar population model. For the most favorable distribution model available (model nr. 15 of [9]), the updated cosmic detection rate for first generation gravitational-wave detectors such as VIRGO, LIGO and GEO can be as high as 1 every 3-5 years at 95% confidence level. Hence, with the discovery of PSR J0737−3039A/B the double-neutron-star coalescence rate estimates enter an astrophysical interesting regime. Within a few years of gravitational-wave detectors operations, it should be possible to directly test these predictions and, in turn, place better constraints on the cosmic population of double-neutron-star binaries.

PROBING PULSAR MAGNETOSPHERE

In pulsar A a ~ 30 second eclipse is seen in the orbital longitude range $89° - 91°$, corresponding to the superior conjunction (i.e. when pulsar A is behind B).

PSR J0737−3039B displays an even more puzzling behavior. The flux density received from B is dramatically changing along the orbit: the signal is clearly visible for two brief periods of about 10 minutes duration each centered on orbital longitudes $210°$ and $280°$. Within these bursts the strength of the emission is such that most pulses are detected individually. In addition, the pulsar often shows weaker emission elsewhere. Observations made with the Green Bank Telescope (GBT; [12]) show that pulsar B's emission is present along almost all the orbit becoming apparently undetectable (eclipsed?) only at inferior conjunction, when pulsar A is between B and the observer. Pulsar B has a really unusual behavior also in pulse profile. In striking contrast with what observed in any other binary pulsar, the pulse profile varies along the orbit. B's strange behavior is unique so far and is likely due to the relativistic wind from the A pulsar penetrating deep into its magnetosphere, as shown in the following.

At the light cylinder radius (delimiting the magnetosphere) of B, the energy density of the relativistic wind from A is about two orders of magnitude greater than that from B, ensuring that the wind from A is penetrating deep into B's magnetosphere. Assuming isotropy, one find infact that the energy densities due to the spin-down luminosity emitted by A and the magnetic field of B are in balance at a distance ~ 0.2 lt-s from B, about 40% of its light cylinder radius. The actual penetration of the wind from A into B's magnetosphere is likely to be a function of the orientation of the rotation and/or magnetic axes of B relative to the direction of the wind and hence it will depend on the precessional and orbital phases of B. This is the most likely explanation for the large flux density and pulse-shape changes of B. The front shock produced in such a violent interaction is probably also the cause of the short eclipse in pulsar A's emission.

The separation of the two pulsars in their orbits (as schematically represented in the top part of figure 3) is typically $\sim 900,000$ km (or 3 lt-s). The large orbital inclination deduced from the measurement of the shape of the Shapiro delay, $s \equiv \sin i$, or from the comparison of the constraints given in the mass-mass diagram by the measured mass ratio, R, and periastron advance, $\dot{\omega}$, implies that, at conjunction, the line-of-sight to one pulsar passes within about 0.15 lt-s of the other (fig. 3, bottom). This is substantially smaller than the 0.45 lt-s radius of J0737−3039B's light cylinder, although much greater than the 0.004 lt-s light cylinder radius of pulsar A. As the pulsars move in their orbits, the line-of-sight from A passes through, and sweeps across, the magnetosphere of B, providing the opportunity to probe its physical conditions. The determination of changes in the radio transmission properties, including the dispersion and rotation measures, will potentially allow the plasma density and magnetic field structure to be probed.

In summary, the close interactions between the two pulsars is for the first time giving the possibility to have a better understanding of pulsars' magnetospheres and emission processes and issues many interesting observational and theoretical challenges: monitoring the changes in the pulse shapes and the flux and polarisation variations along the orbit at many different frequencies will be very important in the aim of assessing the nature of the puzzling behavior of PSR J0737−3039A and PSR J0737−3039B and for better exploiting the J0737−3039 system as a test bed for relativistic gravity.

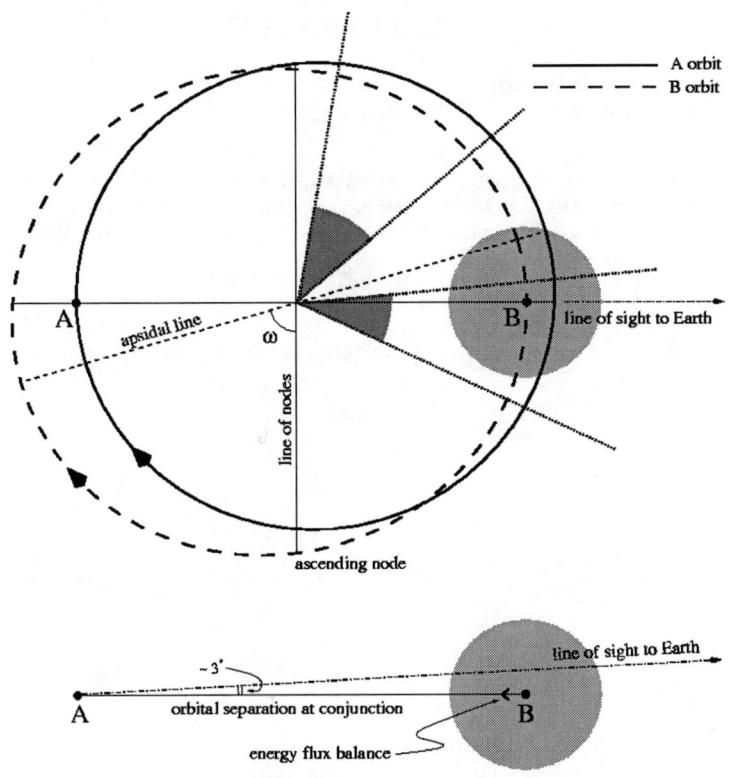

FIGURE 3. The physical configuration of the binary system, at conjunction, as on 19 August 2003 (MJD 52870), showing the relative sizes of the two orbits and B's magnetosphere. Top: view from above the orbital plane with the Earth to the right. The shaded segments indicate orbital phases where B is detected strongly. Bottom: view from the side, showing the passage of the line-of-sight from A to the Earth through the magnetosphere of B. The approximate position of the pressure balance between the relativistic wind from A and the magnetic field of B is indicated.

ACKNOWLEDGMENTS

The Parkes radio telescope is part of the Australia Telescope which is funded by the Commonwealth of Australia for operation as a National Facility managed by CSIRO. The National Radio Astronomy Observatory is facility of the National Science Foundation operated under cooperative agreement by Associated Universities, Inc. MB, NDA and AP received support from the Italian Ministry of University and Research (MIUR) under the national program *Cofin 2003*. DRL is a University Research Fellow funded by the Royal Society. FC acknowledges support from NSF grant AST-02-05853 and a NRAO travel grant.

REFERENCES

1. Burgay, M., et al., *Nature*, **426**, 531 (2003)
2. Kaspi, V. M., et al., *ApJ*, **601**, L179 (2000)
3. Lyne, A. G., et al., *Science*, **303**, 1089 (2004)
4. Damour, T., and Deruelle, N., *Ann.Inst.H. Poincaré*, **44**, 263 (1986)
5. Damour, T., and Taylor, J. H., *Phys. Rev. D*, **45**, 1840 (1992)
6. Taylor, J. H., and Weisberg, J. M., *ApJ*, **345**, 434 (1989)
7. Stairs, I. H., Thorsett, S. E., Taylor, J. H., and Wolszczan, A., *ApJ*, **581**, 501 (2002)
8. Phinney, E.S., *ApJ*, **380**, L17 (1991)
9. Kim, C., Kalogera, V., and Lorimer, D. R., *ApJ*, **584**, 98 (2003)
10. Taylor, J. H., and Cordes, J. M., *ApJ*, **411**, 674 (1993)
11. Kalogera, V., et al., *ApJ*, **601**, L179 (2004)
12. Ransom, S., et al., *Binary Radio Pulsars*, Eds. Rasio and Stairs (2004)

POSTER SESSION

Phase Resolved Blue Spectroscopy of SS433

A.D. Barnes*, P.A. Charles†*, J.S. Clark**, R. Cornelisse* and C. Knigge*

School of Physics and Astronomy, University of Southampton, Southampton, UK
†South African Astronomical Observatory
**Department of Physics and Astronomy, University College London, UK

Abstract. Despite more than 20 years of study, we still possess only limited constraints on the fundamental properties of SS433. With the exception of the 13d period, orbital parameters are ill-constrained, and hence the nature of both compact object, a neutron star or black hole, or the mass donor are still unknown. However, newly acquired high resolution blue spectra, taken when the precessing disc is closest to edge-on, provide the strongest evidence yet for a massive, A4 supergiant donor, in addition to expanding the phase constraints for further observations of the donor.

INTRODUCTION

The bizarre object SS433 is a galactic X-ray binary ($P_{orb} \simeq 13$d) at the centre of the supernova remnant W50, and the first relativistic jet source discovered in the Galaxy. These jets show a 162.5d precession period, which was initially revealed in the radial velocity curves of the strong, blue- and red-shifted Balmer emission lines [1]. The 162.5d modulation is also present in the optical flux, a result of the changing orientation of the precessing accretion disc, the large amplitude of the precessional variation indicating that the disc is the dominant source in the optical band. Recent radio observations have also detected extended emission in a direction that is perpendicular to the main jet outflow [2], likely evidence for a disc-like outflow of matter [3].

Until recently there has been no spectroscopic detection of any feature that is clearly associated with the companion star, and suggestions for its spectral type range from early B to Wolf-Rayet. This is due to the strong and broad emission lines [1] that obliterate any signature of the donor star in SS433 that may be present. One of the key observational difficulties has been that all previous high resolution optical spectroscopy have been performed at longer wavelengths due to the high reddening ($E_{B-V} = 2.1$), i.e. the region where signatures of hot stars are weakest.

TRACES OF HOT STARS

Recognising the absence of high S/N, high resolution blue spectra of SS433, we have assembled such observations over a wide range of precessional and orbital phases for several years. The primary aim was to exploit the new technique of X-ray irradiation of the secondary which can excite sharp Bowen fluorescence features, as observed in a variety of other X-ray binaries [4],[5]. Unfortunately, while the Bowen feature is indeed

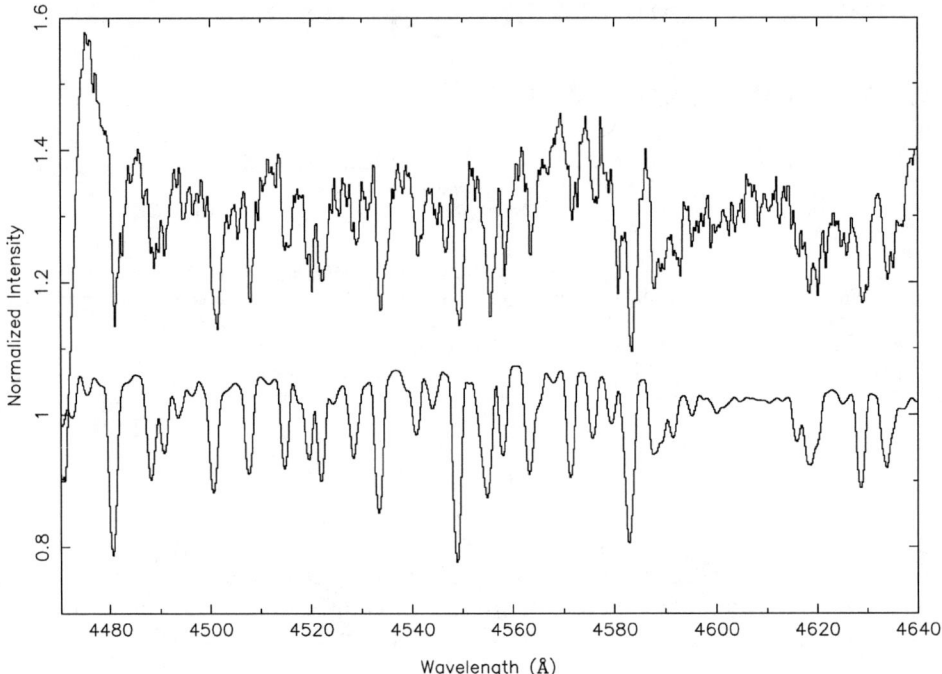

FIGURE 1. A comparison of the SS433 spectrum (top) from June 29 (when $\Psi_{prec} \sim 0.7$, $\phi_{orb} \sim 0.8$) and the A4 supergiant HD9233 (bottom). The spectrum of SS433 has been passed through a filter in order to remove the low-order variations caused by a jet line passing through the region of interest.

strong in SS433, no components clearly attributable to the region of the secondary can be identified.

Hillwig et al. [6] proposed that the *only* time to detect evidence of the donor is when it passes between the X-ray source and the observer at a precessional phase when the disc is most open to the observer ($\Psi_{prec} \sim \phi_{orb} \sim 0$). During this configuration of phases, light from the donor should be least "obscured" by the extended circumbinary disc. Using this criterion, certain features corresponding to a mid-A supergiant were unambiguously identified, with velocities which are consistent with the expected motion of the donor star within the system. However, their observations encompass too small a range of phases to allow for a full kinematical analysis, and hence no precise constraints on fundamental system parameters using this method could be derived by Hillwig et al.

However, our observations reveal that it is possible to observe the signature of the donor outside of the rather limiting criterion by Hillwig et al. Fig. 1 is the prime example (observed when $\Psi_{prec} \sim 0.7$, $\phi_{orb} \sim 0.8$, the disc contribution to the optical flux is at a minimum and the secondary is providing some 70-80% of the signal), with the match between the spectrum of SS433 (top) and the A4I comparison star being quite

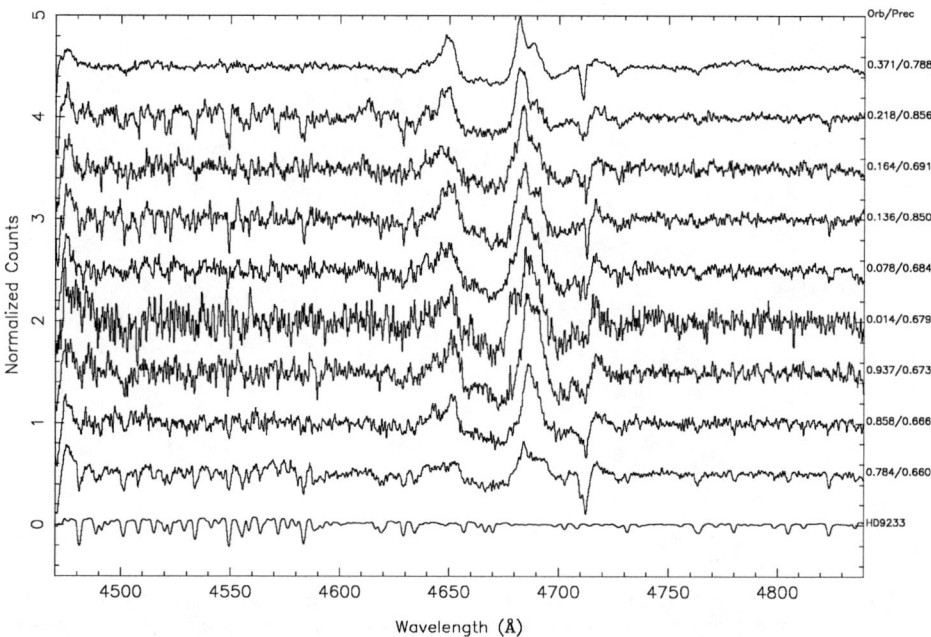

FIGURE 2. Blue spectra of SS433 at different orbital phases (orbital and precessional phase are shown next to each spectrum). For comparison, at the bottom is shown an A4 supergiant.

remarkable in its clarity. Cross-correlating the sharp features of this spectrum with the comparison star produces a velocity measurement entirely consistent with the earlier results of Hillwig *et al.*

Whilst subsequent observations initially appeared promising, these sharp features typical of those seen in late B/A type supergiants become weaker over several nights (Fig. 2). An analysis of the velocities obtained from these subsequent data does not tie in with the idea that these absorption lines originate exclusively upon the secondary star (Fig. 3).

These anomalous velocities can only make sense if the secondary becomes masked at certain times by some dense material, for example an equatorial disc wind, support for which has already been obtained via the radio detection of extended emission in a direction that is perpendicular to the main jet outflow [2]. However, it appears that at a range of orbital and precessional phases (not merely $\Psi_{prec} \sim \phi_{orb} \sim 0$) the secondary star will be detectable and thus a more confident estimate of the system parameters should be possible in the near future.

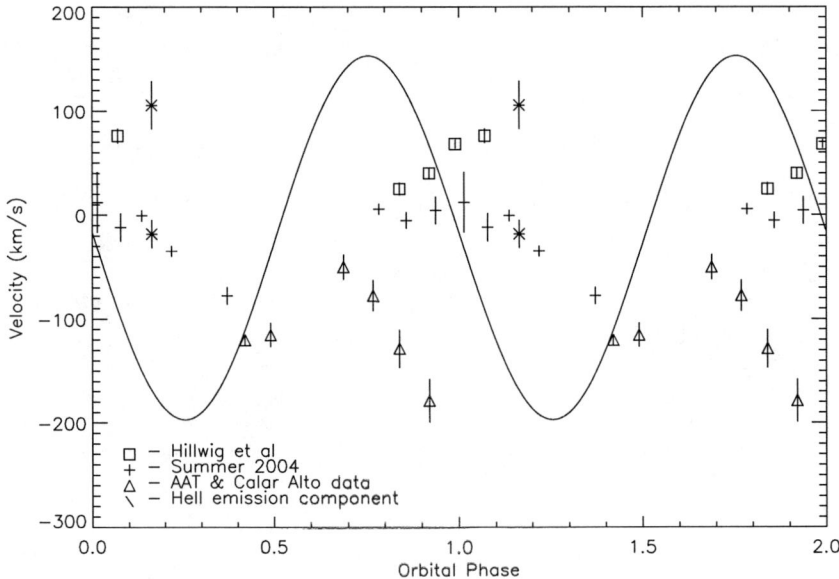

FIGURE 3. Absorption line velocity measurements comparing the results reported by Hillwig et al. (squares) [6] with our earlier (triangles) and latest data (crosses). The solid line shows the HeII emission velocities which are presumed to trace the motion of the accretion disc/compact object (Fabrika and Bychkova [7]). The data is plotted over two orbital cycles for clarity.

REFERENCES

1. Margon, B., *ARAA*, **22**, 507–536 (1984).
2. Blundell, K. M., Mioduszewski, A. J., Muxlow, T. W. B., Podsiadlowski, P., and Rupen, M. P., *ApJ*, **562**, L79–L82 (2001).
3. Zwitter, T., Calvani, M., and D'Odorico, S., *A&A*, **251**, 92–102 (1991).
4. Steeghs, D., and Casares, J., *ApJ*, **568**, 273–278 (2002).
5. Hynes, R. I., Steeghs, D., Casares, J., Charles, P. A., and O'Brien, K., *ApJ*, **583**, L95–L98 (2003).
6. Hillwig, T. C., Gies, D. R., Huang, W., McSwain, M. V., Stark, M. A., van der Meer, A., and Kaper, L., *ApJ*, **615**, 422–431 (2004).
7. Fabrika, S. N., and Bychkova, L. V., *A&A*, **240**, L5–L7 (1990).

Irradiation Effects in Compact Binaries

Martin E. Beer* and Philipp Podsiadlowski[†]

*Department of Physics and Astronomy, University of Leicester, Leicester LE1 7RH, UK.
[†]Department of Astrophysics, University of Oxford, Oxford OX1 3RH, UK.

Abstract. Irradiation of the secondaries in close binary systems affects their appearance and can drastically change their internal structure and hence long-term evolution. We have developed a three-dimensional fluid dynamics code for modelling these effects. We give particular emphasis to the role of circulation driven by the external heating and the radiative surface stress. We present results of self-consistent calculations for the circulation in irradiated systems and show how the inclusion of these is vital to the understanding and interpretation of any system where external irradiation of the secondary is significant.

MOTIVATION

Irradiation of the secondaries in close binary systems affects their appearance. X-ray heating of the stellar surface dramatically changes the observed light curve, e.g., Nova Sco 94 [1]. Heating also affects the observed amplitude of the radial velocity curve as observed in SS Cyg [2]. The amplitude of the radial velocity curve increases with irradiation as the lines used for measuring the radial velocity are no longer observed from the irradiated region [2]. The heating will also distort the radial velocity curve from a pure sinusoid as shown for Nova Sco 94 [3]. The heating also affects regions not directly irradiated. In modelling of Nova Sco 94 heating of a region greater than the directly irradiated surface is required to explain the observed radial velocity amplitude [4]. Understanding the effects of irradiation in these systems is crucial to the accurate determination of system parameters from their observed properties.

Irradiation can also change the secondary's internal structure and hence long-term evolution. Mass-transfer rates appear to be about an order of magnitude larger than the standard model predicts in LMXBs. Ruderman et al. [5] concluded that an irradiation driven wind could maintain the higher rates. Alternatively, is irradiation driven expansion the missing ingredient?

IRRADIATION EFFECTS ON THE STELLAR STRUCTURE

Apart from heating the outer layers of the star and raising the surface temperature, irradiation affects the stellar structure in several fundamental ways. The radiation pressure force itself acting on the surface layer of the star deforms the stellar surface [6]. Heating of the surface layers causes temperature gradients along equipotential surfaces. This drives circulation from the heated surface to the cooler regions [7]. If the surface is taken to be fixed then the circulation currents are found to be highly supersonic [8]. The depth of the circulation pattern can be large compared to the pressure scale height and viscous heating could heat these deeper layers. The radiation pressure force causes a surface stress which drives horizontal motion, a process known as 'Ekman' pumping. Irradiation changes the surface boundary conditions of the star which can lead to long-term structural changes since it effects the efficiency by which internal energy is radiated away at the surface. This could lead to significant expansion of the secondary [9]. Ritter et al. [10] and Hameury [11] have both considered the case of anisotropic irradiation.

METHOD

The irradiation induced circulation has been modelled previously but only in simplified models. The main results of this work are that a circulation pattern develops which reaches the unilluminated side of the star and that the tangential component of the circulation is supersonic when the surface is not deformed. The problem of irradiation-driven circulation requires both three dimensions and accurate treatment of surface geometries and boundary conditions. We have developed a general three-dimensional hydrodynamics code for modelling this circulation [12]. The code is based on methods developed for ocean circulation, has a surface normalised grid and includes the effect of the Coriolis force. With the code we can investigate processes which occur on a dynamic timescale. We currently have a polytropic equation of state and do not solve the thermodynamic equation (although a future extension will). Although the present models are not fully realistic this simplified model will still allow a quantitative determination of the three-dimensional circulation systems. We have time on the UKAFF computing facility on which we are running the code.

THREE-DIMENSIONAL SIMULATIONS

We have modelled a secondary of $1\,M_\odot$ and $2\,R_\odot$ in a binary with a separation of $5\,R_\odot$ and an irradiating flux corresponding to the Eddington limit of a neutron star. Turbulent Reynolds stresses were included in the model and we found high turbulent viscosity coefficients were required in order to find a stable solution. We found that the circulation reaches a steady-state with the largest velocities on the directly irradiated region of order the sound speed. Figure 1 shows the directly irradiated face of the secondary and the steady-state circulation pattern. Figure 1 also shows the divergence of the velocity field where lighter shading indicates expansion and darker regions compression. Shadowing of the equatorial regions by an accretion disc was assumed. The circulation pattern goes completely around the secondary but Coriolis forces cause the circulation to go around the pole rather than over it.

Comparing this to the simulations of others (e.g., [13]) we see that the extent of the circulation current is the same. The magnitude of the circulation, however, is subsonic unlike the previous work. This is because the surface is free to adjust and so can deform to compensate. Only a small deformation is necessary in this case to make the circulation subsonic.

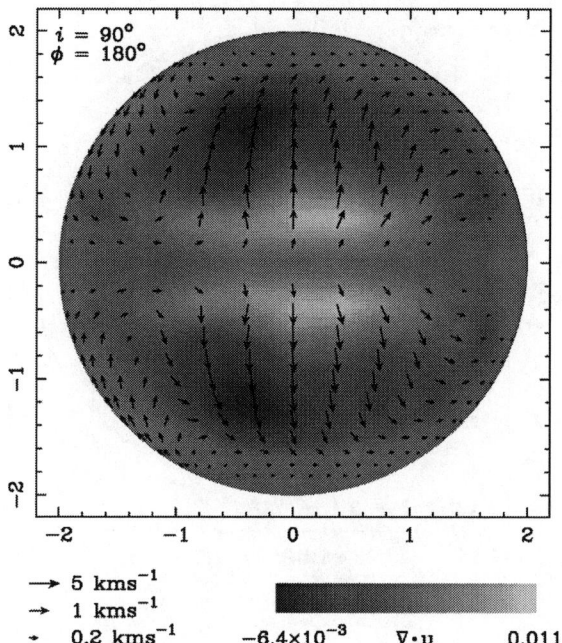

FIGURE 1. The divergence of the velocity on the surface of the directly irradiated portion of the secondary. Accretion disc shadowing of the equatorial regions has been assumed and the velocity is subsonic.

TEMPERATURE EFFECTS

In a two-dimensional simulation we have also considered the effect the circulation has on the temperature distribution compared to the unirradiated case. We calculate this by considering the advection of the heated material and allowing cooling in the non-directly irradiated regions. We find a significant portion of the backside of the secondary can be heated with the circulation penetrating deeply into the secondary (see Figure 2).

CONCLUSIONS

Irradiation dramatically changes the observable properties of close binaries. Understanding irradiation effects is crucial to the reliable determination of system parameters. Irradiation drives circulation currents in the upper layers of the secondaries. Understanding how much expansion of the secondary is induced by irradiation could lead to developments in the modelling of mass-transfer rates. In three-dimensional simulations the circulation pattern extends completely around the star although the velocity in the unilluminated hemisphere is small compared to the illuminated hemisphere. A high viscosity is required to allow a solution to be found. Thermodynamic effects need to be included to find the full solution.

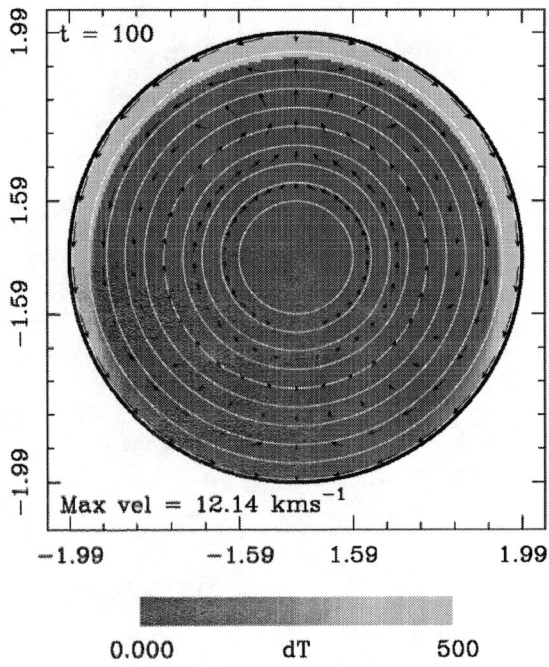

FIGURE 2. The heating of the surface layers in a two-dimensional simulation.

ACKNOWLEDGMENTS

MEB gratefully acknowledges a UK Astrophysical Fluids Facility (UKAFF) fellowship and the UKAFF facility on which some of the calculations reported here were performed.

REFERENCES

1. J. A. Orosz and C. D. Bailyn, *ApJ*, **477**, 876–896 (1997).
2. F. V. Hessman, E. L. Robinson, R. E. Nather and E.-H. Zhang, *ApJ*, **286**, 747–759 (1984).
3. S. N. Phillips, T. Shahbaz and Ph. Podsiadlowski, *MNRAS*, **304**, 839–844 (1999).
4. T. Shahbaz, P. Groot, S. N. Phillips, J. Casares, P. A. Charles and J. van Paradijs, *MNRAS*, **314**, 747–752 (2000).
5. M. Ruderman, J. Shaham and M. Tavani, *ApJ*, **336**, 507–518 (1989).
6. S. N. Phillips and Ph. Podsiadlowski, *MNRAS*, **337**, 431–444 (2002).
7. R. Kippenhahn and H.-C. Thomas, *A&A*, **75**, 281–290 (1979).
8. H. Kırbıyık, *MNRAS*, **200**, 907–923 (1982).
9. Ph. Podsiadlowski, *Nature*, **350**, 136–138 (1991).
10. H. Ritter, Z.-Y. Zhang and U. Kolb, *A&A*, **360**, 959–990 (2000).
11. J.-M. Hameury, *A&A*, **305**, 468–474 (1996).
12. M. E. Beer and Ph. Podsiadlowski, *MNRAS*, **335**, 358–368 (2002).
13. T. J. Martin and S. C. Davey, *MNRAS*, **275**, 31–42 (1995).

Magnetic Pumping in accretion disk coronae

R. Belmont* and M. Tagger*,†

*Service d'Astrophysique, CNRS FRE 2591, CEA Saclay, 91191 Gif-sur-Yvette, France
† Fédération de recherche AstroParticules et Cosmologie

Abstract. Most microquasar models involve a hot plasma in a corona above the disk or at the base of the jet. The Accretion-Ejection Instability (AEI) occurring in magnetized disks leads to the growth of a spiral density wave which can explain low frequency QPOs. It has already been shown to be very efficient in extracting accretion energy from the disk and emitting it upward in the corona as Alfvén waves. Here we present a simple mechanism which also allows the AEI to excite coronal ions. This heating is due to magnetic pumping, i.e. a resonant process occurring as the magnetic field lines emerging from the disk are periodically compressed by the spiral wave. We show how it acts on a collisionless population of ions, trapped above the disk by the joint action of gravity and magnetic stresses. We discuss the efficiency of this mechanism in heating coronal particles.

Keywords: Binary and multiple stars:X-ray binaries — Waves, oscillations, and instabilities in plasmas and intense beams: Macroinstabilities — Plasma properties: Plasma kinetic equations — Plasma dynamics and flow: Gyrokinetics

1. INTRODUCTION

Several different features seem to indicate that accretion disks of microquasars and AGNs are embedded in a gas of hot plasma called corona. The structure and heating mechanism of the corona remain unclear but many models assume that the corona can extend above the disk plane [10],[5]. So far, the main models do not include magnetic field at all [4],[14] or they have a magnetized, but very inhomogeneous corona, with magnetic loops [7],[11] where a large fraction of the gravitational energy is assumed to be released in the corona by reconnection events [8]. They have strong magnetic fields ($\beta \sim 1$) localized in small regions but the net vertical magnetic flux through the disk vanishes. We use here a different geometry, grounded on the results of jet models.

Microquasar jets are though to be very collimated. So far, the most favored models that can explain this property in a consistent way have been MHD models of jets [1],[13],[6]. Their results give some constraints on the magnetic field strength and topology:

- They require the magnetic field to be strong enough to redirect upward the matter radially accreting and to load it at the base of the jet. Namely, the magnetic pressure has to be in equipartition with the gas pressure: $\beta \sim 1$
- Then, in order to accelerate the jet and collimate it efficiently, they constrain the magnetic field to be large scaled and structured with a strong poloidal component, at least close to the disk. Contrary to the models mentioned above, the magnetic field varies on long time- and length scales.

In the two first sections, we present two interesting elements that can be derived from these conditions for the magnetic field. On one hand, coronal particles can be trapped in an oscillating motion along the poloidal field lines. And on the other hand, the Accretion-Ejection Instability (AEI) develops in the disk and leads to the growth of a spiral MHD wave. Then we show that a resonant process can occur between the oscillating motion and the periodic perturbation by the spiral wave. This mechanism, well known in plasma physics, is called magnetic pumping. In last section, we discuss the efficiency of magnetic pumping to heat coronal ions.

2. MODEL

The corona density is supposed to be very low, therefore the time between two collisions might be long in comparison with other time scales. This means that kinetic effects have to be included in the analysis. In the kinetic approach the usual fluid quantities as density, velocity, pressure, temperature... have to be replaced by the distribution function $\mathscr{F}(\vec{x},\vec{v},t)$ that gives the number of particles at time t and position $\vec{x}$ with the velocity $\vec{v}$. Without collision, the evolution of the system is then fully determined by the Vlasov equation which stands for the usual fluid ones (continuity, Euler, energy...) and reads:

$$\partial_t \mathscr{F} + \vec{v}.\vec{\nabla}\mathscr{F} + \Gamma.\partial_{\vec{v}}\mathscr{F} = 0$$

where Γ is the total specific force applied to the particles. This equation cannot be analytically solved in the most general case, and further assumptions have to be made. One is to consider that all time scales of the problem are longer than the cyclotron period. In this limit, it is known that the magnetic moment $\mu = v_\perp^2/2B$ is conserved and that particles can be represented by their guiding center. The parallel motion of the guiding center is then governed by the classical forces, but to keep information about the cyclotron motion and its effects, we have to add a new parallel force called the mirror force: $F = \mu \partial_\parallel B$. The mirror force is thus a repulsive force that pushes charged particles away from strongly magnetized regions. This force is a kinetic effect and has not been used in previous models.

We use here a pertubation approach and consider the MHD wave of the AEI as the small perturbation added to the equilibrium described in introduction.

2.1. Equilibrium Periodic Motions

The first step in this perturbation method is to determine the equilibrium properties. To simplify, let us assume that the magnetic field lines are straight, oblique and purely poloidal. In this geometry, the magnetic field decreases with radius and altitude and the mirror force has to be taken into account. As the disk itself is much denser than the corona, the effect of coronal currents on the field line topology can be neglected. The magnetic field is imposed by the disk, rotates with it and coronal particles have to move along the field lines. It means that all the perpendicular forces act together to keep particles on their field line and we can consider the particles as beads that can freely move along them [9]. The parallel motion is then governed by the projections of gravity, centrifugal force and mirror force. This analysis leads to two different particle behaviors with respect to θ the angle of the field lines with the vertical axis. Two domains are found, separated by a critical angle $\theta_c = 30^o$:

- for $\theta > \theta_c$, the centrifugal force is stronger than gravity and the mirror force helps particles to escape. As a consequence, all particles are ejected as in the MHD case.
- For $\theta < \theta_c$, the mirror force acts against gravity and we find that some coronal particles can oscillate around an equilibrium position above the disk. If their magnetic energy is strong, this position can be high enough so that the particles freely oscillate without crossing the disk where they would otherwise collide dense matter. Particles can thus be trapped in a so called periodic bounce motion.

We will assume that the Vlasov equation, coupled with Maxwell equations and possibly radiative transfer equations has been solved for the equilibrium, giving a consistent solution for all fluid profiles (density, temperature...) and a periodic kinetic behavior of trapped particles. We will now focus on the effect of the perturbation on the equilibrium.

2.2. Perturbation: Periodic excitation

In the conditions described previously for magnetic topology and strength, namely a poloidal magnetic field in equipartition with the gas pressure, we know that the AEI can develop in the disk and perturb the equilibrium described in the last section. The AEI leads to the growth of a MHD wave which rotates through the disk. It was worked out by Tagger and Pellat [15] for a disk in vacuum and is now one of the best candidates to explain the low frequency QPOs [12]. This MHD wave couples the hydrodynamic and magnetic properties of the fluid and appears as a spiral wave, both in density and magnetic intensity: where the matter is denser, the magnetic field is stronger. Numerical simulations show that, in the disk, the perturbed magnetic field is about $b/B \sim 0.1$ [3]. This perturbation extends far above the disk and can therefore affect the behavior of coronal particles. For radii smaller than the corotation radius, the spiral wave rotates more slowly than the gas. In the frame moving with the gas, coronal particles experience a periodic compression of the field lines and therefore a periodic perturbation of the mirror force with the frequency $\tilde{\omega}$.

2.3. Magnetic Pumping

The latter periodic force excites a periodic equilibrium motion. Where the frequencies are equal: $\omega_b \sim \tilde{\omega}$, a resonance occurs. When the bounce motion is due to the mirror force, this resonance is the magnetic equivalent of Landau damping and is called transit-time damping or magnetic pumping. Indeed it pumps energy from the magnetic perturbation to the excited particles. The first result is that the electrons oscillate much too fast to interact with the wave. As $\omega_B^{e-} >> \tilde{\omega}$, there is no resonance for the electrons and they cannot be heated directly by magnetic pumping. On the contrary, coronal ions move slowly enough for the pumping to become efficient and find the heating time scale:

$$\Omega_K \tau = \left(\frac{\omega_B}{\Omega_K}\right)^2 \left(\frac{h_c}{r}\right)^{-2} \left(\frac{b_0}{B^0}\right)^{-2}$$

where h_c is the corona height and b_0 is the magnetic perturbation in the disk. The reference time is the keplerian period and we see that the stronger the perturbation, and the hotter the corona, the quicker the heating and so the more efficient the pumping.

3. DISCUSSION AND CONCLUSION

Going further and giving reliable figures for a consistent coronal model is a very difficult task. The corona is indeed very poorly constrained in the literature. Our goal is not to get global results yet. This would indeed need a full model with a precise description of many other features as cooling mechanisms... Here, we rather do a first attempt to estimate the typical heating time scale understand the significance of magnetic pumping.

For typical microquasars, we have $b/B \sim 0.1$, $T = 50 keV$, $h_c/r \sim .1$. The frequency ratio varies between 0 and and 1 respectively for vertical and critical field lines. This means that for critical lines, the pumping can be infinitely efficient. By choosing an angle of 25°, we find $\tau \sim 1s$. This time scale is comparable with the time for ions to collision the electrons and give their energy, but it is much smaller than the Compton cooling time. Some assumptions made to simplify the analytic resolution could increase the efficiency, when relaxed. This has to be checked, but this kinetic heating is likely to be unable to fight the Compton cooling.

Nevertheless, we can mention that if coronal ions are not collisional, the electron collision time is about the rotation time. As a consequence the kinetic approach becomes less valid for coronal electrons. Such a plasma, between the collisional and the non collisional regime, is known to experience the Braginskii viscosity [2]. For microquasar coronae, this viscosity is quite strong and could result in a high dissipation rate and thus a corona heating. This will be discussed in a future publication.

REFERENCES

1. R. Blandford, and D. G. Payne, *MNRAS*, **199**, 883 (1982)
2. S. I. Braginskii, *Rev. Plasma Phys.*, **1**, 205 (1965)
3. S. E. Caunt, and M. Tagger, *A&A*, **367**, 1095C (2001)
4. D. K. Chakrabarti, L. Titarchuk, D. Kasanas, and K. Ebisawa, *APSS*, **120**, 163 (1996)
5. M. J. Church, and M. Baluncińska-Church, *MNRAS*, **348**, 955 (2004)
6. J. Ferreira, *A&A*, **319**, 340 (1997)
7. A. A. Galeev, R. Rossner, and G. S. Vaiana, *ApJ*, **229**, 318 (1979)
8. F. Haardt, and L. Maraschi, *ApJ*, **380**, L51 (1991)
9. R. N. Henriksen, and D. R. Rayburn, *MNRAS*, **152**, 323 (1971)
10. J. Malzac, A. M. Beloborodov, and J. Poutanen, *MNRAS*, **326**, 417 (2001)
11. A. Merloni, and A. C. Fabian, *MNRAS*, **332**, 165 (2002)
12. J. Rodrigùez, P. Varnière, M. Tagger, and Ph. Durouchoux, *A&A*, **387**, 487R (2002)
13. G. Pelletier, and L. G. Pudritz, *ApJ*, **484**, 794 (1980)
14. A. Różańska, and B. Czerny, *Acta Astron.*, **46**, 223 (1996)
15. M. Tagger, and R. Pellat, *A&A*, **349**, 1003 (1999)

High Resolution Spectroscopy of 4U 1728-34 from a Simultaneous Chandra-RXTE Observation

A. D'Aí*, R. Iaria*, T. Di Salvo*, G. Lavagetto*, N. R. Robba*, L. Burderi[†], M. Mendez** and M. van der Klis[‡]

*Dipartimento di Scienze Fisiche ed Astronomiche, Università di Palermo, via Archirafi n.36, 90123 Palermo, Italy.
[†]Osservatorio Astronomico di Roma, Via Frascati 33, 00040 Monteporzio Catone (Roma), Italy.
**SRON National Institute for Space Research, Sorbonnelaan 2, 3584 CA Utrecht, the Netherlands
[‡]Astronomical Institute "Anton Pannekoek," University of Amsterdam and Center for High-Energy Astrophysics, Kruislaan 403, NL 1098 SJ Amsterdam, the Netherlands

Abstract. We report on a simultaneous Chandra and RossiXTE observation of the LMXB atoll bursting source 4U 1728-34 between 2002 March 3rd and 2002 March 5th. We fitted the 1.2-35 keV continuum with a blackbody spectrum plus a Comptonized component. Absorption edges associated with Fe I and Fe XXV at ~ 7.1 keV and ~ 9 keV respectively, are present in our spectrum. Moreover, an overabundance of Si is highly required for a satisfactory fit. We found no evidence in our fits of broad, or narrow Fe Kα lines, between 6 and 7 keV, contrarily to what previous spectral studies pointed out. We tested our study by reanalysing a previous BeppoSAX observation of 4U 1728-34, finding a general agreement with our new spectral model.

Keywords: accretion discs – stars: individual: 4U 1728-34 — stars: neutron stars — X-ray: stars — X-ray: spectrum — X-ray: general
PACS: 97.80.Jp

INTRODUCTION

4U 1728-34 is a well known prototype of the class of the bursting atoll sources ([4]). This was one of the first sources to display kHz QPO in its power spectrum (PSD) and the first one to display the burst oscillations, around 363 Hz, during some type I X-ray bursts ([7]). Its temporal behaviour has been recently extensively studied ([2], [10]) using a large set of RXTE observations, spanning more than three years. Spectral studies have been carried out in the past with the use of EXOSAT ([11]), RXTE ([6]) and BeppoSAX ([1], [6]) satellites; although the mechanism underneath the continuum emission was quite clear, it was disputed the nature of local features, especially a broad gaussian line at 6.2-6.7 keV. Its distance is estimated between 4.1 kpc and 5.1 kpc ([3]; [1]).

SPECTRAL ANALYSIS AND DISCUSSION

4U 1728-34 was observed by Chandra on 2002 March 4 for a total collecting time of ~ 30 ksec. We used the Chandra High Energy Transmission Grating Spectometer

(HETS) to perform a high resolution spectroscopic analysis. The data were collected in the Timed Exposure Mode, and a subarray was adopted (q = 1 and n = 400, see the Chandra Proposer's Observatory Guide at http://cxc.harvard.edu/proposer/POG/) in order to mitigate the effects of photon pile-up for first order spectra. Consequently the frame time was 1.44 s, the High Energy Grating (HEG) spectrum is cut below 1.6 keV and the Medium Energy Grating (MEG) spectrum below 1.2 keV. No systematic error was added to the data.

The RXTE observation started on 2002 March 3 03:27:12 and ended on 2002 March 5 13:00:00. For the spectral analysis we used only the Proportional Counter Unit 2 and 3 data in Standard2 configuration (with 16 s time resolution and 129 energy channels).

Seven type-I bursts were revealed in the PCA lightcurve and two bursts in the Chandra lightcurve. As our primary concern is to focus on the continuous emission of the source we discarded data around each burst, for a time length of 160 sec.

As concerns the Chandra HETG data, we considered the four first-order dispersed spectra, namely the two HEG spectra and the two MEG spectra on the opposite sides of the zeroth order. We averaged HEG+1 (MEG+1) and HEG-1 (MEG-1) spectra in a single spectrum, after having tested their reciprocal consistency. The used energy range is 1.6-10 keV for the HEG spectrum and 1.2-5 keV for the MEG spectrum. For all the fits we took into account an instrumental feature at 2.07 keV for bright sources (described by [5]) and fit it with an inverse edge (with optical depth $\tau \simeq -0.1$). The HEG and MEG spectra were binned in order to have at least 300 counts for each bin. This, however, still ensures a high number of channels (about 1000) and good spectral resolution throughout the entire covered energy band.

Concerning the RXTE PCA data the standard selection criteria for obtaining the Good Time Intervals were applied. We restricted the spectral analysis to the temporal interval during which RXTE operated simultaneously with Chandra. We limited the energy range to 3.5-35 keV, and applied a 2% systematic error for channels below 25 keV and 2.5% for channels above 25 keV.

Relative normalizations of the three instruments, except for HEG which was fixed to a reference value of 1, were left as free parameters in all the fits performed.

We tried a series of models to simultaneously fit HEG, MEG and RXTE spectra. We found the best-fit model to consist of a soft emission, described by a blackbody of temperature $\simeq 0.52$ keV, plus a Comptonized component (CompTT in XSPEC, [8]), where the seed photon temperature kT_0 is $\simeq 1.3$ keV, the electron temperature kT_e is $\simeq 7.4$ keV and, finally, the optical depth τ associated to the spherical corona is $\simeq 6.2$. Both components are photoelectrically absorbed by an equivalent hydrogen column $N_H \simeq 2.3 \times 10^{22}$ cm^{-2}. The associated χ^2/dof abtained for this fit is 1235/1034. We noted an absorption edge, probably associated to neutral Si around 1.84 keV in the MEG and HEG spectra. To fit this edge we substituted the component *phabs* in XSPEC with the component *vphabs*, which allows us to vary the abundances of single elements with respect to the solar abundances. We found that leaving the Si abundance free improves the fits significantly; Si resulted overabundant by a factor ~ 2 with respect to the solar abundance (χ^2/dof value obtained for this fit is 1066/1031). We did not find any evidence of broad or narrow Fe Kα emission lines in the 5-7 keV energy band. However, we found two absorption edges at 7.03 keV ($\tau \simeq 0.11$) and at 9.0 keV ($\tau \simeq 0.16$), respectively. The addition of these edges improves the fit significantly compared to the simple model described above (from 1066/1031 to 974/1027 χ^2/dof). To test the consistency of our model we reanalyzed a previous BeppoSAX observation performed between August 23 and 24, 1998 (see [1]). A broad gaussian line is no longer statistically required if we introduce two absorption edges at energies above 7 keV. We found a first edge at $\simeq 7.4$ keV ($\tau \simeq 0.08$) and a second edge at $\simeq 8.7$ keV ($\tau \simeq 0.06$). We also found evidence in the BeppoSAX spectrum of another edge at lower energies, which we identified as a Kα edge of Neon IX at 1.196 keV. We introduced this feature also in the Chandra-RXTE model (fixing the energy and the optical depth to the BeppoSAX values). We found a general agreement with the Chandra-RXTE spectrum, obtaining χ^2/dof = 218/178 (instead of χ^2/dof = 236/178 obtained for the model adopted in [1]). In table 1 we present the results of the two fits of the BeppoSAX and Chandra-RXTE datasets for comparison.

We confirm that the best-fit continuum model for this source is composed of two components: a blackbody emission which probably comes from the inner edge of an accretion disk around the compact object, and a Comptonized component comes from a hot corona surrounding the system. The simultaneous Chandra/RXTE spectrum allows us to definitively exclude a scenario, previously hypothesized, where a broad gaussian line (with width between 0.5 keV and 1 keV) was present. We derived, however, an upper limit to the flux of an emission line, setting the values of energy and width to the values found in [1], which is 1.7×10^{-11} erg cm^{-2} sec^{-1} (corresponding to an equivalent width of 28 eV). On the other hand the iron line region is well fitted by two absorption edges at energies at ~ 7 and ~ 9 keV, associated with Fe I and Fe XXV, respectively (see e.g.[9]); we checked that this model is also in agreement with a previous BeppoSAX observation. We observed a shift in the energy of the Fe I edge from 7.1 keV during the Chandra-RXTE observation to 7.4 keV (compatible with Kα edges of moderately ionized iron, Fe IX to Fe XVI) during the BeppoSAX observation, while the energy of the Fe XXV edge is consistent with being the same in both observations. Finally we found, both in the Chandra-RXTE and BeppoSAX spectra, a Si overabundance by a factor $\sim 2-2.5$ with respect to the solar abundance and the values obtained from the fit are consistently in agreement.

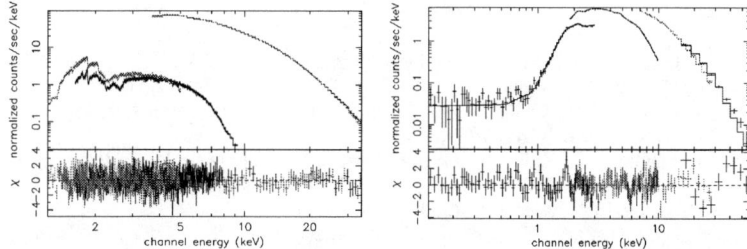

FIGURE 1. Spectra of 4U 1728-34 shown together with the best-fit model (see Table 1). Left panel: 1.2–35.0 keV spectrum obtained from the simultaneous Chandra and RXTE dataset. Right panel: 0.12–60 keV spectrum obtained from the fit of the BeppoSAX dataset. In smaller panels: residuals in unit of σ with respect to the best-fit model.

TABLE 1. Best fit parameters for the two datasets of 4U 1728-34, obtained from a Beppo SAX observation (0.12-60 keV energy band) and a joint CHANDRA-RXTE observation (1.2-35 keV energy band). The photoelectric absorption is indicated as N_H. The continuum emission consists of a thermal blackbody emission (bbody) and a Comptonized spectrum modeled by compTT. kT_{BB} and N_{BB} are, respectively, the blackbody temperature and normalization in units of L_{39}/D_{10}^2, where L_{39} is the luminosity in units of 10^{39} ergs/s and D_{10} is the distance in units of 10 kpc. kT_0, kT_e and τ indicate the seed-photon temperature, the electron temperature and the optical depth of the Comptonizing cloud around the neutron star. N_{Comptt} is the normalization of the Comptt model in XSPEC v.11.2.0 units. Unabsorbed luminosities of the bbody component and of the CompTT component are reported assuming a distance from the source of 5.1 kpc ([1]). For the component Edge, E_{edge} denotes the position of the edge and τ the optical depth. Uncertainties are at 90% confidence level for a single parameter.

Component	Parameter (Unity)	BeppoSAX	Chandra - RXTE
		Values	
vpha	N_H (10^{22} cm^{-2})	$2.17^{+0.14}_{-0.11}$	$2.31^{+0.06}_{-0.07}$
vpha	Si (solar units)	$2.5^{+0.5}_{-0.5}$	$2.24^{+0.13}_{-0.13}$
edge	E_{edge} (keV)	$1.196(fixed)$	$1.196(fixed)$
edge	τ	$0.47^{+0.24}_{-0.3}$	$0.47(fixed)$
edge	E_{edge} (keV)	$7.41^{+0.15}_{-0.14}$	$7.03^{+0.08}_{-0.06}$
edge	τ (10^{-2})	8^{+2}_{-2}	11^{+3}_{-4}
edge	E_{edge} (keV)	$8.73^{+0.26}_{-0.24}$	$9.0^{+0.3}_{-0.4}$
edge	τ (10^{-2})	6^{+3}_{-3}	16^{+4}_{-4}
bbody	kT (keV)	$0.573^{+0.002}_{-0.024}$	$0.516^{+0.002}_{-0.014}$
bbody	N_{BB} (10^{-3})	$21.5^{+1.6}_{-1.3}$	$10.3^{+0.4}_{-0.4}$
bbody	Luminosity (10^{36} erg cm^{-2} sec^{-1})	5.7	0.87
CompTT	kT_0 (keV)	$1.53^{+0.06}_{-0.07}$	$1.33^{+0.05}_{-0.05}$
CompTT	kT_e (keV)	$6.4^{+1.3}_{-0.4}$	$7.4^{+0.5}_{-0.4}$
CompTT	τ	$4.8^{+0.8}_{-1.0}$	$6.2^{+0.4}_{-0.6}$
CompTT	N_{CompTT} (10^{-2})	$6.8^{+1.3}_{-1.4}$	$4.4^{+0.3}_{-0.6}$
CompTT	Luminosity (10^{36} erg cm^{-2} sec^{-1})	10.2	2.6
χ^2/dof		218/178	975/1029

REFERENCES

1. Di Salvo T., Iaria R., Burderi L., Robba N.R., 2000, ApJ, 542, 1034
2. Di Salvo T., Mendez M., van der Klis M., Ford E., Robba N.R., 2001, ApJ, 546, 1107
3. Galloway D.K., Psaltis D., Chakrabarty D., Muno M.P., 2003, ApJ, 590, 999
4. Hasinger, G., & van der Klis, M., 1989, A&A, 225, 79
5. Miller J.M., Fabian A.C., Wijnands R., Remillard R.A., Wojdowski P., Schulz N.S., Di Matteo T., Marshall H.L., Canizares C.R., Pooley D., Lewin W.H.G., 2002, ApJ, 578, 450
6. Piraino S., Santangelo A., Kaaret P., 2000, A&A, 360, L35
7. Strohmayer T. E., Zhang W., Swank J. H., Smale A., Titarchuck L., Day C., Lee U., 1996, ApJ, 469, L9
8. Titarchuck L., 1994, ApJ, 434, 570
9. Turner T.J., Done C., Mushotzky R., Madejski G., Kunieda H., 1992, ApJ, 391, 102
10. van Straaten S., van der Klis M., Di Salvo T., Belloni T., 2002, ApJ, 568, 912
11. White N.E., Peacock A., Hasinger G., Mason K. O., Manzo G., Taylor B.G., Branduardi-Raymont G., 1986, MNRAS, 218, 129

Single Stars and Supernovae from Wolf-Rayet Secondaries

Lynnette Dray* and Christopher Tout[†]

Department of Physics and Astronomy, University of Leicester, LE1 7RH, UK
[†]*Institute of Astronomy, Madingley Road, Cambridge, CB3 0HA, UK*

Abstract. We investigate the population of single Wolf-Rayet (WR) stars in the Milky Way and Magellanic Clouds which may have resulted from massive binary evolution. Observationally, the binary fraction amongst Wolf-Rayet stars is much smaller than that of O stars, their direct progenitors, suggesting that many single WR stars might once have been part of a binary system. We present a grid of stellar models aimed at studying the evolution of the secondary in the case that the supernova explosion of the primary unbinds the binary to see if a suitable population may be produced in this manner. Whilst transferring a significant amount of mass between the stars in such a system is difficult to achieve – a contact binary often resulting instead – the accretion of He-enriched matter onto those secondaries which avoid contact changes their subsequent evolution in a more complex manner than simple rejuvenation and increases the liklihood that they will subsequently undergo a WR phase. If the initial binary fraction is high, a significant population of single WR stars and type Ib/c supernovae could be produced by this seemingly unlikely route.

INTRODUCTION

Wolf-Rayet stars are the hydrogen-depleted descendants of massive O stars, stripped first of their H envelopes by mass loss (WNL and WNE stars) and, if the mass loss continues, eventually ending up as bare CO cores (WC stars) which undergo type Ib/c SNe. Whilst an extra source of mass removal, such as Roche lobe overflow (RLOF) in a binary system, is one way of making a WR star, the observed populations at solar metallicity can be accounted for by stellar winds alone [10]. However at lower metallicities the line-driven wind decreases substantially. RLOF is little-affected by metallicity so the number of WR stars which are formed by binary interaction should remain roughly level with Z and the proportion of initially-single WR stars decrease [9].

This scenario is not borne out by observation. Measured binary fractions in the SMC, LMC and Milky Way are around 40 per cent [6, 14] or less [1] and show no strong metallicity trend. Furthermore, the Galactic O star initial binary fraction is much larger than the WR binary fraction, potentially more than 75 per cent [11]. As O stars are the progenitors of WR stars and mass transfer enhances the possibility of a star entering the WR phase, this suggests that there may be a population of single WR stars which were originally in binaries. Possible routes are the unbinding of binaries by SN kicks or dynamical interactions, or by collision- or contact- induced mergers. The models presented here explore the types of single stars expected from the former route.

SIMULATIONS

Simulations were carried out using an up-to-date version of the Eggleton stellar evolution code [3, 4, 12], including enhanced CO opacities [5] and time-dependent thermohaline mixing. Empirical mass-loss rates were used as by Dray & Tout [2, NL models] with the O star rates of Vink, de Koter & Lamers [15] included when appropriate. For AM loss in the wind and Bondi-Hoyle accretion we used prescriptions from Hurley, Tout & Pols [8]. Rotation is not currently considered, although the empirical mass loss rates naturally include an average amount of rotationally-enhanced mass-loss.

We evolved a core 5 by 7 by 20 grid in q (0.5 – 0.9), P (2 – 200 days) and M_1 (10 – 60$M_\odot$) for a range of metallicities ($Z = Z_\odot$, $Z_\odot/2$, $Z_\odot/5$). As the amount of matter which may be accreted during RLOF is a subject of some debate, as is the size of the kick imparted by a supernova in a binary system, we also consider the cases of conservative RLOF/RLOF in which only half of the matter is accreted and no/large SN kick (where 'large' is enough to unbind the system). For the non-conservative models we extend the q range down to 0.3.

SINGLE STARS AND THE AVOIDANCE OF CONTACT

For the models presented here, accretion of He-enriched matter from the primary significantly lowers the mass limit for the secondary to enter all the WR phases. At $Z = 0.004$ with the mass-loss rates given above a WC phase occurs for single stars with initial mass above $38M_\odot$. However, WC secondaries may be formed by systems whose initial *combined* mass is less than this value, even in the case of non-conservative evolution. If SN kicks are on the whole large, an observable population of 'single' WR secondaries is therefore favoured by the IMF at low Z. These stars could be identifiable by their relatively low masses in comparison with stars of single origin, runaway velocity status and probable rapid rotation.

However, many massive close binaries evolve into contact [16]. Whilst mass loss changes the pre-RLOF parameters in a way which may influence the subsequent outcome (M_1 decreases, q and P increase) the number of systems which escape contact is still small (Fig. 1). The subsequent evolution is uncertain, but probably involves mass being lost rather than transferred to the secondary, significantly reducing the chance of it undergoing a WR phase. A merger is also a possibility.

For an initial parameter distribution $N_p \propto 1/P$, $N_q \propto 1$, 26% of conservative binaries with a $10M_\odot$ primary and $P_i \leq 100$d at solar metallicity avoid contact, decreasing to 13% for a $15M_\odot$ primary. In this case merger systems are likely to play a significant role in the population statistics [eg. 13].

SUPERNOVAE

At low metallicity only the most massive single stars end their lives in a helium-depleted state. Many may then undergo direct collapse to a black hole rather than SNe Ic [7]. If

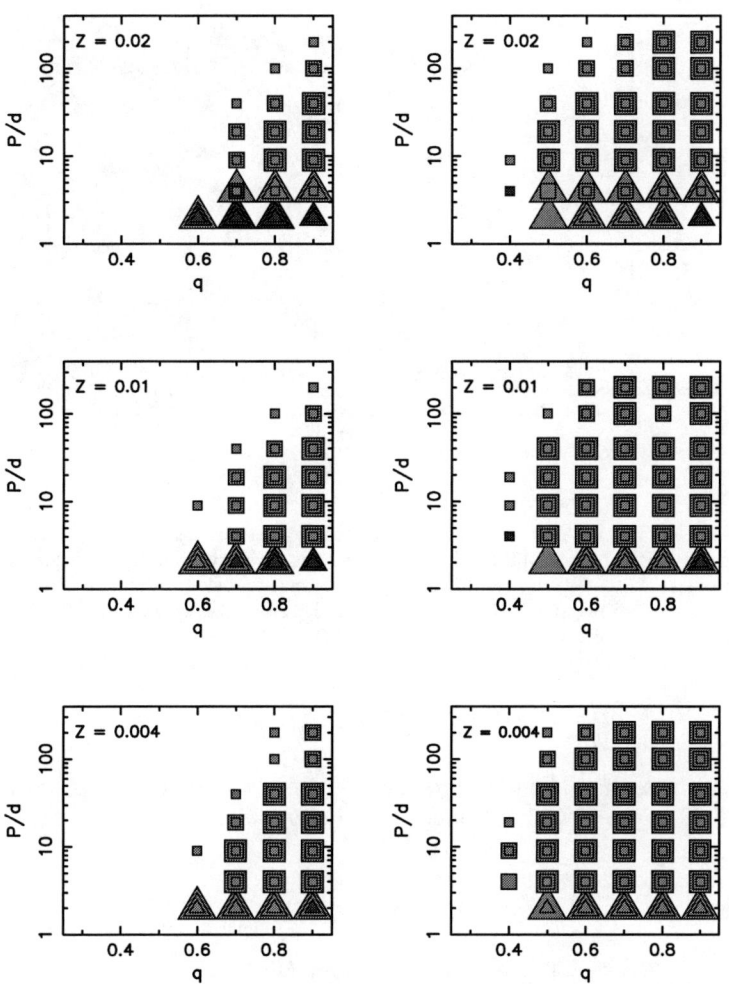

FIGURE 1. Eventual fate of some systems with left: conservative and right: non-conservative RLOF. Only systems which avoid contact are shown. Dark and light grey points illustrate respectively systems with and without reverse mass transfer and triangles and squares case A and case B systems. 10, 11 and 12 solar mass primaries are indicated by points of increasing size. It can be seen that the contact-free parameter space decreases significantly with a small increase in primary mass over this mass range.

WC stars are made from lower-mass stars by binary interaction, they will be represented disproportionately in the SN Ic statistics. At $Z = 0.004$, for a threshold He core mass for direct BH formation of $15M_\odot$ [7] and a Salpeter IMF, over half the single stars which end their lives with exposed or nearly-exposed CO cores are still massive enough to undergo direct collapse rather than SN Ic. The lower-mass population of WR secondaries is little-affected by this problem (Fig. 2). Another population of SNe Ic may result from mergers and a third from the minority of WR primaries which have low final surface He.

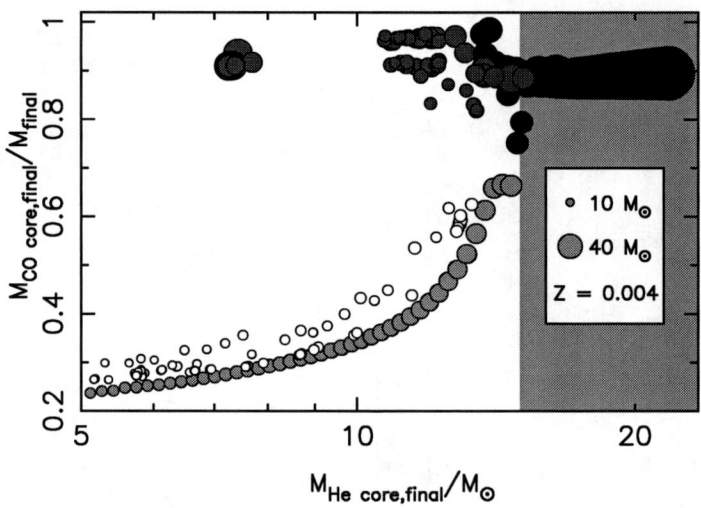

FIGURE 2. Final He core mass (for WC stars total mass) against closeness-to-surface of CO core for single stars (non-WR:light grey, WR:black) and secondaries of massive binaries with conservative RLOF which escape contact (non-WR:white, WR:dark grey). Points are scaled with initial mass (initial primary mass for binaries). The grey area indicates the region in He core mass where direct BH formation rather than a SN may occur.

ACKNOWLEDGMENTS

LMD thanks Peter Eggleton for providing and explaining his code, and PPARC for funding. CAT thanks Churchill College for a Fellowship.

REFERENCES

1. Bartzakos, P., Moffat, A. F. J., & Niemela, V. S. 2001, MNRAS, 324, 33
2. Dray, L. M., & Tout, C. A. 2003, MNRAS, 341, 299
3. Eggleton, P. P. 1971, MNRAS, 151, 351
4. Eggleton, P. P., & Kiseleva-Eggleton, L. 2002, ApJ, 575, 461
5. Eldridge, J. J, & Tout, C. A. 2004, MNRAS, 348, 201
6. Foellmi, C., Moffat, A. F. J., & Guerrero, M. A. 2003, MNRAS, 338, 1025
7. Fryer, C. L. 1999, AJ, 522, 413
8. Hurley, J. R., Tout, C. A., & Pols, O. R. 2002, MNRAS, 329, 897
9. Maeder, A., & Meynet, G. 1994, A&A, 287, 803
10. Maeder, A., & Meynet, G. 2002, A&A, 361, 101
11. Mason, B. D., Gies, D. R., Hartkopf, W. I., Bagnuolo, W. G. Jr., ten Brummelaar, T., & McAlister, H. A. 1998, AJ, 115, 821
12. Pols, O. R., Schröder, K.-P., Hurley, J. R., Tout, C. A., & Eggleton, P. P. 1998, MNRAS, 298, 525.
13. Vanbeveren, D., Van Bever, J., & De Donder, E. 1997, A&A, 317, 487
14. van der Hucht, K. A. 2001, New Astronomy Reviews, 45, 135.
15. Vink, J. S., de Koter, A., & Lamers, H. J. G. L. M. 2001, A&A, 369, 574
16. Wellstein, S. Langer, N., & Braun, H. 2001, A&A, 369, 939

The early spectral evolution of nova Sgr 2004

A. Ederoclite[*,†], E.Mason[*], M.Della Valle[**], R. Gilmozzi[*] and R.E.Williams[‡]

[*]*European Southern Observatory, Alonso de Cordova 3107, Vitacura, Santiago (Chile)*
[†]*Dip Astronomia – Universitá di Trieste, Via Tiepolo 11, Trieste (Italia)*
[**]*INAF - Osservatorio di Arcetri, Largo E.Fermi 5, Firenze, (Italia)*
[‡]*Space Telescope Science Institute, 3700, San Martin Drive, Baltimore, MD, USA*

Abstract.
We secured high resolution spectra of nova Sgr 2004 during its early decline. The spectra were taken within three months after the discovery of the nova with FEROS at the 2.2m at the ESO-La Silla Observatory. Here, we present preliminary analysis and results.

Keywords: Novae: Sgr2004 – Spectroscopy
PACS: 95.55.Qf, 95.75.Fg, 97.30.Qt, 97.80.Gm

INTRODUCTION

Nova Sgr 2004 was independently discovered on Mar 17.3 UT by Nishimura and Liller [8]. Spectroscopic confirmation was given on the following night by Della Valle et al. [3]. The nova had coordinates R.A. = $18^h19^m32^s.29$, Dec. = $-28°36'35".7$ (gal. coord. $l = 3°56'34".7$ $b = -6°18'43".4$). No object brighter than R = 17.5 mag was present at the nova position in the red frame of the Digital Sky Survey [8].

LIGHT CURVE

We derived a light curve (see Fig. 1) from the V broad band observations reported in the literature ([8],[3],[12], and [10]). The observed maximum (V=8.0 mag) occurred on Mar 17.19 UT.

From the light curve we estimate a t_2 and a t_3 of 10 and 18 days, respectively, which characterize nova Sgr 2004 as a fast nova (see [9] and [6]).

SPECTROSCOPY

Spectroscopic observations have been carried out during the Classical Novae Target of Opportunity campaign ongoing at the 2.2m+FEROS (Fiber-fed Extended Range Optical Spectrograph) at the ESO - La Silla Observatory. The spectra cover the wavelength range $\lambda\lambda 3800-9000$ with a resolution R=48000. They have been reduced with the FEROS Data Reduction System under MIDAS and flux-calibrated and analyzed with IRAF. We collected 7 spectra over a 3 months period from maximum light to about V $\sim$ 14 mag.

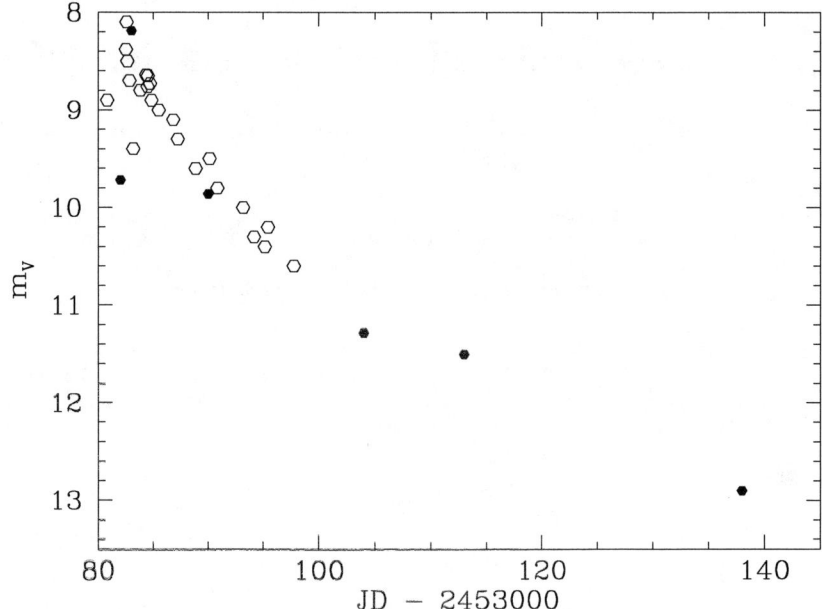

FIGURE 1. V-band light curve of nova Sgr 2004. Empty symbols are the data as from IAUCs and filled symbols are our spectra convolved with a V-filter.

The spectrum taken at maximum brightness (see Fig. 2, top panel) clearly shows the presence of 2 absorption systems: the "Principal" and the "Diffuse enhanced"([9]), which have velocities of -1400 Km/s and -850 Km/s, respectively. The spectrum is dominated by strong FeII (multiplets 42, 49, 55, and 74), OI (34 and 1) and CaII (2) emission lines. This characterizes the nova as a typical "FeII" type object according to the Cerro Tololo classification ([14] and [15]).

One month after maximum light (see spectrum in the mid panel of Fig. 2) the P-Cyg profiles have disappeared together with the FeII lines. The nova has entered the auroral phase, according to [14] and [15]. Forbidden (e.g. [NIII], [OI], [OIII]) and high excitation (e.g. NII and HeI) lines were arising. The OIλ8446 reached its maximum intensity. The FWHM of the Balmer lines has increased to the value of $\sim$2100 Km/s (about twice the value observed at the maximum light).

Two months after maximum (bottom panel of Fig. 2), high ionization elements have completely replaced the low ionization ones and dominate the spectrum. The strongest non Balmer permitted emission is NIII (the 4640 blend is now clearly resolved in NIII λ 4640 and HeII λ 4686); while the strongest forbidden transition is the nebular doublet [OIII]$\lambda\lambda$4959,5007. OIλ8446 and λ7775 have almost disappeared.

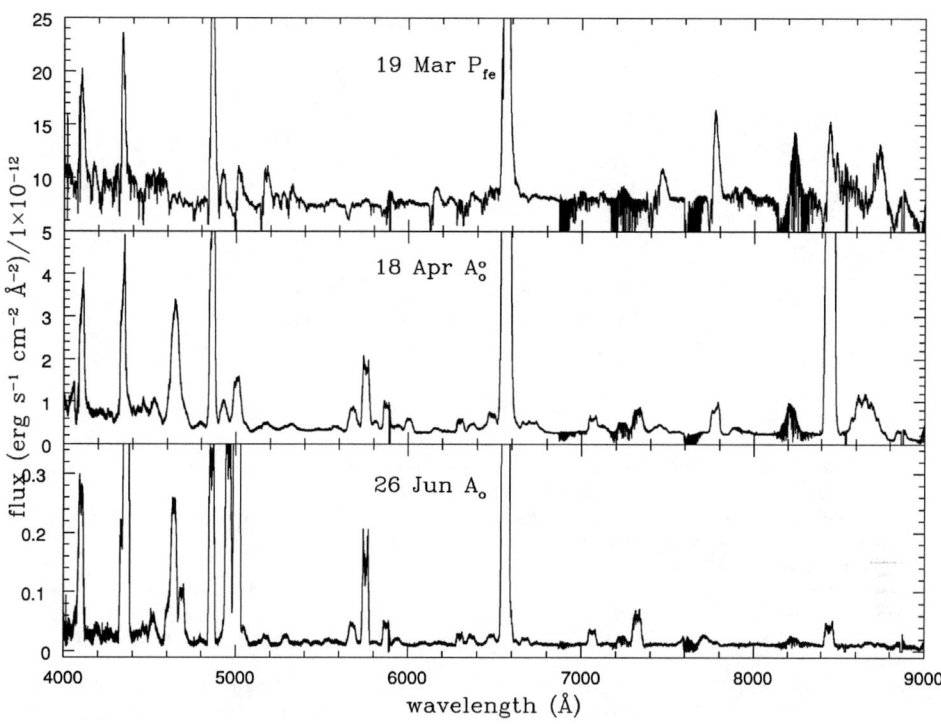

FIGURE 2. Significant spectra: Mar19 (almost at maximum), Apr17 (one month after maximum) and 26Jun (in the auroral phase).

REDDENING AND DISTANCE

Distance determination is strongly dependent on the interstellar absorption. The (B-V) color of the nova at maximum is ~ 0.66. This leads to $E(B-V) = 0.43$ (see [11]). We also used the interstellar K1 $\lambda 7699$ line (see [7]) obtaining $E(B-V) = 0.45$. Hence, averaging the two values for the $E(B-V)$ given above and assuming R=3.2, we derived $A_V = 1.41$. Using this value in the Maximum Magnitude vs. Rate of Decline (MMRD) relation by [5] and in the revised Buscombe - De Vaucouleurs relation [2] we determine the distance of the nova to be $d = 11.4$ kpc and 8.3 kpc, respectively. We assume an average distance of 9.9 kpc, which implies that nova Sgr 2004 is at $z \sim -1$ kpc below the galactic plane. We summarize in table 1 the properties of nova Sgr 2004.

TABLE 1. Summary of the properties of nova Sgr 2004.

t_0	Mar17.19 2004 UT
t_2	10 days
t_3	21 days
m_v (max)	8.0 mag
M_v (max)	-8.5 mag
d	9.9 kpc
m_{15}	10.5 mag
A_v	1.6 mag
Spectral evolution	$P_{fe}, P^o_{fe}, A^o_o, A_o$
Spectral class	FeII

REFERENCES

1. Barbon,R., Benetti,S., Cappellaro,E., Rosino,L. and Turatto,M. 1990, A&A, 237, 79
2. Capaccioli,M., Della Valle, M., Rosino,L., D'Onofrio,M. 1989, AJ, 97, 1622
3. Della Valle,M., Ederoclite,A., Schmidtobreick,L., Germany,L., Dall,T., Saviane,I. 2004, IAUC, 8307
4. DellaValle, M., Pasquini,L., Daou,D., Williams,R.E. 2002, A&A, 390, 155
5. Della Valle,M., Livio,M. 1995, ApJ, 452, 704
6. Della Valle,M., Livio,M. 1998, ApJ, 506, 818
7. Munari, U., Zwitter, T. 1997, A&A, 318, 269
8. Nishimura,H., Nakano, S., Liller,W., West,D., Waagen,E., Royer,R., Bedient,J., Pearce,A. 2004, IAUC, 8306
9. Payne-Gaposchkin, C. 1957 The Galactic Novae (North-Holland Publishing Company, Amsterdam)
10. Soma,M., Shanklin,J. D., Lehky, M., Pearce, A., Shida,R.Y., Schmeer,P., West,J.D., Yoshida,S. 2004, IAUC, 8316
11. van den Bergh, S. & Younger, P.F., 1987, A&AS, 70, 125
12. Yamaoka,H., Itagaki,K., Schaefer,B. E., Liller,W., West,J.D., Gilmore,A.C. 2004 IAUC 8310
13. West,D. 2004 IAUC 8306
14. Williams, R. E., Hamuy, M., Phillips, M. M., Heathcote, S. R., Wells, L., Navarrete, M. 1991, ApJ, 376, 721
15. Williams, R. E., Phillips, M. M., Hamuy, M. 1994, ApJS, 90, 297
16. Williams,R.E. 1994 ApJ, 426, 279

A multiple mass-ejection by the symbiotic prototype Z And during its 2000-03 outburst

A. Skopal*, L. Errico[†], A.A. Vittone[†], S. Tamura**, M. Otsuka**, M. Wolf[‡] and V.G. Elkin[§]

*Astronomical Institute, Slovak Academy of Sciences, 059 60 Tatranská Lomnica, Slovakia
[†]INAF, Osservatorio Astronomico di Capodimonte, via Moiariello 16, I-80131 Napoli, Italy
**Astronomical Institute, Tohoku University, Sendai 980-8578, Japan
[‡]Astronomical Institute, Charles University Prague, CZ-18000 Praha 8, V Holešovičkách 2, Czech republic
[§]Special Astrophysical Observatory, Nizhnij Arkhyz 357147, Karachaevo-Cherkesia, Russia

Abstract. We present optical high and low-resolution spectroscopy and $UBVR$ photometry of the symbiotic prototype Z And from its recent 2000-03 outburst. Our spectra revealed a high-velocity mass-outflow from the star. Structure of the [O III] 5007 profile and a cascade profile of the light curve (LC) on the rise to the maximum suggest a multiple mass-ejection by the hot component of Z And. Our models of the SED confirmed this type of activity.

Keywords: Binaries: symbiotic – Star: individual: Z And
PACS: 97.80.Gm ; 98.38.Fs

INTRODUCTION

Z And is considered as a prototype of the class of symbiotic stars. The binary composes of a late-type M4.5 III giant and a white dwarf accreting from the giant's wind on the 758-day orbit [e.g. 3]. More than 100 years of monitoring Z And (first records from 1887) demonstrated an eruptive character of its LC. On 2000 September 1st, Z And entered a major outburst [5]. In 2002 August, at the position of the inferior conjunction of the giant, a narrow minimum developed in the LC. Skopal [4] associated this effect to the eclipse of the active object by the giant and determined the orbital inclination of the system to $76° - 90°$.

The aim of this contribution is to present and analyze our spectroscopic and photometric observations connected with the mass outflow during the 2000-03 major outburst of Z And.

OBSERVATIONS

High-dispersion spectroscopy was secured at the Asiago Astrophysical Observatory, Okayama Astrophysical Observatory and the Ondřejov Observatory. Low-dispersion spectroscopy was secured at the Terskol Observatory. Photometric observations were performed in the standard Johnson $UBVR$ system at the observatories of the Astronomical Institute of the Slovak Academy of Sciences.

Top panels of Fig. 1 show the $UBVR$ LCs of Z And covering its recent outburst. The rise to maximum was characterized by three rapid increases in the star's brightness and two plateaus, best seen in the U band. The initial rise (September 1st to 7th) was probably caused by an increase of the hot star luminosity, which can be associated to expansion of a shell surrounding the central star, which thus makes it cooler (e.g. the Raman scattered 6830 Å feature disappeared on our spectra from the maximum). The following short-term decline in the $U-B$ index at constant $B-V$ and $V-R$ indices suggested a decrease of the shell's temperature. Then during the 'totality' of the $U-B$ minimum all the colour indices were practically constant indicating thus a constant temperature. Therefore the second increase in the star's brightness, which occurred at constant indices (Fig. 1), resulted from an increase in the luminosity at a relevant expansion of the shell. Finally, the third brightening from the middle of November to the maximum (i.e. the ascending branch of the $U-B$ minimum) indicated an increase of the hot star temperature and its luminosity. Our modeling the SED confirmed this interpretation (see below). Qualitatively, similar behaviour in colour indices is being observed also at early stages of classical novae, prior to the maximum [e.g. Fig. 3 of 2].

Our spectra from the maximum revealed a high-velocity mass-outflow from the star. It was indicated mainly by: (i) Absorption components of the P-Cygni type of the profile pronounced in the He I lines and hydrogen lines of the Paschen series. (ii) The wings of the Hβ and Hα profiles, which extended to about $\pm 1000\,\mathrm{km\,s^{-1}}$ and $\pm 2000\,\mathrm{km\,s^{-1}}$, respectively. The main emission core was modulated by a violet shifted absorption created in the neutral gas at the front of the ejected material. (iii) The shift of the main emission core of the He II 4686 Å line to about $-90\,\mathrm{km\,s^{-1}}$. This means that a significant fraction of the He II emission arised in the hot outflowing gas from the central star at its vicinity. (iv) The profile of the nebular [O III] 5007 Å line, which showed asymmetrically extended core/wings with terminal velocities of about -300 and $+250\,\mathrm{km\,s^{-1}}$. In addition, small emission components at $\pm 400\,\mathrm{km\,s^{-1}}$ accompanied the main profile. A sequence of profiles observed on 10, 11 and 12th December 2000 showed creation of an extra emission component on the violet side of the main emission core. This evolution probably resulted from creation of a mass filament. Some examples are depicted in Fig. 1.

A MULTIPLE MASS-EJECTION

Multiple mass-ejection by Z And on the rise to its maximum (2000 September 1st to 2000 December 15th) is suggested by behaviour in the colour indices and confirmed spectroscopically by the structure of the [O III] 5007 Å profile (Fig. 1).

To understand the nature of this behaviour we reconstructed the SED during the time of the last plateau in the LC and at the maximum. We modeled our optical/near-IR observations and the FUSE points of the continuum [6] by the same way as the 1985-maximum, for which the ultraviolet spectra were available [4]. The result is shown in the bottom panels of Fig. 1. The change in the SED from the plateau to the maximum was connected with an increase of both the hot stellar luminosity and the nebular emission.

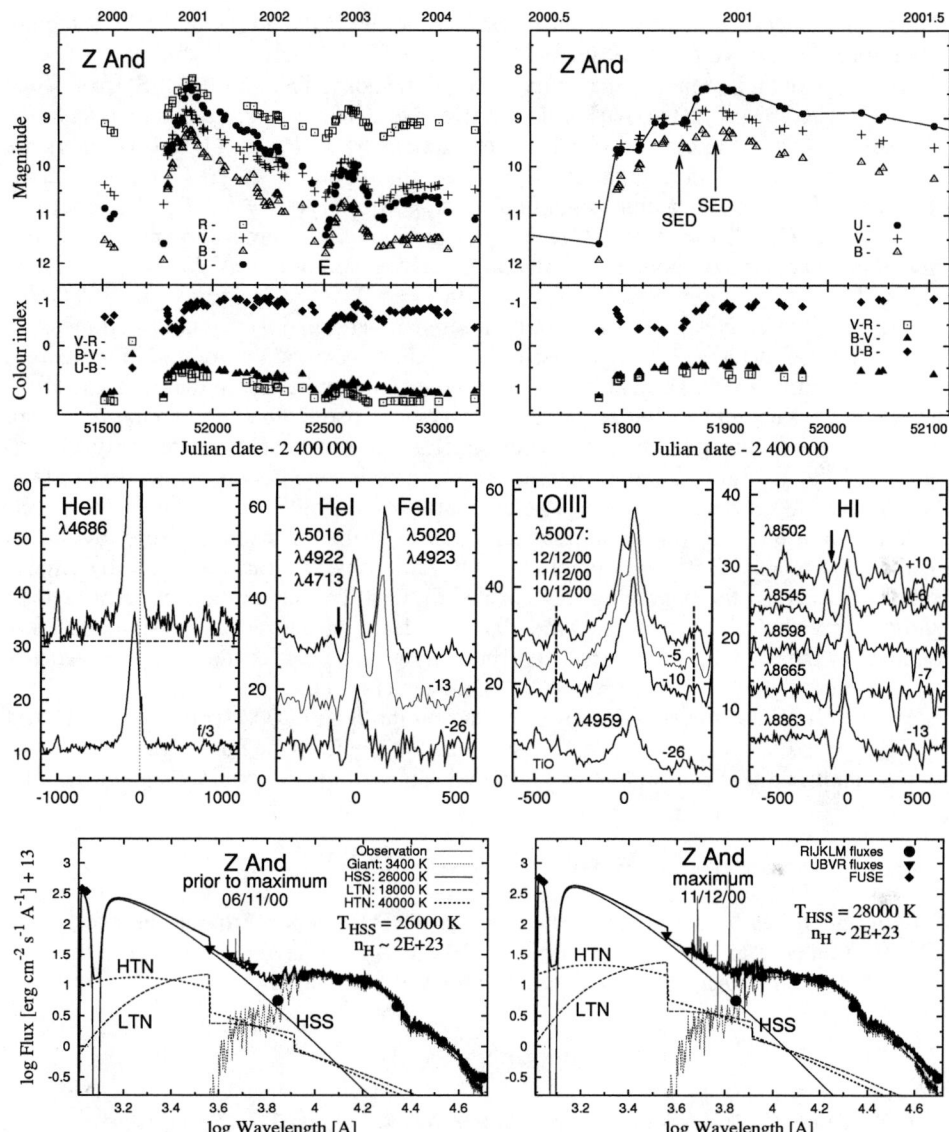

FIGURE 1. Top: The $UBVR$ photometry from the major 2000-03 outburst. Left panel shows the overall evolution in the LC. The eclipse is denoted by E. The right panel shows in detail the beginning of the outburst. Middle: Selected line profiles from the maximum of the star's brightness. Absorption components of the P-Cygni type of profiles are marked by arrows. Fluxes are in units of $10^{-13}\,\mathrm{erg\,cm^{-2}\,s^{-1}\,Å^{-1}}$ and radial velocities in $\mathrm{km\,s^{-1}}$. Bottom: The SED prior to maximum, at the time of the last plateau stage (left) and at the maximum of the Z And brightness (right). Positions are denoted in the top right panel. Electron temperatures of the hot and low temperature nebula (HTN and LTN, respectively) and the n_H parameter were adopted from the 1985 model [4].

These components of radiation contributed more to the U band than in the BVR region. Therefore we observed a larger increase of the star's brightness in U during this transition. In spite of an increase of the temperature of the hot stellar source (HSS), from $T_{HSS} = 26\,000$ K to $T_{HSS} = 28\,000$ K (Fig. 1), its effective radius also increased by a factor of 1.1, from $3.1\,(d/1.5\,\text{kpc})\,R_\odot$ on 06/11/00 to $3.5\,(d/1.5\,\text{kpc})\,R_\odot$ on 11/12/00 as it results from the corresponding angular radii given by the model of the SED. These quantities of the HSS then determine its luminosity, $L_{HSS}(06/11/00) = 4000\,(d/1.5\,\text{kpc})^2\,L_\odot$ and $L_{HSS}(11/12/00) = 6800\,(d/1.5\,\text{kpc})^2\,L_\odot$. Finally, the emission measure determined by nebular components of radiation in our solution was $EM(06/11/00) = 3.2\,10^{60}\,(d/1.5\,\text{kpc})^2\,\text{cm}^{-3}$ and $EM(11/12/00) = 5.2\,10^{60}\,(d/1.5\,\text{kpc})^2\,\text{cm}^{-3}$. The increase of the effective hot star radius, its luminosity and mainly the emission measure during the rise of the star's brightness to its maximum can be interpreted in terms of a mass ejection by the hot object in Z And. Such event supplies more material (i.e. emitters) into the surrounding ionized area and thus gives rise to an extra nebular emission. To the contrary, during the plateau stages the SED did not change considerably, as suggested by the stability of colour indices. This constrains the hot star to be stable in the luminosity and temperature and also no comparable mass outflow could be in the effect, because more emitters would produce an increase of the emission measure. Evolution in the SED is also in a good agreement with radio observations presented by Brocksopp et al. [1]. At the beginning stage of the outburst the hot component became to be significantly cooler, which caused the observed decay in the radio emission. On 2001 September the MERLIN map of Z And showed the transient jet-like extension when a small increase in the star's brightness was detected (Fig. 1).

We conclude that the star's brightening during the major 2000-03 outburst of Z And was connected with an increase in the mass-outflow rate from the hot active star. The rise to the maximum was caused by three different stages of the hot component expansion.

ACKNOWLEDGMENTS

This research has been supported through the NATO Science Programme and by the Slovak Academy of Sciences grant No. 2/4014/4. AS acknowledges the hospitality of the INAF, Osservatorio Astronomico di Capodimonte in Naples.

REFERENCES

1. C. Brocksopp, J.L. Sokoloski, C. Kaiser, et al., *Mon. Not. R. Astr. Soc.* **347**, 430–436 (2004)
2. D. Chochol, and T. Pribulla, *Contr. Astron. Obs. Skalnaté Pleso* **27**, 53–69 (1997)
3. J. Mikołajewska, and S.J. Kenyon, *Astron. J.* **112**, 1659–1669 (1996)
4. A. Skopal, *Astron. Astrophys.* **401**, L17–L20 (2003)
5. A. Skopal, D. Chochol, T. Pribulla, and M. Vaňko, *Inf. Bull. Var. Stars* No. **5005** (2000)
6. J.L. Sokoloski, S.J. Kenyon, A.K.H. Kong, et al., in *The Physics of Cataclysmic Variables and Related Objects*, ASP Conf. Ser. **261**, edited by B.T. Gänsicke, K. Beuermann, and K. Reinsch, San Francisco, 2002, p. 667

What happens when a hot star shines on a cool one?

Katrina Exter[*], Travis Barman[†], Don Pollacco[**], Vladislav Pustynski[‡], Steve Bell[§] and Izold Pustylnik[¶]

[*]IAC, c/ Vía Láctea, La Laguna, E38200 Tenerife, Spain
[†]UCLA, Astronomy & Astrophysics, 475 Portola Plaza, Box 951547, Los Angeles, CA 90095-1547, USA
[**]APS, Physics, Queen's University Belfast, Belfast BT7 1NN, UK
[‡]Tallinn Technical University, Ehitajate tee 5, 19086, Tallinn, Estonia
[§]HMNO, Space Science & Technology Dep't, Rutherford Appleton Lab., Chilton, Didcot, UK
[¶]Tartu Observatory, 61602, Tõravere, Estonia

Abstract. In this, we discuss our on-going work observing the effects, and modeling the physical processes, of the irradiation of a cool star by its hot companion in close-binary systems. In particular, we present here some data and results on the post common-envelope systems EC 11575−1845 and UU Sge.

Keywords: Close binary stars – Irradiation (modeling, observations)
PACS: 97.10.Ex, 97.80.Fk

WHAT HAPPENS TO THE ATMOSPHERE?

Imagine a 100000 K white dwarf located a few solar radii distant from a 3000 K red dwarf; the radiation field from the white dwarf will have a strong effect on the inward facing hemisphere of the cool star. If the orbit is synchronous – that is, the two stars always face each other with the same hemisphere – then there could be a drastic difference in the conditions in the outer atmosphere of the red dwarf on its two sides. We have been observing and modeling this irradiation in a sample of post common-envelope binary stars, particularly EC 11575−1845, UU Sge, and the similar systems V664 Cas, VW Pyx and V477 Lyr.

The irradiation of the red dwarf results in a reversed temperature profile above the photosphere on the heated side, where now the hottest material is located further out from, rather than closer in to the star. The convective-radiative boundary may be moved inwards and some radiation may penetrate through, heating up the convective zone. In Fig. 1 we plot our modeled temperature profiles for UU Sge and EC 11575−1845. Compare the 40000 K peak for UU Sge to the ~4000 K temperature on the unaffected side of the star! A temperature gradient will also be set up on the surface of the star, from the sub-stellar point to the day/night-side boundary; one therefore can refer to the heated hemisphere as a 'hot spot'. For more information on the physical processes in this hot spot, we refer you to Barman, Hauschildt & Allard (2004). Other activities such as circulation currents may be set up, transporting heat around the star (see the poster by Beer et al. in this conference proceedings).

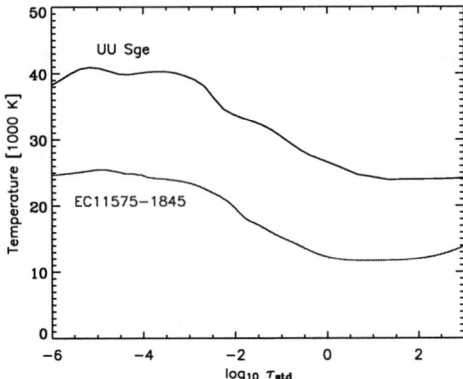

FIGURE 1. Modeled temperature profiles in the irradiated hemispheres of UU Sge and EC 11575−1845. Note that the photospheric temperature (at log $\tau \sim 0$) of the undisturbed side of these red dwarfs is 3000-4000 K.

OBSERVATIONAL CONSEQUENCES

The irradiation of the red dwarf will radically alter the light curve and spectrum at phases when this hemisphere is to view, *i.e.* phases around 0.5 (although if the hot spot is large enough, and the inclination < 90 degrees, it may always be partially to view). The spectrum will be dominated by emission lines and a warm continuum from the ionised gas, as is shown by our observed spectrum and the model for EC 11575−1845 and UU Sge in Fig. 2. In addition, the concentration of neutral ions will be reduced; this together with the additional continuum contribution will alter the photospheric absorption line spectrum one would otherwise normally observe.

The NLTE gas of this irradiated hemisphere cannot be modeled simply as a hotter star or extra black-body component, it will require tailored models for the spectra and light curves. In the light curves, a 'reflection effect' is set up, this arising from the extra continuum of the hot spot and visible at the orbital phases around 0.5. This is shown by our observations and model for UU Sge in Fig. 3. Not visible on the scale of this lightcurve, but present, is also an extra contribution to the normally sinusoidal reflection effect; this arising in extremely heated systems such as UU Sge (and pointed out by Ferguson & James, 1994, for BE UMa; Pustylnik & Pustynski, 1999 for V477 Lyr).

WHAT NEXT?

Our models of the irradiation of red dwarf companion stars will be used to constrain the binary and stellar parameters, as well as to understand the irradiation effects themselves. The spectral models shown here are currently fit to the phase 0.5 spectra only, however, there are orbitally-linked modulations to the spectra with phase – changes to the line fluxes and line profiles (especially for H I) – which must also be explained by the models.

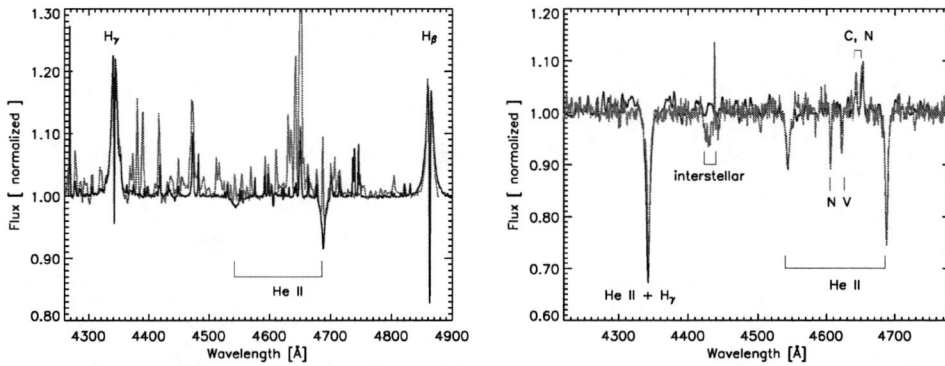

FIGURE 2. Observed and modeled spectrum for EC 11575−1845 (left) and UU Sge (right) at an inclination of 90 degrees and at phase 0.5, when the hot spot is facing us (note that EC 11575−1845 in reality has a much lower inclination). The observed spectrum is the darker. The differences between the spectra of the two systems are mostly due to the different sizes of their stellar components – for UU Sge both are the larger – as both red dwarfs are cool, 3-4000 K, and both sdO stars hot, 100000 K and 90000 K respectively. The fit to EC 11575−1845 will improve when the temperature gradient along the surface of the irradiated face is considered, but note the good fit to the H I profiles. The spectra have been continuum-divided and are plotted in arbitrary units.

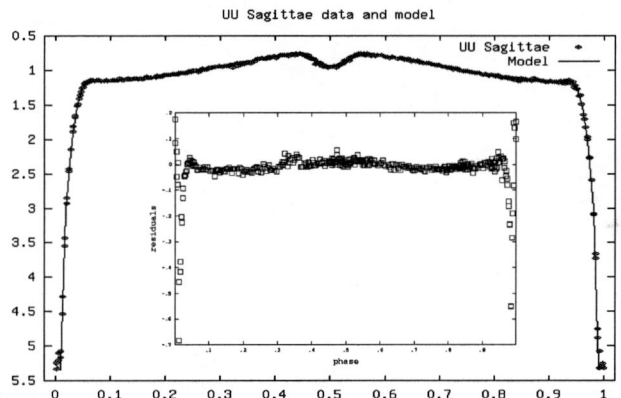

FIGURE 3. The light curve and fit (plus residuals), from irradiation models, for UU Sge.

As our results are likely to evolve with time, we therefore refer the reader to our current (eg. Exter, Pollacco, Bell, 2003) and future (Exter et al, submitted; Barman et al, in progress) publications on these systems for more information.

REFERENCES

- Barman T.S., Hauschildt P.H., Allard, F., 2004, ApJ, 614, 338
- Exter K.M., Pollacco D.L., Bell, S.A., 2003, MNRAS, 341, 1349
- Ferguson D.H., James T.A., 1994, ApJS, 94, 723
- Pustylnik, I., Pustynski, V., 1999, *11th European Workshop on White Dwarfs*, ASP Conference Series, 169, 289

Optical counterpart of the XTE J0929-314 in quiescence: constraints on the magnetic field.

M. Monelli*, G. Fiorentino*, L. Burderi*, F. D'Antona*, N. Robba[†] and V. Testa*

*INAF-Osservatorio Astronomico di Roma, via Frascati 33, 00040, Monte Porzio Catone, Italy
[†]Dipartimento di Fisica, Universitá di Palermo, P.zza del Parlamento 1, 90134 Palermo, Italy

Abstract. We present VLT observations of the optical counterpart of th X-ray millisecond pulsar XTE J0929-314 in quiescence. We detected a very faint candidate in agreement with the position given by radio and X observation. From the observed optical flux we inferred un upper limit to the magnetic field of the system.

Keywords: <Enter Keywords here>
PACS: 90

INTRODUCTION

It is commonly accepted that Low Mass X-ray Binaries (LMXB) consist of a neutron star (NS), generally with a weak magnetic field (B $\sim 10^8$-10^9 Gauss), accreting matter from a low-mass (≤ 1 $M_\odot$) companion. A special subgroup of these sources, are the NS soft X-ray transient (SXT), that are usually found in a quiescence state. XTE J0929-314 is a such kind of source and its X-ray luminosity is $\approx 8 \pm 3 \times 10^{31}$ ergs/s [10]. On occasions this source has shown outbursts, during which its luminosity is increased to $\sim 10^{36} - 10^{38}$ ergs/s. Wijnands & van der Klis ([11]) proposed the idea that NSs in LMXBs were spinning at millisecond periods confirmed by the discovery of coherent X-ray pulsations at $P_{spin} \sim 2.5$ ms in SAX J1808.4-3658. Recently, Galloway et al [4] found that also XTE J0929-314 is a transient low mass X-ray binary (LMXB) first detected in 2002 April by the All Sky Monitor on board the Rossi X-Ray Timing Explorer. The endpoint of a such kind of source is a Millisecond Pulsar (MSP), in particular XTE J0929-314 is in short-period binaries having a binary period of ~ 43 minutes. This short period places this source in the class of ultracompact binaries, defined as having an orbital period ≤ 80 minutes.

These sources belong to a new class of astronomical objects, Millisecond X-ray Pulsars (MSXPs), that could constitute the long searched for bridge between the accretion-powered (X-ray pulsators) and the rotation-powered (Millisecond Radio Pulsar, hereinafter MSP) NS sources. In fact, in the so-called reclycling scenario the X-ray pulsators are the progenitor of MSP [1].

In this work, we are interested to the optical counterpart of XTE J0929-314. Greenhill et al. [5] found a possible candidate for the optical counterpart in the outburst phase, with a apparent magnitude value in V-band of about 18.8 mag. In analogy to the work presented by Burderi et al. [2], we present here the identification of a possible optical counterpart candidate of the source in quiescence phase, that allowed us to constraint the magnetic field and then to reveal a magneto-dipole emission through combined optical and X-ray observations.

OBSERVATION AND DATA REDUCTION

Optical V, R, and I band data have been collected between December 2003 and January 2004 with FORS1@VLT in standard resolution mode. The integration times used were 300s. 20 images were obtained in each band, under excellent seeing conditions ($\leq 0.6''$). The data were prereduced with the ESO pipeline, while the photometric reduction has been made using the DAOPHOTII/ALLFRAME [8, 9] This package performs PSF-fitting photometry simultaneously on all the images available, trying to fit all the sources listed in a reference catalogue. As a first step, every single image has been independently analized, and individual PSF was derived for each frame. Accurate coordinate transformations were obtained using DAOMATCH/DAOMASTER, and then all the images were coadded in a single median image with MONTAGE2. The reference catalogue was obtained by searching for stellar sources in this median image, and ALLFRAME was run with a list of ≈ 27000 stars. Weighted median image were estimated, and the calibration was performed using standard fields (PG1323, SA98) observed during all the runs. The astrometric solution was estimated by the GSC2 catalogue as reference.

Identification of the optical counterpart

In Fig. 1, we show the position of the candidate optical counterpart (small cross). The three circle illustrate the error box on the position of the source from previous works. Our detection is in good agreement with the radio estimate (smallest circle, radius $=0.3''$) The optical (during iutburst) and X-ray position error boxes are $0.4''$ and $0.5''$, respectevely. We have also estimated the probability to have misidentified the source with another field stars, as well as with a background galaxy. By comparing with Galactic models, we estimated that the probability to find two stars within a box of radius $=0.3''$ is $\approx 0.03\%$. Moreover, we estimated the probability to detect a background galaxy within the same area. By adopting deep counts (Grazian 2004, Priv. Comm.) we estimated a probability of $\approx 1\%$.

DISCUSSION AND CONCLUSIONS

Assuming a distance modulus of 5 Kpc and a Hydrogen column density of about 7.6×10^{20} cm-2 (Juett et al. [6]), we estimate from Predhel et al. [7] [N_H=5.6x10^{21}E(B-V)cm^{-2})] a reddening value of 0.14. Then, using extinction laws by Cardelli et al [3] we obtained the extinction coefficients and the absolute magnitude in the optical bands reported in Tab. 1. The measured magnitudes are fainter of about 9 mag than previous optical detected source according to X-ray transient nature of XTE J0929-314. We estimated the bolometric magnitude of the source as

TABLE 1. Quiescent optical de-reddened absolute magnitude in three different bands.

App. Mag.	reddening	Abs. Mag.
V $\approx 27.58 \pm 0.12$	0.42	$M_V \approx 13.66$
R $\approx 27.21 \pm 0.13$	0.36	$M_R \approx 13.55$
I $\approx 27.23 \pm 0.22$	0.26	$M_I \approx 13.47$

By studying the quiescent optical conterpart of SAX J1808.4-3658, Burderi et al. [2] found an optical luminosity of about 10^{31} ergs/s, too high for low mass (≤ 0.14 $M_\odot$) companion star. If this luminosity is produced by an accretion disk, it would require an X-ray luminosity $L_X \sim 10^{34}$ ergs/s, far in excess of the typical

FIGURE 1. Position of the candidate optical counterpart of the XTE 0929-314

quiescent $L_X \sim 10^{31}$ ergs/s. The authors conclude that it can be accounted for by partial reprocessing of the bolometric luminosity of an active magneto-dipole rotator.

Using the absolute magnitudes obtained in previous section, we can give a lower limit for L_{Bol} of 10^{30} ergs/s. Taking into account that reprocessed from the binary companion is about 3% of neutron star luminosity [D'Antona 1996], we conclude that L_{NS} found for this source ($\sim 3.33 \times 10^{32}$) is too large for an accretion disk. In the end, using the Larmor's formula we can estimate for the magnetic field a value of 2.15×10^8 Gauss.

REFERENCES

1. Bhattacharya & van den Heuvel, 1991, PhR, 203, 1
2. L. Burderi, T. Di Salvo, F. D'Antona, N. R. Robba and V. Testa, *A&A*, Publisher Name, Publisher City, 2000, pp. 212–213.

3. J. A. Cardelli, G. C. Clayton and J. S. Mathis, *ApJ*, **345**, 245 (1989).
D'Antona 1996. F. D'Antona, in *epbs.conf*, 1996, **477** pp. 287D.
4. D. K. Galloway et al., *ApJ*, 2002, **576** pp. L137.
5. J. G. Greenhill, A. B. Giles and K. M. Hill, *IAUC*, 2002, **576** n.7889.
6. A. M. Juett, D. K. Galloway, D. Chakrabarty, *ApJ*, 2003, **587** pp. 754.
7. P. Predhel, J. H. M. M. Schmitt, *A&A*, 1995, **293** pp. 889.
8. Stetson, P.B, *PASP*, 1987, **99**, 191
9. P. B. Stetson, *PASP*, 1994, **106** pp. 250.
10. R. Wijnands, *ApJ*, 2003, **588** pp. 425.
11. R. Wijnands, and M. van der Klis *ApJ*, 1998, **507** pp. 63.

Non-axisymmetric Structure of Accretion Disks around the Neutron Star in Be/X-ray Binaries

Kimitake Hayasaki[*,†] and Atsuo T. Okazaki[**]

[*]*Department of Applied Physics, Graduate School of Engineering, Hokkaido University, Kitaku N13W8, Sapporo 060-8628, Japan*
[†]*Centre for Astrophysics and Supercomputing, Swinburne University of Technology, Hawthorn Victoria 3122, Australia*
[**]*Faculty of Engineering, Hokkai-Gakuen University, Toyohira-ku, Sapporo 062-8605, Japan*

Abstract. A non-axisymmetric structure of accretion disks around the neutron star in Be/X-ray binaries is studied, analyzing the results from 3D SPH simulations performed by [1]. It is found that a ram pressure due to the phase-dependent mass transfer from the Be-star disk excites a one-armed, trailing spiral structure in the accretion disk around the neutron star. The spiral wave has a transient nature; it is excited around the periastron, when the material is transferred from the Be disk, and gradually damped afterwards. We also find that the orbital phase-dependence of the mass-accretion rate is mainly caused by the inward propagation of the spiral wave excited on the accretion disks.

Keywords: Accretion and accretion disks, X-ray binaries, Neutron stars, Circumstellar shells, clouds, and expanding envelopes; circumstellar masers
PACS: 97.10.Gz, 97.80.Jp, 97.60.Jd, 97.10.Fy

INTRODUCTION

The majority of the high-mass X-ray binaries have been identified as the Be/X-ray binaries. These systems generally consist of a neutron star and a Be star with a cool ($\sim 10^4 K$) equatorial disk, which is geometrically thin and nearly Keplerian. Be/X-ray binaries are distributed over a wide range of orbital periods ($10\,\mathrm{d} \leq P_{\mathrm{orb}} \leq 300\,\mathrm{d}$) and eccentricities ($e \leq 0.9$).

Most of the Be/X-ray binaries show only transient activity in the X-ray emission and are termed Be/X-ray transients. Be/X-ray transients show periodical (Type I) outbursts, which are separated by the orbital period and have the lumiocity of $L_X = 10^{36-37} \mathrm{erg\,s^{-1}}$, and giant (Type II) outbursts of $L_X \geq 10^{37} \mathrm{erg\,s^{-1}}$ with no orbital modulation. These outbursts have features that strongly suggest the presence of an accretion disk around the neutron star.

Recently, Hayasaki and Okazaki [1] studied the accretion flow around the neutron star in a Be/X-ray binary with a short period ($P_{\mathrm{orb}} = 24.3\,\mathrm{d}$) and a moderate eccentricity ($e = 0.34$), using a 3D SPH code and the imported data by [2]. They found that a time-dependent accretion disk is formed around the neutron star. They also discussed the evolution of the azimuthally-averaged structure of the disk, in which a one-armed spiral structure is seen. This gives a new point of view in Be/X-ray binaries, that is, the systems have double circumsteller disks which interacts mainly via the mass transfer [see Fig 1].

In this paper, we show that the ram pressure due to the material transferred from the Be disk around periastron temporarily excites the one-armed spiral wave in the accretion

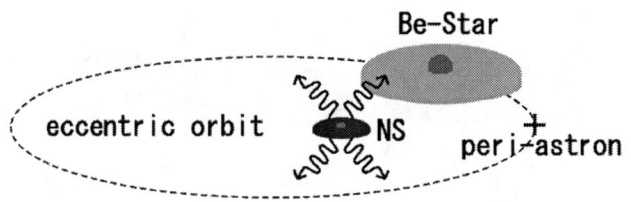

Source: M. Okuno

FIGURE 1. A new schmatic diagram in a typical Be/X-ray binary. The system has double circumstellar disks (the Be disk around the Be star and the accretion disk around the neutron star) which interact mainly via the mass transfer.

disk around the neutron star in Be/X-ray binaries.

ONE-ARMED SPIRAL WAVE EXCITED BY A RAM PRESSURE

Our simulations were performed by using the same 3D SPH code as in [1], which was based on a version originally developed by Benz ([4]; [5]) and later by [6]. In order to inspect the effect of the ram pressure on the accretion disk, we compare results from model 1 in [1] (hereafter, model A) with those from a simulation (hereafter, model B) in which the mass transfer from the Be disk is artificially stopped for one orbital period. Except for this difference, two simulations have the same model parameters: The orbital period P_{orb} is 24.3 d, the eccentricity e is 0.34, and the Be disk is coplanar with the orbital plane. The inner radius of the simulation region r_{in} is $3.0 \times 10^{-3} a$, where a is the semi-major axis of the binary. The polytropic equation of state with the exponent $\Gamma = 1.2$ is adopted. The Shakura-Sunyaev viscosity parameter $\alpha_{SS} = 0.1$ throughout the disk.

Fig. 2 gives a sequence of snapshots of the accretion disk around the neutron star for $7 \leq t \leq 8$ in model A, where the unit of time is the orbital period P_{orb}. The left panels show the contour maps of the surface density, whereas the non-axisymmetric components of the surface density and the velocity field are shown in the right panels. Annotated in each left panel are the time in units of P_{orb} and the mode strength S_1, a measure of the amplitude of the one-armed spiral wave, which is defined by using the azimuth Fourier decomposition of the surface density distribution, details of which are described by [3, Sec 2.2]. It is noted from the figure that the one-armed, trailing spiral is excited at periastron and is gradually damped towards the next periastron. The disk is topologically changing from circular to eccentric with the development of the spiral wave, and then the process reverses to move from eccentric to circular with the decay of the wave during one orbital period.

For comparison purpose, we present the results for model B, in which the mass transfer is artificially turned off for $7 \leq t \leq 8$. Fig. 3 shows the surface density (the left panel) and the non-axisymmetric components of the surface density and the velocity

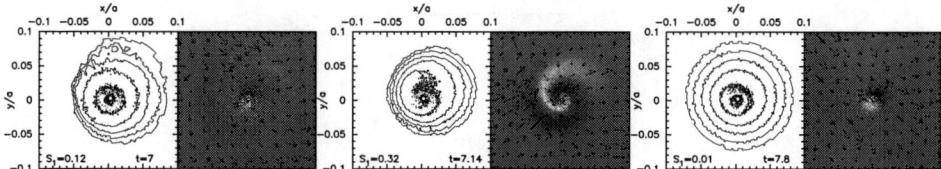

FIGURE 2. Snapshots of the accretion disk for model A. The left panels show the surface density in a range of three orders of magnitude in the logarithmic scale, while the right panels show the non-axisymmetric components of the surface density (gray-scale plot) and the velocity field (arrows) in the linear scale. In the right panels, the region in gray (white) denotes the region with positive (negative) density enhancement. The periastron is in the x-direction and the disk rotates counterclockwise. Annotated in each left panel are the time in units of P_{orb} and the mode strength S_1.

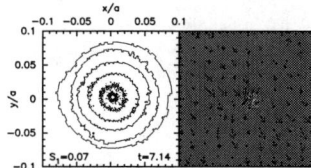

FIGURE 3. Same as Fig. 2, but for model B.

field (the right panel) at the time corresponding to the middle panel of Fig. 2. The format of the figure is the same as that of Fig. 2. It should be noted that the disk deformation due to the one-armed mode is not seen in model B. The disk is more circular and has a larger radius in model B than in model A. This strongly suggests that the excitation of the one-armed spiral structure in the accretion disk is induced by the ram pressure from the material transferred from the Be disk at periastron.

Phase dependence of the mass-accretion rate

After the accretion disk is developed ($t \geq 5$), the mass-accretion rate has double peaks per orbit, a relatively-narrow, low peak at periastron and a broad, high peak afterwards [see Fig. 15(a) of [1]]. While the first low peak at periastron could be artificial, being related to the presence of the inner simulation boundary, the origin of the second high peak was not clear. Below we show that the one-armed spiral wave is responsible for the second peak in the mass-accretion rate.

Fig. 4 shows the time dependence of the mass-accretion rate for $7 \leq t \leq 8$. The thick line denotes the mass-accretion rate in model A, in which the mass transfer from the Be disk is taken into account. For comparison, the mass-accretion rate in model B, in which the mass transfer from the Be disk is artificially turned off at $t = 7$, is also shown by the thin line. The difference between the accretion rate profiles for these two models is striking. The accretion rate in model B monotonically decreases over one orbital period, whereas that of model A shows a broad peak centred at $t \sim 7.32 - 7.35$, which corresponds to the second peak found in [1].

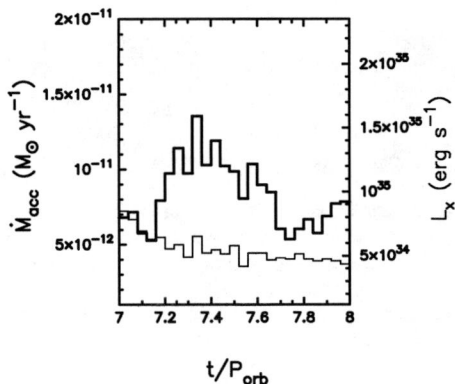

FIGURE 4. Time dependence of the mass-accretion rate for $7 \leq t \leq 8$. The thick and thin lines are for model A and model B, respectively. The right axis shows the X-ray luminocity corresponding to the mass-accretion rate.

Although it is obvious that the above peak is caused by the mass transfer from the Be disk, the mass-transfer rate has a narrow peak at periastron as shown in Fig 2 of [1]. This lag of the peak position on orbital phase between the mass-accretion rate and the mass-transfer rate results from the inward propagation of the wave from the disk outer radius to the inner simulation boundary.

ACKNOWLEDGMENTS

This work has been supported by Grant-in-Aid for the 21st Century COE Scientific Research Programme on "Topological Science and Technology" from the Ministry of Education, Culture, Sport, Science and Technology of Japan (MECSST) and in part by Nukazawa Science Fundation.

REFERENCES

1. Hayasaki, K., and Okazaki, A.T., *MNRAS*, **350**, 971, (2004).
2. Okazaki, A.T., Bate, M.R., Ogilvie, G.I., and Pringle, J.E., *MNRAS*, **337**, 967, (2002).
3. Hayasaki, K., and Okazaki, A.T., submitted to *MNRAS*, (2004).
4. Benz, W. "Smoothed Particle Hydrodynamics - a Review" in *The Numerical Modelling of Nonlinear Stellar Pulsations Problems and Prospects*, edited by J. R. Buchler, Kluwer, Dordrecht, 1990, pp. 269.
5. Benz, B., Bowers, L.R., Cameron, A.G.W., and Press, H.W., *ApJ*, **348**, 647, (1990).
6. Bate, M.R., Bonnell, I.A., and Price, N.M., *MNRAS*, **285**, 33, (1995).

Modeling of Gas Flow Structure in Symbiotic Star Z And in Quiescent and Active States

E.Yu. Kilpio*, D.V. Bisikalo*, A.A. Boyarchuk* and O.A. Kuznetsov*,†

Institute of Astronomy, 48 Pyatnitskaya str., 119017 Moscow, Russia
†*Keldysh Institute of Applied Mathematics, 4 Miusskaya sq., 125047 Moscow, Russia*

Abstract.
2D modeling of gas flow structure in a binary system using the modified model on a fine grid has been carried out for the parameters of the classical symbiotic star Z And. The calculations have confirmed the mechanism of quiescent to active state change as a consequence of the transition from the disk accretion to the accretion from the flow proposed in our previous works [1]. The simulations of the flow structure for the active state of the system have been also carried out. The thermonuclear runaway on the accretor has been modeled by the pressure jump on the accretor's surface. It has been shown that the introduction of the pressure jump leads to the formation of the structure of two shocks and the contact discontinuity in the space between components of the system. The modeled changes of parameters caused by the formation of these shocks are found to be in agreement with the observed brightness changes.

Keywords: Symbiotic stars – Numerical modeling – Outburst – Accretion
PACS: 97.80.Üd, 97.10.Gz

Symbiotic stars are binary systems consisting of a red giant, a white dwarf and surrounding nebulosity. These systems are characterized by non-periodic nova-like outbursts. One of the classical representatives of this class is Z And. The last outburst in this system took place in 2000-2003. During an outburst the system's brightness increases by approximately 3 magnitudes in tens of days, then it begins to decrease and it takes a few hundreds of days for magnitude to reach quiescent values. It is common knowledge now that Z And is a detached binary system and mass exchange is driven by stellar wind from the red giant. In order to explain the transition from quiescent to active state various mechanisms were proposed but there is still a lot of indefiniteness. In case of Z And, the nova-like activity is usually explained by changes of the accretion rate on the white dwarf [2]. Previous calculations allowed us to propose a new mechanism that could provide a significant increase of the accretion rate [1]. The results have shown that the observed value of the giant's wind velocity (25-40km/s) lies near the border dividing different accretion regimes. Namely, if the wind velocity is greater than ≈ 35 km/s the steady accretion disk takes place while for greater values the cone shock forms. So we can suppose that in quiescent state the accretion disk takes place in the system but rather minor variation of the giant's wind velocity can result in the change of the accretion regime. The process of transition between these two cases was studied in details [1] and it was found out that the abrupt jump in the accretion rate takes place during the disk destruction process. This jump in the accretion rate could lead to the change of the burning regime in the accretor's boundary layers and thus to the outburst. In order to study the system's behaviour after the accretion rate jump we have modeled the thermonuclear runaway (TNR) from the accretor by introducing the pressure jump on its surface and carried out the study of the outburst development.

The gas flow structure in the equatorial plane of the system has been considered by means of 2D gasdynamic model. Zero point of the coordinate system was in the center of the donor, x-axis was directed along the line connecting centers of the components, y axis - against the donor's orbital motion. The flow was described by the system of Euler's equations in the reference frame rotating with the angular velocity of the binary system Ω. The adopted force potential differs from the standard Roche potential by the additional term corresponding to the force responsible for donor's wind acceleration. This force is used in parametric form with the parameter value providing the agreement with observations. To complete the system the ideal gas equation of state was used with the ratio of the specific heats $\gamma=5/3$. To solve the system of equations the TVD-type Roe scheme [4] with the restrictions of fluxes in the Oscher form [5] was used. The modified model uses the non-uniform rectangular grid consisting of 679×589 nodes. The considered domain is a square $[-A \ldots 2A] \times [-3/2A \ldots 3/2A]$ with excluded circles of radii equal to the ones of the components and centers in the centers of components. The free outflow boundary conditions ($u=0$, $p=0$) were accepted on the outer border. The situation before the outburst was calculated assuming free outflow boundary conditions on the accretor. The outburst was modeled by introducing the pressure jump on the accretor's surface. The parameters of Z And were taken mainly from [3].

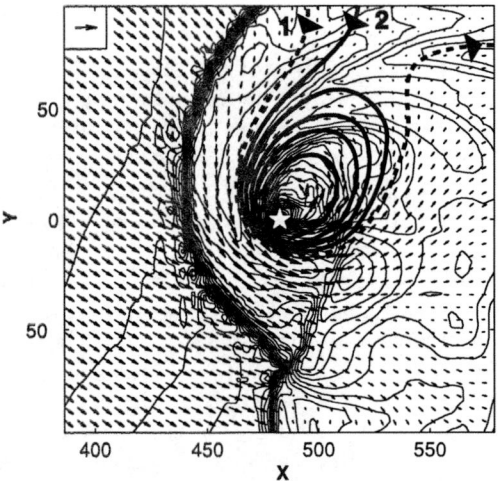

FIGURE 1. Density isolines and velocity vectors in the area near the accretor for the quiescent case (giant's wind velocity v_{wind}=25km/s). Flow lines show the presence of the accretion disk. The radius of the disk approximately equals 50 $R_\odot$.

The results have confirmed the mechanism providing transition from quiescent to active state proposed in our previous works [1] . Namely, it has been confirmed that in quiescent state the steady accretion disk exists in the system (see Fig. 1) but rather minor variations of the giant's wind velocity can lead to the transition from disk accretion to accretion from the flow followed by the abrupt jump of the accretion rate (in dozens times).

After outburst begins the wind from the hot component influences the structure of the nebulosity and as a result the system of two shocks and the contact discontinuity forms in the space between the components. In Fig. 2 the situation for the moment of approximately 100 days after the outburst beginning (maximum brightness on the light curve) is presented.

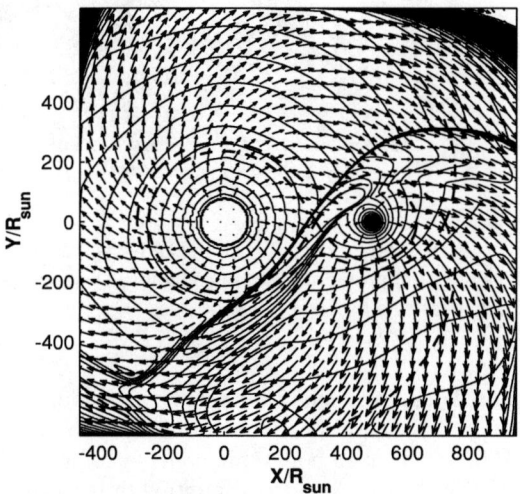

FIGURE 2. Density fields and velocity vectors for the active state. The solution corresponds to approximately 100 days after introducing the pressure jump. Distances are given in solar radii. In the right panel the area near shocks is shown in details.

We can see that each of the flows (from the donor and from the accretor) splits into two oppositely directed flows moving along the tangential discontinuity. In Fig. 3 the area near shocks on the x-axis is shown in details.

In order to estimate how the changes in the flow structure can influence the observed brightness of the system we assumed that brightness changes are proportional to changes in energy losses in the system. We supposed that the energy loss per unit volume and per unit time is $\rho^2 \Lambda(T)$, where $\Lambda(T)$ is the cooling function (see *e.g.* [6]). We calculated the integral energy loss in the whole computational domain for different moments of the outburst development and compared it to the one for the pre-outburst state.

$$Q = \frac{\Sigma \rho^2 \Lambda(T)_{outburst}}{\Sigma \rho^2 \Lambda(T)_{quiescence}} \quad (1)$$

We have found that $Q_{max} \approx 12$ (or $\sim 2^m.7$) while the typical amplitude of Z And outbursts is $(2^m - 3^m)$. So the parameters selected can provide the values of brightness change close to the observed ones.

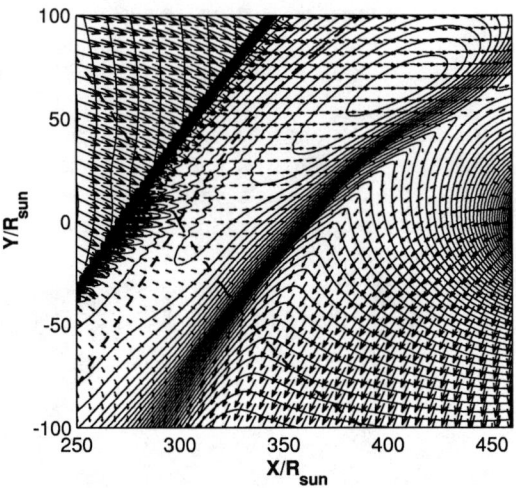

FIGURE 3. Density fields and velocity vectors for the active state. The solution corresponds to approximately 100 days after the pressure jump introducing. Distances are given in solar radii. In the right panel the area near shocks is shown in details.

It should be also noted that TNR can be modeled by introducing not only the pressure jump but the jump in ρv^2 as well. Some indications for high-velocity winds (velocities are of the order of 1000 km/s) from hot components in some symbiotics (see *e.g.* [7]) exist. So if we model the TNR as the in ρv^2 jump with the value of $v \approx 1000$ we can obtain a solution similar to that shown in the Figs. 2-3 but with much greater velocities. In particular, velocities along the tangential discontinuity will be of the order of 1000 km/s. In case of Z And there were some observations (*e.g.* [8]) that could count in favour of this variant.

Acknowledgements: The work was supported by RFBR (NN 02-02-16088, 02-02-17642, 03-01-00311, 03-02-16622), Presidium RAS Programs, Science Schools Support Program (N 162.2003.2) and by Federal Programme "Astronomy".

REFERENCES

1. Bisikalo, D., Boyarchuk, A, Kilpio, E., Kuznetsov, O.: 2002, *Astron. Reports* **46**, 1022.
2. Mikolajewska, J., Kenyon, S. J.: 1992, *MNRAS* **256**, 177.
3. Fernandez-Castro, T., Cassatella, A., Gimenez, A., Viotti, R.: 1988, *ApJ* **324**, 1016
4. Roe, P.L.: 1986, in *Ann. Rev. Fluid Mech.*, **18**, 37.
5. Chakravarthy, S., Osher, S.: 1985, *AIAA Pap* N 85-0363 .
6. Cox, D.P., Daltabuit, E.: 1971, *ApJ* **167**, 113.
7. Vogel, M., Nussbaumer, H.: 1994, *A&A* **284**, 145.
8. Tomov, N., Tomova, M., Taranova, O..: 2004, *ASP Conf. Ser.* "Cataclysmic variables and Related Objects" in press.

Fe II emission lines of RR Tel during an obscuration event

D. Kotnik-Karuza*, M Friedjung[†], K. Exter**, F.P. Keenan** and D.L. Pollacco**

*Physics Dept, University of Rijeka, 51000 Rijeka, Croatia
[†]Institut d'Astrophysique, 98 bis Boulevard Arago, 75014 Paris, France
**Dept of Pure and Applied Physics, Queens University, Belfast, Northern Ireland

Abstract. A study of the behaviour of the permitted Fe II and forbidden [Fe II] emission lines during a dust obscuration episode, suggests a larger flux decrease for the permitted than for the forbidden lines. No other correlation with line properties have been found. Possible interpretations are discussed.

Keywords: Binary stars – Accretion disk – Gas-dynamical modelling
PACS: 97.80.Űd, 98.62.Mw

INTRODUCTION

The symbiotic nova RR Tel is an interacting binary containing a Mira variable, which loses mass via a strong wind, and a compact component, believed to be a white dwarf, that should accrete a part of that wind. The single large amplitude outburst observed up to now (in 1944), involved a rise of 7 magnitudes in the photographic region in less than one year, an almost constant brightness for 5 years and a much slower decline. Specialists explain such outbursts by the onset of thermonuclear burning of the accreting white dwarf.

The Mira component, like normal Miras, is surrounded by dust, whose absorption dominates in the near infra-red. As for only a small proportion of normal Miras, dust obscuration events with increased absorption are observed in that spectral region. Like in other symbiotic Miras such events are much more difficult to detect in the optical.

In this work we present new information about the behaviour of emission lines during a dust obscuration episode of RR Tel.

THE SPECTRA AND THE RESULTS

Flux calibrated spectra at two epochs taken with the Anglo-Australian telescope are compared. One described by Crawford et al. 1999, covering the region from 3100-9800 Å, was obtained on July 22 1996 with a resolution of about 50000 and was flux calibrated on August 2 1996. The other was taken in July 2000 with almost twice the spectral resolution and was flux calibrated with two other spectra, including one taken with the HST in October 2000.

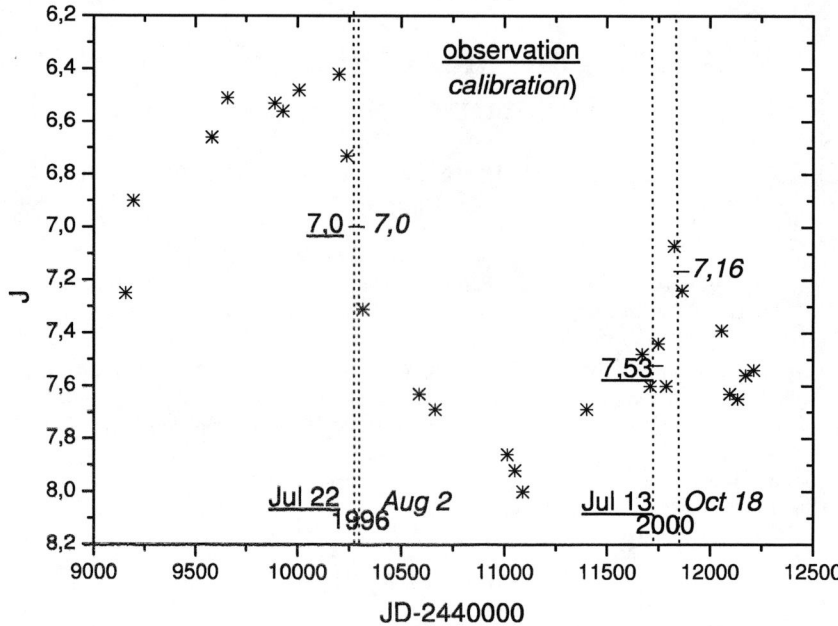

FIGURE 1. J light curve corrected for mira pulsations in the time interval in which the two Fe II line emission spectra were taken: during a dust obscuration episode (2000) and out of it (1996)

A dust obscuration event was taking place when the spectra of 2000 were taken. Correcting for the mira pulsations, a fading of only 0.16 magnitudes in J is obtained between the times of the calibration spectra, because of a temporary J brightening when the calibration of the later spectrum was obtained (Fig. 1). The observed visual fading between the same two dates was 0.48 mag (Fig. 2).

According to Rieke and Lebofsky (1985), the fading of 0.16 in J would correspond to a fading of 0.57 mag in V on the same line of sight or of -0.24 in the logarithm if most of the light in V was due to the cool component.

Log ratios of the Fe II and [Fe II] emission line fluxes of the two spectra, plotted against wavelength, are shown in Fig 3. Lines with wavelengths below 3490 Å were not taken into account, as the Crawford calibration in that region is highly uncertain. The measured line of multiplet 73 at a large wavelength and a few other weak extremely discordant lines were also eliminated.

Only the difference between the permitted and forbidden lines appears to be significant, no significant correlation being found between other line properties and log ratio. The best correlation between the log flux ratio of the forbidden lines and log $(gf\lambda)$ still has a probability of 0.25 and would moreover be difficult to understand for optically thin forbidden lines. The mean flux ratios are -0.845 ± 0.14 for Fe II and -0.658 ± 0.14 for [Fe II] with corresponding non significant slopes of the attempted correlations of $4.28\ 10^{-5} \pm 2.61\ 10^{-5}$ and $-0.28\ 10^{-5} \pm 2.49\ 10^{-5}$.

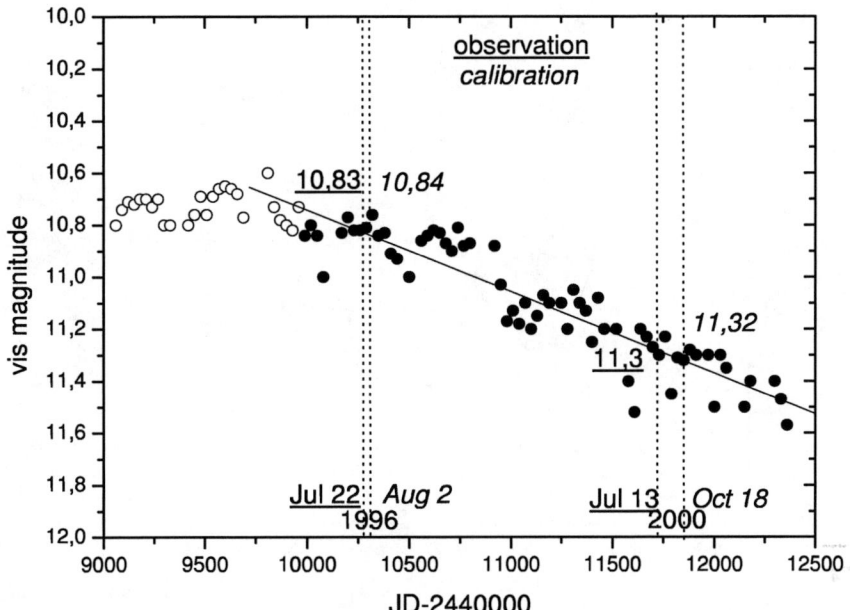

FIGURE 2. Visual light curve in the time interval in which the two Fe II line emission spectra were taken: during a dust obscuration episode (2000) and out of it (1996)

DISCUSSION

Two types of model might be considered. One involves spherical symmetry of the extra dust absorption in the Mira wind, while the other assumes the presence of a cloud. In both cases the apparent radius of the line emitting region found by the SAC (Self-Absorption Curve) method, might be expected to decrease during a dust obscuration episode (Kotnik-Karuza et al. 2002, Kotnik-Karuza et al. 2003). In both cases the presumably less absorbed forbidden line region will be larger.

The log optical absorption of -0.24, corresponding to the J absorption, is less than that of the Fe II lines. One might think that this disagreement is due to the difficulty of finding the exact J magnitudes when the spectra were taken, so the spherically symmetric model, with fewer grains above the forbidden line region than the permitted line one, is not then necessarily contradicted. We must, nevertheless, emphasize that the low velocity of the Mira wind should lead to the line formation regions being occulted by dust significantly later than the Mira itself, so direct comparison of the J and line flux fadings is presumably not justified.

The lack of a clear wavelength dependence of the extra absorption of the optical lines, unlike the behaviour shown in the near infra-red, is however a problem and might suggest rather the presence of an optically thick cloud occulting much of the line emitting regions. More work is required to better understand what is going on.

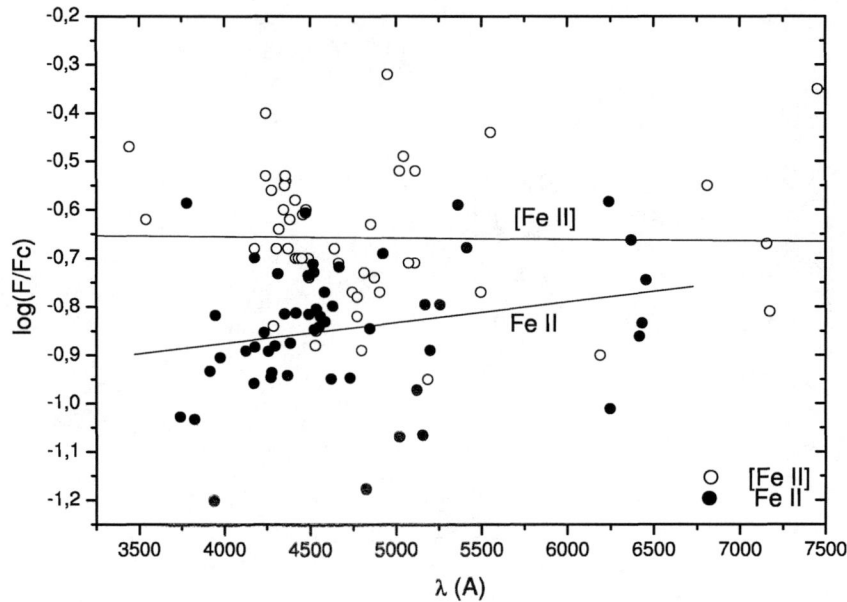

FIGURE 3. Fading of Fe II and [Fe II] log line fluxes from 1996 to 2000

ACKNOWLEDGMENTS

We are grateful to Patricia Whitelock for providing us the IR observations of RR Tel, as well as to Janet Mattei and Rebecca Pollock who made the visual magnitudes of the star available to us.

REFERENCES

1. Crawford, F.L., McKenna, F.C., Keenan, F.P., Aller, L.H., Feibelman, W.A., Ryan, S.G, *A&AS*, 139, 135-140(1999)
2. Kotnik-Karuza, D., Friedjung, M., Selvelli, P.L., *A&A*, 381, 507-516(2002)
3. Kotnik-Karuza, D., Friedjung, M., Exter, K., Keenan.F.P., Pollacco, D.L., "New Results Concerning the Fe II Lines of RR Tel, in *Symbiotic stars probing stellar evolution*, edited by R.L.M.Corradi et al., ASP Conference Series 303, 2003, pp. 136-140
4. Rieke, G.H., and Lebofsky, M.J., *ApJ*, 288, 618-621(1985)

LS 5039 / RX J1826.2-1450: A Young Pulsar?

Andrea Martocchia*, Christian Motch* and Ignacio Negueruela[†]

*CNRS / Observatoire Astronomique de Strasbourg, 11 rue de l'Université, F–67000 Strasbourg, France
[†]Departamento de Física, Ingeniería de Sistemas y Teoría de la Señal, Escuela Politécnica Superior, University of Alicante, Ap. 99, E–03080 Alicante, Spain

Abstract. Recent XMM-*Newton* and *Chandra* observations of the high mass X-ray binary LS 5039 / RX J1826.2-1450 caught the source in a faint X-ray state. In contrast with previous *Rossi*-XTE observations, we fail to detect any evidence of iron line emission. We also fail to detect X-ray pulsations. The X-ray spectrum can be well fitted by a simple powerlaw, slightly harder than in previous observations, and does not require the presence of any additional disk or blackbody component. XMM-*Newton* data imply a X-ray photoelectric absorption consistent with optical reddening ($N_H \sim 7 \times 10^{21}$ cm^{-2}), indicating that no strong local absorption occurs. Among the possible source emission mechanisms and hypotheses on the nature of the compact object, a young pulsar scenario may be the most appropriate.

Keywords: Infall and accretion — Pulsars — X-ray binaries
PACS: 97.60.Gb, 98.35.Mp, 97.80.Jp,

THE SOURCE

LS 5039 / RX J1826.2-1450 is a massive X-ray binary (HMXB) identified in the *ROSAT* all-sky survey by Motch et al. (1997) [1]. Optical follow-up observations led to the detection of a bright V = 11.2, O6.5V((f)) star as a counterpart [2]. The dynamical parameters of this source, as well as the hardness of its X-ray emission, are consistent with the compact object being either a neutron star (NS) or a black hole (BH). It may be accreting directly from the companion's wind, but the presence of an accretion disk has not been excluded yet. The optical photometric variability is very small (< 0.01 mag [3]) and optical colours yield $E(B-V) \sim 1.26$ [1], which implies $N_H \sim 7.2 \times 10^{21}$ cm^{-2}. The source distance has been estimated at about 3.1 kpc ([1], [4]). The binary parameters have been inferred spectroscopically, resulting in a period $P_{orb} = 4.4267 \pm 0.0005$ d and a high eccentricity $e = 0.48 \pm 0.06$ [5]. The mass function is quite low ($f(m) = 0.0017 \pm 0.0005 M_\odot$), yielding a lowest acceptable inclination $i > 9°$ and a compact object mass $M_{co} < 8 M_\odot$. O6.5V((f)) stars typically have $M \sim 36 M_\odot$ and $R \sim 10 R_\odot$ [6].

VLBA observations at milliarcsecond scales revealed that, in this source, persistent radio emission originates from bipolar jets, emerging for at least 6 milli-arcseconds from a central core (see [7], [8], and references therein): therefore, the source is usually referred to as one of the few known *microquasars*, and is one of the very few radio-

TABLE 1. X-ray unabsorbed flux in the 0.3–10 keV band (in units of 10^{-12} erg cm^{-2} s^{-1}), orbital phase and spectral powerlaw index of LS 5059 in the observations performed with different satellites. Orbital phases are computed according to the revised ephemeris [5]; the associated errors range from 2 to 5% (from [3]).

Observation	$F_{0.3-10keV}$	Orbital Phase	Γ
RXTE 08/02/98 (I)	~ 40	~ 0.91	1.95 ± 0.02
RXTE 08/02/98 (II)	~ 40	~ 0.10	1.95 ± 0.02
RXTE 16/02/98	~ 40	~ 0.88	1.95 ± 0.02
ASCA 04/10/99	~ 13	0.22–0.38	~ 1.5
SAX 08/10/00	~ 4.9	0.75–0.96	$\lesssim 1.8$
Chandra 10/09/02	~ 8.1	~ 0.35	$1.15^{+0.23}_{-0.21}$
XMM 08/03/03	~ 10.3	~ 0.79	$1.56^{+0.02}_{-0.05}$
XMM 27/03/03	~ 9.7	~ 0.21	$1.49^{+0.05}_{-0.04}$

emitting HMXBs.[1] The source may have been detected also in the γ-ray band ($E > 100$ MeV): an association with the *EGRET* source 3EG J1824-1514 has been proposed ([7], [8]).

Interestingly, LS 5039 is a runaway system, escaping from the Galactic plane with a total systemic velocity of ~ 150 km s^{-1} and a perpendicular component greater than 100 km s^{-1} with respect to the plane itself ([4], [9]).

LS 5039 IN THE X-RAYS

A summary of the X-Ray "history" of RX J1826.2-1450, after its first detection by *ROSAT* [1], is given in Table 1. The references for the *Rossi*-XTE and *Beppo*SAX observations are [10] and [11], respectively; however, we checked PDS data ourselves to better constrain Γ in the *Beppo*SAX observation. We also analyzed *ASCA* SIS and GIS data, obtaining the results which are reported here. Finally, we analyzed for the first time recent *Chandra* and XMM-*Newton* data [3].

A broad emission line was seen in the 1998 *Rossi*-XTE PCA dataset at $E_0 \sim 6.6$ keV [10], i.e. compatible with a slightly blue-shifted, fluorescent neutral iron feature, or with FeXXV emission at rest velocity. In principle, such a detection could be used to test assumptions on the accretion/ejection flows, and to discriminate among possible scenarios for the central compact object; however, while the PCA energy resolution is insufficient to distinguish between the possible models for the line, its presence could not be confirmed by any of the observations performed afterwards.

LS 5039 was observed with *Chandra* ACIS on 2002, September 10, for about 10 ks (06:44:00–10:05:10 UT) in Faint Mode, through the High Energy Transmission Grating

---

[1] Together with Cyg X-1, SS 433, Cyg X-3, CI Cam, LS I +61°303. However, the latter is the only one to have the same γ-ray properties as LS 5039 / RX J1826.2-1450.

(HETG). We analysed the zeroth-order image, with slightly more than 1000 source counts registered, corresponding – when a simple powerlaw continuum is assumed – to an unabsorbed flux level of $\sim 6.8 \times 10^{-12}$ erg cm^{-2} s^{-1} in the 2 to 10 keV band. This is slightly more than in the 2000 *Beppo*SAX pointing, which therefore corresponded to the lowest observed state of this source. However, the spectrum is the hardest in the *Chandra* observation ($\Gamma \sim 1.1$, see Table).

We tried to perform data fits also by adding an iron line and thermal disk emission. Although the signal-to-noise ratio of the observation is rather poor, we can set some constraints on individual spectral components: in particular, a narrow ($\sigma = 20$ eV) or broad ($\sigma = 390$ eV) Fe line is always compatible (within the 90 % confidence level) with having a null EW, and the F-test shows that its introduction never gives a significantly better fit; parameters cannot be really constrained in the powerlaw-plus-blackbody model. The statistics are also too poor to extract any useful information on short-time variability and/or X-ray pulsations.

More recently, in 2003, two pointings of LS 5039 were performed by XMM-*Newton*. The first one took place on March 8 (07:33:27–10:28:38 UT), the second on March 27 (20:53:47–23:48:59 UT). The EPIC observations were performed with Medium filters on; the observing modes were Prime Small Window and Prime Partial W2, for the pn and MOS cameras respectively. Too few counts were collected in the RGS cameras to get any useful information. We performed fitting of the data from the three instruments together, with the same models used for the *Chandra* data. A simple powerlaw provides a good fit to the combined EPIC data for both observations ([3]). The powerlaw index Γ is now ~ 1.5, as typical of most X-ray binaries in the hard/low state. with other Again, a (narrow or broad) Fe line has a null EW within the 90% confidence level. There is no evidence of edges or fluorescent lines at lower energies, either, which would have given information on the possible cold surrounding medium (e.g. the stellar wind). On the other hand, we could better constrain the hydrogen column density: we get $N_H = 0.70 \pm 0.05 \times 10^{22}$ cm^{-2} in powerlaw models, i.e. no evidence for any intrinsic absorption. By adding a disk multi-blackbody component over the powerlaw we obtain negligible normalisations for the thermal emission ($L_{\mathrm{diskbb}}/L_{\mathrm{total}} < 0.09$ and 0.20, in the two observations respectively). We therefore conclude that there is no evidence of disk emission. Similarly, the addition of a simple blackbody on the top of the powerlaw component does not improve the fit in a statistically significant manner. For a complete description of the spectral modelling and detailed results see [3].

In order to search for pulsations, we applied the Z_1^2 test on the events collected by EPIC pn in a narrow circle centred on the source: we found no significant peaks in the frequency range 0.001–83 Hz.[2] As far as the short time scale variability is concerned, the lightcurves only show slight variation during each pointing, with no evidence of aperiodic random fluctuations typical of wind accretors:[3] we tried with different binnings (in the range 10–1000 s) and found fluctuations of $\sim 20\%$ [3].

---

[2] It must be stressed that there has never been evidence of any periodicities in X-ray lightcurves of this source.

[3] The latter usually display flaring with flux changes of a factor ~ 5 over timescales of a few hundred seconds.

WHAT IS LS 5039 ACTUALLY?

Is RX J1826.2-1450 an accreting compact object? We have seen that *Chandra* and XMM-*Newton* data permit the exclusion of any large disk (thermal, reflected or line) spectral components at a confidence level $> 90\%$. The presence of a large disk, filling the Roche lobe, is in fact excluded by the orbital parameters [12] as well as by the modest source luminosity [9].

Therefore this could be one of the rare cases in which jets form without a well-formed disk.

The compact object may be accreting directly from the companion's wind. Assuming a simple spherical model, the entire X-ray flux range (a factor ~ 10) spanned by the source since its discovery could be explained either by differences in the *orbital phases* from one observation to the other, or by long term changes in the average *wind density*. However, there is no evidence of aperiodic random fluctuations typical of wind accretors, and there is no evidence of intrinsic absorption.

Is RX J1826.2-1450 a young pulsar, interacting with the O star wind? The radio-plus-X-plus-γ luminosity of RX J1826.2-1450 is comparable to the spin-down power of young radio pulsars. Possibly related cases are those of LS I +61°303 and/or PSR 1259-63, in which shock-powered emission between the relativistic wind of the pulsar and the Be star wind can account for the high energy emission (see e.g. [13]). Thus, RX J1826.2-1450 may be a young pulsar, interacting with the wind of the optically visible companion LS 5039. The young pulsar hypothesis is also consistent with the estimated age of the system [4].

REFERENCES

1. C. Motch, F. Haberl, K. Dennerl, M. Pakull, E. Janot-Pacheco, A&A, **323**, 853 (1997)
2. J.S. Clark, P. Reig, S.P. Goodwin, et al., A&A, **376**, 476 (2001)
3. A. Martocchia, C. Motch, and I. Negueruela, A&A, in press, astro-ph/0409608 (2005)
4. M. Ribó, J.M. Paredes, G.E. Romero, et al., A&A, **384**, 954 (2002)
5. M.V. McSwain, D.R. Gies, W. Huang, P.J. Wiita, D.W. Wingert, and L. Kaper, ApJ, **600**, 927 (2004)
6. I.D. Howarth and R.K. Prinja, ApJ Suppl., **69**, 527 (1989)
7. J.M. Paredes, J. Martí, M. Ribó and M. Massi, Science, **288**, 2340 (2000)
8. J.M. Paredes, M. Ribó, E. Ros, J. Martí, and M. Massi, A&A, **393**, L99 (2002)
9. M.V. McSwain and D.R. Gies, ApJ 568, L27 (2002)
10. M. Ribó, P. Reig, J. Martí and J.M. Paredes, A&A, **347**, 518 (1999)
11. P. Reig, M. Ribó, J.M. Paredes, J. Martí, A&A, **405**, 285 (2003)
12. J. Martí, P. Luque-Escamilla, J.L. Garrido, J.M. Paredes, and R. Zamanov, 2004, A&A, **418**, 271 (2004)
13. M. Tavani and J. Arons, ApJ, **477**, 439 (1997)

A refined method for measuring jet speeds

James Miller-Jones[*,†], Katherine Blundell[*] and Peter Duffy[**]

[*]*Astrophysics, University of Oxford, Keble Road, Oxford, OX1 3RH, UK*
[†]*Astronomical Institute 'Anton Pannekoek', University of Amsterdam, Kruislaan 403, 1098 SJ, Amsterdam, The Netherlands*
[**]*Department of Mathematical Physics, University College Dublin, Dublin 4, Ireland*

Abstract. The flux density ratio of corresponding jet knots on opposite sides of the nucleus is often used in quasar and microquasar systems to determine the product of the jet speed β and the cosine of the angle between the jet axis and the line of sight, $\cos\theta$. For this determination to be accurate, and owing to the changing intensities of the jet knots with time, this flux density ratio must be measured when the knots are at equal angular separation from the core. We present a refined formalism which enables the flux density ratio of corresponding knots from a single image to be used to determine the value $\beta\cos\theta$ in the case of adiabatically-expanding, optically-thin, synchrotron-emitting jet knots.

Keywords: Synchrotron radiation, relativity, jets
PACS: 98.38.Fs,98.62.Nx

INTRODUCTION

Relativistic jets in both microquasars and their extragalactic analogues, the quasars, are often seen to consist of a series of discrete knots moving outwards from a central compact object. The flux density ratio of corresponding knots on opposite sides of the core may be used to determine the product $\beta\cos\theta$, where β is the jet speed, and θ is the angle the jet axis makes with the line of sight, thus:

$$\frac{S_{\text{app}}}{S_{\text{rec}}} = \left(\frac{1+\beta\cos\theta}{1-\beta\cos\theta}\right)^{k+\alpha}, \qquad (1)$$

where α is the spectral index of the emission (defined by $S_\nu \propto \nu^{-\alpha}$, where S_ν is the flux density at frequency ν) and $k=3$ for a jet composed of discrete ejecta.

However, the luminosities $L(t)$ of the jet knots change with time, as adiabatic expansion and synchrotron losses take effect, so the true flux density ratio of corresponding knots on either side of the core is in fact

$$\frac{S_{\text{app}}}{S_{\text{rec}}} = \left(\frac{1+\beta\cos\theta}{1-\beta\cos\theta}\right)^{k+\alpha} \frac{L(t_{\text{app}})}{L(t_{\text{rec}})}, \qquad (2)$$

where t_{app} and t_{rec} are the times in the observer's frame when light leaves the approaching and receding knots respectively in order to arrive at the telescope at the same time. Unless the jet axis is perpendicular to the line of sight, the light-travel time between approaching and receding knots will mean that we see the receding knots at an earlier time, i.e. when they were more compact and hence intrinsically brighter (but dimmed by the Doppler deboosting effect taken into account by the original formalism) compared with the approaching knots seen at the same telescope time. To compensate for this effect, Mirabel and Rodríguez [1] proposed that the flux densities of the approaching and receding components should be measured at equal angular separations from the core. However, this requires good enough temporal sampling to be able to interpolate the flux densities back to equal angular separations.

NEW FORMALISM FOR THE CASE OF SYNCHROTRON EMISSION AND ADIABATIC EXPANSION

For the case of symmetric approaching and receding jets, after ejection at $t = 0$, the epochs in the observer's frame at which photons leave corresponding points of the approaching and receding plasmons in order to arrive at the observer at the same time are related by

$$\frac{t_{app}}{t_{rec}} = \frac{1+\beta\cos\theta}{1-\beta\cos\theta}. \tag{3}$$

The total synchrotron emissivity from a single, optically thin jet knot scales as [e.g. 2]

$$J(\nu) \propto B^{3/2} N(\gamma) \gamma^2 \nu^{-1/2}, \tag{4}$$

where B is the magnetic field strength, γ is the Lorentz factor of an individual electron assumed to be radiating at a single frequency

$$\nu = \left(\frac{\gamma^2 eB}{2\pi m_e}\right), \tag{5}$$

and $N(\gamma)$ is the total number of electrons in the plasmon with energies in the range $(\gamma, \gamma+d\gamma)$. Assuming a power-law electron energy spectrum with electron index p, $N(\gamma) \propto \gamma^{-p}$, and a tangled magnetic field, $B \propto R^{-1}$, the plasmon emissivity has a simple dependence on frequency and size given by

$$J(\nu) \propto \nu^{(1-p)/2} R^{(1-3p)/2}. \tag{6}$$

The ratio of flux densities as seen by the observer is then

$$\frac{S_{app}}{S_{rec}} = \left(\frac{R(t_{app})}{R(t_{rec})}\right)^{(1-3p)/2} \left(\frac{1+\beta\cos\theta}{1-\beta\cos\theta}\right)^{k+(p-1)/2}. \tag{7}$$

In the case of linear expansion, $R \propto t$, and using Equation 3, Equation 7 becomes

$$\frac{S_{app}}{S_{rec}} = \left(\frac{1+\beta\cos\theta}{1-\beta\cos\theta}\right)^{k-p}. \tag{8}$$

This is the flux density ratio observed at a given instant by the telescope as opposed to the interpolated flux density ratio at equal angular separation. While the formula is very similar to the original formalism, the exponent is very different.

Caveats

If the spectrum of the source contains a break or turnover, and the approaching and receding knots are not both in the optically-thin regime, then the above formalism will need to be modified further. A detailed discussion of the case of sharp spectral breaks is given, together with a more detailed presentation of this new formalism, in Miller-Jones et al. [3].

Furthermore, measuring the spectral index α (from which the electron index p may be deduced) for an individual knot may be difficult. It is often easier [e.g. 4, 5] to measure the integrated spectrum of the source, and while this may not differ from the spectrum of an individual knot in the jet-dominated case, it will introduce error if there is a significant flat-spectrum core component to the emission.

Care should also be taken in applying this formalism if the expansion mode of the plasmons changes prior to the observation from which the flux density ratio is derived. In such a case, for example a transition from slowed to free expansion [6], the time decay of the flux density would change (steepen with time in this case). In order to constrain the value of $\beta \cos \theta$ in this way, the flux densities of the approaching and receding knots would then have to be measured when the knots were both in the same expansion regime. Unless the transition radius were known, this would require actually measuring (rather than interpolating) the flux densities at equal angular separation from the core.

If there is significant deceleration of the expanding plasmons due to interaction with surrounding material [e.g. 7], then $(R/R_0) \propto (t/t_0)^\eta$, where $\eta < 1$, and Equation 8 then requires modification. Further related caveats are presented by Fender [8], who found that when using proper motions to place limits on the bulk Lorentz factors of jets, any Lorentz factors thus derived are strictly only *lower limits*.

APPLICATION TO REAL SYSTEMS

This formalism may be applied to the radio images of the 2001 outburst of Cygnus X-3 presented by Miller-Jones et al. [4]. Photometric VLA observations determined the overall spectral index of the source to be $\alpha = 0.60 \pm 0.05$, and the measured flux density ratios in the 5- and 15-GHz images varied between 1.14 ± 0.19 and 3.09 ± 0.14. Using the value of $\beta \cos \theta = 0.62 \pm 0.11$ determined from the precession model fitting and the observed proper motions of the jet knots, the predicted flux density ratio is 3.2 ± 1.0. By way of comparison, the original formalism (Equation 1) predicts a flux density ratio of ~ 185, demonstrating the superiority of our new method.

Our formalism can also be applied to the observations of GRS 1915+105 made by Mirabel and Rodríguez [9]. We take the flux density ratio of the observed knots once they had clearly separated from one another and from the nucleus, and we only compare corresponding pairs of ejecta. From their derived values of $\beta \cos \theta = 0.323 \pm 0.016$ and $\alpha = 0.84 \pm 0.03$, we predict a flux density ratio of 1.24 ± 0.05. For the later epochs (1994 April 16, 23 and 30), the measured flux density ratios are 2.33, 2.63 and 1.80 respectively, slightly greater than we predict, but of the right order.

Another case where this method could have been of use are the observations of Ribó et al. [10]. They imaged six microquasar candidates at 5 GHz with the European VLBI

Network (EVN) and the Multi-Element Radio-Linked Interferometer Network (MER-LIN). A two-sided jet was detected in one of the candidates (1RXS J001442.2+580201), but since only a single image was made, and since the components were not at equal angular separations from the core, only a lower limit of $\beta\cos\theta > 0.09 \pm 0.04$ for the inner pair of ejecta could be derived from the flux density ratio of the respective knots. The overall spectral index α was known to be -0.20 ± 0.05 from previous VLA observations [5]. Our method allows an actual value to be derived for $\beta\cos\theta = 0.187$ for the innermost ejecta (an accurate assessment of the uncertainty on this value cannot be made, as Ribó et al. did not state the errors on the flux densities of the components). This compares favourably with their value of $\beta\cos\theta > 0.20 \pm 0.02$ obtained from a proper motion analysis of the outermost pair of ejecta (assuming a constant jet speed and inclination angle). These data demonstrate the advantages of our new method over the equal angular separation method in cases of limited temporal sampling.

CONCLUSIONS

We have considered the evolution of adiabatically-expanding, optically-thin, synchrotron-emitting jet knots in quasars and microquasars. We have developed an expression for the flux density ratio of corresponding knots located on opposite sides of the core, observed at the same time in the observer's frame, for the case of linear expansion of the jet knots. This allows us to constrain the value of $\beta\cos\theta$ from a single image of the jet.

ACKNOWLEDGMENTS

Reproduced by kind permission of the AAS. This is a summary of the original article [3]. J.C.A.M.-J. thanks the UK Particle Physics and Astronomy Research Council for a Studentship. K.M.B. thanks the Royal Society for a University Research Fellowship. K.M.B. and P.D. acknowledge a joint British Council/Enterprise Ireland exchange grant.

REFERENCES

1. I. F. Mirabel, and L. F. Rodríguez, *Ann. Rev. A&A.*, **37**, 409–443 (1999).
2. M. S. Longair, *High energy astrophysics. Vol.2: Stars, the galaxy and the interstellar medium*, Cambridge: Cambridge University Press, 1994, 2nd ed., 1994.
3. J. C. A. Miller-Jones, K. M. Blundell, and P. Duffy, *ApJ Letters*, **603**, L21–L24 (2004).
4. J. C. A. Miller-Jones, K. M. Blundell, M. P. Rupen, A. J. Mioduszewski, P. Duffy, and A. J. Beasley, *ApJ*, **600**, 368–389 (2004).
5. J. M. Paredes, M. Ribó, and J. Martí, *A&A*, **394**, 193–203 (2002).
6. R. M. Hjellming, and K. J. Johnston, *ApJ*, **328**, 600–609 (1988).
7. R. M. Hjellming, and X. Han, "Radio Properties of X-ray Binaries," in *X-Ray Binaries*, edited by W. H. G. Lewin, J. van Paradijs, and E. P. J. van den Heuvel, Cambridge Astrophysics Series 26, Cambridge University Press, Cambridge, MA, 1995, pp. 308–330.
8. R. P. Fender, *MNRAS*, **340**, 1353–1358 (2003).
9. I. F. Mirabel, and L. F. Rodríguez, *Nature*, **371**, 46–48 (1994).
10. M. Ribó, E. Ros, J. M. Paredes, M. Massi, and J. Martí, *A&A*, **394**, 983–991 (2002).

Time-Delayed transfer functions simulations for LMXBs

T. Muñoz-Darias*, I. G. Martínez-País[†] and J. Casares[†]

*Instituto de Astrofísica de Canarias, 38200 La Laguna, Tenerife, Spain ; tmd@iac.es
[†]Instituto de Astrofísica de Canarias, 38200 La Laguna, Tenerife, Spain

Abstract. Recent works (Steeghs & Casares 2002, Casares et al. 2003, Hynes et al. 2003) have demonstrated that Bowen flourescence is a very efficient tracer of the companion star in LMXBs. We present a numerical code to simulate time-delayed transfer functions in LMXBs, specific to the case of reprocessing in emission lines. The code is also able to obtain geometrical and binary parameters by fitting observed (X-ray + optical) light curves using simulated annealing methods. In this work we present the geometrical model for the companion star and the analytical model for the disc and show synthetic time-delay transfer functions for different orbital phases and system parameters.

Keywords: accretion, accretion discs, X-rays:binaries, binaries:close, X-rays:stars
PACS: 90

INTRODUCTION

Optical emission in low mass X-ray binaries (LMXBs, hereafter) mainly arises from X-ray reprocessing in different binary sites. Therefore, the observed optical variability should be delayed due to light travel time difference, with respect to the source X-ray emission. Echo tomography is a very powerful technique which allows one to probe the accretion geometry in active LMXBs by studying time delay as a function of the orbital phase. Previous attempts using broad band photometry and spectroscopy (Hynes et al. 2003 , O'Brien 2001) have shown that the accretion disc is the dominant reprocessing site in the continuum. On the other hand, narrow emission lines originating from the irradiated donor star have been discovered in Sco X-1 (Steeghs & Casares 2002) and others LMXBs (Casares et al. 2004). These narrow features are strongest in the Bowen blend, a set of resonant emission lines, mainly CIII and NIII $\lambda 4640$ Å. In particular, the NIII line are powered by flourescence resonance, which require seed photons of HeII Lyα. The Roche lobe shaped donor star intercepts the energetic photons from the inner accretion disc resulting in the observed optical emission lines from its surface. Therefore, these features are expected to show delayed variability with respect to the irradiating X-ray emission, which can be used to constrain the size of the binary. Since it is clear that Bowen blend emission arises from the companion star, we have built time-delayed transfer functions specific for reprocessing in spectral lines. These functions are based on a geometrical model for the companion star and analytic solutions for the accretion disc.

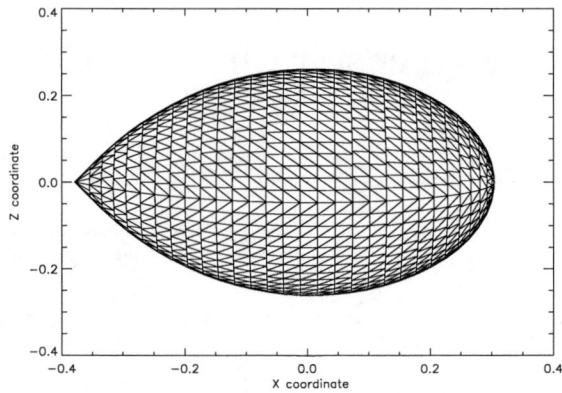

FIGURE 1. Roche lobe filling star for a mass ratio q = 0.3 and orbital phase of 0.25

GEOMETRICAL MODEL FOR THE COMPANION STAR

The model developed is based on the Roche potential for the secondary star. In order to carry out the numerical integration we have divided the shaped surface in small triangular tiles of equal area which cover completely the Roche lobe filling star (see figure 1). The mass ratio q (M_2/M_1) is the free parameter which defines the shape of the companion. On the other hand, the inclination and the orbital phase determine in which moment the different regions are visible by the observer. Eclipses and shadowing effects by a cilindrical accretion disc are also included in the model.

BUILDING TIME-DELAYED TRANSFER FUNCTIONS

We have considered time-delayed transfer functions as the response given by the system to a normalized X-ray emission. As reprocessing time scales for spectral lines are much lower than 1 second the system response can be aproximated by a mirror function whose efficiency only depends on the albedo and the X, Y, Z coordinates of the region considered. For the case of the companion star it is necessary to define a delay-bin (given by the size of the tiles) and to sum up the contributions of the irradiated regions which are visible by the observer. This is repeated for all the possible delays, resulting in time-delayed transfer functions for the companion star which depends on the orbital phase, the inclination, the mass ratio q and the disc geometry. In order to compute the Transfer Function for the disc, we have approximated this by an axially symmetric structure whose vertical cilindrical coordinate, z, depends on the radial coordinate as r^β. The geometry of the disc is characterized by β and both, the inner and outer disc radii. In this way, the transfer function can be obtained analytically taking into account the effects of albedo and the local projection factors. The exponent β is fixed to 9/7, for the case of a standard irradiated disc (Vrtilik et al, 1990).

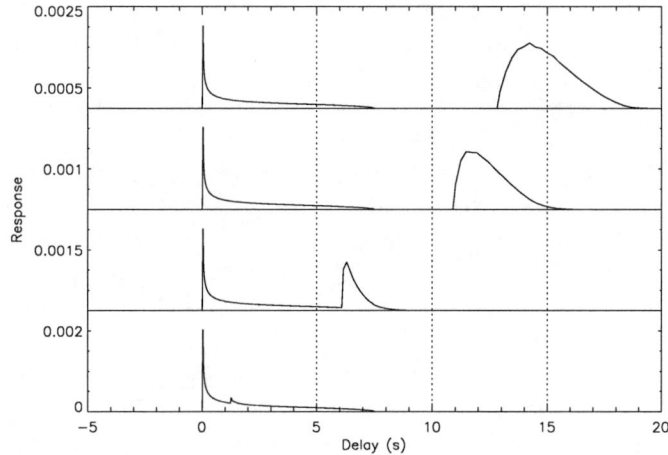

FIGURE 2. Time-delayed transfer functions for i = 80 and system parameters of Sco X-1 (i.e. q= 0.3 and P= 18.9 h). For phases close to 0.5 the delay is maximum (15 s) and the response of the companion star is comparable to the accretion disc contribution.

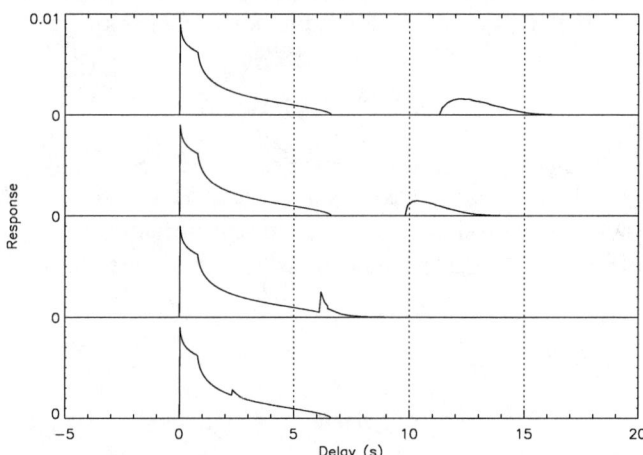

FIGURE 3. Same as in figure 2 but for i = 50. For lower inclinations the disc contribution dominates the transfer function. In this case, for phases close to 0.5, the maximun delay is 12.5 s which is lower than for i = 80 because it scales with cos (i).

After combining the disc and the companion star contributions we get the time-delayed transfer function for the entire system. As we see in figures 2, 3 and 4 the function is strongly dependent on both the orbital phase and system inclination. Therefore,

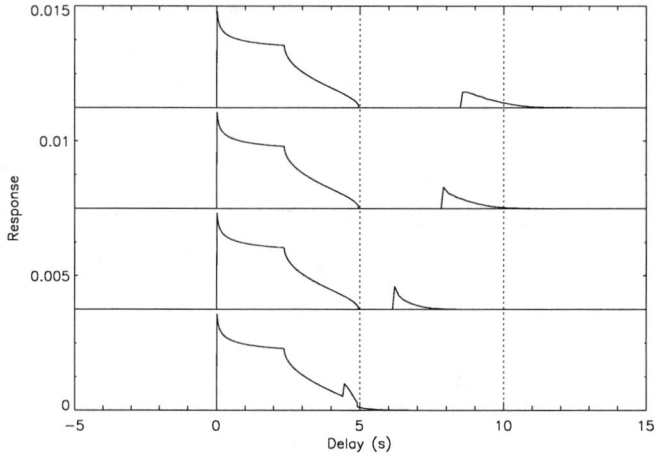

FIGURE 4. Same as in figure 2 but for i=20. The profile of the disc transfer functions is remarkably different in this case because the inner disc is fully visible.

it is possible to compare these simulations with real data (simultaneous X-ray/optical narrow band photometry) in order to constrain the system parametres.

CONCLUSIONS

1. As expected, our Time-delayed transfer functions are strongly dependent on the inclination, the mass ratio (q) and the disc parameters (radius and β coefficient)
2. The transfer function is dominated by the disc contribution for $i \leq 60$
3. The profile of the disc transfer function is very sensitive to β for very low inclinations

REFERENCES

Casares, J., Steeghs, D., Hynes, R. I., & Charles, P. A. 2003, ApJ, 590, 1041

Casares, J., Steeghs, D., Hynes, R. I., Charles, P. A., Cornelisse R., & O'Brien, K. 2004, Rev.Mex AA,30,21

Hynes, R. I., Steeghs D., Casares, J., Charles, P. A., & O'Brien, K. 2003, ApJ, 583, L95

Steeghs, D. & Casares, J., 2002, ApJ, 568, 273

O'Brien, K. 2001 in ASP Conf. Series, (astro-ph/0110267)

Vrtilek, S. D., Raymond, J. C., Garcia, M. R., Verbunt, F., Hasinger, G., & Kurster, M. 1990, A&A, 235, 162

BeppoSAX observations of the X-ray binary pulsar GX 1+4

S. Naik*, P. J. Callanan* and B. Paul[†]

*Department of Physics, University College Cork, Ireland
[†]Tata Institute of Fundamental Research, Mumbai, India

Abstract. We present here the timing and spectral properties of BeppoSAX observations of the binary X-ray pulsar GX 1+4 carried out in August 1996, March 1997, and August 2000. In the middle of the August 2000 observation, the source was in a rare low intensity state that lasted for about 30 hours. Though the source does not show pulsations in the soft X-ray band (1.0-5.5 keV) during the extended low state, pulsations are detected in 5.5-10.0 keV energy band of the MECS detector and in hard X-ray energy bands (15-150 keV) of the PDS instrument. Broad-band (1.0-150 keV) pulse averaged spectroscopy reveals that the best-fit model comprises of a Comptonized continuum along with an iron K_α emission line. A strong iron K_β emission line is detected for the first time in GX 1+4 during the extended low state of 2000 observation. The optical depth and temperature of the Comptonizing plasma are found to be identical during the high and low intensity states whereas the hydrogen column density and the temperature of the seed photons are higher during the low state.

Keywords: stars : neutron — Pulsars : individual (GX 1+4) — X-rays : stars
PACS: 01.30.Cc, 95.85.Nv, 97.60.Gb

INTRODUCTION

The luminous accretion-powered X-ray pulsar GX 1+4 is one of the brightest and hardest X-ray sources in the sky. The neutron star in the binary system is a slow pulsar with a spin period of about 2 minutes. It is one of the brightest and hardest X-ray sources in the sky with a large rate of change of pulse period. The pulsar exhibited spin-up behavior in 1970s which followed by a spin-down activity after an extended low state detected in early 1980s ([1]). *BATSE* monitoring of the source, since 1991, confirmed the spin down trend with occasional torque reversal events ([2]). The pulsar occasionally shows a low state of a few hours duration during which, the pulsations are absent ([3]). This has been interpreted as due to centrifugal prohibition of accretion, also known as 'propeller effect'. The spectral fitting to the phase averaged RXTE data shows that the GX 1+4 energy spectrum is best fitted by a model consisting of a Comptonization continuum component and a Gaussian component, attenuated by the neutral absorption column density ([4]). Pulse-phase resolved spectroscopy reveals that the dip features in the pulse profiles are due to the eclipses of the emitting region by the accretion column ([5]).

Here in this paper, we describe the temporal and spectral properties of GX 1+4 in 1.0–150.0 keV energy band using three BeppoSAX observations. In subsequent sections we give details of the three BeppoSAX observations, the results obtained from the timing and spectral analysis, followed by a discussion of the results.

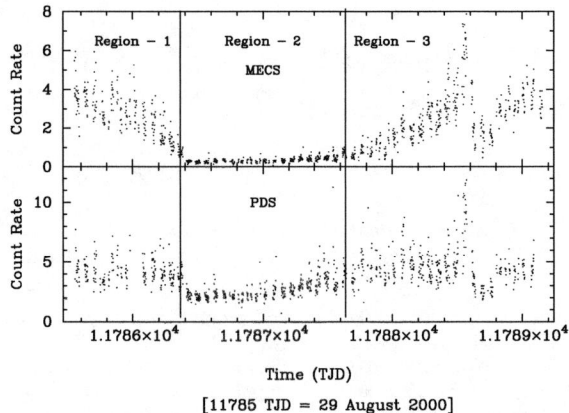

FIGURE 1. The MECS and PDS light curves with time bin intervals same as the spin period of the neutron star obtained from the 2000 BeppoSAX observation of GX 1+4.

OBSERVATION, ANALYSIS AND RESULTS

The observations of GX 1+4 were carried out with the BeppoSAX satellite in 1996 August 18 06:11 UT to August 19 03:38 UT (with 38.6 ks of MECS and 17.6 ks of PDS exposures), 1997 March 25 22:43 UT to March 26 16:08 UT (with 13 ks of LECS, 31.5 ks of MECS and 13.5 ks of PDS exposures), and 2000 August 29 12:36 UT to September 02 03:38 UT (with 56.5 ks of LECS, 132 ks of MECS, and 58.5 ks of PDS exposures). Time resolution of the instruments during these observations was 15.25 μs.

Timing Analysis

Standard procedure was applied to all the observations for data selection and extraction of background subtracted light curves and spectra by selecting appropriate source and background regions of suitable radius. Data from the MECS and PDS detectors were used for the timing analysis as GX 1+4 suffers heavily from absorption at soft X-rays by the intervening cold material. Light curves with time resolution of 1 s were extracted from barycenter corrected event files. An extended low state of $\sim$ 30 hr duration was seen in the light curves (region-2 of Figure 1) of 2000 observation. Pulse folding and chi-square maximization technique was applied to all the light curves and the pulse period was determined to be 124.404(3) s, 126.018(8) s and 134.9256(10) s during the 1996 August 18, 1997 March 25, and 2000 August 29 observations respectively. The pulse profiles, obtained from the MECS and PDS data of the three observations are shown in Figure 2. To get a detailed picture of the pulsation properties during 2000 observation when the pulse fraction was low ($\sim$ 30%), we divided the entire light curve into three regions (Figure 1). Pulsations were detected in the MECS light curves of regions 1 and 3 but not in region 2. However, pulsations were detected in all three regions in hard X-ray energy range.

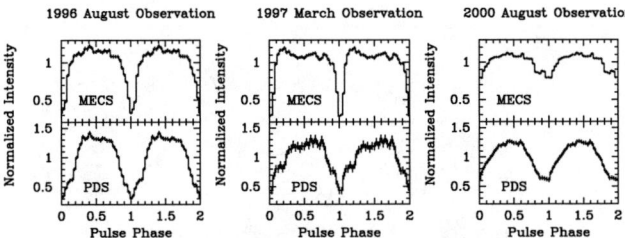

FIGURE 2. The MECS and PDS pulse profiles of the source, of the three observations, are shown in top and bottom panels with 32 phase bins per pulse respectively. Two pulses are shown for clarity.

Energy resolved pulse profiles of GX 1+4 show a clear difference in the shape of profiles in the soft and hard X-ray energy bands. The dip is narrow in soft X-rays and gradually becomes broader with energy. Though the pulse profiles in different energy bands of region 1 and 3 of 2000 August observation are similar to those seen in 1996 observation, pulsations are absent in 1–5.5 keV energy band of region 2 of 2000 observation. A phase difference of about 0.5 is also noticed between the pulse profiles in 5.5–10 keV (a double peaked profile with a pulse fraction of a mere $\sim 13\%$) and higher energy ranges. The light curve above 100 keV is mainly background dominated and pulsations were not detected in 100–250 keV energy band.

Spectral Analysis

The LECS and MECS spectra of GX 1+4 were extracted from regions of radii 6" and 4" respectively centered on the object for the three BeppoSAX observations. Background spectra for both LECS and MECS instruments were extracted from appropriate source-free regions of the field of view by selecting annular regions around the source. Standard procedures were applied to extract background subtracted PDS spectra from the event files of all three observations. The latest response matrices were used for spectral fitting. Events were selected in the energy ranges 1.0–4.0 keV for LECS, 1.65–10.0 keV for MECS and 15.0–150 keV for PDS where the instrument responses are well determined. Combined spectra from the LECS, MECS and PDS detectors, after appropriate background subtraction, were fitted simultaneously. All the spectral parameters, other than the relative normalization, were tied to be the same for all detectors.

The source spectra when fitted to a model consisting of a power law continuum with a Gaussian function for the iron line emission, and interstellar absorption (N_H) yielded poor values of reduced χ^2. Following this, we fitted the spectra with a Comptonized continuum model along with a Gaussian function, interstellar absorption and an absorption edge at ~ 30 keV and obtained a better value of reduced χ^2. The presence of a line like feature at ~ 7.1 keV in the residuals of the spectral fitting of all observations (as shown in Figure 3 for the low state of 2000 observation) allowed us to add another Gaussian function at 7.1 keV to the spectral model for the iron K_β emission line. The residuals of the spectral fitting of the extended low state, after addition of two Gaussian functions for iron K_α and K_β emission lines are shown in the bottom panel of Figure 3. The emission

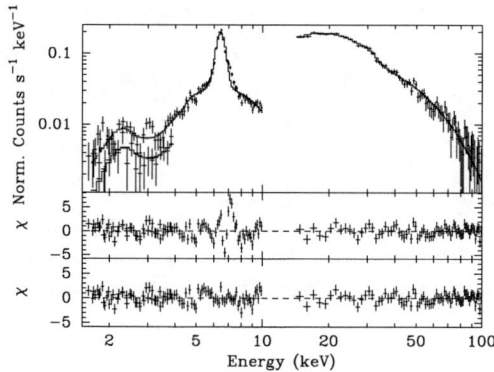

FIGURE 3. Energy spectra of GX 1+4 measured with BeppoSAX during the region 2 of 2000 August observation along with the fitted model (without iron K_β emission line) and the residuals (middle panel). The presence of iron K_β line is evident from the line like structure at ~ 7.1 keV in the residual. Residuals obtained after including a Gaussian function at 7.1 keV in the model are shown in the bottom panel.

TABLE 1. Spectral parameters for GX 1+4 during different intensity states

Parameter	1996 August	1997 March	2000 August Region–1	2000 August Region–2	2000 August Region–3
N_H [a]	21±2	1.1±0.1	$1.37^{+0.05}_{-0.15}$	$14.8^{+1.0}_{-3.4}$	2.4±0.1
kT_{Br} (keV)	$0.36^{+0.12}_{-0.11}$	———	———	$0.26^{+0.06}_{-0.04}$	———
T_0 (keV)	$1.48^{+0.07}_{-0.06}$	$1.28^{+0.04}_{-0.05}$	$1.62^{+0.04}_{-0.02}$	$3.5^{+0.6}_{-0.4}$	1.73±0.03
kT_{Co} (keV)	$13^{+0.4}_{-0.5}$	$12.7^{+1.2}_{-1.1}$	$15.2^{+0.5}_{-1.6}$	$12.6^{+1.9}_{-0.9}$	14.3±0.6
τ	$3.04^{+0.19}_{-0.14}$	$3^{+0.4}_{-0.3}$	$2.38^{+0.38}_{-0.09}$	$3.1^{+0.3}_{-0.5}$	2.55±0.13
Fe^1_c (keV)	6.42±0.02	$6.45^{+0.03}_{-0.02}$	6.49±0.03	6.47±0.01	6.45±0.02
W^1_0 (keV)	0.22±0.01	0.19±0.02	0.24±0.03	$3.0^{+0.6}_{-0.3}$	0.33±0.03
Fe^2_c (keV)	$6.99^{+0.30}_{-0.05}$	$7.06^{+0.10}_{-0.07}$	7.1[†]	$7.10^{+0.02}_{-0.05}$	$7.01^{+0.08}_{-0.09}$
W^2_0 (keV)	0.06±0.01	0.06±0.02	0.02±0.02	$0.55^{+0.14}_{-0.09}$	$0.08^{+0.04}_{-0.02}$
E_{edge} (keV)	33^{+1}_{-2}	30±3	33^{+2}_{-3}	32^{+13}_{-8}	35±2
Reduced χ^2	1.18	1.08	1.0	1.27	1.23
2–10 keV Flux[b]	$2.4^{+0.3}_{-0.2}$	$3.2^{+0.5}_{-0.3}$	$2.07^{+0.34}_{-0.08}$	$0.29^{+0.08}_{-0.05}$	$2.01^{+0.13}_{-0.15}$
10–100 keV Flux[b]	$14.9^{+1.1}_{-0.9}$	$7.3^{+0.8}_{-1.1}$	$10.1^{+1.6}_{-0.3}$	$5.1^{+1.1}_{-0.9}$	$11.2^{+0.7}_{-0.8}$

* Fe^1_c : K_α line energy, W^1_0 : K_α line equivalent width, Fe^2_c : K_β line energy, W^2_0 : K_β line equivalent width, [a] : 10^{22} atoms cm^{-2}, [b] : 10^{-10} ergs cm^{-2} s^{-1}

line at ~ 7.1 keV, with an equivalent width in the range of 25–80 eV is present in all BeppoSAX spectra of GX 1+4 except region 2 of the 2000 observation where it is very strong with an equivalent width of about ~ 0.55 keV. A soft X-ray excess was detected in the residuals of the spectra from the 1996 and region 2 of the 2000 data. Addition of a bremsstrahlung component to the spectral model improved the fit to some extent. The spectral parameters along with the reduced χ^2 obtained are given in Table 1.

DISCUSSION

On several occasions, GX 1+4 was found to show a decrease in X-ray luminosity by an order of magnitude compared to the flux before and after such a state. During the low flux episodes detected with the RXTE-PCA which last for at least a few hours, the pulsations were either absent, or below the detection limit. This has been interpreted as due to onset of a centrifugal barrier or the propeller effect. With a decrease in the mass accretion rate, when the inner radius of the accretion disk recedes beyond the co-rotation radius of the neutron star, this effect may set in. During the low state reported here, the absorbed X-ray flux in 2–10 keV band is comparable to the earlier episodes when pulsations were below the detection level of the RXTE. However, we found that the total X-ray luminosity in 10–100 keV energy band during this state is more than an order of magnitude higher than that in the 2–10 keV band and pulsations are clearly detected in the hard X-rays.

The spectral analysis of the BeppoSAX data showed that in the low state, the medium energy (2–10 keV) X-ray flux is reduced by an order of magnitude, mostly due to an increase in the absorption. In comparison, the hard X-ray flux is reduced by a factor of two only. It was also pointed out from RXTE-PCA data that the X-ray spectrum is harder in the low state [3]. Therefore, there can be significant flux in the hard X-ray band, where the effective area of the RXTE-PCA detectors fall rapidly. It is possible that the low states observed in 1996 [3] and 2002 [6] are similar to the one presented here. The short exposures and decreasing hard X-ray effective area of the RXTE-PCA did not allow detection of the hard X-ray pulses in RXTE-PCA observations during the low states. In such a scenario, the propeller regime for GX 1+4 may occur at still lower mass accretion rate than that those reported earlier.

The broad band X-ray spectrum in both high and low states of GX 1+4 is found to agree well with a Comptonization continuum model. An intense iron emission line at 6.4 keV is clearly seen in both the high and low intensity spectra. In addition, a strong emission line at 7.1 keV was also detected in the low state spectrum, which was very weak during the high intensity states. The K_α line flux was about 3–5 times larger than the K_β line flux in all the observations and is consistent with the ratio expected for fluorescence emission from neutral iron. Fluorescence emission lines with very high equivalent width similar to those found in GX 1+4 in the low state are seen in some persistent X-ray binary pulsars like Her X-1, LMC X-4 and SMC X-1 during their low-intensity phases of the super orbital period ([7, 8, 9, 10]). In these systems, the low-states arise due to increased absorption of the X-ray continuum by processing warped inner accretion disk, resulting in a very large equivalent width. However, unlike these systems, in GX 1+4 the extended low state is observed intermittently, without any hint of a periodicity. The difference in the duration and absence of periodicity in the occurrence of the low states of GX 1+4 rules out a similar mechanism for this source.

Though both the soft and hard X-ray flux decreased during the extended low state, the spectral parameters other than the absorption column density (N_H), are comparable with those during the high intensity states just before and after the low state. The N_H is found to be an order of magnitude higher during the low state. An increase in absorption may not be sole reason for the low state as the hard X-ray flux also decreased by about 50%, and the temperature of the seed photons increased during the low state by a factor of about two. However, the decrease in the hard X-ray flux may arise due to Thompson scattering from the line of sight if the absorbing material is partially ionized and the actual material in the line of sight is more than that measured with the simple absorber model. In the Comptonization model for X-ray pulsar spectra, the seed photons are likely to be produced from a heated surface of the neutron star. If there is a spherical component of accretion, it may cause heating of the surface and these seed photons undergo Compton upscattering in the accretion column. Higher temperature of the seed photons during the low state also indicates that the low state is not only due to an increased absorption along the line of sight. There must have been some additional change in the accretion process, for example, and enhancement in the spherical component of mass accretion. A detailed work is presented elsewhere [11].

REFERENCES

1. Makishima, K., Ohashi, T., Sakao, T., Dotani, T., Inoue, H., Koyama, H., Makino, F., Mitsuda, K., Nagase, F., Thomas, H.D., Turner, M.J.L., Kii, T., and Tawara, Y., *Nature*, **333**, 746 (1988).
2. Chakrabarty, D., Bildsten, L., Finger, M.H., Grunsfeld, J.M., Koh, D.T., Nelson, R.W., Prince, B.A., Vaughan, T.A., and Wilson, R.B., *ApJ*, **481**, L101–L105 (1997).
3. Cui, W., *ApJ*, **482**, L163–L166 (1997).
4. Galloway, D.K., *ApJ*, **543**, L137–L140 (2000).
5. Galloway, D.K., Giles, A.B., Greenhill, J.G., and Storey, M.C., *MNRAS*, **311**, 755–761 (2000).
6. Cui, C., and Smith, B., *ApJ*, **602**, 320–326 (2004).
7. Naik, S., and Paul, B., *A&A*, **401**, 265–270 (2003).
8. Naik, S., and Paul, B., *ApJ*, **600**, 351–357 (2004).
9. Naik, S., and Paul, B., *A&A*, **418**, 655–661 (2004).
10. Vrtilek, S.D., Raymond, J.C., Boroson, B., Kallman, T., Quaintrell, H., and McCray, R., *ApJ*, **563**, L139–L142 (2001).
11. Naik, S., Paul, B., and Callanan, P.J. *ApJ*, **Accepted, astro-ph/0409587** (2005).

Forced oscillations in accretion disks and kHz QPOs

Jérôme Pétri

Astronomical Institute, University of Utrecht - The Netherlands

Abstract. We propose a new model to explain the high frequency QPOs (kHz-QPOs) observed in LMXBs. The idea consists to add a rotating asymmetric component in either the gravitational or magnetic field of the accreting star. We show that the linear stability analysis of a disk evolving in such a structure predicts the existence of 3 instability types: a corotation resonance, a parametric resonance and a resonance due to a driving force, similar to the Lindblad resonance. These results are generalized by performing 2D non-linear simulations using a pseudo-spectral method. Lastly, we discuss the observational consequences of these instabilities on the light curves emitted by the disk, allowing later a confrontation with the observations.

Keywords: Accretion, accretion disks – MHD – Instabilities – Methods: analytical – Methods: numerical – Stars: neutron
PACS: 97.10.Gz

INTRODUCTION

To date, QPOs have been observed in about twenty LMXBs sources containing an accreting neutron star. Among these systems, the high-frequency QPOs which mainly show up by pairs, possess strong similarities in their shape and in their frequencies ranging from 300 Hz to about 1300 Hz (see van der Klis [6] for a review).

The very good agreement in the correlation of low and high frequencies QPOs spanned over more than 6 order of magnitude leads us to the conclusion that the physical mechanism responsible for the oscillations should be the same for the neutron star systems, the black hole candidates and the cataclysmic variables (Warner et al. [8]). Indeed, the presence or the absence of a solid surface, a magnetic field or an event horizon play no relevant role in the production of the X-ray variability (Wijnands [9]). In this paper we propose a new instability related to the evolution of an accretion disk in an asymmetric rotating gravitational or magnetic field.

HYDRODYNAMICAL DISK

We start the study with an accretion disk evolving in a gravitational potential possessing a rotating asymmetric quadrupolar perturbation. The asymmetry could be explained by the deformation of the star's spherical shape due to its rotation or by some magnetic stress in the star's interior. By performing a linear analysis, we have shown that this perturbation induces oscillations in the disk which are related to three kind of instabilities, namely :

- a corotation resonance in the region where the disk's local orbital rotation rate Ω is equal to the rotation speed of the star Ω_* : $\Omega = \Omega_*$;
- a resonance due to a driving force because of the asymmetric part of the gravitational field when $2|\Omega - \Omega_*| = \kappa_{r/z}$, where $\kappa_{r/z}$ are respectively the radial and vertical epicyclic frequencies. This is simply the Lindblad resonance condition well known in the context of the theory of spiral wave density in galactic dynamics ;
- a parametric resonance due to the periodic variation of $\kappa_{r/z}$, and related to Mathieu's equation (Morse and Feshbach [3]). The resonance occurs when $|\Omega - \Omega_*| = \frac{\kappa_{r/z}}{n}$, where n is an integer.

For weak gravitational perturbations, the parametric resonance is irrelevant. Only the corotation and Lindblad resonances are of interest. In this regime, the density perturbation is made of two kind of fluctuations which are different in nature. First, the free waves solutions which cannot propagate between the inner and outer Lindblad resonance will give rise to a spectrum of discrete eigenvalues whose precise value depends on the prescribed boundary conditions. Second, a non-wavelike disturbance due to the gravitational perturbation will propagate in the disk at the star's rotation rate.

We extended these results to the long time evolution of the accretion disk by running non linear 2D simulations using a pseudo-spectral method in the Fourier-Tchebyshev space. The final picture of the density perturbation in the disk, after more than 1000 orbital revolutions of the innermost orbit is shown on the left picture of Fig. 1. The weak gravitational perturbation is 10^{-3} less than the spherical symmetric component. Therefore, the corotation resonance well defined around $r = 40$ which possesses a weak linear growth rate is still not visible at the end of the simulation. However, the Lindblad resonances due to the forcing are clearly apparent. Nevertheless, the main result of this analysis is the persistence of significant density perturbations in the disk with very high coherence and lifetime, (Pétri, [4]).

MHD DISK

In the case of a magnetized rotating star, the main asymmetric contribution to the potential is due to the non aligned dipolar magnetic field. This situation is well suited for LMXBs containing a neutron star. The neutron star is assumed to possess a rotating dipolar magnetic field inclined with respect to the star's rotation axis. Following the same analysis as in the previous section, we showed that the same resonances arise, but for an $m = 1$ asymmetric field instead of an $m = 2$ component as for the HD case. This leads us again to a corotation resonance, $\Omega = \Omega_*$, the Lindblad resonances, when $|\Omega - \Omega_*| = \kappa_{r/z}$, and a parametric resonance when $|\Omega - \Omega_*| = 2\frac{\kappa_{r/z}}{n}$. In the most general case of an azimuthal perturbation of mode m, the term $|\Omega - \Omega_*|$ in the above equations should be replaced by $m|\Omega - \Omega_*|$.

Here again, we extended these results by performing 2D MHD simulations. The final picture of the density perturbation in the disk is shown on the right picture of Fig. 1. However, in order to facilitate the comparison with the HD case we used again a $m = 2$

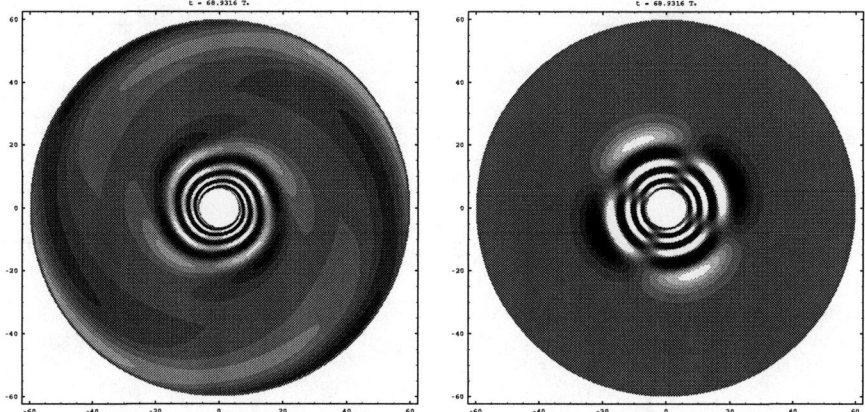

FIGURE 1. Density perturbation in the accretion disk at the final time of the simulation. The disk evolves in a Newtonian potential, with a gravitational perturbation on the left (HD disk) and a magnetic perturbation on the right (MHD disk).

mode. Qualitatively, we can draw the same conclusions as in the hydrodynamical case. In a real accretion disk, clumps of matter can be generated most easily in the regions of strong asymmetric density perturbation which means very close to the inner boundary. These inhomogeneities will rotate at the local pseudo-keplerian speed Ω_{in} which can be interpreted as the highest kHz QPO. In addition, due to the modulation of the density by the magnetic field rotating at Ω_*, the clump's size and radiation intensity will suffer a beat phenomenon at a frequency $\Omega_{in} - \Omega_*$ which can be interpreted as the lowest kHz QPO, (Pétri, [5]). This simplified picture should be improved by adding the viscosity and an inward radial drift of the inhomogeneities and therefore accounting for the QPOs frequencies shift in accordance with the accretion rate. This beat phenomenon can also be applied to a density wave propagating between the inner edge of the disk and the inner Lindblad resonance. One needs only to replace the orbital frequency of the blob by the speed pattern of the density wave.

POWER SPECTRUM DENSITY

Finally, in order to make a connection between these simulations and the observations of QPOs in LMXBs, we computed the power spectrum density of the accretion disk by taking into account the motion of a photon in the Kerr spacetime. We therefore allow for the gravitational and Doppler redshift and as well as for the deflection of the light ray. The disk is assumed to be made of individual punctual particles emitting isotropically in their rest frame. The intensity of each source is proportional to the local density.

Examples of power spectrum densities are shown in Fig. 2 for a HD and a MHD disk. In the case of an hydrodynamical disk, on the left panel, the dominant frequencies are in the range $[v_*, 4v_*]$. In the case of an MHD disk, on the right panel, the QPOs frequencies ranges in the interval $[7v_*, 10v_*]$. These frequencies are very close to that

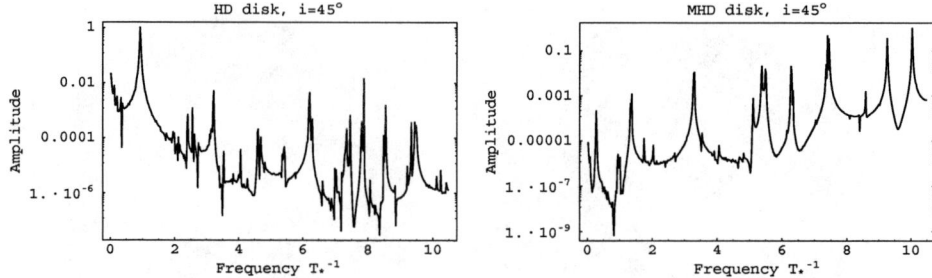

FIGURE 2. Power spectrum density of the accretion disk. The intensity plotted on a logarithm scale have been normalized to unity for the strongest peak and the inclination of the line of sight is equal to $i = 45^o$. The neutron star spin is $\nu_* = T_*^{-1} = 100$ Hz.

of the innermost stable circular orbit and can account for the kHz-QPOs in LMXBs containing a magnetized neutron star.

CONCLUSION

In this paper, we have explored the consequences of a rotating asymmetric gravitational or magnetic field on the evolution of a thin accretion disk. We have shown that the disk becomes unstable and reaches a new quasi-stationary state in which some small scale perturbations in the density emanate at some radii. These perturbations rotate at the local rotation speed of the disk and are concentrated on a very narrow radial extension.

We believe that this model can also be applied to the LMXBs containing a black hole. Indeed, it seems that the observations of the QPOs in black hole candidates is correlated with the detection of jets emanating from these systems, (McClintock & Remillard [2]). The Blandford-Znajek process (Blandford & Znajek [1]) converting the rotational energy of the black hole into the launching of the jet should also link the accretion disk and the black hole via the magnetic field. Therefore, the model presented in this paper could give some explanations for QPOs seen in the BHCs (Wang et al. [7]).

REFERENCES

1. Blandford R. D. and Znajek R. L., 1977, MNRAS, 179, 433
2. McClintock J. E. and Remillard R. A., 2003, astroph/0306213
3. Morse and Feshbach, 1953, Methods of Theoretical Physics, New-York: McGraw Hill
4. Pétri J., 2004a, submitted to A&A
5. Pétri J., 2004b, submitted to A&A
6. van der Klis M., 2000, ARA&A, 38, 717
7. Wang D. and Ma, R. and Lei, W. and Yao, G., 2003, MNRAS, 344, 473
8. Warner B. et al., 2003, MNRAS, 344, 119
9. Wijnands R., 2001, Advances in Space Research, 28, 469

Gamma ray burst progenitors

Jelena Petrovic* and Norbert Langer*

*Sterrenkundig Instituut Utrecht, The Netherlands

Abstract. We follow the evolution of binary systems that include rotational processes for both stars. Neglecting magnetic fields, we show that the cores of massive stars can maintain a high specific angular momentum ($j \sim 10^{17}$ cm^2 s^{-1}) when evolved with the assumption that mean molecular weight gradient suppress rotational mixing processes. We also present models that include magnetic fields generated by differential rotation (Spruit 2002) and we consider the internal angular momentum transport by magnetic torques. We show that the magnetic coupling of core and envelope after the accreting star ends core hydrogen burning leads to slower rotation ($j \sim 10^{15-16}$ cm^2 s^{-1}) than in the non-magnetic case.

Keywords: stars:evolution, stars:rotation, stars:magnetic fields, stars:binaries:close, gamma rays:bursts
PACS: 97.10.Cv,97.10.Kc,97.80.Fk,98.70.Rz

INTRODUCTION

The most widely used model for GRB production in the context of black hole formation in a massive star is the so called collapsar model [1].

A collapsar is a massive ($M \gtrsim 35$-$40 M_\odot$), [2] rotating star whose core collapses to form a black hole [3, 4]. If the collapsing core has enough angular momentum ($j \geq 3 \cdot 10^{16}$ cm^2 s^{-1}), [4] an accretion disk is formed around the black hole. The accretion of the rest of the core at accretion rates up to $0.1 M_\odot$ s^{-1} by the newly-formed black hole is thought to be capable of producing a collimated highly relativistic outflow. This releases large amounts of energy ($\sim 10^{51}$ erg s^{-1}) some of which is deposited in the low density rotation axis of the star. In case the star has no hydrogen envelope, i.e., has a light crossing time which is less or comparable to the duration of the central accretion (about 10s), a GRB accompanied by a Type Ib/c supernova may be produced. The collapsar models for gamma-ray bursts thus needs three essential ingredients: a massive core, loss of the hydrogen envelope, and sufficient angular momentum to form an accretion disk.

A star evolving in a binary system and accreting matter from the companion, increases its surface angular momentum. If this angular momentum can be transported efficiently through the stellar interior, the star may evolve into a red supergiant that has a rapidly spinning core with sufficient specific angular momentum to produce a collapsar.

ROTATING BINARY SYSTEMS WITHOUT MAGNETIC FIELD

To check if accretion can add enough angular momentum to the core, we modeled the evolution of a rotating binary system with initial masses of $M_{1,in}=56 M_\odot$ and $M_{2,in}=33 M_\odot$ and an initial orbital period of $p_{in}=6$ days. The binary system quickly syn-

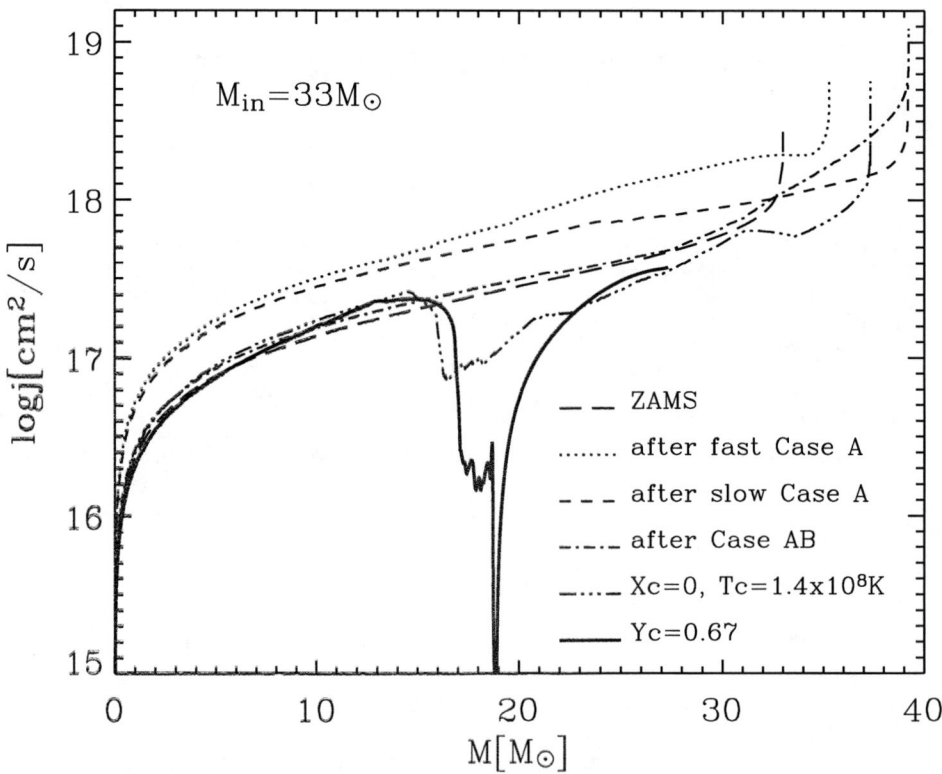

FIGURE 1. Specific angular momentum profiles of the secondary star on the hydrogen ZAMS (long dashed line), after fast (dotted line) and slow (short dashed line) Case A mass transfer, after Case AB mass transfer (dash-dotted line), when helium ignites in the core (three dots-dashed line) and when the central helium abundance is 67% (solid line).

chronizes during the main sequence evolution. Due to this synchronization, both stars lose angular momentum and their initial surface rotational velocities are 92 km s^{-1} for the primary and 64 km s^{-1} for the secondary which is much slower than the typical values for single stars of these masses (200 km s^{-1}), [5]. This means that stars in binary systems lose a significant amount of angular momentum due to synchronization. The angular momentum loss increases with the initial orbital period.

Fig. 1 shows specific angular momentum profiles of the secondary at different points of its evolution. The specific angular momentum of the secondary increases significantly due to fast Case A mass transfer (Fig. 1, dotted line). After this, the secondary loses angular momentum due to stellar wind mass loss, but also gains certain amount through slow Case A and Case AB mass transfer (Fig. 1, dashed and dot-dashed line). The result is that the core has a larger specific angular momentum when central helium burning starts than at the beginning of hydrogen core burning. After core hydrogen exhaustion, the secondary evolves into a red supergiant, the core contracts and the envelope expands.

This leads to a spin-up of the core and a spin-down of the envelope. The specific angular momentum of the core at $3\,M_\odot$ is $\sim 5.5\cdot 10^{16}$ cm^2 s^{-1} (Fig. 1, three dot-dashed line). The envelope is convective and slowly rotating (~ 0.02 km s^{-1}). The core is rigidly rotating with maximum rotational velocity of ~ 100 km s^{-1}. The core and the envelope are separated by layers that have a high μ-gradient. Angular momentum is not efficiently transported through these layers, so the core is not slowed down by the slow rotation of the envelope. When a third of the central helium supply is exhausted, the core (at $3\,M_\odot$) has a specific angular momentum of $\sim 5\cdot 10^{16}$ cm^2 s^{-1}. If we assume that the angular momentum of the core decreases further during helium core burning with the same rate, specific angular momentum of the core at the moment of helium exhaustion is expected to be $\sim 4\cdot 10^{16}$ cm^2 s^{-1}. Since there is no significant angular momentum loss from the core during core carbon burning, we can conclude that this star has enough angular momentum to produce a collapsar and, in the case that the hydrogen envelope is lost during red supergiant phase, a gamma-ray burst.

ROTATING BINARY SYSTEMS WITH MAGNETIC FIELD

We model the evolution of the same binary system including angular momentum transport by magnetic torques, using the improved dynamo model of Spruit, [6]

Fig. 2 shows specific angular momentum profiles of the secondary at different phases of evolution. The specific angular momentum of the secondary increases significantly due to the fast Case A mass transfer (Fig. 2, dotted line). Angular momentum is transported more efficiently through the stellar interior compared to the non-magnetic model, since the incurred magnetic torques are a few orders of magnitude more efficient in angular momentum transport than the rotational instabilities. Comparing the specific angular momentum of the non-magnetic (Fig. 1) and magnetic model (Fig. 2), we notice that during fast Case A the angular momentum of the magnetic star increases more than that of the corresponding non-magnetic star ($2\cdot 10^{17}$ cm^2 s^{-1} for magnetic and $1.25\cdot 10^{17}$ cm^2 s^{-1} for non-magnetic star, $\sim 10^4$ yrs after fast Case A, at $3\,M_\odot$).

The accretion stops when the secondary still has almost 50% of the hydrogen to burn in the core. Angular momentum is efficiently transported from the stellar core to the surface and the μ-gradient can not stop it as in the case of the non-magnetic star. During further main sequence evolution, the stellar core loses significant angular momentum and when hydrogen core burning stops, the specific angular momentum at $3\,M_\odot$ is $2\cdot 10^{16}$ cm^2 s^{-1}. Before helium ignites in the core, the specific angular momentum decreases to $6\cdot 10^{15}$ cm^2 s^{-1}.

CONCLUSIONS

We conclude that our binary models without magnetic field can reproduce stellar cores with a high enough specific angular momentum ($j \geq 3\cdot 10^{16}$ cm^2 s^{-1}) to produce a collapsar and a GRB.

If magnetic field is taken into consideration, however, GRBs at near solar metallicity need to be produced in rather exotic binary channels, or the magnetic effects are

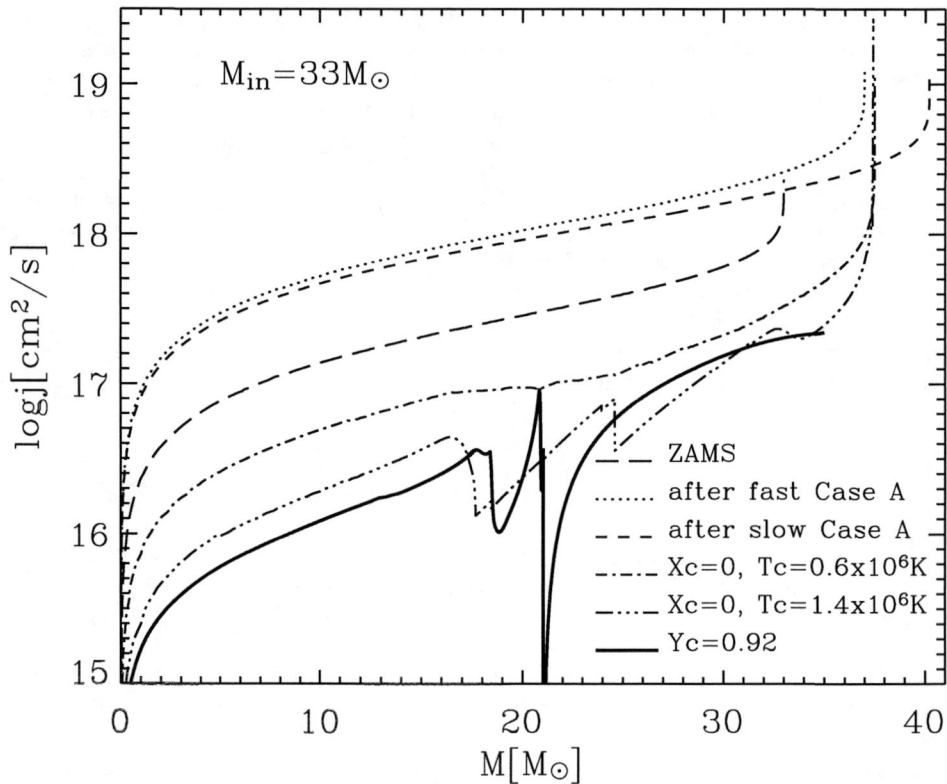

FIGURE 2. Specific angular momentum profiles of the secondary star with magnetic fields on the hydrogen ZAMS (long dashed line), after fast (dotted line) and slow (short dashed line) Case A mass transfer, when all hydrogen is exhausted in the center (dot-dashed line), when helium ignites (three dots-dashed line) and and when the central helium abundance is 92% (solid line).

overestimated in our current models.

REFERENCES

1. Woosley, S. E., 1993, BAAS, 25 894
2. Freyer, C. L., 1999, ApJ, 522, 413
3. Woosley, S. E., 1993, ApJ, 405, 273
4. MacFadyen, A. I. & Woosley, S. E., 1999, ApJ, 524, 262
5. Hager, A., Langer, N, & Woosley, S.E., 2000, ApJ, 528, 368
6. Spruit, H. C., 2002, A&A, 381, 923

The light curve of the companion to PSR B1957+20

Mark Reynolds*, Paul Callanan*, Andy Fruchter[†], Manuel Torres**, Martin Beer[‡] and Rachel Gibbons[§]

*Physics department, University College Cork, Ireland
[†]Space Telescope Science Institute, 3700 San Martin Drive, Baltimore, MD 21218, USA
**Center for Astrophysics, 60 Garden Street, Cambridge, MA 02138, USA
[‡]Department of Physics and Astronomy, University of Leicester, Leicester LE1 7RH, England
[§]University of Maryland, College Park, MD 20742, USA

Abstract. We present a new analysis of the light curve for the secondary star in the PSR B1957+20 system. Combining previous data and new data points at minimum from the *Hubble Space Telescope*, we have 100% coverage in the R-band. We also have a number of new K_s band data points which we use to constrain the IR magnitude of the system. We model this with the Eclipsing Light Curve code. From our best fit model we are able to constrain the system inclination to $63.9 \pm 2.4°$ for pulsar masses ranging from 1.35 – 1.9 $M_\odot$. The pulsar mass is unconstrained. We also find that the secondary is *not* filling its Roche lobe which has important consequences for evolutionary models of this system. The temperature of the un-irradiated side of the companion is in agreement with previous estimates.

Keywords: binaries: eclipsing – pulsars: individual (PSR B1957+20) – stars: neutron, low mass
PACS: 01.30.Cc, 97.60.Hn, 97.80.Gb

INTRODUCTION

The binary millisecond pulsar *PSR B1957+20* [1] is the original and one of the best studied members of its class. It consists of a 1.6ms radio pulsar orbiting a companion of mass no less than $0.022 M_\odot$, in a binary of orbital period 9.17 hours. For 10% of this orbit, the radio emission from the pulsar is eclipsed: the eclipsing region is considerably larger than the Roche lobe of the companion star, suggesting a wind of material from the secondary, due to ablation by the impinging pulsar radiation [1]. In this work we present a new analysis of the optical and IR properties of this system, allowing us to tightly constrain the orbital inclination for the first time.

DATA

The optical data consists of B, V and R band images taken with the *William Herschel Telescope* (*WHT*) at La Palma, (see [2] and references therein for details of the observations), along with two pairs of R & I-band data points taken at minimum with the *Hubble Space Telescope* (*HST*). These data provide us with R-band coverage throughout a complete orbital cycle, as well as colour information during eclipse, for the first time.

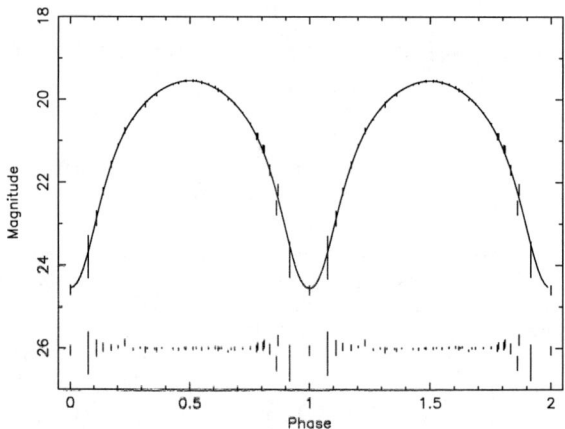

FIGURE 1. The best fit to the combined R-band data with residuals. The residuals are to scale on the 26th magnitude. Two orbital phases are displayed for added clarity. The pulsar mass is 1.40 $M_\odot$. The best fit inclination is $i = 63.3 \pm 1.5\,^\circ$ with a $\chi_\nu^2 = 1.03$

We also obtained a number of K_s band[1] observations of the PSR B1957+20 system, using the IR imager *PANIC* at the Magellan telescope in Chile.

THE ELC MODEL

To facilitate modeling of the light curve we used the ELC code: see [3] and references therein for a complete description of the workings of the code. The ELC code is ideally suited for this type of system as it incorporates the NEXTGEN low temperature model atmosphere tables ($T_{min} = 2200K$), which are critical for systems like *PSR B1957+20*, with a companion of temperature ~ 3000K, as the deviations from a blackbody are almost certainly non-negligible at such temperatures [3]. The ELC code also allows one to fit light curves on a one by one or simultaneous basis. In our case this allowed us to fit the BVRI & K band light curves simultaneously.

RESULTS

Inclination

We modeled the system for a number of possible pulsar masses ranging from 1.35 – 1.9 $M_\odot$, given the fact that all the evolutionary scenarios of this system require the pulsar to have accreted some amount of mass from the secondary. For a given pulsar

---

[1] Based on observations using the Magellan 6.5m telescope at Las Campanas Observatory of the Carnegie Institution

TABLE 1. Optical and IR magnitudes of the companion to *PSR B1957+20* at maximum and during eclipse.

	B	V	R	I	K
Max	21.08 ± 0.05	20.16 ± 0.05	19.53 ± 0.05	18.64 ± 0.05	17.8 ± 0.1
Min	27.7 ± 0.1	25.9 ± 0.1	24.6 ± 0.1	22.52 ± 0.05	19.9 ± 0.1

mass the inclination was constrained to within $3°$, i.e. for a pulsar of mass 1.40 $M_\odot$ $i = 63.3° \pm 1.5°$ and overall for the above range of pulsar masses we find the inclination of the system to be $i = 63.9° \pm 2.4°$ at the 2σ level.

Pulsar Mass

The value of χ_ν^2 only increased from 1.03 to 1.08 as the mass of the pulsar was increased from 1.35 – 1.9 $M_\odot$: hence our models were unable to constrain the pulsar mass.

Roche lobe filling factor

At no point in our attempts to model this system were we able to obtain an acceptable fit for a secondary filling its Roche lobe. Indeed for a blackbody model the filling factor was only $f = 0.75$. For our models using the NEXTGEN model atmospheres the value of f was approximately constant, $0.81 \geq f \geq 0.83$ (2σ level), as we varied the mass of the pulsar between 1.35 and 1.9 $M_\odot$. The model was extremely sensitive to the Roche lobe filling fraction as a change in f by even ± 0.02 was enough to cause the χ_ν^2 to increase from ~ 1.0 to ~ 1.5 (for a primary of mass 1.4 $M_\odot$). Hence we take this as demonstrating that the secondary is currently *not* filling its Roche lobe.

Temperature of the Secondary at min/max

We obtained a temperature of $T = 2935 \pm 120$ K, 2σ level, as the effective temperature of the un-illuminated side of the companion star for pulsar masses in the range 1.35 – 1.90 $M_\odot$. For individual pulsar masses the 2σ error was only ± 100 K i.e. for a pulsar of mass 1.40 $M_\odot$ an effective temperature of $T = 2930 \pm 100$ was obtained at the 2σ level. The corresponding temperatures at maximum are $T = 7250 \pm 120$ K and $T = 7245 \pm 100$ K respectively, both at the 2σ level.

Modeling the Temperature Gradient

It is clear that the light curve of *PSR 1957+20* requires a large temperature gradient to be sustained between the heated and cool hemispheres of the companion star. To test if this is physically sustainable, we decided to model the heat flow along the surface of the secondary in more detail. A two-dimensional model of the irradiation of *PSR 1957+20* was simulated using the a modified version of the code described in [4]. It was found that the heated material extended beyond the directly irradiated region but that not all of the unilluminated portion of the star was heated. Consequently a large temperature gradient between the illuminated and unilluminated sides still exists.

CONCLUSIONS

- A model of a strongly irradiated secondary is consistent with the observations.
- We have constrained the inclination to $i = 63.9 \pm 2.4\,^{\circ}$ at the 2σ level.
- The temperature of the secondary at minimum is in agreement with previous estimates and the observed temperature gradient is physically sustanible
- We find that the secondary is under filling its Roche lobe by about 20%

ACKNOWLEDGMENTS

We thank Jerry Orosz for kindly providing us with the ELC code. This research made extensive use of the *SIMBAD* database, operated at CDS, Strasbourg, France and NASA's Astrophysics Data System.

REFERENCES

1. Fruchter A.S., Stinebring D.R., Taylor J.H., 1988, Nature, 333, 237
2. Callanan P.J., van Paradijs J., Rengelink R., 1995, ApJ, 439, 928
3. Orosz J.A., Hauschildt P.H., 2000, A&A 364, 265.
4. Beer M.E., Podsiadlowski Ph., 2002, MNRAS, 335, 358

Spectroscopic Analysis of the Companion to the Binary MSP PSR J1740−5340 in NGC 6397

E. Sabbi*, F.R. Ferraro*, R. Gratton[†], A. Bragaglia**, A. Possenti[‡] and N. D'Amico[‡]

*Dipartimento d'Astronomia dell'Università di Bologna–Italy
[†]Osservatorio Astronomico di Padova–Italy
**Osservatorio Astronomico di Bologna–Italy
[‡]Osservatorio Astronomico di Cagliari–Italy

Abstract.
By means of high–resolution spectra, we have measured radial velocities of the companion (hereafter COM J1740−5340) to the eclipsing millisecond pulsar PSR J1740−5340 in the galactic globular cluster NGC 6397. The radial velocity curve enables us to derive the mass ratio ($M_{PSR}/M_{COM} = 5.85 \pm 0.13$).

The derived abundances are fully compatible with those of normal unperturbed stars in NGC 6397, with the exception of a few elements. The lack of C, in particular, suggests that the star has been peeled down to regions where incomplete CNO burning occurs, supporting a scenario where the companion is a turn–off star which has lost most of its mass.

A detailed analysis of the Hα lines reveals that optically thin hydrogen gas resided outside the Roche lobe of COM J1740−5340. The line morphology suggests the presence of a steam of material going from the companion toward the neutron star; this material never reachs the neutron star surface, being driven back by the pulsar radiation, far beyond COM J1740−5340.

The unexpected detection of strong He I absorption lines implies the existence of regions at $T > 10,000 \, \text{K}$, which are significantly warmer than the rest of the star, and reveals the existence of a region on the companion surface, heated by the millisecond pulsar flux.

Keywords: Globular Clusters: individual: NGC 6397 – Stars: evolution, abundances – Pulsars: individual: PSR J1740-5340 – Techniques: spectroscopic
PACS: 98.20.Gm, 97.30.-d, 97.80.-d, 39.30.+w, 78.47.+p

INTRODUCTION

COM J1740−5340 is the first example of a MSP companion whose light curve is dominated by ellipsoidal variations, suggestive of a tidally distorted star, which almost completely fills (and is still overflowing) its Roche lobe (Ferraro et al. 2001, ApJ, 561, L93).

Interestingly, this binary system constitute the optically brightest dual-line binary hosting a MSP, thus being a prime target for an accurate determination of both the binary parameters and the mass of a spun-up pulsar, while yielding the possibility of study the possible evolutionary path. Hence, we have planned a spectroscopic campaign using the ESO telescopes at Paranal in Chile.

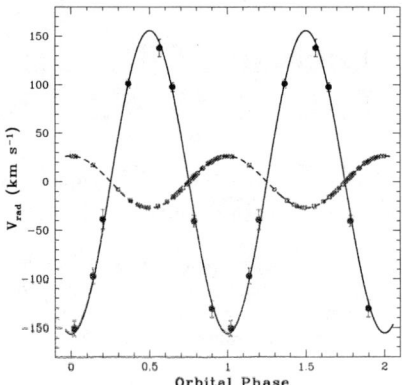

FIGURE 1. The *large dots* are the radial velocity determinations for COM J1740−5340. The *solid* line represents the best-fit sinusoidal curve. The *small open squares* are the radial velocity determinations for PSR J1740−5340 derived from timing measurements and the radio ephemeris (D'Amico et al. 2001, ApJ, 561, L89). The *dashed* line represents the fitted velocity curve of the pulsar. The best fit yields a radial-velocity amplitude $K = 155.8 \pm 3.6 \mathrm{km\,s^{-1}}$ (1σ-error) with $\chi^2_\nu = 0.99$. Inspection of this plot shows that the absolute phase of the RV data matches that of the radio pulsar orbital motion very well: the radial velocities are null at inferior and superior conjunctions which, (according to the convention of D'Amico et al. 2001), correspond to orbital phases 0.25 and 0.75 respectively, confirming unambiguously that PSR J1740−5340 and COM J1740−5340 orbit each other. The mass function $f(M_{\mathrm{COM}})$ of COM J1740−5340 can be easily computed from the radial velocity amplitude K and the orbital period $P_{\mathrm{orb}} = 1.35405939 \pm (5 \times 10^{-8})$ days, and thus we can comupte the mass ratio $q = 5.85 \pm 0.13$.

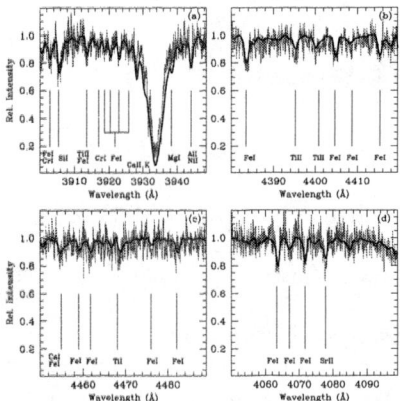

FIGURE 2. Comparison between the COM J1740−5340 spectrum (grey line) and NGC 6397 subgiants template (black line), obtained by avraging, and broadening to account for rotation, three normal NGC 6397 subgiant stars. We show the spectral regions near lines of CaII (panel (a)), TiII (panel (b)) TiI (panel (c)) and SrII (panel (d)). In all these regions many FeI lines are also present. These plots show that the abundance is fully compatible with that of normal, single stars in NGC 6397.

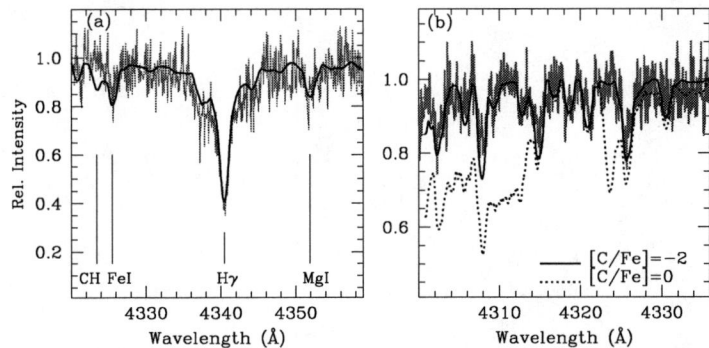

FIGURE 3. Evaluation of the carbon abundance from the CH band. Panel (a) shows a clear depletion in the region of the CH band in the COM J1740−5340 spectrum (grey line) with respect to the NGC 6397 subgiants template (black line). Panel (b) shows the result of spectral synthesis, demonstrating that C is strongly under-abundant in the COM J1740−5340 atmosphere. The strong C depletion seems to indicate that COM J1740−5340 is not a perturbed low mass main sequence star, but had instead a larger mass and has been peeled down to the present $\sim 0.3\,M_\odot$ (see also the models by Burderi et al. 2002, ApJ, 574, 325, and Erga & Sarna 2003, A&A, 399, 237).

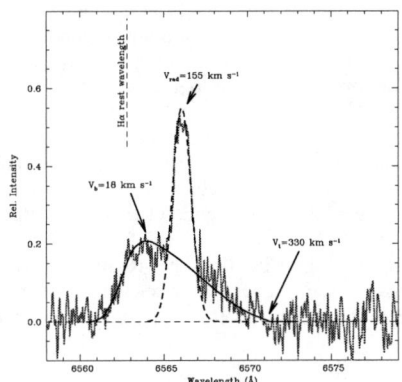

FIGURE 4. Subtracted spectrum of COM J1740−5340 (see Sabbi et al. 2003, ApJ, 589, L41 for description of the adopted procedure) in the Hα region at orbital phase $\phi = 0.56$. No wavelength shift has been applied to the spectrum. The *vertical dashed line* marks the Hα line rest wavelength and the *horizontal dashed line* is the continuum level. Two components can be seen: the narrower and brighter (modeled by the *heavy dashed line*) is emitted by the stellar chromosphere, while the broader and fainter (modeled by the *heavy solid line*) shows a peak (here named V_b) at smaller velocities than the star, but extending farther, with the velocity of its tail (V_t) greater than $300\,\text{km}\,\text{s}^{-1}$. The evident asymmetric shape of the secondary broad component of the Hα emission precludes an origin in a Keplerian disk surrounding PSR J1740−5340, but suggests the presence of a cometary-like tail.

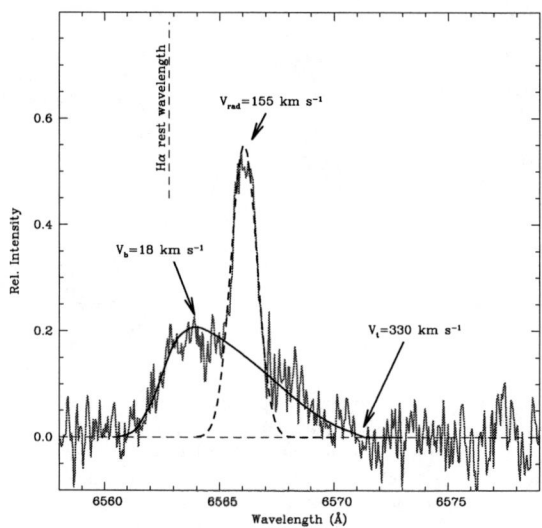

FIGURE 5. The heavy solid lines over-plotted on the observed spectra are empirical profiles obtained by combining the typical rotation broadening profiles with the He I line intensity. An arbitrary shift in intensity is applied to the spectra for the sake of clarity. The detailed inspection of the rotation profile and of the orbital modulation of the He I absorption lines strongly supports the existence of a hot "barbecue-like" strip elongated across the stellar photo-sphere facing the pulsar. The high temperatures required for producing the He I absorption line could be due to a shocked filament standing, e.g., at the boundary between the companion surface and the gas swept back toward the companion by the pulsar flux, or it is possible that the heated strip is due to the effects of the pulsar irradiation on the companion surface. In this case, the shape of the heated region can be accounted for only by a highly anisotropic pulsar emission pattern confined (at least in the direction orthogonal to the binary orbital plane) within a surprisingly small angle of ~ 0.4 degrees. Since such a narrow pencil (or highly flattened fan) beam would probably illuminate the Com J1740−5340 surface only for a portion of the orbit, the tiny width of the heated region would be preserved if the cooling time of the irradiated gas is much shorter than the orbital time scale ~ 1 day. Under basic assumptions (as supposing a perfect gas and an atmospheric density of $10^{-7}\,\mathrm{g\,cm^{-3}}$), we estimate that only a few hundreds seconds are necessary to cool the gas from $10,000\,\mathrm{K}$ to $5,500\,\mathrm{K}$

Radius and temperature evolution of the white dwarf in AS 296 during the 1988-1994 outburst

A.Siviero* and U.Munari*

*INAF - Astronomical Observatory of Padova, Asiago, Italy

Abstract. Symbiotic stars undergo periodic TNR outbursts in which no mass is lost and the accreting white dwarf expands to the dimensions on a A-F giant/supergiant for several months or years before returning to the quiescent state. Several systems display eclipses that, coupled with an orbital solution, allow to determine the absolute values of the white dwarf mass, radius and temperature, and how them evolve during a outburst cycle. In this note we present preliminary results for the symbiotic system AS 296.

Keywords: Stars: binary and variable – Emission spectra: atoms
PACS: 97.80.Űd, 97.30.Űb, 32.30.Űr, 32.50.+d

INTRODUCTION

Symbiotic stars are interacting binaries in which material is transferred from a cool RGB/AGB star to a very hot ($T_{\text{eff}} \geq 10^5$K) and luminous ($L \geq 10^3$ L$_\odot$) white dwarf. The wind from the cool giant is copiously photoioninzed by the hard radiation field of the WD giving rise to a very rich and intense emission line spectrum. Simultaneous presence of a TiO molecular absorption spectrum and high ionization permitted and forbidden emission lines is the hallmark of symbiotic stars. Several of them are detected as Super Soft X-Ray Sources and as such they are belived to experience stable H-burning of the accreted material on the surface of the white dwarf [10]. Many more symbiotic stars would be observable among SSXRS would not be for the self-absorption caused by the local absorption of X-rays by the large amount of circumstellar material. Symbiotic stars frequently undergo powerful outbursts, during which the erupting WD overshines the cool giant/supergiant companion by several magnitudes in the B band. As a rule, these outbursts last for a few years and are not accompanied by appreciable mass-loss from the system. At maximum the WD expands to mimic a A-F giant/supergiant [1] and the ionization of the circumstellar material drops, to regain later as the WD shrinks and warms back during decline. These outbursts are generaly interpreted as TNR events of the non-degenerate type. There are several symbiotic stars that show eclipses [3], and they offer unique opportunities to probe the structure and physical evolution of the WD during the outburst, being able to provide radii and temperatures in absolute units [7]. One such system is AS 296 which 6-yr long decline from outburst maximum in 1988 was interupted by three *total* eclipses of the outbursting component. This contribution aims to outline a preliminary investigation of the eclipses in AS 296, with a more complete one involving full photometric and spectroscopic coverage in the optical and ultraviolet (IUE) being targeted for a future paper.

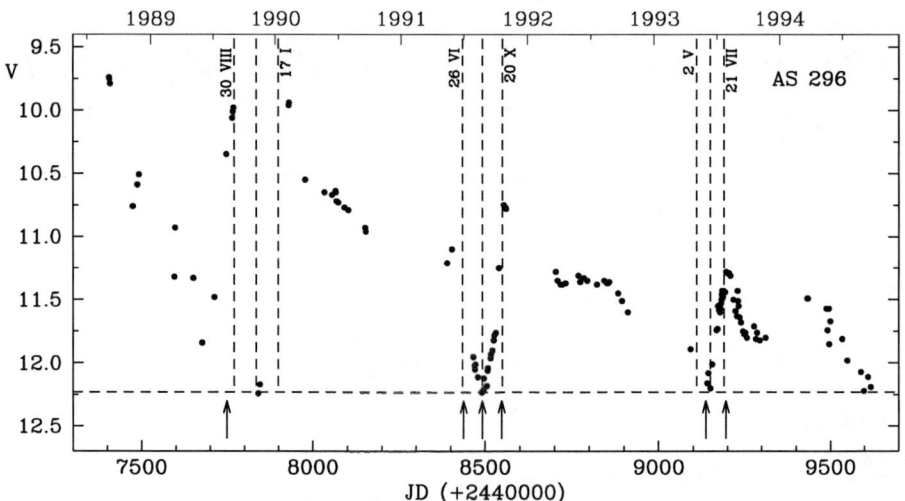

FIGURE 1. The V-band light curve of AS 296 folded with the ephemeris $T_0 = 2448492 + 658 \times E$ derived by [5]. Vertical dashed lines are placed at the center of every total eclipse and to the first and fourth contact. An orizontal dashed lines indicate the constant value of the magnitude at minimum. The arrows mark the time at the low resolution spectra modeled in Table 1 with the CLOUDY photoionization code.

WD RADIUS EVOLUTION DURING THE AS 296 OUTBURST

The symbiotic star AS 296 (= FG Ser = SS73 148) underwent a large outburst in 1988 which lasted till 1994, with a complex multi-maxima decline. It was interupted by three sharp and total eclipses, that allowed [6] to derive in 658 days the orbital period. At the center of each eclipses, the brightness of the system always dropped to $V=12.2$ mag, which corresponds to the value for the M giant alone, which account for the whole V-band brightness of the system in quiescence (*cfr.* Figure 1). The cool giant has a M5III classification, and its intrinsic brightness has remained well constant during quiescence and outburst phases. Early investigation of the outburst was presented by [4], while more complete studies have been reported by [5] and [6]. Partial spectroscopic monitoring of the outburst has been discussed by [12], and [9] has derived the spectroscopic orbit of the M giant in the system. The aim of this work is to derive a preliminary set of values for the radii and photoionizing temperature of the white dwarf during outburst evolution. To obtain the radii we model the width of the three eclipses; during the first one, a year after the first rise to maximum, the hot component was still very expanded and from the profile of the total eclipse we can argue that the radius of the cool component set an upper limit to the radius of the outbursting one. For the purpose of this work we assume that the two stars have the same dimensions during the first total eclipse. Taking the orbital separation $a = 1.95$ AU from [9], from the length of the eclipse (from the first to the last contact) we obtain for the two stars $R_{hot} = R_{cool} = 130$ $R_\odot$. The duration of the second and third eclipse is shorter compared to the first one, as clearly visible on the light curve

TABLE 1. Radii for the cool and hot components of AS 296 as derived from eclipse analysis under the assumption that at the first of the three such events the radii of them were equal (and therefore the orbital inclination is exactly $i=90°$). The WD temperatures are those coming from a photo-ionization modeling of the observed emission line spectrum around the eclipse epochs. The last two columns provide the spectral classification according to [11] for the given radii and temperatures.

		R ($R_\odot$)	$T_{\text{eff.}}$	L (10^3 $L_\odot$)	Sp. Type	Lum. Class
Cool comp.		130 ± 14	3500		M5	III
Hot comp.	First eclipse	130 ± 14	6500	27	F5	Iab
	Second eclipse	90 ± 14	14000	270	B4	Ia
	Third eclipse	30 ± 18	29000	570	O9	Ia

of Figure 1. Since the radius of the cool component is costant this indicates a progressive decrease of the radius of the hot component, which is simultaneously increasing its effective temperature as indicated by the growing ionization of the circusmtellar gas. The radius of the outbursting WD at the second and third eclipse as derived by modeling the light curve of Figure 1 is given in Table 1. The growth in size of the radiation bounded region appears to be the main reason behind the absence of a flat bottom phase in the second and third eclipses, in spite of the sensible decrease in radius of the occulted star.

PHOTOIONIZING TEMPERATURE OF THE HOT SOURCE

The outburst of AS 296 has been intensively monitored with the spectrographs of the Asiago Observatory, in both low and high resolution. In this paper, instead of using the observed photometric color distribution, we estimate the temperature of the hot star by fitting the emission line spectrum observed at dates close to eclipse phases (*cfr.* arrows in Figure 1) with the CLOUDY [2] photoionization code. Provision must be given to the fact that the recombination time scale in the outer gaseous regions is of the same order of the cooling and warming up of the central ionizing source as traced by the multi-maxima in the light curve. To model the spetra, they must be reddening corrected. There are several estimates of the reddening affecting AS 296 (*cfr.* [5] [12]), here we measure it from the equivalent width of interstellar lines following [8] calibrations. The NaI interstellar lines are quite saturated and therefore useful only in indicating a high reddening, while the KI lines are still on the linear increase and therefore well suited for the task. The 7699 Å KI lines appear superimposed on a photospheric 7697 Å SI line from the M5III component. To disentangle the contributions of the two to the equivalent width of the observed line, we estimated the 7697 Å SI contribution from a synthetic Kurucz spectrum that we have computed on purpose for the same spectral resolution and for the best fit parameters T_{eff}=3500 K, [Fe/H]=-0.5 and $\log g$=0.75 and solar chemical partition. The best fit and observed spectra are displayed in Figure 2. The pruned equivalent width of interstellar 7699 Å KI is found to be 0.216 Å, corresponding to E_{B-V}=0.85 mag. The spectroscopical estimate of the reddening allow us to obtain a value for the distance. For a radius of 130 $R_\odot$ and a temperature of 3500 K, and adopting a bolometric correction B.C.=-3.1 from [11], the absolute magnitude of the cool giant in AS 296 is M_V=-0.55, in line with expectations for a slightly metal poor M5III

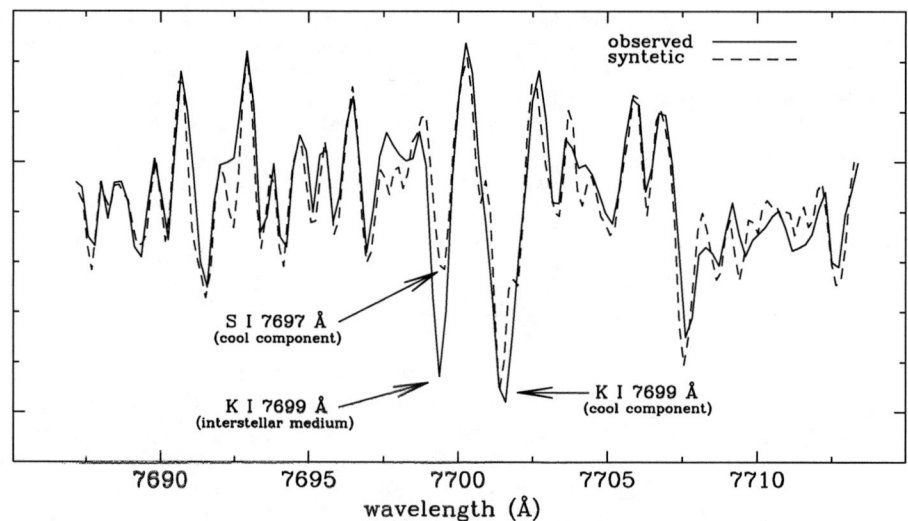

FIGURE 2. Interstellar reddening from 7699 K I line. The AS 296 spectrum has been obtained with the 1.82m telescope of the Padova Astronomical Observatory, operating in Asiago (Italy), with an Echelle spectrograph at resolving power R = 20 000; the syntetic spectrum is for the same resolution, T_{eff}=3500 K, [Fe/H]=−0.5, log g=0.75 and solar chemical partition.

giant. From the above derived E_{B-V}=0.85 mag, the observed value of V=12.22 mag for the cool giant during total eclipses corresponds to a distance of 1.0 kpc, in agreement with values estimated by [9]. The photoionization temperature derived by fitting of the observed emission lines in optical spectra is given in Table 1.

REFERENCES

1. Ciatti, F., The Nature of Symbiotic Stars, IAU Colloq. 70, ed. Reidel, D., 95, 61 (1982)
2. Ferland, G.J., ARA&A 41, 517 (2003)
3. Mikołajewska, J., Symbiotic Stars Probing Stellar Evolution, ASP Conf. Ser., eds. Corradi, R.L.M., Mikołajewska, J., Mahoney, T.J., 303, 9 (2003)
4. Munari, U., Whitelock, P., MNRAS 239, 273 (1989)
5. Munari, U., Whitelock, P., et al., AJ 104, 262 (1992)
6. Munari, U., Yudin, B.F., Kolotilov, E.A., Gilmore, A.C., AJ 109, 1740 (1995)
7. Munari, U., Physical Processes in Symbiotic Binaries and Related Systems, ed. Mikołajewska J., Copernicus Foundation for Polish Astron., 37 (1997)
8. Munari, u., Zwitter, T., A&A 318, 269 (1997)
9. Mürset, U., Dumm, T., Isenegger, S., Nussbaumer, H., Schild, H., Schmid, H.M., Schmutz, W., A&A 353, 952 (2000)
10. Sokoloski, J.L., Symbiotic Stars Probing Stellar Evolution, ASP Conf. Ser., eds. Corradi, R.L.M., Mikołajewska, J., Mahoney, T.J., 303, 202 (2003)
11. Straižys, V., Kuriliene, G., Ap&SS 80, 353 (1981)
12. Wallerstein, G., Gilroy, K.K., Willson, L.A., Garnavich, P., PASP 105, 859 (1993)

Numerical Simulations of the Thermal Instability Collapse in Radiation Pressure Dominated Disks

V. Teresi*, D. Molteni[†] and E. Toscano[†]

Dipartimento di Fisica e Tecnologie Relative, Università di Palermo, Viale delle Scienze, Palermo, 90128, Italy
[†]*<common address for author1, author2 and author3>*

Abstract. We show that accretion disks, both in the subcritical and supercritical accretion rate regime, may exhibit significant amplitude luminosity oscillations. The luminosity time behavior has been obtained by performing a set of time-dependent 2D SPH simulations of accretion disks with different values of α and accretion rate. An explanation of this luminosity behavior is proposed in terms of limit-cycle instability: the disk oscillates between a radiation pressure dominated configuration (with a high luminosity value) and a gas pressure dominated one (with a low luminosity value). The origin of this instability is the difference between the heat produced by viscosity and the energy emitted as radiation from the disk surface (the well-known thermal instability mechanism). We support this hypothesis showing that the limit-cycle behavior produces a sequence of collapsing and refilling states of the innermost disk region.

INTRODUCTION

This work continues our studies on the occurrence of the Shakura and Sunyaev instability [4] in the α-disks when the radiation pressure dominates, i.e. in the so-called A zone. Recently some authors have investigated this problem through the numerical approach. Szuszkiewicz and Miller [6, 7, 8] found that an accretion disc model with low or high viscosity shows a limit-cycle behavior with a period of the cycle of about 780 s. Nayakshin, Rappaport & Melia [3] and Janiuk, Czerny & Siemiginowska [1] used a limit-cycle model to explain the luminosity variability of the micro-quasar GRS 1915+105. Teresi, Molteni & Toscano [9] have clearly shown, with 2D simulations, that, at intermediate accretion rates, accretion discs with A zone suffer a collapse but after a rather long time they show a flaring activity with an intervening refilling phase of the A zone.

We point out that our simulations differ from the Szuszkiewicz and Miller ones since we produce real 2D disks with true vertical motion. Therefore no 'ad hoc' prescription is required to include physical vertical effects, as it is, instead, in the Szuszkiewicz and Miller simulations. Our results suggest that the Shakura-Sunyaev model can be used to explain the luminosity variability shown by many sources.

THE PHYSICAL MODEL

The time dependent equations describing the physics of accretion disks are well known. We used the lagrangean form of them in a cylindrical reference system and in the approximation of local thermal equilibrium (LTE) between gas and radiation [2]. They include:

mass conservation

$$\frac{D\rho}{Dt} = -\rho \, \text{div} \, \vec{v} \tag{1}$$

radial momentum conservation

$$\rho \frac{Dv_r}{Dt} = -\rho \frac{\lambda^2}{r^3} + \rho g_r + (\text{div} \, \vec{\vec{\sigma}})_r + f_r \tag{2}$$

vertical momentum equation

$$\frac{Dv_z}{Dt} = -\frac{1}{\rho}\frac{dP}{dz} - g_z + \frac{f_z}{\rho} \tag{3}$$

energy equation

$$\frac{D}{Dt}\left(\frac{E_{rad}}{\rho} + \varepsilon + \frac{1}{2}v^2\right) =$$

$$= \vec{v} \cdot \vec{g} - \frac{(P_{rad} + P_{gas})}{\rho}\nabla \vec{v} + \vec{v} \cdot \frac{\vec{f}}{\rho} + \frac{1}{\rho}\nabla\left[\vec{v}\vec{\vec{\sigma}}\right] \tag{4}$$

where σ_{ij}, $\vec{\vec{\sigma}}$, is the viscosity stress tensor and $\vec{g}$ is the body force per unit mass (acceleration).

angular momentum equation

$$\frac{D\Omega}{Dt} = -2\Omega\frac{v_r}{r} + \frac{1}{\rho}\frac{\partial}{\partial z}(v\rho\frac{\partial \Omega}{\partial z}) + \frac{1}{\rho}\frac{1}{r^3}\frac{\partial}{\partial r}(r^3 v\rho\frac{\partial \Omega}{\partial r}) \tag{5}$$

Here Ω is the local angular velocity, $\frac{D}{Dt}$ is the comoving derivative, E_{rad} is the radiation energy per unit volume, $\vec{f}$ is the radiation force per unit volume, λ is the angular momentum per unit mass, $E = \varepsilon + \frac{E_{rad}}{\rho}$ is the total internal energy per unit mass, including gas and radiation terms.

THE SIMULATIONS PERFORMED

We performed several simulations, the ones commented here had the following parameter values:
a) $\alpha = 0.1$, $\dot{M} = 0.15\dot{M}_E$, domain $R_1 - R_2 = 3R_g - 100R_g$, $h = 0.25R_g$;
b) $\alpha = 0.1$, $\dot{M} = 2\dot{M}_E$, domain $R_1 - R_2 = 3R_g - 200R_g$, $h = 0.5R_g$;
For all cases the central black hole mass is $M = 10 \, M_\odot$. The reference unit we use for the luminosity is the theoretical value.

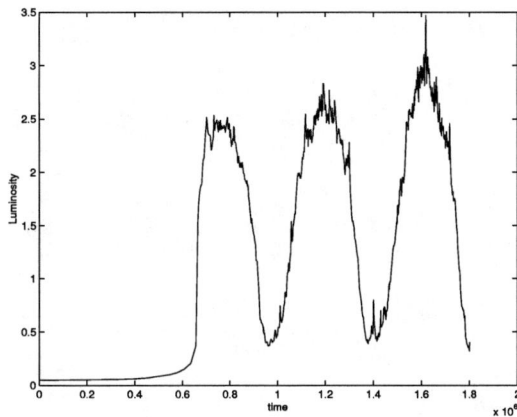

FIGURE 1. The time behavior of the disc luminosity is shown. On the x axis the time values in units of R_g/c are represented. On the y axis the luminosity values in units of $L_{theor} = 0.06\dot{M}c^2 = 2.05 \cdot 10^{39}\ erg\ sec^{-1}$ are represented.

When the instability arises the disc undergoes a collapse phase. Then, because of accretion, the collapsed zone is refilled. The disc luminosity oscillates between the two states. In case 'a' the descending phase of the single oscillation is exponential-like. For the case 'b' the luminosity oscillation is shown in fig. 1.

Here there is no exponential-like behavior as there is in the 'a' case, probably because the energy density of the considered disc region is not uniform throughout the region itself.

DISCUSSION AND CONCLUSIONS

Here we want to discuss the time features of the limit-cycle behavior obtained. There are two time-scales that affect the limit-cycle phenomenon: the thermal time-scale, that determines the development rate of the thermal instability, and the viscous time, that is connected to the A zone refilling after the collapse due to the instability. It is easy to see that, in case 'a', the theoretical viscous time-scales are too large compared to the 'experimental' luminosity cycle-time, whereas the thermal time-scales are too small. We present a table of these typical time-scales.

r	$t_{visc}(LS)$	$t_{visc}(HS)$	t_{therm}
$3R_g$	108.9 s	7.00 s	$4.84 \cdot 10^{-3}$ s
$7R_g$	2717 s	6.79 s	$2.218 \cdot 10^{-2}$ s
$13R_g$	25545 s	63.86 s	$6.046 \cdot 10^{-2}$ s

We propose the explanation that the additional viscosity produced by the 2D turbulent flow and the process of the radial diffusion of radiation reduce the characteristic time

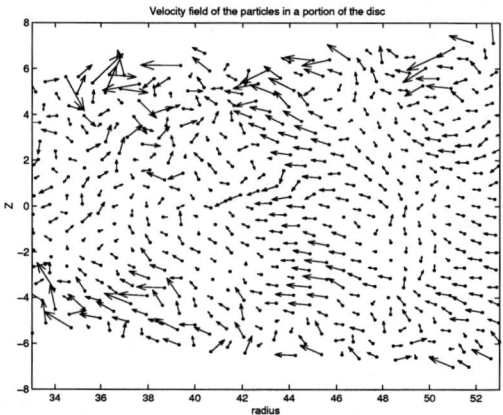

FIGURE 2. The gas velocity field in the radial range between $33R_g$ and $53R_g$ is shown for case 'b'. The disc state is the high one. On the x axis the r values in units of R_g are represented. On the y axis the z values in the same units are represented. The arrows represent the velocity vectors, with their lengths proportional to the speed values.

of the transition LS-HS. As regards the transition HS-LS, we form the hypothesis that the instability development time is increased by convection [5]. The presence of a significant convective motion of the disc gas is one of the most peculiar features of a real 2D simulation. We have found this kind of gas flow both in case 'a' and in the 'b' one. As an example we present in fig. 2 the gas velocity field in a given disc region for the high state of case 'b'.

In conclusion, we can say that the time-scale of the instability and the shape of the light curve depend on the accretion rate in the sense that lower accretion rates produce shorter time-scales. Moreover, the 2D time-scales are shorter (and the oscillation frequencies higher) than the 1D ones.

REFERENCES

Janiuk A., Czerny B., Siemiginowska A., 2002, ApJ, 576, 908J
Mihalas D., Klein R.I., 1982, Jou. Comp. Phys., 46, 97
Nayakshin S., Rappaport S., Melia F., 2000, ApJ, 535, 798N
Shakura N.I., Sunyaev R.A., 1976, MNRAS, 175, 613
Shakura N.I., Sunyaev R.A., Zilitinkevich S.S., 1978, A&A, 62, 179
Szuszkiewicz E., Miller J.C., 1997, MNRAS, 287, 165
Szuszkiewicz E., Miller J.C., 1998, MNRAS, 298, 888
Szuszkiewicz E., Miller J.C., 2001, MNRAS, 328, 36
Teresi V., Molteni D., Toscano E., 2004, MNRAS, 348, 361

On the mass distribution of neutron stars in HMXBs

A. van der Meer*, L. Kaper*, M.H van Kerkwijk[†] and E.P.J. van den Heuvel*

*Astronomical Institute "Anton Pannekoek", University of Amsterdam, Kruislaan 403, NL-1098 SJ Amsterdam, Netherlands
[†]Department of Astronomy and Astrophysics, University of Toronto, 60 St George Street, Toronto, ON M5S 3H8, Canada

Abstract. We present the results of a monitoring campaign of three eclipsing high-mass X-ray binaries (HMXBs: SMC X−1, LMC X−4 and Cen X−3). High-resolution VLT/UVES spectra are used to measure the radial velocities of these systems with high accuracy. We show that the subsequent mass determination of the neutron stars in these systems is significantly improved and discuss the implications of this result.

Keywords: X-ray binaries
PACS: 97.80.Jp

INTRODUCTION

The detailed supernova mechanism producing a neutron star is poorly understood, but it is likely that the many neutrinos that are produced during the formation of the (proto-) neutron star in the center of the collapsing star play an important role [4]. Model calculations show that Type II supernovae (massive, single stars) give a bimodal neutron-star mass distribution, with peaks at 1.28 and 1.73 $M_\odot$, while Type Ib supernovae (such as produced by stars in binaries, which are stripped of their envelopes) will produce neutron stars within a small range around 1.32 $M_\odot$ [17]. Currently, the most massive neutron star is the X-ray pulsar Vela X−1 [2, 12] with a mass of 1.86 ± 0.16 $M_\odot$. It would be consistent with the second peak in this distribution.

Neutron stars are detected as single objects or in binary systems of different evolutionary origin. We focus on the initially more massive binary systems: high-mass X-ray binaries (HMXBs) consisting of a massive OB supergiant and a neutron star or a black hole (for a review see [8]). In five of these systems containing an eclipsing X-ray pulsar the mass of the neutron star has been determined. The masses of all but one (Vela X−1) are consistent with being equal to 1.4 $M_\odot$. However, most spectroscopic observations used for these mass determinations yield too large uncertainties to measure a significant mass difference (see [20]).

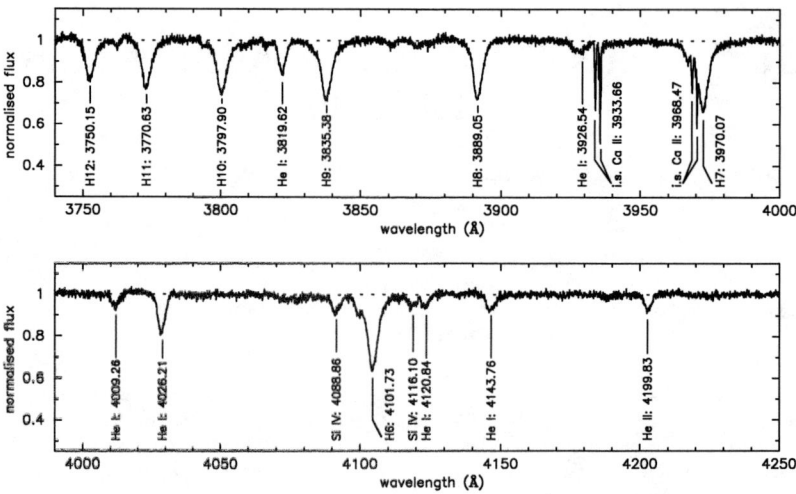

FIGURE 1. Spectrum of SMC X−1 in the wavelength range 3750–4250 Å, obtained by the blue arm of the UVES spectrograph at orbital phase $\phi \sim 0.51$. Many H and He lines are detected as well as a few metal lines. The spectrum has a S/N ~ 45.

OBSERVATIONS

Here we present new, more accurate determinations of the mass of the neutron star in three of these systems, i.e. SMC X−1, LMC X−4 and Cen X−3 using the high-resolution Ultraviolet and Visual Echelle Spectrograph UVES on the *Very Large Telescope* (VLT). These systems are in a phase of Roche-lobe overflow, have well determined, circular orbits (P_{orb} of a few days), and an optical counterpart of $V \simeq 14$ mag, i.e. well within reach of VLT/UVES.

Cen X−3 is located in our own galaxy and was identified as the first X-ray pulsar [5, 6]. The system consists of an O6-7 II-III star [1] with a neutron star in a 2.09 day orbit [11]. SMC X−1 is located in the Small Magellanic Cloud (SMC). The system hosts a B0.5 Ib supergiant [7] with a neutron star in a 3.89 day orbit [11]. LMC X−4 is located in the Large Magellanic Cloud (LMC) and consists of a ~ 20 $M_\odot$ O8 III star and a neutron star in a 1.41 day orbit [10].

We have obtained a dozen spectra of all three systems with VLT/UVES in service mode in the period October 2001 to March 2002. The spectra have a resolving power of $\sim 40\,000$, sufficient to measure line positions with an accuracy of a few km s^{-1} and cover the wavelength range 3600–6600 Å. Figure 1 shows part of the spectrum of SMC X−1. The spectra of LMC X−4 and Cen X−3 are of similar quality. For an extensive description of the full dataset, we refer to [19].

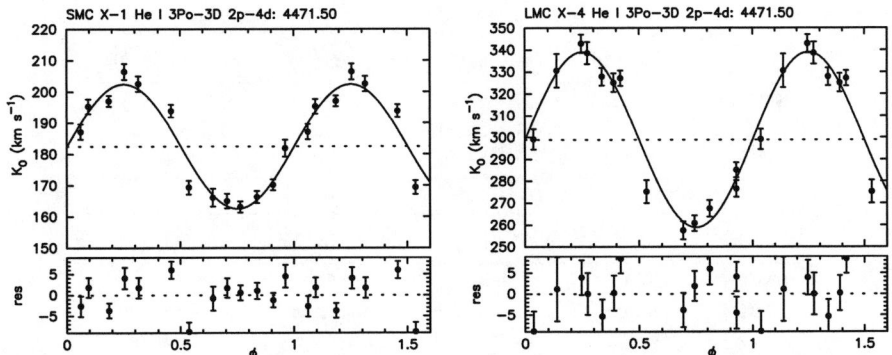

FIGURE 2. For each selected line that is clearly separated, that has a unique identification and is not affected by the stellar wind of the OB-supergiant a radial velocity can be determined. Here the velocity amplitudes of the He I line at 4471.50 Å with its fitted radial velocity curve is shown for SMC X−1 and LMC X−4. The lower panels show the residuals of the fit to the datapoints.

SPECTRAL ANALYSIS

To obtain a radial velocity measurement, often the complete spectra are cross correlated with a template spectrum. In our case the spectra are of such high quality that a radial velocity can be determined for each line separately. The advantage of such a strategy is that it is possible to map the influence of possible distortions due to X-ray heating, gravitational darkening and geometry of the line formation region. Here we select three lines, i.e. H I at 3797.90 Å, He I at 4026.21 Å and He I at 4471.50 Å, that are not affected by the stellar wind and that can be clearly separated from other lines. In [19] we present a velocity moment analysis to select the lines with objective criteria. To determine the line centre we fit the lines with a gaussian profile. In this way we obtain a radial velocity curve, which we fit with a sinusoidal profile (i.e. a circular orbit).

The X-ray pulsar's orbital period is accurately known from pulse time delay measurements. To determine the orbital phase of the system we use the ephemeris of [21], [10] and [11] for SMC X−1, LMC X−4 and Cen X−3, respectively. The orbits are circular. The fit parameters that remain are the amplitude of the radial velocity curve, K_O, and the γ velocity of the system, v_γ. A few example radial velocity curves are shown in Fig. 2 for the He I line at 4471.50 Å.

To determine the mass of the neutron stars in our sample we follow the approach of [13]. The results and literature values are listed in Table 1.

CONCLUSIONS

We show that with our VLT/UVES observations the mass measurements of the neutron stars in HMXBs are significantly improved. The masses are not all higher than 1.4 $M_\odot$, as is the case for Vela X−1. The mass of SMC X−1, which is 1.05 ± 0.09 $M_\odot$, is actually the lowest neutron star mass measured so far. Therefore, we conclude that a mass

TABLE 1. List of all the parameters obtained from literature. All errors are 1 σ.

system	P (days)	$a_x \sin i$ (lt s)	Θ_e	K_X (km s^{-1})
SMC X–1	3.89229090(43)	53.4876(4)	29 ± 2	299.595 ± 0.002
LMC X–4	1.40839776(26)	26.343(16)	27 ± 3	407.8 ± 0.3
Cen X–3	2.08713845(5)	39.56(7)	34 ± 3	413.2 ± 0.7

system	K_O (km s^{-1})	i (°)	M_O (M$_\odot$)	M_X (M$_\odot$)
SMC X–1	20.3 ± 0.9	68 ± 3	15.5 ± 1.5	1.05 ± 0.09
LMC X–4	34.3 ± 1.2	65 ± 4	15.6 ± 1.8	1.31 ± 0.14
Cen X–3	26.0 ± 2.5	73 ± 10	19.7 ± 4.3	1.24 ± 0.24

distribution of neutron stars is present in HMXBs. Possible distortions by gravitational darkening and X-ray heating, that can influence the value of the radial velocity amplitude of the optical companion are discussed in [19].

The most accurate neutron star masses have been derived for the binary radio pulsars. Most of these pulsars have a mass that is consistent with a small range near 1.35 M$_\odot$ [16]. Currently, the sample contains more systems and it is clear that a broader range from $1.25 - 1.44$ M$_\odot$ exists [15]. The secondary formed neutron star in these systems tends to be the less massive of the two [18]. On the other hand, our result shows that the primary formed neutron star can have a low mass as well. A mass distribution of neutron stars in binary radio pulsars as well as in HMXBs sets an important constraint on the formation mechanism of neutron stars.

REFERENCES

1. Ash, T. D. C., Reynolds, A. P., Roche, P., et al., MNRAS, **307**, 357 (1999)
2. Barziv, O., Kaper, L., Van Kerkwijk, M. H., Telting, J. H., and Van Paradijs, J., A&A, **377**, 925 (2001)
3. Brown, G. E., and Bethe, H. A., ApJ, **423**, 659 (1994)
4. Burrows, A., Nature, **403**, 727 (2000)
5. Chodil, G., Mark, H., Rodrigues, R., et al., Physical Review Letters, **19**, 681 (1967)
6. Giacconi, R., Gursky, H., Kellogg, E., Schreier, E., and Tananbaum, H., ApJ, **167**, L67 (1971)
7. Hutchings, J. B., Cowley, A. P., Osmer, P. S., and Crampton, D., ApJ, **217**, 186 (1977)
8. Kaper, L., in ASSL Vol. 264: The Influence of Binaries on Stellar Population Studies, 125 (2001)
9. Lattimer, J. M., and Prakash, M., Science, **304**, 536 (2004)
10. Levine, A. M., Rappaport, S. A., and Zojcheski, G., ApJ, **541**, 194 (2000)
11. Nagase, F., Corbet, R. H. D., Day, C. S. R., et al., ApJ, **396**, 147 (1992)
12. Quaintrell, H., Norton, A. J., Ash, T. D. C., et al., A&A, **401**, 313 (2003)
13. Rappaport, S. A., and Joss, P. C., in Accretion-Driven Stellar X-ray Sources, 1 (1983)
14. Srinivasan, G., in Black Holes in Binaries and Galactic Nuclei, 45 (2001)
15. Stairs, I. H., Science, **304**, 547 (2004)
16. Thorsett, S. E., and Chakrabarty, D., ApJ, **512**, 288 (1999)
17. Timmes, F. X., Woosley, S. E., & Weaver, T. A., ApJ, **457**, 834 (1996)
18. Van den Heuvel, E. P. J., astro-ph/0407451 (2004)
19. Van der Meer, A., Kaper, L., Van Kerkwijk, M. H., et al., to be submitted (2005)
20. Van Kerkwijk, M. H., Van Paradijs, J., & Zuiderwijk, E. J., A&A, **303**, 497 (1995)
21. Wojdowski, P., Clark, G. W., Levine, A. M., Woo, J. W., and Zhang, S. N., ApJ, **502**, 253 (1998)

Creating ultra-compact binaries through stable mass transfer

M.V. van der Sluys*, F. Verbunt* and O.R. Pols*

Astronomical Institute, Post box 80.000, 3508 TA Utrecht, the Netherlands

Abstract. A binary in which a slightly evolved star starts mass transfer to a neutron star can evolve towards ultra-short orbital periods under the influence of magnetic braking. This is called magnetic capture. We investigate in detail for which initial orbital periods and initial donor masses binaries evolve to periods less than 30–40 minutes within the Hubble time. We show that only small ranges of initial periods and masses lead to ultra-short periods, and that for those only a small time interval is spent at ultra-short periods. Consequently, only a very small fraction of any population of X-ray binaries is expected to be observed at ultra-short period at any time. If 2 to 6 of the 13 bright X-ray sources in globular clusters have an ultra-short period, as suggested by recent observations, their formation cannot be explained by the magnetic capture model.

Keywords: Globular clusters, X-ray sources
PACS: 98.20.Gm, 97.80.Jp

INTRODUCTION

About half of the bright X-ray sources in the galactic globular clusters possibly are binaries with ultra-short orbital periods ($\lesssim 40$ min). Two of the five periods known are 11.4 min (in NGC 6624) and 20.6 min (in NGC 6712). The 11.4 min system has a negative period derivative. This high fraction of ultra-short periods is in marked contrast to bright X-ray sources in the galactic disk, where such periods are less common (See Table 3 of [1]).

One of the scenarios to explain the ultra-short periods starts from a binary with a neutron star and a main-sequence star. For a small range of initial orbital periods, strong magnetic braking can shrink the orbit sufficiently that the system evolves to a minimum period in the ultra-short range. This way, an orbital period shorter than 11 min can be reached [2]. At 11.4 min, the period derivative may be negative or positive, depending on whether the system evolves to the period minimum, or has already rebounded.

We determine which initial systems can evolve to ultra-short periods within the age of the globular clusters and what the probability is to observe these systems as X-ray sources. An article about this research has been published in [3].

THE EVOLUTION CODE

We calculate our models using the STARS binary stellar evolution code, developed by Eggleton [4], but with updated physics [5]. The primary is treated as a point mass with an initial mass of $1.4 M_\odot$. Sources of angular momentum loss are gravitational waves,

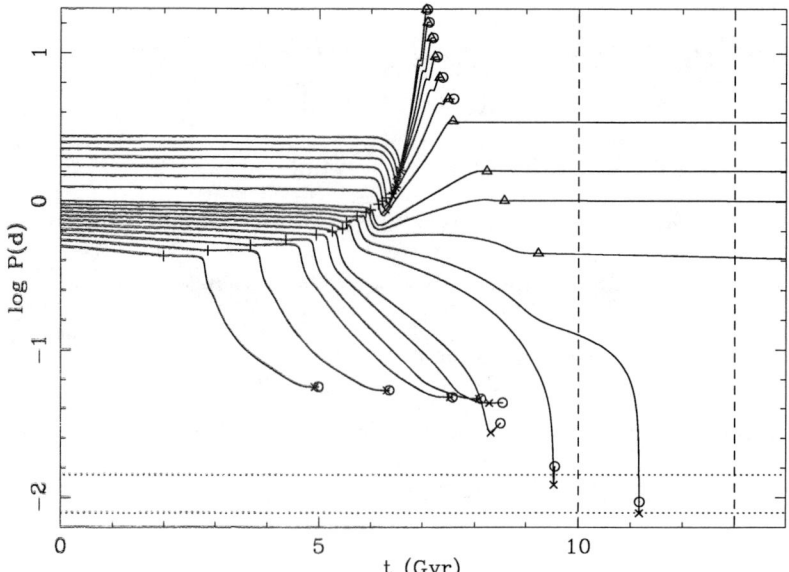

FIGURE 1. Time-period (t-P) tracks for $1.1 M_\odot$. Initial periods are spaced 0.05 d below 1 d and 0.25 d above that. The symbols show: start of mass transfer +; period minimum ×; end of mass transfer △; the the last model ○. The dotted lines are at 11.4 and 20.6 min.

partially conservative mass transfer, and magnetic braking according to [6]:

$$\frac{dJ_{MB}}{dt} = -3.8 \times 10^{-30} M_2 R_2^4 \omega^3 \, \text{dyn cm}, \quad (1)$$

where M_2, R_2 and ω are the mass, radius and angular rotational velocity of the secondary respectively. Tidal effects keep the spin synchronised to the orbit and magnetic braking effectively removes angular momentum from the orbit.

RESULTS

We calculated a grid of models for $Z = 0.01$ (the metallicity of NGC 6624), with initial masses between 0.7 and $1.5 M_\odot$ and initial periods between 0.35 and 3 d. The grid is refined in period around the bifurcation period between converging and diverging systems. We consider models that converge after the Hubble time as diverging.

Figure 1 shows that orbits with low initial period converge to about 70 min, and orbits with high orbital period diverge to several days. A small range in between leads to ultra-compact systems. It is clear that an initial period must be picked carefully to find such a short period minimum.

To determine the probability of observing an ultra-compact binary formed this way, we perform statistics on the t-P tracks. We choose a random initial period from a flat distribution in $\log P$ and determine the t-P track of a system with that initial period by

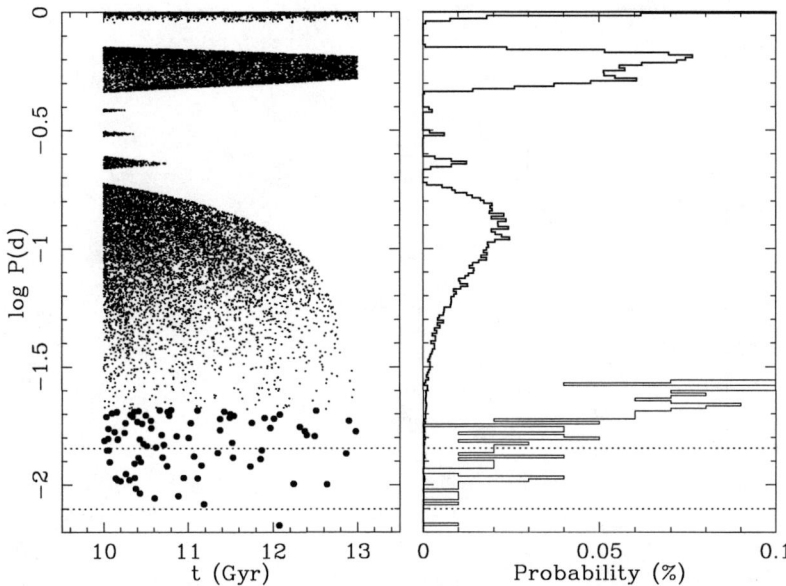

FIGURE 2. Left panel (a): accepted systems for $1.1M_\odot$. Each dot represents one system. The spikes at $\log P \approx -0.5$ are artefacts due to the interpolation. Dots below 30 min are larger for more clarity. Right panel (b): Histogram of the data obtained by summing (a) over the time. The thin line shows the short-period tail, with the probability 100 times enlarged.

interpolation. Once the track is known, we choose a random moment in time between 10 and 13 Gyr (the dashed lines in Fig. 1) and determine the orbital period at that moment. If a system has passed its period minimum, or has no mass transfer at that moment, we reject it. For $1.1M_\odot$, 10.5% of the 10^6 probes is accepted and shown in Fig. 2a. Of these, 86 systems have an orbital period less than 30 min (the large dots in the figure). Figure 2b shows the corresponding period distribution.

We have calculated similar grids for $Z = 0.002$ and $Z = 0.02$ and performed the same statistics as for $Z = 0.01$. The period distributions for each mass have been added using a flat mass distribution. A Salpeter mass distribution leads to little difference, especially for the ultra-short period regime. We show the results for the two of these metallicities in Fig. 3.

Figure 3 shows firstly that there are no big differences in the expected fraction of observable ultra-compact X-ray binaries. Though the exact numbers are uncertain due to the low number of accepted ultra-compact systems, we expect that of a population of 10^7 binaries with initial periods between roughly 0.5 and 2.5 d, 1 to 10 systems have a period of 11.4 min and 10 to 100 systems have a 20.6 min orbital period and emit X-rays.

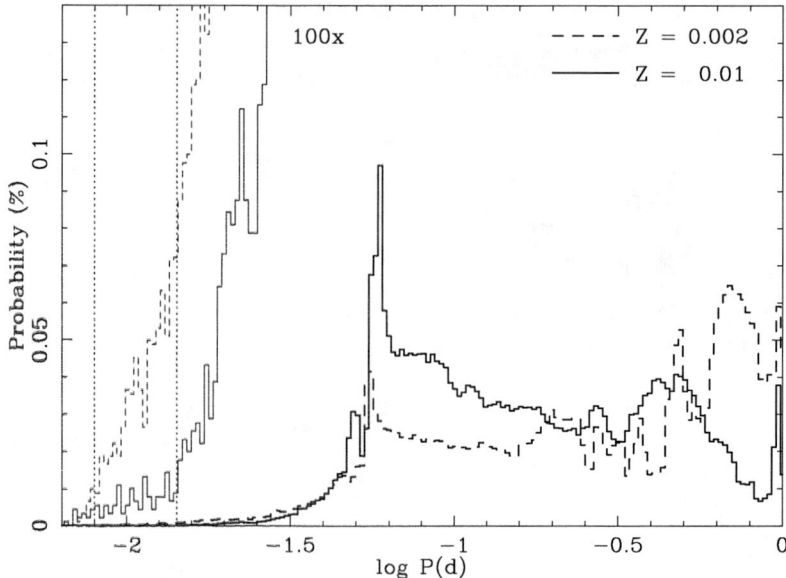

FIGURE 3. Period distributions for $Z = 0.002$ and $Z = 0.01$. The thin lines on the left are enlarged 100 times in the vertical direction.

DISCUSSION

We confirm that magnetic capture can lead to ultra-compact X-ray binaries within a Hubble time. In order to find a binary system that will have its minimum period in the ultra-short period regime, one has to carefully select an initial period, just under the bifurcation period for that system. The systems that reach an ultra-short period remain there for a relatively short time, as can be seen from the steep tracks in Fig. 1. These factors combined make it rather unlikely that such a system can be observed. The metallicity of the stars only has a small influence.

Alternative scenarios to create ultra-compact binaries only allow positive period derivatives for these binaries. For a description and references see Verbunt (these proceedings) and [3].

REFERENCES

1. P. Charles, and M. Coe, "A catalogue of low-mass X-ray binaries," in *Compact stellar X-ray sources*, edited by W. Lewin, and M. van der Klis, Cambridge U.P., Cambridge, 2004.
2. P. Podsiadlowski, S. Rappaport, and E. D. Pfahl, *ApJ*, **565**, 1107–1133 (2002).
3. M. V. van der Sluys, F. Verbunt, and O. R. Pols, *A&A*, **431**, 647–658 (2005).
4. P. P. Eggleton, and L. Kiseleva-Eggleton, *ApJ*, **575**, 461–473 (2002).
5. O. R. Pols, C. A. Tout, P. P. Eggleton, and Z. Han, *MNRAS*, **274**, 964–974 (1995).
6. S. Rappaport, P. C. Joss, and F. Verbunt, *ApJ*, **275**, 713–731 (1983).

X-ray Modulation from Non-Axisymmetric Structures in Accretion Disk

P. Varnière*, E. Blackman* and M. Muno[†]

*Department of Physics & Astronomy, Rochester University, Rochester NY 14627-0171
[†]UCLA

Abstract. Non-axisymmetric accretion disks around compact objects can lead to X-ray variability. Low-frequency Quasi-Periodic Oscillations (LFQPOs), observed in X-ray binaries, have not yet been explained, but could in fact be the result of non-axisymmetry due to spiral waves or blobs in the disk. Here we provide a unified analytic formula that characterizes the non-axisymmetric disk height profile for both spiral waves and blobs. We then study the influence of these structures on the observed flux by relating the height to the local disk temperature and include the important shadowing effects. We predict the observed timing properties and apply the results to LFQPOs, inferring the key features of the non-axisymmetric structure required to match the LFQPO observations.

Keywords: X-ray binaries – observation
PACS: 95.75.-z, 95.85.Nv, 97.80.Jp

INTRODUCTION

X-ray emission from accreting black holes in binary stellar systems varies on time scales ranging from milliseconds to years. Variability on time scales longer than days appears to be driven by changes in the accretion rate onto the black hole, and is often manifested as transient outbursts. At the shortest time scales, quasi-periodic oscillations (QPOs) are observed in the X-ray emission. The highest frequency QPOs (> 100Hz) are consistent with those expected from general relativistic orbits near the innermost stable orbit around the black hole, and they are likely caused by inhomogeneities in the inner accretion flow [3, 4]. However, the cause of low frequency QPO (LFQPO) 0.1–20 Hz is still a mystery. This motivates the work herein.

Two different conceptual paradigms for LFQPOs have been proposed in this regard: (1)the centrifugal pressure supported boundary layer model [1]. In this model, the LFQPO would represent a global radial oscillation that modifies the emitted flux. (2) the accretion-ejection instability model [5, 7]. A spiral shock forms in an accretion disc threaded by a vertical magnetic field. The LFQPO would result from the orbit of this non-axisymmetric structure in the disc.

In this paper we focus on the paradigm exemplified by the latter type of model, namely the production of LFQPOs by the orbit of a non-axisymmetric structure. We investigate the specific question of whether a stable pattern in an accretion disc can reproduce the observed characteristics of LFQPO simply by its orbit around the central engine. In Sec. 2 we give analytical formulae that can be used to model arbitrarily shaped blobs and spirals as non-axisymmetric disc features. In Sec. 3 we compute the emission from an accretion disc with such non-axisymmetric structures and discuss the influence of the shape of the particular non-axisymmetric structures on the LFQPO RMS amplitude. We also discuss some observational consequences of having a non-axisymmetric structure at the origin of the LFQPO.

THE FORMALISM

Blobs and spirals are two useful categories of non-axisymmetric structures. Here we do not detail the formation of these non-axisymmetric structures but focus on providing an analytic framework that characterizes their shape for practical use and allows observational implications for flux modulation to be quantified. We parametrized the disc thickness by:

$$h(r,\vartheta) = h_o(r) + \tilde{\gamma}\left(\frac{r_c}{r}\right)^\beta e^{-0.5\left(\frac{r-r_s}{\delta}\right)^2}$$

where $h_o(r)$ is the axisymmetric unperturbed disk thickness, r_c is the point where the perturbation is the thickest, $\tilde{\gamma}$ is the perturbation thickness at r_c, β is the measure of how fast the disk thickness decreases from the maximum, δ parametrizes the radial extent of the structure on the disk, r_s is the equation of the spiral/blob on the disk ($r_s = r_c e^{-\alpha(r)\vartheta}$) and α is the opening angle of the spiral.

RESULTS

Having obtained an expression for the disc height, we compute the temperature using the approximation that $c_s = h\Omega$. In order to determine the observational effect of a non-axisymmetric disc thickness, we compute the flux as a function of azimuth, assuming that the spectrum at each point is a blackbody.

First notice the lack of influence of the range of $\phi > 2\pi$ in Table 1. For each ϕ, most of the modulation comes from the most inward "bump" in $h(r)$.

Impact of $\tilde{\gamma} = \gamma h_o(r_c)$

The parameter $\tilde{\gamma}$ measures the maximum thickness of the perturbation. Its effect is evident in comparing cases #3, #4, #5. The greater $\tilde{\gamma}$, all else being equal, the stronger the modulation. This is because a thicker structure more strongly shadows the inner disc.

Because $\tilde{\gamma} \equiv \gamma h_o(r_c)$ is the maximum height of the spiral arm, a change in h_o (the zeroth order disc thickness) also increases the modulation. By comparing cases #14 and #15 we see that changing h_o by a factor of two is not exactly the same as changing $\tilde{\gamma}$ by a factor of two because increasing $\tilde{\gamma}$ also increases the difference between the maximum height of the perturbation and h_o.

Impact of β

β measures how fast the the maximum height along the spiral decreases with radius. Its effect is revealed in cases #3, #6, #7. The RMS amplitude of the modulation increases with β for an $\alpha > 0$ spiral. A rapidly decreasing height perturbation with radius means that the height also rapidly decreases along the spiral. This causes a stronger modulation

#	h_o	r_c	α	γ	β	δ	# turn	rms	max
#1	0.01	2.	0.05	0.5	1	0.1	3	2.3%	$\sim 1.28\pi$
#2	0.01	2.	0.05	0.5	1	0.1	2	2.2%	$\sim 1.28\pi$
#3	0.01	2.	0.05	0.5	1	0.1	1	2.2%	$\sim 1.28\pi$
#4	0.01	2.	0.05	0.3	1	0.1	1	0.9%	$\sim 1.28\pi$
#5	0.01	2.	0.05	0.7	1	0.1	1	3.7%	$\sim 1.28\pi$
#6	0.01	2.	0.05	0.5	2	0.1	1	4.3%	$\sim 1.25\pi$
#7	0.01	2.	0.05	0.5	3	0.1	1	7.3%	$\sim 1.25\pi$
#8	0.01	2.	0.05	0.5	2	0.2	1	5.6%	$\sim 1.25\pi$
#9	0.01	1.5	0.05	0.5	2	0.1	1	3.5%	$\sim 1.875\pi$
#10	0.01	2.5	0.05	0.5	2	0.1	1	3.7%	$\sim 1.25\pi$
#11	0.01	3.	0.05	0.5	2	0.1	1	3.2%	$\sim 1.25\pi$
#12	0.01	2.	0.03	0.5	2	0.1	1	1.7%	$\sim 1.28\pi$
#13	0.01	2.	0.1	0.5	2	0.1	1	12.3%	$\sim 1.875\pi$
#14	0.02	2.	0.05	0.5	2	0.1	1	8.6%	$\sim 1.25\pi$
#15	0.01	2.	0.05	1.	2	0.1	1	9.2%	$\sim 1.25\pi$

by producing a stronger azimuthal variation. This is particularly important for tight spirals : were it not for a large β, little non-axisymmetry would otherwise arise.

Impact of the spiral

The simulations relevant for studying the influence of r_c (initial radius where the perturbation begins) are #6, #9, #10 and #11. We see a maximum in the RMS amplitude for $r_c \sim 2\, r_{in}$, where r_{in} is the inner disc radius: when the spiral is too near r_{in}, it obscures less of the inner disc which is the most luminous part. On the other hand, if the spiral is too far out, it can only obscure a less luminous outer region giving a smaller modulation. The combination of these effects is an intermediate r_c for maximal modulation. Since the position of the spiral is chosen based on the LFQPO frequency, a correlation between the RMS amplitude and the frequency of the QPO is expected.

The influence of the parameter α, which determines the opening of the spiral wave can be seen from simulations #6, #12 and #13. There we see that the more open the spiral, the higher the RMS amplitude. This is because a more open spiral means a more non-axisymmetric thickness profile, leading to a higher RMS amplitude of modulation.

Finally, consider the parameter δ, which measures the width of the spiral or blob at its base. From cases #6 and #8 we see that a larger δ implies a larger RMS amplitude. For our choice of a highly inclined system, this trend results because a larger δ makes more of the region inner to the peak of the perturbation more perpendicular to the line of sight. This produces a larger observed flux when the observer is looking at the disk from an azimuth for which the line of sight intersects the inner part of the perturbation. But as the disk rotates, the outer edge of the perturbation comes into view, and the shadowing of the inner region occurs similarly for large or small δ. The contrast in flux (and thus the RMS amplitude of the modulation) is thus larger for larger δ, explaining the trend.

LINK WITH OBSERVATIONS

Because the modulation comes from shadowing, the disc inclination angle θ is important in determining the maximum shadowing and thus the maximum RMS amplitude that a given choice of parameters can produce. Motivated by a comparison with the LFQPOs of GRS 1915+105 (see *e.g.* the review [2]), we again focus on the parameter choices of simulation #8 and vary the inclination angle to obtain different values of the RMS amplitude. As expected, the more edge-on the view, the higher the RMS amplitude. Present observational data are insufficient to definitively confirm or contradict the predicted behavior. Such a trend could explain the absence of LFQPO in Cyg X-1, as the RMS amplitude expected from the inferred inclination angle is very small. More objects with a wider range of inclination angles are needed.

CONCLUSION

The key effect of the asymmetry from our blob or spiral structure that produces the LFQPO is the shadowing of the disk's inner region by the structure. The RMS amplitude of the modulation increases as the height contrast between the spiral or blob becomes sharper. We considered only symmetric height profiles of the structure, but it is the steepness of the slope in the inner edge of the structure that is most important–the sharper the rise, the stronger the shadowing for a fixed maximum height.

We also found that there is an optimal radius where a given non-axisymmetric structure is most effective at producing the LFQPO RMS amplitude: If the structure is too close to the central engine, then too little of the most luminous part of the disk would be shadowed. On the other hand, if the structure is too far from the center, then it is ineffective at shadowing the the most luminous part. This will have to be compared with observations.

REFERENCES

1. Chakrabarti, S. K., & Manickam, S. G. 2000, ApJ, **531**, L41
2. McClintock J. E. & Remillard, R. A., "Compact Stellar X-ray Sources," eds. W.H.G. Lewin and M. van der Klis, Cambridge University Press.
3. Remillard, R. A., Muno, M. P., McClintock, J. E., Orosz, J. A. 2002, ApJ, **580**, 1030
4. Stella, L. & Vietri, M. 1998, ApJ, **492**, L59
5. Tagger, M., and Pellat, R., 1999, A&A, **349**, 1003 (**TP99**)
6. Tomsick, J.A. and Kaaret, P., 2001, ApJ, **548**, 401.
7. Varnière, P. and Tagger. M., 2002, A&A, **394**, 329-338.

Time and spectral changes of GRS 1915+105 in the ρ class

G. Ventura*, E. Massaro*, T. Mineo[†], G. Cusumano[†], M. Litterio**, M. Feroci**, P. Casella[‡] and G. Matt[§]

Department of Physics, Univ. La Sapienza, Roma, Italy
[†]*INAF, IASF, Palermo, Italy*
**INAF, IASF, Roma, Italy*
[‡]*Astronomical Observatory of Roma, M. Porzio Catone, Italy*
[§]*Department of Physics, Univ. Roma Tre, Roma, Italy*

Abstract. We report the results of the temporal and spectral analysis of the longest *BeppoSAX* pointing (about 770 ks) of the microquasar GRS 1915+105, performed in October 2000. The source was mainly observed in the variability class ρ characterized by series of sharp pulses with a typical recurrence time of about 50 s. This behaviour was generally stable but, occasionally, it changed to a more irregular state. A spectral analysis of the 1-100 keV emission showed a small but significant change of the disk temperature between the regular and irregular ρ-mode

Keywords: X-ray sources, Microquasar: GRS 1915+105
PACS: 95.85.Nv, 97.10.Gz, 97.80.Jp

INTRODUCTION

The X-ray emission of GRS 1915+105 is characterized by very different time and spectral behaviors grouped by Belloni et al. (2000) in 12 classes. In particular, the ρ class consists of quasi–periodic 'flares' recurring on a time scale of 1–2 minutes. Their typical shape shows a nearly exponential rising branch followed by a short maximum and a quite fast decay. In this contribution we present some preliminary results of a long wide band X-ray observation of GRS 1915+105 performed with the BeppoSAX satellite from 20 to 29 October 2000. We can define two modes of the ρ class from the MECS (2–10 keV) and PDS (13–200 keV) light curves, characterized by a limited but significant change of the mean spectral parameters.

THE LIGHT CURVES

During the observation of October 2000 the source behavior was for a large fraction of the time in the ρ class alternating stables and irregulars phases. In the final part of the pointing a transition to the ν class was observed. We divided the entire observation in segments each one approximately corresponding to a satellite orbit. Each segment was studied using the Discrete Fourier Transform (DFT) and the wavelet power spectra (WPS), however here we present the results on only two segments of a duration of 2000 s, indicated as A and B. Their MECS light curves, integrated in a time bin of 0.5 s, are plotted in the Fig. 1a, b, respectively. In the segment A pulses repeat regularly with a

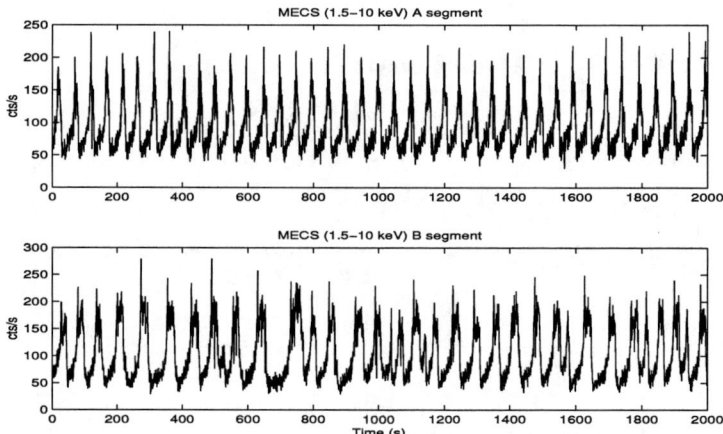

FIGURE 1. MECS light curves of the time segments A and B showing the *regular* and *irregular* modes of the ρ variability class.

stable pattern, while those of B have a variable recurrence time and show also different widths.

THE 2 MODES OF THE ρ CLASS

The DFT power spectra of the A time series in the MECS and PDS energy bands show a single prominent feature at the frequency of 20.28 mHz (corresponding to a recurrence time of 49.30 s). In the spectra of the B series there are two peaks at 14.11 mHz (70.87 s) and 16.56 mHz (60.37 s). WPS, computed by means of the Morlet wavelet transform, show that the A data set is characterized by an uninterrupted line of the maxima, centred at the period value of 49 s. Another line, but with a smaller power, is also present at ∼24.5 s (the first harmonic). Both lines have a small amplitude oscillating behavior over a time scale of a few cycles because of the not precisely periodic recurrence of the pulses. WPS spectra of the B segments have a maximum line showing meanderings and interruptions in the interval 50–120 s. In the last part an isolated maximum appears at ∼40 s. (see Fig. 2).

The use of DFT and WPS allows us to define a criterion, based on the number of peaks and the structure of the highest power line, to distinguish two *modes* of the ρ variability class: segments with spectra having a single prominent peak in the DFT power spectrum and a narrow and uniform strip in the WPS correspond to that we called *regular ρ mode*; segments with two or more high peaks in the DFT spectrum and WPS showing a winding path with bifurcations and/or isolated maxima are typical of the *irregular ρ mode*.

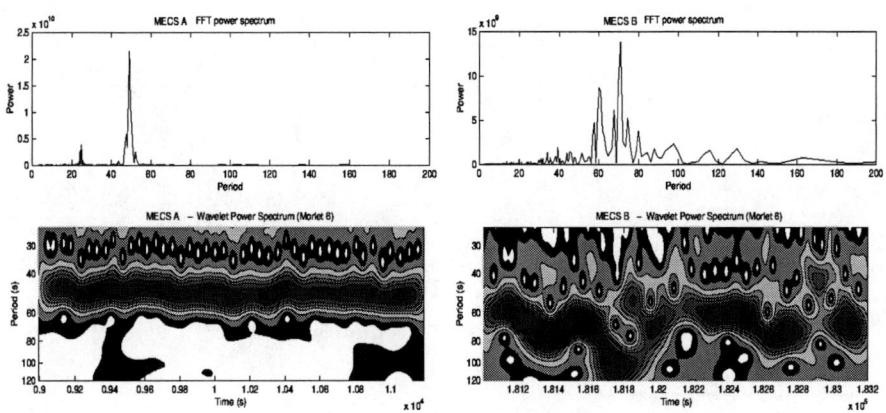

FIGURE 2. DFT and WPS power spectra for MECS A (left panels) and B (right panels) data series.

CHAOTIC BEHAVIOR ANALYSIS

For a more complete description of the two modes of the ρ class we investigated if the source behavior can be considered deterministic or chaotic calculating the *correlation integral* of Grassberger and Procaccia (1983):

$$C_M(R) = [N(N-1)]^{-1} \Sigma_i \Sigma_j H(R - |x_i - x_j|) \quad (1)$$

where x_i and x_j are the data vector in a M dimensional phase space, computed at the instants t_i and t_j, respectively, H is the Heaviside function and R is a distance. The correlation dimension $D_2(M)$ can be estimated from the log-derivative of $C_M(R)$ with respect to R. We found that for the A data set $D_2(M)$ converges rapidly to a value close to 2. For the B series $D_2(M)$ doesn't converge to a finite value and increases above 5 (Fig. 3). We conclude then that while the *regular* ρ mode can be described by a small number of dynamical equation, the *irregular* ρ mode is somewhat of *intermediate* between a stochastic and a deterministic behavior.

SPECTRAL ANALYSIS

Spectral analysis was performed on the entire data sets to verify whether the mean emission changes in the two modes. We considered the events in MECS and PDS energy bands and used a two component model: *i*) a multitemperature blackbody accretion disk (*diskbb*) and *ii*) a logarithmic parabolic law $F(E) = K_h E^{-(\Gamma + \beta Log E)}$ for the high energy emission. Interstellar absorption from an equivalent hydrogen column density was fixed at $N_H = 5 \times 10^{22}$ cm^{-2} and the MECS/PDS intercalibration factor was limited in the range (0.82–0.88). Best fit values of the spectral parameters and their 1σ errors for the two data sets A and B are given in Table 1.

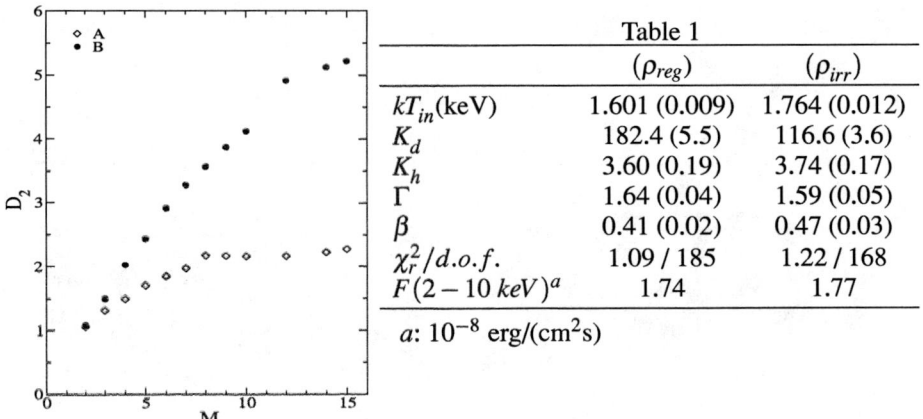

FIGURE 3. Correlation dimensions for the A and B data sets. Table with the best fit spectral parameters

	(ρ_{reg})	(ρ_{irr})
kT_{in}(keV)	1.601 (0.009)	1.764 (0.012)
K_d	182.4 (5.5)	116.6 (3.6)
K_h	3.60 (0.19)	3.74 (0.17)
Γ	1.64 (0.04)	1.59 (0.05)
β	0.41 (0.02)	0.47 (0.03)
$\chi_r^2/d.o.f.$	1.09 / 185	1.22 / 168
$F(2-10\ keV)^a$	1.74	1.77

a: 10^{-8} erg/(cm^{2}s)

Note that the high energy component is unchanged while the B data have a disk temperature higher than A of ∼10%. Although small this difference is significant because when the spectral model for A series is applied to B data the resulting χ^2 is 6.90.

CONCLUSIONS

Our analysis of the long BeppoSAX observation of October 2000 of GRS 1915+105 showed that the ρ variability class (Belloni et al. 2000) presents two modes. In the *regular* mode the distribution of the recurrence times of pulses has a small variance, power spectra have a single prominent peak and the WPSs show a narrow feature. In the *irregular* mode the distributions of recurrence times have a large variance, DFT spectra have two or more prominent peaks and the WPSs show features with meanderings, bifurcations and isolated maxima.

We also found that the mean emission spectrum of the ρ_{irr} mode has a higher inner disk temperature than ρ_{reg}. It is interesting that the source luminosity does not differ between the two data sets. In fact, the ratio between *diskbb* normalisation factors is nearly equal to inverse of the fourth power of the temperature ratio, also shown by the equal values of the 2– 10 keV fluxes. It is unclear what is the physical process triggering the mode change: the change of the dimension of the time series from about 2 to a higher values suggests that non-linear effects should be important in the system physics.

REFERENCES

1. Belloni T., Klein-Wolt M. et al. *Astron. Astrophys.*, **355**, 271 (2000).
2. Grassberger P., Procaccia I. *Physica*, **9D**, 189 (1983)

Chandra Localizations of LMXBs: IR Counterparts and their Properties

Stefanie Wachter[*], Joseph W. Wellhouse[†] and Reba M. Bandyopadhyay[**]

[*]*Spitzer Science Center, Caltech*
[†]*Harvey Mudd College*
[**]*Oxford University*

Abstract. We present new Chandra observations of the low mass X-ray binaries (LMXBs) X1624−490, X1702−429, and X1715−321 and the search for their Infrared (IR) counterparts. We also report on early results from our dedicated IR survey of LMXBs. The goal of this program is to investigate whether IR counterparts can be identified through unique IR colors and to trace the origin of the IR emission in these systems.

Keywords: low mass X-ray binaries, infrared counterparts
PACS: 97.80.Jp

INFRARED PROPERTIES OF LMXBs

Traditionally, LMXBs have been studied in the optical and UV part of the spectrum. In order to explore the IR properties of LMXBs and to investigate the most heavily absorbed sources in the Galactic Bulge, we are undertaking a dedicated IR survey of all LMXBs. In addition to our own observations, we have also searched the literature for published IR magnitudes for these sources. For the brightest LMXBs in fields with moderate crowding, we extracted J, H, and K magnitudes from the 2MASS database. Selected early results from our survey are summarized in Table 1 below. Most of the observations were obtained with the 1.5m telescope at CTIO. Photometry was performed with DAOPHOT II and standardized magnitudes were derived through comparison with 2MASS.

Figure 1 shows the position of the individual LMXBs in the IR color-color diagram (filled circles). Open circles indicate multiple measurements of the same sources. Also shown are the main sequence and giant branch tracks. The intrinsic variability of the LMXBs limits the predictive power of the IR colors (see e.g. Sco X−1). A few sources reveal the contribution of a giant mass donor. X1608−52 and X1636−536 appear to show very unusual colors. These are some of the faintest sources we measured and require deeper observations to confirm our photometry.

Figure 2 shows the positions of the individual LMXBs in the IR color-magnitude diagram (note that apparent, not absolute, K magnitudes are plotted). The symbols are the same as used in Figure 1. For comparison, we also include the location of field stars from a representative Galactic Bulge field (small filled circles). The different branches visible in the color-magnitude diagram distinguish different types of stars. The first branch, roughly up to $J - K = 1.8$, corresponds to nearby main sequence stars, while the clump of stars around $J - K = 2.0 - 2.5$ represents a superposition of giant stars

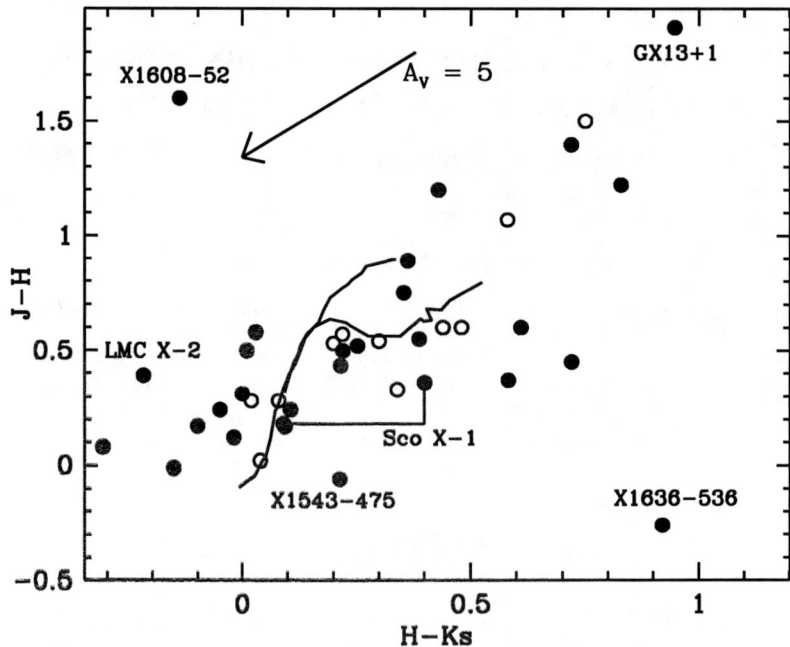

FIGURE 1. IR color-color diagram for LMXBs. For details, please see text.

with different values of extinction and distance. The LMXBs appear to preferentially cluster in an almost vertical strip around $J - K = 0$. GX 13+1 stands out as a remarkably red source.

CHANDRA LOCALIZATIONS

X1624−490 and X1702−429

We observed X1624−490 on 2002 May 30 and X1702−429 on 2003 June 19 with the Chandra HRC-I for 1 ksec each. In the X1624−490 data set, a single bright source is detected at the center of the 30'×30' field. The best position is 16:28:02.825 −49:11:54.61 (J2000) with the nominal 0.6" positional uncertainty. In the X1702−429 data set, the X-ray binary is the only source detected. Our best localization gives 17:06:15.314 −43:02:08.69 (J2000).

We also obtained deep Ks band observations of each source at the ESO NTT with SOFI and the CTIO 4m telescope with ISPI, respectively. A single, faint ($Ks = 18.3 \pm 0.1$) source is visible inside the Chandra error circle of X1624−490, and we propose this source as its IR counterpart. For X1702−429, a $Ks = 16.5 \pm 0.07$ source is visible

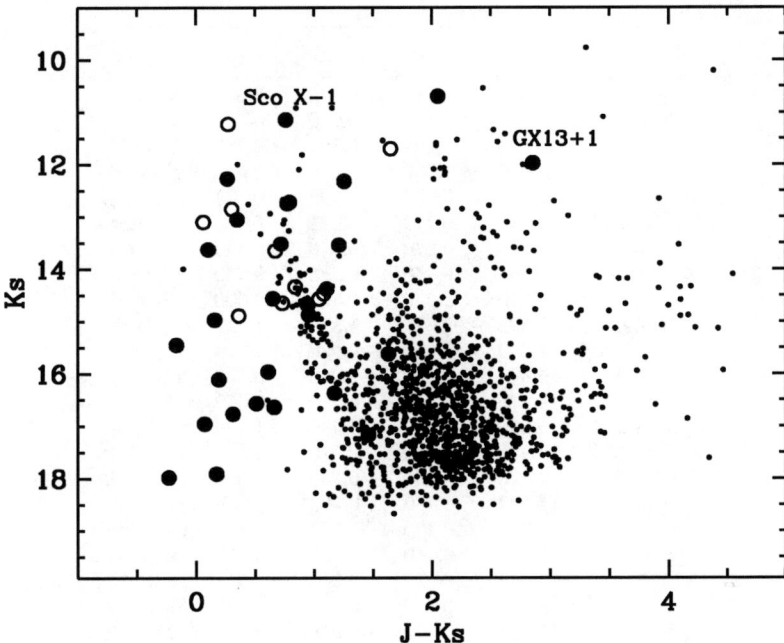

FIGURE 2. IR color-magnitude diagram for LMXBs. For details, please see text.

at the edge of the Chandra error circle. The brightness of both counterpart candidates is comparable to that of other low mass X-ray binary IR counterparts when corrected for extinction and distance. For details, please refer to Wachter et al. 2005, ApJ, in press.

X1715−321

X1715−321 is a poorly studied burster and transient at a distance of 5-7 kpc. We obtained a 1 ksec HRC-I observation of the source in an effort to detect its quiescent counterpart. No source was detected in the observation, placing an upper limit of 2.8×10^{-14} ergs cm^{-2} s^{-1} for the quiescent flux from this source.

ACKNOWLEDGMENTS

The research described in these pproceedings was carried out, in part, at the Jet Propulsion Laboratory, California Institute of Technology, and was sponsored by the National Aeronautics and Space Administration. We made use of data products from the 2 Micron All Sky Survey, which is a joint project of the University of Massachusetts and the In-

TABLE 1. IR Observations of LMXBs

Source	K	J−K	H−K	J−H	Ref.
LMC X−2	17.910	0.170	−0.220	0.390	this work
X0614+091	16.370	1.170	0.720	0.450	this work
X0620−003	14.383	1.105	0.354	0.751	2MASS
	14.470	1.080	0.480	0.600	this work
X0748−676	16.960	0.070	−0.100	0.170	this work
X0921−630	13.518	0.719	0.221	0.498	2MASS
	13.650	0.670	0.340	0.330	Lit.
Cen X−4	14.663	0.938	0.388	0.550	2MASS
	14.570	1.040	0.440	0.600	this work
Cir X−1	10.692	2.051	0.829	1.222	2MASS
	11.700	1.650	0.580	1.070	Lit.
X1543−475	14.890	0.360	0.080	0.280	this work
	14.970	0.155	0.214	−0.059	2MASS
X1550−564	15.620	1.630	0.430	1.200	this work
X1556−605	17.980	−0.230	−0.310	0.080	this work
X1608−522	17.170	1.460	−0.140	1.600	this work
Sco X−1	11.147	0.760	0.400	0.360	2MASS
	11.230	0.270	0.090	0.180	this work
X1636−536	16.640	0.660	0.920	−0.260	this work
X1655−40	12.744	0.772	0.253	0.519	2MASS
	12.720	0.790	0.220	0.570	this work
Her X−1	13.628	0.101	−0.019	0.120	2MASS
	13.100	0.060	0.040	0.020	this work
X1658−298	16.570	0.510	0.010	0.500	this work
GX 349+2	14.563	0.650	0.216	0.434	2MASS
	14.650	0.730	0.200	0.530	this work
	14.340	0.840	0.300	0.540	this work
GX 9+9	16.110	0.190	−0.050	0.240	this work
GX 1+4	7.979	2.116	0.719	1.397	2MASS
	8.060	2.250	0.750	1.500	Lit.
X1735−444	16.770	0.310	0.000	0.310	this work
GX 5−1	13.540	1.210	0.610	0.600	Lit.
GX 13+1	11.974	2.855	0.947	1.908	2MASS
J1819.3-2525	12.270	0.262	0.094	0.168	2MASS
	12.850	0.300	0.020	0.280	this work
X1822−371	15.450	−0.165	−0.153	−0.012	2MASS
Aql X−1	15.960	0.610	0.030	0.580	this work
X2023+338	12.321	1.254	0.363	0.891	2MASS
X2129+470	14.873	0.953	0.582	0.371	2MASS
Cyg X−2	13.049	0.347	0.106	0.241	2MASS

frared Processing and Analysis Center/California Institute of Technology, funded by the National Aeronautics and Space Administration and the National Science Foundation. It also utilized NASA's Astrophysics Data System Abstract Service and the SIMBAD database operated by CDS, Strasbourg, France. SW was supported by Chandra award GO2-3044X. SW and JW acknowledge support through Chandra grant GO3-4036X.

Discovering Interacting Binaries with Hα Surveys

Andrew Witham*, Christian Knigge*, Janet Drew[†], Paul Groot**, Robert Greimel[‡] and Quentin Parker[§,¶]

*School of Physics & Astronomy, University of Southampton, Highfield, Southampton, SO17 1BJ, UK
[†]Astrophysics Group, Department of Physics, Imperial College London, Exhibition Road, London SW7 2AZ, UK
**Department of Astrophysics, University of Nijmegen, P.O. Box 9010, NL - 6500 GL Nijmegen, NL
[‡]Isaac Newton Group, Apartado de Correos, 321, 38700 Santa Cruz de La Palma, Canary Islands, ESP
[§]Department of Physics, Macquarie University, NSW 2109, Australia
[¶]Anglo-Australian Observatory, PO Box 296, Epping NSW 1710, Australia

Abstract. A deep (R $\sim$ 19.5) photographic Hα Survey of the southern Galactic Plane was recently completed using the UK Schmidt Telescope at the AAO. In addition, we have recently started a similar, CCD-based survey of the northern Galactic Plane using the Wide Field Camera on the INT. Both surveys aim to provide information on many types of emission line objects, such as planetary nebulae, luminous blue variables and interacting binaries.

Here, we focus specifically on the ability of Hα emission line surveys to discover cataclysmic variables (CVs). Follow-up observations have already begun, and we present initial spectra of a candidate CV discovered by these surveys. We also present results from analyzing the properties of known CVs in the Southern Survey. By calculating the recovery rate of these objects, we estimate the efficiency of Hα-based searches in finding CVs.

Keywords: Hα, Galactic Surveys, Interacting Binaries, Cataclysmic Variables
PACS: 97.80.Gm, 95.80.Kr, 97.30.Eh, 97.80.Jp

INTRODUCTION

Binary evolution theory predicts a large (but currently undetected) population of short-period, faint cataclysmic cariables (CVs) (see for example Kolb 1993 [1]; and Howell, Rappaport, & Politano 1997 [2]). Such systems should exhibit particularly strong Balmer emission (Patterson 1984 [3]), making Hα surveys an ideal way to find them, if they exist. Previous Hα surveys (see for example Kohoutek and Wehmeyer 1997 [4]; and Gaustad et al 2001 [5]) have become incomplete at bright magnitudes or have been of low spatial resolution. Below, we briefly describe two new photometric Hα surveys which are able to push to deeper magnitudes with good resolution and will allow for a significant population of new Hα emitters to be discovered, including many Cataclysmic Variables.

Hα SURVEYS

The galactic plane is being observed in $H\alpha$ by two Surveys: (i) The INT/WFC Photometric $H\alpha$ Survey of the Northern Galactic Plane (IPHAS), and (ii) the AAO/UKST Southern $H\alpha$ Survey (SHS). IPHAS is a CCD survey which combines photometry in the bands $H\alpha$, r and i from images obtained with the Wide Field Camera (WFC) on the Isaac Newton Telescope (INT). The WFC provides a field of view of 34'x34' and the resolution of the accepted photometry is $\leq 1.7"$. At the time of writing, IPHAS is $\sim 1/3$ complete and when finished will cover the whole of the Northern galactic plane with latitudes $|b| < \sim 5°$ down to ~ 19.5 mags in r and i. Further information can be found at http://astro.ic.ac.uk/Research/Halpha/North and in a forthcoming paper by Drew et al (in preparation).

The SHS is a photographic survey which, using the UK Schmidt Telescope (UKST), has provided images of 233 galactic fields at four degree centres. Fields were observed in two bands ($H\alpha$ and short red $\simeq$ R) with a spatial resolution of $\sim 1"$. This has resulted in a survey that is complete to ~ 19.5 mags in R and has an area of coverage defined by $-75 < Dec < +2.5$ and $|b| <\sim 10°$. The data can be accessed via the SuperCOSMOS website at http://www-wfau.roe.ac.uk/sss/halpha and additional information can be found in Parker & Phillipps (1998) [6]

SPECTROSCOPIC FOLLOW-UP OF IPHAS

IPHAS has the advantage of having far fewer spurious $H\alpha$ emitters than the SHS because of the photographic nature of the SHS. Defining selection criteria for follow-up is therefore less problematic with IPHAS. Initial selection of potential new CVs for follow-up was done on the basis that we expect many to be $H\alpha$ emitters and to show a large $H\alpha$ excess. These CVs can thus be expected to lie above the main stellar locus in colour-colour plots of r-$H\alpha$ vs r-i.

Spectroscopic follow-up of IPHAS is in progress and has already lead to the discovery of three new CV candidates. Figure 1 illustrates the selection criteria used to pick objects for follow-up, while Figure 2 shows the resulting spectrum of one of the new CV candidates. Classification of the new CV candidates will be done using time-resolved spectroscopy and photometry.

RECOVERY RATE OF KNOWN CVS IN THE SOUTH

We have analyzed the properties of the CVs contained in the Ritter & Kolb (2003) [7] catalog that fall within the SHS area. Using R-$H\alpha$ vs R band colour-magnitude plots we classify as "emitters" those CVs that would have been selected for spectroscopic follow-up with 6df i.e. 5 objects per square degree. The result is that 52% of the sample would be recovered by the Southern Survey. Furthermore, if consideration is limited to objects with good photometry in the survey, the recovery rate is even higher at 62%. The histogram in Figure 3 shows how the recovery rate depends on the CV orbital period. Somewhat surprisingly, there is no obvious bias, in favour of detecting faint, short period

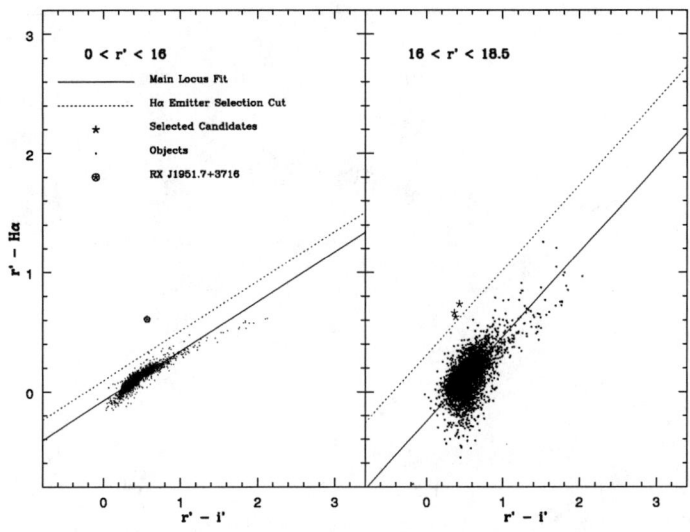

FIGURE 1. Illustration of our selection strategy. The plots compromise two r magnitude bins containing the objects from one IPHAS field. A Hα excess of 0.1mag corresponds to an equivalent width of $\sim$10Å. Solid lines are linear fits to the stellar loci, the dotted lines indicate the selection cut. The known CV RX J1951.7+3716 is present within this field and would be selected as it lies above the cut.

CVs.

The fact that many of the CVs (including several novae) appear blue as well as having a Hα excess can be used to define more stringent selection cuts; however this could have the negative effect of introducing a further selection bias into the CV populations discovered with Hα surveys. A more thorough discussion of the recovery rate of the known population of CVs in both IPHAS and the SHS will presented by Witham *et al* (in preparation). Overall the recovery rate is very encouraging and the surveys have a great chance of finding many new CVs.

REFERENCES

1. Kolb, U., 1993, A&A, 271, 149
2. Howell, S.B., Rappaport, S., Politano, M., 1997, MNRAS, 287, 929
3. Patterson, J., 1984, ApJS, 54, 443
4. Kohoutek, L., Wehmeyer R., 1997, Catalogue of Stars in the Northern Milky Way Having H-alpha in Emission, Abhandl. Sternw. XI
5. Gaustad, J.E., McCullough, P.R., Rosing, W., Van Buren, D., 2001, PASP, 113, 1326
6. Parker Q., Phillipps S., 1998, PASA, 15, 28
7. Ritter H., Kolb U., 2003, A&A, 404, 301

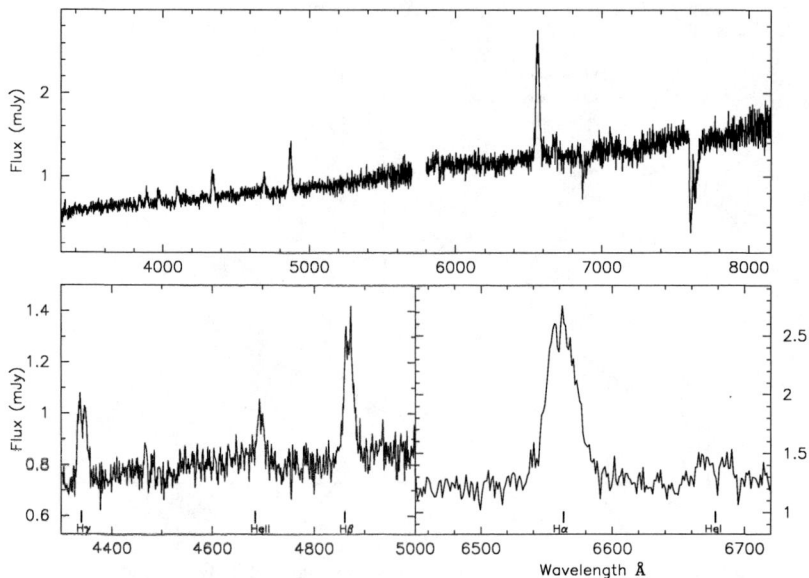

FIGURE 2. WHT/ISIS spectra of a newly discovered CV candidate. The dispersion is 0.86Å/pixel in the blue and 1.49Å/pixel in the red. Based on the broad and strong Hα emission line and the presence of HeII emission, this is a CV candidate.

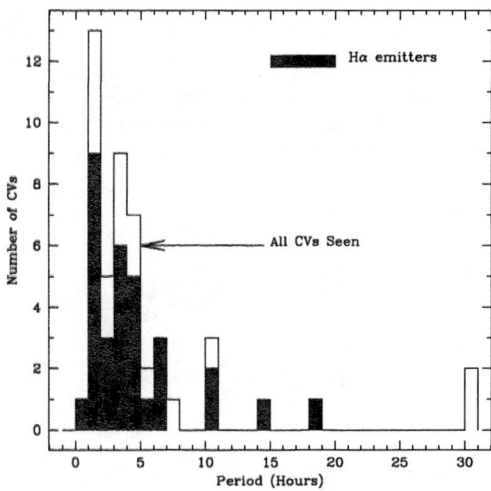

FIGURE 3. The period distribution of known CVs classified as emitters (black) overlaid on the distribution of the known CVs with good photometry in the survey (white).

Spectroscopy and Near-Infrared Photometry of the Helium Nova V445 Puppis

Patrick A. Woudt[*] and Danny Steeghs[†]

[*]*Department of Astronomy, University of Cape Town, Rondebosch 7700, South Africa*
[†]*Center for Astrophysics, MS-67, 60 Garden Street, Cambridge, MA 01238, USA*

Abstract. V445 Pup (Nova Puppis 2000) is the first observed example of a helium nova (Kato & Hachisu 2003). Here we present Magellan spectroscopy of V445 Pup, obtained three years after outburst ($V = 19.91$ mag). The optical spectrum consists of He I, [O I], [O II], [O III] emission lines; no hydrogen is present. The emission lines show multiple velocity components (around $\pm$ 850 km s^{-1} and $\pm$ 1600 km s^{-1}), associated with an expanding and clearly asymmetric nova shell. Images of V445 Pup obtained under good seeing conditions show this object to be elongated and extended. The expanding nova shell should allow an accurate distance determination of V445 Pup through high-resolution imaging and integral field unit spectroscopy. We have continuously monitored the near-infrared (J, H, K_s) secular evolution since 2002 March with the Infrared Survey Facility (IRSF) from the Sutherland site of the South African Astronomical Observatory (SAAO). Nearly four years of the outburst, V445 Pup is still covered by an optically thick dust shell (2004 September 29: $E(J - K_s) \sim 4.6 \pm 0.1$ mag).

Keywords: Novae – Infrared photometry – Spectroscopy
PACS: 97.30.Qt, 97.80.Gm, 95.85.-e, 95.55.Qf, 95.75.Fg

V445 PUPPIS AS A HELIUM NOVA

The last nova of the previous millenium (Nova Puppis 2000 = V445 Pup) is unlike any nova observed to date. It is hydrogen-deficient (Wagner, Foltz & Starrfield 2001; Ashok & Banerjee 2003), enriched in helium and carbon, and the initially formed optically thin dust shell has developed into an optically thick dust shell (Henden, Wagner & Starrfield 2001; Lynch et al. 2001; Lynch et al. 2004), which currently still obscures the nova nearly four years after its outburst. Ashok & Banerjee (2003) and Kato & Hachisu (2003) suggested that V445 Pup might be the first observed helium nova. Such helium novae were theoretically predicted (Kato, Saio & Hachisu 1989; Iben & Tutukov 1991), but have never before been observed. In these systems, the 'nova' outburst is caused by a helium shell flash on the white dwarf which has been accreting from a helium-rich companion star.

Kato & Hachisu (2003) modelled the optical light curve of V445 Pup and deduced various model-dependent parameters for the system: mass of the primary (M_1) $\gtrsim$ 1.33 $M_\odot$, a mass-transfer rate ($\dot{M}$) of several times 10^{-7} $M_\odot$ yr^{-1}, and an estimated nova recurrence time t_r of $\sim$70 yr (based on the ignition mass of several times 10^{-5} $M_\odot$). The model suggests that V445 Pup is a likely candidate progenitor of a type Ia supernova, although it could end up via accretion-induced collapse as a neutron star (Kato & Hachisu 2003).

The nature of the underlying binary still remains unknown. There are three likely

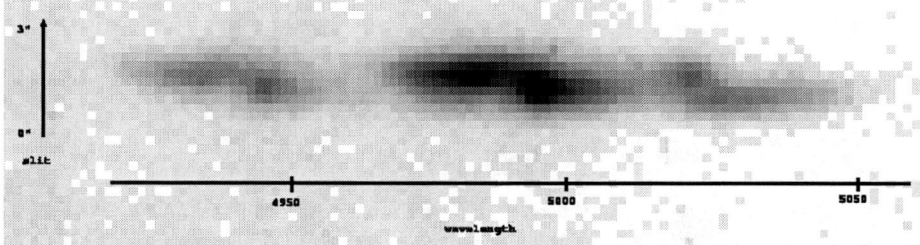

FIGURE 1. The 2-dimensional structure of the [O III] emission line complex around 5000 Å of V445 Pup taken with Magellan (+ IMACS) optical spectroscopy on 2003 December 17. The vertical axis correspond to the spatial direction (the length of the arrow equals 3″). The 2-dimensional velocity structure of the shell is barely resolved along the slit.

possibilities for the parent population of these helium novae, distinguished by the nature of the mass donor star: I. Double white dwarfs with short orbital periods ($\sim 5-10$ min) and $\dot{M}$ comparable to those derived by Kato & Hachisu (2003). These are ultra-compact helium-transferring AM CVn stars (see Nelemans, Yungelson & Portegies Zwart (2004) for a review). II. CVs with helium donors (Iben & Tutukov 1991). III. A high $\dot{M}$ hydrogen accretor (probably a supersoft source, but note that V445 Pup has not been detected in the ROSAT all-sky survey). Each of these populations have a distinctly different range in absolute magnitudes. Since the pre-outburst magnitude of V445 Pup and the Galactic reddening towards V445 Pup are well-known (Ashok & Banerjee 2003), an accurately determined distance (and hence pre-outburst absolute magnitude) of V445 Pup will allow one to determine the nature of the underlying binary.

Magellan spectroscopy

Deep optical imaging with the 6.5-m Magellan telescope in good seeing conditions (FWHM $\sim 0.6''$) revealed V445 Pup to be a source of extended and elongated emission. Optical spectroscopy (3600 – 9000 Å) of V445 Pup was obtained at the same time using the imaging spectrograph IMACS on Magellan on 2003 December 17. The spectrum is shown in figure 1 of Woudt & Steeghs (2005); it contains He I, [O I], [O II] and [O III] emission lines and no hydrogen is present, suggesting extreme He/H abundances (e.g. EW(5876Å)/EW(Hβ) > 50). Most importantly, the [O II] and [O III] emission lines show a complex structure consisting of multiple velocity components – redshifted and blueshifted – at $\pm$ 850 km s^{-1} and $\pm$ 1600 km s^{-1} (see the 2-dimensional structure of the [O III] emission line complex around 5000 Å shown in Figure 1). Figure 1 shows that one side of the nova shell is predominantly blueshifted, whereas the other side is largely redshifted. This is indicative of an expanding asymmetric nova shell and essentially offers the opportunity to determine the expansion parallax of V445 Pup through high angular-resolution imaging in combination with long-slit or integral field unit spectroscopy.

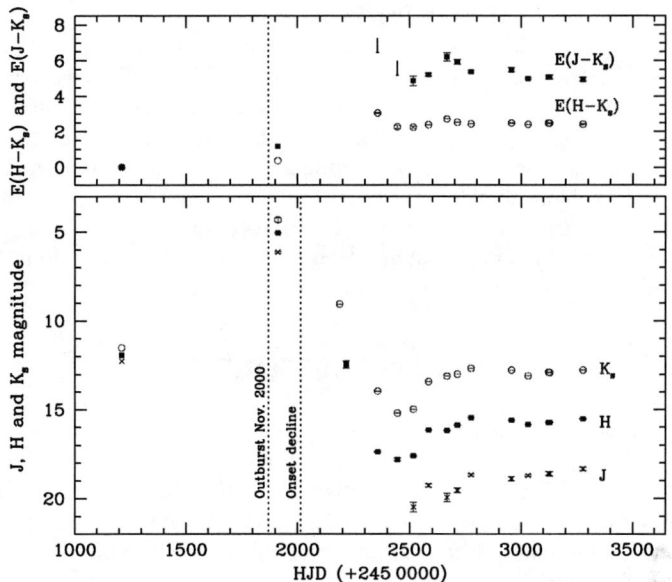

FIGURE 2. The near-infrared (J, H, K_s) light curves of V445 Pup (lower panel). The upper panel shows the evolution of the post-outburst colour excess $E(J-K_s)$ and $E(H-K_s)$.

Near-Infrared Photometry

We have continuously followed the secular evolution of V445 Puppis in the near-infrared (J, H, K_s) since 2002 March. The observations were done with the Infrared Survey Facility (IRSF) – a 1.4-m telescope – and the SIRIUS camera. The latter has an $7.7' \times 7.7'$ field of view, a pixel scale of $0.45''$ pixel^{-1} and simultaneous J, H, and K_s imaging capability (Nagashima et al. 1999; Nagayama et al. 2003). The photometry was calibrated using standard stars from Persson et al. (1998), covering a substantial range in airmass (1.0 – 2.1), and a wide range in colours (0.3 mag $\leq (J-K_s) \leq$ 3.0 mag, and 0.2 mag $\leq (J-H) \leq$ 2.1 mag). The resulting photometry agrees very well with 2MASS; a comparison of eight stars in the vicinity of V445 Pup (local relative standards) show that our measurements are within the 2MASS errors, i.e. differences are $\leq$ 0.05 mag.

The near-infrared light curves are shown in the lower panel of Figure 2. The light curves include one observation prior to outburst (a 2MASS measurement in 1999 February: J, H, K_s = 12.27, 11.94 and 11.52 mag, respectively) and two subsequent observations reported by Ashok & Banerjee (2003). Our observations commenced on HJD 245 2356. Initially V445 Pup was too faint to be detected in the J-band (fainter than $J \sim 20.5$ mag), with typical integration times of 15 minutes (three sets of ten 30-s exposures) on target. The first two observations with the IRSF showed a continued dimmening (Woudt 2002) in both the H- and K_s-band. Subsequently V445 Pup is graduately brightening, albeit very slowly.

The extreme red colour of V445 Pup is immediately obvious from the near-infrared photometry. In the upper panel of Figure 2 the evolution of the colour excess is plotted for both $E(J-K_s)$ and $E(H-K_s)$. On 2003 January 25, V445 Pup measured $(J-K_s)$ = 6.8 ± 0.2 mag, which translates into a post-outburst colour excess of $E(J-K_s)$ = 6.1 ± 0.2 mag when compared to the pre-outburst 2MASS observation. This very large colour excess is due to a massive optically thick dust shell (Lynch et al. 2001; Lynch et al. 2004). Note that the extremely red colour of V445 Pup could introduce a small systematic uncertainty in our near-infrared photometric calibration as none of the Persson et al. (1998) standard stars observed have such a red colour. However, we estimate this effect, if present, to be less than ~ 0.1 mag.

CONCLUSIONS

A tentative estimate for the radius of the nova shell in 2005 (following Krautter et al. (2002) and assuming $v_{exp} \sim 1600$ km s^{-1} and the distance d to be 700 pc), is around 2–2.5 arcsec. We plan to use adaptive optics imaging (on the VLT) and integral field unit spectroscopy to determine the expansion parallax of V445 Pup. The resulting distance determination should be accurate enough ($\sim 10\%$) to determine the nature of the underlying binary of V445 Puppis.

ACKNOWLEDGMENTS

We kindly acknowledge Prof. Nagata for scheduling the IRSF observations of V445 Pup and thank the observers at the IRSF (Mr. Baba, Kadowaki, Nakaya, Nishiyama, Kato, Matsunaga, and Drs. Tanabe and Ita) for the service mode observations.

REFERENCES

1. Ashok, N.M., and Banerjee, D.P.K., A&A, **409**, 1007 (2003).
2. Henden, A.A., Wagner, R.M., and Starrfield, S.G., IAU Circ., **7730** (2001).
3. Iben, I., Jr., and Tutukov, A.V., ApJ, **370**, 615 (1991).
4. Kato, M., and Hachisu, I., ApJ Lett., **598**, L107 (2003).
5. Kato, M., Saio, H., and Hachisu, I., ApJ, **340**, 509 (1989).
6. Krautter, J., Woodward, C.E., Schuster, M.T., et al., AJ, **124**, 2888 (2002).
7. Lynch, D.K., Rudy, R.J., Venturini, C.C., et al., AJ, **128**, 2962 (2004).
8. Lynch, D.K., Russell, R.W., and Sitko, M.L., AJ Lett., **122**, L3313 (2001).
9. Nagashima, C., Nagayama, T., Nakajima, Y., et al., in *Star Formation 1999*, edited by T. Nakamoto, Nobeyama Radio Observatory, 397 (1999).
10. Nagayama, T., Nagashima, C., Nakajima, Y., et al., Proc. SPIE, **4841**, 459 (2003).
11. Nelemans, G., Yungelson, L.R., and Portegies Zwart, S.F., MNRAS, **349**, 181 (2004).
12. Persson, S.E., Murphy, D.C., Krzeminski, W., et al., AJ, **116**, 2475 (1998).
13. Wagner, R.M., Foltz, C.B., and Starrfield, S.G., IAU Circ., **7556** (2001).
14. Woudt, P.A., IAU Circ., **7955** (2002).
15. Woudt, P.A., and Steeghs, D., in *The Astrophysics of Cataclysmic Variables and Related Objects*, edited by J.M. Hameury and J.P. Lasota, ASP Conf. Ser., in press (2005) (astro-ph/0409525).

Stabilization of helium shell burning by rotation in accreting white dwarfs

S.-C. Yoon and N. Langer

Astronomical Institute, Utrecht University, Princetonplein 5, 3584 CC, Utrecht, The Netherlands

Abstract. The currently favored scenario for the progenitor evolution of Type Ia supernovae (SNe Ia) presumes that white dwarfs in close binary systems grow to the Chandrasekhar limit via mass accretion from their non-degenerate companions. However, the accreted hydrogen and/or helium usually participate thermally unstable or even violent nuclear reactions in a geometrically confined region, due to the compactness of the white dwarf. Since shell flashes induced by the thermal instability might induce significant loss of mass, efficient mass increase of white dwarfs by hydrogen and/or helium accretion has been seriously questioned. A good understanding of the stability of thermonuclear shell sources is therefore crucial in order to investigate the evolution of accreting white dwarfs as SNe Ia progenitors. Here, we present a quantitative criterion for the thermal stability of thermonuclear shell sources, and discuss the effects of rotation on the stability of helium shell burning in helium accreting CO white dwarfs with $\dot{M} \approx 10^{-7}...10^{-6}$ $M_\odot$ yr^{-1}. In particular, we show that, if the effects of rotation are properly considered, helium shell sources are significantly stabilized, which might increase the likelihood for accreting white dwarfs to grow to the Chandrasekhar limit.

Keywords: Nuclear reactions – Stars: white dwarf – Stars: rotation – Supernovae: Type Ia
PACS: 97.10.Cv; 97.10.Kc

INTRODUCTION

Many important astrophysical phenomena are related to the thermonuclear shell burning in accreting white dwarfs. Nova outbursts are induced by explosive hydrogen shell flash in highly degenerate hydrogen envelopes of white dwarfs which accretes hydrogen at a rate of $\dot{M} \approx 10^{-10}...10^{-8}$ $M_\odot$ yr^{-1}. Steady burning of hydrogen and/or helium in accreting white dwarfs with $\dot{M} \approx 10^{-7}...10^{-6}$ $M_\odot$ yr^{-1} may be, on the other hand, relevant to the so-called super-soft X-ray sources (van den Heuvel et al. [2]). In particular, growth of accreting white dwarfs to the Chandrasekhar limit via such steady nuclear burning of accreted hydrogen and/or helium will lead to one of the most energetic events in the universe: Type Ia supernova (SN Ia).

Recent binary star evolution model by Yoon & Langer [7], who considered the evolution of a binary system consisting of a CO white dwarf and a helium giant, confirms that white dwarfs can reach the Chandrasekhar limit, at least, by helium accretion. However, accretion rates which allow steady nuclear shell burning in white dwarfs are limited to a very narrow range, for both hydrogen and helium. Further, steady burning of hydrogen usually leads to unstable helium shell flashes (Cassisi et al. [1]; Kato & Hachisu [3]). It is hence much debated whether hydrogen and/or helium accreting white dwarfs could be a major source for SNe Ia.

To have a better understanding on the issue, here we provide a quantitative criterion

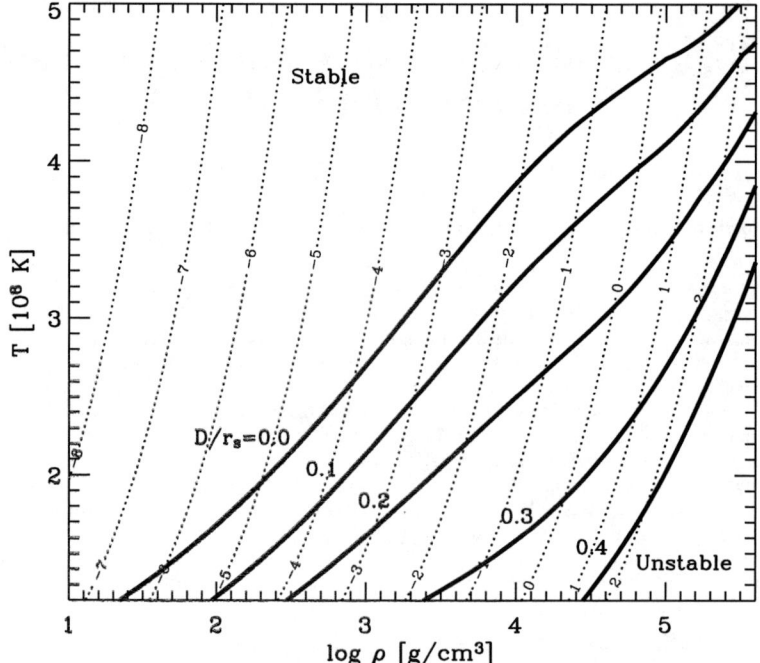

FIGURE 1. Stability conditions for a helium shell source in the density – temperature plane. The solid lines separate the thermally unstable region from the stable region, for 5 different relative shell source thicknesses (i.e., $D/r_s = 0.0, 0.1, 0.2, 0.3$ and 0.4). The dotted contour lines denote the degeneracy parameter ($:= \psi/kT$). $X_{He} = 0.662$ and $X_C = 0.286$ have been assumed. From Yoon, Langer & van der Sluys [9]

for the stability of thermonuclear helium shell sources. We also present evolutionary models of helium accreting CO white dwarfs, where the effects of the centrifugal force on the white dwarf structure, and rotationally induced chemical mixing are taken into account.

STABILITY OF THERMONUCLEAR SHELL SOURCES

Thermonuclear reaction rates are extremely sensitive to temperature and small increase in temperature may drastically enhance the energy generation. In the core of non-degenerate stars, however, increased energy output is consumed mostly for expansion work instead of increasing internal energy, and thermal stability is thus ensured (e.g. Kippenhahn & Weigert [4]).

In white dwarfs, on the other hand, accreted hydrogen and/or helium participate nuclear burning in a very dense and geometrically confined region in the white dwarf envelope. Since both electron degeneracy and strong geometrical confinement tend to keep local pressure in the shell source constant, nuclear reactions in the shell – especially in the helium burning shell – are prone to the thermal instability (cf. Schwarzschild & Härm [5]; Weigert [6]; Kippenhahn & Weigert [4]). In other words, a shell source becomes more susceptible to the thermal instability if it is more degenerate and geometrically thinner. Another factor to influence the stability of shell sources is temperature. With higher temperature, the role of radiation pressure becomes important, and sensitivity of nuclear reactions to a change in temperature becomes weaker, which favors thermal stability.

Yoon, Langer & van der Sluys [9] summarize conditions for the stability of helium shell sources, in terms of density, temperature, and relative thickness of the shell source, as in Fig. 1. In conclusion, the less dense, hotter, and thicker a shell source is, the more stable it becomes. See Yoon, Langer & van der Sluys [9] for more detailed discussion.

STABILIZING EFFECTS OF ROTATION

White dwarfs accrete matter through the Keplerian disk, and believed to be spun up by the angular momentum gained from the accreted matter. Such a spin-up process might have two important effects on the behavior of helium shell burning in accreting white dwarfs.

Firstly, the centrifugal force will lift up the helium burning layers to reduce electron degeneracy, compared to the non-rotating case. Secondly, rotationally induced hydrodynamic instabilities in the spun-up layers will induce mixing of chemical elements. i.e., accreted helium will be mixed into the carbon-oxygen core, and helium burning layers will be accordingly extended. Both effects – decrease of degeneracy and widening of the shell source – favor thermal stability. Moreover, simulations of helium accreting white dwarfs by Yoon, Langer & Scheithauer [10], who considered the above mentioned effects, show that the helium shell source becomes somewhat hotter with rotation, mainly due to the increased $^{12}C(\alpha,\gamma)^{16}O$ reaction.

Fig. 2 clearly shows these stabilizing effects of rotation in helium accreting white dwarf models. In non-rotating models with $\dot{M} = 5 \times 10^{-7}$ $M_\odot$ yr^{-1}, thermal pulses induced by unstable helium shell burning derive the white dwarf envelope to reach the Eddington limit. In the corresponding rotating model, very weak thermal pulses appear soon after the mass accretion, but steady burning follows soon. With $\dot{M} = 10^{-6}$ $M_\odot$ yr^{-1}, helium shell burning remains stable even when the white dwarf grows to ~ 1.5 $M_\odot$. This result implies that mass accumulation of accreting white dwarfs might be much more efficient than believed before, in support of the currently favored scenario for the SNe Ia progenitor evolution (i.e., single degenerate Chandrasekhar mass scenario). Readers are referred to Yoon, Langer & Scheithauer [10] for more comprehensive discussion.

CONCLUDING REMARKS

The above discussion indicates that rotation might play such a crucial role in the evolution of accreting white dwarfs that its effects should not be neglected in the study of SNe Ia progenitors. Apart from the stabilization of helium shell burning which may in-

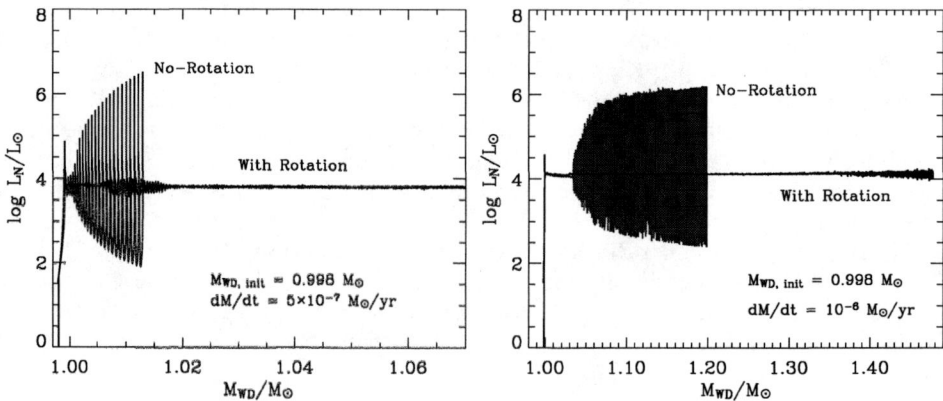

FIGURE 2. Evolution of the nuclear luminosity due to helium burning in helium accreting white dwarf models for $\dot{M} = 5 \times 10^{-7}$ $M_\odot$ yr^{-1} (left panel) and $\dot{M} = 10^{-6}$ $M_\odot$ yr^{-1} (right panel). The initial mass of the white dwarf is 0.998 $M_\odot$. The results with rotation and without rotation are given by the black and gray lines, respectively. From Yoon, Langer & Scheithauer [10].

crease the mass accumulation efficiency of accreting white dwarfs, rotation may involve another important aspect. Differential rotation inside the white dwarf as a result of mass and angular momentum accretion might lead to SN Ia explosion at super-Chandrasekhar masses (Yoon & Langer [8]). Future work, including multi-dimensional studies on the effects of rotation (cf. Yoon & Langer 2005, in preparation), will reveal whether consideration of rotation would give better explanations of various aspects – such as production rate and diversity – of SNe Ia.

ACKNOWLEDGMENTS

This work has been supported by the Netherlands Organization for Scientific Research (NWO).

REFERENCES

1. Cassisi, S., Castellani, V., and Tornambé, *ApJ*, **459**, pp. 298–306 (1996)
2. van den Heuvel, E.P.J., Bhattacharya, D., Nomoto, K., and Rappaport, S.A., *A&A*, **262**, pp. 97–105 (1992)
3. Kato, M., and Hachisu, I., *ApJ Lett.*, **613**, pp. L129–L132 (2004)
4. Kippenhahn, R., and Weigert, A., *Stellar Structure and Evolution*, Springer-Verlag, 1992
5. Schwarzschild, M., and Härm, R., *ApJ*, **142**, pp. 855–867 (1965)
6. Weigert, A., *Z. Astrophys.*, **64**, pp. 395–425 (1966)
7. Yoon, S.-C., and Langer, N., *A&A*, **412**, pp. L53–L56 (2003)
8. Yoon, S.-C., and Langer, N., *A&A*, **419**, pp. 623–644 (2004)
9. Yoon, S.-C., Langer, N. and van der Sluys, M. *A&A*, **425**, pp. 207–216 (2004)
10. Yoon, S.-C., Langer, N. and Scheithauer, S. *A&A*, **425**, pp. 217–228 (2004)

White dwarfs with jets as non-relativistic analogues of quasars and microquasars?

R. Zamanov*, M.F.Bode*, P.Marziani[†], R.J. Davis**, S.P.S. Eyres[‡], A. Gomboc*, J. Porter * and A. Skopal [§]

*Astrophysics Research Institute, Liverpool John Moores University, UK
[†]Osservatorio Astronomico di Padova, INAF, Padova, Italy
**Jodrell Bank Observatory, University of Manchester, UK
[‡]Centre for Astrophysics, University of Central Lancashire, UK
[§]Astronomical Institute, Slovak Academy of Sciences, Slovakia

Abstract. We explore the similarities between accreting white dwarfs (CH Cyg and MWC 560) and the much more energetic jet sources - quasars and microquasars. To-date we have identified several common attributes: (1) they exhibit collimated outflows (jets); (2) the jets are precessing; (3) these two symbiotic stars exhibit quasar-like emission line spectra; (4) there is a disk-jet connection like that observed in microquasars. Additionally they may have a similar energy source (extraction of rotational energy from the accreting object). Study of the low energy analogues could have important implications for our understanding of their higher energy cousins.

Keywords: Symbiotic stars – Quasars: emission lines – Stars: individual (CH Cygni, MWC 560)
PACS: 98.54.Aj, 98.62.Ra

EMISSION LINE SIMILARITIES

As illustrated in Zamanov and Marziani (2002), there are striking similarities between the optical spectra of Active Galactic Nuclei (AGNs) and two accreting white dwarfs. Almost every emission line visible in the AGN spectrum of I Zw 1 shows a corresponding feature in the spectra of CH Cyg and MWC 560. The similarity between the UV spectrum of CH Cyg and I Zw 1 is demonstrated in Fig.1.

In AGN, hydrogen and FeII emission lines are emitted from the so-called Broad Line Region. This region is thought to lie within <1 pc of the central black hole. Its structure is still poorly understood. The clear similarity between the emission lines suggests that we are observing a scaled down version of the quasar Broad Line Region in galactic objects like MWC560 and CH Cyg (see also Zamanov and Marziani, 2002).

DISK-JET CONNECTION

Comparison of the flickering behaviour of T CrB (February 28, 1995) and MWC 560 (March 05, 1990) is shown on Fig. 2. In the case of MWC 560, the short term variability (on a time scale of minutes) is missing. Only smooth, hour-timescale variations are present. This indicates disruption of the inner part of the accretion disk during the time of jet ejection. A disruption of the inner disk (disk-jet) connection is also observed in CH Cyg (Sokoloski and Kenyon 2003). The behaviour is closely analogous to that of

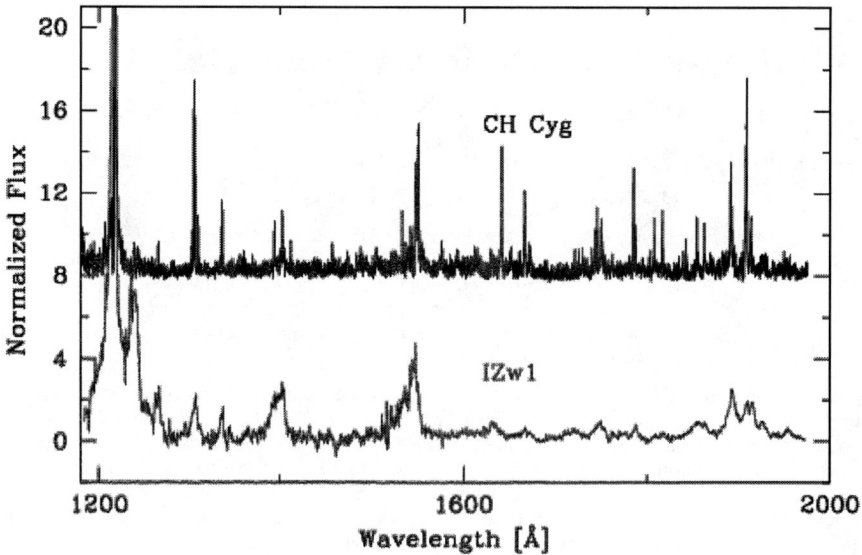

FIGURE 1. The UV spectra of symbiotic star CH Cyg and Narrow Line Seyfert 1 galaxy I Zw 1. Clear similarity in the emission lines is visible.

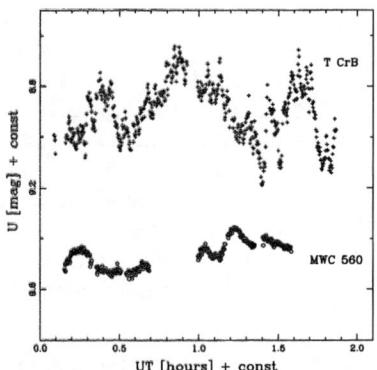

FIGURE 2. The flickering behaviour of MWC 560 and T CrB.

the microquasar GRS1915+105. This supports the view that there may be a common mechanism for jets in quasars, microquasars and symbiotic stars (see also Livio, Pringle and King 2003).

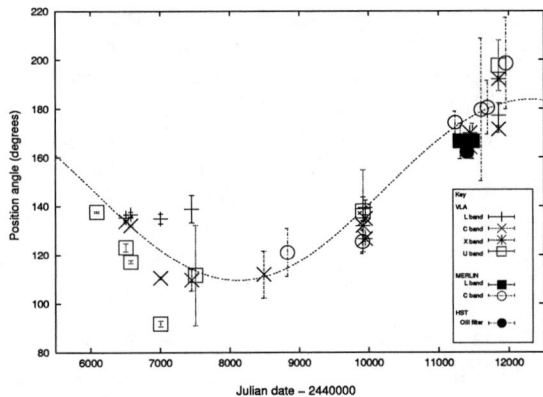

FIGURE 3. Variation of the position angle of the extended emission in the central region of CH Cyg between 1985 and 2000, along with a curve representing the precessing jet model, adapted from that for the high-velocity jets of SS 433.

PRECESSING JETS

The best known precessing jets in astrophysics are probably those of SS 433. In recent years precessing jets have been identified in two symbiotic stars: CH Cyg (on the basis of radio imaging by Crocker et al. 2002) and MWC 560 (from optical spectroscopy by Iijima 2002). In both cases the model of the jets of the microquasar SS 433 has been adopted to fit the evolution of the morphology of the outflows (using velocities appropriate for white dwarfs).

ENERGY SOURCE OF JETS

It is worth noting that interacting binaries, where a white dwarf accretes material from the wind of a red giant (usually classified as symbiotic stars), are strongly variable objects. We show above that the spectra of CH Cyg and MWC 560 are similar to low-redshift quasars around the times when jet activity is detected (CH Cyg - July 1984, MWC 560 - November 1990).

Jets are detected in systems quite different from those harboring black holes (for a review see Livio, 2001): young stellar objects (velocity $v \sim 200$ km s^{-1}), planetary nebulae ($v \sim 200$-1000 km s-1), supersoft X-ray sources ($v \sim 1000$ km s^{-1}). The jet velocities observed in the accreting white dwarfs (we call them *"nanoquasars"*) are ~ 1000 km s^{-1} in CH Cyg (Taylor et al. 1986) and 1000-6000 km s^{-1} in MWC 560 (Tomov et al. 1992). They are consistent with an overall picture in which the jet velocity is of the same order as the escape velocity from the accretor (Livio 2001).

The luminosities of MWC 560 and CH Cyg are considerably less than the Eddington limit. The mass accretion rate is about $M_{acc} \sim 0.05$ M_{Edd}. At such mass accretion rates the most probable jet energy source involves extraction of rotational energy from the compact object. In the case of nano-quasars the extraction is probably occurring via the propeller action of a magnetic white dwarf (Mikolajewski et al. 1996). The most probable source of jet formation in quasars is the extraction of energy and angular momentum via the Blandford and Znajek (1977) mechanism. In this sense the jets in the "nanoquasars" probably represent a low energy (non-relativistic) analogue of the jets in quasars and microquasars. They involve a similar energy source - the extraction of rotational energy from the central compact object.

CONCLUSIONS

We think it is appropriate to call the two accreting white dwarfs discussed here "nanoquasars" because they represent the very low energy analogue of quasars and microquasars. The name is chosen by analogy with the quasar and microquasar denominations, and also because $νανος$ (ancient greek) = nano (ital.) = dwarf(engl.).

We suggest that the "nanoquasars" could be an important link in our understanding of a broad range of accreting sources. They could help us to create a unified picture of accreting objects from cataclysmic variables and stellar-mass black holes up to the most powerful quasars.

References
Blandford, R., and Znajek, R., 1977, MNRAS 179, 433
Crocker, M.M., et al., 2002, MNRAS 335, 1100
Iijima, T., 2002, A&A 391, 617
Livio, M., 2001, ASP Conf. Ser., v.224, p.225
Livio, M., Pringle, J.E., and King, A.R., 2003, ApJ 593, 184
Mikolajewski, M., Milkolajewska, J., and Tomov, T., 1996, IAUS 165, 451
Sokoloski, J., and Kenyon, S., 2003, ApJ 584, 1021
Taylor, A.R., Seaquist, E.R., and Mattei, J.A., 1986, Nature 319, 38
Tomov, T., Zamanov, R., Kolev, D., et al. 1992, MNRAS 258, 23
Zamanov, R., and Marziani, P., 2002, ApJ Lett. 571, L77

INTERNATIONAL CONFERENCE
INTERACTING BINARIES: ACCRETION, EVOLUTION AND OUTCOMES
Cefalù, Sicily, July 4-10, 2004

List of Participants:

L.A. Antonelli
Astronomical Obs. of Rome
Via di Frascati, 33
00040 Monteporzio Catone, Rome (Italy)
antonelli@mporzio.astro.it

Monika Balucinska-Church
University of Birmingham
School of Physics & Astronomy
Edgbaston, Birmingham B15 2TT (UK)
mbc@star.sr.bham.ac.uk

Reba Bandyopadhyay
Oxford University
Dept. of Astrophysics
Keble Road, Oxford OX1 3RH (UK)
rmb@astro.ox.ac.uk

Robin Barnard
The Open University,
Physics & Astronomy
Walton Hall, Milton Keynes MK7 6AA (UK)
R.Barnard@open.ac.uk

Andrew Barnes
Univ. of Southampton
Astronomy Group
Southampton SO17 1BJ (UK)
adb@astro.soton.ac.uk

Cees Bassa
Astronomical Institute
University of Utrecht
P. O. Box 80 000
3508 TA Utrecht (NL)
C.G.Bassa@astro.uu.nl

Giacomo Beccari
INAF-Astronomical Obs. Bologna (Italy)
Via Zamboni, P.O. Box 596
40126 Bologna (Italy)
giacomo.beccari@bo.astro.it

Martin Beer
University of Leicester
Dept. of Physics & Astronomy
Leicester LE1 7RH (UK)
martin.beer@astro.le.ac.uk

Tomaso Belloni
INAF-Astr. Obs. of Brera
Via E. Bianchi, 46,
23807 Merate (Lecco - Italy)
belloni@merate.mi.astro.it

Renaud Belmont
Service d'Astrophysique
CEA, Saclay
Orme des Merisiers Bat. 709
91191 Gif sur Yvette (France)
belmont@cea.fr

Dmitry Bisikalo
Institute of Astronomy-RAS
48, Pyatnitskaya Str.
Moscow (Russia)
bisikalo@inasan.rssi.ru

Ignazio Bombaci
Physics Dept.
University of Pisa
Via Buonarroti, 2
56127 Pisa (Italy)
bombaci@df.unipi.it

Erin Bonning
Obs. of Paris in Meudon
5, place Jules Janssen
92195 Meudon Cedex (France)
Erin.Bonning@obspm.fr

Enrico Bozzo
Astronomical Obs. of Rome
Via di Frascati, 33
00040 Monteporzio Catone, Rome (Italy)
bozzo@mporzio.astro.it

Hale Bradt
M.I.T. – Room 37-587
Cambridge MA 02139-4307
Massachusetts (USA)
bradt@mit.edu

Eduardo Bravo
Univ. Politecnica de Catalunya,
Av. Diagonal 647
08028 Barcelona (Spain)
eduardo.bravo@upc.es

Roberto Buonanno
Astronomical Obs. of Rome
Via di Frascati, 33
00040 Monteporzio Catone, Rome (Italy)
buonanno@mporzio.astro.it

Luciano Burderi
Astronomical Obs. of Rome
Via di Frascati, 33
00040 Monteporzio Catone, Rome (Italy)
burderi@mporzio.astro.it

Marta Burgay
INAF- Astr. Obs. of Cagliari
Loc. Poggio dei Pini, Strada 54
09012 Capoterra
Cagliari (Italy)
burgay@ca.astro..it

Annalisa Calamida
Astronomical Obs. of Rome
Via di Frascati, 33
00040 Monteporzio Catone, Rome (Italy)
calamida@mporzio.astro.it

Paul Callanan
University College Cork
Dept. of Physics
Cork (Ireland)
paulc@ucc.ie

Sergio Campana
INAF-OAB-Astr. Obs. of Brera
Via Bianchi, 46
23807 Merate - Lecco (Italy)
campana@merate.mi.astro.it

Jorge Casares
IAC –Inst. de Astrofisica de Canarias
c/ Via Lactea s/n
38200 La Laguna, Tenerife (Spain)
jcv@ll.iac.es

Piergiorgio Casella
Astronomical Obs. of Merate
Via Bianchi, 46
23807 Merate, Lecco (Italy)
casella@merate.mi.astro.it

Santi Cassisi
INAF-Astr. Obs. of Teramo
Via M. Maggini
64100 Teramo (Italy)
cassisi@te.astro.it

Deepto Chakrabarty
M.I.T.
70 Vassar Street, Room 37-626A
Cambridge, MA 02139 (USA)
deepto@space.mit.edu

Phil Charles
South African Astr. Obs.
P.O. Box 9
Observatory, 7935 (South Africa)
pac@saao.ac.za

Cristina Chiappini
INAF-OAT, Astr. Obs. of Trieste
Via G. B. Tiepolo, 11
34131 Trieste (Italy)
chiappini@ts.astro.it

Guido Chincarini
INAF-Astr. Obs. of Brera
Via E. Bianchi, 46,
23807 Merate (Lecco - Italy)
guido@merate.mi.astro.it

Monica Colpi
Univ. Milano-Bicocca
Dept. of Physics
Piazza della Scienza, 3
20126 Milano (Italy)
colpi@mib.infn.it

Virginia Corless
M.I.T.
Dept. of Physics, 37-626A
Cambridge, MA 02139 (USA)
vcorless@mit.edu

Remon Cornelisse
Univ. of Southampton
School of Physics & Astronomy
Highfield Campus S)17 1BJ
Southampton (UK)
cornelis@astro.soton.ac.uk

Stefano Covino
INAF-Astr. Obs. of Brera
Via E. Bianchi, 46,
23807 Merate (Lecco - Italy)
covino@mi.astro.it

Mark Cropper
MSSL-UCL,
Mullard-Space Laboratory
Holmbury St. Mary
Dorking, Surrey, RH5 6NT (UK)
msc@mssl.ucl.ac.uk

Francesca D'Antona
Astronomical Obs. of Rome
Via di Frascati, 33
00040 Monteporzio Catone, Rome (Italy)
dantona@mporzio.astro.it

John Danziger
INAF-OAT, Astr. Obs. of Trieste
Via G. B. Tiepolo, 11
34131 Trieste (Italy)
danziger@ts.astro.it

Teodoro Munoz Darias
IAC – Inst. de Astrofisica de Canarias
Juana Blanca Ii portal 1 piso 1A
C. P. 38202
La Laguna - S/C Tenerife (Spain)
tmd@iac.es

Massimo Della Valle
INAF-Astr. Obs. of Arcetri
Largo E. Fermi, 5
50125 Firenze(Italy)
massimo@arcetri.astro.it

Elisa Di Carlo
INAF-OACT-Astr. Obs. of Teramo
Via M. Maggini
64100 Teramo (Italy)
dicarlo@te.astro.it

Tiziana Di Salvo
University of Palermo
Dept. of Physics
Via Archirafi, 36
90123 Palermo (Italy)
disalvo@gifco.fisica.unipa.it

Mauro Dolci
INAF-OACT, Astr. Obs. of Teramo
Via M. Maggini
64100 Teramo (Italy)
dolci@te.astro.it

Inma Dominguez
Univ. of Granada
Dept. of Theoretical Physics
18071 Granada (Spain)
inma@ugr.es

Lynnette Dray
Univ. of Leicester
Theoretical Physics Group
Dept. of Physics & Astronomy
Leicester LE1 7RH (UK)
Lynnette.Dray@astro.le.ac.uk

Mohammad T. Edalati
Physics Dept. – School of Sciences
Univ. of Ferdowsi
Mashhad (Iran)
tedalati@yahoo.com

Alessandro Ederoclite
Univ. Trieste/ESO Santiago
Alonso de Cordova 3107
Casilla 19001
Santiago 19 (Chile)
aederocl@eso.org

Luigi Errico
INAF-OA
Astr. Obs, of Capodimonte
Via Moiariello, 16
80131 Napoli (Italy)
errico@na.astro.it

Katrina Exter
IAC –Inst. de Astrofisica de Canarias
c/ Via Lactea s/n
38200 La Laguna, Tenerife (Spain)
Katrina@ll.iac.es

Maurizio Falanga
Service d'Astrophysique
CEA, Saclay
Orme des Merisiers Bat. 709
91191 Gif sur Yvette (France)
mfalanga@cea.fr

Francesco Ferraro
University of Bologna
Dept. of Astronomy
Via Ranzani, 1
40126 Bologna (Italy)
ferraro@bo.astro.it

Giuliana Fiorentino
INAF-OAR, Astr. Obs. of Rome
Via di Frascati, 33
00040 Monteporzio Catone - Rome (Italy)
giuliana@mporzio.astro.it

Simona Gagliardi
INAF-OACT, Astr. Obs. of Teramo
Via M. Maggini
64100 Teramo (Italy)
gagliardi@te.astro.it

Elena Gallo
Univ. of Amsterdam
Kruislaan 403
1098 SJ Amsterdam (NL)
egallo@science.uva.nl

Andreja Gomboc
Dept. of Physics
University of Ljubljana
Jadranska 19
1000 Ljubljana (Slovenia)
andreja.gomboc@fmf.uni-lj.si

Jonay I. Gonzalez-Hernandez
IAC- Inst. de Astrofisica de Canarias
c/ Via Lactea s/n
La Laguna -Tenerife (Spain)
jonay@ll.iac.es

Josh Grindlay
CFA-Harvard-Smithsonian Center for
Astrophysics
60 Garden Str.
Cambridge, MA 02138 (USA)
josh@cfa.harvard.edu

Kimitake Hayasaki
Hokkaido University,
Physics Division
Kitaku N108W8, Sapporo 060-0810 (Japan)
kimi@astro1.sci.hokudai.ac.jp

Craig Heinke
Northwestern Univ.
Dept. of Physics & Astronomy
2131 Tech Drive
Evanston IL 60208 (USA)
cheinke@cfa.harvard.edu

Jordi Isern
IEEC, Despatx 201 – Edifici Nexus
C/ Gran Capità, 2-4
08034 Barcelona (Spain)
isern@ieec.fcr.es

Gianluca Israel
INAF-OAR, Astronomical Obs. of Rome
Via di Frascati, 33
00040 Monteporzio Catone - Rome (Italy)
gianluca@mporzio.astro.it

Natalia Ivanova
Northwestern University
Physics & Astronomy
2145 Sheridan Rd,
Evanston, ILL 60208 (USA)
nata@northwestern.edu

Vassiliky Kalogera
Northwestern University,
Dearborn Observatory
2131 Tech Drive
Evanston-ILL 60208 (USA)
vicky@northwestern.edu

Margarita Karovska
CFA-Harvard University
60 Garden Str.
Cambridge, MA 02138 (USA)
karovska@cfa.harvard.edu

Elena Kilpio
Institute of Astronomy-RAS
48 Pyatnitskaya Str.
119017 Moscow (Russia)
lena@inasan.rssi.ru

Andrew King
University of Leicester
Theoretical Astrophysics Group
Leicester LE1 7RH (UK)
ark@star.le.ac.uk

Marcus Kirsch
VILSPA-ESA
Apartado – P.O. Box 50727
28080 Madrid (Spain)
mkirsch@xmm.vilspa.esa.es

Wlodek Kluzniak
Zielona Gora Univ./Copernicus Astr. Center,
Bartycka, 18
00-716 Warsaw (Poland)
wlodek@camk.edu.pl

Dubravka Kotnik-Karuza
Univ. of Rijeka
Omladinska, 14
HR-51000 Rijeka (Croatia)
kotnik@mapef.pefri.hr

Attay Kovetz
Tel Aviv University
School of Physics & Astronomy
Ramat Aviv, Tel Aviv 69978 (Israel)
attay@etoile.tau.ac.il

Erik Kuulkers
ESA/ESTEC,
Science Operations & Data Systems Division
Keplerlaan, 1
2201 AZ Noordwijk (NL)
ekuulker@rssd.esa.int

Oleg Kuznetsov
Kelysh Inst. of Applied Mathematics
4, Miusskaya sq.
Moscow 125047 (Russia)
kuznecov@spp.keldysh.ru

Norbert Langer
University of Utrecht
Princetonplein, 5
NL-3584 CC Utrecht (NL)
N.Langer@astro.uu.nl

Thierry Lanz
University of Maryland
College Park, MD 20742-2421 (USA)
tlanz@umd.edu

Giuseppe Lavagetto
University of Palermo
Physics Dept.
Via Archirafi, 136
90123 Palermo (Italy)
lavaget@fisica.unipa.it

William Lee
Institute of Astronomy UNAM,
Apdo Postal 70-264
Mexico DF 04510
wlee@astroscu.unam.mx

Walter Lewin
M.I.T. - Room 37-627
Cambridge, MA 02139 (USA)
lewin@mit.edu

Andrew Lyne
University of Manchester
Jodrell Bank Observatory
Macclesfield, Cheshire SK11 9DL (UK)
agl@jb.man.ac.uk

Dipankar Maitra
Yale University
Dept. of Astronomy
260 Whitney Ave.
New Haven, CT 06511 (USA)
maitra@astro.yale.edu

Craig Markwardt
Univ. Maryland/GSFC-NASA
Code 662 – Greenbelt, MD 20771 (USA)
craigm@lheamail.gsfc.nasa.gov

Andrea Martocchia
Astronomical Obs. of Strasbourg
11, rue de l'Universitè
F-67000 Strasbourg (France)
martok@isaac.u-strasbg.fr

Giorgio Matt
University Roma Tre
Physics Dept.
Via della Vasca Navale, 84
00146 Roma (Italy)
matt@fis.uniroma3.it

Francesca Matteucci
University of Trieste
Astronomy Dept.
Via G.B. Tiepolo, 11
34124 Trieste (Italy)
matteucci@ts.astro.it

Maria Teresa Menna
INAF-OAR, Astron. Obs. of Rome
Via di Frascati, 33
00040 Monteporzio Catone - Rome (Italy)
menna@mporzio.astro.it

James Miller-Jones
Astronomical Inst. « A. Pannekoek »
University of Amsterdam
Kruislaan 403
1098 SJ Amsterdam (NL)
jmiller@science.uva.nl

Teresa Mineo
IASF-CNR,
Via La Malfa, 153
90123 Palermo (Italy)
teresa.mineo@pa.iasf.cnr.it

Felix Mirabel
ESO Representative in Chile
Head of the office of Sience
Alonso de Cordova, 3907
Santiago 19 (CHILE)
fmirabel@cea.fr

Ulisse Munari
INAF-Astr. Obs. of Padova
Via dell'Osservatorio, 8
36012 Asiago – Vicenza (Italy)
munari@pd.astro.it

Sachindra Naik
Univ. College Cork
Physics Dept.
Cork (Ireland)
sachi@ucc.ie

Gijs Nelemans
Radbond Univ. Nijmegen,
Dept. of Astrophysics
P.O. Box 9010
NL-6500 Nijmegen (NL)
nelemans@astro.ru.nl

Marina Orio
INAF-Astron. Obs. of Torino
Strada Osservatorio, 20
10025 Pino Torinese, Torino (Italy)
orio@to.astro.it

Alessandro Papitto
INAF-Astr. Obs. of Rome
Via di Frascati, 33
00040 Monteporzio Catone - Rome (Italy)
papitto@mporzio.astro.it

Jerome Petri
Max-Planck Inst. für Kernphysik
Saupfercheckweg, 1
Postfach 10 39 80
69029 Heidelberg (Germany)
j.petri@mpi-hd.mpg.de

Jelena Petrovic
Univ. of Utrecht
Princetonplein, 5
3584 CC Utrecht (NL)
petrovic@astro.uu.nl

Luciano Piersanti
INAF-OACT, Astr. Obs. of Teramo
Via M. Maggini
64100 Teramo (Italy)
piersanti@astrte.te.astro.it

Tsvi Piran
Racah Inst. for Physics
Hebrew University
Jerusalem (Israel)
tsvi@phys.huji.ac.il

Philipp Podsiadlowski
Oxford University
Astrophysics Dept.
The Denys Wilkinson Building
Keble Road - Oxford OX1 3RH (UK)
podsi@astro.ox.ac.uk

Silvia Poggi
University of Trento
Via Arioste, 579/16
45022 Bagnolo di Po, Rovigo (Italy)
poggi@science.unitn.it

Dina Prialnik
Tel Aviv University
Dept. of Geophysics & Planetary Sciences
Ramat Aviv, Tel Aviv 69978 (Israel)
dina@planet.tau.ac.il

Saul Rappaport
M.I.T., 37-602b
Cambridge, MA 02139. (USA)
sar@mit.edu

Nanda Rea
INAF-OAR, Astr. Obs. of Rome
Via di Frascati, 33
00040 Monteporzio Catone, Rome (Italy)
rea@mporzio.astro.it

Ron Remillard
M.I.T., Center for Space Research
Room 37-595
Cambridge, MA 02139 (USA)
rr@space.mit.edu

Mark Reynolds
Univ. College Cork
Room 104D, Dept. of Physics
Cork (Ireland)
kramreynolds@hotmail.com

Hans Ritter
M.P.A.
Karl-Schwarzschild-Str., 1
Postfach 1317
D-85741 Garching (Germany)
hsr@mpa-garching.mpg.de

Elena Sabbi
University of Bologna
Dept. of Astronomy
Via Ranzani, 1
40126 Bologna (Italy)
elena.sabbi@bo.astro.it

Patrizia Santolamazza
ASDC-ASI,
Via Galileo Galilei
00044 Frascati - Rome (Italy)
santolamazza@asdc.asi.it

Klaus Schenker
University of Leicester
Dept. of Physics & Astronomy
University Road - Leicester LE1 7RH (UK)
kjs@astro.le.ac.uk

Gabriele Schoenherr
XMM-Vilspa, ESA
Apartado – P.O. Box 50727
28080 Madrid (Spain)
gschoen@xmm.vilspa.esa.es

Stephen I. Chun Shih
University of Southampton
53A Highfield Lane
Southampton SO17 1BJ (UK)
icshih@astro.soton.ac.uk

Alessandro Siviero
INAF-Astr. Obs. of Padova
Via dell'Osservatorio, 8
I-36012 Asiago - Vicenza (Italy)
siviero@pd.astro.it

Luigi Stella
INAF- Astr. Obs. of Rome
Via di Frascati, 33
00040 Monteporzio Catone - Rome (Italy)
stella@mporzio.astro.it

Oscar Straniero
INAF-Astr. Obs. of Teramo
Via M. Maggini
64100 Teramo (Italy)
straniero@te.astro.it

Vincenzo Teresi
University of Palermo
Dept. of Physics & Technologies
Viale delle Scienze - Edificio 18
90128 Palermo (Italy)
vteresi@unipa.it

Vincenzo Testa
INAF-Astr. Obs. of Rome
Via di Frascati, 33
00040 Monteporzio Catone - Rome (Italy)
testa@mporzio.astro.it

Andrea Tiengo
IASF-CNR,
Via Bassini, 15
20133 Milano (Italy)
tiengo@mi.iasf.mi.cnr.it

Amedeo Tornambè
INAF-Astr. Obs. of Teramo
Via M. Maggini
64100 Teramo (Italy)
tornambe@te.astro.it

Gaghik Tovmassian
UNAM
Apartado Postal 877
22800 Ensenada, B.C. (Mexico)
gag@astrosen.unam.mx

Gaetano Valentini
INAF-Astr. Obs. of Teramo
Via M. Maggini
64100 Teramo (Italy)
valentini@te.astro.it

Dany Vanbeveren
University of Brussels
Astrophysical Institute
Pleinlaan 2
B-1050 (Belgium)
dvbevere@vub.ac.be

Michiel van der Klis
University of Amsterdam
Sterrenkundig Instituut
Kruislaan 403
1098 SJ Amsterdam (NL)
michiel@science.uva.nl

Arjen van der Meer
University of Amsterdam
Sterrenkundig Instituut - Kruislaan 403
1098 SJ Amsterdam (NL)
ameer@science.uva.nl

Marc van der Sluys
University of Utrecht
Astronomical Institute
P.O. Box 80000
NL-3508 TA Utrecht (NL)
sluys@astro.uu.nl

Walter Van Rensbergen
University of Brussels
Martin Vannier
E.S.O.
Casilla 19001
Alonso de Cordova 3107
Vitacura/Santiago (Chile)
mvannier@eso.org

Peggy Varniere
University of Rochester,
Dept. of Physics & Astronomy
NY 14627-0171 (USA)
pvarni@pas.rochester.edu

Paolo Ventura
INAF-Astr. Obs. of Rome
Via di Frascati, 33
00040 Monteporzio Catone - Rome (Italy)
ventura@mporzio.astro.it

Frank Verbunt
University of Utrecht
Astronomical Institute
P.O. Box 80000
NL-3508 TA Utrecht (NL)
F.W.M.Verbunt@astro.uu.nl

Alberto Angelo Vittone
INAF-Astr. Obs. of Capodimonte
Via Moiariello, 16
80131 Napoli (Italy)
vittone@na.astro.it

Stefanie Wachter
Spitzer Science Center/Caltech M>S 220-6
1200 E. California Bvd.
Pasadena, CA 91125 (USA)
wachter@ipac.caltech.edu

Brian Warner
University of Cape Town
Dept. of Astronomy
Rondebosch 7700 - Capetown (South Africa)
warner@physci.uct.ac.za

Natalie Webb
CESR
9 Avenue du Colonel Roche
31028 Toulouse (France)
Natalie.Webb@cesr.fr

Andrew Witham
University of Southampton
Flat 4E St. Margaret's House
Pleinlaan, 2
B-1050 Brussels (Belgium)
wvanrens@vub.ac.be
6-8 Hulse Road
Southampton SO15 2JX (UK)
arw@phys.soton.ac.uk
Patrick Woudt
Univ. of Cape Town
Dept. of Astronomy
Rondebosch 7700 - Capetown (South Africa)
pwoudt@circinus.ast.uct.ac.za

Graham Wynn
Univ. Leicester
Dept. of Physics & Astronomy
University Road
Leicester LE1 7RH (UK)
graham.wynn@astro.le.ac.uk

Sung-Chul Yoon
Astr. Inst. « A. Pannekoek »
Kruislaan 403
NL-1098 SJ Amsterdam (NL)
scyoon@science.uva.nl

Lev R. Yungelson
INASAN-RAS
48 Pyatnitskaya Str.
119017 Moscow (Russia)
lry@inasan.rsi.ru

Silvia Zane
MSSL-Mullard-Space Laboratory
Holmbury St. Mary
Dorking, Surrey, RH5 6NT (UK)
sz@mssl.ucl.ac.uk

Author Index

A

Antonelli, L. A., 173
Audard, M., 313

B

Badenes, C., 453
Bailyn, C. D., 249
Bałucińska-Church, M., 339
Bandyopadhyay, R. M., 410, 639
Barman, T., 561
Barnard, R., 219
Barnes, A. D., 533
Barret, D., 359
Bauer, F. E., 410
Beer, M. E., 386, 537, 607
Bell, S., 561
Belloni, T., 197, 225
Belmont, R., 541
Bisikalo, D. V., 295, 573
Blackman, E., 631
Blundell, K. M., 410, 585
Bode, M. F., 181, 655
Bogdanov, S., 40
Bombaci, I., 132
Bono, G., 61
Boyarchuk, A. A., 295, 573
Boyd, P. T., 507
Bragaglia, A., 611
Bravo, E., 453, 497
Buonanno, R., 61
Burderi, L., 95, 110, 116, 371, 545, 565
Burgay, M., 523

C

Calamida, A., 61
Callanan, P. J., 593, 607
Camilo, F., 40, 523
Campana, S., 81, 307
Carter, D., 181
Casares, J., 81, 365, 589
Casella, P., 225, 635
Chakrabarty, D., 71

Charles, P., 81
Charles, P. A., 533
Chiappini, C., 476
Chincarini, G., 163
Church, M. J., 339
Clark, J. S., 533
Cohn, H. N., 40
Colpi, M., 205
Cornelisse, R., 365, 533
Corsi, C. E., 61
Covino, S., 144, 307
Cusumano, G., 635

D

D'Aí, A., 545
Dall'Ora, M., 61
Dall'Osso, S., 307
D'Amico, N., 523, 611
D'Antona, F., 46, 95, 116, 565
Danziger, J., 491
D'Avanzo, P., 81
Davis, R. J., 655
Della Valle, M., 150, 553
De Loore, C., 301
den Herder, J. W., 507
Devecchi, B., 205
Dhillon, V. S., 365
Di Salvo, T., 95, 110, 116, 371, 545
Domínguez, I., 497
Drew, J., 643
Duffy, P., 585

E

Ederoclite, A., 553
Edmonds, P. D., 40
Elkin, V. G., 557
Errico, L., 557
Espaillat, C., 287
Exter, K., 561, 577
Eyres, S. P. S., 655

F

Fender, R., 189
Feroci, M., 635
Ferraro, F. R., 103, 611
Fiore, F., 371
Fiorentino, G., 565
Freire, P. C., 40, 523
Freyhammer, L. M., 61
Friedjung, M., 577
Fruchter, A., 607
Fugazza, D., 307

G

Gagliardi, S., 497
Gallo, E., 189
García-Berro, E., 463
García-Senz, D., 453
Ghisellini, G., 144
Gibbons, R., 607
Gilmozzi, R., 553
Gomboc, A., 181, 655
González Hernández, J. I., 416
Gratton, R., 611
Greimel, R., 643
Grindlay, J. E., 13, 40
Groot, P., 643
Guainazzi, M., 513
Guerrero, J., 463
Guidorzi, C., 181

H

Hailey, C. J., 507
Han, Z., 386
Haswell, C. A., 219
Hayasaki, K., 569
Heinke, C. O., 40
Henden, A., 331
Henninger, M., 241
Homan, J., 225
Hynes, R., 81

I

Iaria, R., 116, 545
Ibarra, A., 513
Isern, J., 463

Israel, G. L., 81, 307
Ivanova, N., 53, 241

J

Jimenez-Bailon, E., 513
Jimenez-Garate, M. A., 507
Jonker, P., 396
Joshi, B. C., 523

K

Kaiser, C., 189
Kalogera, V., 241
Kaper, L., 623
Karovska, M., 265
Kaygorodov, P. V., 295
Keenan, F. P., 577
Kilpio, E. Y., 573
Knigge, C., 533, 643
Kohmura, T., 519
Kolb, U., 219
Kotnik-Karuza, D., 577
Kovetz, A., 319
Kramer, M., 523
Kuulkers, E., 402
Kuznetsov, O. A., 271, 295, 573

L

Langer, N., 603, 651
Lanz, T., 313
Lavagetto, G., 116, 545
Lazzati, D., 144
Lee, W. H., 138
Leibowitz, E., 471
Levin, T., 241
Lewin, W. H. G., 23
Litterio, M., 635
Lobo, J. A., 463
Lorén-Aguilar, P., 463
Lorimer, D., 523
Lugger, P. M., 40
Lyne, A., 523

M

Maitra, D., 249
Malesani, D., 144
Manchester, D., 523
Mangano, V., 307
Mapelli, M., 205
Marconi, G., 81, 307
Marsh, T. R., 365
Martínez-País, I. G., 365, 589
Martocchia, A., 581
Marziani, P., 655
Mason, E., 553
Massaro, E., 635
Matsuda, T., 295
Matt, G., 513, 635
McLaughlin, M., 523
Mendez, M., 545
Menna, M. T., 110
Mereghetti, S., 307
Miller-Jones, J. C. A., 410, 585
Mineo, T., 635
Molteni, D., 619
Monelli, M., 61, 565
Monfardini, A., 181
Motch, C., 581
Munari, U., 307, 331, 615
Mundell, C. G., 181
Muno, M., 631
Muñoz-Darias, T., 365, 589
Munteanu, A., 61

N

Naik, S., 593
Negueruela, I., 581
Nelemans, G., 396
Newsam, A. M., 181

O

Okazaki, A. T., 569
Olive, J.-F., 359
Orio, M., 471
Osborne, J. P., 219
Otsuka, M., 557

P

Paerels, F., 313
Papitto, A., 110
Parker, Q., 643
Patruno, A., 205
Patterson, J., 287
Paul, B., 593
Perna, R., 434
Pétri, J., 599
Petrovic, J., 603
Pfahl, E., 386
Piersanti, L., 482, 497
Piran, T., 123
Podsiadlowski, P., 386, 410, 422, 537
Pollacco, D. L., 561, 577
Porter, J., 655
Possenti, A., 205, 523, 611
Prialnik, D., 319
Pustylnik, I., 561
Pustynski, V., 561

R

Ramsay, G., 507
Rappaport, S., 386, 410, 422
Rasio, F. A., 53
Rauch, T., 471
Remillard, R. A., 231
Reynolds, M., 607
Ritter, H., 377
Robba, N. R., 95, 116, 545, 565
Rossi, E., 144

S

Sabbi, E., 611
Santolamazza, P., 371
Sarkissian, J., 523
Siviero, A., 615
Skopal, A., 557, 655
Smith, R. J., 181
Steeghs, D., 365, 647
Steele, I. A., 181
Stella, L., 81, 110, 225, 307, 434
Still, M., 507
Straniero, O., 497

T

Tagger, M., 541
Tamura, S., 557
Telis, G. A., 313
Tepedelenlioglu, E., 471
Teresi, V., 619
Testa, V., 565
Tornambé, A., 482, 497
Torres, M., 607
Toscano, E., 619
Tovmassian, G. H., 257

V

Vanbeveren, D., 301, 445
van den Heuvel, E. P. J., 623
van der Klis, M., 345, 545
van der Meer, A., 623
van der Sluys, M. V., 627
van Kerkwijk, M. H., 623
Van Rensbergen, W., 301
Varnière, P., 213, 631
Ventura, G., 635
Verbunt, F., 23, 30, 627
Vittone, A. A., 557

W

Wachter, S., 639
Wang, Q. D., 410
Warner, B., 277, 287
Webb, N. A., 359
Wellhouse, J. W., 639
Willems, B., 241
Williams, R. E., 553
Witham, A., 643
Wolf, M., 557
Woudt, P. A., 277, 287, 647

Y

Yoon, S.-C., 651
Yungelson, L. R., 1

Z

Zamanov, R., 655
Zane, S., 507